TREATISE ON INVERTEBRATE PALEONTOLOGY

Prepared under Sponsorship of
The Geological Society of America, Inc.

The Paleontological Society *The Society of Economic Paleontologists and Mineralogists*
The Palaeontographical Society *The Palaeontological Association*

RAYMOND C. MOORE
Founder

RAYMOND C. MOORE and CURT TEICHERT
Editors

JACK D. KEIM, LAVON McCORMICK, ROGER B. WILLIAMS
Assistant Editors

Part T
ECHINODERMATA 2

VOLUME 2

By GEORGES UBAGHS, R. C. MOORE, H. WIENBERG RASMUSSEN, N. GARY LANE, ALBERT BREIMER, H. L. STRIMPLE, J. C. BROWER, RUSSELL M. JEFFORDS, JAMES SPRINKLE, R. E. PECK, D. B. MACURDA, JR., D. L. MEYER, MICHEL ROUX, HERTHA SIEVERTS-DORECK, R. O. FAY, and R. A. ROBISON

THE GEOLOGICAL SOCIETY OF AMERICA, INC.
and
THE UNIVERSITY OF KANSAS
BOULDER, COLORADO, and LAWRENCE, KANSAS
1978

Library of Congress Catalogue Card
Number: 53-12913
I.S.B.N. 0-8137-3021-X
TIVPA 1-1027 (1978)

Text Composed by
THE UNIVERSITY OF KANSAS PRINTING SERVICE
Lawrence, Kansas

Illustrations and Offset Lithography
THE MERIDEN GRAVURE COMPANY
Meriden, Connecticut

Binding
TAPLEY-RUTTER COMPANY
Moonachie, New Jersey

Published 1978

Distributed by the Geological Society of America, Inc., 3300 Penrose Place, Boulder, Colo., 80301, to which all communications should be addressed.

PART T
ECHINODERMATA 2
CRINOIDEA

By Georges Ubaghs, R. C. Moore, H. Wienberg Rasmussen, N. Gary Lane, Albert Breimer, H. L. Strimple, J. C. Brower, Russell M. Jeffords, James Sprinkle, R. E. Peck, D. B. Macurda, Jr., D. L. Meyer, Michel Roux, Hertha Sieverts-Doreck, R. O. Fay, and R. A. Robison

VOLUME 2

SYSTEMATIC DESCRIPTIONS

By R. C. Moore, H. Wienberg Rasmussen, N. Gary Lane, Georges Ubaghs, H. L. Strimple, R. E. Peck, James Sprinkle, R. O. Fay, and Hertha Sieverts-Doreck

CONTENTS

[Volume 2, p. *T*403-*T*812; Volume 3, p. *T*813-*T*1027]

CRINOIDEA

Class CRINOIDEA Miller, 1821

[Crinoidea MILLER, 1821, p. 7 (family)] [=Stylastritae MARTIN, 1809; Stylasteritae GOLDFUSS, 1826-33; Asterencrinidae DE BLAINVILLE, 1834-37; Pinnigrada FORBES, 1841; Pinnastella AUSTIN & AUSTIN, 1842, p. 5; Brachiata BURMEISTER, 1856; Encrines PICTET, 1857, p. 278 (order); Actinoidea ROEMER, 1852-54, p. 224 (suborder); Eucrinoidea ZITTEL, 1879, p. 341] [Diagnosis by GEORGES UBAGHS]

Crinozoa provided with true arms; pentameral symmetry well developed; theca divided into aboral cup and adoral tegmen, comprising five radial plates from which invariably the aboral skeleton of the arms starts; radial growth pattern concentrated on arms, which are directed away from theca; column ordinarily well developed, lost in postlarval stage in some forms. *M. Cam.; L.Ord.-Holo.*

ECHMATOCRINEA

By JAMES SPRINKLE and RAYMOND C. MOORE
[University of Texas at Austin; University of Kansas]

Subclass ECHMATOCRINEA Sprinkle & Moore, new subclass

Primitive crinoids with an irregularly plated cup, plate circlets unrecognizable except for possible thin radials(?) directly below arms; no stem, cup attached to objects on substrate by long or short, irregularly plated holdfast; eight to 10 short, uniserial, atomous arms attached to rim of cup, 10 to 12 heavily plated brachials per arm bearing smaller soft appendages (?tube feet) attached alternately to each side of brachials; no tegmen structures known. *M.Cam. (Bathyuriscus-Elrathina Zone).*

The subclass Echmatocrinea is based on a single genus and species known from only five partially complete specimens and a single brachial plate from the Middle Cambrian Burgess Shale of western Canada. Unfortunately, none of these specimens is well enough preserved to show the original plating arrangement of the cup and holdfast. At least four specimens were found by C. D. WALCOTT between 1910 and 1917, but owing to their rarity, incompleteness, and relatively poor preservation, he did not describe or name them. The best preserved specimen was obtained by the Geological Survey of Canada in the summer of 1967 and has been chosen as holotype of *Echmatocrinus brachiatus* SPRINKLE (1973a).

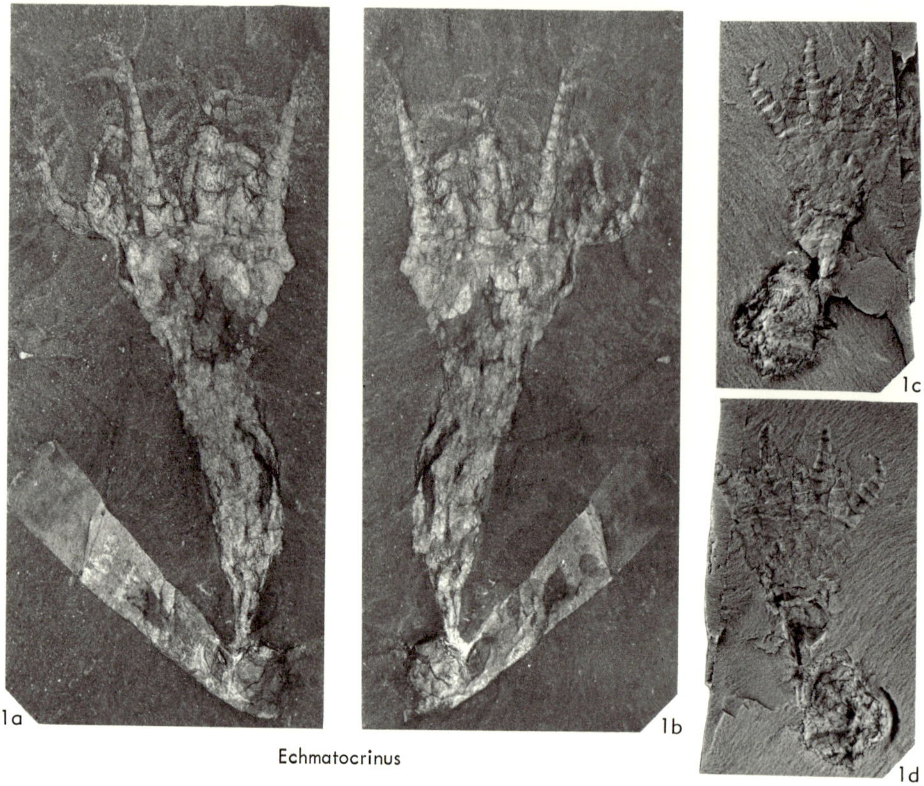

FIG. 219. Echmatocrinidae (p. T407) (Sprinkle, 1973a; from Special Publication of the Museum of Comparative Zoology, Harvard University (1973), courtesy of the Museum of Comparative Zoology).

Echmatocrinus is now the earliest crinoid (and crinozoan) known from the fossil record, predating the next oldest occurrence in the Early Ordovician (Tremadocian-Arenigian) of France and England by approximately 30 to 40 million years. It is also the most primitive known crinoid in terms of its cup, attachment appendage, and arm morphology. Except for its heavily plated atomous arms, which are definitely crinoid-like, *Echmatocrinus* resembles several Cambrian eocrinoids that also have irregular calyx plating and an attachment holdfast.

Echmatocrinus is probably most closely related to some early members of the crinoid subclass Inadunata, especially in its simple arms, but cannot be assigned to this subclass because of its irregular cup plating and presence of a holdfast. It differs considerably from members of the subclasses Camerata and Flexibilia, which have much more complex arms and specialized cup and tegmen plating. Therefore, it is here assigned to the new subclass Echmatocrinea.

The largest and best preserved specimen of *Echmatocrinus* has numerous soft appendages attached to the arms (Fig. 219; 220). These secondary appendages are 4 to 6 mm. long; show no plate sutures, relief, or reticulate ornament like that on the arms; are severely folded over on themselves; in several places are apparently attached to opposite sides of the arm, one per brachial plate; and are visible only when this specimen is immersed in a liquid. These soft appendages most likely represent tube feet from the crinoid's water vascular system, although they might possibly be unplated pinnules instead. If this first interpretation is correct, then these soft appendages are the earliest preserved echinoderm tube feet known from the fossil record.

The two best preserved specimens show

a considerable difference in the size and length of their holdfasts (Fig. 219,*1a-d*). This might represent a growth difference, related to the size of the crinoid, or perhaps the holdfast plates were imbricately arranged, so that *Echmatocrinus* could vary the height of its cup above the sea floor by extending or telescoping these holdfast plates. Unfortunately, neither of these holdfasts is well enough preserved to determine whether the plating was actually adjacent or imbricate, although the former seems more likely because of similarities to the cup plating.

The presence of a holdfast in *Echmatocrinus* implies that crinoids underwent a holdfast-to-stem transition similar (but independent) to that shown by Middle Cambrian eocrinoids (SPRINKLE, 1973a). However, crinoids passed through a different intermediate stage characterized by multipart columnals, such as those in the genus *Aethocrinus* described by UBAGHS (1969), and did not evolve a true stem with one-piece columnals until the Early Ordovician. The development of a columnal-bearing stem was probably one of the main features that led to the great diversity and success of Paleozoic crinoids. *Echmatocrinus* represents an earlier pre-stem stage of crinoid evolution characterized by apparent low diversity and primitive morphology.

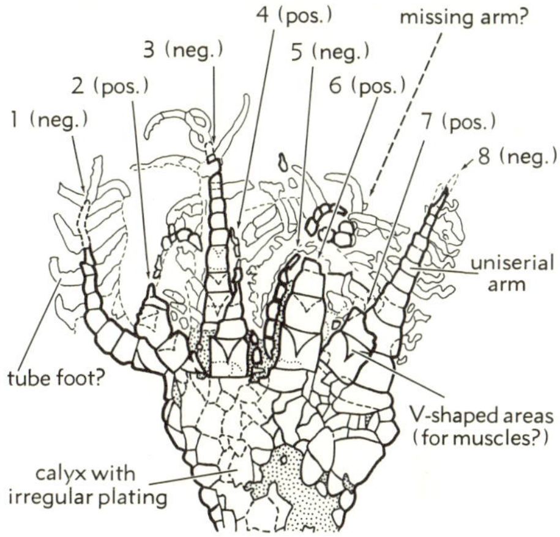

FIG. 220. *Echmatocrinus brachiatus* SPRINKLE, holotype, M.Cam., Can.; drawing showing cup, arm, and ?tube foot morphology. Arms numbered and labeled with type of relief (positive or negative). Note irregular but imperfectly preserved cup plating, elongate cup plates (?radials) beneath several arms, V-shaped areas (?for muscles) on lower brachial plates, alternate branching of the ?tube feet (one per brachial plate) on arms *1* and *3*, and alternating arm relief except between arms *6* and *7*, where an additional arm (now missing) from the rear of the cup may have been located (after Sprinkle, 1973a; from Special Publication of the Museum of Comparative Zoology, Harvard University (1973), courtesy of the Museum of Comparative Zoology).

Order ECHMATOCRINIDA Sprinkle & Moore, new order

Characters of subclass. *M.Cam.(Bathyuriscus-Elrathina Z.)*.

Family ECHMATOCRINIDAE Sprinkle, 1973

[Echmatocrinidae SPRINKLE, 1973a, p. 177]

Characters of subclass. *M.Cam.(Bathyuriscus-Elrathina Z.)*.

Echmatocrinus SPRINKLE, 1973, p. 177 [**E. brachiatus*; OD]. Cup moderately large, maximum cup diameter just below arms and greater than cup height, plated with numerous small to medium-sized adjacent plates not organized into obvious circlets, perhaps as many as 140 cup plates present, several arms mounted on thin, elongate plates, which may be radials(?); no stem present, but instead base of cup grades into tapering, cylindrical, irregularly-plated holdfast between 1 to 2 times height of cup, maximum holdfast diameter about one-fourth to one-half length; holdfast plates appear to be slightly larger but otherwise similar to those in cup, several holdfasts attached to objects (worm tubes, hyolithids) apparently lying on sea floor; 8 to 10 short, uniserial, atomous, heavily-plated arms present, 10 to 12 brachials per arm, smaller soft appendages (?tube feet) attached ?alternately to each brachial, dark V-shaped areas perhaps for muscle attachment on aboral side of proximal brachials; location of mouth, anal pyramid or tube, and other tegmen structures unknown. *M.Cam.(Bathyuriscus-Elrathina Z., Burgess Shale)*, Can.(B.C.).——FIG. 219,*1*; 220. **E. brachiatus*; 219,*1a,b*, counterparts of best specimen (immersed in water); note long holdfast attached to *Selkirkia* worm tube and ?tube feet attached to arms, ×1; 219,*1c,d*, counterparts of smaller specimen (coated with ammonium chloride) showing positive and negative relief of arms and much shorter holdfast, ×0.9; 220, drawing of cup and arms of best specimen showing plate sutures and ?tube feet (Sprinkle, 1973a).

CAMERATA

By Georges Ubaghs
[Université de Liège]
[With contributions by N. Gary Lane, R. C. Moore, A. L. Bowsher, and H. L. Strimple]

INTRODUCTION

The Camerata comprise one of the three main groups of Paleozoic crinoids. In the system that has been adopted in the *Treatise,* they are considered as a subclass, taxonomically equivalent to the subclasses Inadunata, Flexibilia, and Articulata. This subclass corresponds to the Cladocrinoidea of Jaekel [originally (1894) introduced as an order, later (1918) raised to subclass rank] and to the orders Adunata, Monocyclica Camerata, and Dicyclica Camerata of Bather's classification (1899b, 1900a).

Camerate crinoids form a unit fairly well defined on morphological grounds. They include 209 genera and two subgenera recognized herein, many of them comprising numerous species. Their known range extends from Early Ordovician to Late Permian time. They are represented by largest number of genera in Upper Silurian rocks and by largest number of species and individuals in Lower Mississippian deposits. Occurrence of relatively few representatives has been reported from the Late Mississippian, Pennsylvanian, and Permian, but seemingly none reached the Mesozoic Era.

The subclass Camerata contains both dicyclic and monocyclic genera, the latter being more than three times as numerous as the former. Throughout their history, these crinoids retained readily discernible characters. One of these is rigid union of the thecal plates, although early representatives and at least one specialized genus *(Scyphocrinites)* have somewhat loose connection between skeletal components of the test. The tegmen—one of their most salient features, from which the name Camerata is derived (Latin, *camera,* chamber)—forms a continuous generally (but not invariably) rigid vaulted ceiling above the sides and bottom of the theca. Except in a very few genera (e.g., *Cyttarocrinus, Lyonicrinus*), it covers the mouth and proximal parts of ambulacral tracts. The anus, commonly raised on a tube, is the only aperture found on the tegmen. The presence of an external hydropore or madreporite, as in some inadunates and articulates, has never been recorded.

Typically, a variable number of arm plates (and in some genera pinnulars) are incorporated in the calyx by means of interprimibrachial, intersecundibrachial, and (less commonly) intertertibrach and interpinnular plates. The number of fixed brachials may be as few as one or two to each ray or as many as 140 in a ray (e.g., *Himerocrinus*); in *Strotocrinus* up to 12 orders of brachials may enter composition of the calyx. The fixed brachials are uniserial, except in a few genera (e.g., *Spyridiocrinus, Cyphocrinus, Abacocrinus, Tryblioc rinus*), in which they show a biserial arrangement distally. On the other hand, in several families (i.e., Nyctocrinidae, Hexacrinitidae, Parahexacrinidae, Dichocrinidae, Acrocrinidae, Haplocrinidae, Platycrinitidae, Paradoxocrinidae) the proximal brachials, though generally more or less suturally connected either with the first interprimibrach (if present) or with interambulacrals, offer virtually all characteristics of free arm plates, whereas generally no interbrachials take part in formation of the calyx. It is such crinoids, provided with a solid tegmen and a calyx typically reduced to the patina (radials, basals) and in some forms a single anal plate, that were included by Bather (1899b, 1900a) in his order Adunata.

In many camerate genera, the posterior *(CD)* interray is very distinctive, the anal plates forming a median series that splits this interradial area and consequently impresses a marked bilateral symmetry on the calyx; in many, however, they are reduced to a single plate or a few supplementary elements, or they may be lacking entirely. The anal plates of camerates are unlike those of the Inadunata and Flexibilia in that none of them can be considered as equivalent to the radianal of these crinoids. They have no particular relation to the *C* ray and show no tendency to migrate obliquely upward in a left posterior (*D*-ray) direction. The proximal anal plate **(primanal)** deserves special attention, for its position in or above the radial circlet is considered to be an extremely important taxonomic character.

The radials of camerates, invariably five in number, are never compound, i.e., divided into an inferradial and a superradial. At least ten free arms are present in all genera and there may be as many as 65 or 70 (e.g., *Strotocrinus*). Wherever only a single free arm seems to occur in each ray, it is because the two normal primary rami resulting from first division have either fused together longitudinally (as in *Melocrinites*) or have developed so unequally that one of them resembles a mere ramule (as in *Cytidocrinus*).

The free arms of camerates may be simple (atomous) or branching. Types of arm branching include 1) isotomy (division into equal rami), probably the primitive type of arm branching, 2) bilateral heterotomy (division into unequal rami with smaller ones disposed alternately on opposite sides), 3) exotomy (division into unequal rami with smaller ones all given off toward outer sides of the ray), and 4) endotomy (like exotomy but with smaller rami all directed toward the midline of the ray). Robust arm trunks carrying relatively slender ramuli characterize genera belonging to various families such as the Rhodocrinitidae (e.g., *Rhipidocrinus*), Lampterocrinidae (e.g., *Lampterocrinus*), Periechocrinidae (e.g., *Thamnocrinus*), Actinocrinitidae (e.g., *Cytidocrinus, Manillocrinus, Steganocrinus*), Polypeltidae (e.g., *Trybliocrinus*), and Platycrinitidae (e.g., *Eucladocrinus*). Marked reduction of arms and, still less, complete atrophy of arms, as in some inadunates, has never been recorded among camerate crinoids.

The arms are uniserial (condition judged to be primitive) or biserial, and almost invariably pinnulate. The brachials may be compound, and provided with two, three, or four (rarely more) pinnules. At points of bifurcation of uniserial arms are axillary brachials, which differ from those of other crinoids by the fact that in some species they carry pinnules. In biserial arms axillaries generally are lacking, the division being effected by splitting of each of the two brachials just below the bifurcation.

The radials, brachials, and pinnulars (except distal pinnulars devoid of an ambulacral groove) are unperforated in nearly all camerate species. As a rule, the free brachials are united by immovable or very slightly movable articulations of zygosynostosial or symplectial type; however, even in such crinoids limited flexibility of the arms probably was achieved by the relatively large number of joints in a given length, for in biserial arms one commonly may count 20 to 25 transverse sutures in a span of 1 cm. on each side of the arm, the maximum recorded number being 60 in 1 cm. In some camerates with free arms directly supported by radials, the arm-bearing facets on the radials are more differentiated than others and the brachials are united by synostosial or pseudosynarthrial articulation, a condition allowing a greater capacity for movements of the arms.

Contrasting with the small amount of differentiation in articulations of brachials with contiguous neighbors, the pinnule-bearing facets on the arms of camerates are rather complicated. They certainly show that in most species the pinnules could move actively so as to spread out widely and fold up closely against the adoral side of the arm. In many species distal portions of the pinnules are composed of perforated pinnulars lacking an adoral groove. Such portions of the pinnules were able to flex inward to form a rooflike cover protecting soft parts of the arms and pinnules. This cover, in at least one genus (*Barrandeocrinus*), was permanent and solid, but generally it could open, as indicated by observing distal pinnulars in both extended and inwardly flexed positions.

All known camerates are stem-bearing, with possibly the single known exception of *Monstrocrinus*, which W. E. Schmidt (1942) reported as showing no trace of a stem insertion on the base of its calyx. No camerate species has been recorded in which the column is divided into parts by longitudinal sutures, that is to say in which the columnals are composed of tetrameres or pentameres. In all other respects, by its structure, composition, presence or absence of cirri, mode of growth, and nature of the holdfast, the stalk of camerates is similar to that of most Paleozoic crinoids. Although some camerates have very distinctive columns, most dissociated columnals and pluricolumnals cannot now be attributed certainly to any particular camerate genus.

Rarely stated is general agreement that main component parts and ossicles of the body of camerate crinoids are homologous to specified parts and ossicles of other crinoids. This has been denied by JAEKEL (1894, 1918), who considered the camerates to be utterly different in origin and morphogenesis from all other crinoids. For him, the most distinctive feature of camerate crinoids is branching of the ambulacra within the theca (hence the name Cladocrinoidea which he gave them) and their carrying of free articulated appendages outside the theca which he called *Finger* (German vernacular term), because with respect to each of the five rays they are given off like fingers of a hand. He termed **costals (costalia),** rather than radials and fixed brachials, the vertical series of calyx plates that support the fingers, which he judged to have been produced by a further evolution of the brachioles of eocrinoids through transformation of their covering ambulacral plates into pinnules. The arms of camerates, however, have precisely the same morphological relations with the theca as the arms of other crinoids. Seemingly, therefore, no valid reason exists for doubting the strictly homologous nature of these appendages and supporting calyx plates in all crinoids. This does not mean, of course, that any plate of a camerate is necessarily homologous with a corresponding plate in other crinoids. For instance, homologies of orals and anals of the camerates are quite uncertain, since these skeletal elements may be very similar in appearance to those of articulates and inadunates and yet be quite distinct in origin.

Subclass CAMERATA Wachsmuth & Springer, 1885

[*nom. transl.* MOORE & LAUDON, 1943a, p. 21 (*ex* suborder Camerata WACHSMUTH & SPRINGER, 1891, p. 345, *nom. correct. pro* Camarata WACHSMUTH & SPRINGER, 1885, p. 6 (228), *nom. imperfect.*)] [=Spheroidae WACHSMUTH, 1877, p. 186; family Sphaeroidocrinidae WACHSMUTH, 1880, p. 22; order Sphaeroidocrinacea NEUMAYR, 1889, p. 462; order Cladocrinoidea JAEKEL, 1894, p. 113; orders Monocyclica Adunata + Monocyclica Camerata + Dicyclica Camerata BATHER, 1899b, p. 921-923; order Camerida DELAGE & HÉROUARD, 1904, p. 376] [Materials for this subclass prepared by GEORGES UBAGHS, with contributions by N. GARY LANE, R. C. MOORE, A. L. BOWSHER, and H. L. STRIMPLE]

Crinoidea characterized by monocyclic or dicyclic base; thecal plates typically united by rigid sutures; proximal brachials generally, though varyingly, incorporated in theca by union with aboral or adoral plates or both; interbrachials ordinarily present; ambulacrals corresponding to fixed brachials either incorporated in tegminal surface or pressed below it; mouth and proximal portion of food grooves (except very rarely) covered by tegminal pavement; anus opening directly through tegmen or at end of anal tube; posterior side of calyx typically differentiated by supplementary plates arranged according to crinoidal plane of symmetry (through midline of *A* radial and *CD* interray) and without any definite relation to the *C* ray; no compound radials, no radianal; rays dividing at least once; free arms uniserial or biserial; brachials mono- or hyperpinnulate, imperforate. *L.Ord.-U. Perm.*

Order DIPLOBATHRIDA Moore & Laudon, 1943

[*nom. correct.* MOORE in MOORE, LALICKER & FISCHER, 1952, p. 614 (*pro* Diplobathra MOORE & LAUDON, 1943a, p. 79)] [=Dicyclica Camerata BATHER, 1899b, p. 923] [Materials for this order prepared by GEORGES UBAGHS]

Calyx dicyclic. *M.Ord.-U.Ord.; M.Sil.-L.Carb.*

Suborder ZYGODIPLOBATHRINA Ubaghs, 1953

[Zygodiplobathrina UBAGHS, 1953, p. 735]

Basals alternating with radials in a circlet of ten plates, all of them in contact with infrabasals. *M.Ord.,L.Dev.*

Family CLEIOCRINIDAE S. A. Miller, 1890

[Cleiocrinidae S. A. MILLER, 1890, p. 323]

Theca high conical or pyriform, test thin and probably pliant; infrabasals five, distinct, in deep concavity and overlapped by radial-basal circlet, which projects downward over stem; rays bifurcating several times in calyx, giving off fixed pinnules (or ramules); fixed brachials extremely

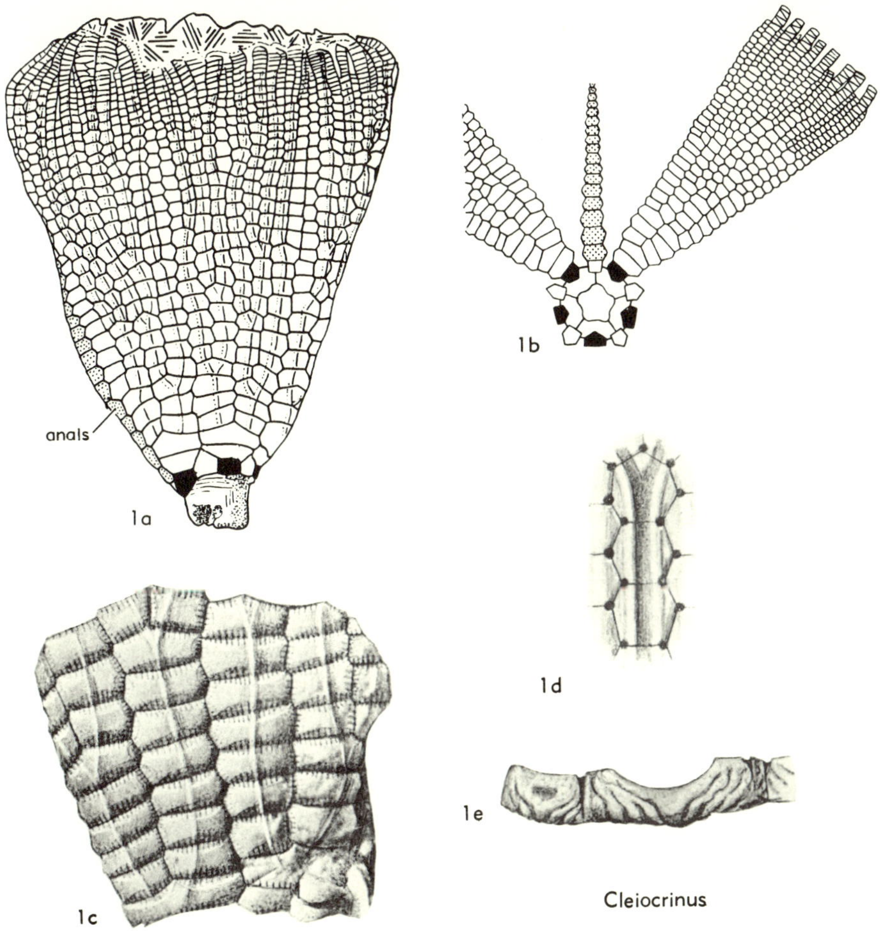

FIG. 221. Cleiocrinidae (p. *T*411).

numerous, contiguous, and interlocking; no interprimibrachs; anals in long uniserial vertical column resting on *CD* basal; numerous pores along suture lines comprising external apertures of sutural canals, which converge into large tunnel-like pores opening to interior of theca at plate corners. Tegmen many-plated, undifferentiated into ambulacral and interambulacral areas. Arms simple, uniserial. Column obtusely pentagonal or nearly round in cross section. *M. Ord.*

Cleiocrinus Billings, 1857, p. 276 [**C. regius*; M] [=*Campanulites* Troost, 1850b, p. 419 *(nom. nud.)* (type, *C. tesselatus*; M)]; *Cleistocrinus* Ludwig, 1906, p. 8 *(nom. null.)*; *Cliocrinus* Hudson, 1911, p. 211 *(nom. van.)*]. Characters of family. *M.Ord.*, N.Am.(Can.-USA).——Fig. 221,*1a-d*. **C. regius*, Trenton., Can.; *1a*, lat. view of type specimen (radials black), ×1.3 (Springer, 1905); *1b*, diagram of base, *C* and *D* rays and *CD* interray (upward) (radials black, anals stippled) (Ubaghs, 1950); *1c*, portion of arm, ×2.25 (Springer, 1911b); *1d*, inner side of fixed brachials, showing broad long. groove and pores at corners of plates, ×5 (Springer, 1911b).——Fig. 221,*1e*. *C. sculptus* Springer, Trenton., Ky.; distal faces of thecal plates, showing sutural pore system, ×5 (Springer, 1911b).

Family SPYRIDIOCRINIDAE Jaekel, 1918

[Spyridiocrinidae Jaekel, 1918, p. 32]

Calyx broad, low, with a more or less deep basal concavity; infrabasals fused; rays bifurcating several times before becoming free; fixed ramules proximally uniserial, dis-

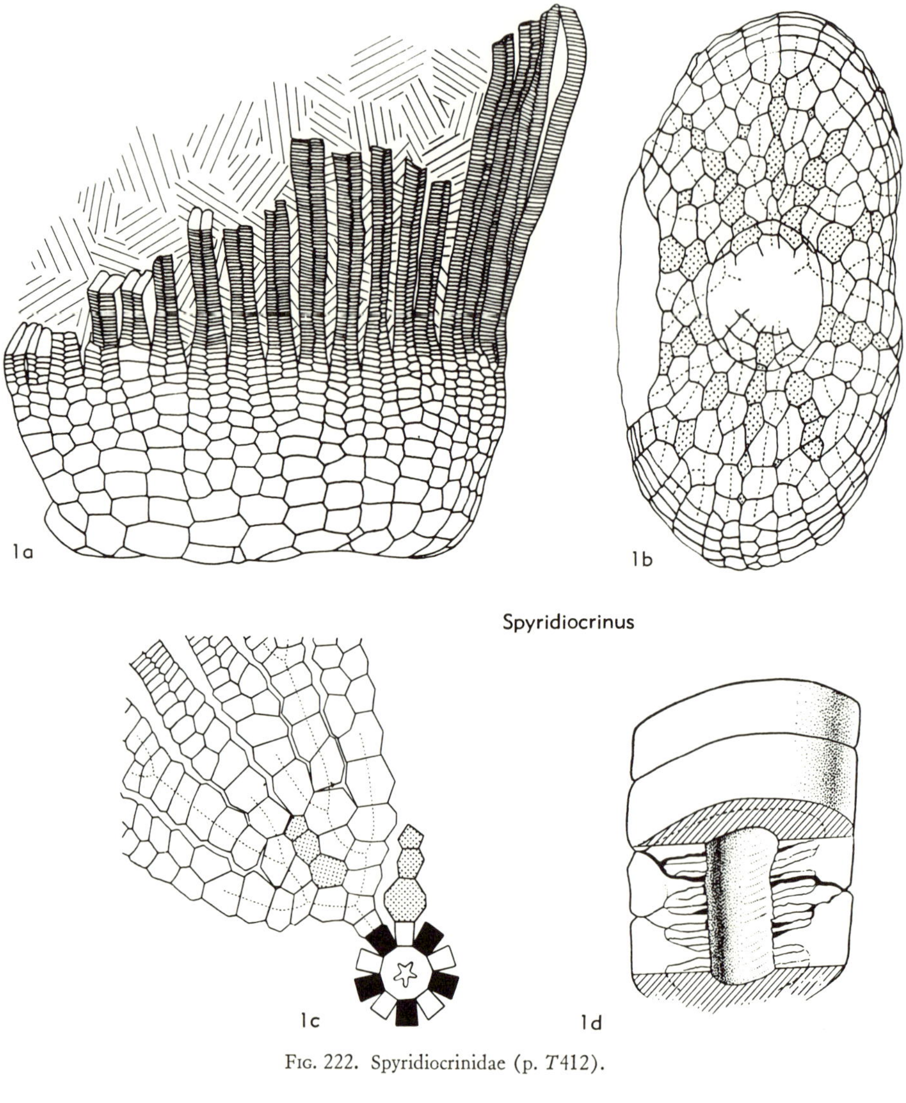

Fig. 222. Spyridiocrinidae (p. T412).

tally biserial; interprimibrachs in short uniserial series resting on basals; intersecundibrachs few, intertertibrachs reduced or lacking; posterior side not differentiated in calyx. Arms simple, biserial. Stems round in cross section, with nudinodals concealing very short internodals, which do not reach periphery, at least in proximal region of stem; axial canal quinquelobate. *L.Dev.*

Spyridiocrinus Oehlert, 1889, p. 786 [**S. cheuxi*; OD] [=*Lahuseniocrinus* Chernyshev, 1893, p. 92 (type, *L. tirlensis*; OD)]. Characters of family. *L.Dev.*, Eu.(France-Boh.-USSR).——Fig. 222,*1*. **S. cheuxi,* France; *1a,b,* lat. and dorsal views of calyx (interbrachials stippled), ×1 (Ubaghs, 1950); *1c,* plate diagram of base, *D* ray, and *CD* interray (radials black, anals and interbrachials stippled; *CD* interray directed upward) (Ubaghs, 1950); *1d,* stem structure at 12 cm. below theca, ×3.3 (Ubaghs, 1950).

Suborder EUDIPLOBATHRINA Ubaghs, 1953

[Eudiplobathrina Ubaghs, 1953, p. 735]

Basals and radials alternating in two distinct circlets. *M.Ord.-U.Ord.; M.Sil.-L.Carb.*

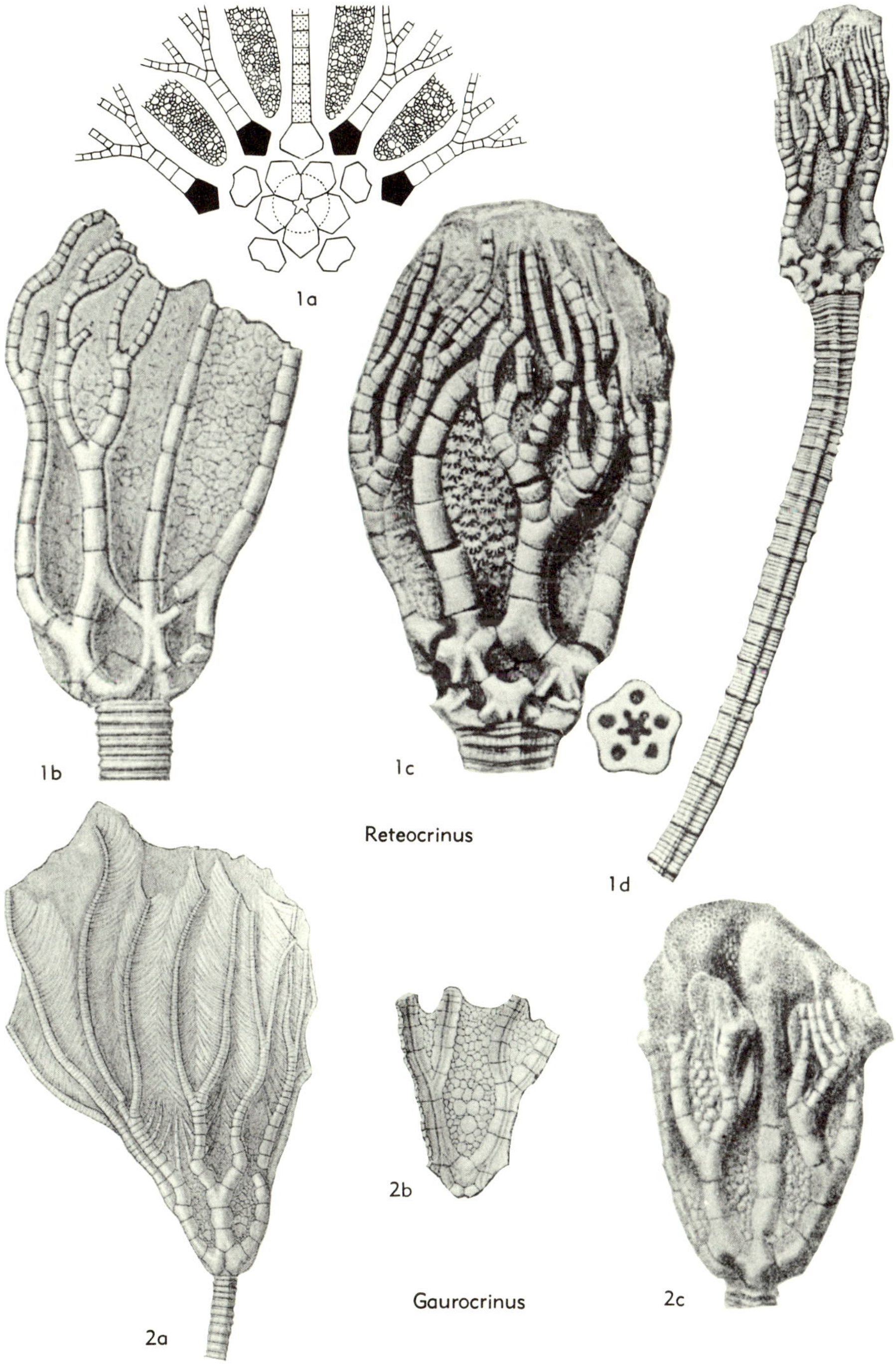

FIG. 223. Retiocrinidae (p. *T*414).

Superfamily RHODOCRINITACEA Roemer, 1855

[*nom. transl. et correct.* UBAGHS, herein (*ex* family Rhodocrinidae ROEMER, 1855, p. 228)] [=Reteocrinicae UBAGHS, 1953, p. 735]

Radials separated from each other by interprimibrachs, which adjoin basals. *M. Ord.-U.Ord.; M.Sil.-L.Carb.*

Family RETEOCRINIDAE Wachsmuth & Springer, 1885

[Reteocrinidae WACHSMUTH & SPRINGER, 1885, p. 91 (313)] [=Gaurocrinidae S. A. MILLER, 1890, p. 321]

Calyx obconical, with marked bilateral symmetry through midline of *A* radial and *CD* interray; thecal walls thin and weak, sustained by ray and anal plates plicated in strong externally rounded ridges; infrabasals five, projecting more or less beyond column; basals large; several secundibrachs incorporated in calyx; interprimibrachs depressed, small, numerous, and irregular; *CD* interray with sagittal uniserial row of very prominent anal plates separating two depressed fields of very diminutive plates. Tegmen low, slightly convex, pliant, composed of minute irregular pieces forming a continuation of interbrachials; no anal tube. Arms uniserial. Stem pentagonal or round in cross section. *M.Ord.-U.Ord.*

Reteocrinus BILLINGS, 1859, p. 63 [**R. stellaris*; M] [=*Retiocrinus* BIGSBY, 1868, p. 23 *(nom. van.)*]. Small spaces between infrabasals, basals, and radials generally present; primibrachs 2 to 6 in each ray. Arms many-branched, irregularly dichotomous; brachials rather long; no pinnules (according to KOLATA, 1975). Cross section of stem round or pentagonal or passing from pentagonal to round distally. *M.Ord.-U.Ord.*, N.Am. (Can.).——FIG. 223,*1a,b*, **R. stellaris*, Trenton., Can.; *1a*, plate diagram of part of calyx (radials black, anal series stippled, directed upward) (Ubaghs, n); *1b*, post. view, ×2 (Wachsmuth & Springer, 1897).——FIG. 223,*1c,d*. *R. alveolatus* MILLER & GURLEY, Trenton., Can.; *1c*, *CD*-interray and *C*-ray views with cross section of stem, ×2, enl. (Springer, 1911b); *1d*, *A*-ray view of calyx and stem, ×1 (Springer, 1911b).

Gaurocrinus S. A. MILLER, 1883, p. 228 [**Glyptocrinus nealli* HALL, 1866, p. 2; OD]. Similar to *Reteocrinus*, but with interbrachials very irregular in form and size. Arms branching isotomously once or twice at rather long intervals; 2 to 3 fixed pinnules on each arm, 1st one considerably stouter than others; free brachials very short, wedge shaped, pinnulate. Stem pentagonal transversely. *U.Ord.*, N.Am.(USA).——FIG. 223,*2*. **G. nealli* (HALL), Richmond., Ohio; *2a*, *A*-ray view of crown and part of attached stem, ×2 (Wachsmuth & Springer, 1897); *2b*, portion of inner side of calyx ×2 (Wachsmuth & Springer, 1897); *2c*, *CD*-interray view of crown, ×4 (Springer, 1905).

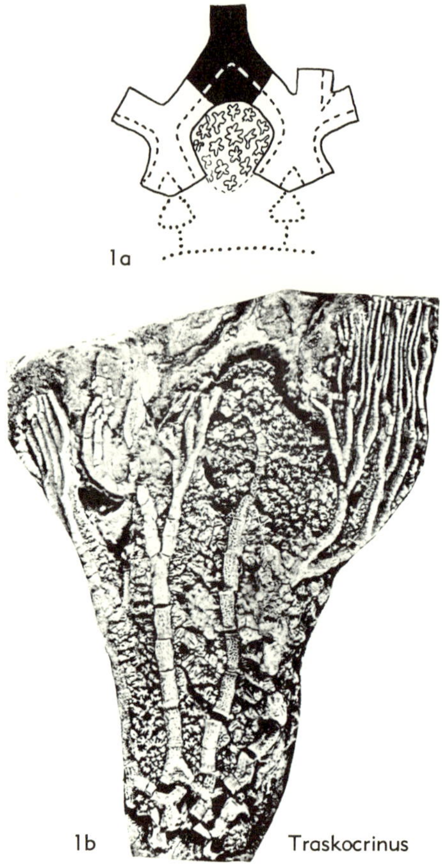

FIG. 224. Retiocrinidae (p. *T* 414). [From Memoir 7, Supplement to Vol. 49, no. 3, *Journal of Paleontology*, 1975; courtesy of the Paleontological Society and Dennis R. Kolata.]

Traskocrinus KOLATA, 1975, p. 46 [**T. mahlburgi*; OD]. Differs from *Reteocrinus* in having large spaces between deeply cleft sides of infrabasals, basals, and radials filled with numerous and irregular minute ossicles similar to interbrachials. *M.Ord.*, N.Am.(USA).——FIG. 224,*1*. **T. mahlburgi*, Platteville Gr., Blackriver., USA(Ill.); *1a*, holotype, *D* radial (in black), two basals and one infrabasal (dotted lines) surrounding spaces occupied by tiny ossicles, ×2.6 (Kolata, 1975, mod.); *1b*, holotype, *CD* interray view of crown, ×0.8 (Kolata, 1975).

Family OPSIOCRINIDAE Kier, 1952

[Opsiocrinidae KIER, 1952, p. 64]

Calyx relatively low, subconical; infra-

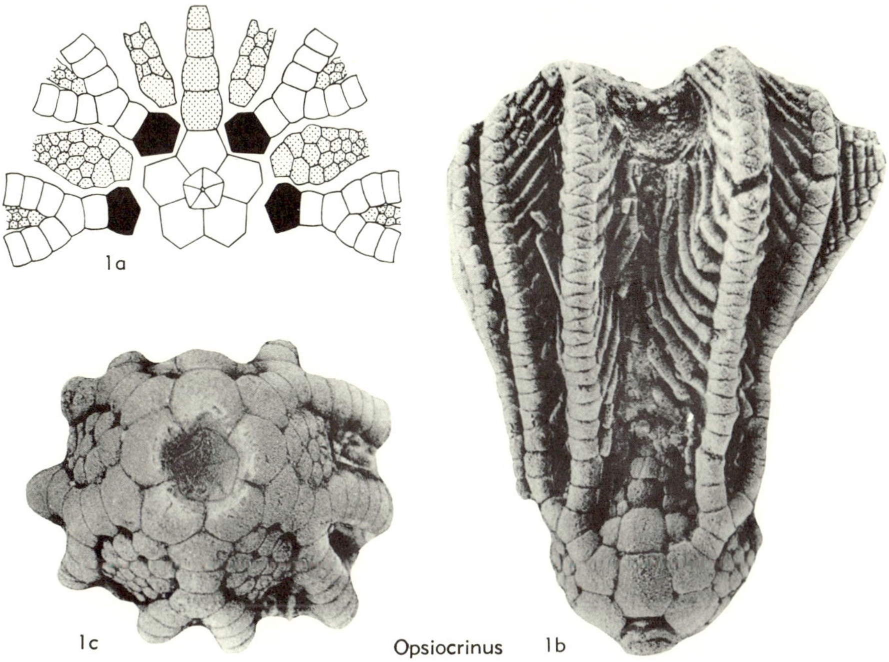

FIG. 225. Opsiocrinidae (p. T415).

basals forming a small pentagon entirely hidden by proximal columnal; interbrachials depressed, small, fairly regular, especially in proximal part of interrays; *CD* interray conspicuous, with median uniserial row of large, subquadrangular anal plates raised above field of tiny, rather irregular plates on each side. Arms ten, free at about level of second secundibrachs, biserial, unbranched distally. Stem star shaped in cross section, heteromorphic, comprising nodals and internodals; axial canal round. *M.Dev.*

Opsiocrinus KIER, 1952, p. 65 [**O. mariana*; OD] [=*Ameliacrinus* KESLING, 1968, p. 159 (type, *A. benderii*; OD)]. Characters of family. *M.Dev.*, USA(Mich.).——FIG. 225,*1*. **O. mariana*, Silica F.; *1a*, plate diagram of part of calyx, arms, and interradial areas (radials black, anals and interbrachials stippled, *CD* interray directed upward) (Kesling & Meyer, 1963); *1b,c*, *CD*-interray views of immature holotype, and dorsal side of another specimen, ×7 (Kesling & Meyer, 1963).

Family ARCHAEOCRINIDAE Moore & Laudon, 1943

[Archaeocrinidae MOORE & LAUDON, 1943a, p. 82]

Calyx more or less elongate; faint or well marked median ray ridges typically present; interprimibrachs and intersecundibrachs regular, numerous, depressed, merging with tegmen; posterior side wider, generally with extra plates, but without distinct anal ridge. Arms uniserial to biserial. Column cylindrical. *M.Ord.-U.Ord.*

Archaeocrinus WACHSMUTH & SPRINGER, 1881, p. 189(363) [**Glyptocrinus lacunosus* BILLINGS, 1857; OD]. Calyx high subconical; infrabasals very small concealed by stem or resting in concavity formed by lower part of rather large basals; median ray ridges faint proximally, but more conspicuous distally; each first interprimibrach followed by 2 plates, primanal by 3 plates. Free arms branching, biserial. [As restricted by the above diagnosis, *Archaeocrinus* comprises only a small compact group of species from the Trentonian of North America.] *M.Ord.*, N.Am. (Can.).——FIG. 226, *1a,b*. **A. lacunosus* (BILLINGS), Trenton., Can.; *1a*, *CD*-interray view of holotype, ×1.5 (Ubaghs, n; Geol. Survey Canada, 1589); *1b*, *C*-ray view of paratype, ×2 (Ubaghs, n; Geol. Survey Canada).——FIG. 226,*1c,d*. *A. microbasalis* (BILLINGS), Trenton., Can.; *1c*, plate diagram of calyx, lower part of arms, and interradial areas (radials black, anals and interbrachials stippled, *CD* interray directed upward) (Ubaghs, n); *1d*, *CD*-interray view of crown and stem, slightly reduced (Wachsmuth & Springer, 1897).

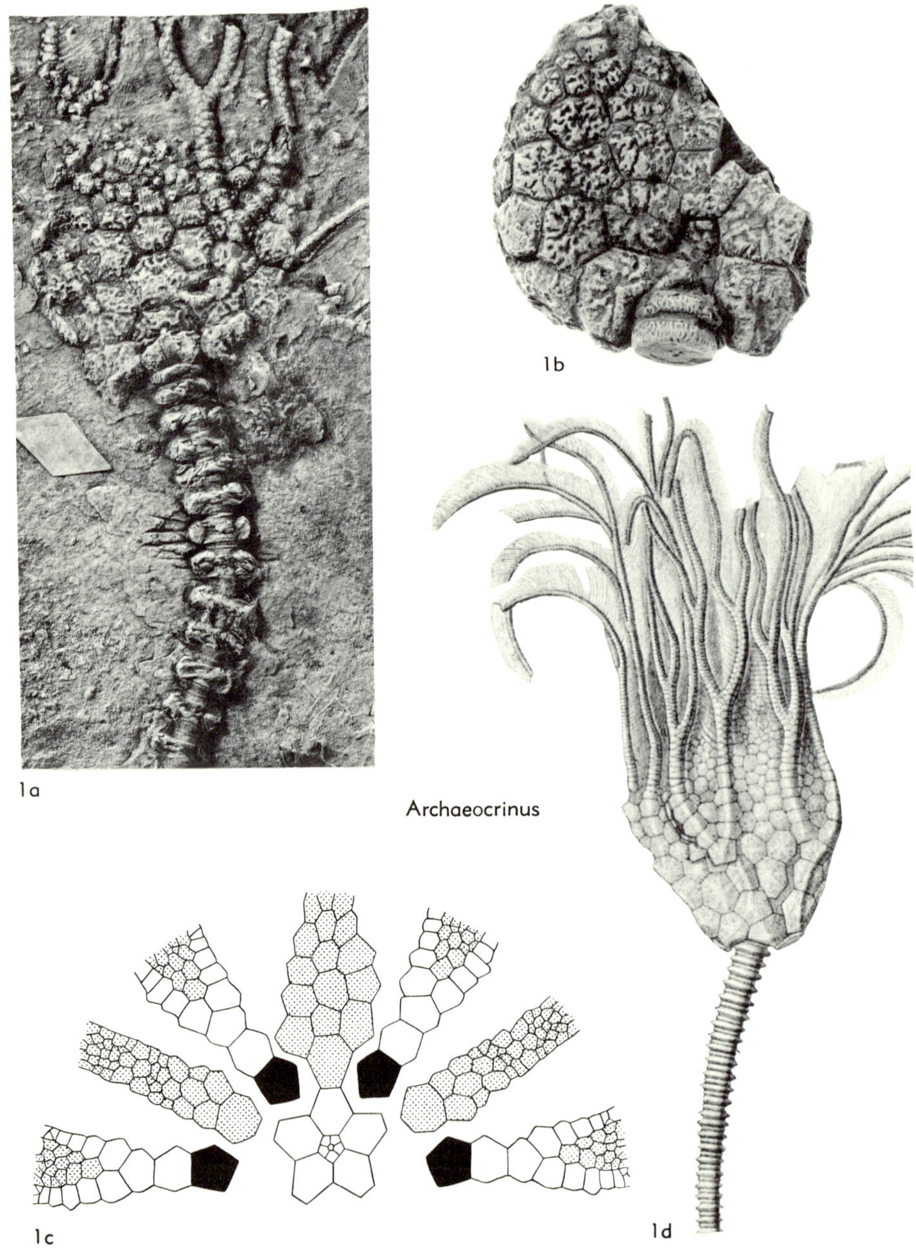

FIG. 226. Archaeocrinidae (p. *T*415).

Balacrinus RAMSBOTTOM, 1961, p. 24 [**Glyptocrinus basalis* M'COY, 1850, p. 289; OD] [=*Coelocrinus* SALTER, in RAMSAY, 1866, p. 283 (*non* MEEK & WORTHEN, 1865b)]. Calyx subglobular to conical; infrabasals visible in side view; median ray ridges narrow but distinct; all interradial areas with usually 3 plates in second row; *CD* interray without extra plates. Arms uniserial; brachials short, cuneate. Stem formed of alternating nodals and internodals; lumen round and large. *M.Ord. (Caradoc.),* G.Brit.——FIG. 227,*1*. **B. basalis* (M'COY); *CD*-interray view of crown and part of stem, ×2 (Ramsbottom, 1961). [=*Coenocrinus* DELAGE & HEROUARD, 1903 *(nom. null.)*.]

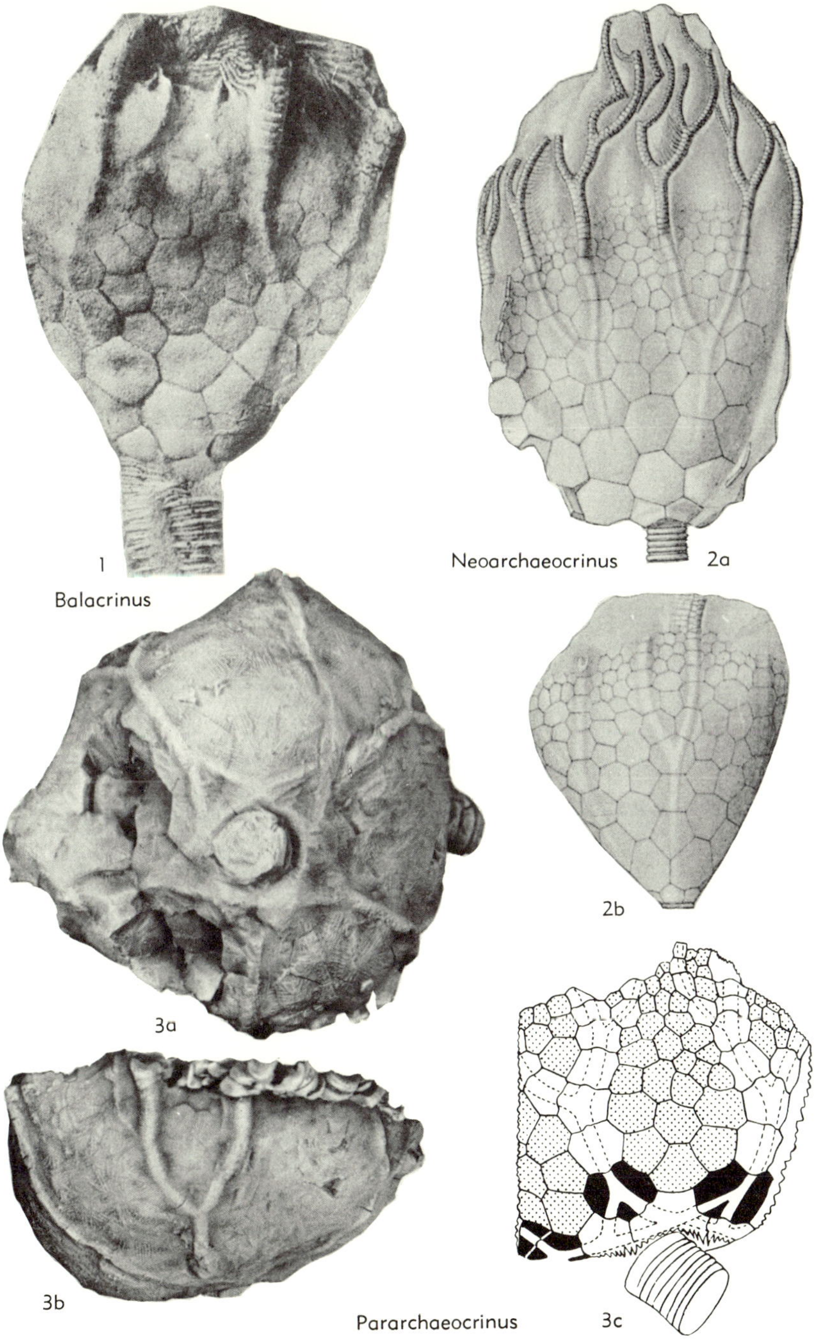

FIG. 227. Archaeocrinidae (p. T416-T418).

Neoarchaeocrinus STRIMPLE & WATKINS, 1955, p. 351 [**Thysanocrinus (Rhodocrinus) pyriformis* BILLINGS, 1857; OD]. Similar to *Archaeocrinus* but calyx conical or pyriform and relatively large with upflaring infrabasals clearly visible in side view. *M.Ord.-U.Ord.,* N.Am.(Can.)-Eu.(Scot.).

——FIG. 227,*2*. **N. pyriformis* (BILLINGS), Trenton., Can.(Ont.); *2a,b*, lat. views of crushed holotype and paratype, ×1 (Wachsmuth & Springer, 1897).

Pararchaeocrinus STRIMPLE & WATKINS, 1955, p. 351 [**P. decoratus*; OD]. Calyx subglobular; basal concavity including infrabasals and proximal 3rd of basals, strong median ray ridges joined to other ridges which form pentagon about basal concavity; interradial areas except *CD* interray with 3 plates in second row; *CD* interray protruding, with 2 or 3 plates in direct contact with posterior basal. Arms uniserial; free brachials very short, cuneate. Stem obtusely pentagonal in outline (at least proximally); lumen large, round. *M.Ord.*, USA(Okla.).——FIG. 227,*3*. **P. decoratus*, Bromide F.; *3a,b*, basal and lat. views of holotype (U.S. Natl. Museum S 4871), ×2 (Kier, n); *3c*, plate diagram of part of calyx, *ED* interray directed upward (radials black, interbrachials stippled) (Strimple & Watkins, 1955).

Rhaphanocrinus WACHSMUTH & SPRINGER, 1885, p. 98(320) [**Glyptocrinus? subnodosus* WALCOTT, 1883; OD]. Calyx conical; infrabasals small, hidden by column; strong median ray ridges; 1st interprimibrachs followed by 2 or 3 plates in next row; primanal supporting 3 plates. Arms 10 to 20, simple, uniserial; brachials externally quadrangular to cuneate. *M.Ord.-U.Ord.*, N.Am.-Eu.(Eng.-Est.).——FIG. 228,*1a*. **R. subnodosus* (WALCOTT), M.Ord.(Trenton.), N.Y.; lat. view of holotype crown with attached stem, ×1 (Wachsmuth & Springer, 1897).——FIG. 228,*1b*. *R. sculptus* (S. A. MILLER), U.Ord.(Richmond.), Ohio; lat. view of crown, ×1 (Wachsmuth & Springer, 1897).

Family ANTHRACOCRINIDAE Strimple & Watkins, 1955

[Anthracocrinidae STRIMPLE & WATKINS, 1955, p. 348]

Calyx with basal concavity involving infrabasals and part of basals; interbrachials regular, separated from tegmen by fixed pinnulars; *CD* interray differing little from others, lacking sagittal series of anal plates; fixed plates of lower pinnules enlarged, those of one ray meeting neighboring ones (or ramules in some rays) above interradial (or interbrachial) areas of theca; fixed brachials and pinnulars forming weblike extension at base of each arm. Arms uniserial to biserial. *M.Ord.*

Anthracocrinus STRIMPLE & WATKINS, 1955, p. 348 [**A. primitivus*; OD]. Calyx tall, with deep basal concavity involving elongate infrabasals and more than proximal half of basals; median ray ridges prominent; interbrachials few, mildly depressed; proximal interprimibrachs large; primanal smaller than proximal interprimibrachs but succeeding plates larger than those found in other interrays. Arms uniserial, typically 3 to each ray, one half-ray in each ray bifurcating on 2nd fixed secundibrach (exceptionally on first); free brachials wedge shaped. Column cylindrical, proximally composed of expanded nudinodals and very thin narrow internodals; lumen large, quinquelobate in outline. *M.Ord.*, USA(Okla.).——FIG. 228,*2*. **A. primitivus*, Bromide F.; *2a,b*, lat. views of holotype (U.S. Natl. Museum S4759) and paratype (USNM S4761), ×3 (Kier, n); *2c,d*, plate diagrams of *CD* interray and normal interray with parts of adjoining rays (radials black, anals and interbrachials stippled, shape of basal concavity indicated by dotted lines) (modified from Strimple & Watkins, 1955).

Deocrinus HUDSON, 1907, p. 121 [**Rhodocrinus asperatus* BILLINGS, 1859, p. 27; OD]. Calyx globose, with rather shallow basal concavity involving small infrabasals and about one third or less of proximal part of basals; no median ray ridges; interprimibrachs numerous, comprising large plate completely surrounded by small ones; *CD* interray with 2 large plates partially surrounded by small plates; 1 or 2 relatively large intersecundibrachs in each ray; fixed secundibrachs relatively numerous. Tegmen composed of many very small plates extending out over ambulacral grooves of lower pinnules and base of arms; anal tube nearly central. Arms 2 to each ray, brachials externally wedge shaped. Stem lumen round. *M.Ord.*, N.Am.(Can.).——FIG. 229,*1*. **D. asperatus* (BILLINGS), Chazyan, Can.; *1a,b*, lat. and dorsal views of holotype crown, ×2 (Ubaghs, n; Geol. Survey Canada, 1019a); *1c*, plate diagram of part of theca (radials black, interbrachials stippled, *CD* interray directed upward) (Hudson, 1907).

Hercocrinus HUDSON, 1907, p. 125 [**H. elegans*; OD]. Calyx globose, more or less flattened at base with narrow basal concavity involving very small infrabasals and less than proximal half of each basal; interradial areas generally differing from each other in arrangement of plates and comprising relatively large plates partly or entirely surrounded by smaller ones; generally 2 interprimibrachs between radials; *CD* interray not wider than others and having about same plate arrangement; fixed secundibrachs 2 or more, each giving off a pinnule partly incorporated in calyx. Tegmen formed by very numerous small plates extending upward onto arm bases; short anal tube subcentral. Arms biserial. Stem circular in cross section, lumen moderately large, pentagonal, *M.Ord.*, USA(N.Y.).——FIG. 229,*2*. **H. elegans*, Chazyan; plate diagram of part of calyx and arms (radials black, interbrachials stippled) (Hudson, 1907).

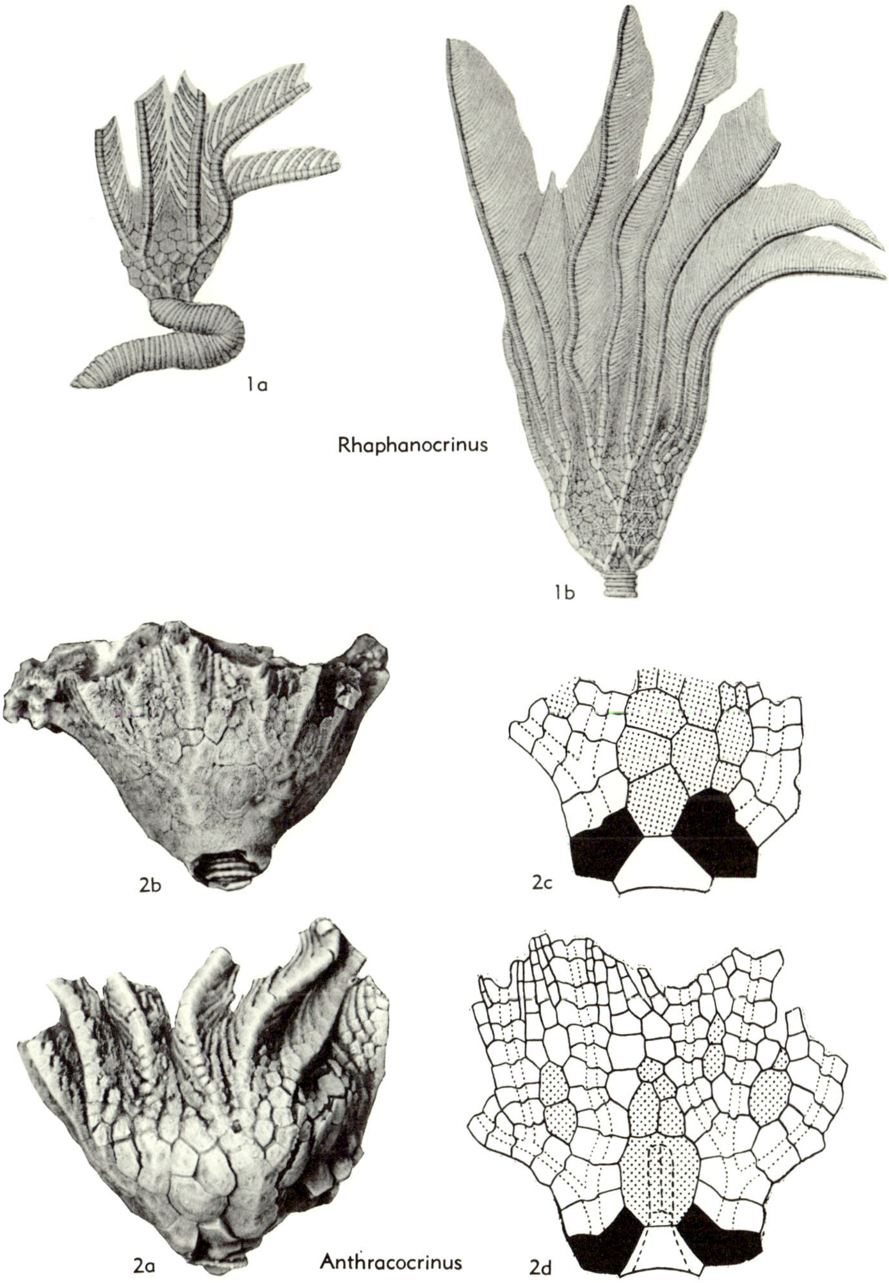

FIG. 228. Archaeocrinidae *(1)*; Anthracocrinidae *(2)* (p. *T*418).

Family ANTHEMOCRINIDAE Jaekel, 1918

[Anthemocrinidae JAEKEL, 1918, p. 40]

Calyx moderately to very low; infrabasals four, unequal; interbrachials few, not connected with tegmen; *CD* interray indistinct or barely differentiated. Tegmen many-plated, with subcentral anal tube. Arms

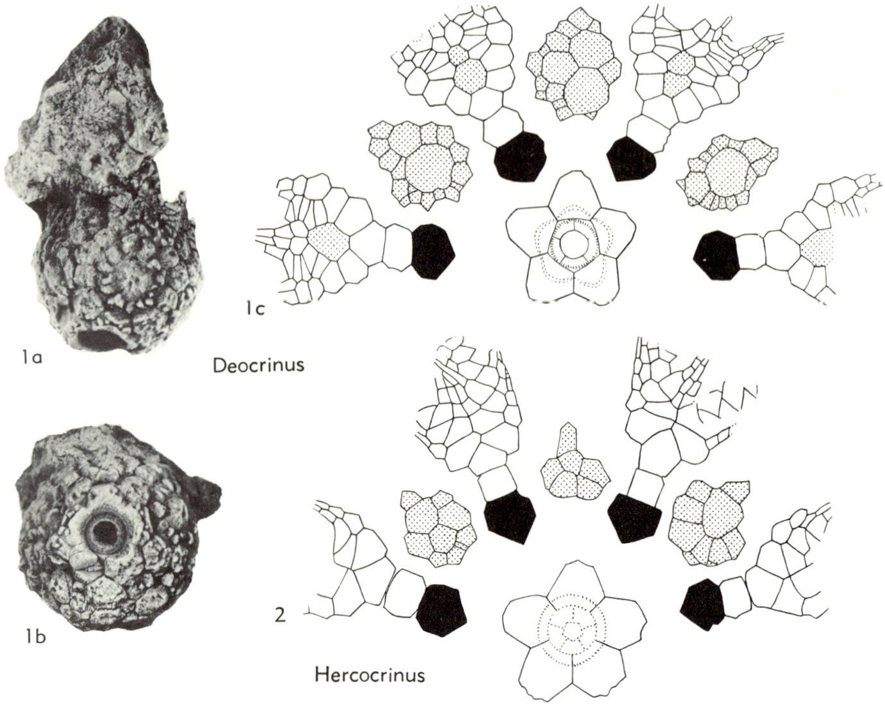

FIG. 229. Anthracocrinidae (p. *T*418).

ten, biserial. Axial canal of stem very small. *M.Sil.-U.Sil.*

Anthemocrinus WACHSMUTH & SPRINGER, 1881, p. 208(382) [**Eucrinus venustus* ANGELIN, 1878, p. 25; OD]. Calyx bowl shaped; fixed primibrachs 1 in each ray; fixed secundibrachs 2 or 3; interprimibrachs 2(1 + 1); intersecundibrachs few. Arms branching at least once, rami undivided or bifurcating isotomously once or twice. *U.Sil.,* Eu. (Sweden).——FIG. 230,*1a*. **A. venustus* (ANGELIN), Gotl.; lat. view of lectotype crown (Stockholm Museum), ×2 (Ubaghs, n).——FIG. 230, *1b,c*. *A. minor* (ANGELIN), Gotl.; lat. and dorsal views of lectotype crown (Stockholm Mus.) (radials black, interbrachials stippled), ×3 (Ubaghs, n).

Wilsonicrinus SPRINGER, 1926, p. 23 [**W. discoideus*; OD]. Calyx broad, wheel shaped, slightly convex aborally; fixed secundibrachs 4 or 5, externally cuneate; single interprimibrach in each interray, *CD* slightly wider than others; arm openings 10, widely separated; fixed pinnulars similar to interprimibrachs, leading to conspicuous interradial pinnule openings at margin between arm openings. *M.Sil.,* USA(Ind.).——FIG. 230,2. **W. discoideus,* Niagaran (Laurel Ls.); *2a,b,* dorsal and ventral views of holotype, ×2 (Springer, 1926a).

Family RHODOCRINITIDAE Roemer, 1855

[*nom. correct.* BASSLER, 1938, p. 25 (*pro* Rhodocrinidae ROEMER, 1855, p. 228)] [=Gilbertsocrinidae JAEKEL, 1918, p. 42]

Calyx generally globose, with excavate or flattened base; median ray ridges weak or typically absent; interprimibrachs regular, well defined, not depressed or only slightly so; posterior side usually widest with or without extra plates but generally lacking an anal ridge. Tegmen many-plated, anal tube ordinarily present. Arms free above secundibrachs, uniserial to biserial; stem typically round in cross section. *M.Ord.-U.Ord.; M.Sil.; L.Dev.-L.Carb.*

Rhodocrinites MILLER, 1821, p. 15, 106 [**R. verus*; SD ROEMER, 1855, p. 241] [=*Rhodocrinus* AGASSIZ, 1836, p. 196 (*nom. van.*)]. Calyx rounded, with sides nearly vertical or diverging upward, plates typically with stellate ornamentation; base flattened or concave; basals relatively small; fixed secundibrachs 1 or more in each ray; interradial areas large, many-plated, merging with tegmen; intersecundibrachs present; *CD* interray wider than others, with extra anal plates. Teg-

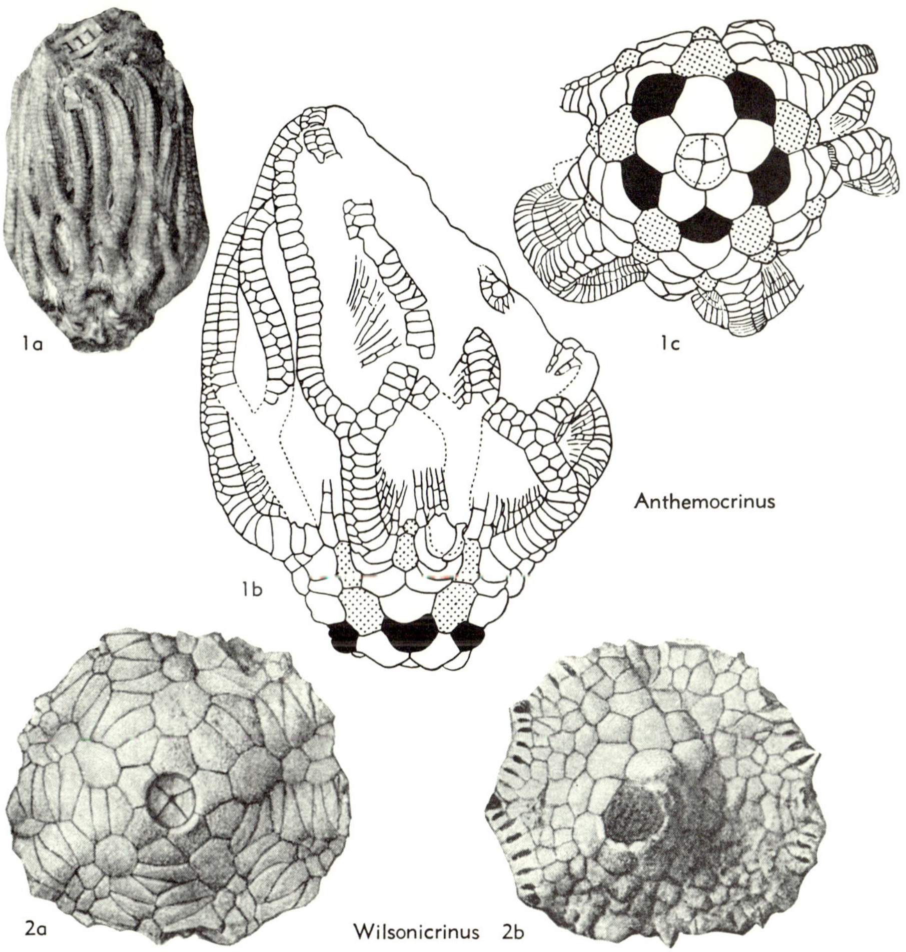

Fig. 230. Anthemocrinidae (p. *T*420).

men rather flat, composed of numerous small plates; anal tube short, excentric. Arms branching isotomously, uniserial to most distal bifurcations but compactly biserial above them. Stem cylindrical, composed of several orders of rather short columnals. *L.Carb.*, Eu.-N.Am.-Asia-?Australia. [In the original description of *Rhodocrinites,* Miller confounded two species under *R. verus,* one from the Silurian of Dudley, Worcestershire, Eng., the other from the Mountain Ls. (L.Carb.) along the River Avon, near Bristol, England. The Silurian form has never been definitely selected as lectotype of *R. verus* by a subsequent author, but has been attributed to *Sagenocrinites expansus* (Phillips) by Wachsmuth & Springer (1881, p. 201[375]). The Carboniferous form, with Miller's specific name was designated as the type species of *Rhodocrinites* by Roemer in 1855 (p. 241).]——Fig. 231, *1a,b.* **R. verus,* Tournais., Eng.; *1a,* lat. view of crown, $\times 2$ (orig. figure by Miller); *1b,* lat. view of theca of another specimen, $\times 1$ (Grenfell, 1876).——Fig. 231,*1c-e. R. kirbyi* (Wachsmuth & Springer), Miss., USA(Iowa); *1c, A*-ray view of crown, $\times 2$ (Wachsmuth & Springer, 1897); *1d, CD*-interray view, $\times 2$ (Wachsmuth & Springer, 1897); *1e,* diagram of part of calyx (radials black, anals and interbrachials stippled (Ubaghs, n).

Acanthocrinus F. A. Roemer, 1850, p. 679 [**A. longispina*; SD Jaekel, 1895, p. 24]. Calyx bowl shaped, with slightly depressed base; faint median ray ridges and long spines on basals, radials, tegmen, and proximal interprimibrachs; slightly depressed interprimibrachs numerous, connecting with tegmen, which is low and composed of small irregular pieces; anus marginal on tegmen and without tube. Arms 10, branching, biserial.

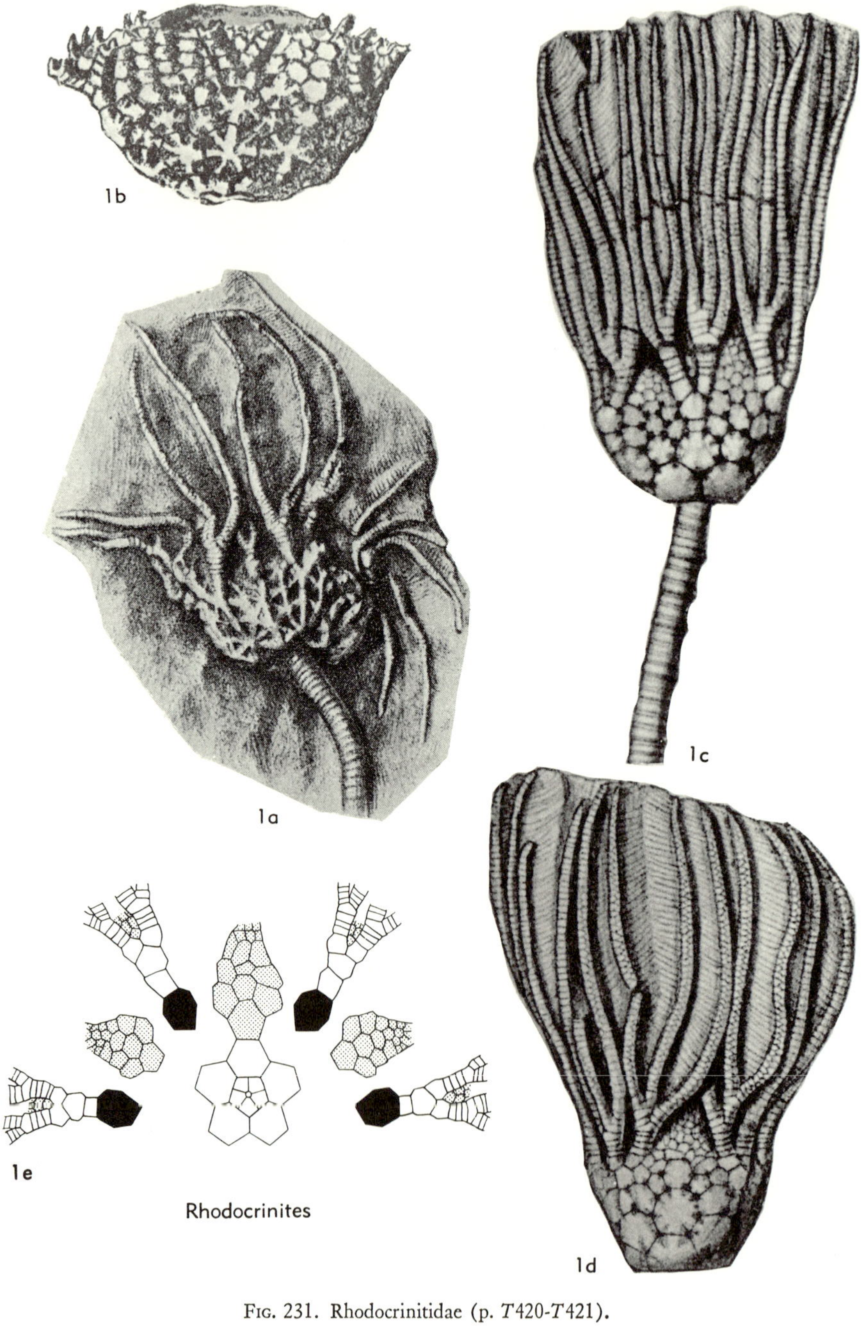

FIG. 231. Rhodocrinitidae (p. T420-T421).

Nudinodals generally prominent and internodals very short. *L.Dev.-M.Dev.*, Eu.-N.Am.-N.Afr.——FIG. 232,*1*. *A. rex* JAEKEL, L.Dev. (Hunsrück-Schiefers), W.Ger.; complete specimen, ×0.5 (Schmidt, 1934).

Atactocrinus WELLER, 1916, p. 239 [**A. wilmingtonensis*; OD]. Calyx bowl shaped; radials separated by 1st interradials in 2 adjoining interrays

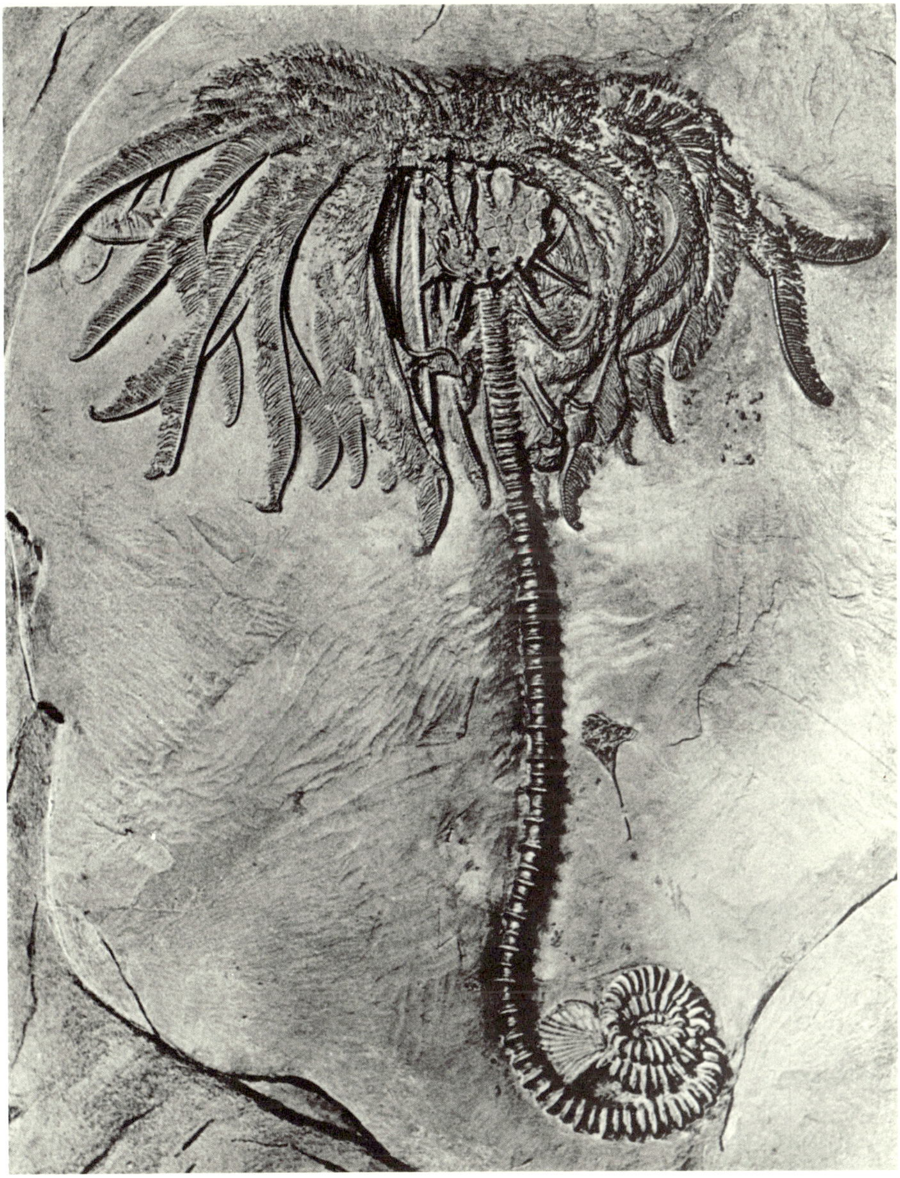

1 Acanthocrinus

FIG. 232. Rhodocrinitidae (p. *T*421-*T*422).

and a 3rd opposite to them but meeting across the 2 remaining interrays; 1st interprimibrach followed by 2 plates; *CD* interray indistinct in cup. Tegmen, arms, and stem unknown. [As suggested by BATHER (1917), this may be an abnormal *Lyriocrinus*.] *U.Ord.(Richmond.)*, USA (Ill.).

Cadiscocrinus KIRK, 1945, p. 346 [**C. southworthi*; OD]. Calyx subcylindrical, with flattened flaring invaginated base, containing small infrabasals and proximal part of very large basals; fixed secundibrachs 2; interprimibrachs 1 or 2 in each interray, that on posterior side largest and bearing median series of anal plates; arms

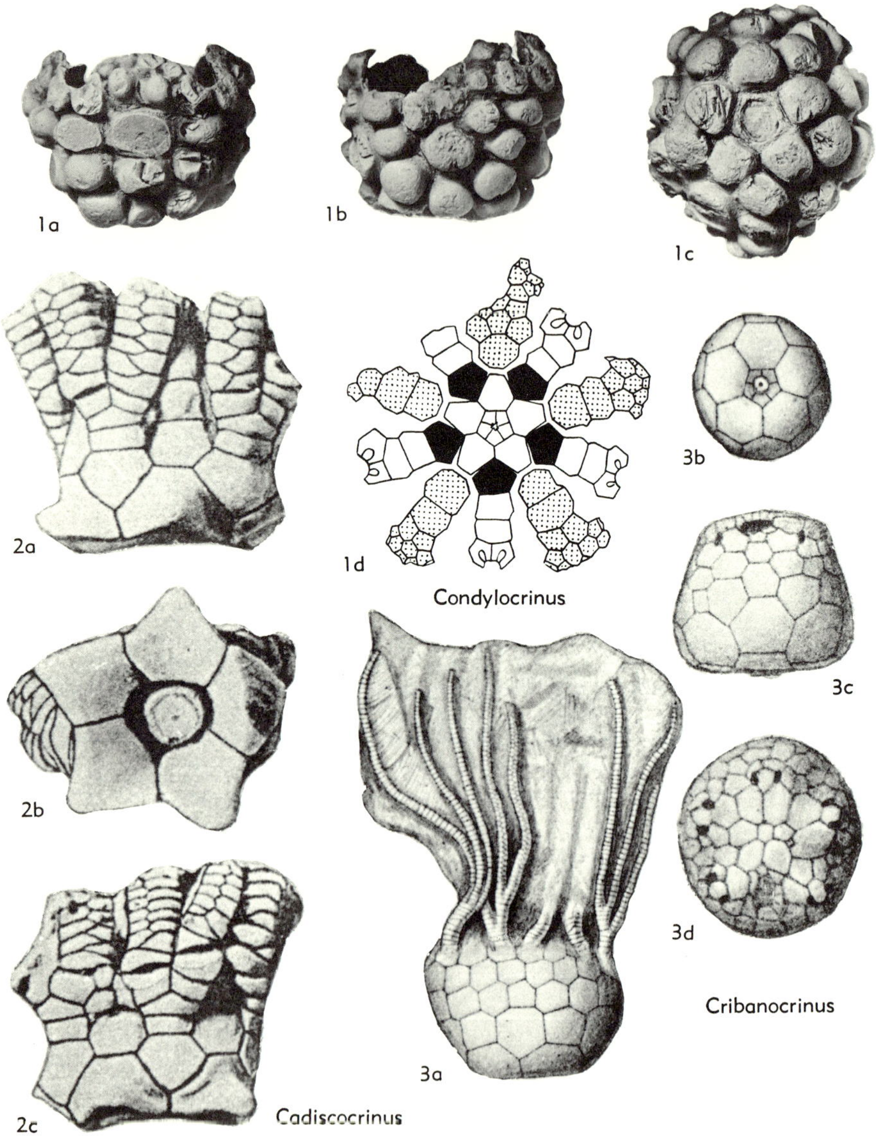

FIG. 233. Rhodocrinitidae (p. T423-T425).

10, biserial. Axial canal of stem round. *M.Dev.* N.Am.(Can.).——FIG. 233,*2*. **C. southworthi*, Arkona Sh.(Hamilton.), Can.(Ont.); *2a-c*, holotype, *CD*-interray, dorsal, and *A*-ray views, ×2 (Kirk, 1945a).

Condylocrinus EICHWALD, 1860, p. 612 [**C. verrucosus*; M]. Calyx subhemispherical; infrabasals small, flat, partly hidden by stem; basals large, widely in contact with 1st interprimibrachs; fixed secundibrachs 1 to each half-ray bearing brachial facet; interprimibrachs running 1, 1, 2, 3 or 4, 1st 2 larger than any other plates of calyx; no intersecundibrachs; posterior side with 3 anal plates in second row, at least 4 in 3rd row and numerous small elements in distal part. Tegmen, arms, and stem unknown; stem insertion round, with small, quinquelobate lumen. *L.Dev.*, USSR (Urals).——FIG. 233,*1*. **C. verrucosus*; *1a-c*, *AB*-interray, *CD*-interray, and dorsal views of theca, ×1.5 (Arendt, 1971c); *1d*, plate diagram (radials black, anals and interbrachials stippled, *CD*-interray directed upward) (Arendt, 1971c, mod.).

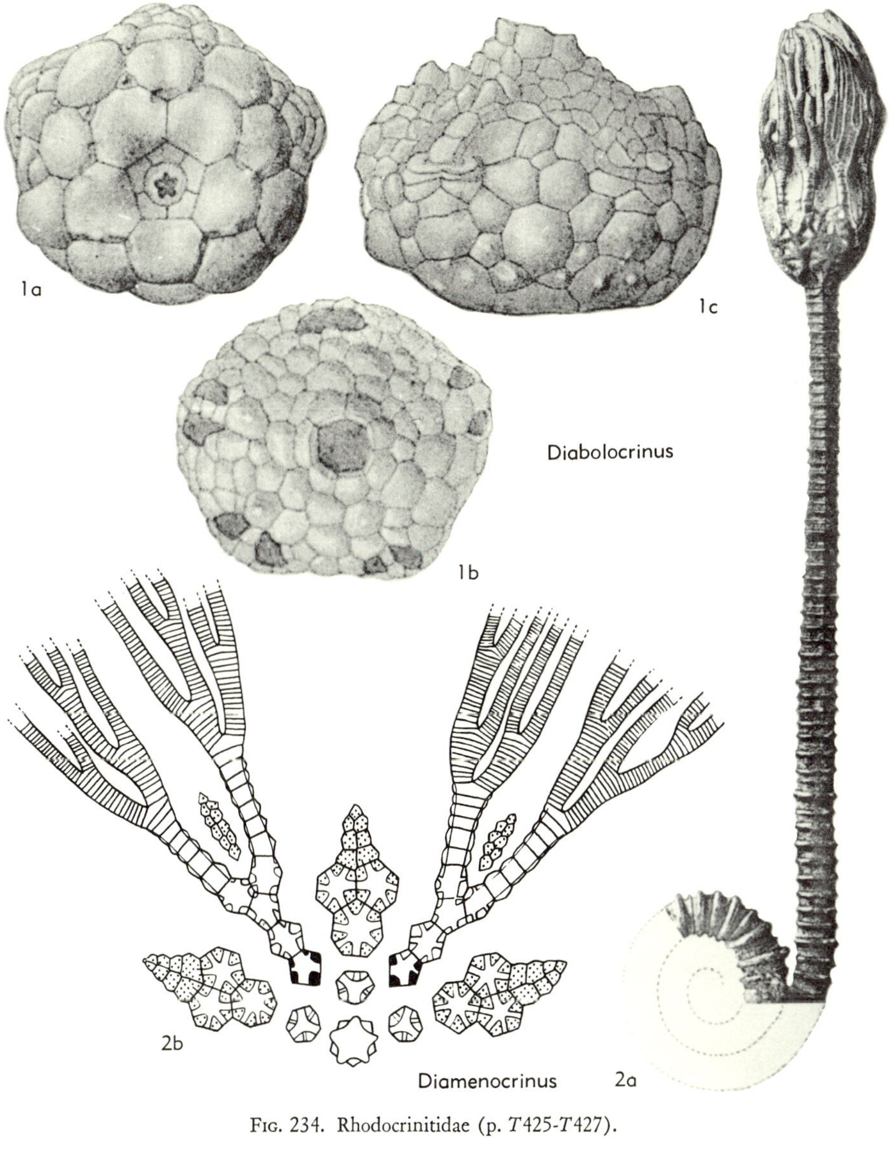

FIG. 234. Rhodocrinitidae (p. T425-T427).

Cribanocrinus KIRK, 1944, p. 13 [**Rhodocrinus wortheni* HALL, 1858b, p. 556; OD]. Similar to *Rhodocrinites* but having subglobose to ovate calyx constricted at arm level; ornamentation lacking or slight; basals and radials large in comparison to primibrachs; fixed secundibrachs 1 or 2 in each half-ray; interradial areas narrow, constricted at arm level; no intertertibrachs; *CD* interray wider than others, rarely with poorly defined median row of plates. Tegmen very small, convex, made up of small plates; small anal protuberance, marginal or excentric. *L.Miss.*, USA(Iowa-N. Mexico); *L.Carb.*, Eu.-E.Australia.——FIG. 233, *3a,b*. **C. wortheni* (HALL), L.Miss.(Burlington Ls.), Iowa; *3a*, lat. view of crown, ×1.5 (Wachsmuth & Springer, 1897); *3b*, dorsal view of calyx, ×1.5 (Wachsmuth & Springer, 1897).——FIG. 233,*3c,d*. *C. urceolatus* (WACHSMUTH & SPRINGER), L.Miss.(Lake Valley Ls.), N.Mexico; *CD* interray and ventral views of theca, ×2 (Wachsmuth & Springer, 1897).

Diabolocrinus WACHSMUTH & SPRINGER, 1897, p. 249 [**D. perplexus*; OD]. Calyx globose, deep, wide basal concavity, involving infrabasals and

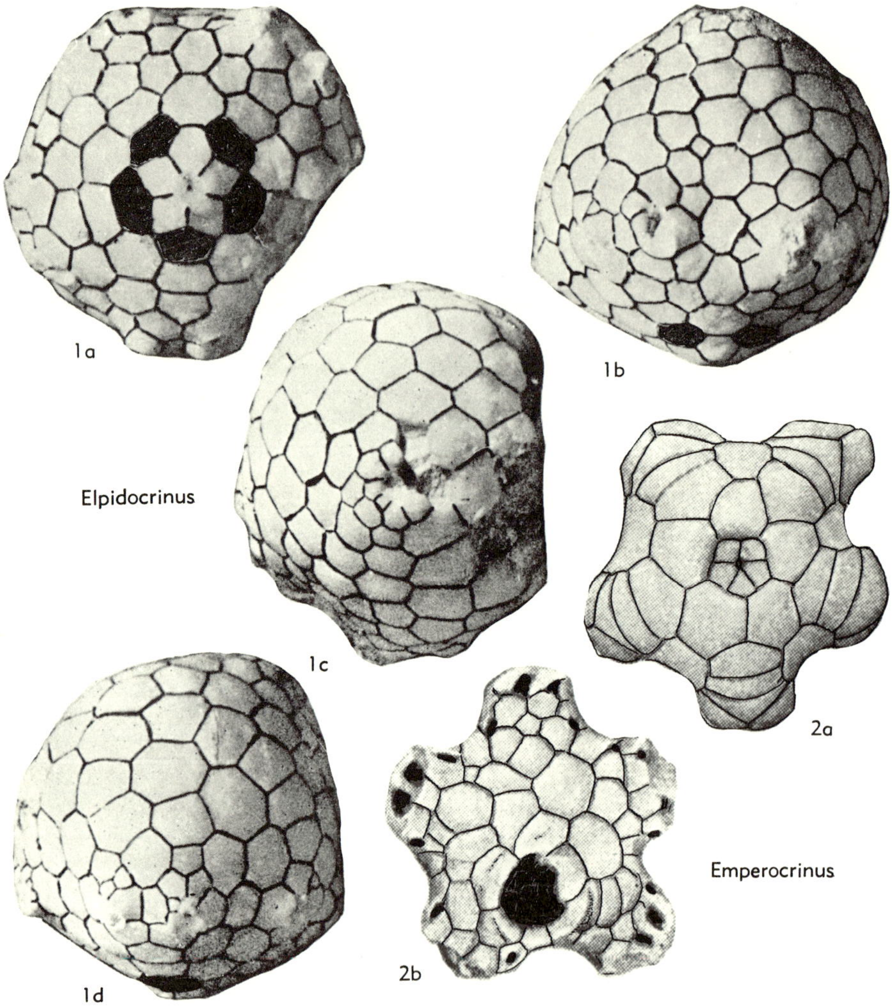

FIG. 235. Rhodocrinitidae (p. T427).

large portions of basals; fixed secundibrachs 1 or 2 in each half-ray; interradial areas wide, connected with tegmen, composed of large plates (1 + 2 + 3) usually surrounded by supplementary plates, which wholly or partly isolate proximal interradials from basals, radials, and primibrachs; *CD* interray wider than others, with 3 plates in 2nd row and 4 or more in 3rd row. Tegmen low, composed of moderate-sized plates with almost central anal tube. Arm openings very large, directed obliquely upward and arranged in groups. Column round or subpentagonal, with large quinquelobate canal. *M.Ord.*, N.Am.-Eu.(Scot.).——FIG. 234,*1a,b*. **D. perplexus*, Chazyan, Tenn.; dorsal and ventral views of theca, ×2 (Wachsmuth & Springer, 1897). ——FIG. 234,*1c*. *D. vesperalis* (WHITE), Chazyan, Tenn.; lat. view, ×2 (Wachsmuth and Springer, 1897).

Diamenocrinus OEHLERT, 1891, p. 836 [**D. jouani*; M]. Calyx subglobose or subcylindrical; plates thick, tumid, generally with stellate ornamentation; some species with each proximal plate bearing fixed spine or node; median ray ridges more or less conspicuous; each half-ray containing 5 to 8 fixed secundibrachs. Tegmen flat, many-plated. Arms biserial, repeatedly branching; brachials very short, cuneate, exteriorly in zigzag. Stem pentagonal, axial canal quinquelobate. *L. Dev.*, Eu.-?N.Afr.-?Sib.——FIG. 234,2. **D.*

jouani, Manche, France; *2a,* almost complete specimen, ×1 (Oehlert, 1891); *2b,* plate diagram of cup and post. arms (Oehlert, 1891).

Elpidocrinus STRIMPLE, 1963, p. 77 [**E. tholiformis*; OD]. Calyx shallow, conical; infrabasals small, not visible in side view and probably entirely covered by proximal columnals; interprimibrachs running 1, 2, 3, connected with tegmen; *CD* interray with 3 anal plates in second row; fixed secundibrachs 2 in each half-ray; arm openings directed outward, paired but with proximal extremity of one intersecundibrach separating them. Tegmen high, made of rather large plates; anal opening almost central but somewhat displaced rearward. Stem insertion round, lumen small and quinquelobate. *M.Sil.,* USA(Okla.).——FIG. 235,*1*. **E. tholiformis*; Niagaran (Henryhouse F.); *1a-d,* dorsal (radials black), *CD*-interray, *E*-ray, and ventral views of holotype, ×2 (Strimple, 1963a).

Emperocrinus MILLER & GURLEY, 1895, p. 42 [**E. indianensis*; OD]. Calyx low, pentagonal in outline in dorsal and ventral views; infrabasals ?3-5, interradial areas large, depressed, connecting with tegmen, each one occupied by single large plate; *CD* interray wider, with 3 anal plates in second row; arm openings 10; fixed pinnule openings few, apparently one to each arm on outer side, close to arm-base. Tegmen composed of relatively large plates; posterior anal tube. Lumen of stem insertion small, quinquelobate. *M.Sil.,* USA(Ind.).——FIG. 235,*2*. **E. indianensis,* Niagaran (Laurel Ls.); *2a,b,* dorsal and ventral views, ×2 (Springer, 1926a).

Gilbertsocrinus PHILLIPS, 1836, p. 207 [**G. calcaratus*; SD BASSLER, 1938, p. 102] [=*Ollacrinus* CUMBERLAND, 1826 *(nom. nud.)*; *Ollacrinites* DE BLAINVILLE, 1834, p. 659 *(nom. van.)*; *Gilbertsocrinites* BRONN, 1848, p. 530 *(nom. van.)*; *Goniasteroidocrinus* LYON & CASSEDAY, 1859, p. 233 (type, *G. tuberosus*; M); *Trematocrinus* HALL, 1860c, p. 70 (type, *T. typus*; OD); *Goniastroidocrinus* MARSCHALL, 1873, p. 400 *(nom. null.)*]. Calyx usually short, subcylindrical, basally excavate; interradial areas large, many-plated; *CD* interradial (primanal) usually not distinct in cup. Tegmen flat or low hemispherical, composed of numerous plates, with 5 interradial depressions in some species, and extended into 10 hollow tubular appendages, which taper gradually to a point; these appendages may remain distinct for their whole length, but generally those of same interray coalesce for some distance and then separate, in some late species branching once or twice; anus subcentral, directly through tegmen. Arms slender, pendent or folding over tegmen, branching, typically biserial. Nodals prominent. *M.Dev.-L.Carb.,* W.Eu.-N.Am.——FIG. 236,*1a-c*. **G. calcaratus,* L.Carb.(Tournais.), Eng.; *1a-c,* dorsal, *C*-ray, and ventral views of calyx, ×1.5 (Wright, 1955-60).——FIG. 236,*1d-f*. *G. tuberosus* (LYON & CASSEDAY), *L.Miss.,* USA(Ind.-Iowa-Ky.); *1d,e,* lat., and ventral views of theca, ×1 (Wachsmuth & Springer, 1897); *1f,* articular face of ossicles belonging to a tubular appendage, ×5 (Van Sant & Lane, 1964).

[The tubular appendages of *Gilbertsocrinus* are a feature that stands alone among crinoids. Each one is composed of a single series of cylindrical ossicles (primitive condition) or of one row of plates aborally and two rows of plates adorally; these interlock like covering plates (Fig. 236,*1d,e*). Where the tubes are joined, the number of elements appears to be doubled. No ventral groove is observed. Each tube is perforated longitudinally by a central canal; when the appendages meet near the tegmen, their individual central canals remain separate. In some species, according to WACHSMUTH & SPRINGER (1897), each central canal connects with the subtegminal ambulacral groove of the adjacent ray within the theca. The correctness of this observation has been questioned by VAN SANT (VAN SANT & LANE, 1964), but confirmed by LANE (personal communication, 1969). This tends to demonstrate that these appendages, while interradial in position, are radial in origin and morphological relation. The tubular extensions of *Gilbertsocrinus* have been erroneously taken for the arms by early workers (HALL, PHILLIPS, LYON, CASSEDAY, ROFE), until the true arms were discovered by MEEK & WORTHEN (1866a, p. 217). Although apparently they do not open to the exterior, they have been generally regarded as analogous to the thecal pores of *Batocrinus* and other genera, even if they did not perform the same function. SPRINGER (1917) thought that they resulted from hypertrophy of proximal pinnules, which approached the cirri in structure and possibly in function. On the other hand, their position and construction have suggested to VAN SANT (VAN SANT & LANE, 1964) that they were modified tegminal extensions, provided with limited mobility and serving directly or indirectly to 1) help support or anchor the theca, 2) protect the delicate arms, and 3) contribute to producing circulation of water around the theca.]

Lyriocrinus HALL, 1852, p. 197 [**Marsupiocrinites? dactylus* HALL, 1843, p. 114; M]. Calyx depressed globose; infrabasals very small, in deep pit; interprimibrachs arranged 1, 2, 1, ordinarily connected with tegmen, but not invariably touching basals; intersecundibrachs none or only 1 in a ray; *CD* interray indistinct or with single anal plate; fixed secundibrachs 2 in each half-ray. Tegmen flat, somewhat depressed in interrays, composed of many small plates, among which some larger ones (orals?) may be differentiated; probably small subcentral anal tube. Arms 10, simple, stout, biserial, rising in straight line with sides of calyx. Stem cylindrical, formed of regularly alternating nudinodals and internodals. *M. Sil.,* N.Am.(USA)-Eu.(Eng.).——FIG. 237,*1*. **L. dactylus* (HALL), Niagaran, USA(N.Y.)-Can.(Ont.); *1a,* oblique view of holotype crown, ×1 (Wachsmuth & Springer, 1897); *1b,* side view of crown showing pinnules, ×1 (Wachsmuth & Springer, 1897).

Maquoketocrinus SLOCOM, in SLOCUM & FOERSTE, 1924, p. 330 [**M. ornatus*; M]. Calyx bowl shaped; infrabasals forming flat pentagonal disc extending beyond column; no fixed secundibrachs; interprimibrachs arranged 1, 2, 3, connecting with tegmen; *CD* interray indistinct. Stem insertion circular in outline, with pentagonal lumen. [RAMSBOTTOM (1961), has suggested that this genus probably represents an unusual *Lyriocrinus*.] *U.Ord.,* USA(Iowa).——FIG. 237,*2*. **M. ornatus,* Maquoketa Sh.(Richmond.), Iowa; *2a,b,* lat. and

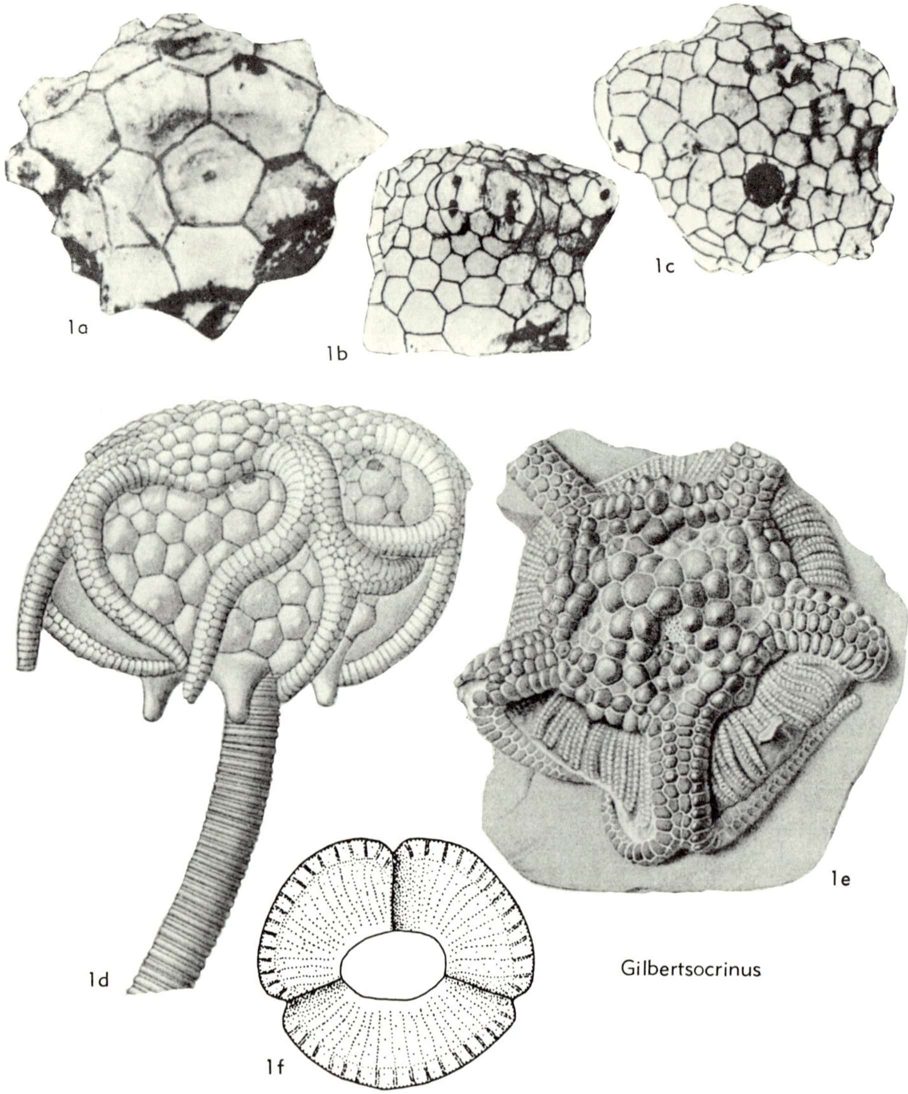

FIG. 236. Rhodocrinitidae (p. T427).

dorsal views of calyx (Field Museum Chicago; Ubaghs, n), ×5; *2c,* plate diagram of calyx (radials black, interbrachials stippled) (Slocom & Foerste, 1924).

Monstrocrinus W. E. SCHMIDT, 1942, p. 213 [**M. securifer*; OD]. Similar to *Acanthocrinus* but bearing gigantic fixed spines on basals, radials, and interprimibrachs of 1st and 2nd orders. No stem(?). *L.Dev.-M.Dev.,* Eu.(Ger.).——FIG. 237, *3.* **M. securifer,* M.Dev.; *3a,b,* part of calyx and isolated spine, ×1 (Schmidt, 1942).

Ophiocrinus SALTER, 1856, p. 223 (*non* CHARLESWORTH, 1865; *non* SEMPER, 1868; *non* ANGELIN, 1878) [**O. stangeri*; M]. Calyx moderately bowl shaped, slightly depressed below; infrabasals flat, small, entirely concealed by stem; basals prominent, overhanging stem; radials penetrating deeply between basals; fixed ray plates prominent, not ridged, including 2 primibrachs per ray, 2 to 3 secundibrachs per half-ray, and 1 to 2 tertibrachs to each arm; interbrachial areas well developed, slightly depressed, in contact with tegmen, composed of small, numerous plates; intersecundibrachs and intertibrachs present; extra plates in

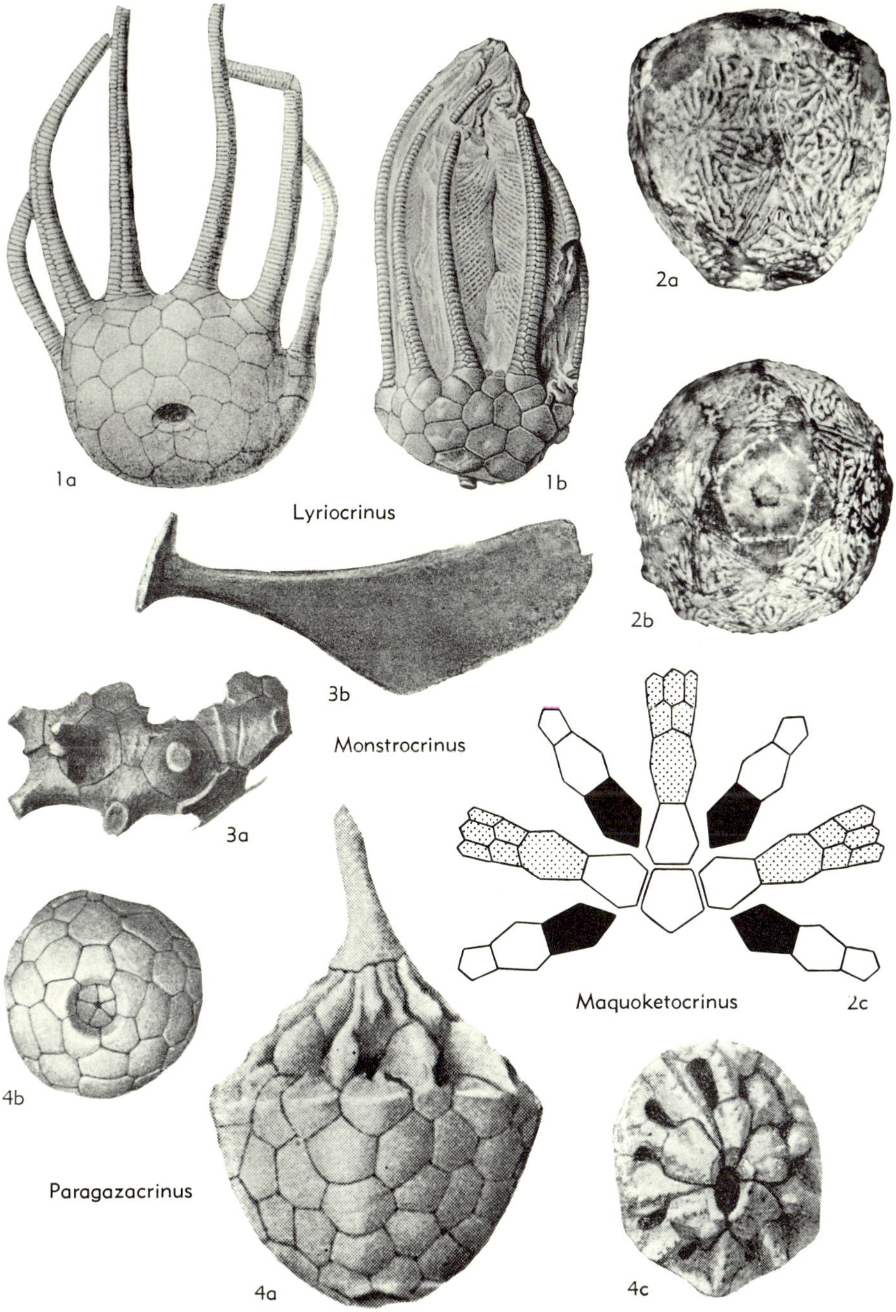

FIG. 237. Rhodocrinitidae (p. *T*427-*T*428, *T*430).

CD interray. Tegmen unknown. Free arms 4 to each ray, stout, simple, uniserial; brachials wedge shaped. Column circular in cross section, heteromorphic. *L.Dev.*, S.Afr.——FIG. 238,2. **O. stangeri*, Bokkeveld Series, Cape Prov.; holotype, Sedgwick Museum Cambridge no. A 3441, latex

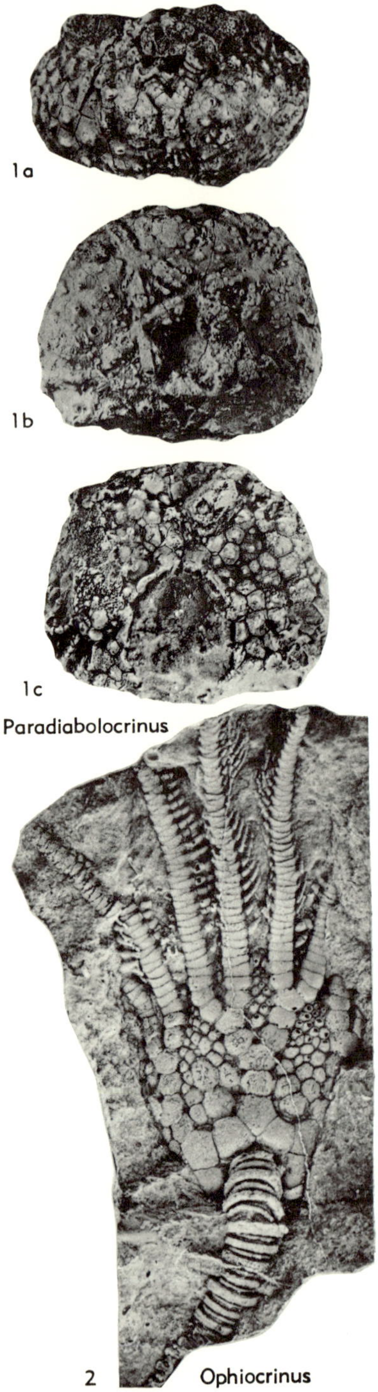

FIG. 238. Rhodocrinitidae (p. T428-T430).

cast of C-ray view, ×1.5 (Ubaghs, n).

Paradiabolocrinus BROWER & VEINUS, 1974, p. 92 [**P. irregularis*; OD]. Cup depressed globular or widely conical, with basal concavity; median-ray ridges and stellate ridges on basals commonly present; fixed secundibrachs 3 to 5 per half-ray; interprimibrachs very numerous (17 to 36) irregularly shaped and placed, connecting with tegmen; intersecundibrachs several; *CD* interray wider and containing greater number of plates than other interrays. Tegmen many-plated, smoothly convex or with ambulacral areas faintly lobate. Two free arms per ray, originating from short lobate protruding arm bases. Free arms and column unknown. *M.Ord.,* USA(Tenn.).——FIG. 238,*1*. **P. irregularis; 1a-c, D*-ray, basal, and tegminal views, ×1 (Brower & Veinus, 1974).

Paragazacrinus SPRINGER, 1926, p. 22 [**P. rotundus*; OD]. Calyx medium bowl shaped; infrabasals small, in vertical-sided pit; interradial areas in contact with tegmen, composed of 4 or 5 relatively large plates; *CD* interray with 1 or 2 anal plates; fixed secundibrach 1 in each half-ray. Tegmen forming pyramid divided into 10 grooves by ridges and surmounted by tall erect spine. Arms 10, resting in tegminal grooves. *M.Sil.,* USA(Ind.).——FIG. 237,*4*. **P. rotundus,* Niagaran (Laurel Ls.); *4a,* lat. view of complete theca, ×3 (Springer, 1926a); *4b,c,* dorsal and tegminal views of theca, central spine on tegmen lacking, ×3 (Springer, 1926a).

Rhipidocrinus BEYRICH in ZITTEL, 1879, p. 376 [**Rhodocrinites crenatus* GOLDFUSS, 1831, p. 212; M] [=*Ripidocrinus* WACHSMUTH & SPRINGER, 1881, p. 205 (379) *(nom. van.)*]. Calyx low, flat-bottomed, globose; basals in lateral contact all around or with variable number of them separated from each other by radials in contact with infrabasals; interprimibrachs 5 to 9 in different interrays separated from tegmen by fixed ramules; *CD* interray indistinct; intersecundibrachs 1 to 5, not connected with tegmen; fixed secundibrachs usually 2 in half-ray, last one axillary and bearing ramule on its outer shoulder and main arm trunk on its inner shoulder, proximal part of both ramule and arm trunk being incorporated in calyx. Tegmen low, composed of numerous small plates which decrease in size toward periphery; anus excentric. Each ray produced into 2 long, heavy, uniserial trunks, carrying biserial, pinnuliferous ramules alternating on opposite sides. Stem cylindrical, comprising thin nudinodals with broadened peripheral rims and thin internodals concealed by nodals; axial canal pentalobate. *M.Dev.,* Eu.(Ger.-Eng.). [This genus stands apart from other *Rhodocrinitidae,* in organization of the calyx base and peculiar arm structure. It is retained in this family until its real affinities are known.]——FIG. 239,*1*. **R. crenatus* (GOLDFUSS), Givet., Ger., *1a,* lat. view of crown, ×1 (Schultze, 1867); *1b,c,* dorsal and ventral

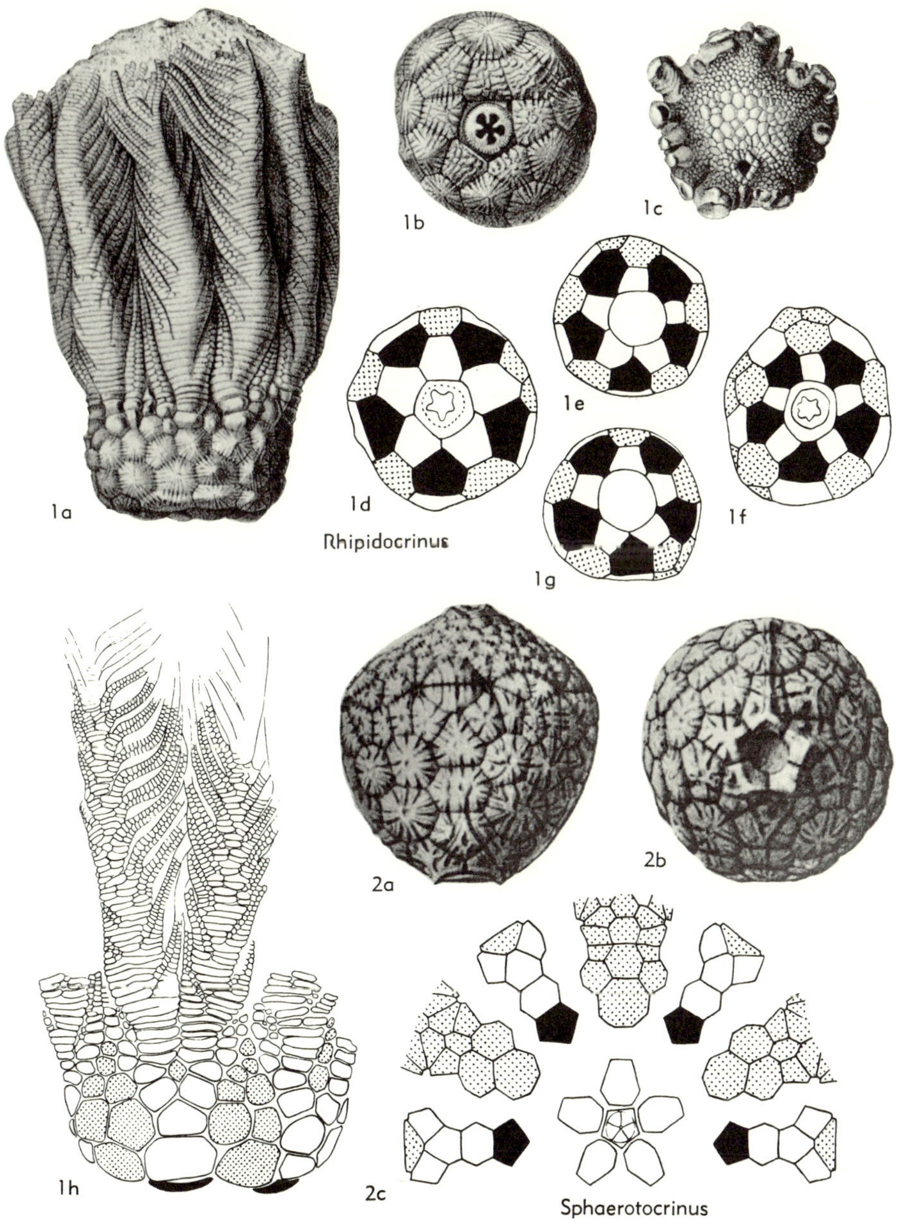

FIG. 239. Rhodocrinitidae (p. *T*430-*T*432).

views of theca, ×1 (Schultze, 1867); *1d-g,* diagrams of various arrangements of plates in base of calyx (radials black, interbrachials stippled) (Breimer, 1960).——FIG. 239,*1h.* *R.* sp. *cf.* *R. perloricatus* W. E. SCHMIDT, Givet., Ger.; *E*-ray view of incompl. crown (radials black, interbrachials stippled), ×1 (Breimer, 1960).

Sphaerotocrinus GOLDRING, 1923, p. 99 [**S. ornatus*; OD]. Calyx subglobose, with shallow basal depression; infrabasals very small, concealed by stem impression; single fixed secundibrach in each half-ray; 1st interprimibrach large, followed in 2nd row by 2 plates and in 3rd row by 3 plates which meet tegmen; *CD* interray with

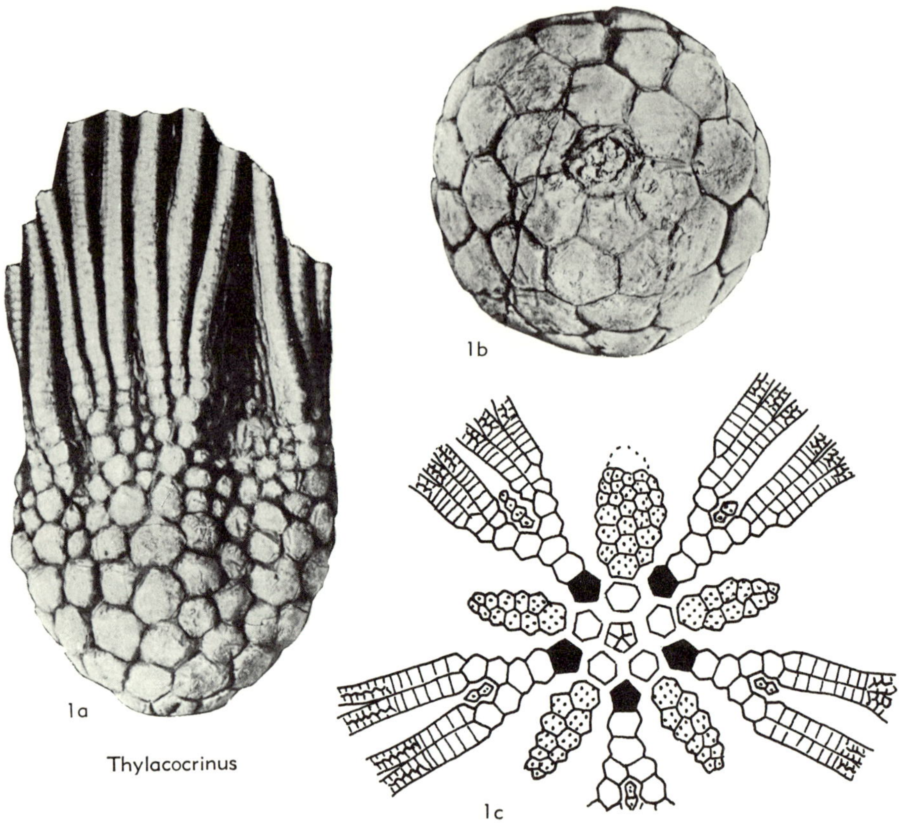

Fig. 240. Rhodocrinitidae (p. *T*432).

median line of anal plates bordered on each side by single row of rather irregular, small plates; triangular intersecundibrachs 1 in each ray; arm openings 10, very small. Tegmen strongly arched, composed of numerous small plates, with central anal tube. Arms unknown, but probably slender and abruptly separated from calyx. Stem insertion circular; axial canal small, round. *L.Dev.*, USA (Pa.).——Fig. 239,*2*. **S. ornatus*, Helderberg.; *2a,b*, *A*-ray and dorsal views of theca, ×4 (Goldring, 1923); *2c*, plate diagram of part of calyx (radials black, and interbrachials stippled) (Goldring, 1923).

Thylacocrinus Oehlert, 1878, p. 6 [**T. vannioti*; M]. Calyx large, medium to high conical or bowl shaped; infrabasals 5, small; fixed secundibrachs 2; fixed tertibrachs 6 to 9; interprimibrachs merging with tegmen, numerous; *CD* interray wider and with a sagittal series of supplementary plates; intersecundibrachs few. Tegmen low convex, composed of a large number of small pieces. Free arms 20 to 24, generally 4 per ray, occasionally 5 or 6 in *C* and *D* rays, biserial, simple. Stem heteromorphic, round or pentagonal in cross section. *L.Dev.-M.Dev.*, Eu.-N.Afr.-N.Am.-?E.Australia.——Fig. 240,*1*. **T. vannioti*, L.Dev., France; *1a*, *D*-ray view of crown, ×1 (Le Menn, 1974); *1b*, basal view of lectotype, ×1 (Le Menn, 1974); *1c*, plate diagram of part of theca and arms (radials black, interbrachials and anals stippled) (Le Menn, 1974, mod.).

Trichinocrinus Moore & Laudon, 1943, p. 265 [**T. terranovicus*; OD]. Calyx bowl shaped; low median ray ridges present; infrabasals in basal pit; large interprimibrachs at least 1 in each interray. Rami 10, simple, uniserial; brachial externally cuneate; pinnules bearing appendages or pinnulets which may represent unusually long, weakly calcified tentacles. Stem transversely round, with thick projecting nodals and thin internodals. *M.Ord.*, Can.(Newf.).——Fig. 241, *1*. **T. terranovicus*; side view of crown, *ca.* ×5 (Moore & Laudon, 1943b).

Family UNCERTAIN

Paulocrinus Springer, 1926, p. 22 [**P. biturbinatus*; OD]. Calyx truncate conical, spreading to horizontally directed arm bases which form projecting rim with fixed pinnulars; infrabasals visible in side view, unequal, 2 of them truncated by radials; radials unequal; primibrachs usually 2, exceptionally only 1 in ray; fixed secundibrachs 2 to 4 in each half-ray; fixed tertibrachs present; arm openings 4 to ray; interbrachials few, separated from tegmen by fixed pinnulars; *CD* interray barely differentiated. Tegmen conical, many-plated, culminating in large excentric anal tube. [This genus, based on a single possibly abnormal specimen, is of uncertain affinity.] *M.Sil.,* USA (Ind.).——FIG. 242,*1*. **P. biturbinatus,* Niagaran (Laurel Ls.); *1a,b, CD* interray and basal views of theca, ×2 (Springer, 1926a); *1c,* plate diagram of calyx base (radials black) (Springer, 1926a).

Siderocrinus WOOD, 1909, p. 104 [**S. ornatus*; M] [=*Siderocrinites* TROOST in WOOD, 1909, p. 104 *(nom. nud.)*]. Deep basal excavation; infrabasals 5, large; basals thick; radials quadrangular, each bearing long spine. Other parts of theca, arms, and stem unknown. *M.Sil.,* USA(Tenn.). [This genus, founded on a single specimen, is insufficiently known for trustworthy recognition.]

Superfamily DIMEROCRINITACEA Zittel, 1879

[*nom. correct.* BROWER, 1973, p. 445 (*pro* Dimerocriniticae UBAGHS, 1953, p. 737, *nom. transl. ex* family Dimerocrinidae ZITTEL, 1879, p. 344)]

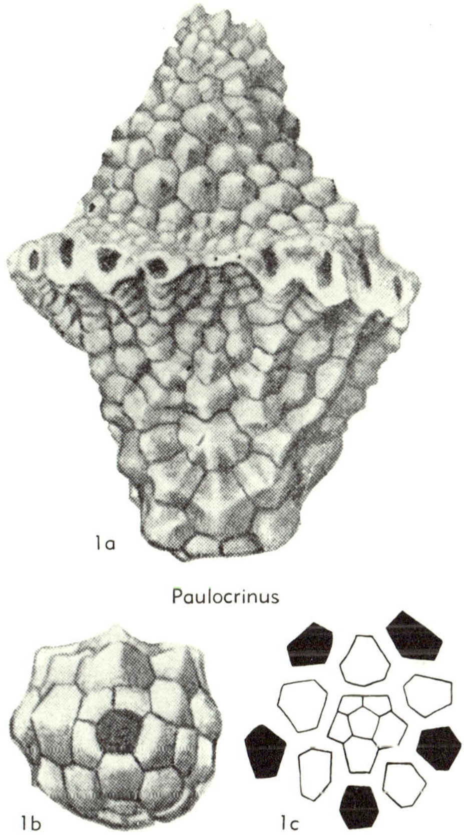

FIG. 242. Family Uncertain (p. *T*432-*T*433).

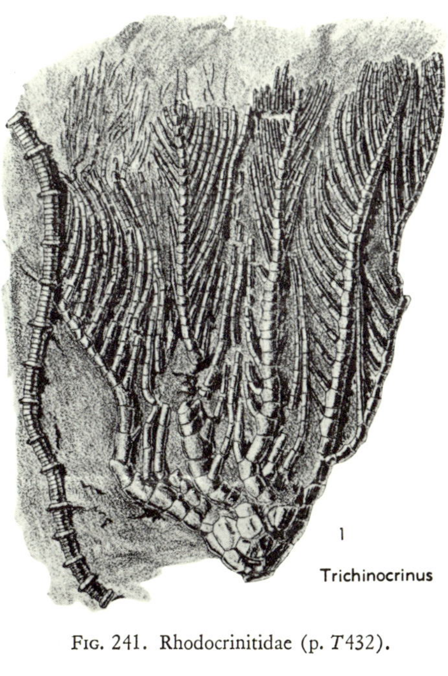

FIG. 241. Rhodocrinitidae (p. *T*432).

Radials adjoining each other except for separation by primanal on posterior side of calyx, primanal resting on posterior basal. *U.Ord.; M.Sil.-U.Dev.*

Family DIMEROCRINITIDAE Zittel, 1879

[*nom. correct.* BASSLER, 1938, p. 25 (*pro* Dimerocrinidae ZITTEL, 1879, p. 344, 367)] [=Eucrinidae ANGELIN, 1878, p. 24; Glyptasteridae WACHSMUTH & SPRINGER, 1885, p. 89 (314); Thysanocrinidae WACHSMUTH & SPRINGER, 1897, p. 188; Ptychocrinidae JAEKEL, 1918, p. 41]

Median ray ridges present in primitive members; interprimibrachials well developed, regular, depressed or not, connecting with tegmen, first one typically followed by two plates; primanal supporting three plates in next higher row; generally with sagittal series of ridged or unridged extra plates, leading to anus in *CD* interray. Arms simple or branching, uniserial, biserial, or with compound brachials. Stem transversely round to substellate. *U.Ord.; M.Sil.-U.Dev.*

Dimerocrinites PHILLIPS in MURCHISON, 1839, p. 674 [**D. decadactylus*; SD ROEMER, 1855, p. 237] [=*Dimerocrinus* MÜLLER, 1843, p. 208 *(nom. van.)*; *Glyptaster* HALL, 1852, p. 187 (type, *G. brachiatus*; M); *Thysanocrinus* HALL, 1852, p. 188 (type, *T. liliiformis*; SD WACHSMUTH & SPRINGER, 1897, p. 191); *Harmocrinus* ANGELIN, 1878, p. 22 (type, *H. longimanus*; M); *Eucrinus* ANGELIN, 1878, p. 24 (type, *E. laevis*; SD WACHSMUTH & SPRINGER, 1881, p. 197 (371)]. Calyx low to medium conical; plates moderately thick to heavy; median ray ridges more or less conspicuous and interray areas depressed; sagittal series of flush or keeled anal plates generally present joining tegmen, which contains distinct orals and ambulacrals but lacks anal tube. Biserial rami 2 to 4 in each ray, generally free above 3rd brachial. Stem circular in section with quinquelobate axial canal. *M.Sil.-M.Dev.*, Eu.-N.Am.——FIG. 243,*1*. **D. decadactylus*, M.Sil.(Wenlock Ls.), Eng.; *1a*, post. view of crown, X2 (Bather, 1900a); *1b*, plate diagram of part of calyx (radials black, anals and interbrachials stippled) (Ubaghs, n).

Ambicocrinus KIRK, 1945, p. 353 [**Thysanocrinus arborescens* TALBOT, 1905; OD]. Calyx somewhat elongate; thecal plates thin, giving an incompetent structure; basals comparatively large, without median ray ridges; interrays not depressed; intersecundibrachs numerous; 3 or more fixed secundibrachs. Free arms compactly biserial, dividing once or twice. Column substellate in section. *L.Dev.*, N.Am.——FIG. 243,*2*. **A. arborescens* (TALBOT), Coeymans Ls., USA(N.Y.); *D*-ray and *CD*-interray view of crown with attached stem, X1.5 (Goldring, 1923).

Cyphocrinus S. A. MILLER, 1892, p. 50 [**C. gorbyi*; M] [=*Hyptiocrinus* WACHSMUTH & SPRINGER, 1892, p. 138 (type, *H. typus*; OD)]. Calyx low, wide, and recurving from narrow obconical base to spreading inverted bowl; fixed pinnule openings 4 to 6 in each interray. Tegmen with central and spiniferous radial dome plates. Arms 10, biserial, apparently stout and pendent (unknown beyond small remnants). *M.Sil.*, USA (Ind.).——FIG. 243,*3*. **C. gorbyi*, Niagaran (Laurel Ls.); *3a*, dorsal view of holotype theca, X1.5 (Wachsmuth & Springer, 1897); *3b,c*, lat. and ventral views of theca, X1.5 (Springer, 1926a).

Eudimerocrinus SPRINGER, 1926, p. 14 [**E. multibrachiatus*; OD]. Like *Dimerocrinites* but with free arms branching several times and column subpentagonal in cross section. *M.Sil.*, USA (Tenn.).——FIG. 243,*4*. **E. multibrachiatus*, Niagaran; *A*-ray view, X1.5 (Springer, 1926a).

Griphocrinus KIRK, 1945, p. 350 [**Rhodocrinus (Acanthocrinus) nodulosus* HALL, 1862, p. 126; OD]. Calyx subturbinate to bowl shaped, composed of thick, commonly nodose, plates; median ray ridges more or less distinct; infrabasals 5 (4 in *G. ovetensis*), small, but visible in side view; basals and radials large; in some specimens, all radials in lateral contact, in others all or variable number of radials separated by 1st interprimibrachs in contact with basals; interrays relatively narrow, with 2 plates in 2nd row and 2 or 3 in 3rd row; *CD* interray wide, with sagittal series of anal plates; intersecundibrachs present; fixed secundibrachs 1, 2 or more. Tegmen low, many-plated, with excentric anal opening; anal tube present in at least one species. Arms biserial, branching dichotomously twice or more. Column large, circular in section, with large quinquelobate lumen. *Dev.*, N.Am.-Eu.-N.Afr.——FIG. 243,*5*. **G. nodulosus* (HALL), M.Dev.(Moscow F.), USA (N.Y.); lat. view of holotype crown, X1 (Goldring, 1923).

Macarocrinus JAEKEL, 1895, p. 36 [**M. springeri*; M]. Calyx cupuliform or widely conical; no median ray ridges; interprimibrachs not depressed. Tegmen seemingly pliant, composed of coarse platelets, irregular in size and arrangement, with possibly oral and radial dome plates in at least some species. Arms uniserial, dichotomously branching 1 to 3 times. Column circular in cross section. *Dev.*, Eu.——FIG. 244,*1*. **M. springeri*, L.Dev.(HunsrückSchiefer), Ger.; *1a*, dorsal view of holotype crown, X1 (Jaekel, 1895); *1b*, plate diagram of calyx and lower part of arms (radials black, interbrachials stippled), X2.5 (Schmidt, 1934).

Pterinocrinus GOLDRING, 1923, p. 86 [**P. quinquenodus*; OD]. Similar to *Dimerocrinites* but having compound bipinnulated brachials. *Dev.*, N.Am.-Eu.——FIG. 244,*2a*. **P. quinquenodus*, U.Dev.(L.Chemung.), USA(N.Y.); lat. view of crown, X2 (Goldring, 1923).——FIG. 244,*2b*. **P. decembrachiatus* BREIMER, L.Dev., Spain; lat. view of crown, X1.5 (Breimer, 1962).

Ptychocrinus WACHSMUTH & SPRINGER, 1885, p. 99 (321) [**Gaurocrinus splendens* S. A. MILLER, 1883b, p. 230; SD WACHSMUTH & SPRINGER, 1897, p. 197]. Calyx obconical, thinly plated, with well-developed median ray ridges; interprimibrachs depressed; distinct anal ridge. Tegmen incompetent, composed of numerous small plates. Arms slender, uniserial, simple or branching. Column round in section. *U.Ord.*, N.Am.(USA).——FIG. 244,*3*. **P. splendens* (S. A. MILLER), Girardeau Ls., USA(Mo.-Ill.); lat. view of crown and attached stem, X3 (Wachsmuth & Springer, 1897).

Family LAMPTEROCRINIDAE Bather, 1899

[Lampterocrinidae BATHER, 1899b, p. 923]

Calyx like that of Dimerocrinitidae in general structure but asymmetrical owing to bulged posterior interray. *M.Sil.*

Lampterocrinus ROEMER, 1860, p. 37 [**L. ten-*

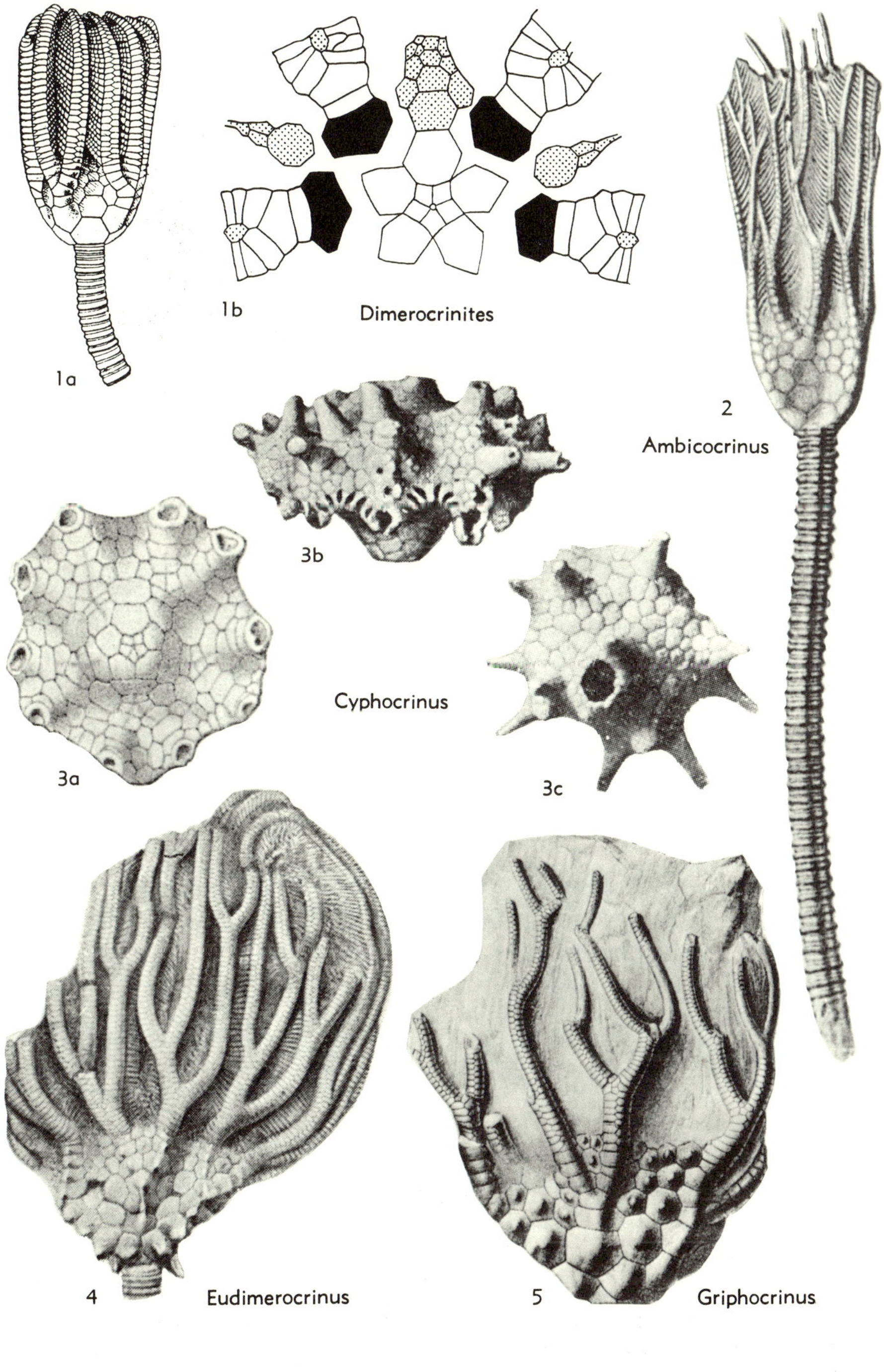

FIG. 243. Dimerocrinitidae (p. *T*434).

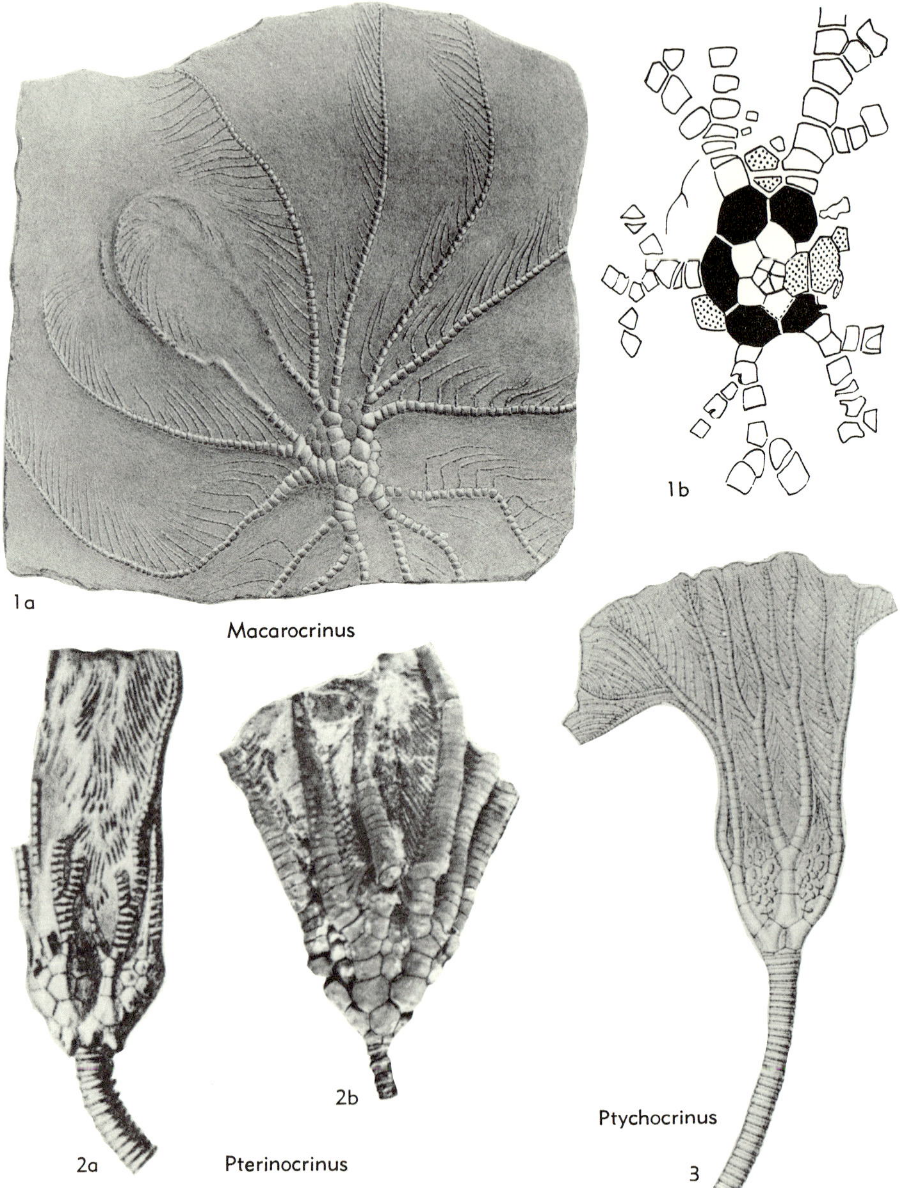

FIG. 244. Dimerocrinitidae (p. T434).

nesseensis; M] [=*Balanocrinites* TROOST, 1850b, p. 419 *(nom. nud.)* (type, *B. sculptus*; M)] [=*Balanocrinus* HALL, 1861d, p. 22 *(non* AGASSIZ, 1845) *(nom. van.)*]. Calyx high bowl shaped; infrabasals and basals rather high spreading; interprimibrachials few. Tegmen composed of numerous convex plates passing into strong subcentral anal tube. Rays produced into 5 tubular arm trunks bearing pinnuliferous ramuli alternately on opposite sides; brachials of ramuli compound and bipinnulate. Column sharply pentagonal in cross section. *M.Sil.*, USA(Tenn.).——FIG. 245,*1*. **L. tennesseensis,* Niagaran (Beech River F.); *1a,b, CD-* and *DE*-interray views of theca, ×1

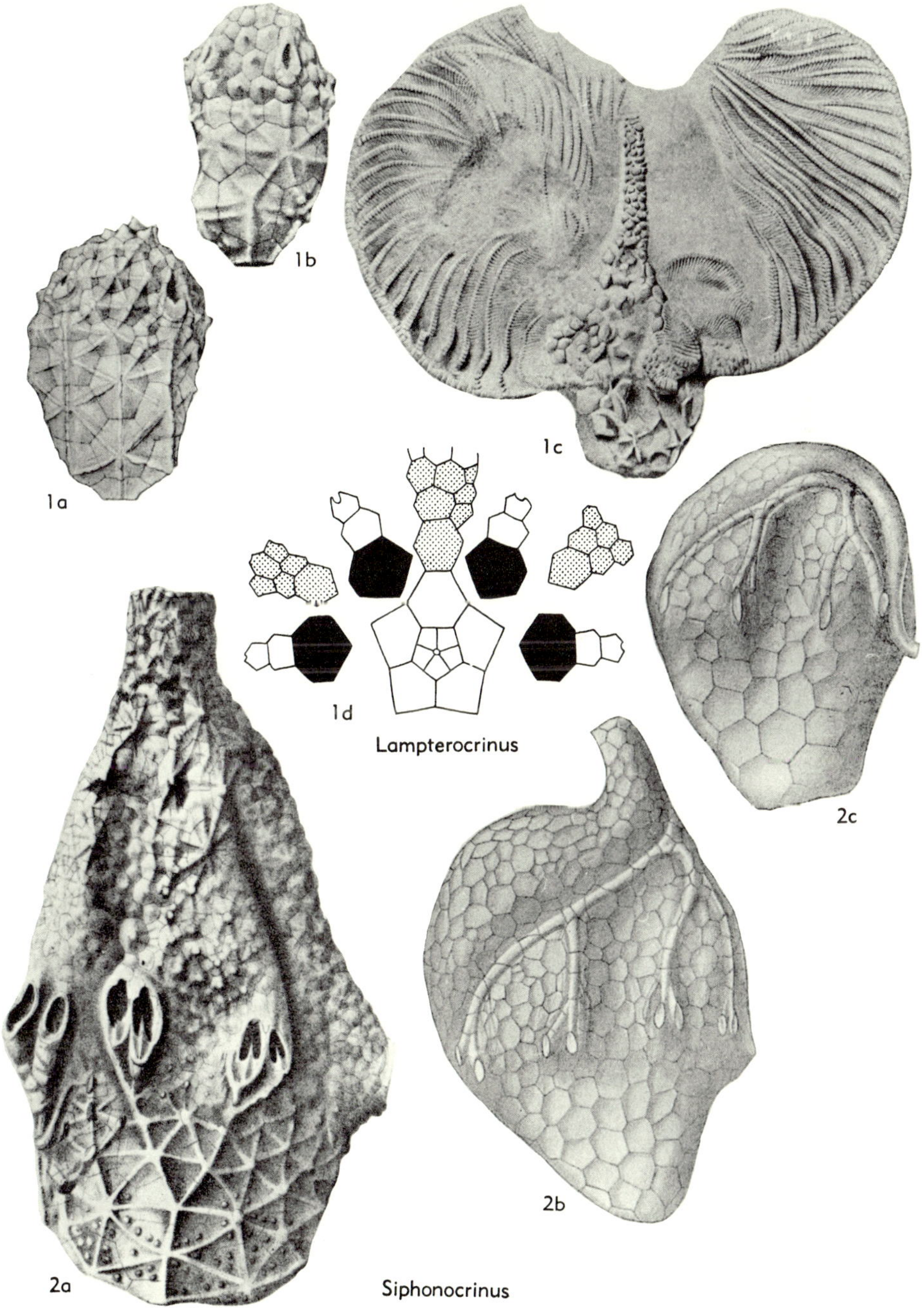

FIG. 245. Lampterocrinidae (p. *T*434-*T*438).

(Springer, 1926a); *1c,* lat. view of crown, ×1 (Springer, 1926a); *1d,* plate diagram of part of theca (radials black; anals and interbrachials stippled) (Ubaghs, n).

Siphonocrinus S. A. MILLER, 1888, p. 263 [**Glyptocrinus nobilis* HALL, 1861d, p. 21; OD] [=*Ochlerocrinus* STRIMPLE, 1963a, p. 87 (type, *Eucalyptocrinus armosus* MCCHESNEY, 1861, p. 95; OD)]. Calyx very asymmetrical, somewhat depressed interradially and lobed at arm level; basals

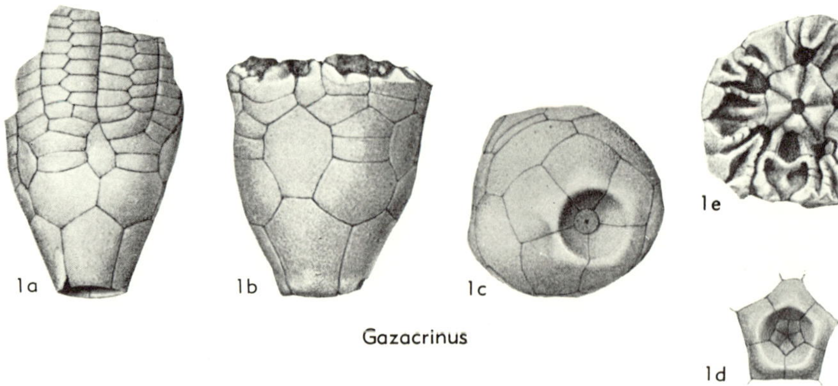

FIG. 246. Gazacrinidae (p. T438).

high, more or less separating radials and meeting 1st interprimibrachs in some individuals; interprimibrachs numerous; intersecundibrachs present. Tegmen very high, with conspicuous protuberance occupying whole *CD* interray and encroaching on *D* ray, distal part of protuberance either terminating in central anal tube or continuing across apex to anterior anal aperture at or even beneath arm bases; subtegminal ambulacral tracts and rectum in open grooves on inner side of tegmen. *M.Sil.*, N.Am.(USA).——FIG. 245,*2a,b*. **S. nobilis* (HALL), Niagaran (Racine Dol.), USA(Wis.-Ill.); *2a*, *E*-ray lat. view of cast from external mold, ×1 (Hall, 1865); *2b*, *BC*-interray lat. view of internal mold, ×1 (Wachsmuth & Springer, 1897).——FIG. 245,*2c*. **S. armosus* (MCCHESNEY), Niagaran (Racine Dol.); *BC*-interray lat. view of internal mold, ×1 (Wachsmuth & Springer, 1897).

Family GAZACRINIDAE S. A. Miller, 1892

[Gazacrinidae S. A. MILLER, 1892a, p. 49]

Infrabasals confined to basal concavity; single interprimibrach in each interray. Tegmen composed of few large interambulacrals crowned with central pyramid of five distinct or ankylosed plates; vertical tegminal ridges may form grooves for reception of arms. Arms ten simple, biserial. Column round in cross section. *M.Sil.*

Gazacrinus S. A. MILLER, 1892, p. 49 [**G. inornatus*; M] [=*Idiocrinus* WACHSMUTH & SPRINGER, 1892 (obj.)]. Characters of family. *M.Sil.*, N.Am. (USA).——FIG. 246,*1*. **G. inornatus*, Niagaran (Laurel Ls.), USA(Ind.); *1a*, *A*-ray view of calyx and lower part of arms, ×2 (Springer, 1926a); *1b*, *CD*-interray view of theca, ×2.5 (Springer, 1926a); *1c*, dorsal view of theca (infrabasals con-

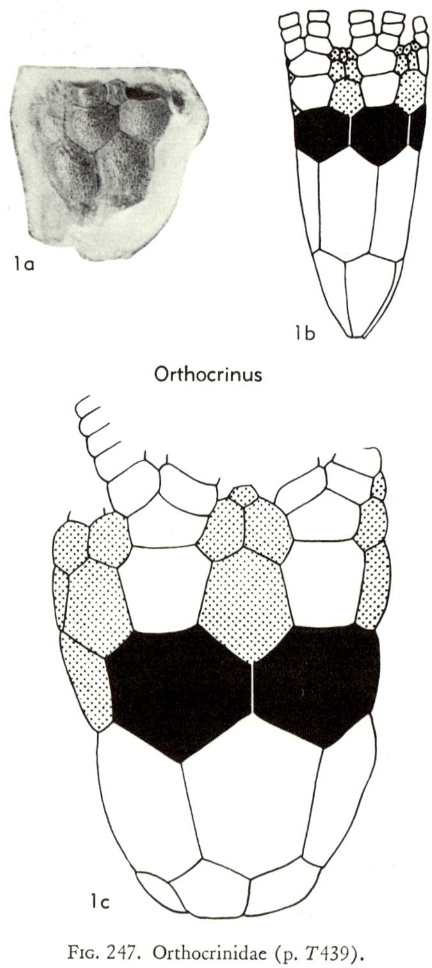

FIG. 247. Orthocrinidae (p. T439).

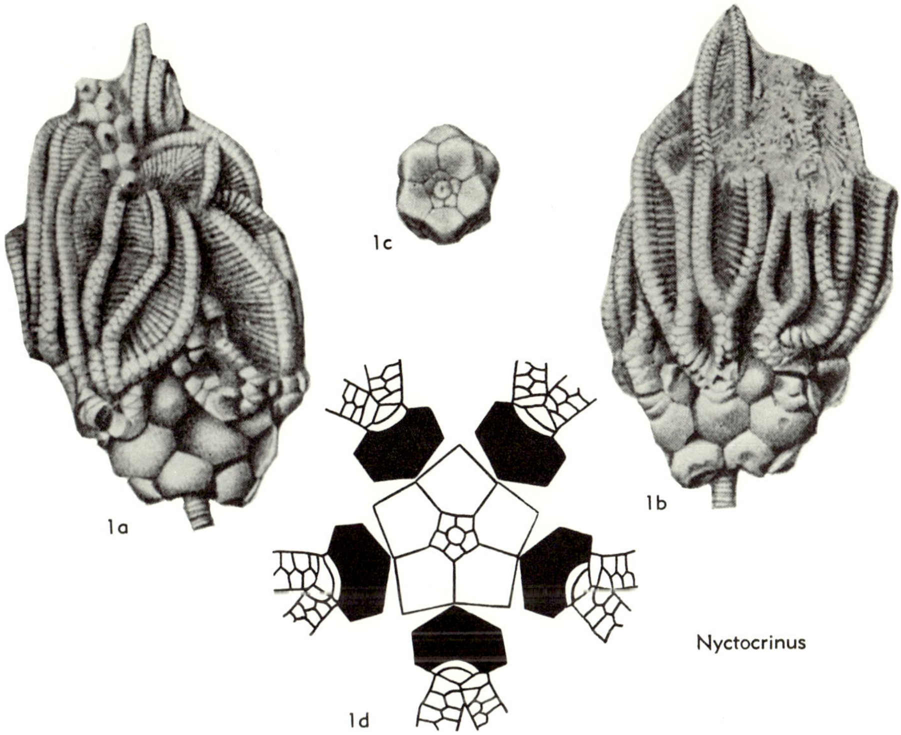

FIG. 248. Nyctocrinidae (p. T440).

cealed by columnal), ×2 (Springer, 1926a); *1d,e,* base and tegmen, ×2 (Springer, 1926a).

Family ORTHOCRINIDAE Jaekel, 1918

[Orthocrinidae JAEKEL, 1918, p. 42]

Calyx more or less elongate; infrabasal and basal circlets very high, occupying about one third to two thirds of calyx height; radials and fixed primibrachs (two in each ray) relatively small; interradial areas reduced, with two plates in second range and three in *CD* interray. Tegmen very low, composed of many minute elements. Arms ten, stout, simple, composed of compound, bipinnulated brachials. Column cylindrical, very delicate, with circular lumen. *L.Dev.-M.Dev.*

Orthocrinus JAEKEL, 1895, p. 29 [**O. simplex*; M]. Characters of family. *L.Dev.-M.Dev.,* W.Eu. ——FIG. 247,*1a.* **O. simplex,* M.Dev., Ger.; lectotype, fragment of calyx, ×1 (Schmidt, 1942). ——FIG. 247,*1b.* *O. elongatus* BREIMER, L.Dev. (U.Ems.), Spain; lat. view of holotype (radials black, interbrachials stippled), ×1 (Breimer, 1962).——FIG. 247,*1c.* *O. robustus* BREIMER, M.Dev., Spain; lat. view of holotype (radials black, interbrachials stippled), ×1 (Breimer, 1962).

Superfamily NYCTOCRINACEA Moore & Laudon, 1943

[*nom. correct.* UBAGHS, herein (*pro* Nyctocrinicae UBAGHS, 1953, p. 737, *nom. transl. ex* family Nyctocrinidae MOORE & LAUDON, 1943a, p. 84)]

Radials adjoining each other all around, forming uninterrupted circlets. *M.Sil.*

Family NYCTOCRINIDAE Moore & Laudon, 1943

[Nyctocrinidae MOORE & LAUDON, 1943a, p. 84]

Calyx low conical, with truncated base; basals large, curving into concavity in which are located five small infrabasals; radials large, with narrow radial facets; primibrachs two in each ray, very short; interprimibrachs and first anal plate above radial circlet and largely confined to tegmen; primanal plate slightly larger than first interprimibrachs; long anal tube. Arms biserial, two in each ray, simple or more generally branching once or twice. Column cylindrical. *M.Sil.*

Nyctocrinus SPRINGER, 1926, p. 54 [**N. magnitubus*; OD]. Characters of family. *M.Sil.*, N.Am. (USA).——FIG. 248,*1*. **N. magnitubus*, Niagaran (Beech River F.), USA(Tenn.); *1a,b*, opposite views of somewhat crushed specimen, ×1.5 (Springer, 1926a); *1c*, dorsal view of calyx base, ×1.5 (Springer, 1926a); *1d*, plate diagram of calyx and lower part of arms (radials black) (Ubaghs, n).

Order MONOBATHRIDA Moore & Laudon, 1943

[*nom. correct.* MOORE, 1952a, p. 614 (*pro* Monobathra MOORE & LAUDON, 1943a, p. 86)] [Materials for this order prepared by GEORGES UBAGHS, with contributions by N. GARY LANE, R. C. MOORE, A. L. BOWSHER, and H. L. STRIMPLE, as indicated in the text]

Calyx monocyclic. *?L.Ord., M.Ord.-U.Perm.*

Suborder COMPSOCRININA Ubaghs, new suborder

[=Tanaocrinina MOORE, 1952a, p. 614]

Basal circlet hexagonal; radials adjoining each other except for separation by primanal on posterior side. *?L.Ord., U.Ord.-U.Perm.*

Superfamily XENOCRINACEA S. A. Miller, 1890

[*nom. correct.* BROWER, 1973, p. 443 (*pro* Xenocrinicae UBAGHS, 1953, p. 738, *nom. transl. ex* Xenocrinidae S. A. MILLER, 1890, p. 318)]

Calyx elongate; fixed brachials numerous; basals four; interbrachials numerous, connecting with interambulacrals; *CD* interray divided by median series of anal plates. Tegmen many-plated, without anal tube. Free arms few per ray. *?L.Ord.; U.Ord.; U.Sil.*

Family XENOCRINIDAE S. A. Miller, 1890

[Xenocrinidae S. A. MILLER, 1890, p. 318]

Median ray and anal ridges conspicuous; radials usually adjoining each other in all interrays except posterior; sutures between some adjacent radials covered by small supplementary plates; fixed secundibrachs two, if there is further division, otherwise three to four to each half ray; few fixed tertibrachs may be present; interbrachial areas depressed, flexible, composed of many small and irregular plates; *CD* interray with radial-like primanal and median anitaxis of prominent convex plates. Free arms 2, 3 or 4 per ray, unbranched, uniserial or composed of cuneiform and interlocking brachials. Stem quadrangular in cross section, with pentagonal axial canal. *U.Ord.*

Xenocrinus S. A. MILLER, 1881, p. 71 [**X. penicillus*; M]. Characters of family. *U.Ord.*, N.Am. (USA)-Eu.(Scot.).——FIG. 249,*1*. **X. penicillus*, Richmond., USA(Ohio); *1a,b*, *CD* interray and *D*-ray views of crowns, ×2 (Wachsmuth & Springer, 1897); *1c*, plate diagram of part of calyx and arms (radials black, anal plates stippled) (Ubaghs, n).

Family TANAOCRINIDAE Bather, 1899

[Tanaocrinidae BATHER, 1899b, p. 921]

Median ray and anal ridges present; radials in full lateral contact except in *CD* interray; interbrachial areas moderately depressed, probably solid, composed of rather numerous and moderately large and regular plates. *?L.Ord., U.Ord.*

Canistrocrinus WACHSMUTH & SPRINGER, 1885, p. 94(316) [**Glyptocrinus richardsoni* WETHERBY, 1880; OD] [=*Tanaocrinus* WACHSMUTH & SPRINGER, 1897, p. 185 (type, *T. typus*);*Canistocrinus* JAMES, 1897, p. 104 (*nom. null.*)]. Basals visible in side view; fixed secundibrachs 5 or 6 in each half-ray; fixed tertibrachs passing to free brachials. Free arms uniserial, 4 to each ray, forking isotomously once or twice at long intervals, with short, cuneate brachials; proximal pinnules partly incorporated in calyx, fixed pinnulars prominent. Column cylindrical. [As first recognized by ULRICH (in MS), the types of *Tanaocrinus* have 4 basals (not 5 as erroneously stated by WACHSMUTH & SPRINGER, 1897, p. 173) and a whole set of features undistinguishable from those of *Canistrocrinus*; consequently *Tanaocrinus* is here treated as a subjective junior synonym of the latter.] *U.Ord.*, N.Am.(USA).——FIG. 250, *1a*. **C. richardsoni* (WETHERBY), Waynesville F., USA(Ohio); lat. view of crown, ×3 (Ubaghs, n). ——FIG. 250,*1b*. *C.* sp., Girardeau, USA(Mo.); *BC* interray view (U.S. Natl. Museum 92768), ×3 (Ubaghs, n).——FIG. 250,*1c*. *C. typus* (WACHSMUTH & SPRINGER), paratype of *Tanaocrinus typus*, Richmond., USA(Ohio); *BC* interray view (U.S. Natl. Museum S. 716), ×3 (Ubaghs, n).

Compsocrinus S. A. MILLER, 1883, p. 219, 223 [**Glyptocrinus harrisi* S. A. MILLER, 1881a, p. 74; OD] [*non Compsocrinus* PERNER in ZELIZKO,

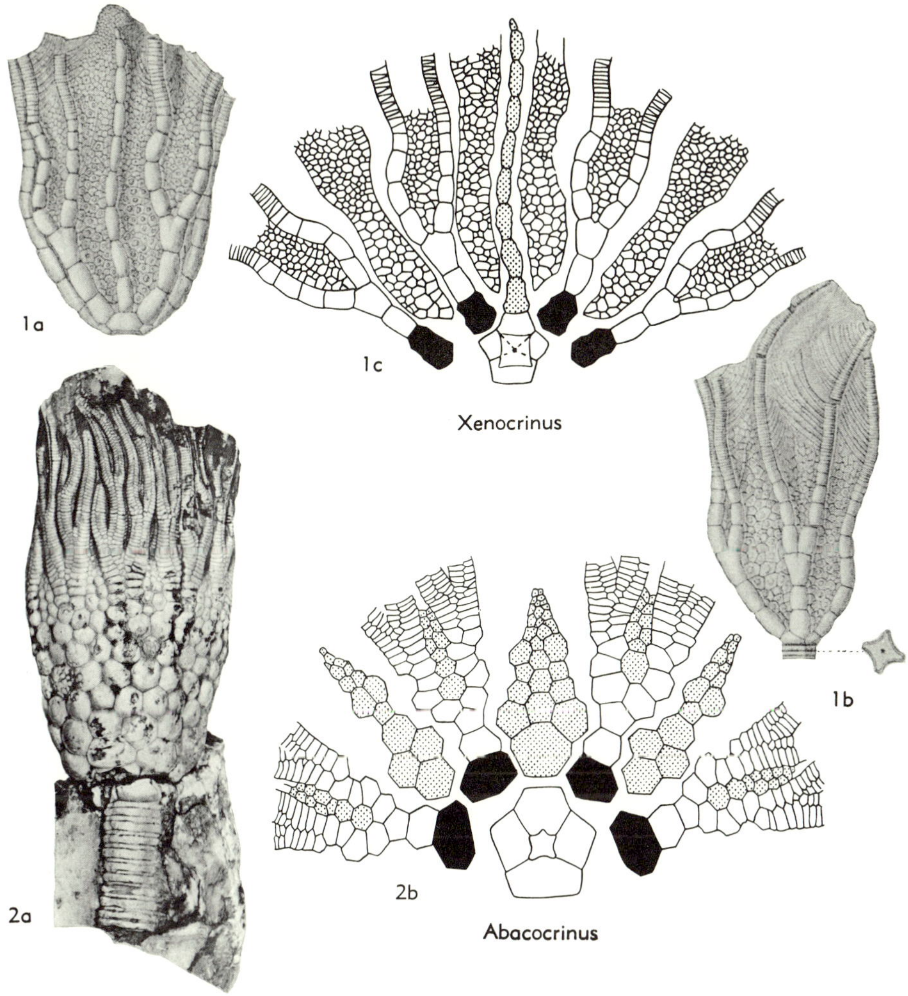

FIG. 249. Xenocrinidae *(1)*; Abacocrinidae *(2)* (p. *T*440, *T*442-*T*443).

1911, p. 6 *(nom. nud.)* (=*Akadocrinus* PROKOP, 1962, Eocrinoidea)]. Basals visible from side; fixed secundibrachs 2 if any further brachitaxis; fixed tertibrachs, if any, 4 or 5 in each ramus, passing to free brachials; all or only proximal interprimibrachs large and regular, according to species. Free arms unbranched, 2, 3, or 4 per ray, uniserial. Column quadrangular or round in cross section. *U.Ord.*, N.Am.(USA).——FIG. 250, *2*. **C. harrisi* (S. A. MILLER), Richmond, USA (Ohio); *CD* interray view, ×2 (Wachsmuth & Springer, 1897).

?**Proexenocrinus** STRIMPLE & MCGINNIS, 1972, p. 72 [**P. inyoensis*; M]. Calyx bowl shaped; basals confined to basal concavity; interbrachials moderately large, mildly depressed, first one in notch at summit of two radials; *CD* interray with median series of plates accompanied by single series of small plates on each side. Free arms 2 per ray, uniserial; first pinnule stout; free brachials slightly wedge shaped. [This genus was placed among the Xenocrinidae by STRIMPLE & MCGINNIS (1972, p. 72); however, uncertainty concerning the structure of the proximal part of its theca up to the radial circlet and ignorance of the number of basal plates render its systematic position doubtful. Because of the relatively large size and regular arrangement of its interbrachials, it is here classified with doubt in the Tanaocrinidae.] *L.Ord.*, USA(Calif.).——FIG. 250,*3*. **P. inyoensis*, Al Rose F.; holotype, *C*-ray view, ×4 (Strimple & McGinnis, 1972).

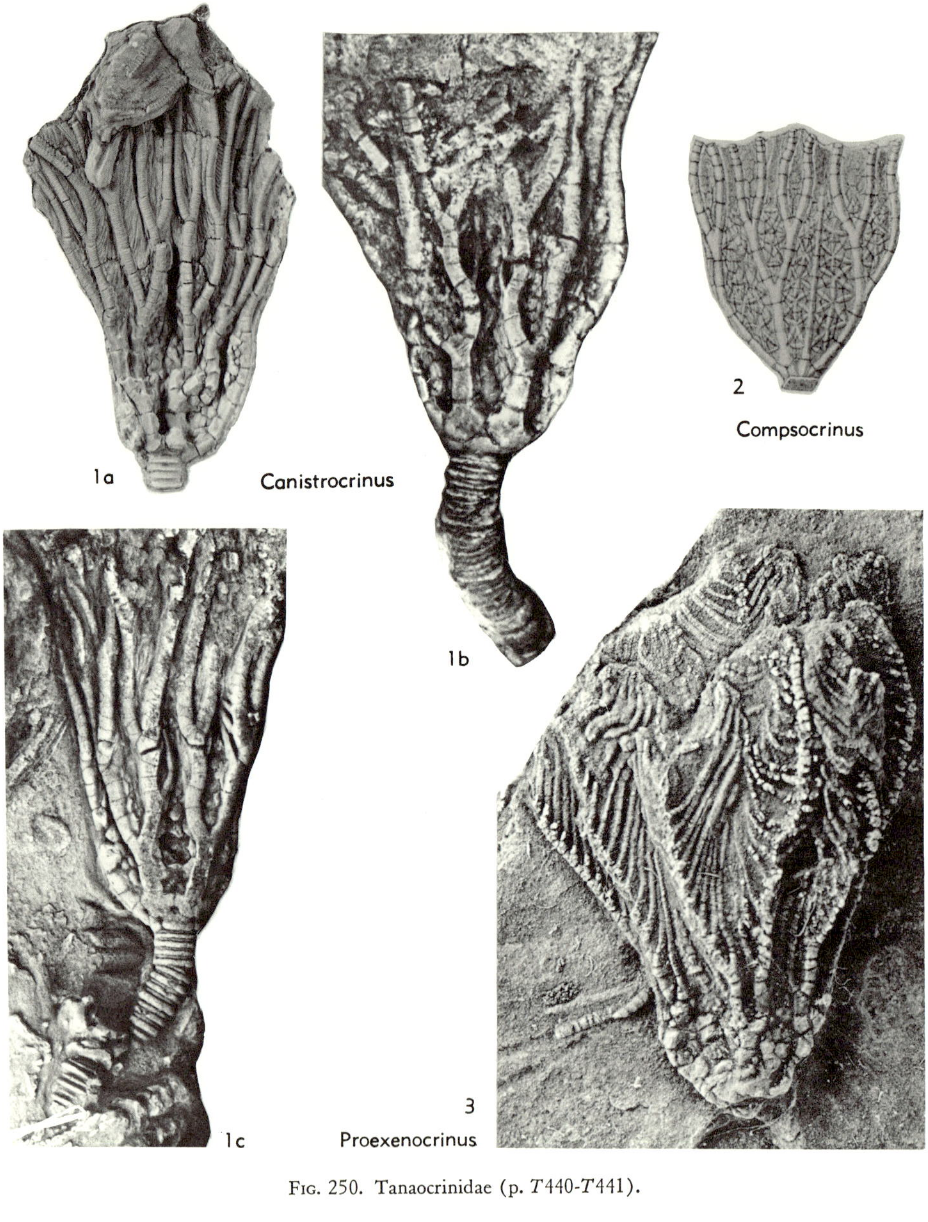

FIG. 250. Tanaocrinidae (p. T440-T441).

Family ABACOCRINIDAE Jaekel, 1918

[Abacocrinidae JAEKEL, 1918, p. 31]

Calyx heavily plated, without distinct ray and anal ridges; fixed secundibrachs numerous, proximally uniserial, distally biserial; interbrachials not depressed, large, stout, and regular; *CD* interray with extra plates. Tegmen solid, with anal pyramid. Each half ray generally dividing on becoming free; free arms 4 per ray, forking isotomously two or three times at wide intervals; proximal pinnules partly incorporated in calyx. Column round in cross section; heteromorphic. *U.Sil.*

Abacocrinus ANGELIN, 1878, p. 19 [**Actinocrinites tesseracontadactylus* GOLDFUSS, 1831, p. 194; SD WACHSMUTH & SPRINGER, 1881, p. 135(309)]. Characters of family. *U.Sil.,* Sweden(Gotl.).——

FIG. 249,*2*. **A. tesseracontadactylus* (GOLDFUSS); *2a*, lat. view of crown with attached part of stem (Nat. Riksmus. Stockholm, Ec 10867), ×1 (Ubaghs, n); *2b*, plate diagram of part of theca and arms (radials black, anals and interbrachials stippled) (Ubaghs, n).

Superfamily PERIECHOCRINACEA Bronn, 1849

[*nom. correct.* UBAGHS, herein (*pro* Periechocrinicae RAMSBOTTOM, 1954, p. 687, *nom. correct. pro* Periechocriniticae UBAGHS, 1953, p. 738, *nom. transl. ex* Periechocrinidae BRONN, 1849, p. 179)] [=sections Periechocrinites + Actinocrinites WACHSMUTH & SPRINGER, 1881, p. 93, 267]

Three equal basals (sutures typically in *B* and *E* rays and *CD* interray) but may be ankylosed; first primibrachs hexagonal, but quadrangular in advanced members; primanal heptagonal, followed by three plates in next higher row, or hexagonal with two plates in second row. *Sil.-U.Perm.*

Family PERIECHOCRINIDAE Bronn, 1849

[Periechocrinidae BRONN, 1849, p. 179] [=Periechocrinitidae BASSLER, 1938, p. 28; Saccocrininae JAEKEL, 1918, p. 34]

First primibrachs typically hexagonal; two to five fixed secundibrachs in each ray and two to seven tertibrachs may be incorporated in calyx; interprimibrachs not depressed or only slightly so, numerous and merging with tegmen; *CD* interray wide, primanal heptagonal, followed by three plates in second row. Tegmen many-plated, with ambulacrals and orals rarely distinct. Arms 10 to 40, biserial or uniserial, simple or branching. Column cylindrical. *Sil.-Miss.*

Periechocrinus MORRIS, 1843, p. 56 [**Periechocrinites costatus* AUSTIN & AUSTIN, 1843 (=*Actinocrinites moniliformis* PHILLIPS in MURCHISON, 1839, p. 673, *non* MILLER, 1821); SD RAMSBOTTOM, 1954, p. 687] [=*Crumenaecrinites* TROOST, 1850a, p. 420 *(nom. nud.)* (type, *C. ovalis*; M); *Geocrinus* D'ORBIGNY, 1850, p. 46 (obj.); *Saccocrinus* HALL, 1852, p. 205 (type, *S. speciosus*; M); *Crumenaecrinus* BATHER in LANKESTER, 1900a, p. 202 *(nom. van.)*; *Periechocrinites* BASSLER, 1938, p. 149 (*non* AUSTIN & AUSTIN, 1843) (obj.); *Periecocrinites* AUSTIN & AUSTIN, 1842, p. 110 *(nom. nud.)*]. [In 1843, AUSTIN & AUSTIN erected a genus *Periechocrinites* and designated *P. articulosus* as its type species, but the type specimen of this species belongs to *Actinocrinites? expansus* PHILLIPS, type species of the genus *Sagenocrinites* (RAMSBOTTOM, 1951b). It thus follows that *Periechocrinites* is a junior synonym of *Sagenocrinites* since their respective type species are subjective synonyms. A new name was then required for the genus hitherto called *Periechocrinites,* and as *Periechocrinus* MORRIS, 1843, was the earliest of several available names, it has been chosen by RAMSBOTTOM (1954) to replace *Periechocrinites.*] Calyx typically elongate, expanding to arm bases; calyx plates thin, and particularly in radial and primibrach series, very long; median ray ridges generally present; basals high; primaxils seven-sided; rays branching once or several times within calyx; interprimibrachs numerous, not depressed, forming 2 series of plates in each interray; *CD* interray with ridged median series of anal plates. Tegmen incompetent, entirely or mostly composed of many minute irregular plates and bearing anal tube. Arms 10 to 40, biserial, slender, simple or heterotomously branched. Heteromorphic stem composed of alternate nodals and internodals. *M.Sil.*, Eu.-N.Am.——FIG. 251,*1*. **P. costatus* (AUSTIN & AUSTIN), Wenlock, Eng.; *1a,b*, drawings of *A*-ray and *CD* interray views of crown (Brit. Museum Nat. Hist. 57249) (radials black, anals and interbrachials stippled), ×0.5 (Ubaghs, n); *1c*, lat. view of complete crown, ×0.75 (Brit. Palaeoz. Fossils, 1964); *1d*, drawing of tegmen (Geol. Survey, London, 89.929), ×1 (Ubaghs, n); *1e*, plate diagram of part of calyx (radials black, anals and interbrachials stippled) (Ubaghs, n).

Aryballocrinus BREIMER, 1962, p. 72 [**Actinocrinus (Megistocrinus) whitei* HALL, 1861a, p. 271; OD]. Calyx globose, composed of thin plates; radial circlet high; faint median ray ridges present in a few species; fixed secundibrachs few; interprimibrachs not depressed, few; *CD* interray very wide. Tegmen low, with anus excentric, little above surface of tegmen. Arms 10, dichotomously branching, biserial, densely pinnulate. Stem stout, composed of equal plates. *L.Carb.*, Eu.; *L.Miss.*, N.Am.——FIG. 252,*1*. **A. whitei* (HALL), Osag. (Burlington Ls.), USA(Iowa); *1a,b*, *A*-ray view of crown with attached stem and side view of calyx, ×1 (Wachsmuth & Springer, 1897).

Athabascacrinus LAUDON, PARKS, & SPRENG, 1952, p. 566 [**A. colemanensis*; OD]. Calyx low bowl shaped, with gradually sloping walls; basals and radials laterally directed; fixed secundibrachs 1 or 2 in each half-ray; interprimibrachs not depressed, rather numerous and large; *CD* interray broad, with primanal larger than radials. Tegmen low, conical, with markedly depressed interambulacral areas, and with single central plate surrounded by 6 slightly smaller ?orals and 4 still smaller plates on posterior side; plates covering ambulacral areas relatively large. Arm bases markedly projecting, facets directed laterally, 2 to 4 according to ray; arms unknown. *L.Miss.*, N.Am.(Can.).——FIG.

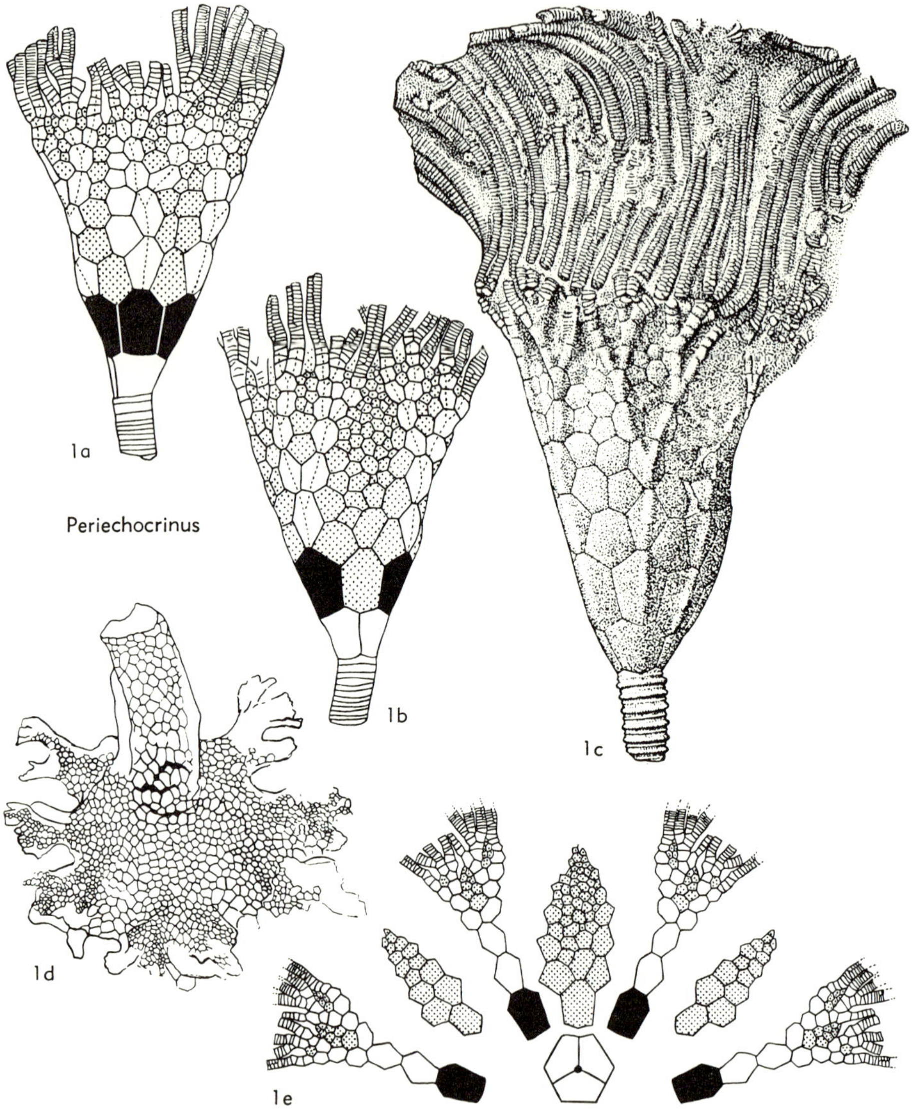

FIG. 251. Periechocrinidae (p. T443).

252,*2*. **A. colemanensis,* Banff F., Can.; *2a-c,* dorsal, ventral, and *A*-ray view of holotype, ×2 (Laudon, Parks & Spreng, 1952).

Beyrichocrinus WAAGEN & JAHN, 1899, p. 12 [**B. humilis*; M]. Calyx bowl shaped, with plates heavy and tumid; base slightly depressed; radials unequal, separated partially or entirely by interprimibrachs; primibrachs larger than radials; fixed secundibrachs 2 in each half-ray; fixed tertibrachs 6 or 7, becoming biserial distally; interprimibrachs numerous, not depressed; intersecundibrachs and intertertibrachs present; *CD* interray slightly wider than others, contracting distally; primanal meeting posterior basal narrowly. Tegmen low, composed of small plates isolating larger ones; anus slightly excentric. Arm openings 20. *L.Dev.,* Eu.(Czech.).——FIG. 252,*3*. **B. humilis,* Koneprusy Ls.; *3a,b,* dorsal and ventral views of theca, ×0.5 (Natl. Museum, Prague; Ubaghs, n).

Corocrinus GOLDRING, 1923, p. 202 [**C. ornatus*; OD]. Calyx turbinate or bowl shaped with stellate or nodose plates; median ray ridges present; primaxils 7-sided; fixed secundibrachs few; interradial areas narrow, distally depressed; *CD* interray widely expanded, with ridged median series of anal plates. Tegmen many-plated, with anus

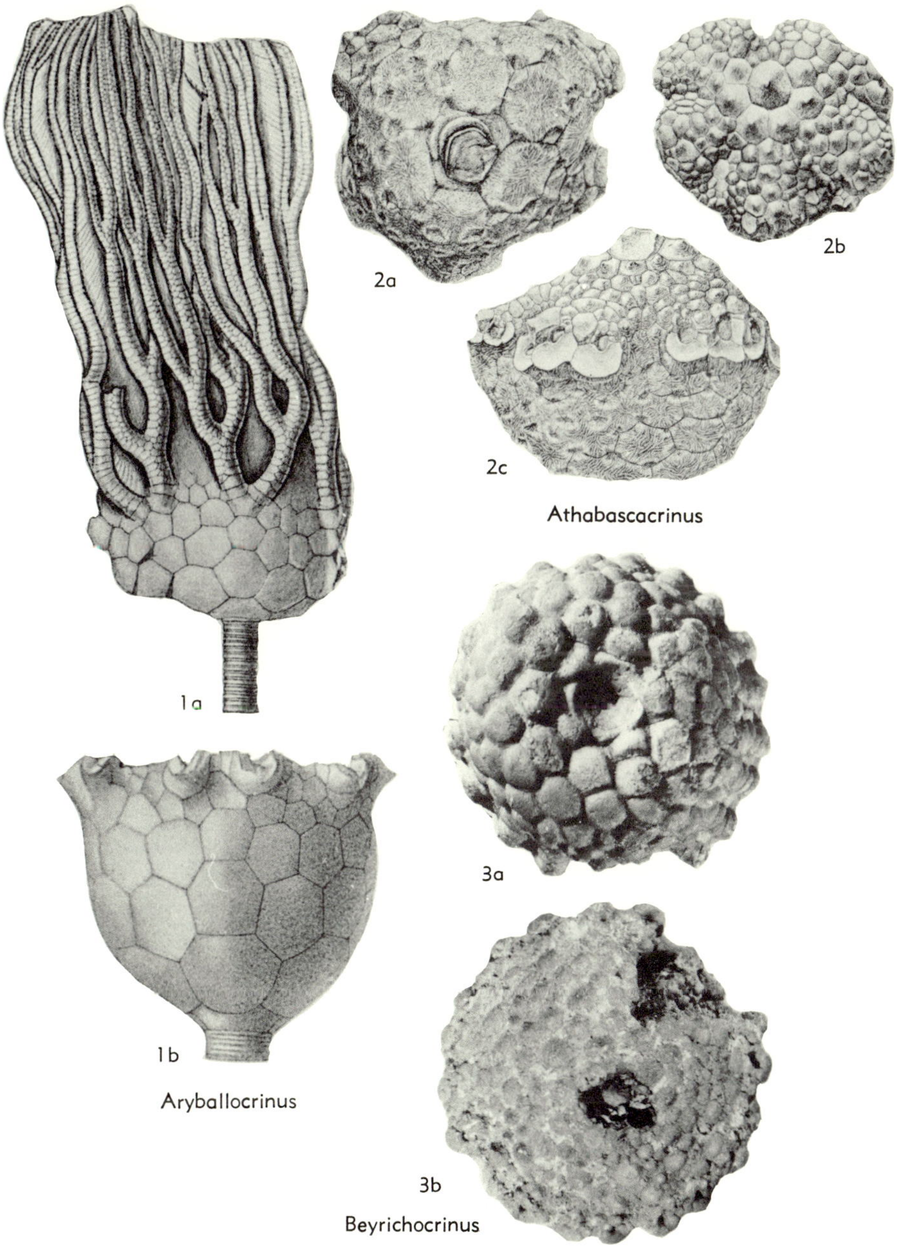

FIG. 252. Periechocrinidae (p. *T*443-*T*444).

at top of protuberance or anal tube. Arms 2 in each ray, bifurcating only once, biserial or uniserial according to species. *L.Dev.-M.Dev.,* N. Am.-Eu.——FIG. 253,*1.* **C. ornatus,* M.Dev., USA(N.Y.); *CD*-interray view of theca with attached part of stem, ×1 (Goldring, 1923).

Gennaeocrinus WACHSMUTH & SPRINGER, 1881, p. 160(334) [**Actinocrinus kentuckiensis* SHUMARD, 1868, p. 345; OD]. Calyx low to high, conical to globose, generally flattened at base, lobed at arm bases; plates thin and nearly smooth, ornamented in low relief, or coarsely ornamented with

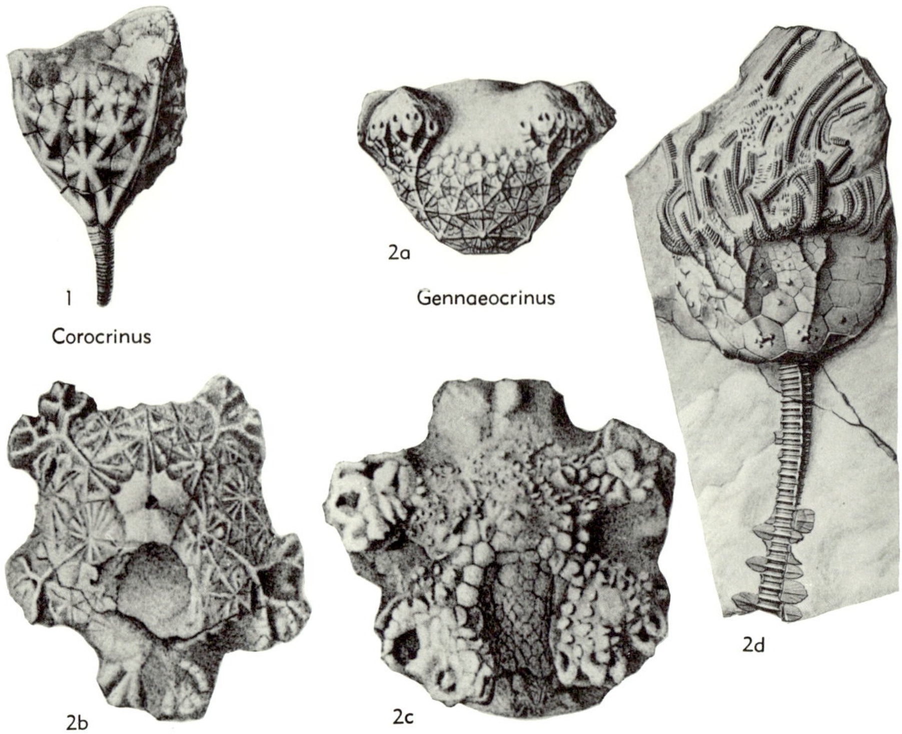

FIG. 253. Periechocrinidae (p. T444-T446).

nodes, spines, radiating ridges, and parallel carinae according to species; median ray ridges generally distinct; primaxils 5-, 6-, or 7-sided; rays heterotomously branching within cup; interprimibrachs numerous, not depressed; *CD* interray wide. Tegmen low, many-plated, interradially depressed, may be spiniferous, with small orals and distinct ambulacrals. Free arms 6 to 8 in each ray, biserial, long, slender, simple or branching once. Column with large quinquelobate axial canal. *L.Dev.-M.Dev.,* N.Am.(USA)-Eu.——FIG. 253,*2a-c.* **G. kentuckiensis* (SHUMARD), M.Dev.; *2a,* N.Y., *CD*-interray view of holotype, ×1 (Goldring, 1923); *2b,c,* Ky., dorsal and ventral views of another theca, ×1 (Wachsmuth & Springer, 1897).——FIG. 253,*2d.* *G. eucharis* (HALL), M.Dev., N.Y.; crown and part of stem, ×0.5 (Goldring, 1923).

Lenneocrinus JAEKEL, 1918, p. 35 [**L. cirratus*; OD]. Calyx conical; plates thin, with prominent radial ornamentation; median ray ridges present; primaxils 7-sided; fixed secundibrachs 2 in each half-ray; interprimibrachs few, depressed distally. Free arms 2 in ray, biserial, branching irregularly dichotomously; pinnules long and slender. Column composed of nodals of 2 orders separated by numerous internodals; long cirri on both orders of nodals. *M.Dev.-U.Dev.,* Eu.(Ger.-Spain).——FIG. 254,*1.* **L. cirratus,* Givet., Ger.; *1a,b,* lat. view of holotype crown with attached cirriferous stem, dorsal view, ×3 (Jaekel, 1918).

Megistocrinus OWEN & SHUMARD, 1852, p. 594 [**Actinocrinus evansii* OWEN & SHUMARD, 1850, p. 68; M] [=*Tylocrinus* WOOD, 1904, p. 66]. Calyx bowl shaped with thick plates and flattened or excavate base; basals may be ankylosed; no median ray ridges; primaxils 5-, 6-, or 7-sided; fixed secundibrachs numerous; higher order of brachials more or less incorporated in cup; interprimibrachs numerous, not depressed; *CD* interray very wide. Tegmen flat to conical, with orals, radial dome-plates and ambulacrals generally differentiated; anus excentric, ordinarily at end of tube. Arms 4 in each ray, or 4 in *A, C,* and *D* rays and 2 (rarely 3) in others, biserial, branching dichotomously. Column stout; axial canal wide, quinquelobate. *L.Dev.-L.Miss.,* N.Am.(mainly)-W.Eu.-New Zealand.——FIG. 255,*1a,b.* **M. evansii* (OWEN & SHUMARD), L.Miss.(Burlington Ls.), USA(Iowa-N.Mexico); lat. and ventral views, ×0.7 (Wachsmuth & Springer, 1897).——FIG. 255,*1c.* *M. nobilis* WACHSMUTH & SPRINGER, L. Miss.(Hampton F.), USA(Iowa); *CD*-interray view, ×0.7 (Wachsmuth & Springer, 1897).

Pithocrinus KIRK, 1945, p. 341 [**P. cooperi*; OD]. Calyx wide globose, with rather thick plates, convex to tumid or spiny, with gonioporoids; base commonly produced into rim; primaxils usually 5-sided; fixed portions of rays commonly curved; interrays relatively small, with interprimibrachs in 2 series; *CD* interray larger than others; interbrachials few. Tegmen convex to highly elevated, composed of many relatively small plates, some of which bear subspinous processes; stout subcentral anal tube. Free arms, 2 to 6 in each ray according to species and even according to ray in individuals, densely biserial, irregularly branching. *L.Dev.-M. Dev.*, Eu.-N.Am.——FIG. 255,*2*. **P. cooperi*, M. Dev.(Traverse Gr.), USA(Mich.); *2a-c*, *C*-ray, *CD*-interray, and dorsal views of holotype, ×1 (Kirk, 1945a).

Pradocrinus DE VERNEUIL, 1850, p. 184 [**P. baylii*; M]. Calyx very high, urn shaped; plates thin, slightly convex, with radial ornamentation; median ray ridges present; primaxils 7-sided; 2 secundibrachs in each half-ray incorporated in calyx; interprimibrachs not depressed, 9 to 10, merging with tegmen; *CD* interray with sagittal ridged series of anal plates. Tegmen low, composed of rather large plates, with central nodose plate; anus excentric, directly through tegmen. Free arms 10, stout, biserial. Column unknown. *L.Dev.*, Eu. ——FIG. 255,*3*. **P. baylii*, L.Ems., Spain; *3a-c*, holotype theca, *CD*-interray, dorsal and ventral views, ×1 (Ubaghs, n).

Pyxidocrinus MÜLLER, 1855, p. 82 [**Actinocrinus prumiensis* ZEILER & WIRTGEN, 1855, p. 81; M]. Calyx conical; thin plated, with gonioporoids; no median ray ridges; primaxils 5-sided; 2 secundibrachs and 2 tertibrachs of each ray incorporated in calyx; interradial areas relatively small, with interprimibrachs in 2 series; interbrachials up to 3 in ray; *CD* interray without median ridge. Tegmen relatively low, composed of irregularly arranged plates and central nodose plate, which may be surrounded by 4 interradial spiny plates; anus excentric, directly through tegmen. Free arms 4 to ray, not certainly known. Column unknown. *L.Dev.-M.Dev.*, Eu.——FIG. 256,*1a,b*. **P. prumiensis* (ZEILER & WIRTGEN), probably M.Dev., Ger.; *CD*-interray and ventral views of holotype, ×3 (Müller, 1855).——FIG. 256,*1c,d*. *P. collensis* BREIMER, L.Ems., Spain; *CD*-interray and ventral views of theca, ×1.5 (Ubaghs, n).

Stamnocrinus BREIMER, 1962, p. 59 [**Saccocrinus? intrastigmatus* W. E. SCHMIDT, 1932, p. 21; OD]. Calyx high turbinate; primaxils 7-sided; interradial areas composed of many plates in 3 series; intersecundibrach areas large. Tegmen with anal tube and spines in ambulacral parts. Free arms and stem unknown. *L.Dev.-M.Dev.*, Eu.-N.Am. ——FIG. 256,*2*. **S. intrastigmatus* (W. E. SCHMIDT), L.Ems., Spain; *2a-c*, *B*-ray, dorsal, and ventral views of theca, ×1 (Breimer, 1962).

Stiptocrinus KIRK, 1946, p. 33 [**S. spinosus*; OD].

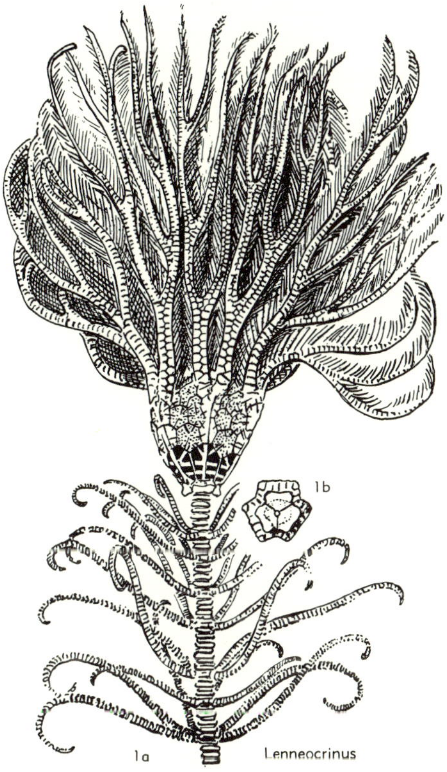

FIG. 254. Periechocrinidae (p. T446).

Cup subturbinate to subglobose, with somewhat depressed interradial areas and pronounced tendency toward lobation; primibrachs usually much smaller than radials; 2 or 3 fixed secundibrachs, other brachs free; intersecundibrachs lacking, or single and enclosed by brachs; *CD* interray much larger than others, with median series of 2 to 3 or more plates; primanal equal in size to radials. Tegmen low, competent, many-plated; anal tube excentric. Free arms 2 per ray, united and uniserial proximally, beyond unknown. Column unknown. *Sil.*, USA.——FIG. 257,*1a*. *S. benedicti* (MILLER), Laurel Ls., Ind.; *CD*-interray view (radials black, anal plates and interbrachials stippled), ×2 (Springer, 1926a).——FIG. 257,*1b*. **S. spinosus*, Laurel Ls., Ind.; *AE*-interray view (radials black, interbrachials stippled), ×2 (Springer, 1926a).——FIG. 257,*1c*. *S. nodosus* (SPRINGER), Decatur Ls., Tenn.; tegmen, ×1.5 (Springer, 1926a).

Thamnocrinus GOLDRING, 1923, p. 239 [**T. springeri*; OD]. Calyx subturbinate with prominent median ray ridges; primaxils 7-sided; ray plates incorporated in to level of secundaxils or 1 or 2 plates beyond; interradial areas flattened, many-plated, uniting main rami up to last bifurcation;

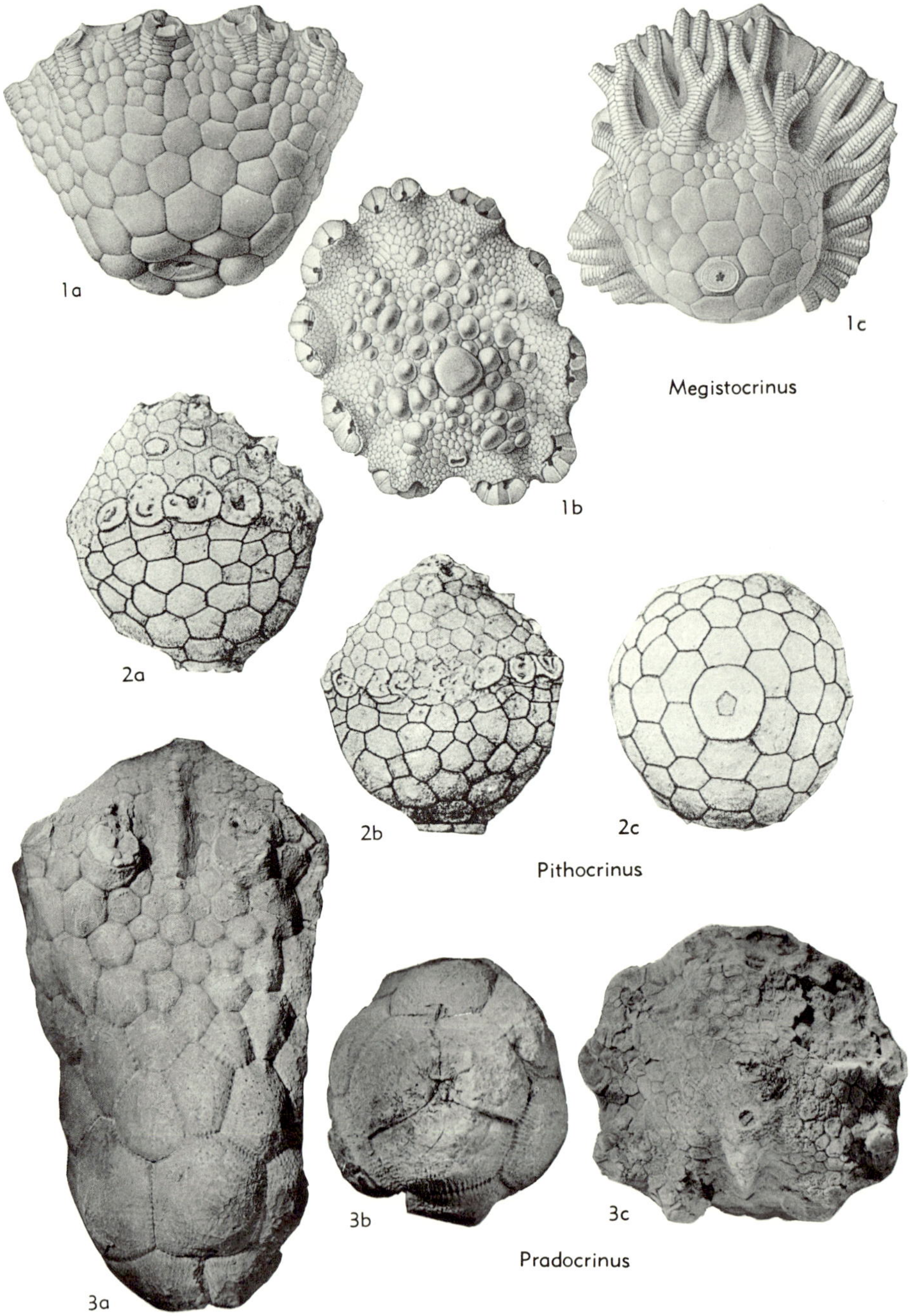

FIG. 255. Periechocrinidae (p. *T*446-*T*447).

median series of anal plates well marked. Tegmen lobed, interradially depressed sharply, spiniferous. Free arms 2 in each ray, short, uniserial, endotomous, bearing undivided biserial long and pinnulate ramuli. Column with large circular axial canal. *M.Dev.*, N.Am.——FIG. 256,*3*. **T. springeri*, Moscow F., USA(N.Y.); *3a,b*, *E*-ray and dorsal views of crowns, ×1 (Goldring, 1923).

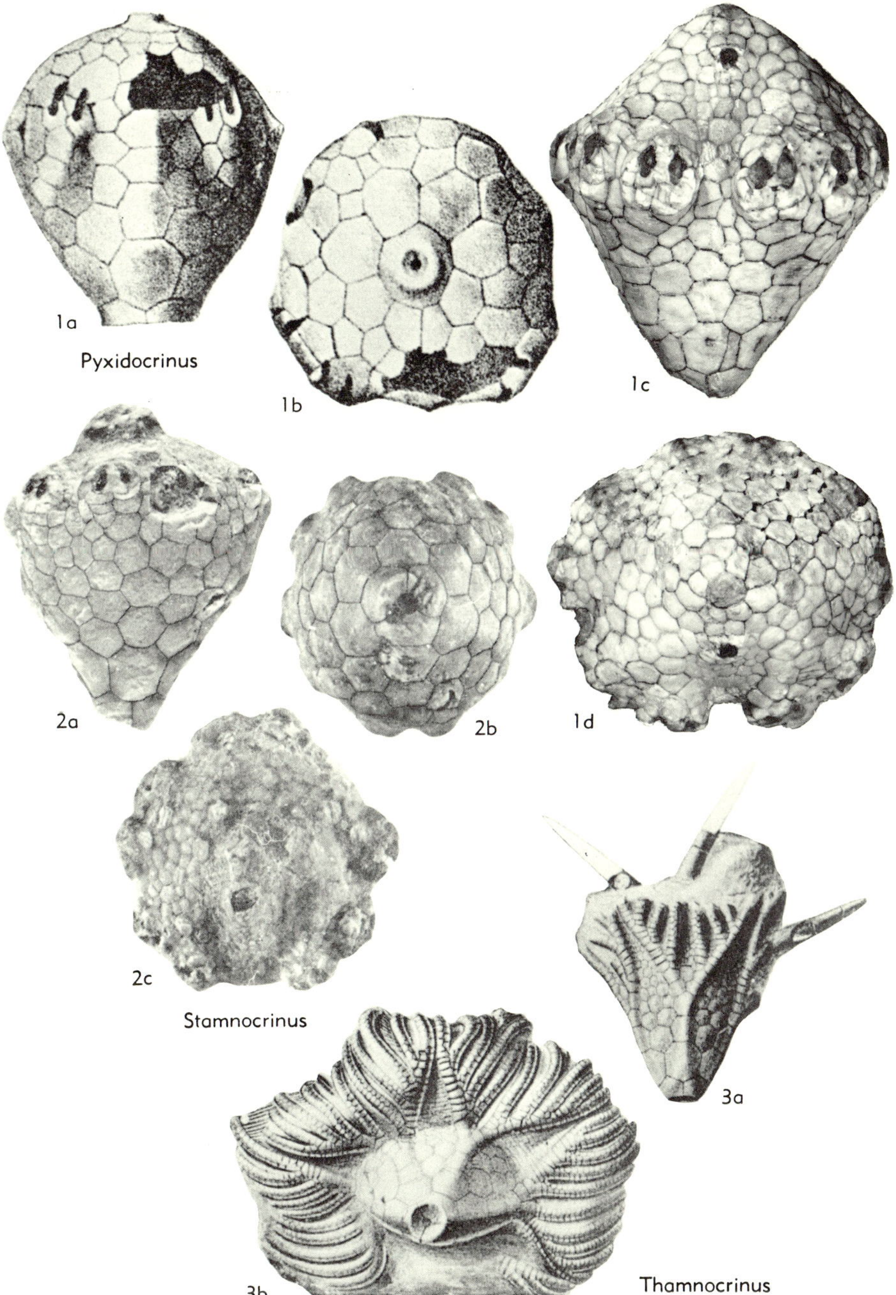

FIG. 256. Periechocrinidae (p. *T*447-*T*448).

Family PARAGARICOCRINIDAE
Moore & Laudon, 1942

[Paragaricocrinidae MOORE & LAUDON, 1942, p. 68]

Calyx low rounded, basally depressed; basals concealed by column; radials laterally directed, commonly forming part of basal

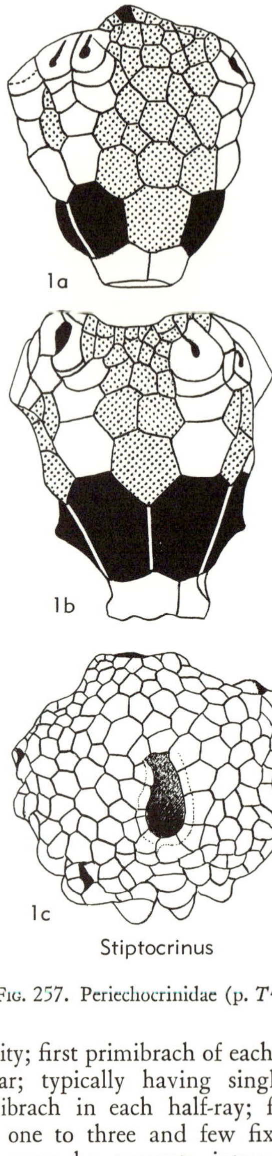

FIG. 257. Periechocrinidae (p. T447).

concavity; first primibrach of each ray quadrangular; typically having single axillary secundibrach in each half-ray; fixed tertibrachs one to three and few fixed quartibrachs may be present; interprimibrachs few, not in contact with tegmen, first one large; *CD* interray wider than others, with two or three anals in second row, connected with tegmen in primitive members, but not in advanced ones. Free arms proximally biserial, more or less in contact all around. *U.Carb., U.Perm.*

Paragaricocrinus YAKOVLEV, 1934, p. 271 [**P. mediterraneus*; OD]. Rays incorporated in calyx up to proximal tertibrachs; generally single interprimibrach in interrays; *CD* interray slightly wider than others, not connecting with tegmen, 2 plates in 2nd row. Tegmen moderately elevated. Free arm bases in lateral contact all around, 8 in each ray. *U.Perm.*, Eu.(Sicily).——FIG. 258,*1*. **P. mediterraneus; 1a,b,* lat. and dorsal views of theca, ×1 (Yakovlev, 1934).

Iberocrinus SIEVERTS-DORECK, 1951, p. 105 [**I. multibrachiatus*; OD]. Rays incorporated in calyx up to tertibrachs and in some rays to proximal quartibrachs; interprimibrachs few, 2 in 2nd row; *CD* interray wide, depressed, largely merging with tegmen, 3 plates in 2nd row. Tegmen conical, with subcentral spine; anal opening excentric. Free arm bases 4 to 5 in each ray, but 6 in *C* and *D* rays. *U.Carb.*, Eu.; *L.Penn.*, ?N.Am. [*Megaliocrinus exotericus* STRIMPLE (1951a), from L.Penn. of Oklahoma, seems nearer to *Iberocrinus* than to *Megaliocrinus*.]——FIG. 258,*2*. **I. multibrachiatus,* Westphalian D, Spain; *2a-d,* lat., *CD*-interray, dorsal and ventral views of holotype theca, ×2 (Sieverts-Doreck, 1951a).

Megaliocrinus MOORE & LAUDON, 1942, p. 68 [**M. aplatus*; OD]. Rays incorporated in calyx up to tertibrachs and in some rays to quartibrachs; interprimibrachs few, 1 or 2 in 2nd row; *CD* interray not depressed, narrowly connected with tegmen, 3 plates in 2nd row. Tegmen high conical, many plated, terminating in a large, blunt point; anus near summit. Free arm bases 5 in each ray, but 6 in *C* and *D* rays, in lateral contact all around except at posterior side. Column unknown. *L.Penn.*, USA(Okla.).——FIG. 258,*3*. **M. aplatus; 3a,b,* dorsal and *CD*-interray views of theca, ×2 (Moore & Laudon, 1942); *3c,* hypotype, calyx from side, ×2.7 (Moore & Strimple, 1973a).

Wannerocrinus MAREZ OYENS, 1940, p. 294 [**W. glans*; OD]. Rays incorporated in calyx up to 3rd tertibrachs; single interprimibrach in each interray; *CD* interray wider than others, not connecting with tegmen, 2 plates in 2nd row. Tegmen dome shaped, with subcentral spine-bearing plate. Free arm bases in lateral contact all around, 4 in each ray. Column unknown. *U.Perm.*, Indon.(Timor).——FIG. 259,*1*. **W. glans; 1a-d, E*-ray, *CD*-interray, dorsal, and ventral views of holotype theca, ×1 (Marez Oyens, 1940b).

Family AMPHORACRINIDAE Bather, 1899

[Amphoracrinidae BATHER, 1899b, p. 922]

Fixed secundibrachs one or two in each half-ray; fixed tertibrachs absent or present; interprimibrachs few, connected with tegmen; primanal typically hexagonal and followed by two plates, rarely heptagonal and followed by three plates; calycal plates generally with granulo-vermicular ornament. Tegmen usually high in relation to calyx; anal tube moderately developed. Proximal region of arms horizontal or projecting

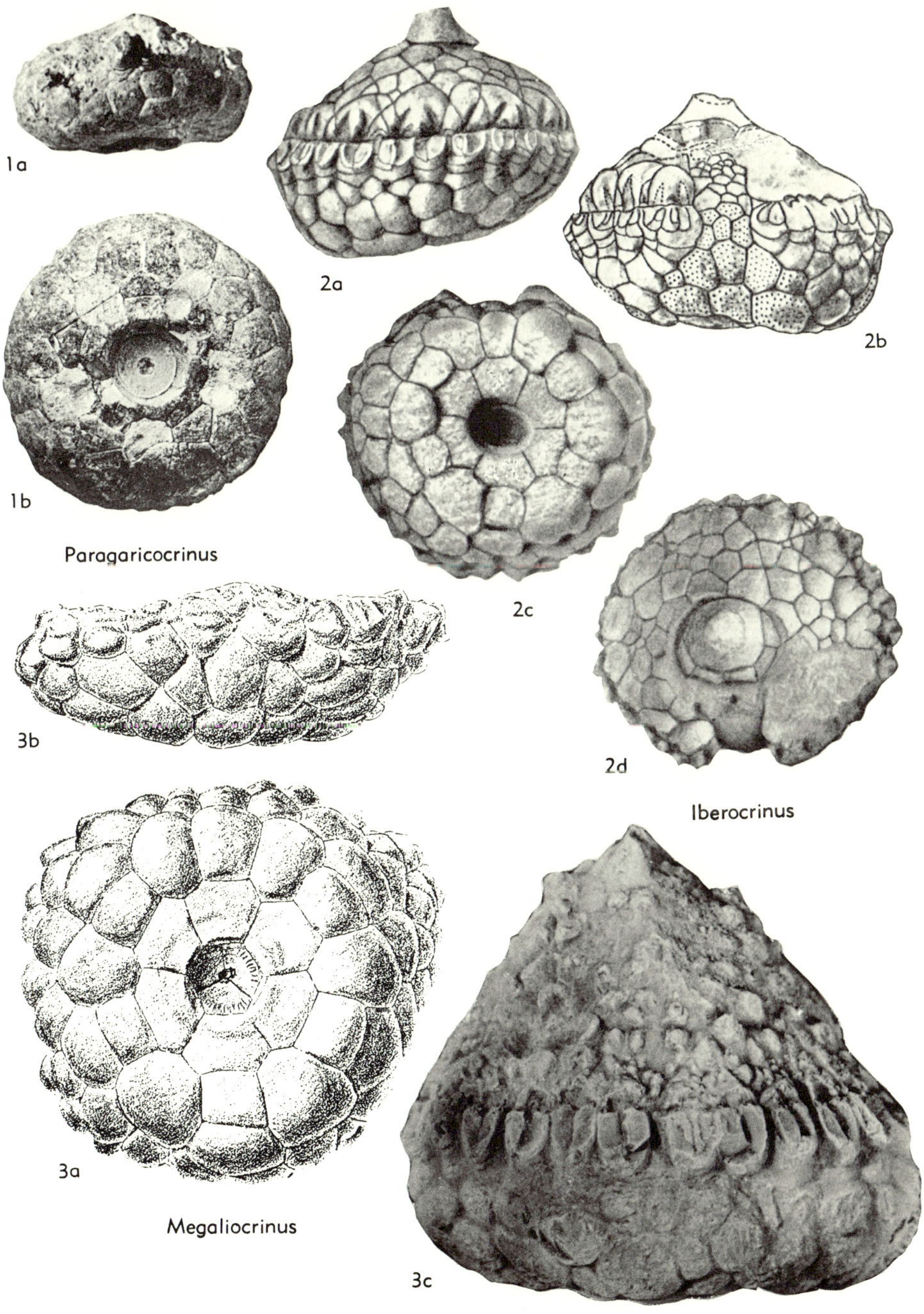

FIG. 258. Paragaricocrinidae (p. *T*450).

downward. *L.Carb.(Tournais.)-U.Carb.(Namur.).*

Amphoracrinus AUSTIN, 1848, p. 292 [**Actinocrinus gilbertsoni* MILLER in PHILLIPS, 1836, p. 206; SD ROEMER, 1855, p. 250]. Calyx high or low, rounded or conical; fixed secundibrachs 2 in each

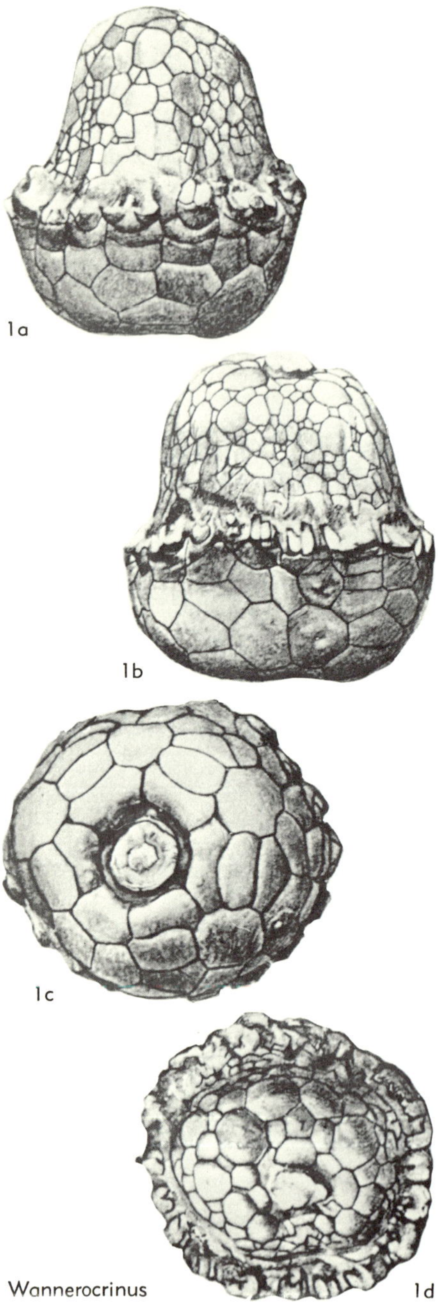

FIG. 259. Paragaricocrinidae (p. T450).

half-ray; primanal typically followed by 2 or rarely 3 plates. Tegmen usually more than calyx in height, but may be equal to or less than this, tapering upward and prominently lobed, generally flat summit occupied by 5 orals, 4 of which form semicircle enclosing larger posterior one; anal tube excentric on distal side of posterior oral. *L.Carb.*, Eu.; ?Japan; *U.Carb.(low.Namur.)*, N.Afr.; *Miss.*, N.Am.——FIG. 260,*1*. **A. gilbertsoni* (MILLER), L.Carb., Eng.; *1a-c, A*-ray, *CD*-interray, and ventral views of neotype theca (anal tube lacking), ×1 (Wright, 1955-60); *1d, BC*-interray view of another specimen, with anal tube partly preserved, ×1 (Wright, 1955-60).

Ectocrinus WRIGHT, 1955, p. 204 [**Actinocrinus (Amphoracrinus?) olla* McCOY, 1849, p. 248; OD]. Calyx rounded; 2 fixed secundibrachs in each half-ray. Tegmen constricted, with steeply sloping or concave sides and flat or slightly concave summit, composed of 5 large oral plates with posterior one largest; anal tube directed sideward immediately under or at short distance below posterior oral. *L.Carb.(Visean)*, Eng.——FIG. 260,*2*. **E. olla* (McCOY); *2a-c, A*-ray, *CD*- interray, and ventral views of lectotype theca, ×1 (Wright, 1955-60).

Pimlicocrinus WRIGHT, 1943, p. 89 [**Amphoracrinus clitheroensis* WRIGHT, 1942, p. 272; OD]. Calyx low to very shallow; low basals may be covered by column; single fixed secundibrach in each half-ray, axillary; 1 or 2 fixed tertibrachs usually present in half-rays. Tegmen moderately to very high in relation to calyx, without conspicuous orals and with subcentral anal tube. *L.Carb.*, Eu.(Eng.).——FIG. 260,*3*. **P. clitheroensis* (WRIGHT), Tournais.; *3a-d, A*-ray, *CD*-interray, dorsal, and ventral views of holotype theca, ×1 (Wright, 1955-60).

Family ACTINOCRINITIDAE Austin & Austin, 1842

[*nom. correct.* BASSLER, 1938, p. 27 (*pro* family Actinocrinoidea AUSTIN & AUSTIN, 1842, p. 109, based on *Actinocrinites* MILLER, 1821, p. 15)] [=Actinocrinidae AUSTIN & AUSTIN, 1843, p. 199; tribu Actinocriniens PICTET, 1857, p. 321 (vernacular)] [Materials for this family prepared by A. L. BOWSHER, revised by GEORGES UBAGHS]

Calyx generally conical, less commonly bowl shaped; lower brachials usually forming rather important part of calyx; interprimibrachs numerous, connected with tegmen or not; primanal hexagonal with two plates next above it; calycal plates generally ornamented by axial folds. Tegmen solid, orals not prominent or distinct, ambulacrals may be differentiated; anus either on tube or piercing tegmen. Free arms bifurcating according to different patterns, in all known species (except in *Dialutocrinus*) becoming biserial just above calyx or forming uniserial trunks with biserial armlets. *L.Carb. (L.Miss.)-U.Carb., ?Perm.*

Actinocrinitidae comprise a highly diversified

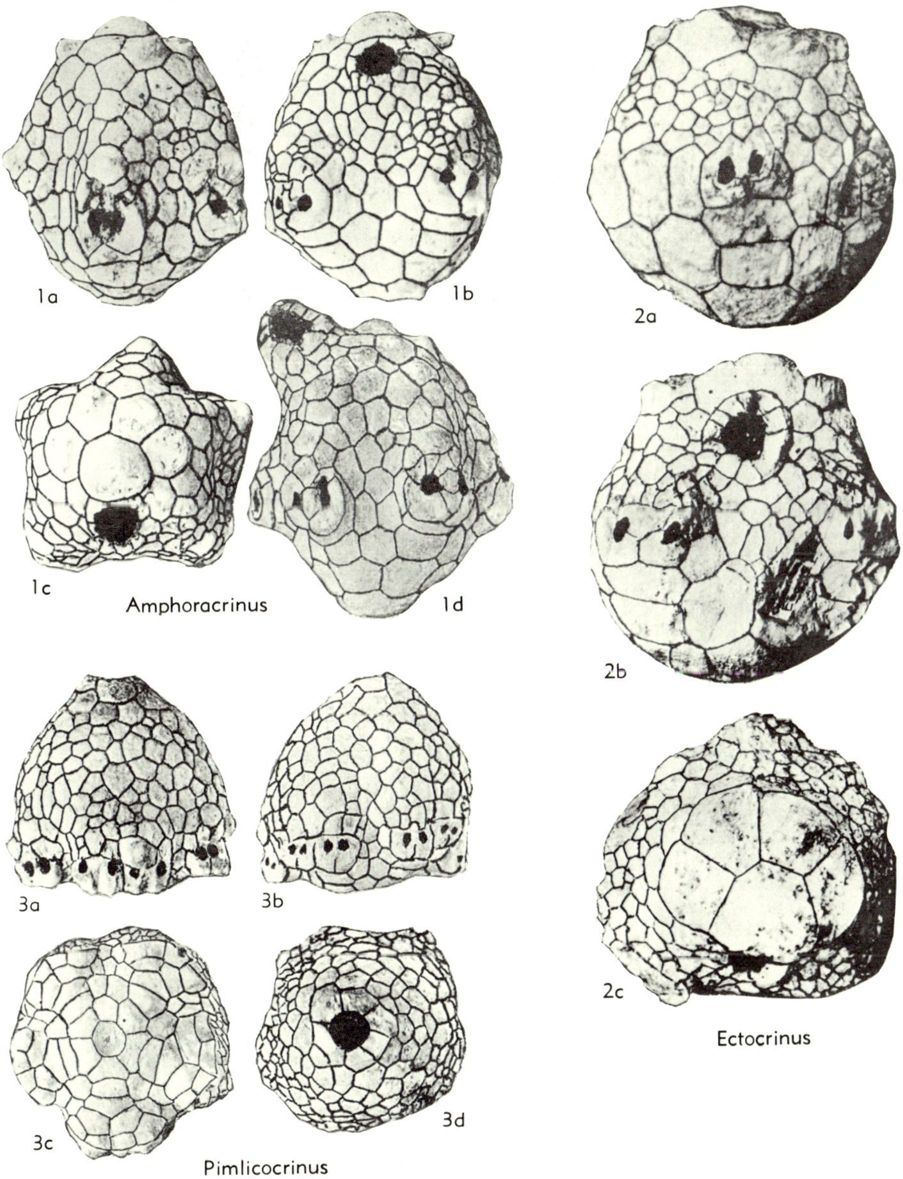

FIG. 260. Amphoracrinidae (p. *T*451-*T*452).

group, mostly of Mississippian age. In a recent review, BROWER (1967) has recognized four sections of subfamily rank based on common morphology and evolutionary trends; however, he did not give them formal names. The same classification is adopted herein, but each section is given a subfamily status and name.

Subfamily ACTINOCRINITINAE Austin & Austin, 1842

[*nom. transl.* UBAGHS, herein (*ex* Actinocrinitidae BASSLER, 1938, p. 27, *pro* Actinocrinoidea AUSTIN & AUSTIN, 1842, p. 109)] [=Aacocrinid section BROWER, 1967, p. 676 (vernacular)] [=Actinocrininae JAEKEL, 1918, p. 36]

Fixed portion of arms strongly grouped,

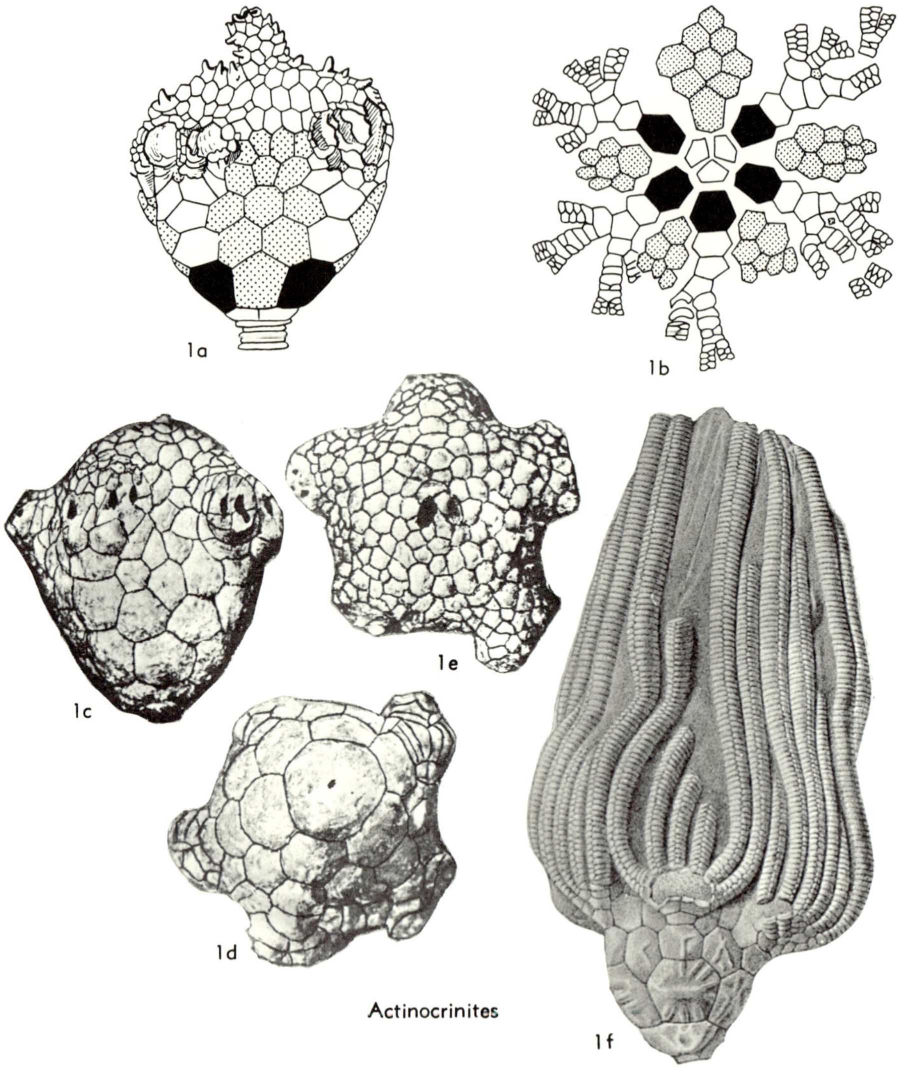

FIG. 261. Actinocrinitidae (Actinocrinitinae) (p. T454-T455).

forming protuberant brachial lobe in each ray. *L.Carb.(L.Miss.)-U.Carb.; ?Perm.*

The dominant evolutionary trend of this group is in development of lobation. This is accomplished by producing outwardly flared fixed brachials and ramifying arms of the rays in various ways. Other evolutionary trends shown by all or part of the group are 1) development of a large-plated tegmen with anal tube, 2) development of externally rectangular first primibrachs, 3) changes in thecal shape, and 4) addition of arms (BROWER, 1967-69).

Actinocrinites MILLER, 1821, p. 15, 94 [**A. triacontadactylus*; SD WACHSMUTH & SPRINGER, 1881, p. 146(320)] [=*Actinocrinus* AGASSIZ, 1836, p. 196 *(nom. van.)*]. Calyx usually elongate conical; fixed brachials relatively small, up to 3rd or 4th tertibrachs outwardly flared and braced by complex arrangement of covering plates and tegminals; brachial lobes widely separated by interprimibrachs. Tegmen somewhat variable in height, many-plated; orals and ambulacrals generally undifferentiated; central or subcentral anal tube prominent. Free arms biserial, 30 or perhaps more (number indeterminant in type and other British species), usually (but not invariably) unbranched and grouped in clusters. *L.Miss.-Penn., ?Perm.,* N.Am.; *L.Carb.-U.Carb., ?Perm.,* Eu.-N.Afr.-E.Australia-?Indon.(Timor). [Species of Permian age assigned to *Actinocrinites* by WANNER (1916, 1924) are highly divergent. Many of them have very massive and protuberant brachial lobes. Their first primibrachs are rectangular and primaxils commonly are pentagonal. In some the primanals are followed by 3 plates. They probably represent a late and specialized development from some unknown actinocrinitid stock.]——FIG. 261,*1a-e.* **A. triacontadactylus,* L.Carb., Eng.; *1a, CD*-interray view of theca (radials black, anals and interbrachials stippled), ×1 (Ubaghs, n, based on Miller, 1821, pl. II, opp. p. 98); *1b,* plate diagram (radials black, anals and interbrachials stippled) (Bowsher, n); *1c-e, CD*-interray, dorsal, and tegminal views of theca, ×1 (Wright, 1955-60).——FIG. 261,*1f. A. multiradiatus* (SHUMARD), L.Miss.(Osag.), USA(Iowa); *A*-ray view of crown, ×1 (Wachsmuth & Springer, 1897).

Aacocrinus BOWSHER, 1955, p. 3 [**A. nododorsatus*; OD]. Calyx conical, wide to elongate, composed of large, more or less tumid plates; distal fixed brachials large, directed semihorizontally. Tegmen low arched, commonly with nodose orals and ambulacrals. Free arms practically unknown, numbering (as judged from ambulacral tracts) 2 to 4 (exceptionally 3) in ray. Stem unknown. *L.Miss.(Kinderhook.),* N.Am.(USA).——FIG. 262,*1.* **A. nododorsatus,* Chouteau Ls., USA(Mo.); *1a-c, CD*-interray, *E*-ray, and ventral views of theca, ×2 (Bowsher, 1955b); *1d-f,* plate diagrams of *CD*-interray, *E*-ray and ventral sides of calyx (radials black, anals and interbrachials stippled), ×1.3 (Bowsher, 1955b).

Abactinocrinus LAUDON & SEVERSON, 1953, p. 528 [**A. rossei*; OD]. Calyx conical, strongly lobate, expanding rapidly upward, and flaring at top of 1st primibrachs into tubelike projections; fixed brachials comprising primibrachs and 1st secundibrachs only; interrays constricted at arm level, giving markedly pentagonal outline in dorsal and ventral views; ambulacral openings 2 in each ray, directed laterally and slightly upward. Tegmen low conical, many-plated, with depressed interambulacral areas and prominent ambulacral tracts; anus excentric, opening directly through tegmen. Free arms and stem unknown. *L.Miss.(Kinderhook.),* N.Am.(USA).——FIG. 262,*2.* **A. rossei,* Lodgepole F., USA(Mont.); *2a-c,* post., *B*-ray, and ventral views of holotype theca, ×2 (Laudon and Severson, 1953).

Blairocrinus S. A. MILLER, 1891, p. 69 [**B. trijugis*; M]. Calyx low, bowl shaped; basals forming flattened disc; cup plates tumid, generally marked by heavy stellate ridges. Tegmen convex, composed of large plates, all or part of them spinose or nodose, orals and ambulacrals differentiated by size and spinosity; anal tube subcentral, stout at base; arm facets 4 in each ray, facing obliquely upward. Free arms and stem unknown. *L.Miss.(Kinderhook.),* N.Am.; *L.Carb.,* Eu.(Belg.).——FIG. 263,*1.* **B. trijugis,* Compton Ls., USA(Mo.); *1a-c, CD*-interray, dorsal, and tegminal views of holotype theca, ×2 (Peck & Keyte, 1938).

Diatorocrinus WRIGHT, 1955, p. 235 [**D. angustus*; OD]. Calyx similar to that of *Actinocrinites*; brachial lobes strongly developed. Tegmen low convex to medium conical, many-plated; anus subcentral or excentric, directly through tegmen. Free arms (at least 4 in each ray) and column unknown. *L.Carb.(low.Visean),* Eu.(Eng.).——FIG. 263,*2.* **D. angustus; 2a,b, DE*-interray and tegminal views of theca, ×1 (Wright, 1955-60).

Sampsonocrinus MILLER & GURLEY, 1895, p. 51 [**S. hemisphericus*; OD] [=*Amphora* CUMBERLAND, 1826, p. 36 (type, *A.* 1st species, unnamed, =*Encrinites loricatus* VON SCHLOTHEIM, 1820)]. Like *Actinocrinites* but with more bowl-shaped and less elongate calyx, relatively larger radials, distal portion of 2nd primibrachs flaring outward and forming part of underside of brachial lobes. Large-plated tegmen with poorly developed anal tube and fewer arm openings (4 in ray) than in *Actinocrinites.* Free arms and stem unknown. *L.Miss.(Kinderhook.),* USA, *L.Carb.,* G.Brit.——FIG. 263,*3.* **S. hemisphericus,* Chouteau Ls., USA (Mo.); *3a-c, BC*-interray, dorsal, and ventral views of holotype, ×1.5 (Peck & Keyte, 1938).

Steganocrinus MEEK & WORTHEN, 1866, p. 195 [**Actinocrinus pentagonaus* HALL, 1858b, p. 577; OD] [=*Shumardocrinus* MILLER & GURLEY, 1895, p. 40 (type, *Actinocrinus concinnus* SHUMARD, 1855; OD)]. Calyx conical, bowl shaped, to highly elongate, strongly lobate; distal portion of 1st primibrachs and primaxils directed almost horizontally, forming base of ray trunk; calycal plates typically stellate, rarely almost smooth or highly tumid; interradial areas wide, connecting with tegmen, which is low arched and generally made up of few fairly large and heavy plates, rarely composed of many small plates; subcentral or eccentric anal tube usually present. Free arms (known only in *S. pentagonus*) consisting of 2 strong, long, tubular, uniserial, and nonpinnulate trunks, composed of high brachials, each of which bears relatively short, stout, biserial and pinnulate ramule on alternate sides of ray trunk. Column unknown. *L.Miss.(Osag.),* N.Am.(USA).——FIG. 263,*4.* **S. pentagonus* (HALL), Burlington Ls., USA(Iowa-Mo.-N.Mexico); *4a,b, CD*-interray and tegminal views of theca, ×1 (Wachsmuth & Springer, 1897); *4c,* dorsal view of incomplete crown with arms preserved, ×1 (Wachsmuth & Springer, 1897).

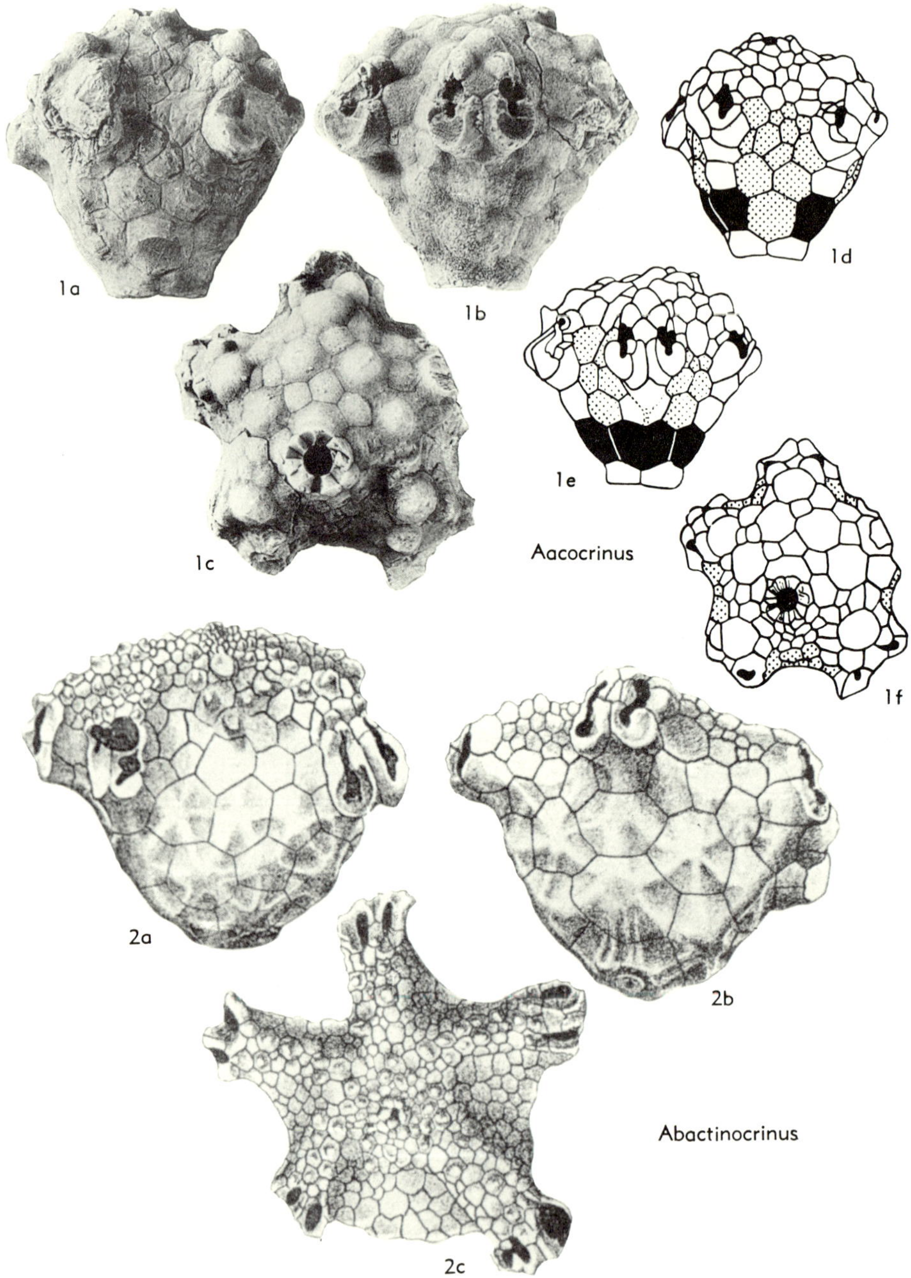

FIG. 262. Actinocrinitidae (Actinocrinitinae) (p. T455).

Subfamily EUMORPHOCRININAE Ubaghs, new subfamily

[=eumorphocrinid section BROWER, 1967, p. 675 (vernacular)]

Fixed portions of arms strongly grouped, but never forming protuberant brachial lobes; fixed brachials relatively numerous; tegmen many-plated. *L.Carb.*

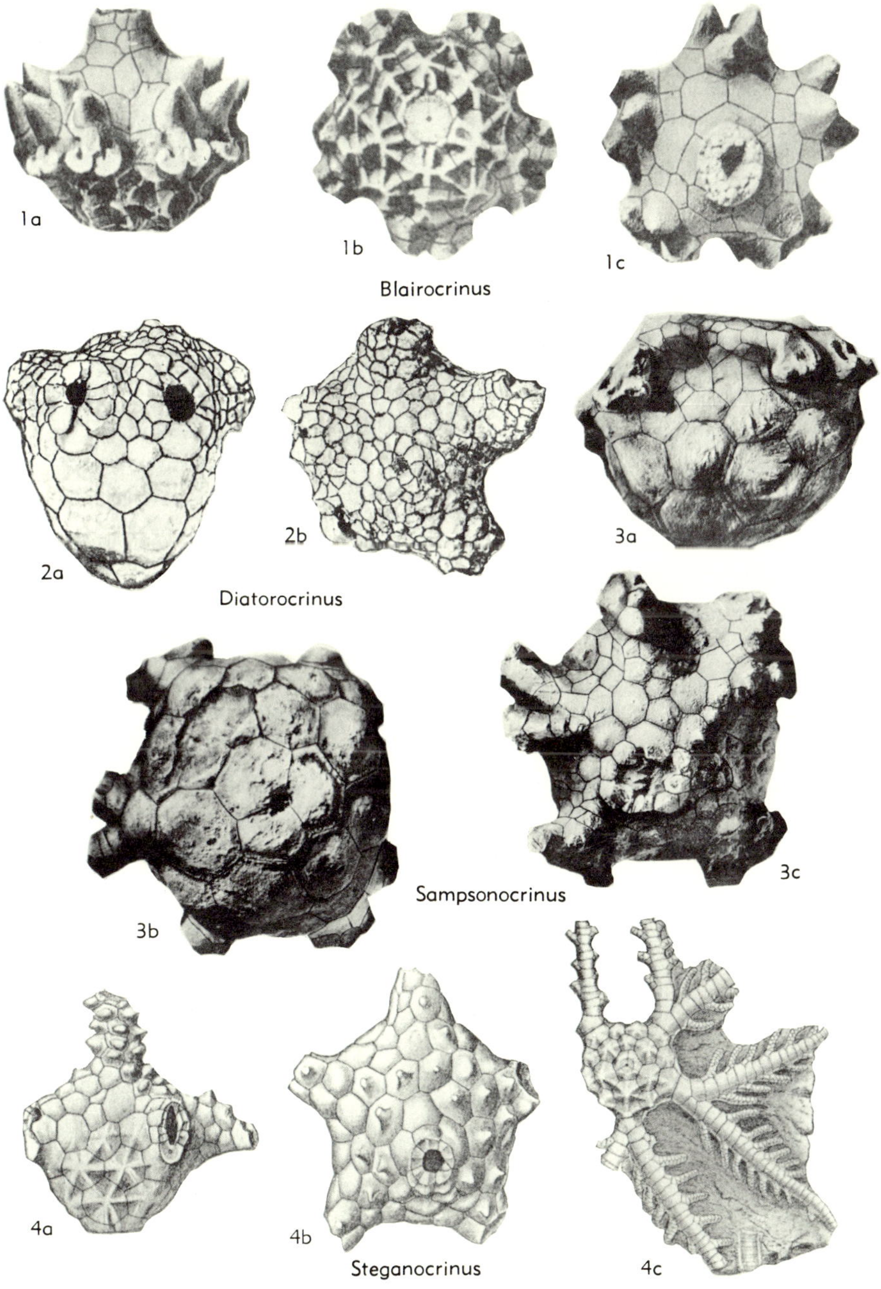

FIG. 263. Actinocrinitidae (Actinocrinitinae) (p. *T*455).

In the nature of their calyx, the crinoids included in this subfamily closely resemble the Periechocrinidae, from which they differ mainly in having two plates in the second range of the *CD* interray. Most of them (particularly *Maligneocrinus,* Kinderhookian) may indeed be direct descendants of some periechocrinid stock; however, *Eumorphocrinus,* the type of the subfamily, being early Visean in age, appears as a relatively late actinocrinitid. It is therefore suggested (LANE, personal communication) that its "primitive" features could have been secondarily acquired from a more advanced, but older, ancestor. If so, the Eumorphocrininae would unite genera of basically similar calyx structure but not necessarily related to each other phylogenetically.

Eumorphocrinus WRIGHT, 1955, p. 232 [**E. erectus*; OD]. Calyx conical or slightly rounded; 2nd secundibrachs fixed and axillary; interprimibrachs numerous, openly connected with tegmen; no brachial lobes, but strongly grouped fixed brachials may project at slight angle from theca; brachial openings 4 in each ray, directed more or less upward. Tegmen low to steeply conical, many-plated, with central anal tube. Free arms unknown above 2nd or 3rd tertibrachs. *L.Carb. (low.Visean),* Eu.(Eng.-N.Eire).——FIG. 264,*1*. **E. erectus; 1a,b, AE*-interray and tegminal views of holotype theca, ×1 (Wright, 1955-60).

Cytidocrinus KIRK, 1944, p. 85 [*pro Cyrtocrinus* KIRK, 1943a, p. 263 (*non* JAEKEL, 1891)] [**Actinocrinus sculptus* HALL, 1858b, p. 582; OD]. Calyx conical, with practically no lobation at arm level; interprimibrachs widely connected with tegmen; each ray divided into small, biserial, pinnuliferous ramus and strong, rigid, uniserial, nonpinnuliferous, tubular arm-trunk which carries long, stout, biserial, pinnuliferous ramuli on alternate sides on each 2nd brachial. Tegmen high, many-plated, with central anal tube. *L.Miss. (Osag.),* N.Am.——FIG. 264,*2*. **C. sculptus* (HALL), USA(Iowa); *2a,* plate diagram (radials black, anals and interbrachials stippled) (Bowsher, n); *2b,c,* lat. view of theca and *CD*-interray view of crown, ×0.7 (Wachsmuth & Springer, 1897).

Maligneocrinus LAUDON, PARKS, & SPRENG, 1952, p. 564 [**M. medicinensis*; OD]. Calyx conical, broadly expanded; plate ornamentation stellate; 1st primibrachs hexagonal, primaxils typically 7-sided; single fixed secundibrach in each half-ray, axillary; 1 fixed tertibrach in each ramus, followed by free arm; brachial openings 4 in ray; interprimibrachs relatively large in 1st 3 ranges and small at arm level; *CD* interray wider than others. Tegmen consisting of many small plates, with interambulacral areas slightly depressed. Free arms and column unknown. *L.Miss.(Kinderhook.),* N. Am.(Can.). [This genus originally placed among Periechocrinidae has been assigned to the Actinocrinitidae by BROWER (1967).]——FIG. 264,*3*. **M. medicinensis,* Banff F., Alberta; *3a,b, A*-ray and dorsal views of holotype theca, ×1 (Laudon, Parks & Spreng, 1952); *3c, CD*-interray view of paratype, ×1 (Laudon, Parks, & Spreng, 1952).

Manillocrinus CAMPBELL & BEIN, 1971, p. 427 [**Cactocrinus? brownei* DUN & BENSON, 1920, p. 342; OD]. Calyx steep conical; plate ornamentation stellate; 1st primibrachs hexagonal, primaxils octagonal; 2nd secundibrachs axillary; interprimibrachs well developed, widely connecting with tegmen; intersecundibrachs present; each ray divided into 4 rami, consisting of 2 small biserial pinnuliferous unbranched arms on outer sides and 2 main uniserial nonpinnuliferous exotomous trunks on inner sides; main arm trunks parallel, bearing biserial undivided pinnuliferous armlets on opposite sides at more or less regular intervals. Stem round, columnals alternating proximally, equidimensional and cirriferous distally. *L.Carb.,* E.Australia.——FIG. 265,*1*. **M. brownei* (DUN & BENSON); *1a,* plate diagram of calyx and *D*-ray arm (radials black, anals and interbrachials stippled) (Campbell & Bein, 1971); *1b, D*-ray view of crown with attached stem, ×1.5 (Campbell & Bein, 1971).

Subfamily CACTOCRININAE
Ubaghs, new subfamily

[=cactocrinid section BROWER, 1967, p. 676]

Fixed portions of arms weakly grouped or ungrouped (except in some species of *Dialutocrinus*); rays never protuberant. Tegmen composed of large undifferentiated plates, anal tube present. *L.Carb.*

Most genera of this subfamily show two dominant evolutionary trends. As additional arms are incorporated into the calyx, distal ranges of interbrachials tend to be resorbed or displaced out of the calyx, whereas the arms, less and less grouped, tend to be arranged in a continuous belt around the calyx.

Cactocrinus WACHSMUTH & SPRINGER, 1897, p. 600 [**Actinocrinus proboscidialis* HALL, 1858b, p. 584; OD]. Calyx conical, wider than high, with arm bases laterally in contact all around; interprimibrachs numerous, not connecting with tegmen. Tegmen high, steeply conical, large-plated, passing into prominent, almost central, anal tube. Free arms biserial, 4 to 8 in each ray, equidistant or nearly so, forming continuous ring; pinnules commonly with sharp-pointed hooks. Column large, with pentangular axial canal. *L.Miss. (Osag.),* N.Am.(USA).——FIG. 266,*1a,b*. **C. proboscidialis* (HALL), low. Burlington Ls., USA (Iowa-Mo.-N.Mexico); *1a, A*-ray view of theca, ×1 (Wachsmuth & Springer, 1897); *1b,* plate diagram of calyx and arms (radials black; interbrachials stippled) (Bowsher, n).——FIG. 266,*1c*. *C. multibrachiatus* (HALL), low. Burlington Ls., USA(Iowa); *A*-ray view of crown, ×1 (Wachsmuth & Springer, 1897).

Cusacrinus BOWSHER, 1955, p. 7 [**Actinocrinus nodobrachiatus* WACHSMUTH & SPRINGER in S. A. MILLER, 1889, p. 219; OD]. Calyx steeply conical; interprimibrachs numerous, 1 or 2 at arm level, connecting with tegmen. Tegmen low arched to

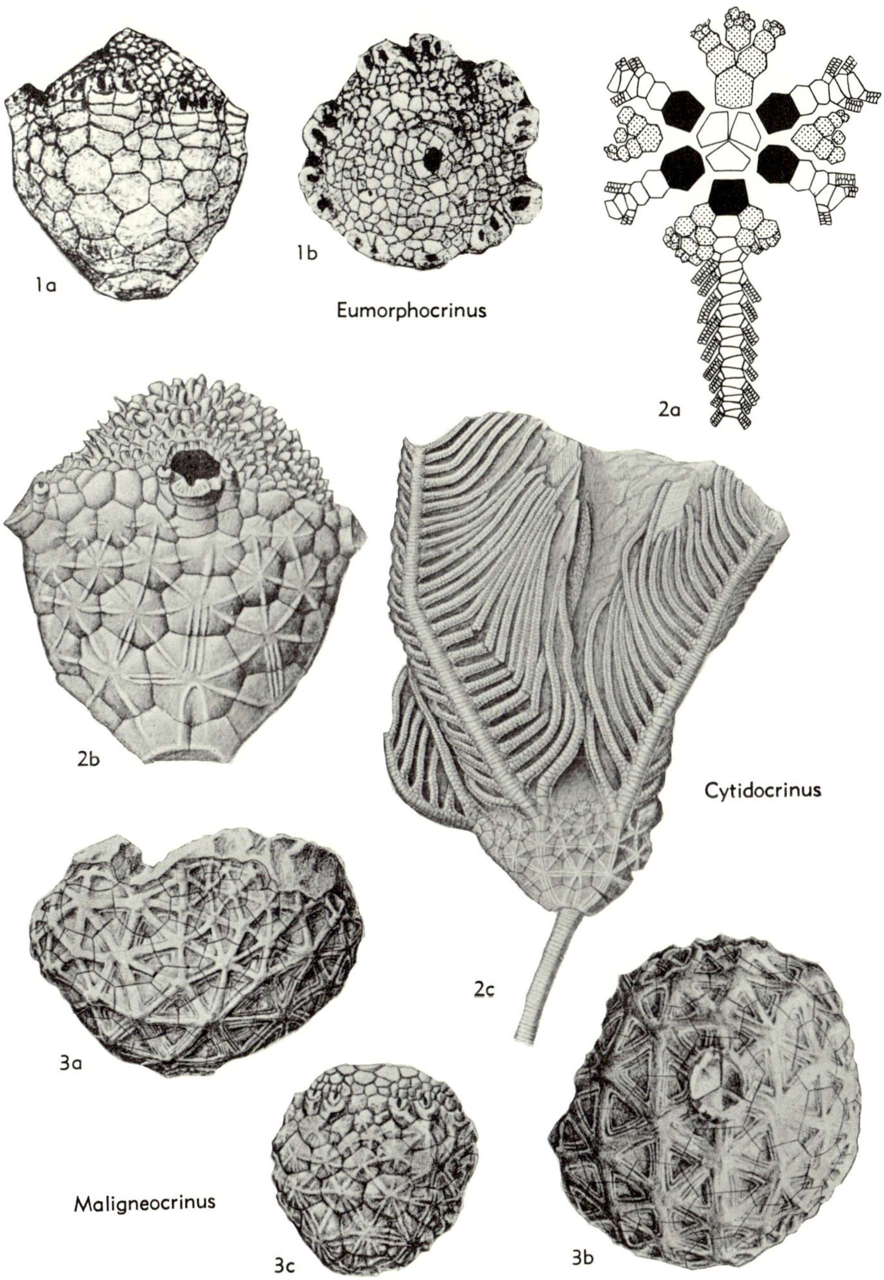

FIG. 264. Actinocrinitidae (Eumorphocrininae) (p. *T*458).

conical, large-plated, bearing prominent central anal tube. Arms biserial, weakly grouped, 6 or 8 (only 5 in some species) in each ray, unbranched. Column relatively large, composed of nudinodals and internodals. *L.Miss.(Kinderhook.-Osag.)*, N.Am.(USA).——FIG. 266,2. **C. nodobrachiatus* (WACHSMUTH & SPRINGER), Kinderhook, USA(Iowa); *2a*, crown showing incurling arms, ×1 (Bowsher, 1955b); *2b-d*, drawings of *A*-ray, *CD*-interray and dorsal sides of theca

FIG. 265. Actinocrinitidae (Eumorphocrininae) (p. T458).

(radials black, anals and interbrachials stippled), ×1.3 (Bowsher, 1955b).

Dialutocrinus WRIGHT, 1955, p. 241 [**D. milleri*; OD]. Calyx conical or rounded; fixed brachials weakly or strongly grouped according to species; interprimibrachs numerous, one, two, or more at arm level; tegmen rounded, rising to anal tube which is long and composed of small or large polygonal plates. Free brachials remaining uniserial for some distance above calyx and not becoming biserial until above last branching; each ray dividing into 6 to 10 or even more free branches, which are long, and in some species characterized by thornlike processes on aboral surface; division of ray as shown by Fig. 267,*1b-d*. Column composed of alternating nudinodals. *L. Carb.(Tournais.)*, Eu.(Eire-Eng.-Belg.).——FIG. 267,*1a*. *D. laevissimus* (AUSTIN & AUSTIN), Eire; ×1 (Wright, 1955-60).——FIG. 267,*1b-d*. Ray structure in **D. milleri*, Eng., Eire *(1b)*, *D. aculeatus* (AUSTIN & AUSTIN), Eire *(1c)*, and *D. laevissimus* (AUSTIN & AUSTIN), Eire *(1d)* (Wright, 1955-60).

Nunnacrinus BOWSHER, 1955, p. 17 [**N. mamillatus*; OD]. Calyx conical; fixed portions of arms weakly grouped and separated by 1 or 2 elongate interprimibrachs. Tegmen large-plated, with central anal tube. Free arms biserial, stout, unbranched, 4 in each ray. *L.Carb.*, Eu.; *L.Miss.*, N.Am.(USA).——FIG. 267,*2*. **N. mamillatus*, Lake Valley F.(Osag.), USA(N.Mexico); *2a-d*, drawings of theca from *A*-ray, *CD*-interray, dorsal, and ventral sides (radials black, anals and interbrachials stippled), ×1.3 (Bowsher, 1955b).

Teleiocrinus WACHSMUTH & SPRINGER, 1881, p. 146(320) [**Actinocrinus umbrosus* HALL, 1858b, p. 590; OD] [=*Calathocrinus* HALL, 1861b, p. 12 (*non* VON MEYER, 1848) (type, *Actinocrinus (Calathocrinus) erodus* HALL, 1861b, p. 12)]. Calyx steeply conical; interprimibrachs numerous, separated from tegmen by fixed brachials, which with interbrachials form flange at arm level. Free arms moderately long, biserial, simple, very numerous (up to 16 in ray), and closely crowded together. Tegmen convex, with strong nodose central anal tube. Column covered with vertical rows of angular processes. *L.Miss.(Osag.)*, N.Am. (USA).——FIG. 268,*1*. **T. umbrosus* (HALL), USA(Iowa); *1a*, lat. view of crown, ×1 (Wachsmuth & Springer, 1897); *1b,c*, *A*-ray and tegminal views of theca, ×1 (Wachsmuth & Springer, 1897); *1d*, plate diagram of calyx and arms (radials black, anals and interbrachials stippled) (Bowsher, n).

Subfamily PHYSETOCRININAE Ubaghs, new subfamily

[=physetocrinid section BROWER, 1967, p. 676 (vernacular)]

Fixed portions of arms usually weakly grouped, not protuberant, tending to be in lateral contact in advanced members; tegminals typically small; no anal tube. *L.Carb.*

Physetocrinines throughout their history retained a small-plated tegmen with no anal tube. Otherwise, their evolutionary trends closely paralleled those of the cactocrinine group. *Teleiocrinus* and *Strotocrinus* are judged to represent end products of similar evolutionary trends in unrelated crinoid lineages (BROWER, 1967).

Physetocrinus MEEK & WORTHEN, 1869, p. 158 [**Actinocrinus ventricosus* HALL, 1858b, p. 595; SD WACHSMUTH & SPRINGER, 1881, p. 157)]. Calyx cup conical to basin shaped, more or less elevated, distinctly lobed at arms level in early species, slightly so in later and more advanced species; interbrachials narrowly connected with tegmen. Tegminals relatively large (primitive con-

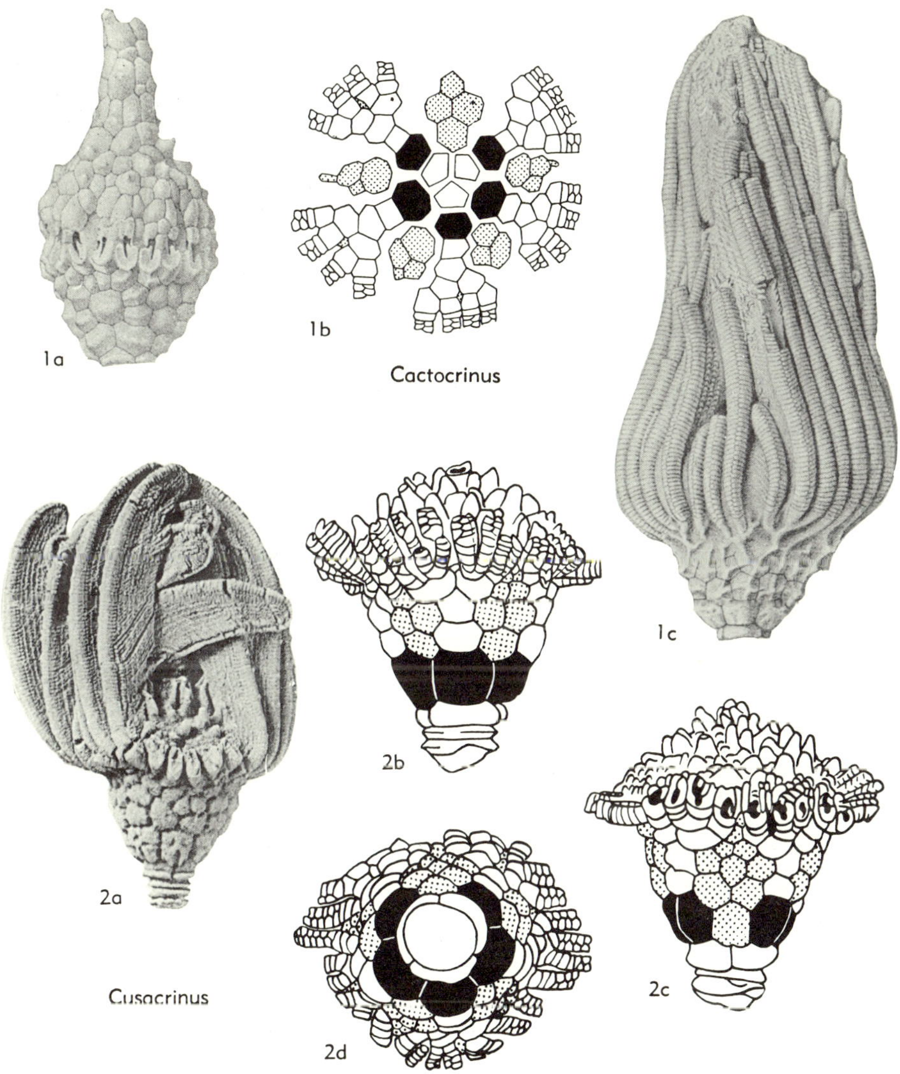

FIG. 266. Actinocrinitidae (Cactocrininae) (p. *T*458-*T*459).

dition) to smaller and smaller (advanced condition). Free arms biserial, simple, 4 in each ray, rather strongly grouped in early forms, slightly grouped in advanced members, but never forming rim around theca; bifurcating series branching on every brachial until full complement of arms is developed. *L.Miss.(Osag.)*, N.Am.; *L.Carb.*, G. Brit.-E.Australia.——FIG. 269,*1*. **P. ventricosus* (HALL), Burlington Ls., USA(Iowa); *1a, CD*-interray views of crown with attached stem, ×1 (Wachsmuth & Springer, 1897); *1b,c, A*-ray and ventral views of theca, ×1 (Wachsmuth & Springer, 1897); *1d,* plate diagram of calyx and arms (radials black, interbrachials stippled) (Bowsher, n).

Strotocrinus MEEK & WORTHEN, 1866, p. 253 [**Actinocrinus perumbrosus* HALL, 1860c, p. 7; SD S. A. MILLER, 1889, p. 284]. Calyx large, steeply conical; fixed brachials, and interbrachials forming very prominent flange at arms level from which arise very numerous (up to 30 in each ray) rather short, unbranched, biserial free arms. Tegmen flat or very slightly convex, composed of many minute irregular plates. *L.Miss.(Osag.)*, N.Am.(USA).——FIG. 270,*1a,b*. **S. perumbrosus* (HALL), USA(Iowa); *A*-ray and tegminal sides

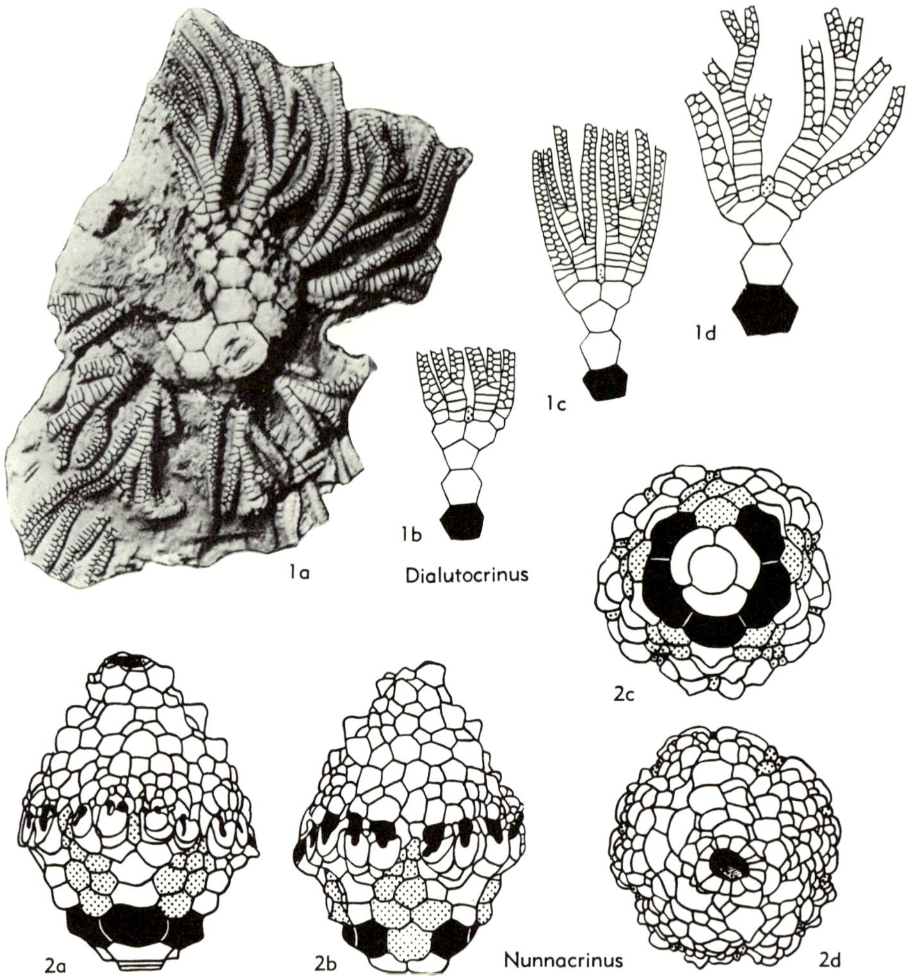

FIG. 267. Actinocrinitidae (Cactocrininae) (p. T460).

of theca (radials black, interbrachials stippled), ×1 (drawn from Wachsmuth & Springer, 1897). ——FIG. 270,*1c*. *S. glyptus* (HALL), USA(Iowa); oblique view of crown from below with attached stem, ×1 (Wachsmuth & Springer, 1897).

Superfamily CARPOCRINACEA de Koninck & Le Hon, 1854

[*nom. transl.* UBAGHS, herein (*ex* Carpocrinidae DE KONINCK & LE HON, 1854, p. 115)]

Basals three, generally equal; first primibrachs ordinarily quadrangular; primanal generally followed by three plates in next higher row. *M.Sil.-U.Sil.; M.Dev.-L.Carb.*

Family CARPOCRINIDAE de Koninck & Le Hon, 1854

[*nom. correct.* UBAGHS, herein, *pro* Carpocrinidées DE KONINCK & LE HON, 1854, p. 115] [=Carpocriniens PICTET, 1857, p. 326 *("tribu") (partim)*; Desmidocrinidae ANGELIN, 1878, p. 5; Habrocrinidae ANGELIN, 1878, p. 3; Barrandeocrinidae ANGELIN, 1878, p. 7; Carpocrinidae ZITTEL, 1879, p. 344; Methabocrinidae UBAGHS, 1958a, p. 52]

Fixed secundibrachs one to three in each half-ray, exceptionally more; interprimibrachs one to many, connecting with or separated from tegmen; *CD* interray primitively and typically with median series of anal plates. Tegmen solid, with orals and ambulacrals ordinarily distinct. Free arms simple, usually two in each ray, uniserial

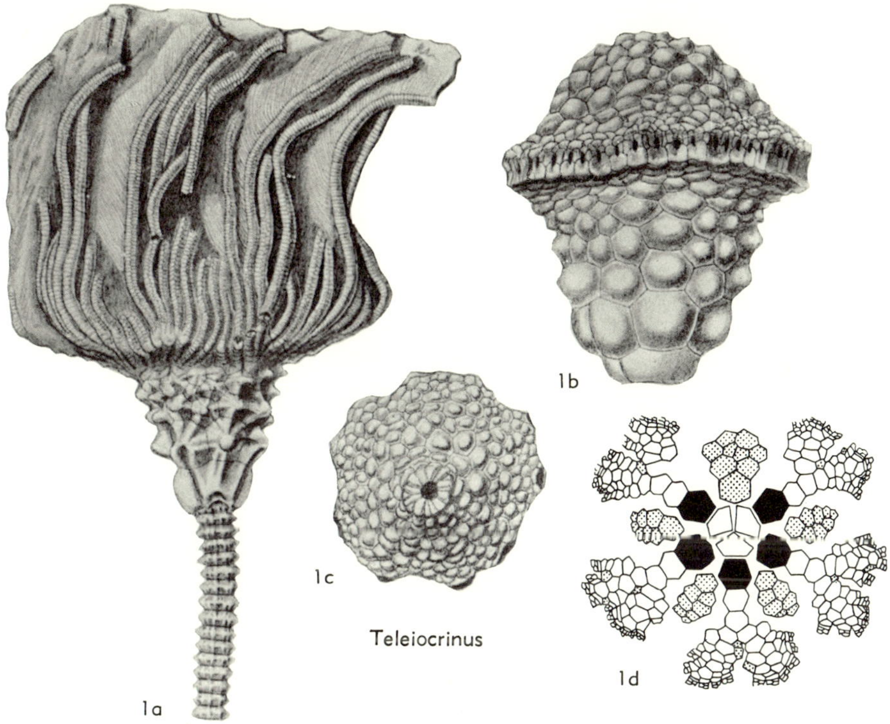

FIG. 268. Actinocrinitidae (Cactocrininae) (p. T460).

or biserial; brachials simple or compound. Column cylindrical. *M.Sil.-U.Sil.; M.Dev.*

Carpocrinus MÜLLER, 1840, p. 33 [**Actinocrinites simplex* PHILLIPS, in MURCHISON, 1839, p. 673; SD ROEMER, 1855, p. 237] [=*Carpiocrinus* AGASSIZ, 1842, p. 3 *(nom. null.); Phoenicocrinites* AUSTIN & AUSTIN, 1843, p. 205 (obj.); *Carpocrinites* GEINITZ, 1846, p. 550 *(nom. van.); Abracrinus* D'ORBIGNY, 1850, p. 47 (obj.); *Habrocrinus* ANGELIN, 1878, p. 3 *(nom. subst. pro Abracrinus); Pionocrinus* ANGELIN, 1878, p. 4 (type, *P. pulchellus*; SD BASSLER, 1938, p. 153); *Leptocrinus* ANGELIN, 1878, p. 3 (type, *L. raridigitatus*; M); *Phoenicocrinus* SPRINGER, 1913, p. 194 *(nom. van.)*]. Calyx conical to cyathiform; basals equal or unequal; 1st primibrachs 4- to 6-sided; interbrachial areas narrow, in contact with tegmen; *CD* interray wide, with prominent median series of anal plates leading to anus. Tegmen with ambulacrals distinct near center; posterior oral well differentiated, others small or undistinct; no anal tube. Free arms 2 to 4 in each ray; outer rami of some species reduced, without pinnules and resembling large proximal pinnules; brachials compound, bi- or hyperpinnulated. *M.Sil.*, Eu.-N.Am.——FIG. 271,*2a*. **C. simplex* (PHILLIPS), Wenlock., Eng.; part of crown, ×1 (Bather, 1900a).——FIG. 271,*2b,c*. *C. ornatus* (ANGELIN), Gotl.; *E*-ray view (Stockholm Mus. Ec. 11299), ×1.5, and *CD*-interray view (Stockholm Mus. no. Ec. 11300), ×1 (Ubaghs, n).

Acacocrinus WACHSMUTH & SPRINGER, 1897, p. 515 [**A. elrodi*; OD]. Calyx short and rotund; basals equal; 1st interprimibrachs large, followed by smaller plates connecting with tegmen; *CD* interray wide, composed of numerous plates arranged generally in 3 longitudinal rows. Arms free from 2nd or 3rd secundibrach, 2 in each ray, simple, uniserial, with wedge-shaped brachials. *M.Sil., M.Dev.*, N.Am.(USA).——FIG. 272,*2*. **A. elrodi*, Niagaran, USA(Ind.); side view of holotype crown, ×2 (Wachsmuth & Springer, 1897).

Barrandeocrinus ANGELIN, 1878, p. 7 [**B. sceptrum*; M]. Crown melon shaped, with permanently pendent arms, calyx cyathiform; basals subequal; radials irregularly shaped; interbrachial areas with 1 followed by 2 plates and indefinitely numerous, variously shaped platelets; primanal large, supporting 3 plates in next row; all interradial (including posterior) and interbrachial areas connecting with tegmen. Tegmen dome shaped, moderately high, with 10 large summit plates.

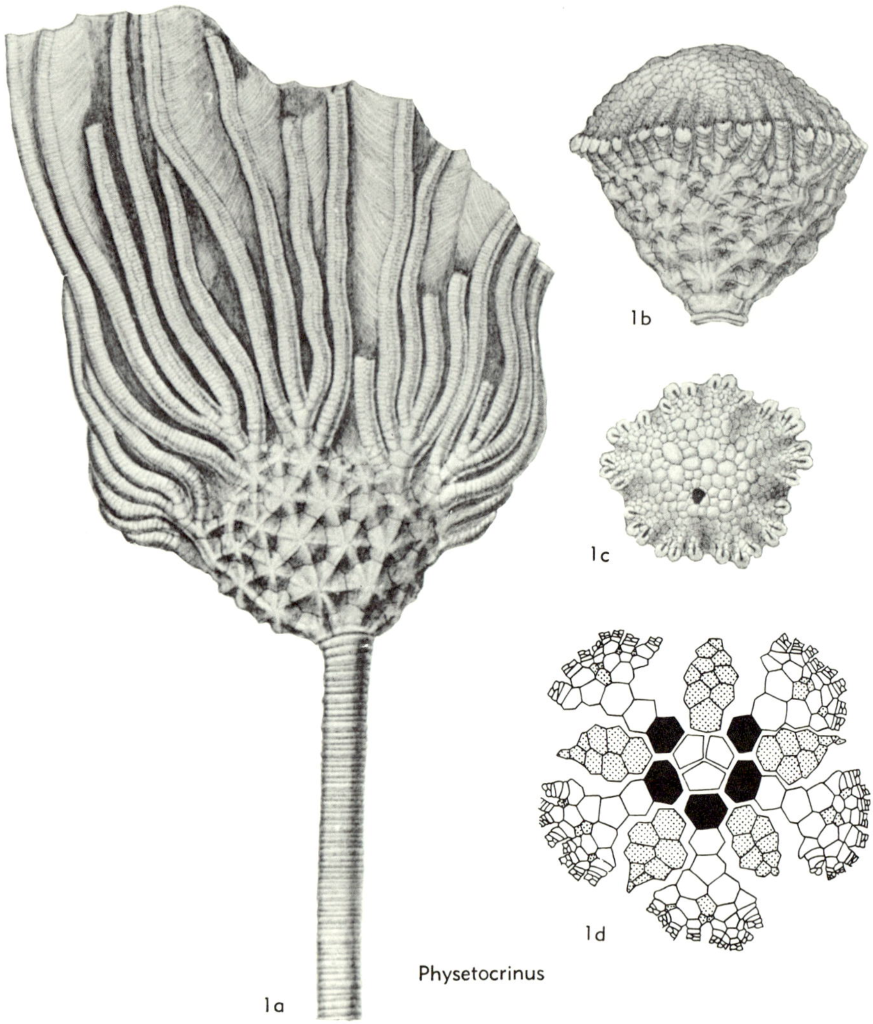

FIG. 269. Actinocrinitidae (Physetocrininae) (p. T460-T461).

Brachial insertions very large, secondarily extended over calyx; free arms 10, biserial, with very short, cuneate brachials; pinnules close-set, folded at midlength and attached together to form a firm external protective pavement enclosing 10 brachial chambers; very distal parts of pinnules free and allowing entrance of seawater into chambers; exhalant common aperture at apex of crown. Nudinodals concealing internodals of stem proximally; columnals all alike distally; radicular cirri present. *U.Sil.*, Eu.(Sweden).——FIG. 273,*1*. **B. sceptrum*, Gotl.; *1a*, lat. view of crown, ×1.5; *1b*, apical view of crown showing exhalant central aperture, ×1.5; *1c*, crown with 2 pendent arms removed, ×1.5; *1d*, lat. view of theca, ×2; *1e*, cross section of crown showing brachial chambers formed by rigid portions of pinnules, ×2; *1f*, plate diagram of calyx (radials black, interbrachials and anals stippled) (all Ubaghs, 1956).

Bohemicocrinus WAAGEN & JAHN, 1899, p. 15 [**B. pulverens*; M] [=*Bohemiocrinus* SHARPE, 1899 (Zool. Rec. 36), p. 15 *(nom. null.)*; *Vletavicrinus* WAAGEN & JAHN, 1899, p. 98 (type, *V. haueri*; M)]. Calyx vase shaped, elongate, basally rounded; basals equal; radials irregular; fixed secundibrachs 7 or 8; interbrachial areas long, narrow, many-plated, separated from tegmen by fixed pinnulars; *CD* interray without median series of anal plates, with only 1 plate resting on primanal. Free arms

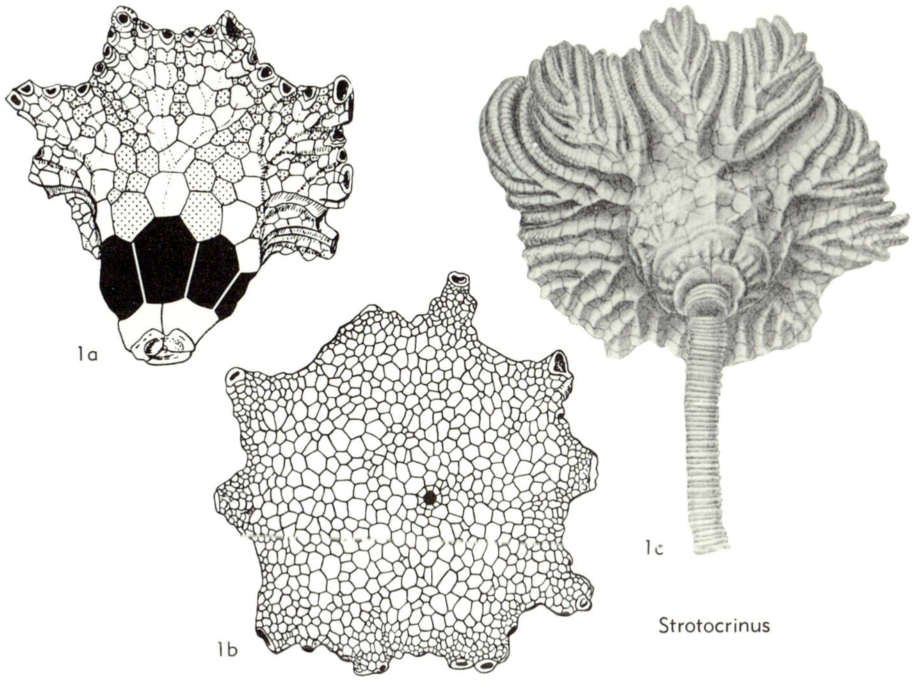

FIG. 270. Actinocrinitidae (Physetocrininae) (p. *T*461-*T*462).

2 in each ray (unknown in *Bohemicocrinus,* but isotomously branching and composed of compound brachials in *Vletavicrinus,* herein taken as synonym). *U.Sil.,* Eu.(Boh.).——FIG. 271,*1*. **B. pulverens; 1a-c,* drawings of theca from *A*-ray, *CD*-interray, and dorsal sides (radials black, anals and interbrachials stippled), ×0.7 (Ubaghs, n); *1d,* lat. view of crown with attached stem (type of *Vletavicrinus haueri,* Vienna Museum 2156), ×1 (Ubaghs, n).

Cylicocrinus S. A. MILLER, 1892, p. 31 [**C. canaliculatus*; M]. Calyx truncate conical; basals generally equal; radials very large; fixed primibrachs small; one large interprimibrach followed by very small plates connecting with tegmen in each interray; *CD* interray wide, 3 plates in 2nd and 3rd rows, connected with smaller plates merging with tegmen. Tegmen convex, composed of ambulacrals and minute interambulacrals. Free arms and stem unknown. *M.Sil.,* N.Am.——FIG. 271,*3*. **C. canaliculatus,* Niagaran (Laurel Ls.), USA(Ind.); *3a,b, A*-ray and *CD*-interray views of calyx, ×1.5 (Springer, 1926a).

Desmidocrinus ANGELIN, 1878, p. 5 [**D. pentadactylus*; SD SPRINGER, 1926a, p. 43]. Calyx conical to cyathiform; basals equal or unequal; fixed tertibrachs present in all species and quartibrachs in some; fixed portions of rays isotomously or heterotomously branched; interbrachial areas composed of few plates, mostly small and very narrowly connected with tegmen; 1st interprimibrachs large; *CD* interray wider than others, with nonridged median series of 4 or 5 anal plates passing to tegmen. Free arms 3 to 12 (ordinarily 3 to 8) in each ray, composed of extremely short brachials arranged in zigzag. Nudinodals concealing internodals at least in proximal portion of stem; axial canal small, pentalobate in cross section. *U.Sil.,* Eu.-?N.Am. [American species referred to this genus differ markedly from typical representatives found in Europe.]——FIG. 272,*1*. **D. pentadactylus,* Sweden (Gotl.); *1a, B*-ray view of complete crown with short part of attached stem (Stockholm Museum Ec. 11366), ×1.5 (Ubaghs, n); *1b, CD*-interray view of incomplete crown (Stockholm Museum Ec. 11377), ×1.5 (Ubaghs, n); *1c,* plate diagram of part of calyx (radials black, anals and interbrachials stippled) (Ubaghs, n).

Methabocrinus JAEKEL, 1918, p. 37 [**M. erraticus*; OD]. Calyx moderately high, truncate conical; basals unequal; fixed secundibrachs (one in each half-ray) in contact all around, except in *CD* interray; single large interprimibrach in each interray; *CD* interray with three plates in row above primanal followed by a few plates in contact with tegmen. Tegmen solid, composed of central posterior oral plate surrounded by four orals, five

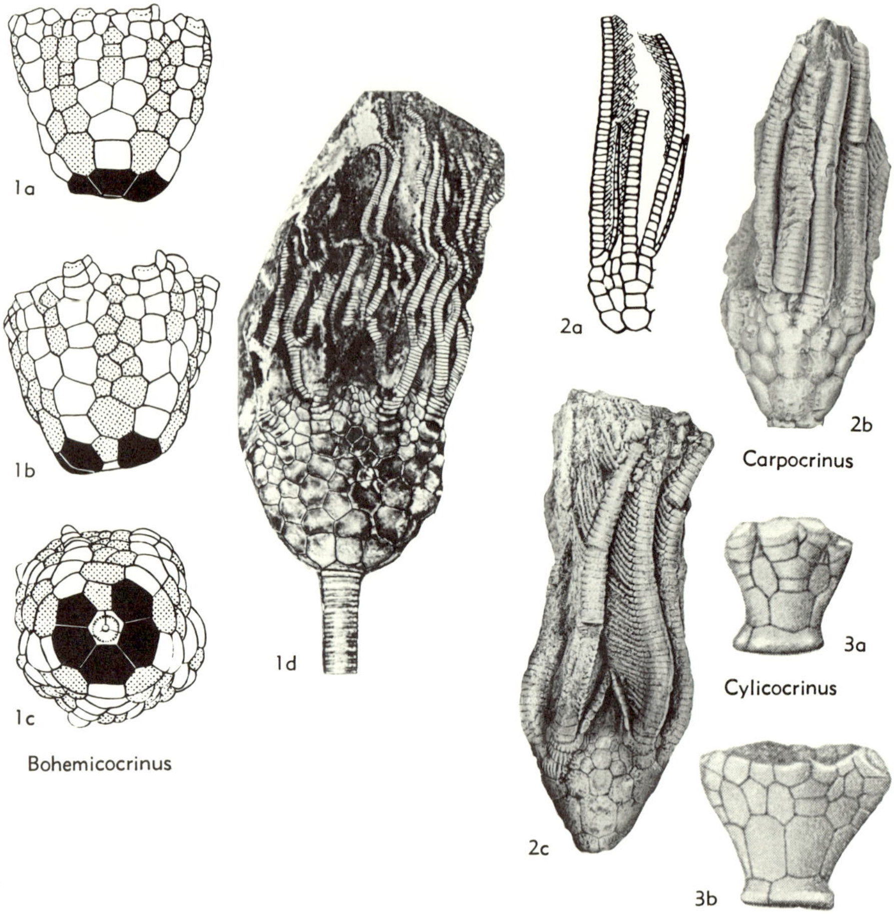

FIG. 271. Carpocrinidae (p. T463-T465).

interambulacrals in exsert position and few smaller elements in anal interray. Free arms unknown, resting in ten vertical compartments of tegmen bounded by partitions. *U.Sil.*, N.Eu.——FIG. 274,*1*. **M. erraticus*, Ger.;*1a-c*, *D*-ray, *CD*-interray, and tegminal views of holotype theca, ×5 (Ubaghs, 1958a); *1d*, plate part of calyx (radials black, primanal and other interbrachial plates stippled), ×5 (Ubaghs, n).

Family BATOCRINIDAE Wachsmuth & Springer, 1881

[*nom. transl.* WACHSMUTH & SPRINGER, 1897, p. 361 (*ex* section Batocrinites WACHSMUTH & SPRINGER, 1881, p. 267 [93])] [=Batocrininae JAEKEL, 1918, p. 35] [Materials for this family prepared by N. GARY LANE]

Theca variable in shape; basals three, equal; one to 15 interprimibrachs in each interray may or may not be in contact with tegmen; primanal in line with radials and succeeded by three anal plates in next range, three to 19 anal plates may or may not be in contact with tegmen; first fixed brachial externally quadrangular, or axillary in advanced genera; second fixed brachial typically axillary; one to five fixed secundibrachs and none to five fixed tertibrachs in each half-ray, latter commonly present in *C* and *D* rays, rarely in *A* ray. Free arms biserial, unbranched, single or paired from each ambulacral opening into theca, 10 to 40 and typically unequal in number in different rays of same individual. Anal opening at end of long, slender central or subcentral anal tube; oral and radial dome plates distinct. *L.Miss.-U.Miss.*

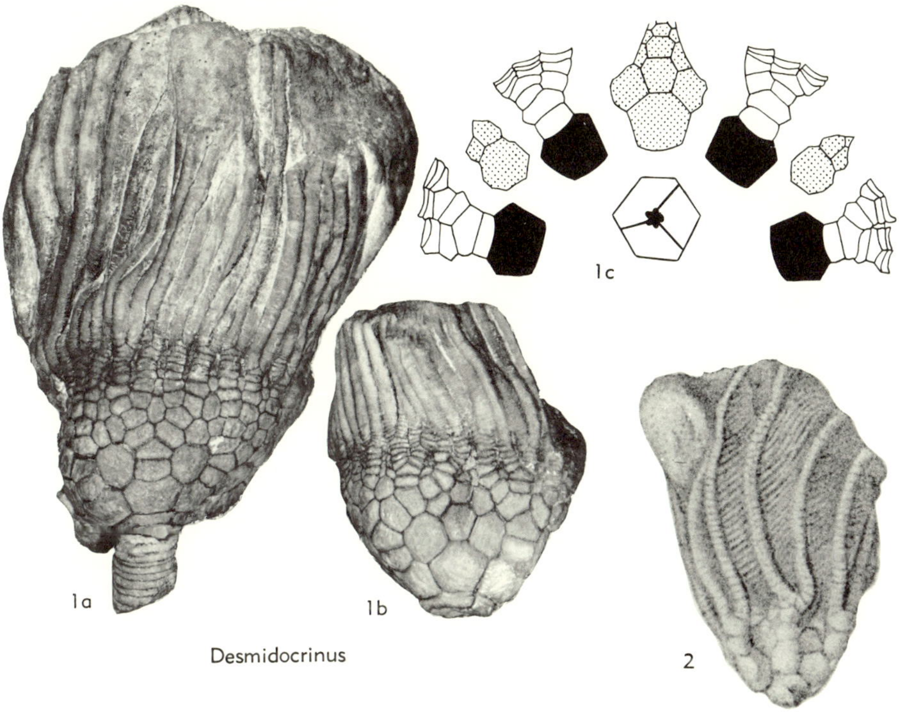

FIG. 272. Carpocrinidae (p. T463, T465).

Batocrinus CASSEDAY, 1854, p. 237 [*B. icosidactylus; SD WACHSMUTH & SPRINGER, 1881, p. 340]. Calyx low, broad, plate sutures distinct or beveled; nodes, granules, or ridges on ray plates; basal circlet low, broad; 1 to 4 interprimibrachs to interray; primanal followed by 3, and 1 to 3 additional, anal plates; all interray areas arched over by fixed brachials and separated from tegmen; arm facets large, horizontally directed, with distinct peripheral transverse and vertical articular ridges. Tegmen high, conical, with prominent spinose or tuberculate plates; anal tube long, central, typically with some spinose plates. Free arms 18 to 22. *U.Miss.(Meramec.)*, USA(Ind.-Ill.-Ky.-Mo.-Tenn.-Va.).——FIG. 275,*6*. *B. icosidactylus*, Ind.; *6a,b*, lat. and dorsal views of theca, ×1 (Wachsmuth & Springer, 1897).

Abatocrinus LANE, 1963, p. 696 [**Actinocrinus turbinatus* HALL, 1858b, p. 587; OD]. Calyx relatively high, sides straight or gently convex; cup plates smooth, convex, or nodose, unornamented; basal and radial circlets relatively high and prominent; 3 to 5 plates in each interray and 6 to 9 anal plates above primanal; all interrays arched by fixed brachials and not in contact with tegmen; arm facets small, slightly concave, with fine radial striations; anal tube stout and central, tegminal plates convex or nodose. Free arms 18 to 26, single, except for rare paired arms in most advanced species; 2 to 4 arms in *A* ray, 4 to 7 in *C* and *D* rays. *L.Miss.(Kinderhook.-Osag.)*, USA(Mo.-Ill.-Ky.-Iowa-Ind.-Tenn.).——FIG. 275, *2*. **A. turbinatus* (HALL), Ia.;*2a,b*, post. and lat. views of theca, lat. views of crown, ×1 (Wachsmuth & Springer, 1897).

Alloprosallocrinus CASSEDAY & LYON, 1862, p. 29 [**A. conicus*; SD WACHSMUTH & SPRINGER, 1881, p. 288] [=*Conocrinites* TROOST, 1850a *(nom. nud.)*; *Conocrinus* SHUMARD, 1868 (*nom. van. pro Conocrinites* TROOST)]. Calyx low, flat, concave centrally, sutures not impressed; 1st fixed primibrachs present or absent, partially eliminated from cup in some specimens; interprimibrachs 1 to 3, 3 anals above primanal, all interrays arched by fixed brachials. Tegmen high, conical, plates nodose; anal tube subcentral or slightly in front of tegmen center. Free arms 10 to 12, *C* or *D* rays may have more than two arms; arm facets large, crescent shaped, each facet composed of parts of 2 fixed brachs. *L.Miss.(Osag.)*, USA(Ind.-Ky.-Tenn.-Ala.).——FIG. 275,*3*. **A. conicus*, Borden Gr., Ky., Ind.; *3a-c*, ant., post., and dorsal views of theca, ×1 (Wachsmuth & Springer, 1897).

Azygocrinus LANE, 1963, p. 698 [**Actinocrinus dodecadactylus* MEEK & WORTHEN, 1861, p. 131;

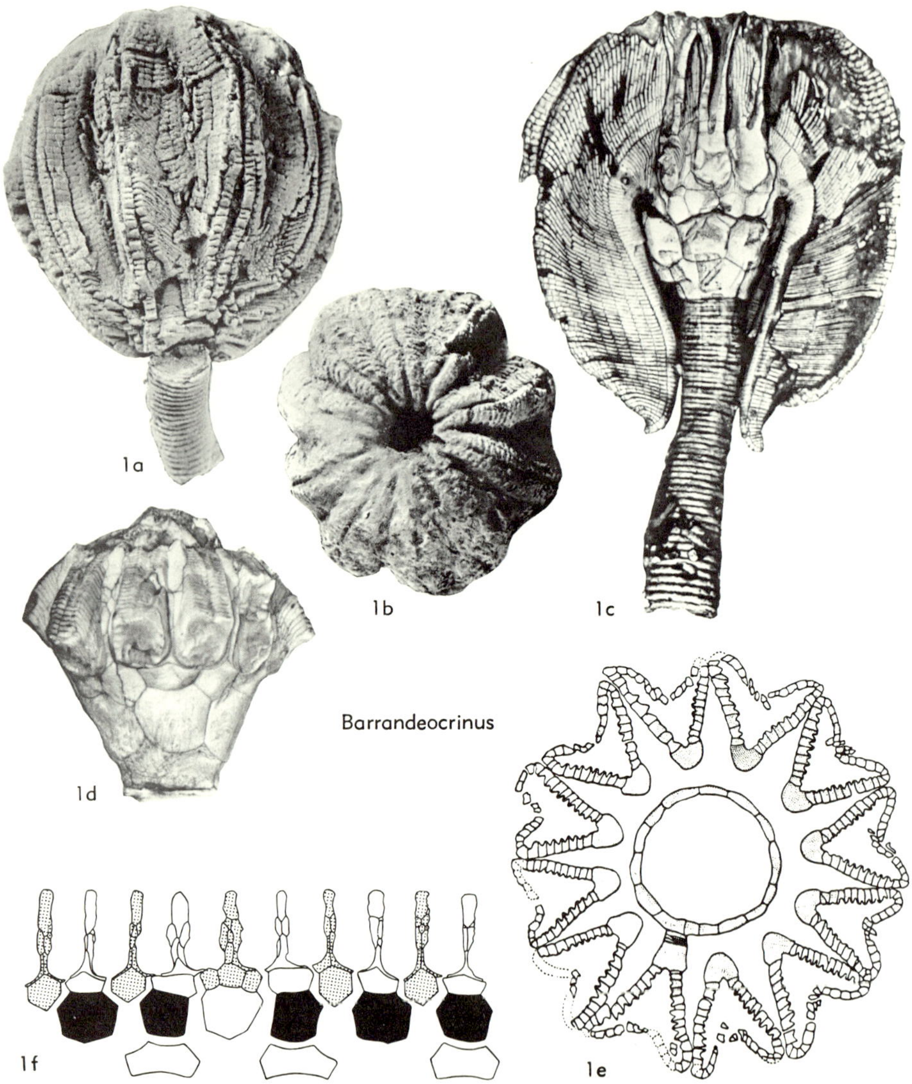

FIG. 273. Carpocrinidae (p. T463-T464).

OD]. Calyx subspherical, with flat smooth plates and indistinct sutures; aboral cup low bowl shaped, basal circlet inconspicuous, not flanged; 1 to 5 interprimibrachs arched by fixed brachials, primanal followed by 3 to 11 anals which may or may not be arched by fixed brachials; arm openings 12 to 23; arm facets small, depressed or flush with surface of fixed brachials. Tegmen low, rounded, plates smooth; anal tube short, situated toward posterior side of tegmen. Arms simple. *L.Miss.(Osag.),* USA(Mo.-Iowa-Ill.).—FIG. 275, *4.* **A. dodecadactylus* (MEEK & WORTHEN), Burlington Ls., Iowa; *4a,b,* post. and ventral views of theca, ×1 (Wachsmuth & Springer, 1897).

Dizygocrinus WACHSMUTH & SPRINGER, 1897, p. 413 [**Actinocrinus indianaensis* LYON & CASSEDAY, 1860, p. 75; OD]. Calyx small, bowl shaped, sutures impressed; ray plates granulose, nodose, or bearing ridges; basal circlet low, broad, extended proximally into rim; 2 to 5 interprimibrachs in each interray, 5 to 12 anals above primanal; anal plates but not regular interprimibrachs commonly in contact with tegmen; arm facets small, semicircular, elevated above surrounding fixed brachials. Tegmen low, with convex sides and nodose plates; anal tube short, slender. Free arms 14 to 20, commonly 16, paired, with 2 biserial arms from each ambulacral opening into calyx; all or only some arms in an individual may be paired; arms slightly spatulate at distal tips. *L.Miss.(Osag.)-U.Miss.(Meramec.),* USA(Ind.-Ill.-Mo.-Ky.-Tenn.).——FIG. 275,*7*. **D. indianaensis* (LYON & CASSEDAY), L.Miss.(Borden Gr.), Ind.; *7a,b,* lat. view of crown and post. view of theca, ×1 (Wachsmuth & Springer, 1897).

Eretmocrinus LYON & CASSEDAY, 1859, p. 241 [**E. magnificus*; M]. Calyx low, broad; basals low, forming projecting rim at base of calyx; 1 to 3 interprimibrachs in each interray; 4 to 7 anals above primanal, not in contact with tegmen, which is asymmetric, bulging, and highest anteriorly, flattened posteriorly; *CD* oral large, central; anal tube short and curved in some species. Free arms 12 to 26; *C* and *D* rays only may have more than 4 arms; distal parts of arms greatly widened and without pinnules. *L.Miss.(Kinderhook.-Osag.),* USA (Iowa-Ill.-Mo.-Ind.-Ky.-N.Mexico-Tenn.).——FIG. 275,*8a*. **E. magnificus,* Tenn.; lat. view of theca, ×1 (Wachsmuth & Springer, 1897).——FIG. 275,*8b*. *E. remibrachiatus expansus* WACHSMUTH & SPRINGER, Osag.(Burlington Ls.), Iowa; dorsal view of crown, ×1 (Wachsmuth & Springer, 1897).

Eutrochocrinus WACHSMUTH & SPRINGER, 1897, p. 408 [**Actinocrinus christyi* SHUMARD, 1855, p. 191; OD]. Calyx narrow proximally, abruptly expanded above level of radials; interprimibrachs 2 to 15 in each interray, may or may not be in contact with tegmen; 9 to 13 anals above primanal, in contact with tegmen; interbrachials commonly present between secundibrachs, and rarely between tertibrachs. Tegmen almost flat peripherally, rising centrally to form stout, central, long anal tube. Free arms short, incurved, 18 to 40, single or paired from each ambulacral opening into calyx. *L.Miss.(Osag.),* USA(Iowa-Ill.-Mo.).——FIG. 276,*1*. **E. christyi* (SHUMARD), Burlington Ls., Iowa; *1a,b,* lat. view of crown and post. view of theca, ×0.7 (Wachsmuth & Springer, 1897).

Globocrinus WELLER, 1920, p. 330 [**Batocrinus unionensis* WORTHEN, 1890, p. 84; OD]. Like *Dizygocrinus* except that primibrachs *1* are anchylosed with primaxils or eliminated from calyx, and commonly 2 rather than 3 anals present next above primanal. Free arms unknown. *U.Miss. (Meramec.-Chester.),* USA(Ill.).——FIG. 275,*5*. **G. unionensis* (WORTHEN), Chester.; *5a,b,* lat. and dorsal views of theca, ×1 (Weller, 1920).

Macrocrinus WACHSMUTH & SPRINGER, 1897, p. 446 [**Actinocrinus konincki* SHUMARD, 1855, p. 194; OD]. Calyx small; calyx high, with straight or slightly concave sides, base broad, flat; basals high, radials and first interprimibrachs nodose; 1 to 4 interprimibrachs in each interray arched by fixed brachials; 3 to 5 anals above primanal, in contact with tegmen; interray areas of cup depressed and ray areas elevated, giving lobed aspect to calyx. Tegmen high, conical; anal tube central, long, and slender. Free arms stout, short, 12 to 20, commonly with fewer than 4 arms in *A, B,* and *E* rays. *L.Miss.(Osag.),* USA(Iowa-Ill.-Ind.-

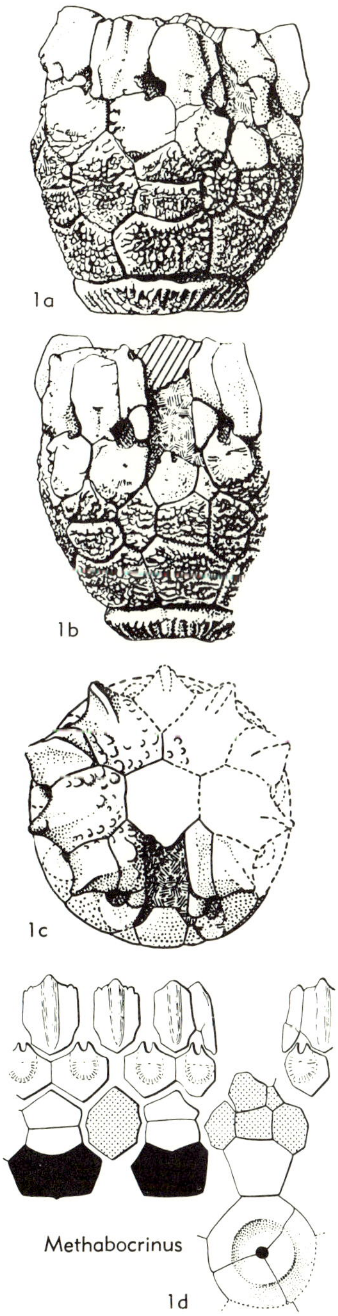

FIG. 274. Carpocrinidae (p. *T*465-*T*466).

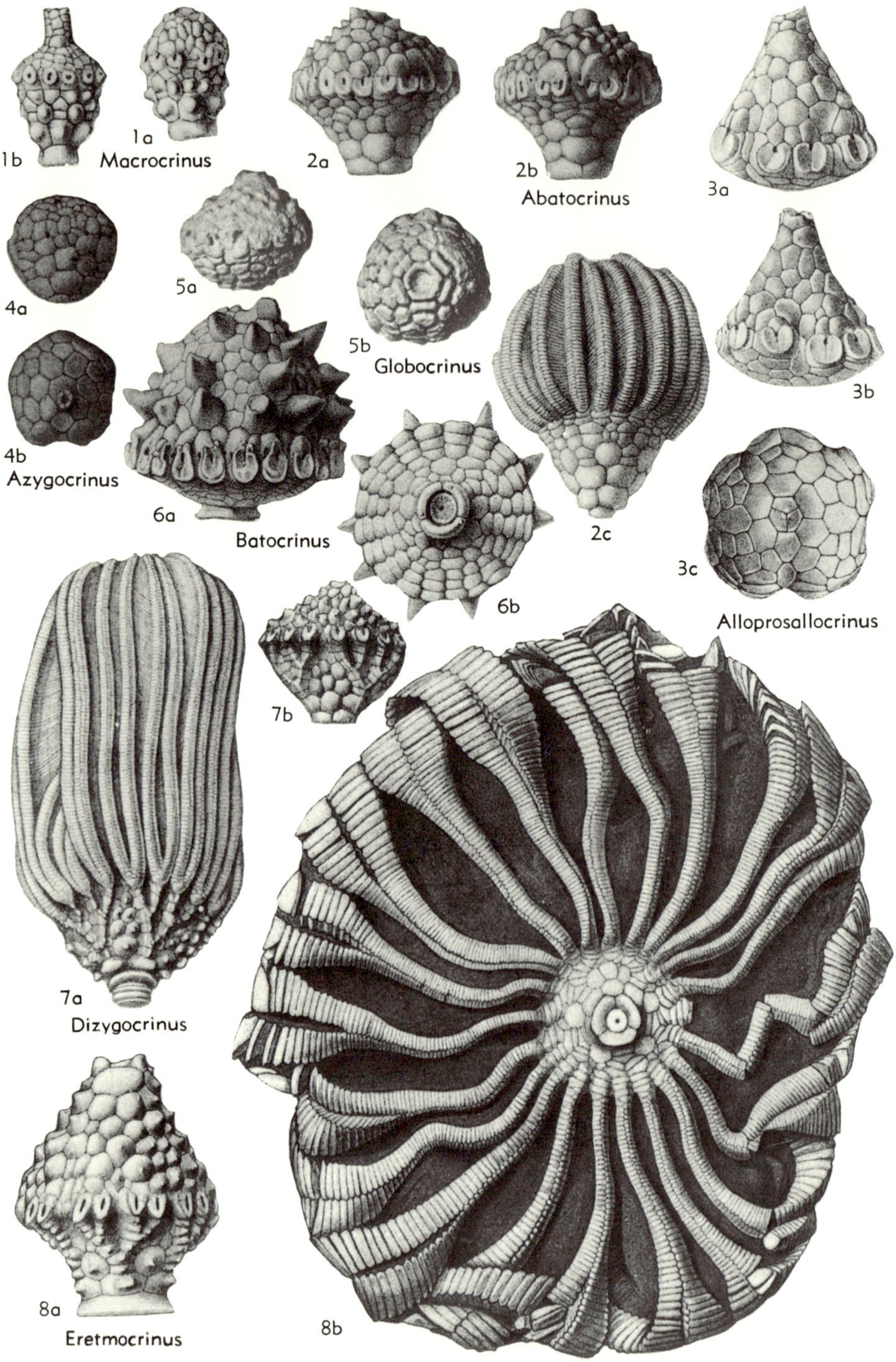

FIG. 275. Batocrinidae (p. T467-T469, T471).

Mo.).——FIG. 275,*1*. **M. koninck*i (SHUMARD), Burlington Ls., Iowa; *1a,b,* post. and ant. views of theca, ×1 (Wachsmuth & Springer, 1897). ——FIG. 276,*2*. *M. jucundus* (MILLER), Borden Gr., Ind.; *2a,b,* ant. view of theca with anal tube, and crown, ×0.7 (Wachsmuth & Springer, 1897).

Uperocrinus MEEK & WORTHEN, 1865, p. 153 [**Actinocrinus pyriformis* SHUMARD, 1855, p. 192; OD] [=*Lobocrinus* WACHSMUTH & SPRINGER, 1897, p. 434 (type, *Actinocrinus Nashvillae* TROOST, 1850a); *Hyperocrinus* BATHER, 1897 (Zool. Rec. 34), p. 117 *(nom. van.)*]. Radial and basal circlets prominent; rays grouped, interrays depressed and in contact with tegmen; interprimibrachs and anals numerous, in several ranges; anal tube stout, large, commonly long and central; arm openings into calyx large, distally directed, without fixed pinnular openings. Free arms 10 to 22. *L.Miss.(Osag.),* USA(Iowa-Ill.-Mo.-Ind.-Ky.-Tenn.).——FIG. 276,*3*. **U. pyriformis* (SHUMARD), Burlington Ls., Iowa; *3a,b,* lat. view of crown and post. view of theca, ×0.7 (Wachsmuth & Springer, 1897).

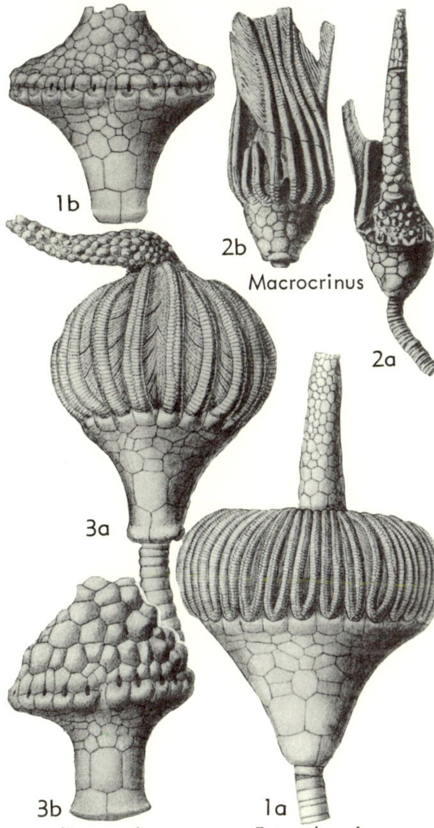

FIG. 276. Batocrinidae (p. *T*469, *T*471).

Family COELOCRINIDAE Bather, 1899

[Coelocrinidae BATHER, 1899b, p. 922] [Materials for this family prepared by N. GARY LANE]

Calyx commonly lobed at arm-base region; interprimibrachs few, commonly three in each interray, in contact with tegmen; *CD* interray wide, with median row of large anals. Anal opening flush with tegmen or raised on low protuberance, but without anal tube; fixed pinnular openings absent. Free arms single or paired, biserial, two to eight in each ray. Column transversely round, with small, pentagonal axial canal. *M.Dev.-L.Miss.(Osag.).*

Coelocrinus MEEK & WORTHEN, 1865, p. 273 [**Actinocrinus (Amphoracrinus) concavus* MEEK & WORTHEN, 1861, p. 132; OD] [=*Sphaerocrinus* MEEK & WORTHEN, 1865b, p. 154 (*non* ROEMER, 1858) (obj.)]. Crown small, base of calyx flat to concave, basals not visible in side view; radials large. Tegmen flat. Arms 12. *L.Miss.(Osag.),* USA(Ill.-Iowa).——FIG. 277,*2*. **C. concavus* (MEEK & WORTHEN), Burlington Ls., Ill.; *2a-c,* dorsal, ventral, and lat. views of theca, ×1 (Wachsmuth & Springer, 1897).

Agaricocrinus HALL, 1858, p. 560 [**A. tuberosus* HALL, 1858b, p. 561; SD MILLER & GURLEY, 1897, p. 11] [=*Agaricronites* TROOST, 1850a *(nom. nud.)*; *Agaricocrinites* TROOST, 1850a *(nom. nud.)*; *Agaricocrinus* AUSTIN, 1851, p. 282 *(nom. nud.)*]. Theca small to large, calyx conical or hemispherical; base flat or concave, basals and commonly radials not visible in side view; interprimibrachs and anals elongate. Orals and radial dome plates nodose, *CD* oral commonly large. Free arms long, ponderous, commonly 10, or with 2 to 4 arms in *C* and *D* rays. *L.Miss.(Kinderhook.-Osag.),* USA(Tenn.-Ky.-Ark.-Iowa-Mo.-Ill.-Ind.).——FIG. 277,*4*. **A. tuberosus,* Keokuk Ls., Iowa; *4a-c, CD*-interray, dorsal, and ventral views of theca, ×1 (Hall, 1858b).

Aorocrinus WACHSMUTH & SPRINGER, 1897, p. 470 [**Dorycrinus immaturus* WACHSMUTH & SPRINGER in MILLER, 1889, p. 240; OD]. Calyx small, conical to biturbinate; basals small, visible in side view; *CD* interray protuberant along median line, *CD* oral generally nodose. Free arms single, strong, grouped, 2 to 4 in each ray; brachials commonly spinose. *M.Dev.-L.Miss.(Osag.),* USA (N.Y.-Mich.-Tenn.-Iowa-Ind.-Mo.-Ohio-Pa.)., ?Eu. ——FIG. 277,*1*. **A. immaturus* (WACHSMUTH & SPRINGER); Hampton F., Iowa; *1a,b,* lat. view of partial crown and stem, and *CD*-interray view of crown, ×1 (Wachsmuth & Springer, 1897).

Dorycrinus C. F. ROEMER, 1854, p. 207 [**D. mississippiensis*; M]. Basals large, calyx distinctly lobed at arm-base regions. Anal opening excentric, opening laterally; *CD* oral, or *CD* oral and 5

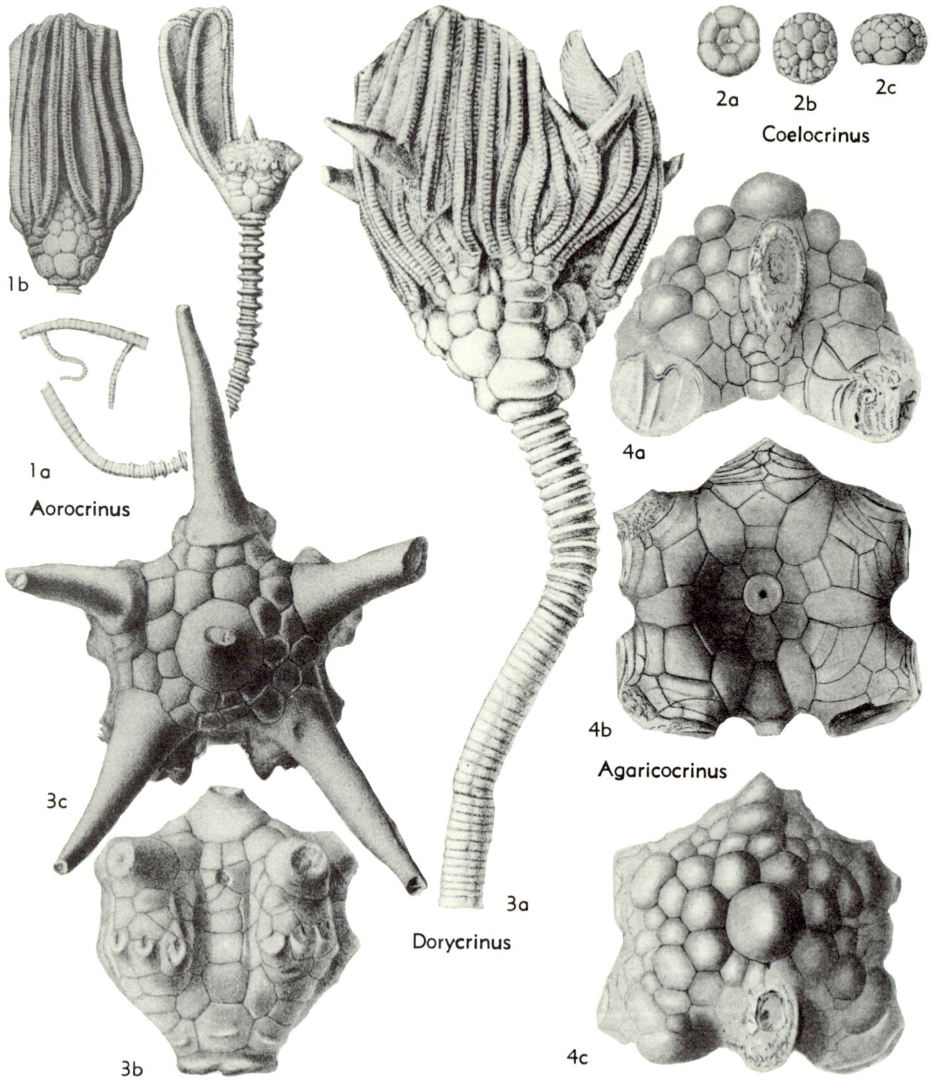

FIG. 277. Coelocrinidae (p. T471-T472).

radial dome plates bearing short to long, smooth or barbed spines. Free arms 16 to 20 pairs, commonly fewer paired arms in *A, B,* and *E* rays than in *C* and *D* rays; small spines on every 2nd to 4th free brachial. *L.Miss.(Kinderhook.-Osag.)*, USA (Iowa-Mo.-N.Mexico-Ind.-Ky.-Tenn.-Ill.).——FIG. 277,*3*. **D. mississippiensis,* Keokuk Ls., Iowa; *3a-c, D*-ray view of crown and stem, *CD*-interray and ventral views of theca, ×1 (Wachsmuth & Springer, 1897).

Family UNCERTAIN

Sunwaptacrinus LAUDON, PARKS, & SPRENG, 1952, p. 569 [**S. brazeauensis*; OD]. Calyx rapidly expanding, flaring abruptly distally, with arm bases forming lobes; basals 3, equal; 1st primibrachs quadrangular; fixed secundibrachs 1 in each half-ray, axillary; fixed tertibrachs at least 2 in each arm; 1st interprimibrachs large, followed by few small plates in contact with tegmen; primanal followed by 3 plates, widely connected to tegmen, which is high, with long stout anal tube. Arms biserial immediately beyond calyx, branching isotomously more than once; 1st bifurcation high above theca. *L. Miss.,* N.Am. (Can.). [This genus was placed by LAUDON, PARKS, & SPRENG (1952) among Batocrinidae, but

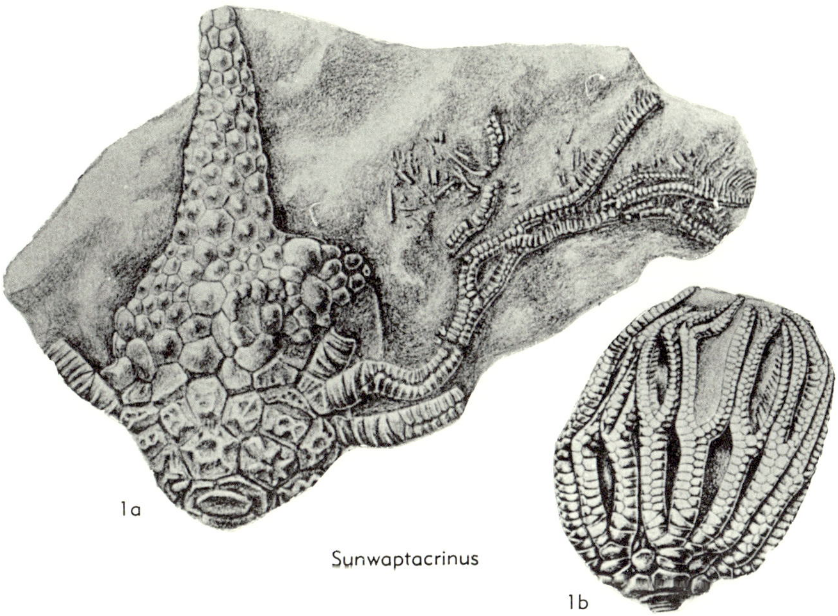

FIG. 278. Family Uncertain (p. *T*472-*T*473).

considered as exhibiting a stage of evolution between Devonian desmidocrinids and typical Mississippian batocrinids. It was excluded by LANE (1963) from Batocrinidae, because of its possession of free arm branching isotomously at least twice. It is here considered as a rather isolated form, with possible, yet undetermined, desmidocrinid affinities.]——FIG. 278,*1*. **S. brazeauensis,* Banff F., Alberta; *1a,* post. view of holotype, ×1.5; *1b,* lat. view of paratype, ×1.5 (Laudon, Parks, & Spreng, 1952).

Superfamily HEXACRINITACEA Wachsmuth & Springer, 1885

[*nom. correct.* STRIMPLE, 1975, p. 27 (*pro* Hexacriniticae UBAGHS, 1953, p. 740, *nom. transl. ex* family Hexacrinidae WACHSMUTH & SPRINGER, 1885, p. 93 [315])]

Calyx essentially confined to basals, radials, and primanals; basals equal, three, two, or fused together; radials large; interprimibrachs typically reduced in size or lacking; radial articular facets for reception of arms generally occupying only small part of radial summit. Arms slightly incorporated in calyx or not at all; primibrach series tending to be markedly reduced. Anal tube small or absent. Stem circular or elliptical in section. *U.Sil.-U.Perm.*

Family HEXACRINITIDAE Wachsmuth & Springer, 1885

[*nom. correct.* BASSLER, 1938, p. 24 (*pro* Hexacrinidae WACHSMUTH & SPRINGER, 1885, p. 93 (315), *nom. transl. ex* section Hexacrinites WACHSMUTH & SPRINGER, 1881, p. 56, 230)]

Calyx medium conical to bowl shaped, with low to moderately high tegmen; basals three, subequal; primibrachs typically not incorporated in calyx but commonly joined with interprimibrach tegminal plates; primanal generally of approximately same size as radials. Tegmen stout, composed of small to medium-sized plates, orals and ambulacrals commonly distinct; anus opening directly through tegmen or at end of short tube. Arms two in each ray, branching. Column circular in cross section. *U. Sil.-U.Dev., ?Penn.*

Hexacrinites AUSTIN & AUSTIN, 1843, p. 19 [**Platycrinus interscapularis* PHILLIPS, 1841, p. 28; M] [=*Hexacrinus* AUSTIN & AUSTIN, 1844, p. 48 (type, *H. melo*; M) (=*Platycrinus interscapularis* PHILLIPS, 1841, p. 28 (obj.); *Triplaricrinites* GOLDFUSS in SCHULTZE, 1866, p. 67 (type, *T. exculptus,* =*Hexacrinus pateraeformis* SCHULTZE, 1866, p. 67); *Triplaricrinus* ZITTEL, 1879, p. 365 (*nom. van.*)]. Primibrachs 1 or 2 in each ray, so reduced in some species as to be concealed by 1st secundibrachs which rest directly on radials; anus subcentral or marginal on tegmen. Free arms (known only in a few species) 2 in each ray, uniserial, simple or bearing ramuli on one or both sides. *U.Sil.-U.Dev., ?U.Carb.(?Penn.),*

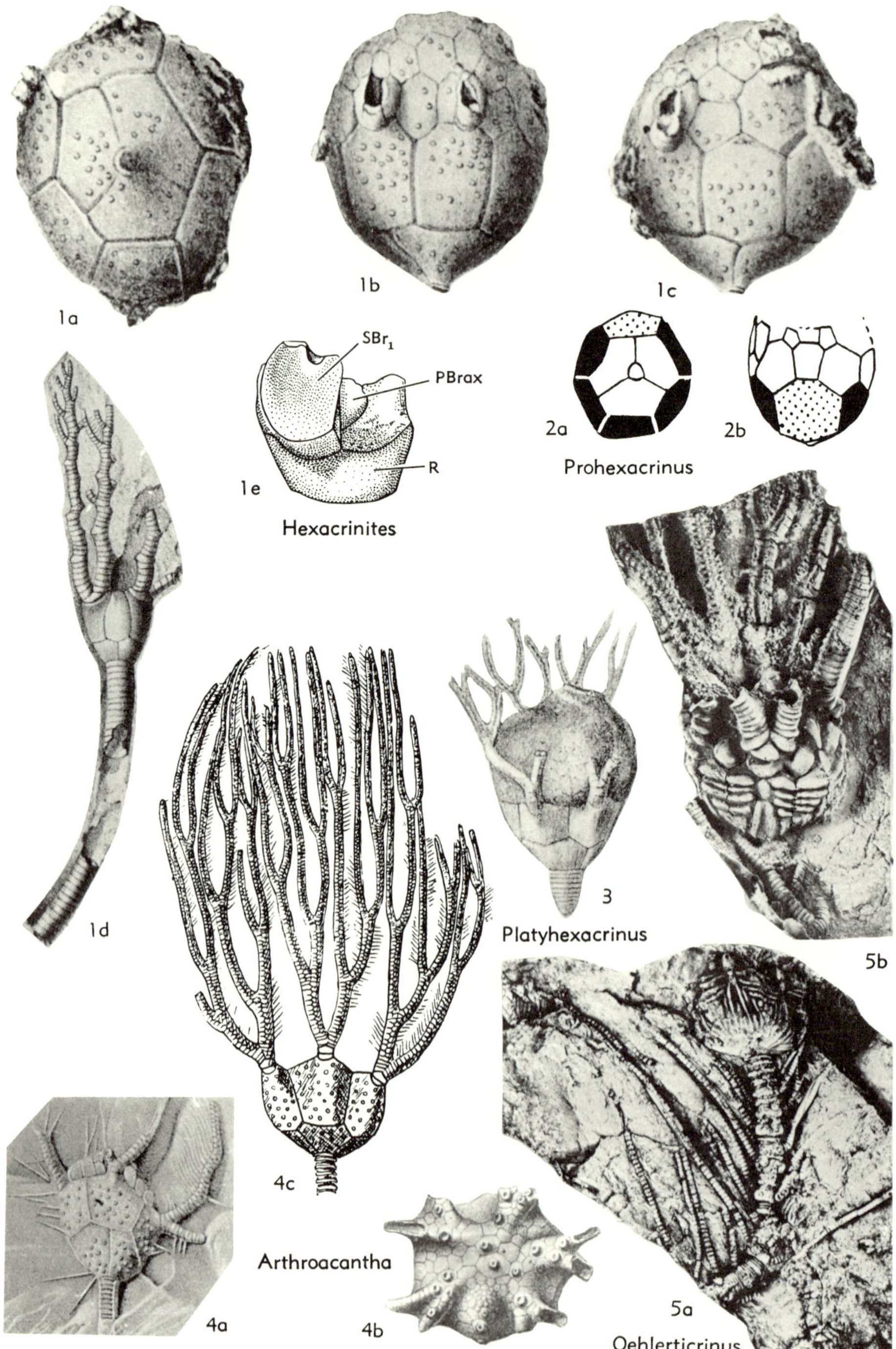

Fig. 279. Hexacrinitidae (p. T473, T475).

W.Eu.-USSR-Sib.-China-N.Am.-?E.Australia-?New Zealand. [*H. carboniferous* (YAKOVLEV & IVANOV) from M.Carb. (Russian stratigraphical column) of USSR is very different from typical species and may belong to another lineage.] ——FIG. 279, *1a-c*. **H. interscapularis* (PHILLIPS), M.Dev., Eng.; aboral, ant. and post. views of theca, ×1 (Austin & Austin, 1843-49).——FIG. 279,*1d*. *H. limbatus* (MÜLLER), M.Dev., Ger.; ×1 (Schultze, 1867).——FIG. 279,*1e*. *H. verrucosus* (DEWALQUE), U.Dev., Belg.; radial *(R)*, and primaxil *(PBrAx)* concealed by first secundibrachs *(SBr_1)*, ×2 (Ubaghs, n).

Arthroacantha WILLIAMS, 1883, p. 84 [**A. ithacensis*; OD] [=*Hystricrinus* HINDE, 1885, p. 158 (type, *H. carpenteri*, p. 162; M); *Arthracantha* HINDE, 1886, p. 273 *(nom. van.)*]. Primibrachs 2, tegmen rather low; interradial tegminal plates 1 to 3 or even more, well developed, joined to proximal brachials; anus excentric on slight swelling; pitted tubercles bearing movable spines on calycal plates and in some species on brachials and tegminal plates. Free arms biserial, isotomously branching. Stem transversely round, nodals typically cirriferous. *L.Dev.-U.Dev.*, W.Eu.-N.Am. (USA).——FIG. 279,*4a*. **A. ithacensis*, U.Dev. (Portage F.), USA(N.Y.); lat. view of theca, ×1 (Wachsmuth & Springer, 1897).——FIG. 279, *4b*. *A. depressa* (WACHSMUTH & SPRINGER), U. Dev.(L.Chemung.), USA(N.Y.); tegmen, ×1 (Goldring, 1923).——FIG. 279,*4c*. *A. schwerdii* (FOLLMANN), L.Dev.(Ems.), Ger.; crown, ×0.7 (Schmidt, 1942).

Oehlerticrinus LE MENN, 1975, p. 243 [**O. seillouensis*; OD]. Primibrachs 2, incorporated in cup; interbrachials rather well developed, one to several per interray; primanal supporting 1 to 4 plates; plates of cup strongly ornamented. Free arms dichotomous or heterotomous, with compound brachials. Column heteromorphic and cirriferous. *L.Dev.-M.Dev.*, Eu.-N.Afr.——FIG. 279, *5a*. **O. seillouensis*, L.Dev., France; calyx with part of column, ×1 (Le Menn, 1975).——FIG. 279,*5b*. *O. lefretensis* LE MENN, M.Dev., France; calyx and part of arms, ×1 (Le Menn, 1975).

Platyhexacrinus W. E. SCHMIDT, 1913, p. 301 [**P. inornatus*; SD W. E. SCHMIDT, 1915, p. 119]. Like *Arthroacantha*, but without movable spines on radials and basals. *L.Dev.-M.Dev.*, Eu.-Sib. ——FIG. 279,*3*. **P. inornatus*, M.Dev.(L.Couvin.), Ger.; ×1 (Schmidt, 1913).

?**Prohexacrinus** YAKOVLEV, 1946, p. 154 [**P. arcticus*; OD]. Primibrachs 2 in each ray, well developed, incorporated in calyx; 1st interprimibrachs large, followed by 2 smaller plates in next range. *?U.Sil.*, USSR. [This genus, based on poorly preserved material, should perhaps be placed among Carpocrinacea.]——FIG. 279,*2*. **P. arcticus*, limestones on Es-to-vis River (N. Urals; *2a,b*, dorsal and post. views of theca (radials black, primanal stippled), ×1 (Yakovlev, 1946a).

Family PARAHEXACRINIDAE Shevchenko, 1967

[Parahexacrinidae SHEVCHENKO, 1967, p. 76]

Theca fungiform to spheroidal; basals three, or fused together into large hexagonal plate, with depressed columnal insertion; radials large, directly supporting either brachial facets or proximal fixed brachials, which are very low and irregular; one intersecundibrach in each ray may be present; primanal generally smaller than radials. Tegmen rather high, dome shaped, consisting of small and large plates with five differentiated orals and ambulacrals commonly distinct; anus directly through tegmen more or less excentric. Arms four to eight in each ray, invariably paired, probably slender, tending to form more or less continuous wreath around tegmen. Stem circular to elliptical in cross section; nudinodals entirely or partially concealing very short internodals. *L.Dev.*

Parahexacrinus SHEVCHENKO, 1967, p. 77 [**P. fungiformis*; OD]. Theca fungiform in outline; basals fused, subhorizontal, barely visible in side view; radials directly carrying 4 pairs of brachial facets. Tegmen as high as calyx or slightly higher, composed of 5 orals, each surrounded by 4 ambulacrals and 1 large interambulacral. Stem subelliptical in transverse section, with dichotomizing crenulae on columnals. *L.Dev.(lower Kshtut beds)*, C.Asia(SE.Tien Shan).——FIG. 280,*1*. **P. fungiformis*; *1a,b*, *A*-ray and *CD*-interray views of holotype theca, ×1.5; *1c*, plate diagram of calyx (radials black, primanal stippled), ×1.5; *1d,e*, side view of columnals, and columnar articular facet, ×2 (all Shevchenko, 1967).

Agathocrinus SHEVCHENKO, 1967, p. 84 [**A. globosus*; OD]. Theca subspheroidal; basals distinct, visible in side view; each radial carrying directly 2 groups of short irregular proximal brachials, separated by intersecundibrachs; 2 paired brachial facets in each ray. Tegmen somewhat higher than 0.3 total height of theca, many-plated; anus lateral. Column circular in cross section, with nondichotomizing crenulae. *L.Dev. (upper Kshtut beds)*, C.Asia(SE.Tien Shan).——FIG. 280, *2*. **A. globosus*; *2a,b*, *D*-ray and *CD*-interray views of theca, ×3; *2c*, plate diagram of calyx (radials black, primanal stippled), ×3; *2d,e*, side view and articulum of columnals, ×3 (all Shevchenko, 1967).

Amonohexacrinus SHEVCHENKO, 1967, p. 82 [**A. adelius*; OD]. Calyx low conical or low bowl shaped, with rounded tegmen; basals fused, visible in side view; radials directly supporting low and

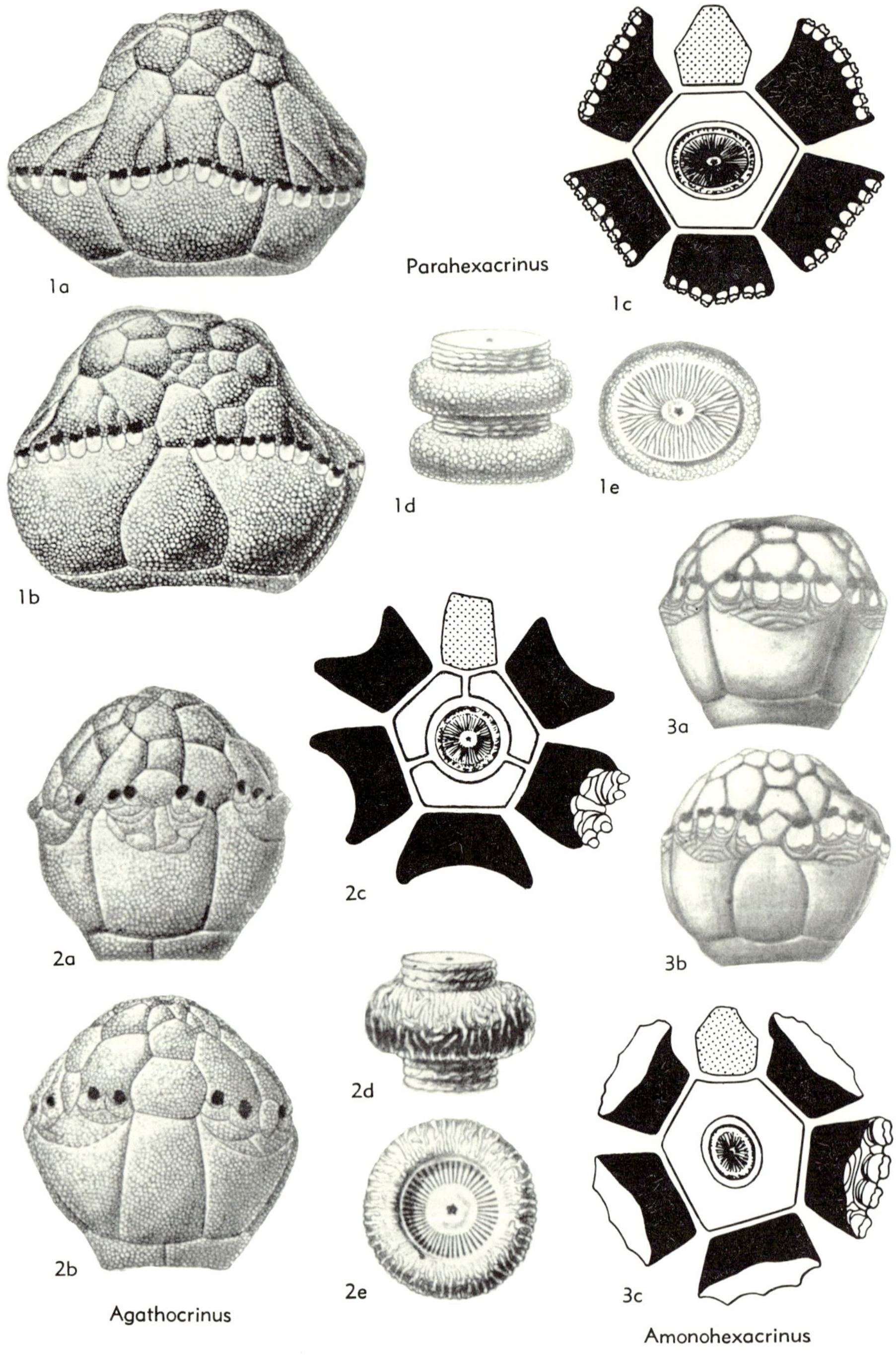

FIG. 280. Parahexacrinidae (p. T475, T477).

irregular brachials; 4 arm facets in each ray, median ones joined into pair. Tegmen (apical part not preserved) comprising peripheral circlet of 20 ambulacrals and 5 interambulacrals. Stem elliptical in cross section, columnals with flabelliform crenulae. *L.Dev.(up. low. Kshtut beds),* C.Asia(SE.Tien Shan).——FIG. 280,*3*. **A. adelius; 3a,b, E*-ray and *CD*-interray views of holotype theca, ×2; *3c,* plate diagram of calyx (radials black, primanal stippled), ×2.2 (all Shevchenko, 1967).

Family DICHOCRINIDAE S. A. Miller, 1889

[Dichocrinidae S. A. MILLER, 1889, p. 214] [=Dichocrininae, Talarocrininae UBAGHS, 1953, p. 741] [Materials for this family prepared by N. GARY LANE]

Calyx small; basals two, equal, with *A-CD* interbasal suture; radials five, separated on posterior side by primanal. Tegmen variable in height and prominence, with hypertrophied wing plates, or secondarily simplified to five orals, in advanced genera. Free arms primitively uniserial and two to each ray, becoming biserial and four to six in each ray in advanced forms. Stem round transversely and straight, or coiled and bilateral. *L.Miss.-U.Perm.*

Dichocrinus MÜNSTER, 1839, p. 31 [**D. radiatus*; M] [=*Dichocrinites* GEINITZ, 1846, p. 560 (obj.) *(nom. van.); Cotyledonocrinus* CASSEDAY & LYON, 1862, p. 56 (type, *C. pentalobus* CASSEDAY & LYON); *Edwarsocrinus* D'ORBIGNY, 1850, p. 157 (type, *Platycrinus ornatus* M'COY, 1844, p. 176); *Edwardsocrinus* D'ORBIGNY, 1852 *(nom. correct.)*]. Calyx small, bullet shaped, composed of 2 equal basals and primanal in line with radials. Tegmen formed of numerous small interambulacrals and ambulacrals, with 5 central orals, *CD* oral enlarged; anal opening in *CD* interray of tegmen on small protuberance or at end of short, conical anal tube. Free arms isotomous, uniserial or biserial, erect or pendent, branching on primibrachs *2*; secundibrachs *2* or *3* axillary in some species or additional branching may occur high above calyx. Column round, straight. *L.Carb.,* Eu. (Eng.-Eire-Belg.-USSR), *L.Miss.,* N.Am.(USA). ——FIG. 281,*1a,b*. *D. crassitestus* WHITE, USA (Iowa); *1a,b,* ant. view of crown and post. view of theca with anal tube, ×1 (Wachsmuth & Springer, 1897).——FIG. 281,*1c,d*. **D. radiatus,* Eng.; ant. and post. views of calyx, ×1 (Wright, 1956a).

Camptocrinus WACHSMUTH & SPRINGER, 1897, p. 779 [**C. myelodactylus*; OD]. Crown like *Dichocrinus*; stem coiled, bilateral, with short columnals round proximally, becoming crescentic or lozenge-shaped distally, with concave side of stem crown at center of coil, bent back on stem typically covered by long round cirri on paired or single nodals, typically confined to 2 rows at or near margins of flattened columnal sides. *L.Miss.-Perm.,* USA-Eu. (Scot.-USSR).-Indon. (Timor).——FIG. 281,*2*. **C. myelodactylus,* L.Miss.(Borden Gr.), USA(Ind.); *2a,b,* lat. view of crown with attached stem and lat. view of stem, ×1; *2c,* lat. view of columnals and cirri, ×3 (Springer, 1926b).

Neodichocrinus WANNER, 1937, p. 84 [**N. nanus*; OD]. Calyx like *Dichocrinus*. Tegmen composed of 5 orals, among which *CD* oral is enlarged; anal opening between primanal and *CD* oral. *U.Perm.,* Indon.(Timor).——FIG. 281,*3*. **N. nanus; 3a,b,* post. and ventral views of theca, ×5.5; *3c,* dorsal view of theca, ×4.8 (Wanner, 1937).

Paradichocrinus SPRINGER, 1926, p. 51 [**Dichocrinus polydactylus* CASSEDAY & LYON, 1862, p. 18; OD]. Calyx similar to *Dichocrinus* but typically larger. Tegmen high, conical, composed of numerous large interambulacrals, ambulacrals, and 5 orals; anal opening near summit of tegmen. Free arms biserial, branching exotomously several times above axillary primibrachs *2*. *L.Miss.(Osag.),* USA(Ind.).——FIG. 281,*6a*. **P. polydactylus* (CASSEDAY & LYON); post. view of theca, ×1 (Springer, 1926b).——FIG. 281,*6b*. *P. planus* SPRINGER; lat. view of crown, ×1 (Springer, 1926b).

Pterotocrinus LYON & CASSEDAY, 1859, p. 301 [**Asterocrinus capitalis* LYON, 1857, p. 472; OD] [=*Codonocrinites* TROOST, 1850a *(nom. nud.); Codonocrinus* SHUMARD, 1868 *(nom. van.); Asterocrinus* LYON, 1857, p. 472 (*non* MÜNSTER, 1839)]. Calyx low, bowl shaped, basally concave in some species; basals large and convex or small and flattened; radials low, strongly convex or flat, deeply excavated and scalloped distally; primanal narrow, high or triangular and confined between *C* and *D* radials. Tegmen high, pyramidal or distally flattened; 5 axillary tegminal ambulacrals hypertrophied, variable in shape, produced into massive, thin bladelike, or forked wing plates; anal opening at summit of tegmen. Free arms short, 4 or 6 in each ray, biserial, confined to area between hypertrophied wing plates, with primibrachs *1* very small, axillary, and commonly not visible, or lacking; secundibrachs *1* axillary, partly or completely supported by radial; abmedial tertibrach *1* in contact with lateral edge of radial, or 2nd lateral axillary secundibrach present in youngest species; single intersecundibrach may be present between adjacent radials. *U.Miss.,* USA (Ky.-Ill.-Ind.).——FIG. 281,*7a*. **P. capitalis* (LYON), Ky.; *D*-ray view of crown, ×1 (Springer, 1926b).——FIG. 281,*7b*. *P. coronarius* (LYON), Ky.; ventral view of crown, ×1 (Springer, 1926b).——FIG. 281,*7c*. *P. bifurcatus* WETHERBY, Ky.; dorsal view of crown, ×1 (Wachsmuth & Springer, 1897).

Stomiocrinus WANNER, 1937, p. 84 [**S. sub-*

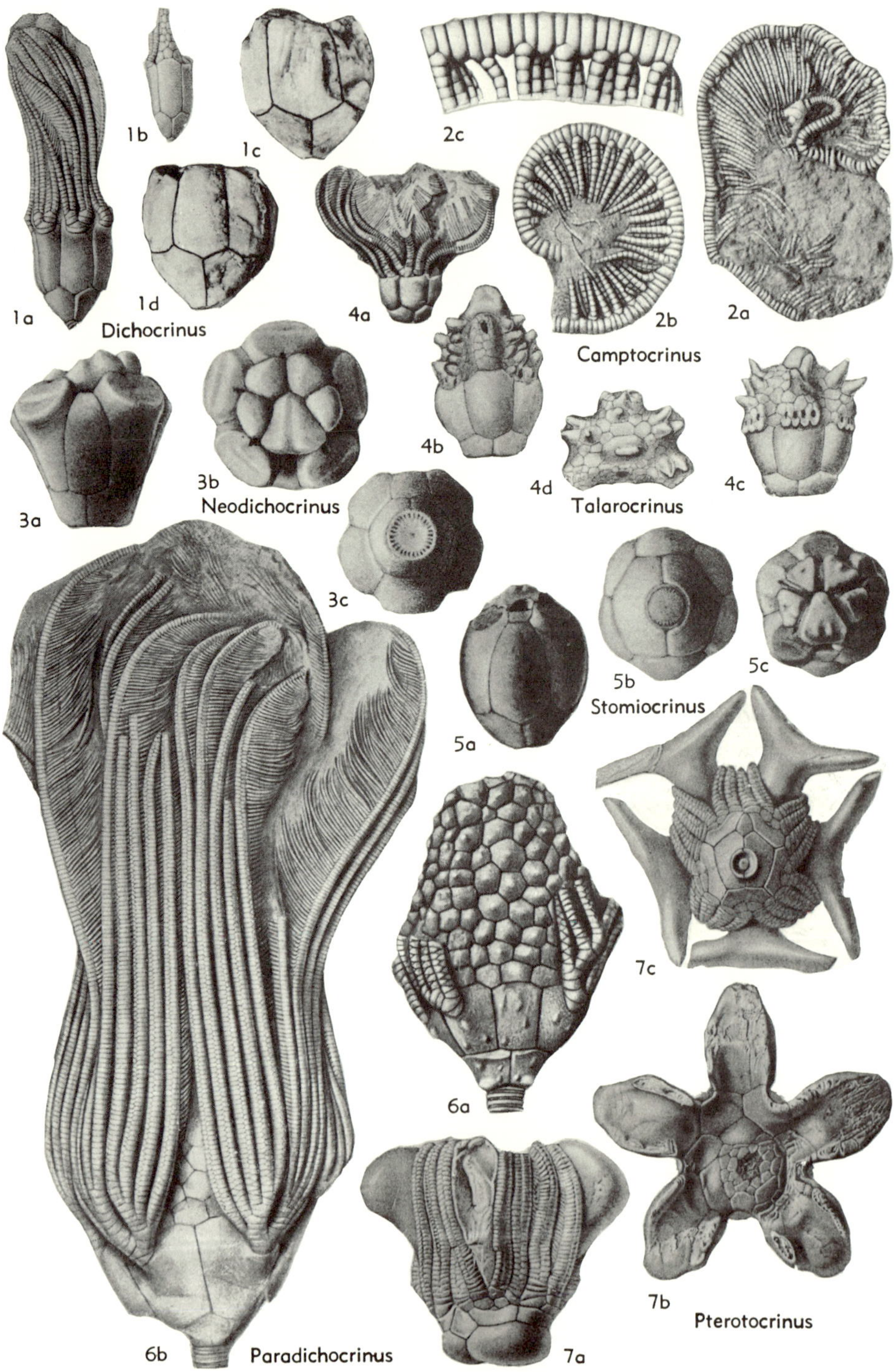

FIG. 281. Dichocrinidae (p. T477, T479).

globosus; OD]. Like *Neodichocrinus* except orals not in contact medially. *U.Perm.*, Indon.(Timor).——FIG. 281,*5*. **S. subglobosus; 5a-c,* post., dorsal, and ventral views of theca, ×2 (Wanner 1937).

Talarocrinus WACHSMUTH & SPRINGER, 1881, p. 259(85) [**Dichocrinus cornigerus* SHUMARD, 1857, p. 72; OD] [=*Daemenocrinites* TROOST, 1850a *(nom. nud.); Daemenocrinus* BATHER, in LANKESTER, 1900a, p. 202 *(nom. nud.); Doliolocrinites* TROOST in HALL, 1858 *(nom. nud.); Doliolocrinus* WACHSMUTH & SPRINGER, 1881, p. 231 (405), *(nom. nud.)*]. Calyx like *Dichocrinus* except plates more massive. Tegmen as high as cup, with large central *CD* oral, anal opening high on tegmen, laterally directed. Arms with primibrachs *1* axillary, very small; secundibrachs *1* in contact with lateral edge of radial facet; arms biserial, 4 in each ray. *U.Miss.*, USA (Ill.-Ind.-Ky.-Mo.-Va.-Ala.).——FIG. 281,*4a*. *T. patei* MILLER & GURLEY, Ky.; lat. view of crown, ×1 (Springer, 1926b).——FIG. 281,*4b-d*. **T. cornigerus* (SHUMARD), Ala.; post., ant., and ventral views of theca, ×1.5 (Wachsmuth & Springer, 1897).

Family ACROCRINIDAE
Wachsmuth & Springer, 1885

[Acrocrinidae WACHSMUTH & SPRINGER, 1885, p. 315] [=Acrocrininae UBAGHS, 1953, p. 741] [Materials for this family prepared by R. C. MOORE and H. L. STRIMPLE]

Crown ovoid to pyriform, with height exceeding greatest width; arms (where known) erect or pendent, biserial, pinnulate; calyx subconical, vaselike, tall ovoid, or subglobular, with two equal basals separated by suture between them oriented anteroposteriorly as in Dichocrinidae; radials five, laterally adjoining one another except *C* and *D* radials, which are separated by primanals; radials and basals not in contact but set apart by few to very many intervening supplementary plates collectively designated as intercalaries; tegmen flat to gently convex, with or without differentiated orals and having few or extremely numerous small plates, anal vent near posterior margin, inconspicuous or located at tip of anal pyramid. Column transversely circular, homeomorphic or heteromorphic, with columnal articular facets peripherally crenulate; axial canal diminutive, circular to distinctly quinquelobate in section. [The outstanding attribute of this family is the presence of intercalaries in calices of all genera. These may be arranged in moderately regular alternating circlets or appear quite irregular, with proximal plates notably smaller than distal ones. The advent of intercalaries in a late Paleozoic group of camerate crinoids suggests evolutionary reversion to cystoidlike structure of the calyx and divergence of genera in significant morphological features indicates unusual plasticity of the stock.] *L.Miss.-U.Penn., up.L.Carb.-M.Carb.*

Subfamily ACROCRININAE
Wachsmuth & Springer, 1885

[*nom. transl.* UBAGHS, 1953, p. 741 (*ex* Acrocrinidae WACHSMUTH & SPRINGER, 1885, p. 315)]

Calyx with very wide, low radials or width of these plates only moderately exceeding height, mainly characterized by gently scalloped articular facets that reach to interradial sutures; primanal in circlet of radials and similar to them in shape, or subpentagonal in outline narrowing upward; basals clearly visible from side; intercalaries numerous, most distal circlet containing 12, 15, or 18 plates, of which seven or eight (including subanals) are located interradally and others radially. Arms erect or pendent. *U.Miss.(Chester.)* or *up.L. Carb.(Visean)-L.Penn.(Morrow.)*.

Acrocrinus YANDELL, 1855, p. 135 [**A. shumardi*; M]. Calyx urn shaped, with usually low, wide radials and primanal; distal articular facet of radials supporting minute axillary primibrachs *1* and 2 secundibrachs *1*, latter followed by axillary secundibrach *2*; most distal circlet of intercalaries including 10 small subradials and 8 in interradial positions, those of *CD* interray comprising 2 small subanals and 2 distinctly larger intercalaries next to them on either side. Arms erect. Stem homeomorphic, composed of very low columnals. *U.Miss.(Chester.)*, N.Am.(USA); *up.L.Carb.(Visean)*, Eu.(Eng.).——FIG. 282,*1*;283,*3*. **A. shumardi*, up.Chester.(Glen Dean Ls.), USA (Grayson Co., Ky.); 282,*1*, diagram showing radial circlet and primanal next above circlet of 18 most distal intercalaries (radials black, distal intercalaries shaded) (Moore & Strimple, 1969); 283,*3a*, lat. view of holotype crown (rays not identified), ×1 (Wachsmuth & Springer, 1897); 283,*3b*, lat. view of small crown, ×1 (Springer, 1926b).

Amphoracrocrinus MOORE & STRIMPLE, 1969, p. 9 [**Acrocrinus amphora* WACHSMUTH & SPRINGER, 1897, p. 808; OD]. Similar to *Acrocrinus* but differing in its notably taller and narrower radials and primanal, in having only 12 plates in most distal circlet of intercalaries (5 subradial, 7 interradial, latter including single subanal), and espe-

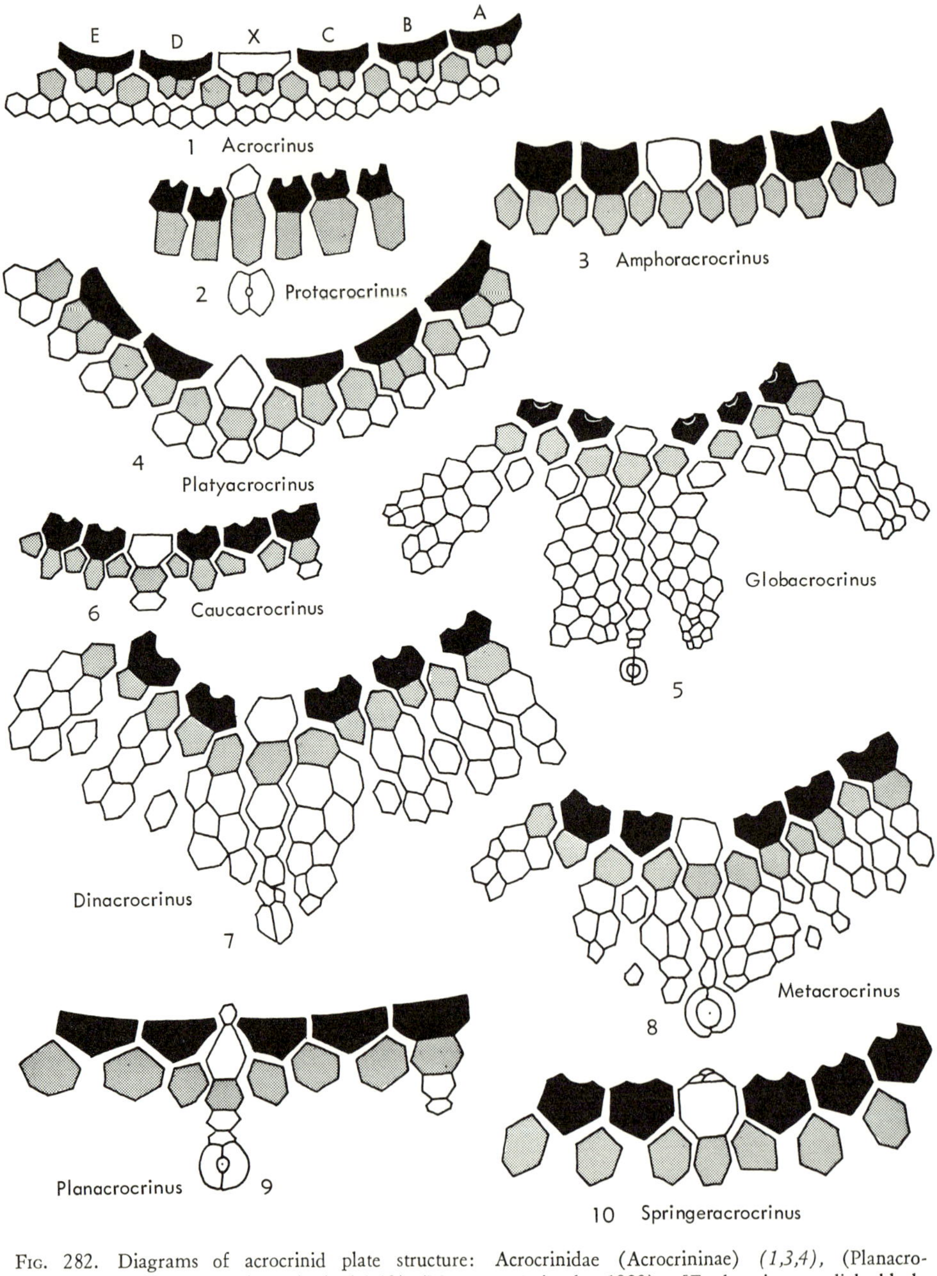

FIG. 282. Diagrams of acrocrinid plate structure: Acrocrinidae (Acrocrininae) *(1,3,4)*, (Planacrocrininae) *(9)*, (Globacrocrininae) *(2,5-8,10)* (Moore & Strimple, 1939). [Explanation: radials black, distal intercalaries shaded, others and bipartite basal circlet unshaded.]

cially in its robust, long, pendent arms. Tegmen formed by extremely numerous small plates without differentiated orals and lacking anal pyramid. *U.Miss.(up.Chester.)*, USA(Ala.).——FIG. 282,*3*; 283,*2*. **A. amphora* (WACHSMUTH & SPRINGER), Gasper F.; 282,*3*, diagram showing radial circlet and primanal next above circlet of 12 most distal intercalaries (radials black, distal intercalaries shaded) (Moore & Strimple, 1969); 283,*2a*, view of tegmen, post. side down, ×4 (Wachsmuth &

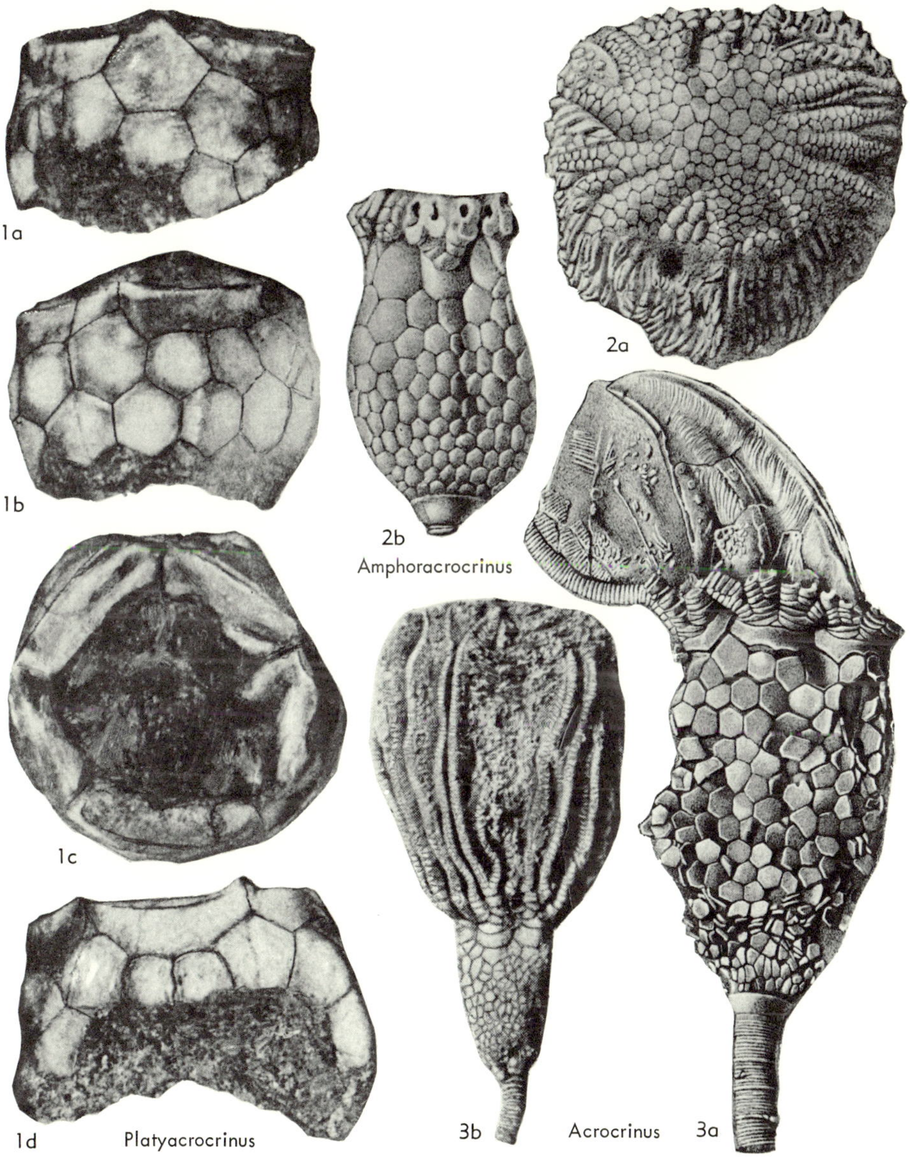

FIG. 283. Acrocrinidae (Acrocrininae) (p. *T*479-*T*482).

Springer, 1897); 283,*2b*, *C*-ray view of calyx, ×2 (Springer, 1926b).

Platyacrocrinus MOORE & STRIMPLE, 1969, p. 10 [**Acrocrinus brentwoodensis* MOORE & PLUMMER, 1938, p. 222; OD]. Radials most similar to those of *Acrocrinus* in being very low and wide; primanal distinctive in its subpentagonal shape, narrower than radials, with distal tip projecting slightly above rim of calyx; intercalaries fewer and relatively larger than in other genera of subfamily, most distal circlet containing 15 plates, of which 8 are subradial (single ones beneath *C* and *D* radials, pairs of plates beneath other radials) and 7 interradial in position, with 3 plates in *CD* interray (including subanal). Tegmen, lower part of calyx, arms, and stem unknown. *L.Penn. (Morrow.)*, USA(NW.Ark.).——FIG. 282,*4*; 283, *1*. **P. brentwoodensis* (MOORE & PLUMMER), Bloyd F.(Brentwood Ls.), near Woolsey, Washington Co., Ark.; 282,*4*, diagram showing radial

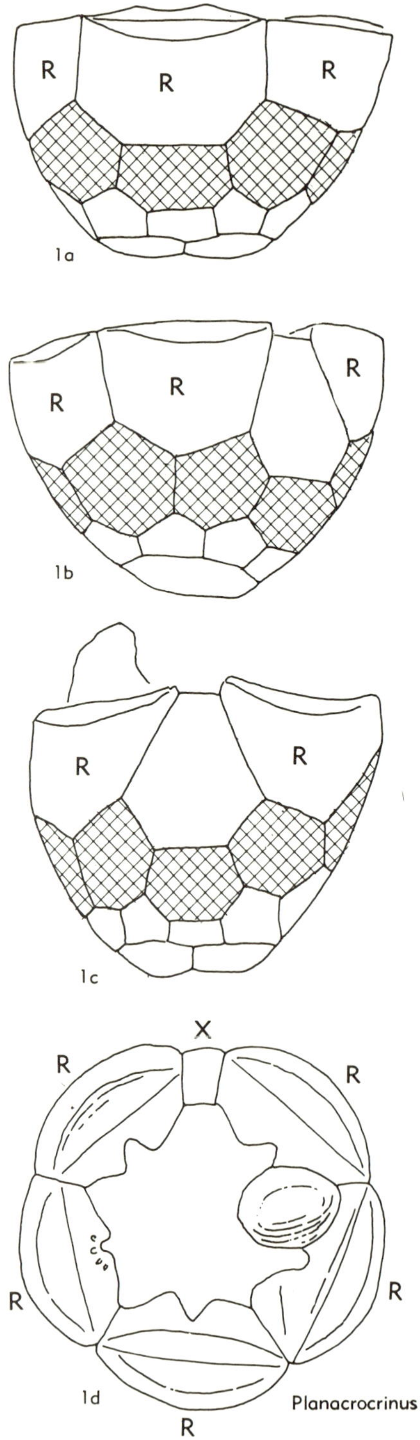

FIG. 284. Acrocrinidae (Planacrocrininae) (p. T482).

circlet and primanal with subjacent 15 intercalaries bilaterally symmetrical in arrangement with respect to *A-CD* axis (radials black, distal intercalaries shaded) (Moore & Strimple, 1969); 283,*1a-d, CD-, A*-ray, ventral, and *B*-ray views of holotype, ×3 (Moore & Plummer, 1938).

Subfamily PLANACROCRININAE Moore & Strimple, new subfamily

Calyx flat at base and summit, bowl shaped to subconical with steep nearly straight sides; basals clearly visible in side view; radials wider than high, with straight articular facets occupying full width of plates and extending well inward from outer margin, resembling facets of such inadunate crinoids as *Delocrinus* and *Cromyocrinus* in presence of ligament fossae, transverse ridge, muscle-attachment areas, and intermuscular notch; *C* and *D* radials separated by hexagonal primanal which is wider in lower third and narrowest at summit slightly below level of radial facets or even with them; circlet of most distal intercalaries comprising eight plates, only *A* radial having one directly beneath it; tegmen and arms unknown. *L.Penn.(Morrow.).*

Planacrocrinus MOORE & STRIMPLE, 1969, p. 15 [**P. ambix*; OD]. Characters of subfamily. *L. Penn.(Morrow., Brentwood Ls.-Wapanucka Ls.),* USA(NE.Okla.-S.Okla.).——FIG. 282,*9*; 284,*1*. **P. ambix,* Wapanucka Ls., S.Okla.; 282,*9,* diagram of plate arrangement in summit part of calyx (radials black, distal intercalaries shaded) (Moore & Strimple, 1969); 284,*1a-d, A-* and *D*-ray, *CD*-interray, and ventral views of holotype (camera lucida drawings, *R,* radials, distal intercalaries cross hatched), ×16.5 (Moore & Strimple, 1969, mod.).——FIG. 285,*1*. *P. conicus* MOORE & STRIMPLE, Brentwood Ls., Greenleaf Lake, NE. Okla.; *1a-c, A-* and *B*-ray, and *DE*-interray views of holotype (camera lucida drawings), ×10; *1d,e,* ventral and *CD*-interray views of same (camera lucida drawings, *R,* radials; distal intercalaries cross hatched), ×21 (after Moore & Strimple, 1969).

Subfamily GLOBACROCRININAE Moore & Strimple, 1969

[Globacrocrininae MOORE & STRIMPLE, 1969, p. 21]

Distinguished from other subfamilies of Acrocrinidae by narrow, strongly arcuate radial articular facets; basals visible or not visible from side; distalmost ring of intercalaries composed of six to 12 plates, subradial intercalaries ranging from none to five, some genera with longitudinal series

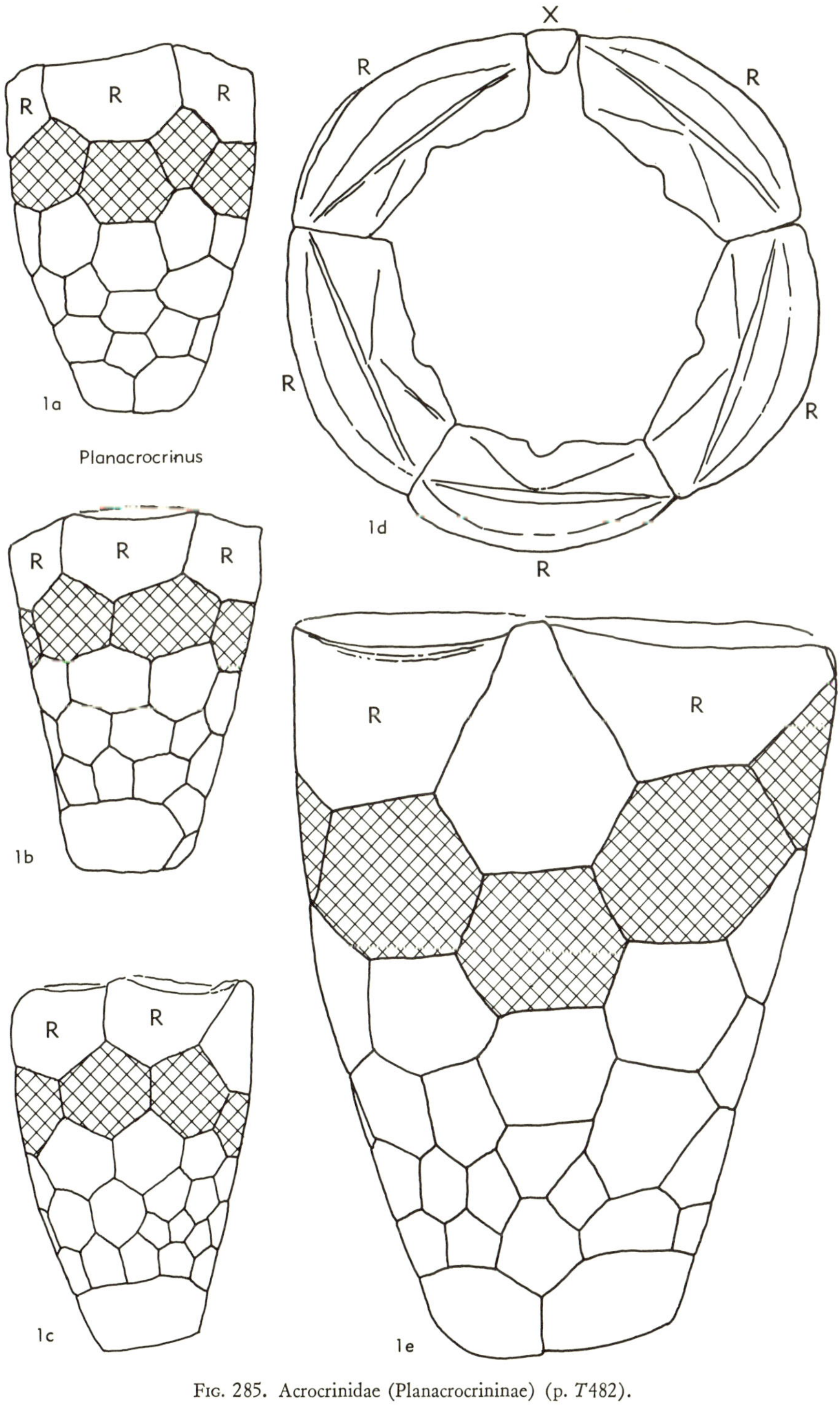

FIG. 285. Acrocrinidae (Planacrocrininae) (p. T482).

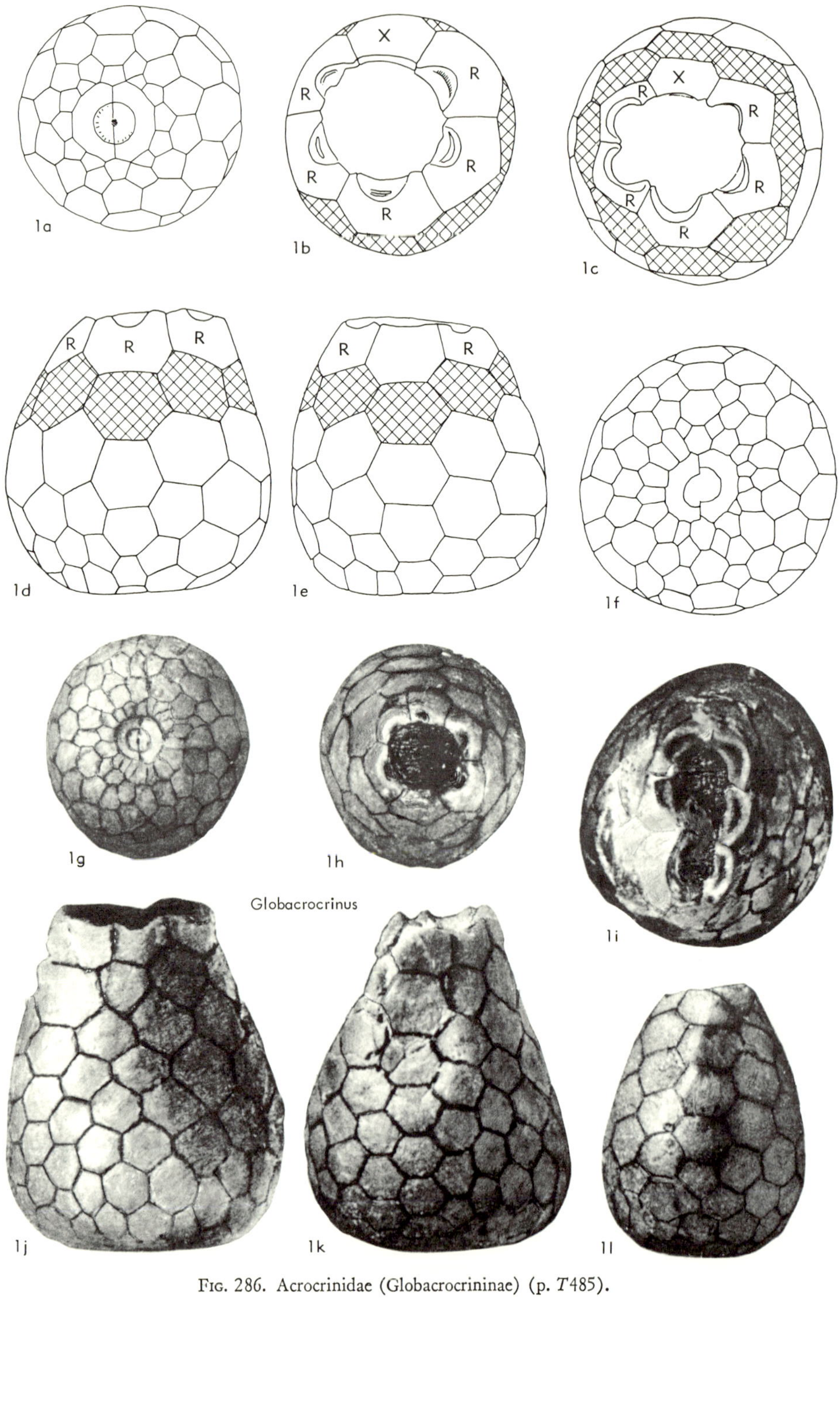

FIG. 286. Acrocrinidae (Globacrocrininae) (p. T485).

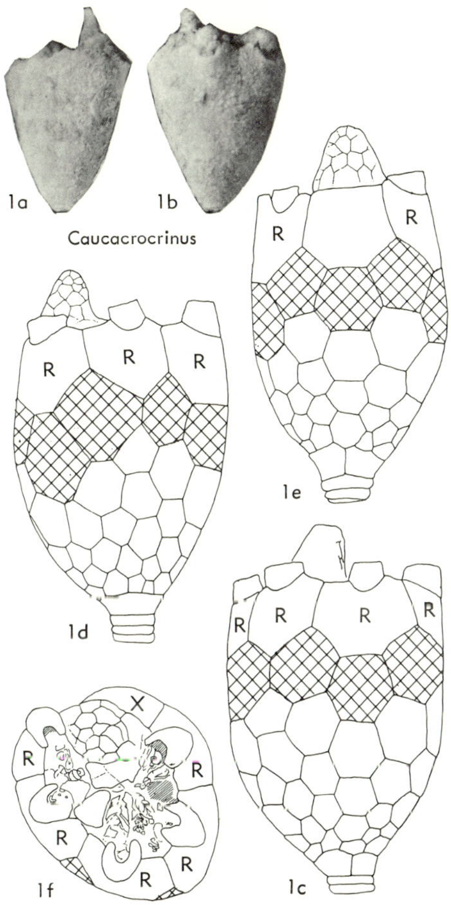

FIG. 287. Acrocrinidae (Globacrocrininae) (p. T485).

of superposed intercalaries beneath one to three radials in addition to primanal series present in all; tegmen with distinct orals, of which posterior one is largest, and commonly showing low anal pyramid near *CD* border. *L.Miss.-U.Penn.(Missour.).*

Globacrocrinus MOORE & STRIMPLE, 1969, p. 21 [**Acrocrinus pirum* MOORE & PLUMMER, 1938, p. 218; OD]. Calyx pear shaped to globose, small to very diminutive basals not visible from side; radials with height and width subequal, little larger than distalmost intercalaries, which form circlet of 8 plates with subradial one confined to *A* ray, (including subanal) interradial in position, longitudinal series of intercalaries beneath *A* radial and primanal. Tegmen with orals and anal pyramid. Stem slightly heteromorphic, at least next to calyx. Arms unknown. *L.Penn.(Morrow.)-M.Penn.(Desmoines.),* USA(Okla.).——FIG. 282,*5*; 286,*1g-l*. **G. pirum* (MOORE & PLUMMER), L.Penn.(Morrow., Brentwood Ls.), USA(Okla.); 282,*5*, diagram showing plate arrangement of calyx with 8 distalmost intercalaries (radials black, distal intercalaries shaded) (Moore & Strimple, 1969); 286,*1g,h,l*, dorsal, ventral and *CD*-interray views of paratype calyx, ×2.7 (Moore & Plummer, 1938); 286,*1i-k*, ventral, *C*-ray, and *CD*-interray views of holotype, ×2.7 (Moore & Plummer, 1938).——FIG. 286,*1a-f*. *G. rotundus* MOORE & STRIMPLE, L.Penn.(Morrow., Wapanucka Ls.), USA(Okla.); *1a,b*, camera lucida drawings of dorsal and ventral sides of paratype, ×10 (Moore & Strimple, 1969); *1c-f*, camera lucida drawings of holotype, ventral, *A-ray*, *CD*-interray, and dorsal views ×5.4 (*R*, radials; distal intercalaries cross hatched) (Moore & Strimple, 1969, mod.).

Caucacrocrinus MOORE & STRIMPLE, 1969, p. 28 [**C. urnula*; OD]. Diminutive calyx urn shaped, basals upflaring, visible from side; distalmost intercalaries with 4 (*A, C, D, E* rays) subradial or exceptionally 5 (including *B* ray), otherwise as in *Globacrocrinus*. *U.Penn.(Missour.),* USA(Ill.).——FIG. 282,*6*; 287,*1*. **C. urnula,* LaSalle Ls., Livingston Co., Ill.; 282,*6*, plate diagram of distal part of calyx (radials black, distal intercalaries shaded) (Moore & Strimple, 1969); 287,*1a,b*, side views of holotype calyx looking toward *A* ray and *CD* interray, ×6.5; 287,*1c-f*, *AB*-interray, *B*-ray, *CD*-interray, and ventral views of holotype (photos and camera lucida drawings, *R*, radials; distal intercalaries cross hatched), ×8.5 (Moore & Strimple, 1969, mod.).

Dinacrocrinus MOORE & STRIMPLE, 1969, p. 36 [**Acrocrinus expansus* STRIMPLE, 1951c, p. 192; OD]. Calyx rotund, truncate at summit of radials, basals not visible from side; differing from *Globacrocrinus* in occurrence of 12 intercalaries next beneath radial circlet, 5 subradial,

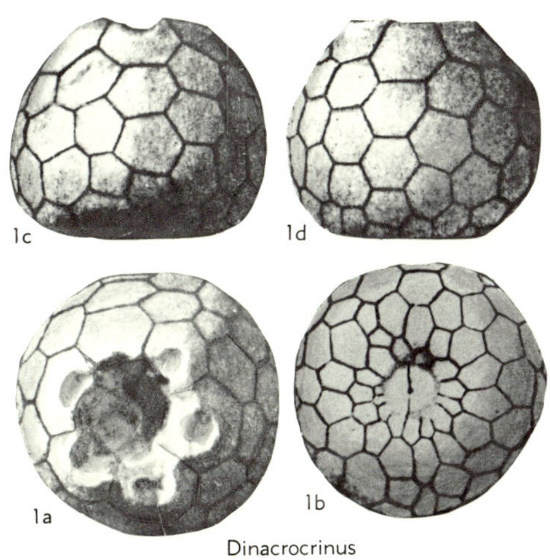

FIG. 288. Acrocrinidae (Globacrocrininae) (p. T485-T486).

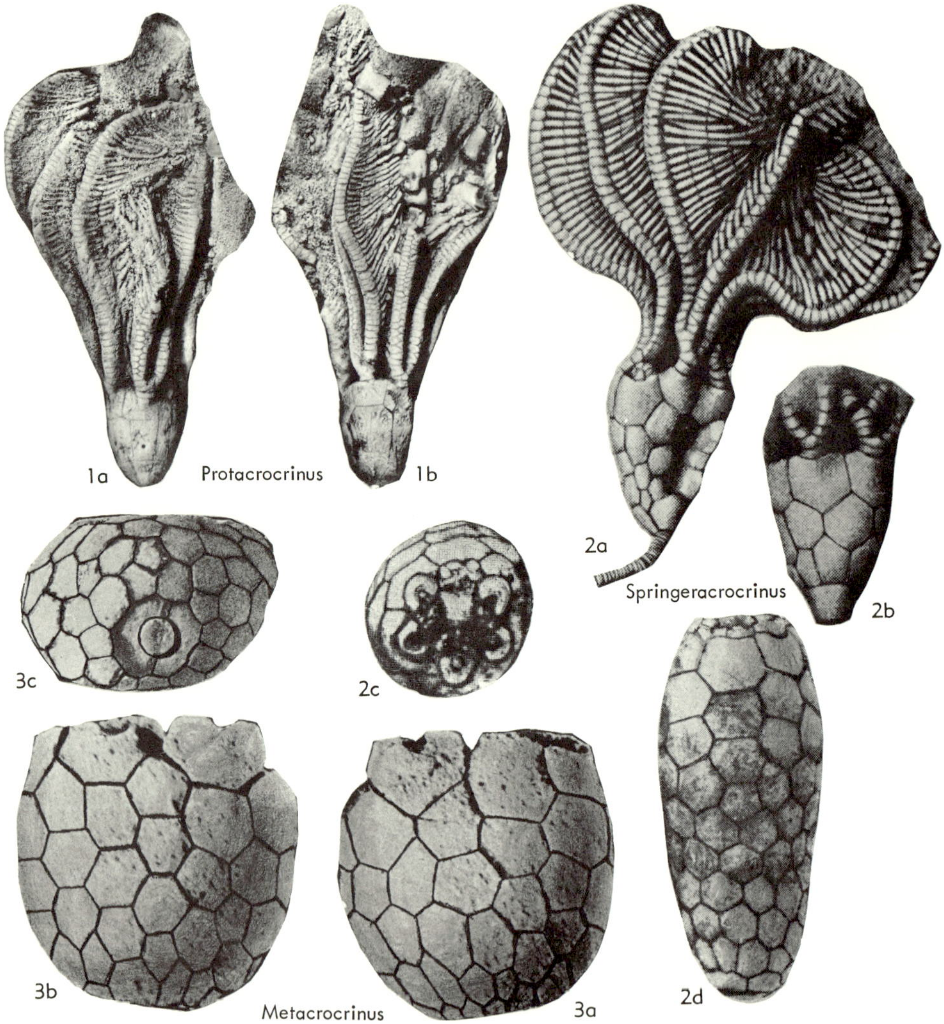

Fig. 289. Acrocrinidae (Globacrocrininae) (p. T486-T487).

with longitudinal series of intercalaries under *A* radial and primanal. *M.Penn.(Desmoines.)-U. Penn.(Missour.)*, USA(Okla.-Ill.).——Fig. 282,7; 288,*1*. **D. expansus* (Strimple), M.Penn.(Desmoines., Oologah Ls.), USA(Okla.); 282,*7*, diagram of calyx plates showing 12 distalmost intercalaries (radials black, distal intercalaries shaded) (Strimple, 1951, mod.); 288,*1a-d*, ventral, dorsal, *CD*-interray, and *A*-ray views of holotype, ×3 (Strimple, 1951e).

Metacrocrinus Moore & Strimple, 1969, p. 35 [**Acrocrinus pumpkensis* Strimple, 1949b, p. 900; OD]. Differs from *Dinacrocrinus* only in absence of intercalary next beneath *D* radial. *M.Penn.(Desmoines.)*, USA(Okla.).——Fig. 282, *8*; 289,*3*. **M. pumpkensis* (Strimple), M.Penn. (Lake Murray F., Pumpkin Creek Ls.), S. of Ardmore, Carter Co.; 282,*8*, diagram of calyx plates (radials black, distal intercalaries shaded) (Strimple, 1949b, mod.); 289,*3a-c*, *D*-ray (slightly left of midline), *EA*-interray, and dorsal views of holotype, ×4 (Strimple, 1949b).

Protacrocrinus Moore & Strimple, 1969, p. 38 [**Acrocrinus primitivus* Laudon & Beane, 1937, p. 252; OD]. Crown pyriform, with 10 biserial arms bearing long slender pinnules. Calyx deep bowl shaped, evenly rounded at base, truncate at top, composed of 5 small radials subequal in height and width, primanal rising slightly above summit of radials, 6 tall intercalaries directly beneath radials and primanal. Stem impression very small, stem unknown. *L.Miss.(Kinderhook.),* USA (Iowa).——FIG. 282,*2*; 289,*1*. **P. primitivus* (LAUDON & BEANE), Hampton F., LeGrand, Tama Co., Iowa; 282, *2*, diagram of calyx plates (radials black, distal intercalaries shaded) (Moore & Strimple, 1969); 289,*1a,b*, views of holotype crown from *D*- and *B*-ray sides, ×2.25 (Moore & Strimple, 1969).

Springeracrocrinus MOORE & STRIMPLE, 1969, p. 41 [**Acrocrinus intermedius* SPRINGER, 1926b, p. 43; OD]. With characters of subfamily but all distal intercalaries located interradially, none directly beneath radial, 1 intercalary beneath primanal, sum of 7 plates in this circlet. *L.Miss.(Osag.)-U.Miss.(Chester.),* USA(Iowa-Ill.); *U.Carb.(Moscov.),* Eu.(USSR).——FIG. 282,*10*; 289,*2a,b*. **S. intermedius* (SPRINGER), low.Chester.(Renault F.), Ill.(Monroe Co.); 282,*10*, diagram of calyx plates (radials black, distal intercalaries shaded) (Moore & Strimple, 1969); 289,*2a,b*, *C*-ray view of crown and *E*-ray side of calyx, both ×3.4 (Springer, 1926b).——FIG. 289,*2c,d*. *S. mjatschkowensis* (YAKOVLEV), U.Carb.(Myachkova F.), USSR (Moscow basin); ventral and *CD*-interray views of calyx, ×4.5 (Yakovlev & Ivanov, 1956).

Suborder GLYPTOCRININA Moore, 1952

[Glyptocrinina MOORE, 1952 in MOORE, LALICKER, & FISCHER, p. 614]

Basal circlet pentagonal; radials adjoining each other all around. *M.Ord.-U.Perm.*

Superfamily GLYPTOCRINACEA Zittel, 1879

[*nom. transl.* UBAGHS, herein (*ex* Glyptocrinidae ZITTEL, 1879, p. 374)]

Calyx conical to subglobose, plates generally with stellate ornamentation; basals five; fixed secundibrachs two to eight or even more in each ray; fixed tertibrachs present in some species; interprimibrachs numerous and well defined; intersecundibrachs numerous, less definite; primanal followed by three plates in next row above; median anal ridge present. Tegmen composed of many small plates; anus opening directly through it. Rays ordinarily not branching beyond tertibrach series; brachials uniserial or biserial. Column cylindrical; axial canal pentagonal. *M.Ord.-U.Sil.*

Family GLYPTOCRINIDAE Zittel, 1879

[Gryptocrinidae ZITTEL, 1879, p. 374]

Characters of superfamily. *M.Ord.-U.Sil.*

Glyptocrinus HALL, 1847, p. 281 [**G. decadactylus*; M] [=*Icosidactylocrinites* OWEN, 1843 *(nom. nud.)*; *Icosidactylocrinus* SHUMARD, 1868, p. 372 *(nom. subst. pro Icosidactylocrinites)*; *Glyptocrinites* CARPENTER, 1884, p. 929 *(nom. van. pro Glyptocrinus)*]. Secundibrachs 2 in each ray; fixed tertibrachs passing to free arms. Arms 20, simple or branching once, uniserial; brachials cuneate externally. *M.Ord.-U.Sil.,* N.Am.——FIG. 290,*1*. **G. decadactylus*, M.Ord., USA; *1a,b*, *E*-ray and *CD*-interray views of slightly crushed crown, ×2 (Ubaghs, n); *1c*, tegmen, ×2 (Wachsmuth & Springer, 1897); *1d*, plate diagram of part of calyx (radials black, interbrachials stippled) (Ubaghs, n).

Periglyptocrinus WACHSMUTH & SPRINGER, 1897, p. 277 [**P. billingsi*; OD]. Similar to *Glyptocrinus*, but basals larger and arms biserial. *M.Ord.,* N.Am.——FIG. 291,*1*. **P. billingsi*; Trenton., Can.; *CD*-interray view of crown with attached stem, ×1 (Wachsmuth & Springer, 1897).

Pycnocrinus S. A. MILLER, 1883, p. 219 [**Glyptocrinus shafferi* S. A. MILLER, 1875, p. 277; OD]. Secundibrachs numerous; fixed pinnules as many as 5 or 6 in each ray, 1st one borne by fixed secundibrachs *2* (outer) and 2nd one by fixed secundibrachs *4* (inner). Arms 10, branching once or twice, uniserial. *M.Ord.-U.Ord.,* N.Am.——FIG. 291,*2a*. **P. shafferi* (S. A. MILLER), U.Ord., USA(Ohio); type crown and stem, ×2 (Wachsmuth & Springer, 1897).——FIG. 291,*2b-d*. *P. dyeri* (MEEK), U.Ord., Cincinnati., USA(Ohio); *2b*, lat. view of almost complete crown with attached stem, ×1 (Wachsmuth & Springer, 1897); *2c*, *AB*-interray view of theca showing fixed pinnules, ×1 (Springer, 1917); *2d*, tegmen, ×1.25 (Springer, 1917).

Superfamily MELOCRINITACEA d'Orbigny, 1852

[*nom. correct.* UBAGHS, herein (*pro* Melocriniticae UBAGHS, 1953, p. 741, *nom. transl. ex* Melocrinidae D'ORBIGNY, 1852, p. 141)]

Calyx generally higher than wide, typically expanding from relatively narrow base to zone of greatest width at level of arms; basals, four; median ray ridges present in primitive members; interprimibrachs numerous, well defined; primanal followed by three plates in next row; tegmen composed of many, generally undifferentiated plates, anus opening directly through it or at end of tube; plate ornamentation largely stellate. Rami two to four in each ray, branching isotomously or heterotomously; brachials uniserial, biserial, or compound. Column cylindrical. *U.Ord.-U.Dev.*

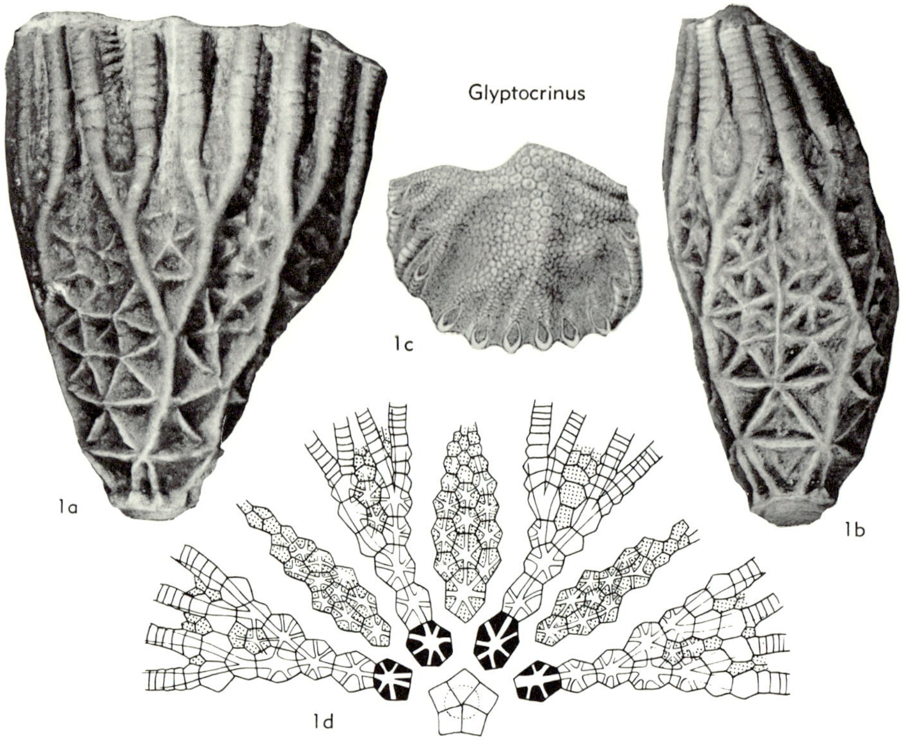

FIG. 290. Glyptocrinidae (p. T487).

Family SCYPHOCRINITIDAE Jaekel, 1918

[*nom. correct.* UBAGHS, 1953, p. 741 (*pro* Scyphocrinidae JAEKEL, 1918, p. 31)] [=Camarocystidae ZITTEL, 1895, p. 154; Camarocystida HAECKEL, 1896a, p. 168]

Secundibrachs numerous, many of them incorporated in calyx; interbrachials separated from tegmen by fixed pinnulars; posterior side slightly differentiated in calyx; rami two in each ray, isotomously branching. *?L.Sil.; U.Sil.-L.Dev.*

In view of the very distinctive morphological features of this family, chiefly represented by its type genus, coupled with mass accumulation of well-preserved crowns and other skeletal parts, including the distinctive bulbous holdfasts known as loboliths or *Camarocrinus,* the very restricted stratigraphic range and wide repartition of these crinoids in Europe, Asia, North Africa, and North America are very noteworthy. In Bohemia, where *Scyphocrinites* was recognized first, this genus is located in the Pridoli Formation (*eβ*2), which extends from the zone of *Pristiograptus ultimus* to the zone of *Monograptus angustidens.* It has about the same range in Germany (Harz, Kellerwald, Thuringia), Bulgaria, and Poland, whereas in North Africa it reaches a little higher level, in the zone of *Monograptus uniformis.* It has also been reported from the Upper Silurian of England (Cornwall), France, Spain (eastern Pyrenees), USSR (Podolia, Khazakstan), Burma, and China. In North America (Tennessee, Missouri, Oklahoma, West Virginia) the recorded occurrence is Silurian (Decatur? and Henryhouse formations) and Lower Devonian (Helderbergian). Interpretation of *Scyphocrinites* as a pelagic crinoid provided with float *(Camarocrinus),* which ranged the globe very widely, appearing and disappearing abruptly within a relatively short time span, accords with geographic, stratigraphic, morphologic, and paleontologic evidence (see HAUDE, 1972). It should receive recognition as an exceptionally precise age indicator useful for correlation of remotely separate marine deposits.

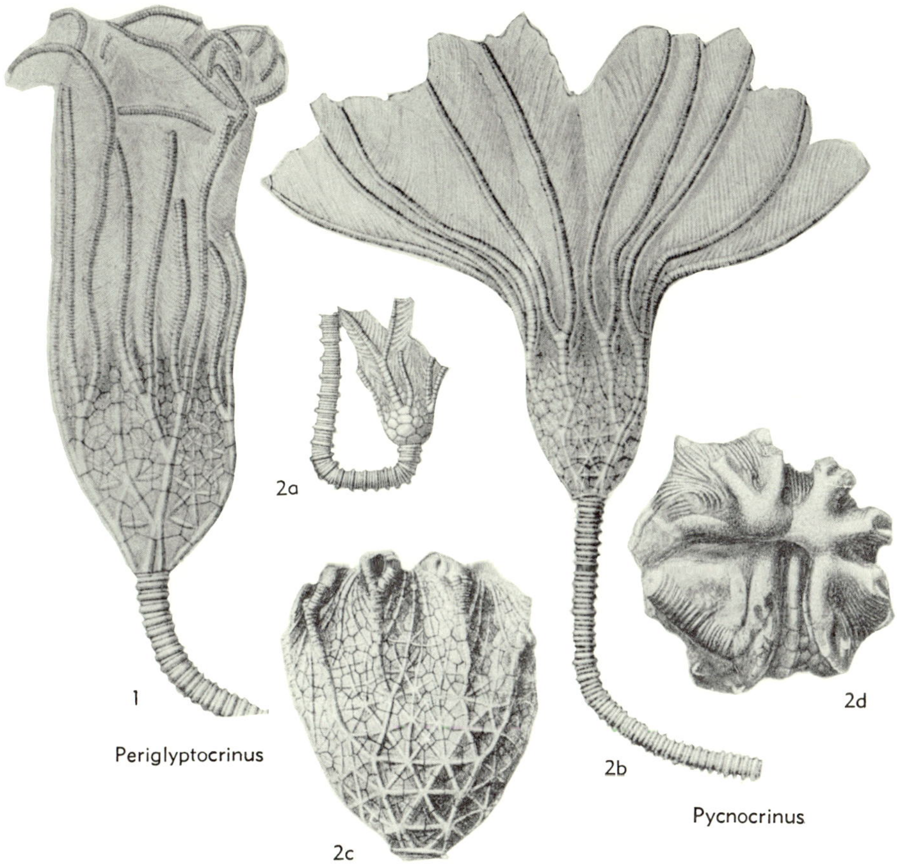

FIG. 291. Glyptocrinidae (p. T487).

Scyphocrinites ZENKER, 1833, p. 25 [**S. elegans*; M] [=*Scyphocrinus* GEINITZ, 1846, p. 549 (*nom. van. pro Scyphocrinites* (obj.) (*non Scyphocrinus* HALL, 1847); *Scyphonocrinites* ROEMER in BRONN, 1855, p. 255 (*nom. null.*); *Siphocrinus* KREJCI, 1860, p. 751 (*nom. null.*); *Lobolithus* BARRANDE in BIGSBY, 1868, p. 192 (*nom. nud.*) (holdfast); *Camarocrinus* HALL, 1879b, p. 205 (type, *C. stellatus* HALL, 1879b; SD MILLER, 1889, p. 230) (holdfast); *Camerocrinus* FRIČ, 1904, p. 277 (*nom. van. pro Camarocrinus*)]. Calyx large, expanding or contracting toward arm bases, with interradial areas depressed or protuberant, largely formed of fixed pinnulars; fixed secundibrachs 10 to 20 or even more, 1st pinnule borne by fixed secundibrach *2* (outer) and 2nd pinnule by fixed secundibrachs *4* (inner). Anal tube strong, subcentral on tegmen. Free arms composed of very short brachials arranged uniserially or biserially. Column ending distally either in small encrusting root or (more typically) in large bulbous chambered structure designated separately as *Lobolithus* or *Camarocrinus* with walls of minute irregular plates, having their origin in bifurcations and modifications of innumerable root-branches; axial canal of stem quinquelobate to sharply quinquestellate. [For description and illustration of the holdfast, see section on morphology of fossil crinoids, p. T88.] *?L.Sil., U.Sil.-L.Dev.*, Eu.-N. Am.(USA)-N.Afr.-Asia.——FIG. 292,*1a-d*. **S. elegans; 1a,* side view of crown (diagram.) from L.Dev.(Bailey Ls.), USA(Mo.), ×0.7 (Springer, 1917); *1b-d,* median secs. of pluricolumnals from proxistele, mesistele, and dististele regions, specimens from L.Dev.(Bailey Ls.), USA(Mo.), ×2 (Springer, 1917).——FIG. 292,*1e*. *S.* sp., U.Sil., N.Afr.; tegmen and anal tube, ×1 (Ubaghs, n).——FIG. 292,*1f*. *S.* sp., L.Dev., USA; plated bulbous holdfast (*Camarocrinus*), ×0.5 (Hall, 1879b).——FIG. 292,*1g*. *S. ulrichi* (SCHUCHERT), L.Dev.(Helderberg.), USA(Okla.), upper part of large bulb (not included in figure) showing distal part of stem and branches merging into plated area, ×1 (Springer, 1917).

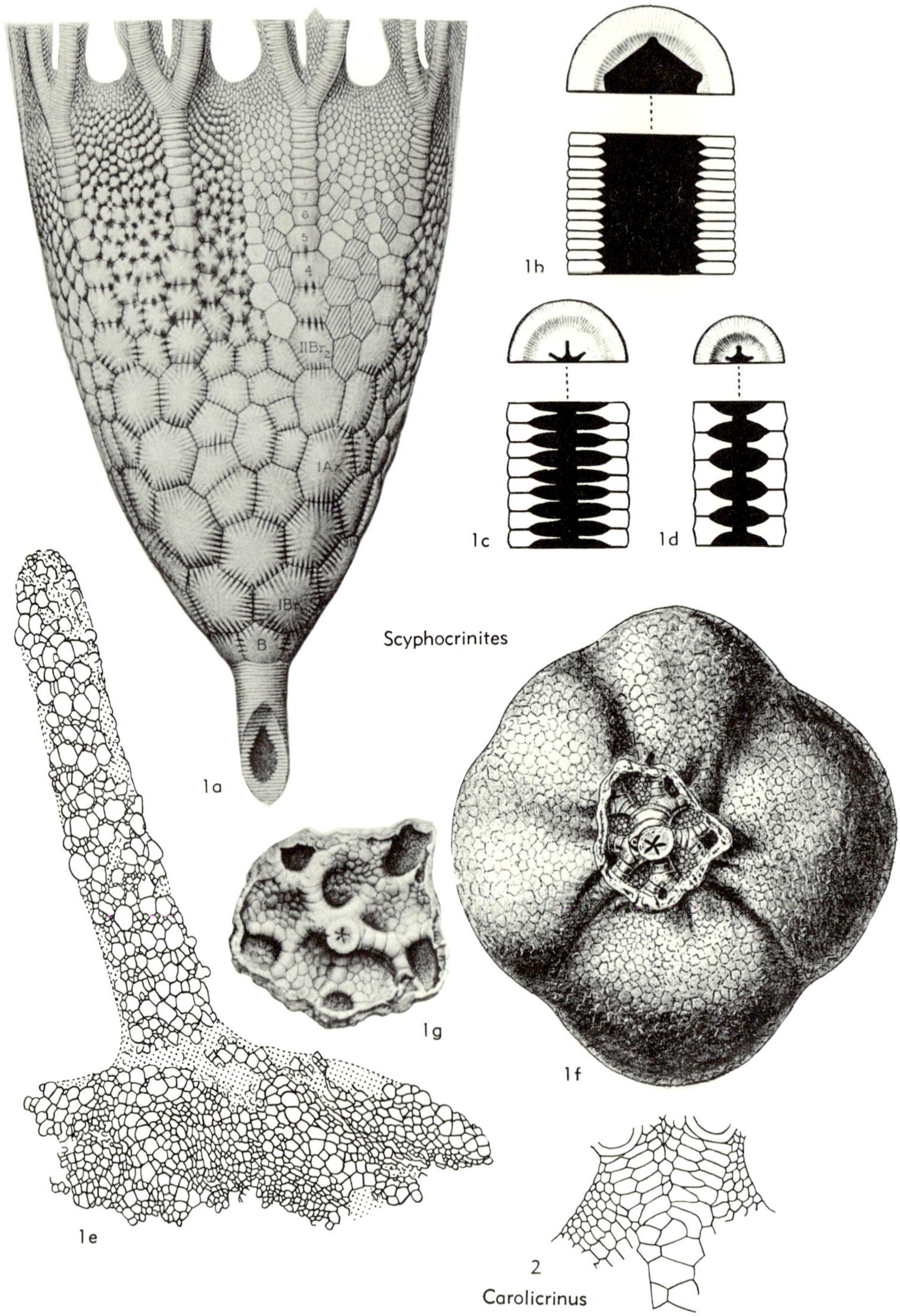

Fig. 292. Scyphocrinitidae (p. T489, T491).

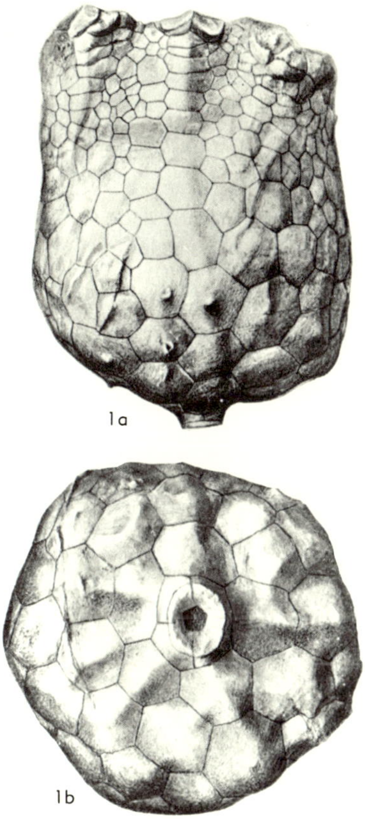

FIG. 293. Scyphocrinitidae (p. T491).

[*Lobolithus* BARRANDE, 1868 (in BIGSBY, Thesaur. Silur., p. 192) has been treated by authors as a generic name, even though not associated by anyone with a named species. BARRANDE (1887, p. 1), ZITTEL (1895, p. 154), WAAGEN & JAHN (1899, p. 189) and other authors have referred to these fossils in French vernacular as *lobolithes* (loboliths in English, *Lobolithen* in German), a term commonly used in Bohemia since the middle of the 19th century. Thus, *Lobolithus* has no standing in zoological nomenclature.]

?**Carolicrinus** WAAGEN & JAHN, 1899, p. 22 [**C. barrandei*; M]. Similar to *Scyphocrinites* but having proximal biserial brachials incorporated in calyx; free arms compactly biserial. Base in type specimen possibly abnormal. [This genus, imperfectly known, may be a mere synonym of *Scyphocrinites*.] *U.Sil.*, Eu.(Boh.).——FIG. 292,*2*. **C. barrandei*, Boh.; biserial fixed brachials and pinnulars, ×0.7 (Waagen & Jahn, 1899).

Liomolgocrinus STRIMPLE, 1963, p. 103 [**L. dissutus*; OD]. Theca ovoid, with rapidly expanding base, widest at about 2nd range of interprimibrachs, and contracting slightly from there to near arm-bases; fixed secundibrachs apparently less than 10 in half-ray; wide interradial areas; fixed pinnules undifferentiated from interbrachials; thecal plates flat or low convex. Arms and stem unknown. *L.Dev.*, USA(Okla.).——FIG. 293,*1*. **L. dissutus*, Helderberg.; *1a,b*, *D*-ray and dorsal views of theca, ×1 (Springer, 1917).

Family PARAMELOCRINIDAE Ubaghs, 1958

[Paramelocrinidae UBAGHS, 1958b, p. 261]

Secundibrachs numerous, few of them incorporated in calyx; interbrachials merging with tegmen; *CD* interray slightly larger than others, with three plates in second range. Arms two to each ray, isotomously branching, biserial. *U.Sil.*

Paramelocrinus UBAGHS, 1958, p. 261 [**P. angelini*; OD]. Characters of family. *U.Sil.*, Sweden(Gotl.).——FIG. 294,*1*. **P. angelini*; *1a*, plate diagram of part of calyx (radials black, anals and interbrachials stippled) (Ubaghs, n); *1b*, lat. view of holotype crown, ×2.7 (Ubaghs, 1958b).

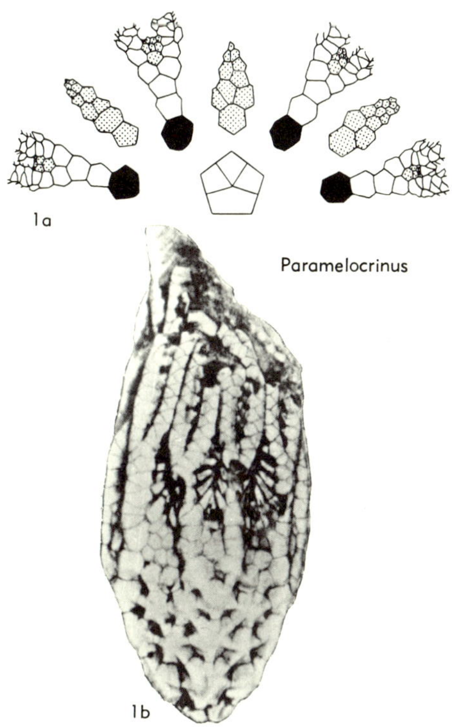

FIG. 294. Paramelocrinidae (p. T491).

Family MELOCRINITIDAE d'Orbigny, 1852

[*nom. correct.* BASSLER, 1938, p. 26 (*pro* Melocrinidae D'ORBIGNY, 1852, p. 141)] [=Ctenocrinidae C. F. ROEMER, 1855, p. 228]

Secundibrachs two or three in each ray, rarely more; interprimibrachs merging with tegmen; presence of intersecundibrachs and intertertibrachs dependent on degree of brachial incorporation in calyx; primanal generally followed by median row of anal plates; anal tube may be present. Free arms four in each ray, undivided or exotomously branching; inner pair of arms primitively distinct, secondarily fusing into stout biserial ramuli-bearing trunk; outer pair progressively reduced and disappearing in ultimate members. Column with sharply defined nodals and internodals. *U.Ord.-U.Dev.*

Melocrinites GOLDFUSS, 1831, p. 197 [**M. hieroglyphicus*; SD C. F. ROEMER, 1855, p. 251] [=*Melocrinus* AGASSIZ, 1836, p. 197 (*nom. van. pro Melocrinites*); *Castanocrinus* C. F. ROEMER, 1855, p. 228, 252 (type, *Melocrinites gibbosus* GOLDFUSS, 1831, p. 211; OD)]. Intersecundibrachs absent or reduced; anus subcentral, directly through tegmen or moderately protruding. One ramuli-bearing trunk in each ray, typically without outer rami. *M.Dev.-U.Dev.*, Eu.-N.Am.-Sib. ——FIG. 295,*1a-e*. **M. hieroglyphicus*, U.Dev. (Frasn.), Stolberg, Ger.; lectotype, Geol.-paleont. Inst. Univ. Bonn(unnumbered) (chosen by UBAGHS, herein, orig. GOLDFUSS, 1831, pl. 60, fig. 1c); *1a*, plate diagram of part of calyx (radials black, anals and interbrachials stippled) (Ubaghs, n); *1b-e*, *A*-ray, *CD*-interray, dorsal, and tegminal views of calyx, ×1.5 (Ubaghs, n).——FIG. 295, *1f*. *M. aequus* SCHMIDT, M.Dev.(Couvin.), Ger.; lat. view of incomplete crown, ×1 (Schmidt, 1942).

Alisocrinus KIRK, 1929, p. 343 [**Mariacrinus warreni* RINGUEBERG, 1888, p. 133; OD]. Median ray ridges present; fixed secundibrachs 2 or rarely 3; fixed tertibrachs may be present; interbrachials well developed. Tegmen, so far as known, low, many-plated, anus excentric. Free arms subequal, undivided, uniserial, pinnulate, typically 4 in each ray; free brachials cuneate. *U.Ord.-M.Sil.*, N.Am. ——FIG. 295,*2*. **A. warreni* (RINGUEBERG), M. Sil.(Niagaran), USA(N.Y.); lat. view of type crown with attached stem, ×1 (Wachsmuth & Springer, 1897).

Ctenocrinus BRONN, 1840, p. 547 [**C. typus*; OD] [=*Ctenocrinites* STEININGER, 1849, p. 22 (*nom. van. pro Ctenocrinus*); *Astrocrinites* CONRAD, 1841, p. 34 (type, *A. pachydactylus*) (*non Astrocrinites* AUSTIN & AUSTIN, 1843); *Mariacrinus* HALL, 1859, p. 104 (type, *M. nobilissimus*); *Astrocrinus* BATHER, 1900a, p. 161 (*nom. van. pro Astrocrinites*) (*non Astrocrinus* MORRIS, 1843); *Clonocrinus* OEHLERT, 1879, p. 3 (type, *C. bigsbyi*) (*non* QUENSTEDT, 1876); *Lindstroemiocrinus* JAEKEL, 1918, p. 31 (type, *Trochocrinites gottlandicus* PANDER in HELMERSEN, 1858, p. 312)]. Median ray ridges more or less developed; secundibrachs generally 2; intersecundibrachs few or absent; excentric anal tube present in some if not all species. Inner arms of each ray partially or entirely united in stout, biserial, ramuli-bearing trunks, without pinnules; outer arms more or less reduced and similar to ramuli, which are numerous, pinnulate, with uniserial, biserial or compound brachials. *U.Sil.-M.Dev.*, Eu.-N.Am.-S.Am.——FIG. 296,*1a*. *C. gottlandicus* (PANDER), U.Sil.(Ludlov.), Sweden(Gotl.); crown with proximally united inner rami, ×2 (Ubaghs, 1958b).——FIG. 296,*1b*. **C. typus*, L.Dev.(Siegen.), Ger.; entirely united arm trunks of lectotype crown, ×1 (Schmidt, 1937).

Promelocrinus JAEKEL, 1902, p. 1067 [**P. anglicus*; M] [=?*Zenkericrinus* WAAGEN & JAHN, 1899, p. 102 (*nom. subst. pro Xenocrinus* JAHN, 1892) (*non* S. A. MILLER, 1881a) (type, *Z. melocrinoides* WAAGEN & JAHN; M); ?*Proctenocrinus* JAEKEL, 1918, p. 32 (*nom. nud.*) (no species named)]. Median ray ridges faint or absent; secundibrachs typically 2. Inner arms of each ray uniserial, generally larger, not pinnuliferous, bearing 1 to several uniserial or biserial pinnuliferous ramuli on outer side; outer rami generally smaller, undivided, uniserial or biserial, pinnuliferous. *M. Sil.-U.Sil.*, Eu.-N.Am.(USA).——FIG. 295,*3a*. **P. anglicus*, M.Sil.(Wenlock.), Eng.; *CD*-interray view of crown, ×1 (Springer, 1926a).——FIG. 295,*3b*. *P. fulminatus* (ANGELIN), U.Sil.(Wenlock.), Sweden; *CD*-interray view of crown, ×2.5 (Ubaghs, 1958a).

Trichotocrinus OLSSON, 1912, p. 27 [**Melocrinus (Trichotocrinus) harrisi*; M]. Main arm trunks bearing 1) pinnulate ramuli and 2) ramuliferous branches at irregular intervals; branches and trunks without pinnules; brachials of branches biserial, those of ramuli biserial or compound; outer rami present or absent. *U.Dev.*, N.Am. (USA).——FIG. 296,*2a*. *T. lutheri* GOLDRING, Portage beds, N.Y.; arm trunk, ×1 (Goldring, 1923).——FIG. 296,*2b*. **T. harrisi*, Portage-Ithaca beds, N.Y.; ray structure (Olsson, 1912a).

Family UNCERTAIN

Cytocrinus ROEMER, 1860, p. 46 [**C. laevis*; M] [=*Turbinocrinus* WACHSMUTH & SPRINGER, 1881, p. 237 (*nom. nud.*); *Turbinicrinites* WOOD (*ex* TROOST MS), 1909, p. 44 (type, *T. verneuili*; OD)]. Calyx conspicuously lobed, having same structure as in *Melocrinites*, but arms apparently extended into 5 uniserial trunks; anal tube excentric. Arms and column unknown. *M.Sil. (Niagaran)*, N.Am.(USA).——FIG. 297,*1*. **C. laevis*, Beech River Ls., Tenn.; *1a,b*, lat. and tegminal views of theca, ×2 (Springer, 1926a).

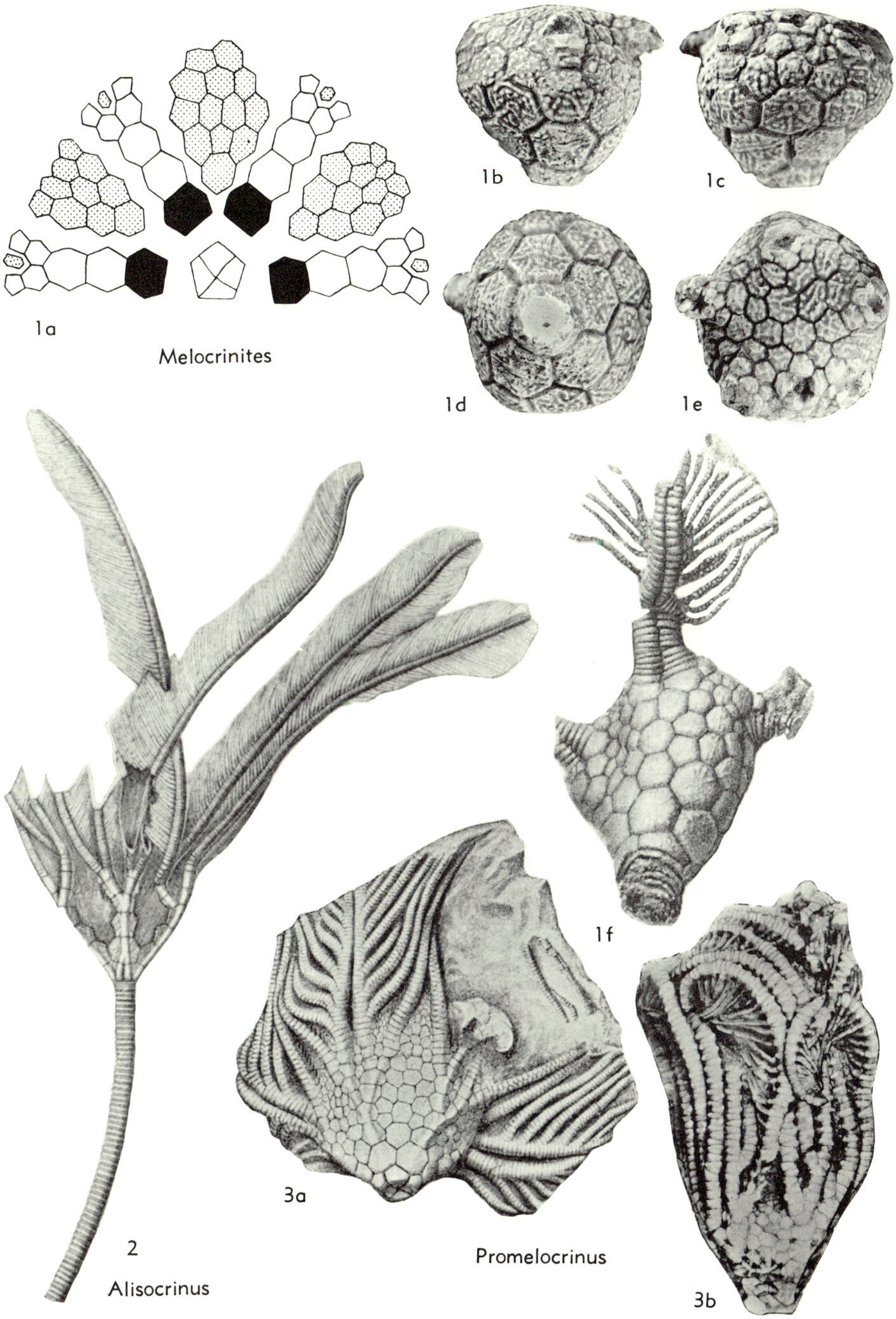

Fig. 295. Melocrinitidae (p. *T*492).

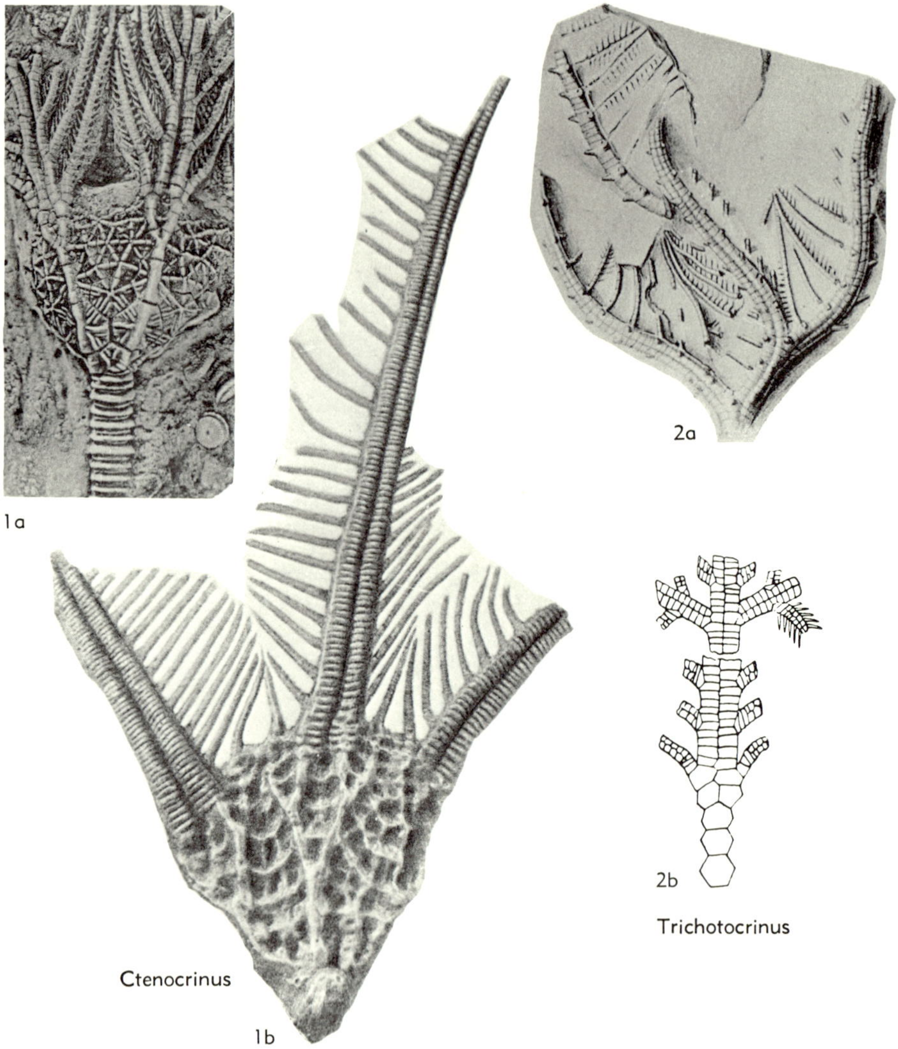

FIG. 296. Melocrinitidae (p. T492).

Superfamily EUCALYPTOCRINITACEA Roemer, 1855

[*nom. transl. et correct.* UBAGHS, herein (*ex* Eucalyptocrinidae ROEMER, 1855, p. 229)] [=Dolatocrinicae UBAGHS, p. 742]

Calyx typically wider than high, broadly rounded below, with exsert or more commonly concave base; unequal basals three, four, or fused together; evenly pentamerous radial circlet composed of five plates commonly wider than high, laterally adjoining one another all around; fixed brachials firmly joined to interbrachials, posterior side of cup barely differentiated from other interrays or not at all. Tegmen stoutly plated, low and rounded or strongly elevated, with or without lateral projections in form of vertical ridges which produce alcoves for reception of free arms; free brachials mostly biserial but may be uniserial. Column circular transversely, generally heteromorphic, some genera with laterally flanged columnals or nodals bearing cirri; attached generally by many-branched radix. *M.Sil.-M.Dev.*

Family CLONOCRINIDAE Bather, 1899

[Clonocrinidae BATHER, 1899b, p. 922] [=Corymbocrinidae ANGELIN, 1878, p. 18]

Basals four; fixed primibrachs two, fixed secundibrachs one or two, and fixed tertibrachs varying in number in each ray; fixed quartibrachs also may be present; interprimibrachs rather few and large; posterior side undifferentiated in cup. Free arms four in each ray, biserial. *U.Sil.-L.Dev.*

Clonocrinus QUENSTEDT, 1876, p. 638 [**C. gothlandicus*; M] [=*Corymbocrinus* ANGELIN, 1878, p. 18 (type, *Eucalyptocrinus polydactylus* M'COY, 1849, p. 249; SD BASSLER, 1938, p. 73)]. Basal concavity including basals and major part of radials; quadrangular fixed primibrachs 1, fixed secundibrachs 2, and more or less numerous fixed tertibrachs in each ray, passing to free arm plates; interprimibrachs in single vertical series of 2 or 3 large plates followed by smaller and irregular elements; intersecundibrachs few. Arms branching isotomously several times; distal pinnulars without oral grooves. Stem transversely round, heteromorphic, with nodals covering internodals at least proximally. *U.Sil., ?L.Dev.*, W.Eu.-N.Am.——FIG. 298,*1a,b*. **C. gothlandicus*, U.Sil., Sweden(Gotl.); *1a*, portion of holotype crown, ×1 (Ubaghs, n); *1b*, lat. view of complete crown (Stockholm Museum Ec. 11843) (Ubaghs, n).——FIG. 298,*1c*. *C.* sp.; vert. sec. of theca and proxistele (radials black, interbrachials stippled) (Ubaghs, n).——FIG. 298,*1d*. *C.* sp., plate diagram of part of calyx (symbols as in Fig. *1c*) (Ubaghs, n).

Technocrinus HALL, 1859, p. 139 [**Mariacrinus (Technocrinus) andrewsi*; SD WACHSMUTH & SPRINGER, 1881, p. 117(291)]. Base convex; 1st primibrachs hexagonal; fixed secundibrachs 1; and fixed tertibrachs 1 or 2 in each half-ray; interprimibrachs in upward sequence 1,2, 1 or 2, 0 or 1; no intersecundibrachs. Arms long, simple. Stem transversely circular. *U.Sil.-L.Dev.*, N.Am. (USA).——FIG. 298,*2*. **T. andrewsi*, L.Dev. (Oriskany), Md.; lat. view of holotype crown with attached stem, ×1 (Wachsmuth & Springer, 1897).

Family EUCALYPTOCRINITIDAE Roemer, 1855

[*nom. correct.* BASSLER, 1938, p. 27 (*pro* Eucalyptocrinidae ROEMER, 1855, p. 229)] [=Polycrinidae D'ORBIGNY, 1852, p. 141 (invalid because containing no type genus *Polycrinus*—*non Polycrinus* JAEKEL, 1918); Polycrinides PICTET, 1857 (vernacular); Calyptocrinidae ANGELIN, 1878, p. 14 (invalid because containing no type genus *Calyptocrinus*); Calyptocrinidées DE LORIOL, 1884, p. 59 (vernacular)]

Calyx typically with narrow, moderately deep basal concavity having sides composed of four basals and proximal parts of radials;

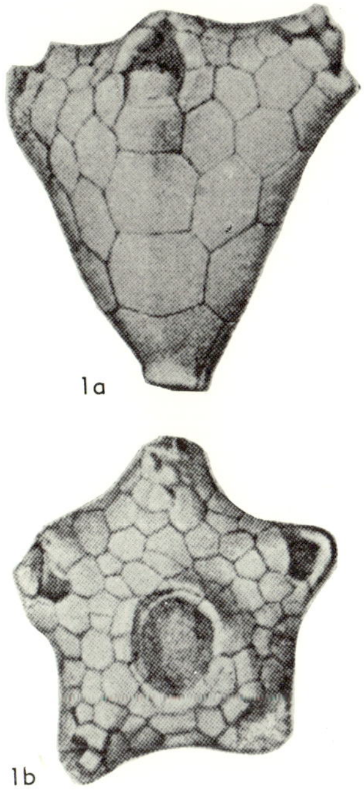

Cytocrinus

FIG. 297. Family Uncertain (p. T492).

fixed primibrachs one or two, fixed secundibrachs two, and fixed tertibrachs one in each ray, followed by free arms; each interray with single large interprimibrach followed by two smaller ones, each ray with single interbrachial between secundibrachs of branched fixed brachs; *CD* interray undifferentiated from others. Tegmen elevated, stoutly built of relatively large plates in circlets or alternating, disposed around central hollow cylinder, produced laterally to form vertical partitions between arm alcoves or lacking these. Transversely circular column with small pentastellate axial canal and radicular holdfast. *M.Sil.-M.Dev.*

Eucalyptocrinites GOLDFUSS, 1831, p. 214 [**E. rosaceus*; M] [=*Eucalyptocrinus* AGASSIZ, 1836, p. 197 *(nom. van.)* (obj.); *Hypanthocrinites* PHILLIPS in MURCHISON, 1839, p. 672 (type, *H. decorus*; M); *Hypanthocrinus* MORRIS, 1843, p. 59 *(nom. van.)*; *Eucalyptrocrinites* GEINITZ, 1846, p. 551 *(nom. null.)*; *Crinocystites* HALL, 1864, p. 317 (type, *C. chrysalis*; M); *Crinocystis* HAECKEL, 1896a, p. 69 *(nom. van.)*]. First primibrachs invariably present in Silurian species, may be absent in Devonian representatives; single interbrachial plate in each ray subequal in shape and size to pairs of distal interprimibrachs in each

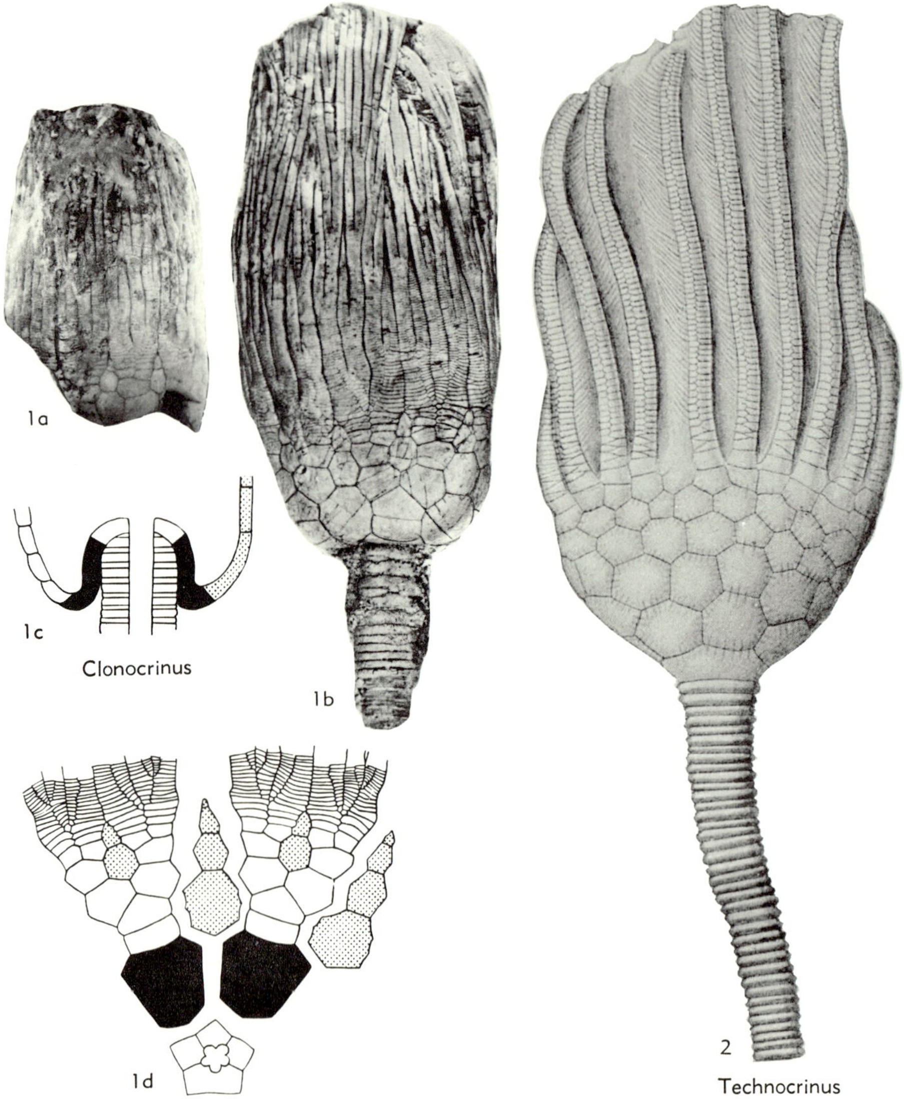

FIG. 298. Clonocrinidae (p. T495).

interray. Tegmen composed of 1) 4 large plates forming roof over viscera (recorded only from *E. crassus*), 2) 10 high, riblike plates resting against 4 lower tegminal plates (which they conceal) and cup, 3) shorter ribbed plates capping higher riblike ones to form apex of tegmen and meeting around central opening, which may be protected by special anal plates or raised on anal tube. *M.Sil.-M.Dev.*, Eu.-N.Am.(USA)-Australia-Asia(Sib.).——FIG. 299,*1a-c*. **E. rosaceus*, M. Dev., Ger.; crown with arm pairs removed in front, viewed from side, summit, and base, ×1 (Schultze, 1867).——FIG. 299,*1d*. *E. proboscidalis* (MILLER), U.Sil.(Cedarville Dol.), USA(Ohio); side view of holotype theca and part of attached stem with articular side columnal (black) in lower corner, ×1 (Foerste, 1920).——FIG. 299,*1e*. *E.* sp., plate diagram of cup (radials black, interbrachials stippled) (Ubaghs, n).——FIG. 299,*1f*. *E. crassus* (HALL), M.Sil.(Niagaran), USA(Ind.); summit view of theca, showing 4 plates which roof viscera (Macurda, 1968).

Calliocrinus D'ORBIGNY, 1850, p. 45 [**Eugeniacrinites? costatus* HISINGER, 1837, p. 90; M] [=*Cal-*

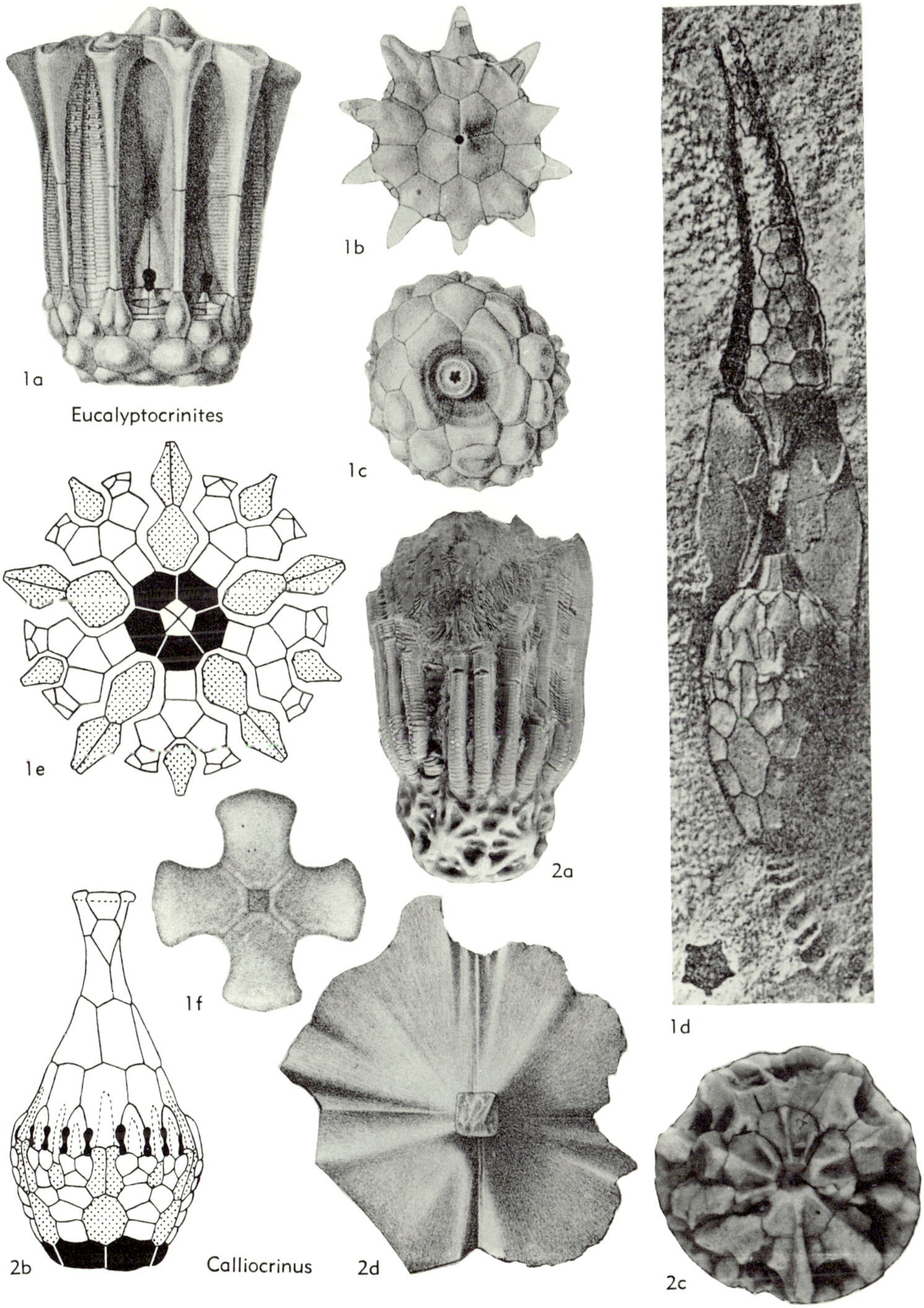

FIG. 299. Eucalyptocrinidae (p. T495-T498).

liocrinites CHAPMAN, 1857, p. 304 *(nom. van.)*; *Cryptodiscus* HALL, 1867b, pl. 11, fig. 8 (type, *C. bilobus* WELLER, 1897, p. 749; SD WELLER); *Callicrinus* ANGELIN, 1878, p. 14 *(nom. van.)*; *Chi-*

cagocrinus WELLER, 1900, p. 126 (type, *C. ornatus*; SD BASSLER, 1938, p. 65)]. Partitions on tegmen between arm pairs restricted to their proximal parts; large spinose processes generally borne by calyx and tegminal plates; summit plate circlet of anal tube may be extended laterally in 4 winglike lobed expansions forming umbrella above arm tips. *U.Sil.-L.Dev.*, NW.Eu.-USSR-N.Am.——FIG. 299, *2a-c*. **C. costatus*, U.Sil., Sweden(Gotl.); *2a*, side view of crown (Stockholm Museum, Ec. 12088), ×1 (Ubaghs, n); *2b*, diagrammatic composition of calyx (radials black, interbrachials stippled), ×1 (Ubaghs, n, adapted from Angelin, 1878); *2c*, plates at distal end of anal tube, ×4 (Angelin, 1878).——FIG. 299,*2d*. *C.* sp., U.Sil., USA(N.Y.); top of umbrellalike summit plates of tegmen (=*Cryptodiscus* HALL), ×1 (Hall, 1865).

Family DOLATOCRINIDAE S. A. Miller, 1890

[Dolatocrinidae MILLER, 1890, p. 34] [=section Dolatocrinites WACHSMUTH & SPRINGER, 1897, p. 264 (vernacular)]

Basals three, unequal or commonly fused; fixed primibrachs one or two and fixed secundibrachs one to seven in each ray; brachitaxes as far as fixed quartibrachs may be incorporated in calyx, which tends to acquire perfect pentameral symmetry; interradials not exceeding three or four ranges, first one large, none to two next above, extra interprimibrachs in *CD* interray of some genera; intersecundibrachs generally and intertibrachs rarely developed. Tegmen stout, with or without anal tube. Free arms simple or branching, composed of biserial or compound brachials. Column cylindrical, relatively large, may bear lateral flanges. *L.Dev.-M.Dev.*

Dolatocrinus LYON, 1857, p. 482 [**D. lacus*; M] [=*Cacabocrinites* TROOST, 1850a, p. 419 (*nom. nud.*); *Cacabocrinus* HALL, 1862, p. 137 (*nom. subst. pro Cacabocrinites* TROOST); *Stereocrinus* BARRIS, 1878, p. 282 (type, *S. triangulatus*; SD S. A. MILLER, 1889, p. 283)]. Calyx broadly rounded, higher than tegmen, with nearly to perfect pentamerous symmetry, plates smooth in some species but tending in most to elaborately sculptured patterns; basals generally fused, slightly exsert and visible from side, plane, or most commonly in moderately deep concavity; fixed brachials including 1 or 2 primibrachs, and 1 or 4 secundibrachs in each half-ray; brachitaxes of tertibrachs, quartibrachs uncommon in most species; fixed pinnulars typically present near summit of cup, with prominent pinnule openings at border of tegmen; first interprimibrach commonly largest plate of calyx; *CD* interray usually not differentiated in calyx by increase in number of plates, but in some species this interray may have larger plates than in others. Tegmen flat or low convex, stoutly built of medium to large, irregularly arranged plates, with or without rounded central prominences; subcentral anal tube generally not preserved in fossils. Free arms mostly 10 to 20, rarely more, composed of biserial brachials. Column strongly heteromorphic, nodals lacking cirri but commonly bearing ridgelike to vertically extended flanges. *L.Dev.-M.Dev.*, N.Am.(USA). ——FIG. 300,*1a-c*. **D. lacus*, L.Dev.(Jeffersonville Ls.), Ky.; *A*-ray, dorsal, and tegminal views of holotype theca, ×1 (Wachsmuth & Springer, 1897).——FIG. 300,*1d*. *D.* sp., plate diagram of calyx (radials black, interbrachials stippled) (Kesling & Mintz, 1963a).——FIG. 300,*1e*. *D. grandis* MILLER & GURLEY, L.Dev.(Onondaga Ls.), Ky.; detail of interray showing 12 fixed pinnule openings, ×2 (Springer, 1921a).——FIG. 300,*1f-h*. *D. exculptus* MOORE & JEFFORDS, M. Dev.(Ludlowville F.), N.Y.; facetal and 2 side views of pluricolumnals with vertical flanges on nodals, ×3.5 (Moore & Jeffords, 1968).

Clarkeocrinus GOLDRING, 1923, p. 179 [**Cacabocrinus troosti* HALL, 1862, p. 138; OD]. Basals ankylosed but sutures between them commonly visible; fixed brachials in each ray comprising 2 primibrachs and 2 secundibrachs; 1st interprimibrach large, reaching well up toward top of primaxil, supporting 2 or 3 plates in next range. Tegmen many-plated, at least in type species with very long and slender nearly central anal tube. Free arms 4 in each ray, forking 2 or 3 times; brachials compound, each bearing 2, 3, or (rarely) 4 to 6 pinnules. Column with conspicuous nodals, each carrying a whorl of 5 very long cirri; axial canal circular. *M.Dev.*, USA(N.Y.)-Eu.(Ger.). ——FIG. 301,*1a*. **C. troosti* (HALL), Moscow F., N.Y.; side view of crown and proximal part of column, ×1 (Goldring, 1923).——FIG. 301,*1b*. *C. spriesterbachi* W. E. SCHMIDT, Couvin., Ger.; crown and proximal part of column, ×1 (W. E. Schmidt, 1942).

Comanthocrinus SPRINGER, 1921, p. 4 [**Stereocrinus indianensis* MILLER & GURLEY, 1897, p. 38; OD]. Calyx very low; basals depressed, azygous one in *ED* interray; fixed 1st primibrach in each ray, more or less reduced and irregular in outline; fixed secundibrachs 3 to 7 in each half-ray; fixed pinnulars present, with pinnule openings at border of tegmen; *CD* interray with extra plate in 2nd row and ridged conspicuous series of tegminal plates extending to base of subcentral anal tube. Tegmen high, strongly bulged posteriorly. Arms simple, 10; brachials compound, hyperpinnulate. Heteromorphic stem divided into well-defined cirrinodals and inconspicuous internodals. *L.Dev.-M.Dev.*, N.Am.(USA).——FIG. 302,*1*. **C. indianensis* (MILLER & GURLEY), M.Dev.(Hamilton), Ky., N.Y.; *1a-c*, dorsal, *DE*-, and *CD*-interray views of theca, ×2 (Springer, 1921a);

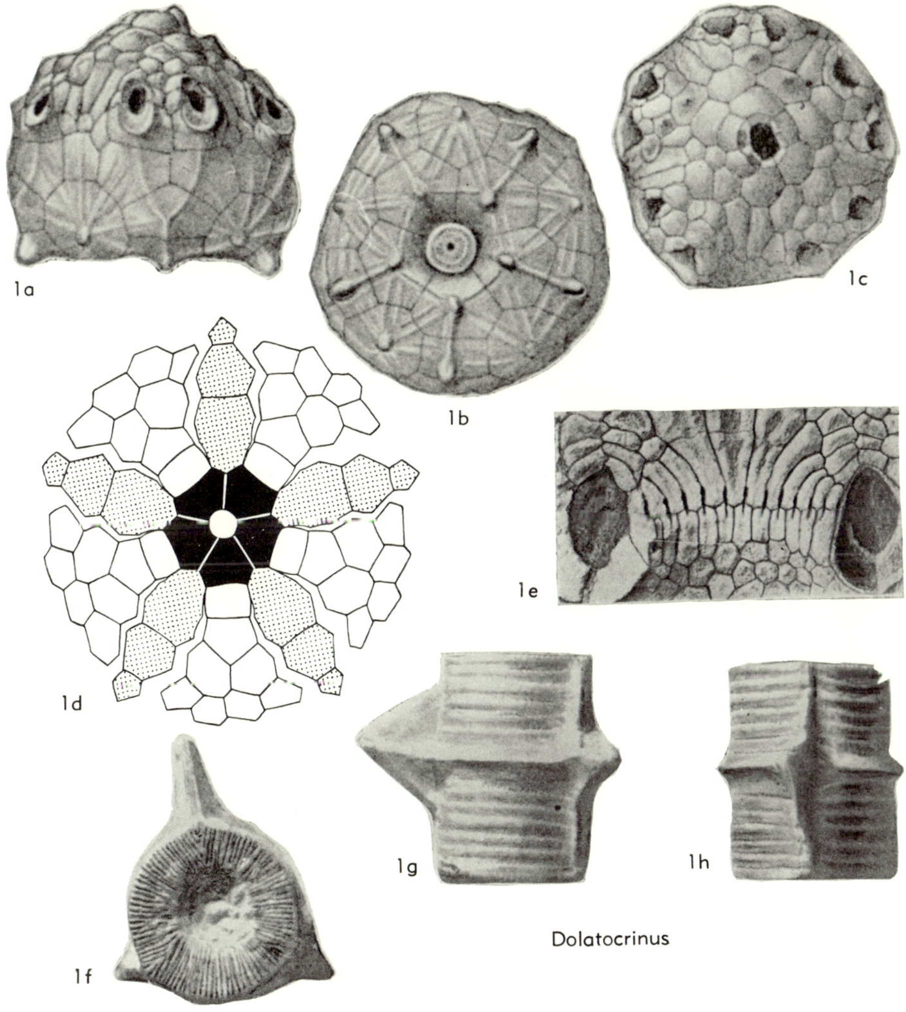

FIG. 300. Dolatocrinidae (p. *T*498).

1d, plate diagram of calyx (radials black, interbrachials stippled), ×2 (Springer, 1921a, mod.).

Craterocrinus GOLDRING, 1923, p. 185 [**C. ruedemanni*; OD]. Calyx depressed, plates ornamented with low radiating wrinkles; basals 3, azygous one in *AB* interray, or fused; fixed primibrachs 2 in each ray; fixed secundibrachs 1 to 3 (usually 2), and fixed tertibrachs 2 to 5 or even more in each half-ray; fixed pinnulars few, with 2 to 4 pinnule openings between arm bases; 1st interprimibrach large, separated from distal small interprimibrachs by 1, 2, or 3 ranks of fixed tertibrachs; *CD* interray not differentiated in calyx. Tegmen low-arched, with interambulacral areas slightly depressed; anal opening subcentral. Free arms biserial, typically 4 in each ray. *L.Dev.,* N.Am.(USA).——FIG. 302,2. **C. ruedemanni,* Onondaga Ls., N.Y.; *2a,b,* dorsal and tegminal views of theca, latter showing scar made by attached parasitic gastropod around anal opening, ×1 (Goldring, 1923); *2c,* plate diagram of calyx (radials black, interbrachials stippled) (Goldring, 1923, mod.).

Family POLYPELTIDAE Angelin, 1878

[Polypeltidae ANGELIN, 1878, p. 27]

Theca large, calyx bowl shaped and many-plated; allometric growth-zone near base of calyx resulting either in inverted

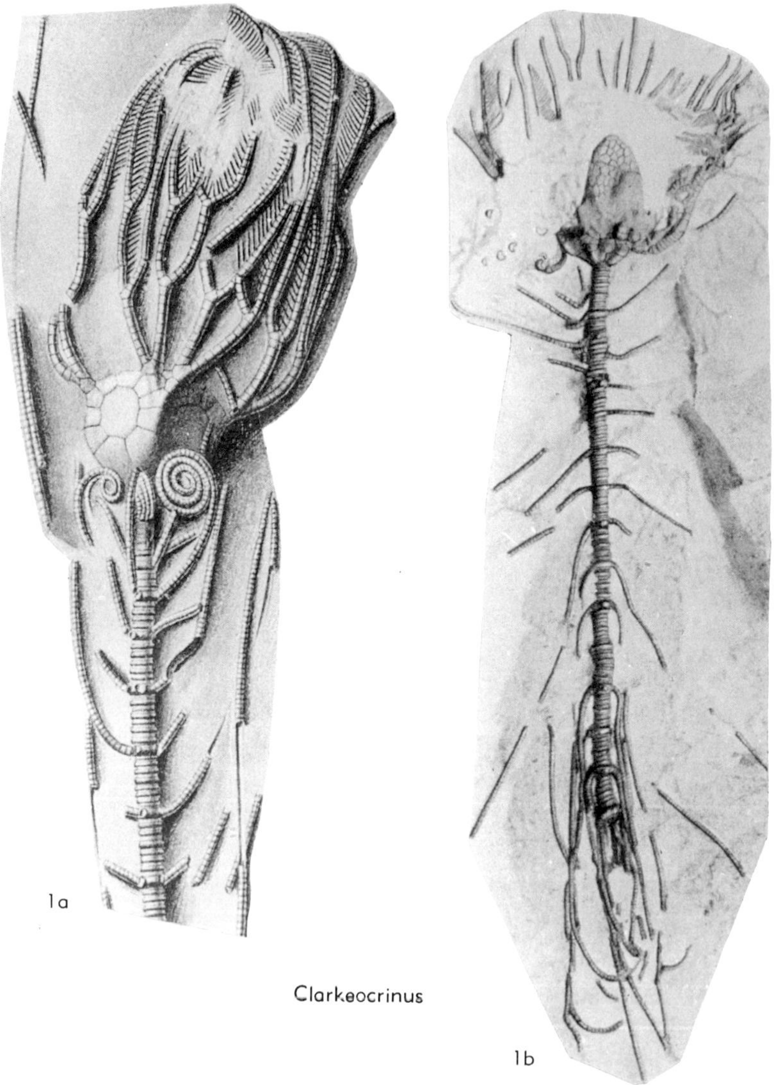

FIG. 301. Dolatocrinidae (p. T498).

position of lower calycal plates or in progressive covering of proximal thecal elements by stem facet; basals three, unequal (azygous one in *AE* interray), either internal only or atrophied (at least in external aspect); large number of ray plates incorporated in calyx; fixed primibrachs one or two in each ray; fixed secundibrachs, if arms divide, two in each half-ray, otherwise numerous; last fixed brachitaxis proximally uniserial, distally biserial; interprimibrachs, intersecundibrachs and intertertibrachs (if present) few to many, connecting with tegmen or not; generally two interprimibrachs in second range and three in posterior interray. Tegmen low, many-plated. Free arms biserial, typically simple or weakly branching. Large cylindrical stem, with quinquelobate axial canal. *U.Sil.-L.Dev.*

Polypeltes ANGELIN, 1878, p. 27 [**P. granulatus*; M]. Calyx pyriform to wide bowl shaped; basals atrophied (at least not visible externally in adult specimens); radials more or less deeply notched by axial canal; fixed primibrachs 2 in each ray;

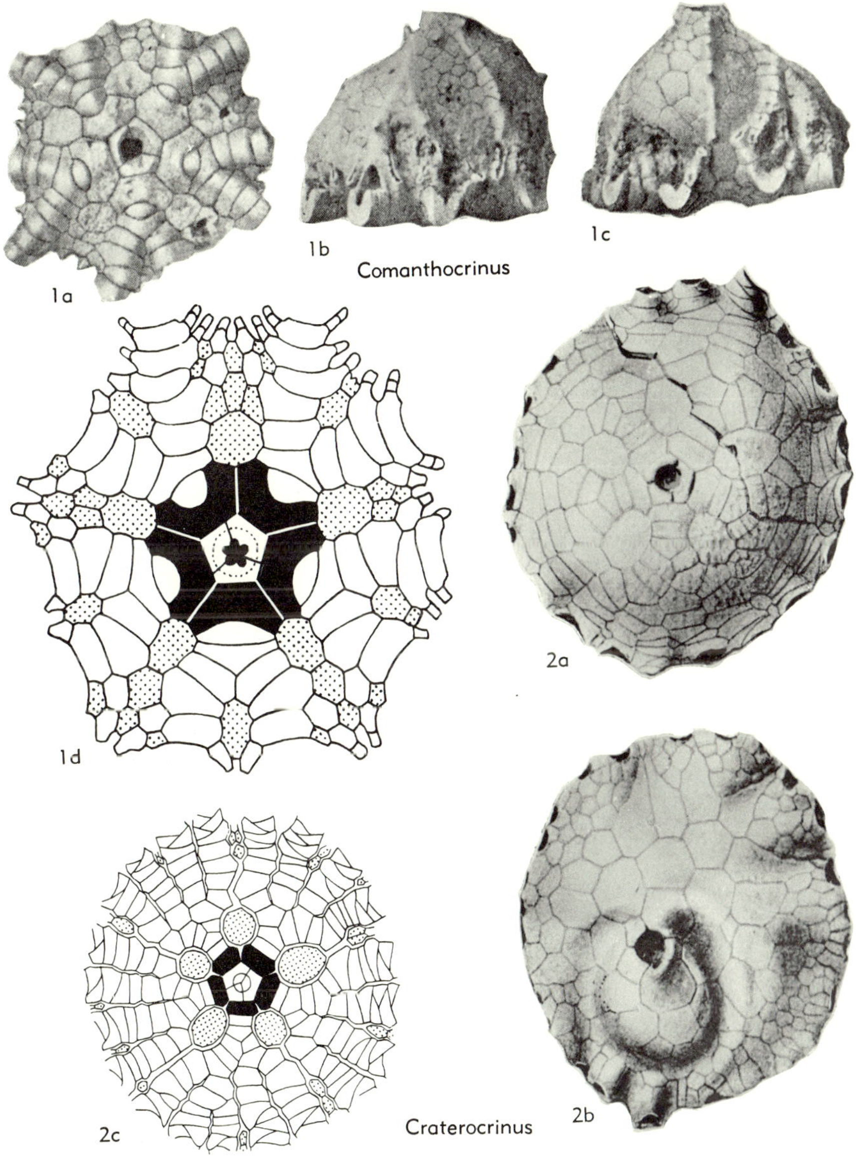

FIG. 302. Dolatocrinidae (p. *T*498-*T*499).

fixed secundibrachs 2 and fixed tertibrachs numerous in each half-ray; proximally uniserial, distally biserial; interprimibrachs, intersecundibrachs and intertertibrachs in narrow vertical series, connecting with tegmen; *CD* interray with few extra plates. Arms 40, heavy, isotomously branching at least once; pinnules short, distal pinnulars without oral groove. Stem transversely round, heteromorphic, with short nodals covering extremely thin interprimibrachs, at least proximally. *U.Sil.*, Sweden(Gotl.).——FIG. 303,*1*. **P. granulatus; 1a*, lat. view of lectotype crown (Stockholm Museum, Ec. 6150), ×1.5 (Ubaghs, 1956b); *1b*, dorsal view of small theca, ×1.5 (Ubaghs, 1956b); *1c*, plate diagram of calyx (radials black, interbrachials stippled) (Ubaghs, 1956b, mod.).

Hadrocrinus LYON, 1869, p. 445 [**H. discus*; SD WACHSMUTH & SPRINGER, 1897, p. 327]. Similar to *Polypeltes* but having only 1 fixed primibrach in each ray, more definite interprimibrachs, inter-

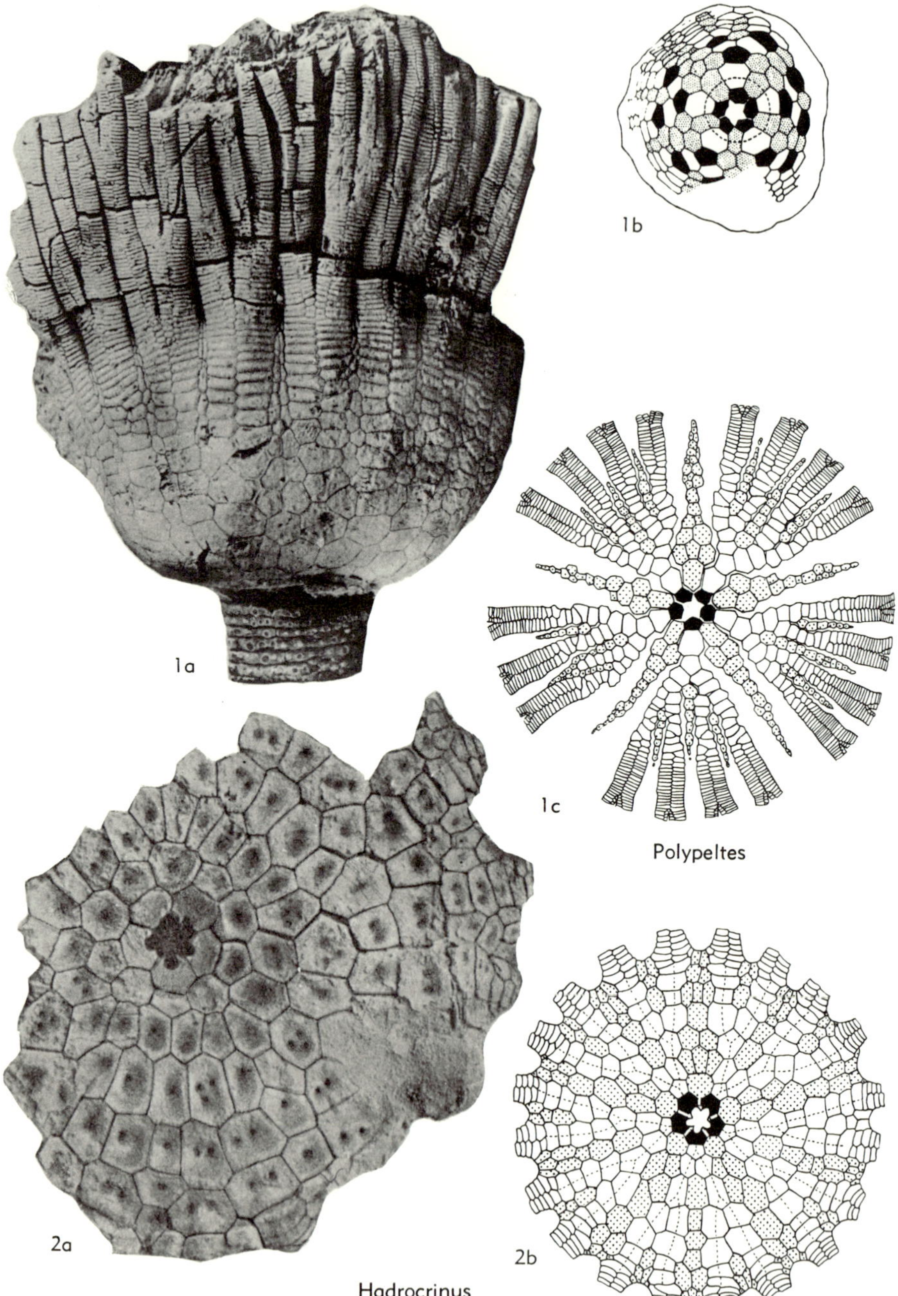

FIG. 303. Polypeltidae (p. *T*500-*T*502).

secundibrachs and intertertibrachs and only 20 arm bases. Free arms and stem unknown. *L.Dev.*, N.Am.(USA).——FIG. 303,*2*. **H. discus*, Jeffersonville Ls., Ky.; *2a*, dorsal view of part of theca, ×0.9 (Springer, 1921a); *2b*, plate diagram of calyx (radials black, interbrachials stippled)

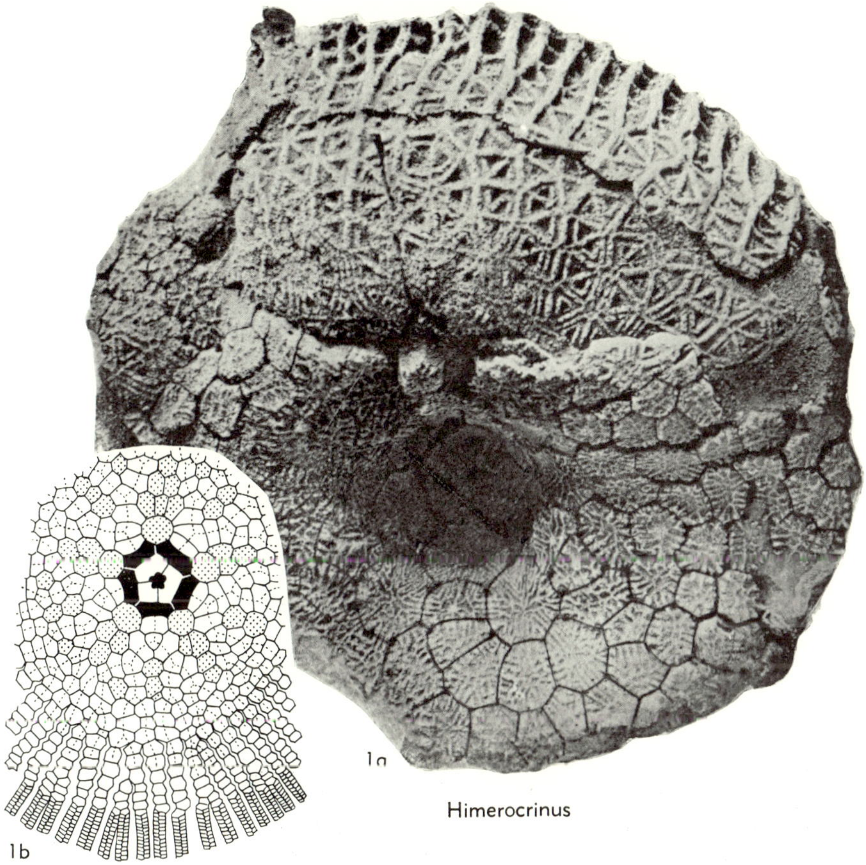

FIG. 304. Polypeltidae (p. *T*503).

(Springer, 1921a, mod.).

Himerocrinus SPRINGER, 1921, p. 12 [**Hadrocrinus plenissimus* LYON, 1869, p. 445; OD]. Calyx broadly rounded; plates with profuse and highly sculptured ornamentation (at least in type species); basals forming inverted cone; azygous basal in *AE* interray; fixed brachials extremely numerous, each ray dividing 3 to 5 (exceptionally 6) times before becoming free, at intervals of 2 plates for 1st 3 bifurcations, and longer higher up; primaxils somewhat reduced in size, and irregular in shape; interprimibrachs few, 2 in 2nd row, but 3 in *CD* interray, not connecting with tegmen; interbrachials few, located in proximal part of theca; gonioporoids present between cup plates. Tegmen low, composed of innumerable elements. Free arms simple, small, biserial, 16 to 17 in each ray. Column with conspicuous nodals characterized by flanged peripheral rim studded with projecting cogs, overhanging and even concealing thin internodals. *L.Dev.*, N.Am.(USA).——FIG. 304,*1*. **H. plenissimus* (LYON), Jeffersonville Ls., Ky.; *1a*, oblique dorsal view of theca, ×0.9 (Springer, 1921a); *1b*, plate diagram of calyx (radials black, interbrachials stippled) (Springer, 1921a, mod.).

Trybliocrinus GEINITZ, 1867, p. 284 [**Glyptocrinus* (in error for *Trybliocrinus*) *flatheanus*; M) (=*Hadrocrinus hispaniae* W. E. SCHMIDT, 1932, p. 7)] [=*Tribliocrinus* SPRINGER, 1913, p. 191 (*nom. null.*)]. Calyx wide bowl shaped; basal region deeply concave, forming inverted cone comprising basals, radials and primibrachs (in advanced growth stages); fixed part of rays composed of 2 primibrachs and numerous secundibrachs; fixed secundibrachs proximally uniserial, distally biserial; interprimibrachs and intersecundibrachs numerous, connecting with tegmen; *CD* interray commonly not differentiated. Tegmen low, composed of prismatic plates, many of them with bulbous expansions forming an underlying structure; anus opening directly through tegmen;

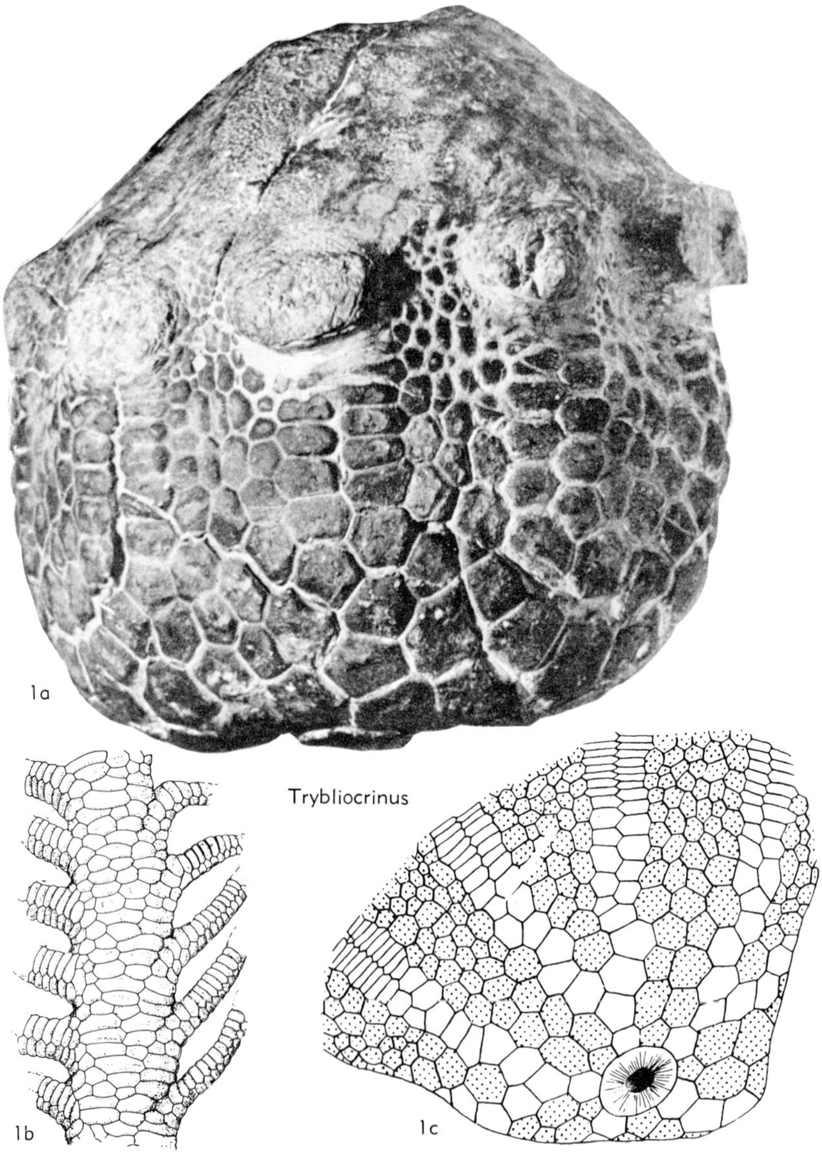

FIG. 305. Polypeltidae (p. T503-T504).

vessel system within tegminal plates. Free arms 10, stout, biserial, with pinnulated biserial ramules alternating along opposite sides; oral sides of arms and ramules roofed by stratified covering plates. Column composed of regular alternation of 1 nodal and 1 to 3 internodals; distal part provided with branching radicular cirri. Coelomic canal system, probably for respiration, present in column as radial canals between columnals, in cup as gonioporoids, in tegmen as lacunae between lower parts of tegminal plates. *L.Dev.*, Eu.(Spain-?Ger.).——FIG. 305,*1*. **T. flatheanus* GEINITZ, L.Ems., Spain; *1a*, lat. view of theca, ×1 (Breimer, 1962); *1b*, portion of arm, aboral view, ×3 (Breimer, 1962); *1c*, oblique lat. view of calyx (interbrachials stippled), ×0.7 (Schmidt, 1932).

Family UNCERTAIN

Abathocrinus STRIMPLE, 1963, p. 109 [**Maria-*

crinus? rotundus Springer, 1926a, p. 29; OD]. Calyx shallow conical; basals diminutive, confined to small concavity, number not known; radials relatively large, in contact with each other all around; fixed primibrachs 2 in each ray; fixed secundibrachs 2 in each half-ray; brachial openings directed outward; interprimibrachs few, 2 in 2nd row, merging with tegminal plates; posterior interray not differentiated in calyx. Tegmen high, rounded, many-plated, pierced by small subcentral anal opening. Free arms and stem unknown. *M.Sil.,* N.Am.(USA).——Fig. 306,*1*. **A. rotundus* (Springer), Niagaran (Decatur Ls.), Tenn.; *1a,b, DE*-interray and dorsal views of theca, ×2 (Springer, 1926a).

Abathocrinus

Fig. 306. Family Uncertain (p. *T*504-*T*505).

Superfamily PATELLIOCRINACEA Angelin, 1878

[*nom. correct.* Brower, 1973, p. 331 (*pro* Patelliocrinicae Ubaghs, 1953, p. 742, *nom. transl. ex* Patelliocrinidae Angelin, 1878, p. 1)]

Glyptocrinina with few brachials incorporated in calyx; radials large; fixed primibrachs two, generally much smaller than radials; first primibrachs quadrangular; fixed secundibrachs one to three in each half-ray; interprimibrachs few, first usually large; primanal supporting typically three plates but tending to become like other proximal interprimibrachs. Free arms simple, uniserial, biserial or with compound brachials. Column cylindrical. *U.Ord.-M.Dev.*

Family STELIDIOCRINIDAE Angelin, 1878

[Stelidiocrinidae Angelin, 1878, p. 21]

Calyx low to medium conical; basals five; interprimibrachs two between each pair of rays, in vertical succession, followed by single interradial tegminal plate; one intersecundibrach in each half-ray followed by one interambulacral on tegmen; primanal supporting three plates directly; posterior side of theca marked off by median series of rather large plates leading to laterally placed anus, protected during life by valvular pyramid of small plates. Tegmen low, solid, composed of 14 large plates. Free arms ten, biserial. Column cylindrical. *U.Sil.*

Stelidiocrinus Angelin, 1878, p. 21 [**S. capitulum*; SD Wachsmuth & Springer, 1881, p. 99(273)] [=*Chytrocrinus* Jaekel, 1918, p. 33 (type, *Stelidiocrinus laevis* Angelin, 1878, p. 21; M)]. Characters of family. *U.Sil.,* Sweden(Gotl.).——Fig. 307,*1a-d*. **S. capitulum; A*-ray, *CD*-interray, dorsal, and ventral views of theca (Stockholm Museum, Ec. 11738) (radials black, interbrachials and anals stippled), ×4 (Ubaghs, n).——Fig. 307,*1e*. *S. laevis* Angelin; *C*-ray view of crown (Stockholm Museum, Ec. 11775), ×3 (Ubaghs, n).

Family PATELLIOCRINIDAE Angelin, 1878

[Patelliocrinidae Angelin, 1878, p. 1]

Calyx conical or subglobose to subcylindrical; basals ordinarily three, unequal, azygous one in *AE* or *ED* interrays; one genus with all basals fused together; *CD* interray usually little or not at all differentiated in calyx; tegmen rarely preserved, probably incompetent, consisting of numerous plates differentiated into ambulacrals and interambulacrals in genera where it is known. Free arms uniserial, biserial, or made of compound brachials. Column typically cylindrical. *U.Ord.-M.Dev.*

Patelliocrinus Angelin, 1878, p. 1 [**P. pachydactylus*; SD Wachsmuth & Springer, 1881, p. 102(276)]. Calyx conical to subglobose; fixed secundibrachs 2 in each half-ray, other secundibrachs free; 1st interprimibrach large, followed

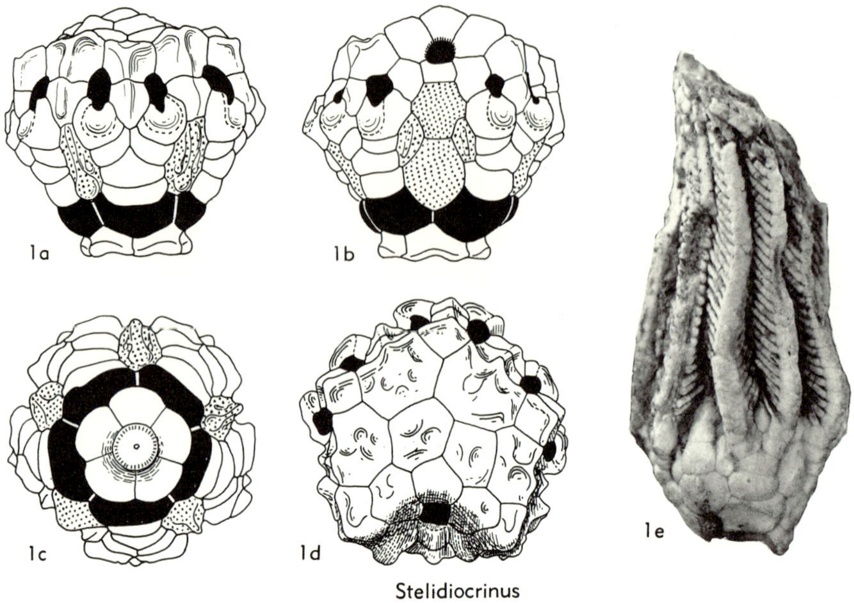

FIG. 307. Stelidiocrinidae (p. T505).

by 1, 2, or several small plates narrowly in contact with tegmen; *CD* interray little or not differentiated in calyx. Free arms 10, simple, biserial; 1st pinnule may be distinctly larger than others. *U.Sil.*, Eu.(Sweden-Eng.)-N.Am.——FIG. 308,*1*. **P. pachydactylus*, Gotl.; *1a*, *AE*-interray view of slightly reconstructed holotype (Stockholm Museum, Ec. 11586), ×2 and *1b*, plate diagram of part of calyx (based on Stockholm Museum, Ec. 11592) (radials black, interbrachials stippled) (Ubaghs, n).

Allocrinus WACHSMUTH & SPRINGER in MILLER, 1889, p. 222 [**A. typus*; OD]. Calyx low bowl shaped, basally excavated; 1st interprimibrach very large followed by 1 or 2 small plates connected with tegmen; *CD* interray little differentiated in calyx or not at all; fixed brachials small, including 1 or 2 secundibrachs. Free arms 10, simple, more or less stout; brachials compound, hyperpinnulated. Columnals rather long, with narrow, pentagular axial canal. *M.Sil.(Niagaran)*, N.Am.(USA).——FIG. 309,*1a*. **A. typus*, Beech River F., Tenn.; lat. view of crown, ×2 (Springer, 1926a).——FIG. 309,*1b,c*. *A. longidactylus* SPRINGER, Beech River F., USA(Tenn.); *CD*-interray and dorsal views of theca, ×2 (Springer, 1926a).

Briarocrinus ANGELIN, 1878, p. 1 [**B. inflatus*; SD JAEKEL, 1918, p. 38] [=*Angelinocrinus* JAEKEL, 1918, p. 38 (type, *Briarocrinus angustus* ANGELIN, 1878, p. 1)]. Calyx high bowl shaped; radials and fixed primibrachs large, unequal, irregular in outline; fixed secundibrachs 2 or 3, 1st 2 in lateral contact in each ray, 3rd ones free or united by single small intersecundibrachs; interbrachials generally 1 in each interray, small, resting on 1st primibrachs or primaxils or even higher according to species, in contact with tegmen or not; primanal generally larger than interprimibrachs, followed by tegminal series of large plates, may rest on adjacent radials. Free arms 10, simple, massive, with compound, hyperpinnulated brachials. *U.Sil.*, Sweden(Gotl.).——FIG. 308,*2a,b*. **B. inflatus*; lat. views of holotype crown (Stockholm Museum, Ec. 11415), (radials black, interbrachials stippled in *2b*), ×1.5 (Ubaghs, n).——FIG. 308,*2c*. *B.* sp.; *C*-ray view of crown (radials black, anals and interbrachials stippled) (Stockholm Museum, Ec. 11417), ×1.5 (Ubaghs, n).

Centriocrinus BATHER, 1899, p. 922 [*nom. subst. pro Centrocrinus* WACHSMUTH & SPRINGER, 1881, p. 104(278) (*non* AUSTIN & AUSTIN, 1843, p. 6)] [**Actinocrinus pentaspinus* LYON, 1869, p. 453; OD]. Calyx subcylindrical, as wide as high, flat below; basal disc forming circular collar around stem, showing no traces of interbasal sutures; radials very large, followed in each ray by 2 relatively small fixed primibrachs and in each half-ray by 2 still smaller fixed secundibrachs; 1st interprimibrach large, with 2 small plates in next row, and very narrow plate connected with tegmen in 3rd row; *CD* interray differentiated by extra anal plate in 2nd range. Free arm bases in each ray close together. Columnal facet very small. All other parts unknown. *M.Dev.*, N.Am.(USA).——FIG. 308,*3*. **C. pentaspinus* (LYON), Jeffersonville Ls., Ky.; *3a,b*, dorsal and lat. views of

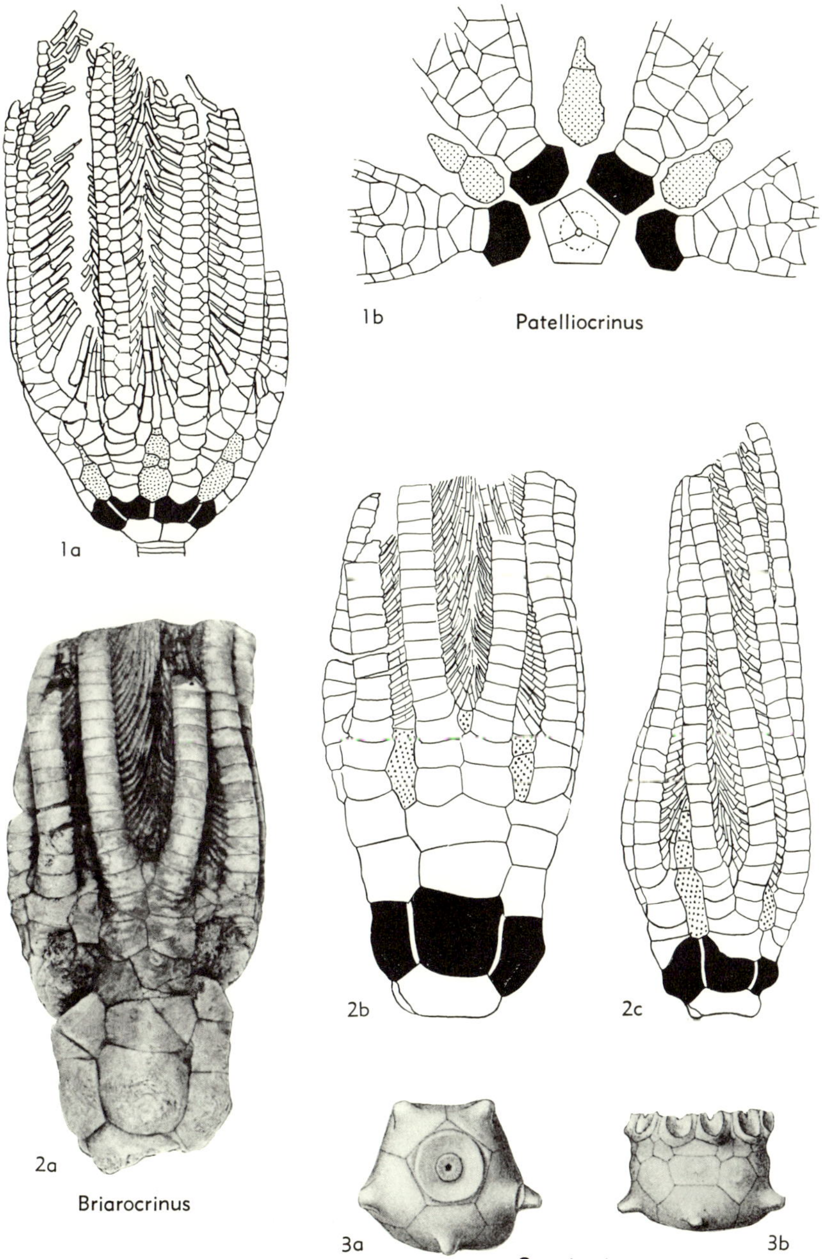

FIG. 308. Patelliocrinidae (p. T505-T506).

holotype, ×1 (Wachsmuth & Springer, 1897).

Eopatelliocrinus BROWER, 1973, p. 332 [*E. scyphogracilis; OD]. Calyx conical; basals 3, erect, forming part of cup wall; *CD* interray wider than others; primanal followed by 3 plates; anal interray ridge present. Free arms 10, uniserial. Column round or pentagonal in cross section. *U.Ord.*, N.Am.(Ill.-Mo.).——FIG. 310,*1*. *E. scyphogracilis*, Girardeau, Mo.; holotype, lat. view, ×3 (interbrachials stippled) (Brower, 1973).

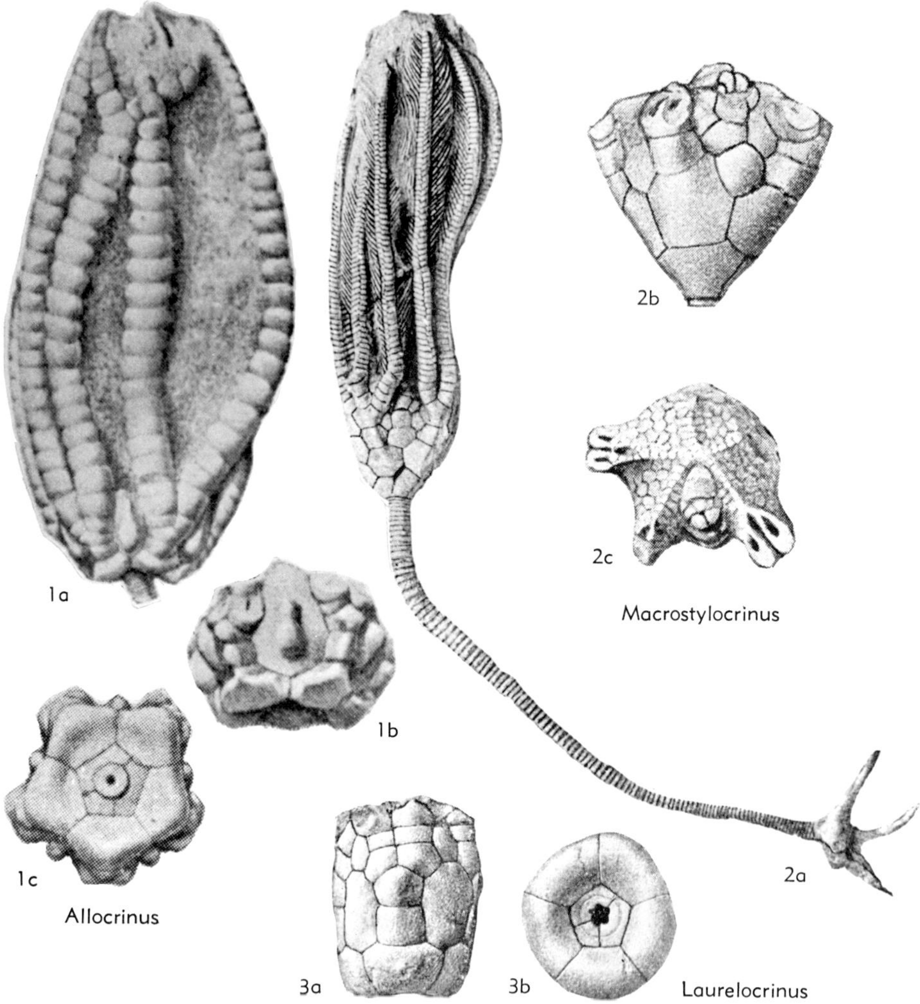

FIG. 309. Patelliocrinidae (p. *T506, T508*).

Laurelocrinus SPRINGER, 1926, p. 32 [**L. paulensis*; OD]. Calyx elongate cylindrical or slightly contracting upward; basals small, forming with radials broad, flattened, or shallow concave disc; radials very large; fixed secundibrachs 3; 1st interprimibrach large and elongate, with or without small plate above it; single intersecundibrach may occur in each ray; posterior side not differentiated in cup. Free arm bases 4 in each ray, but free arms, tegmen, and column unknown. Columnal facet circular, with large lumen. *M.Sil.(Niagaran)*, N.Am.(USA).——FIG. 309,*3*. **L. paulensis*, Laurel Ls., Ind.; *3a,b*, lat. and dorsal views of theca, ×1 (Springer, 1926a).

Macrostylocrinus HALL, 1852, p. 203 [**M. ornatus*; M] [=*Loveniocrinus* JAEKEL, 1918, p. 37 (type, *L. gotlandicus*; M)]. Calyx conical to subglobose; fixed secundibrachs 1 or 2; interprimibrachs few, connected with tegmen, 1st one large; *CD* interray much wider than others; primanal followed by 3 to 5 plates in next range. Tegmen low, composed of small irregular pieces, with narrow ambulacral tracks and more or less distinct orals; anus marginal. Free arms generally 10, biserial and simple, recumbent and as many as 23 to 25 in a Devonian species. Column with small axial canal; whorls of cirri may be present. [In one or perhaps two rays of the type and only known specimen of *Loveniocrinus* one of the proximal pinnules is replaced by an arm, bringing to 3 the number of arms in this ray; such difference is judged not sufficient to warrant a generic distinction.] *U.Ord.*-

L.Dev., Eu.-N.Am.(USA).——Fig. 309,*2a*. **M. ornatus*, M.Sil.(Niagaran), N.Y.; *CD*-interray view of complete crown with attached stem and holdfast, ×1 (Springer, 1926a).——Fig. 309,*2b,c*. *M. granulosus* Hall, M.Sil.(Niagaran), Tenn.; post. and tegminal views of theca, ×3 (Springer, 1926a).

Superfamily PLATYCRINITACEA Austin & Austin, 1842

[*nom. correct.* Brower, 1973, p. 397 (*pro* Platycriniticae Ubaghs, 1953, p. 742, *nom. transl. ex* family Platycrinoidea Austin & Austin, 1842, p. 109)]

Calyx tending to be confined to lowermost two plate circlets (patina); basals three, unequal (small one normally in *AE* interray), or fused together; brachials and interbrachials generally little represented in calyx, proximal brachials tending to stand out clearly from radials although joined firmly to calyx, tegmen, or both by interradially situated plates; posterior side slightly or not at all differentiated in calyx. *?U.Ord., Sil.-Perm.*

Family MARSUPIOCRINIDAE Bronn, 1855

[Marsupiocrinidae Bronn, 1855, p. 22] [=family Marsupiocrinoidea Austin & Austin, 1842, p. 109 (invalid family-group name based on nominal genus which is a junior homonym); Marsupiocrinidae Austin & Austin, 1843, p. 198 (rejected for same reason); Cypellocrininae Bather, 1899b, p. 922; Marsipocrininae Bather, 1900a, p. 156; Marsipocrinidae Jaekel, 1918, p. 38]

Calyx flat to low bowl shaped; basals 3, unequal, or fused; fixed primibrachs one in each ray, small, trigonal, axillary; first fixed secundibrachs on either side of it resting on radials, with one to several following brachials, wholly incorporated in calyx; fixed secundibrachs, if arms divide further, one or two; single large interbrachial between adjoining rays, succeeded by smaller plates which may be restricted to tegmen. Tegmen flat to low conical, many-plated, with or without well-defined ambulacrals, and directly pierced by subcentral anus. Free arms 2 to 4 to each ray, simple, stout, biserial. Column circular in section, with nudinodals and interradials; axial canal large, pentagonal or quinquelobate. *M.Sil.-L.Dev.*

Marsupiocrinus Morris, 1843, p. 54 [**Marsupiocrinites coelatus* Phillips in Murchison, 1839, p. 672; M] [=*Marsupiocrinites* Phillips in Murchison, 1839, p. 672 (*non* de Blainville, 1830) (type, *M. coelatus*; M); *Cupellaecrinites* Troost,

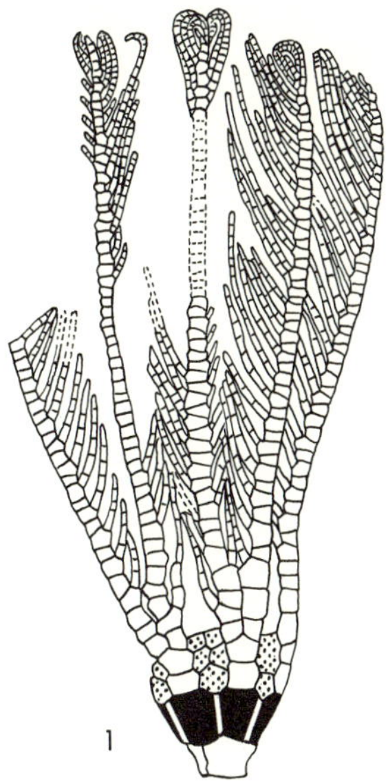

1

Cupelliocrinus

Fig. 310. Patelliocrinidae (p. T507).

1850a, p. 419 *(nom. nud.)* (type, *C. buchii*; SD Bassler, 1938, p. 78); *Cupellaecrinus* Shumard, 1865, p. 361 *(nom. van.)*; *Cupellaecrinus* Meek & Worthen, 1866a, p. 172 *(nom. van.)*; *Cypellocrinus* Bather, 1899b, p. 921 *(nom. van.)* (*non* Bather in Lankester, 1900, p. 177); *Marsipocrinus* Bather, 1889b, p. 173 *(nom. van.)*]. Characters of family. *M.Sil.-L.Dev.*, Eu.-N.Am.

M. (Marsupiocrinus). Four arms per ray, no basal rim, broad basal invagination involving both basals and radials, gently convex base, and greatly elongated interbrachials. *M.Sil.-L.Dev.*, Eu.-N. Am.—Fig. 311,*1*. **M. (M.) coelatus* (Phillips), M.Sil.(Wenlock.), Eu.(Eng.); *1a,b*, lat. and dorsal views of crown, ×1 (Springer, 1926a); *1c*, plate diagram of part of calyx (radials black, interbrachials stippled) (Ubaghs, n).

M. (Amarsupiocrinus) Frest, 1975, p. 569 [**Marsipocrinus striatissimus* Springer, 1926a, p. 61; OD]. Generally two arms per ray, pronounced basal rim surrounding column facet, nearly planar base, and subquadrate interbrachials. *M.Sil.*, N.Am.(USA).——Fig. 311,*2*. **M. (A.) striatissimus* (Springer), M.Sil.(Niagaran), Tenn.; *2a,b*, aboral and lat. views of calyx, ×0.67; *2c*, aboral side of crown, ×0.67 (Springer, 1926a).

Family HAPALOCRINIDAE Jaekel, 1895

[Hapalocrinidae Jaekel, 1895, p. 94] [=Coccocrinidae Zittel, 1895, p. 124 *(partim)*; Coccocrininae Bather, 1899b, p. 922; Thallocrinidae, Agriocrinidae Jaekel, 1918, p. 91]

Primibrachs two in each ray, exceptionally fewer or more, generally well developed;

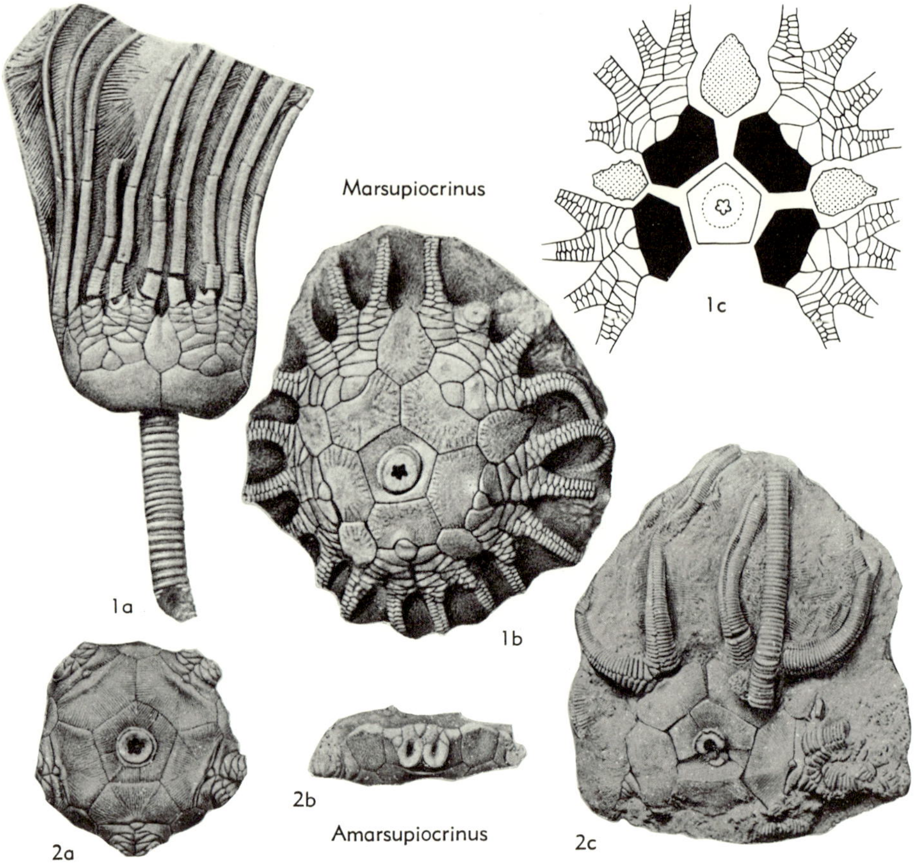

FIG. 311. Marsupiocrinidae (p. T509).

secundibrachs more than two in each half-ray. Tegmen mainly composed of five orals; ambulacral tracts ordinarily distinct, but tending to become covered by orals which themselves tend to remain in apposed apical position; no differentiated axillary ambulacrals and no interambulacrals; mouth subtegminal or suprategminal; anal tube absent or incipient. Free arms generally two in each ray, simple or forking, proximally uniserial, then biserial or composed of compound brachials. Column typically cylindrical; axial canal small and round. *?U.Ord., Sil.-M.Dev.; ?U.Perm.*

Hapalocrinus JAEKEL, 1895, p. 95 [**H. elegans*; M] [=*Agriocrinus* JAEKEL, 1895, p. 103 (type, *A. frechi*; SD BASSLER, 1938, p. 36)]. Primibrachs 2, exceptionally 3 or 5; arms uniserial, forking once or more; brachials quadrangular, generally elongate, with pinnule on each or every other one; tegmen unknown. *L.Dev.*, Eu.——FIG. 312,*1*. **H. elegans*, Bundenbacher Schiefer, Ger.; oblique dorsal view of crown, ×0.7 (Jaekel, 1895).

Amblacrinus D'ORBIGNY, 1849, p. 104 [**Platycrinus rosaceus* ROEMER, 1844, p. 63; SD BASSLER, 1938, p. 38] [=*Coccocrinus* MÜLLER in ZEILER & WIRTGEN, 1855, p. 20 (obj.)]. Primibrachs 2, 1st one slightly incorporated in calyx; interprimibrachs 1 in each interray, pentagonal, posterior one larger than others. Tegmen composed of 5 large orals, with little or no differentiation of posterior oral; ambulacra suprategminal in narrow grooves along interoral sutures; mouth subtegminal; anus between posterior oral and posterior interprimibrach. *M.Dev.*, Eu.——FIG. 313,*1*. **A. rosaceus* (ROEMER), Couvin., Ger.; *1a-c*, lat., dorsal, and tegminal views of theca, ×1 (Springer, 1926a).

Bogotacrinus W. E. SCHMIDT, 1937, p. 5 [**B. scheibei*; OD]. Primibrachs 3 in each ray, primaxils bearing bilaterally heterotomous rami; brachials compound and generally bipinnulated. *Sil.*

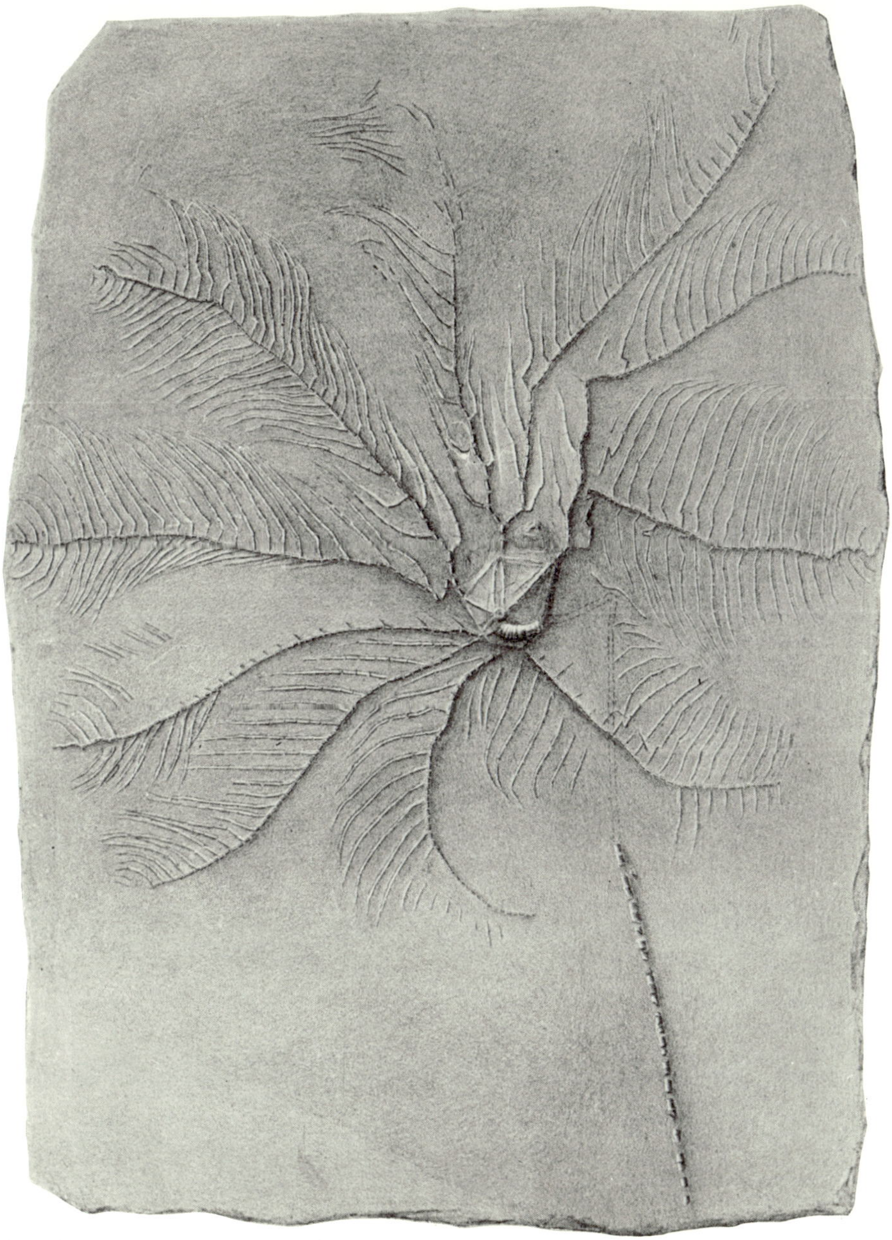

Fig. 312. Hapalocrinidae (p. T510).

or *L.Dev.*, S.Am.(Colom.).——Fig. 313,*2*. **B. scheibei*; arms of ray (radial black, pinnules omitted), ×1.3 (Schmidt, 1937).

Cantharocrinus Breimer, 1962, p. 117 [**C. minor*; OD]. Small basal in *AE* or *CD* interray; primibrachs 2, 1st one slightly incorporated in calyx; single hexagonal interprimibrach in each interray, but 3 in posterior interray. Tegmen composed of 5 triangular orals, posterior one largest; ambulacral tracts covered by alternating platelets, not reaching

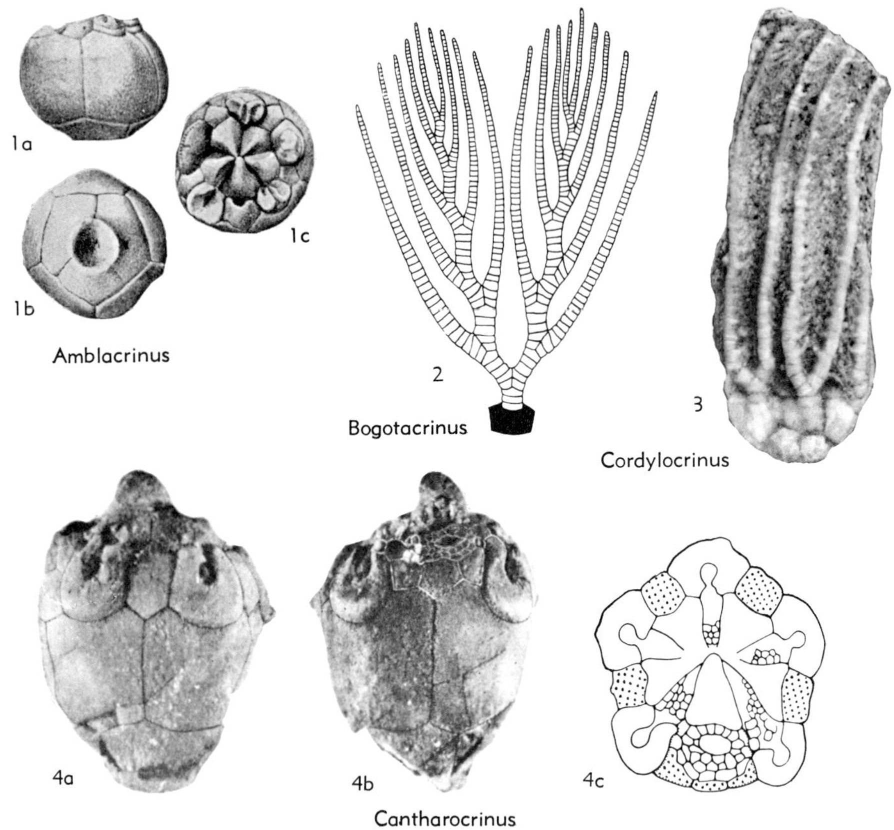

FIG. 313. Hapalocrinidae (p. T510-T512).

center; mouth subtegminal; anus surrounded by many small plates. Arms and stem unknown. *L.Dev.-M.Dev.*, Eu.(Spain).——FIG. 313,*4*. **C. minor,* L. Ems., León Prov.; *4a-c, AB-, CD*-interray, and tegminal views of holotype theca, ×4 (interbrachials stippled) (Breimer, 1962).

Clematocrinus JAEKEL, 1898a, p. 47 [**Actinocrinites? retiarius* PHILLIPS in MURCHISON, 1839, p. 674; M]. Primibrachs 2, joined to calyx; 1st interprimibrachs relatively large. Free arms simple, 10 to 20; brachials cuneate in zigzag or biserial. Stem with whorls of cirri. *M.Sil.,* Eu.-N.Am.-?Australia.——FIG. 314,*1*. **C. retiarius* (PHILLIPS), Wenlock Ls., Eng.; lat. view of complete crown with attached stem, ×3 (interbrachials stippled) (Bather, 1900a, mod.).

Cordylocrinus ANGELIN, 1878, p. 3 [**C. comtus*; M]. Primibrachs 2 in each ray; arms incorporated in calyx up to 2nd secundibrachs (at least in type species); 1st interprimibrachs large. Tegmen, known only in *C. plumosus* (HALL), many-plated and bearing anal cone. Arms simple, with compound, bipinnulate brachials. Column with whorls of cirri (at least in some species). *U.Sil.-L.Dev.,* Eu.-N.Am. [The characters shown by the tegmen of *C. plumosus* (HALL) do not fit well with the family diagnosis. One has to be cautious about reference of this species to *Cordylocrinus*.]——FIG. 313,*3*. **C. comtus,* U.Sil., Sweden(Gotl.); lat. view of crown, ×3 (Ubaghs, n).

Culicocrinus MÜLLER in ZEILER & WIRTGEN, 1855, p. 23 [**Platycrinus nodosus* MÜLLER in ZEILER & WIRTGEN, 1855, p. 15; M] [=*Protoeuryale* MÜLLER in ZEILER & WIRTGEN, 1855, p. 24 (type, *P. confluentina*; M); *Cylicocrinus* BATHER in LANKESTER, 1900a, p. 156 *(nom. van. pro Culicocrinus)*]. Primibrachs 2 and proximal secundibrachs incorporated in calyx by 5 large interprimibrachs. Tegmen composed of 5 orals, generally ornamented by protuberances; posterior oral differentiated; anus in notch of posterior oral. Free arms biserial, forking once or twice. *?U.Ord.,* N.Am.; *L.Dev.,* W.Eu. [The position of this genus among Hapalocrinidae is somewhat uncertain.]——FIG. 314,*2*. **C. nodosus* (MÜLLER), U.Ems., Ger.; *2a,* lat. view of theca and part of

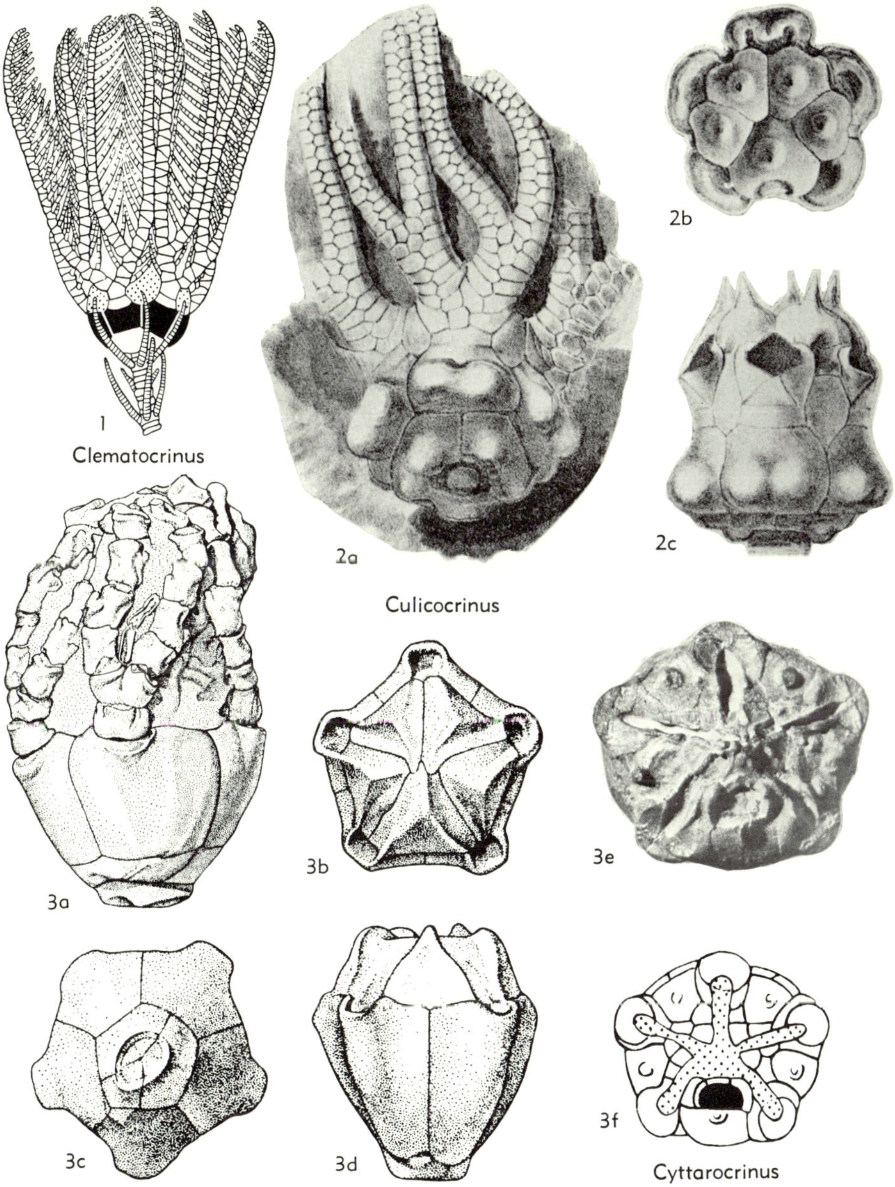

FIG. 314. Hapalocrinidae (p. T512-T514).

free arms, ×2 (Müller, in Zeiler & Wirtgen, 1855); *2b,c,* theca in tegminal and lat. views, ×3 (Müller, in Zeiler & Wirtgen, 1855).

Cyttarocrinus Goldring, 1923, p. 265 [**Platycrinus eriensis* Hall, 1862; OD]. Small basal in *AE* or *AB* interray; primibrachs 1 or 2; interprimibrachs comparatively large, largest in *CD* interray. Tegmen (in *C. jewetti* Goldring) composed of 5 orals and 2 interambulacrals in each interray, bordering each interprimibrach; ambulacra and ?mouth suprategminal; anus large, bordered above by posterior oral, narrow interambulacrals on each side, and below by posterior interprimibrach. Free arms simple, uniserial. [This diagnosis combines features of both *C. eriensis* and *C. jewetti*; however, it is far from proved that these species are congeneric, as admitted by Goldring (1923). The type species is based on an incomplete speci-

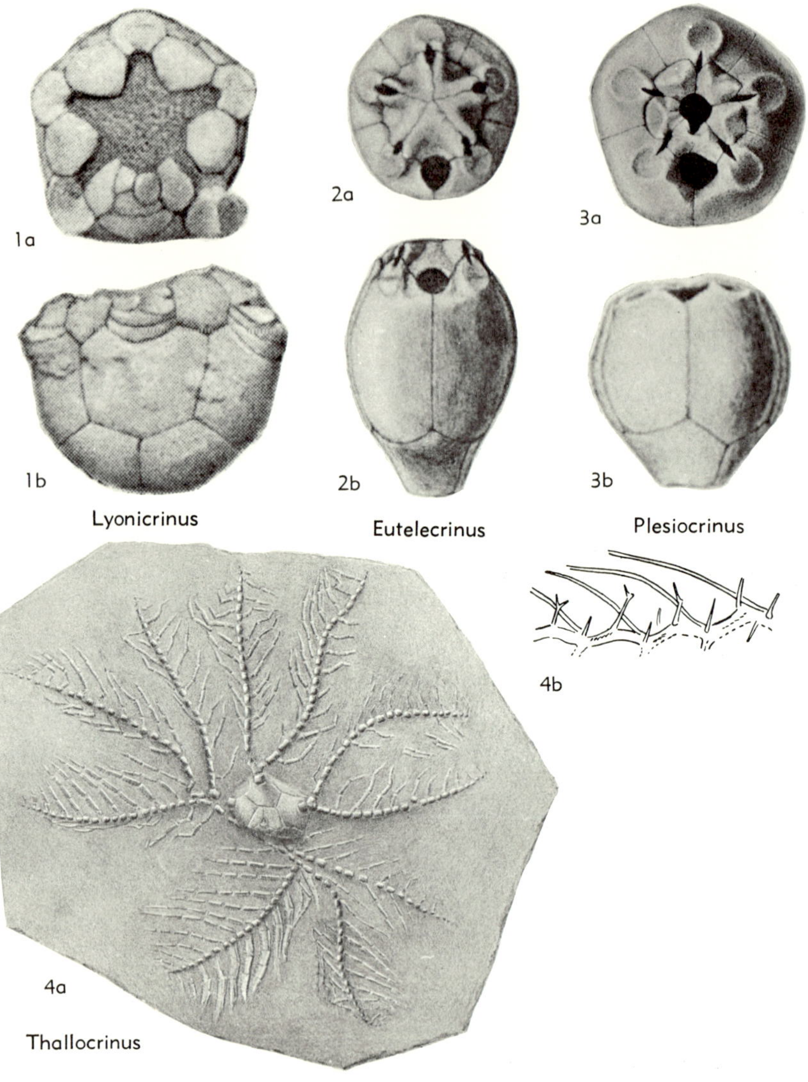

FIG. 315. Hapalocrinidae (p. T514-T515).

men, lacking a stem and right half of the crown; neither tegmen nor stem insertion are shown. An ontogenetic series assumed to belong to *C. eriensis* and described by KOENIG (1965) suggests that the tegmen of *Cyttarocrinus,* as in *Amblacrinus,* is composed of 5 orals, with interoral sutures at bottom of shallow troughs; in these immature crinoids the stem cicatrix is elliptical. If this assumption be correct, *C. jewetti,* which closely resembles *Lyonicrinus bacca,* ought probably to be transferred to the latter genus and the diagnosis of *Cyttarocrinus* changed accordingly.] *L.Dev.-M.Dev.,* Eu.-N.Am.(USA).——FIG. 314,*3a*. **C. eriensis* (HALL), Hamilton Sh., N.Y.; *D*-ray view of holotype crown, ×4 (Koenig, 1965).——FIG. 314,*3b-d*. *C.* sp. *cf. C. eriensis* (HALL), Moscow F., USA(N.Y.); ventral, dorsal, and lat. views of immature specimens, ×15 (Koenig, 1965).——FIG. 314,*3e,f*. *C. jewetti* GOLDRING, Hamilton Gr., USA(N.Y.); tegmen, photograph and diagram, ×4 (mouth and ambulacral areas stippled) (photo courtesy Buffalo Museum; diagram, mod. from Koenig, 1965).

?**Eutelecrinus** WANNER, 1916, p. 23 [**E. piriformis*; OD]. Theca distally constricted; tegmen composed of 5 large orals. *U.Perm.,* Indon.(Timor.).

[Genetic relations of this genus with Silurian and Devonian hapalocrinids are uncertain.]——FIG. 315,*2*. **E. piriformis; 2a,b,* tegminal and *CD*-interray views of theca, X2 (Wanner, 1937).

Lyonicrinus SPRINGER, 1926, p. 48 [**Coccocrinus bacca* ROEMER, 1860, p. 51; OD]. Small basal in *DE* interray; primibrachs 2, joined to adjacent interprimibrachs or interambulacrals. Tegmen composed of 4 plates interambulacral in position, separated by open clefts and surrounding central open space (peristome), which during life may have been partly occupied by a posterior plate and covered by superoral plates; ambulacra suprategminal, flanked by adambulacrals and probably closed by covering plates; 1 large posterior plate bordered by 2 small ones supporting short plated anal protuberance. Free arms 10, simple, uniserial. *M.Sil.*, N.Am.(USA).——FIG. 315,*1*. **L. bacca* (ROEMER), Niagaran (Beech River F.), Tenn.; *1a,b,* tegminal and *B*-ray views of theca, X3 (Springer, 1926a).

?**Plesiocrinus** WANNER, 1937, p. 74 [**P. piriformis*; OD]. Like *Eutelecrinus* but with central tegminal opening between orals, probably covered during life by superoral plates. *U.Perm.*, Indon.(Timor.). [Genetic relations of this genus with Silurian and Devonian hapalocrinids are uncertain.]——FIG. 315,*3*. **P. piriformis; 3a,b,* tegminal and *CD*-interray views of holotype theca, X2 (Wanner, 1937).

Thallocrinus JAEKEL, 1895, p. 106 [**T. hauchecornei*; M]. Similar to *Hapalocrinus* but with movable spines on ventral side of free arms. *L.Dev.*, Eu.(Ger.).——FIG. 315,*4a*. **T. hauchecornei,* Bundenbacher Schiefer; type, X1 (W. E. Schmidt, 1932).——FIG. 315,*4b*. *T. procerus* W. E. SCHMIDT, Bundenbacher Schiefer; portion of arm with spines, enl. (W. E. Schmidt, 1934).

Family PLATYCRINITIDAE Austin & Austin, 1842

[*nom. correct.* BASSLER, 1938, p. 23 (*pro* family Platycrinoidea AUSTIN & AUSTIN, 1842, p. 109)] [=Platycrinidae AUSTIN & AUSTIN, 1843, p. 199] [Materials for this family prepared by N. GARY LANE]

Calyx confined to patina of five radials, and three unequal basals which may be fused; posterior side commonly not differentiated; orals distinct; interambulacrals and ambulacrals important tegminal elements but absent in youngest genera; first primibrachs typically axillary. Arms biserial, pinnulate. Column transversely round proximally, elliptical and twisted distally, with synarthrial articulation, fulcral ridge following long diameter of columnals; axial canal small, round. *?U.Sil., Dev.-Perm.*

Platycrinites MILLER, 1821, p. 73 [**P. laevis*; SD MEEK & WORTHEN, 1865c, p. 160] [=*Platycrinus* AGASSIZ, 1836, p. 197 (obj.) *(nom. van.)*; *Centrocrinus* AUSTIN & AUSTIN, 1843, p. 6 (type, *Platycrinites gigas* GILBERTSON, in PHILLIPS, 1836, p. 204); *Atocrinus* M'COY in GRIFFITH, 1842, p. 11 (type, *A. milleri*)]. Calyx bowl shaped; basals 3, unequal, small one typically in *AE* interray, or fused with others; radials large, with narrow articular facets. Tegmen flat or pyramidal, composed of numerous plates, orals distinct; anal opening excentric, directly through tegmen or at end of short or long anal tube. Arms with primibrachs *1* or rarely *2* axillary, with 1 to 5 higher, heterotomous divisions, becoming biserial above distal axillary of branches; proximal brachials like free arm plates but fixed to tegmen by 1 to 3 interradials. *?U.Sil.,L.Dev.-U.Perm.*, Eu.-N.Am.-Indon.(Timor.).——FIG. 316,*2a,b*. **P. laevis,* L. Carb., Eng.; lat. views of crown and partial crown with anal tube, X0.9 (Wright, 1955).——FIG. 316,*2c-e*. *P. burlingtonensis* (OWEN & SHUMARD), L.Miss.(Osag., Burlington Ls.), USA(Iowa); lat. views of crown and theca, ventral view of theca, X0.9 (Wachsmuth & Springer, 1897).——FIG. 317,*4*. *P. saffordi* (HALL), L.Miss.(Osag., Burlington Ls.), USA(Iowa); *4a,b,* lat. views of crown and theca, X0.7 (Wachsmuth & Springer, 1897).

[Revision of M'COY's species of Irish crinoids by G. SEVASTOPULO, Univ. of Dublin, has resulted in transfer of *Edwardsocrinus* D'ORBIGNY, 1849, from *Platycrinites* to *Dichocrinus* (G. SEVASTOPULO, pers. comm.).]

Brahmacrinus SOLLAS, 1900, p. 264 [**B. ponderosus*; OD]. Calyx as in *Platycrinites*; primibrachs and secundibrachs in sutural contact with tegminal interambulacrals and interbrachials. Tegmen with subcentral anal tube. Arms free above 1st tertibrachs. *L.Carb.(Tournais.)*. Eng.——FIG. 317,*5*. **B. ponderosus; 5a,b,* post. and *BC*-interray views of theca, X0.7 (Sollas, 1900).

Eucladocrinus MEEK, 1872, p. 373 [**Platycrinites (Eucladocrinus) montanaesis* MEEK, 1872a, p. 373 *(nom. imperf. pro montanaensis)*; M]. Calyx like *Platycrinites*. Arms consisting of 2 main uniserial rami in each ray, covered by large ambulacral plates, forming tubular extensions of thecal cavity and giving off on alternate sides short, biserial, pinnulate ramuli. *L.Miss.(Osag.)*, USA(Iowa-Ill.-Mont.).——FIG. 316,*3*. **E. montanaensis,* Lodgepole Ls., Mont.; lat. view of crown, X0.9 (Wachsmuth & Springer, 1897).——FIG. 317,*3*. *E. pleuroviminus* (WHITE), Burlington Ls., Iowa; lat. view of crown, showing strongly elliptical twisted stem, X0.7 (Wachsmuth & Springer, 1897).

Neoplatycrinus WANNER, 1916, p. 32 [**N. dilatatus*; OD]. Aboral cup and lateral anal opening as in *Pleurocrinus*. Tegmen with 5 orals and few interambulacral plates, lacking interambulacrals; anal opening between *CD* oral and *C* and *D* radials. Arms unknown. *U.Perm.*, Indon.(Timor).——FIG. 317,2. **N. dilatatus; 2a-c,* post., post., and ventral views of theca, X0.7; *2d,* ventral view of theca, X1 (Wanner, 1937).

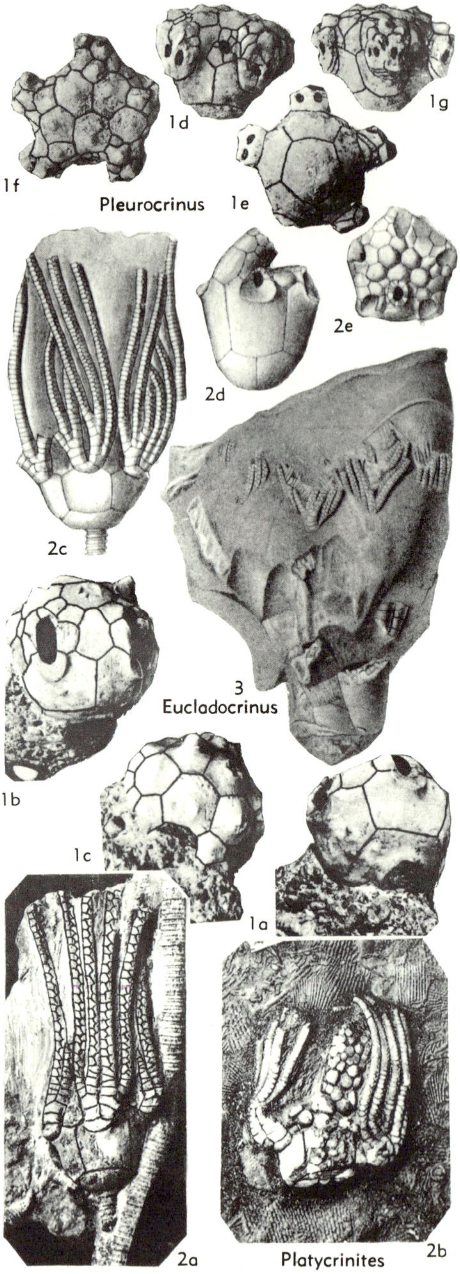

FIG. 316. Platycrinitidae (p. T515-T516).

Oenochoacrinus BREIMER, 1962, p. 124 [*O. *princeps*; OD]. Small basal in *CD* or *AE* position; posterior interray wider than others. Tegmen composed of 5 central orals, of which *CD* oral is largest, and 5 large axillary ambulacral plates in radial position; anal opening bounded below by 3 small plates. *L.Dev.-M.Dev.*, Spain.——FIG. 317,*6*. **O. princeps*, L.Dev.; *6a-d*, post., dorsal, ventral, and ant. views of theca, ×2 (Breimer, 1962).

Plemnocrinus KIRK, 1946, p. 435 [**P. beebei*; OD]. Calyx low, basin shaped or flat; radials widest medially, distally constricted into protruding neck. Arms with hypertrophied primaxil forming with 2 fused secundibrachs in each half-ray short trunk that bears 2 uniserial, heterotomously branching rami; ramuli biserial, pinnulate. *L.Miss.(Osag.)*, USA(Iowa-Ill.-Mo.).——FIG. 317,*1*. **P. beebei*, Burlington Ls., Iowa; *1a-c*, lat. and dorsal views of crown, ventral view of theca, ×0.7 (Kirk, 1946d).

Pleurocrinus AUSTIN & AUSTIN, 1843, p. 6 [**Platycrinites mucronatus* AUSTIN in AUSTIN & AUSTIN, 1843, p. 22; OD (conditional) AUSTIN & AUSTIN, 1843, p. 25] [=*Medusacrinus* AUSTIN, 1875, p. 91 (type, *Platycrinites mucronatus* AUSTIN & AUSTIN, 1843; SD BASSLER, 1938, p. 130)] [*Pleurocrinus* was originally proposed by AUSTIN & AUSTIN as a "division" of the genus *Platycrinites* and therefore originally had subgeneric rank. These authors apparently believed that a "division" did not need a type species, because they state (1843, p. 25), "Did we consider it expedient to elevate it to generic distinction we should take the *mucronatus* [*Platycrinites mucronatus*] as the type of our proposed genus *Pleurocrinus*." This statement is taken as a clear indication that *Platycrinites mucronatus* is to be considered the type species of *Pleurocrinus*.] Like *Platycrinites* except anal opening just above *C* and *D* radials is laterally directed, with or without intervening small plates between opening and radials. *L.Carb.-U.Perm.*, N.Am.-Eu.-Indon.(Timor).——FIG. 316,*1a-c*. **P. mucronatus* (AUSTIN), L.Carb., Eng.; oblique, lat., and ventral views of theca, ×0.9 (Wright, 1956).——FIG. 316,*1d-g*. *P. coronatus* (GOLDFUSS), L.Carb., Eng.; post., dorsal, ventral, and ant. views of theca, ×0.9 (Wright, 1956a).

ORDER, SUBORDER, SUPERFAMILY, and FAMILY UNCERTAIN

Becharocrinus TERMIER & TERMIER, 1956, p. 316 [**Triacrinus paradoxus* TERMIER & TERMIER, 1950; OD] [*nom. subst. pro Triacrinus* TERMIER & TERMIER, 1950, p. 84 (*non* MÜNSTER, 1839)]. Based on an apparently abnormal trirayed specimen; probably actinocrinitid. *U.Carb.(low.Westphal.)*, N.Afr.(Alg.).——FIG. 318,*1*. **B. paradoxus* (TERMIER & TERMIER); *1a,b*, lat. and tegminal views, ×2 (Termier & Termier, 1950).

Coronocrinus HALL, 1859, p. 124 [**C. polydactylus*;

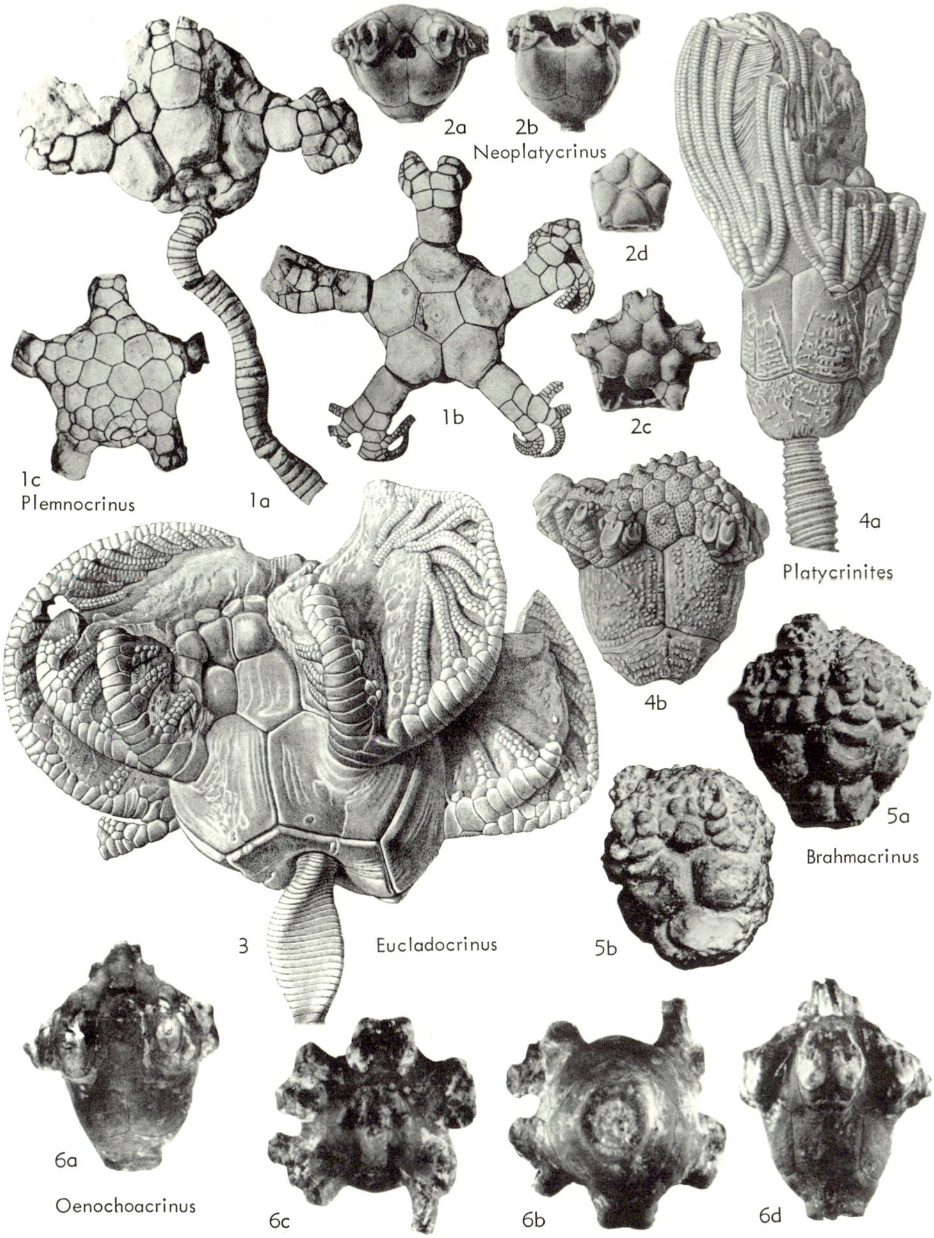

FIG. 317. Platycrinitidae (p. *T515-T516*).

M]. Based on fragment of large, very broad and many-plated theca; tegmen slightly depressed or elevated, composed of numerous small plates; arms (only bases of some preserved) numerous, biserial, proceeding abruptly from calyx margin; column and base unknown. [According to GOLDRING (1923), may be same as *Himerocrinus*]. *L.Dev.*, USA(N.Y.).——FIG. 318,2. **C. polydactylus*, Helderberg.; lat. view of holotype, ×1 (Goldring, 1923).

Craspedocrinus DAHMER, 1921, p. 292 [**C. schmidti*; M]. Genus based on very small poorly preserved specimens; presence of infrabasals uncertain; median ray and anal ridges distinct;

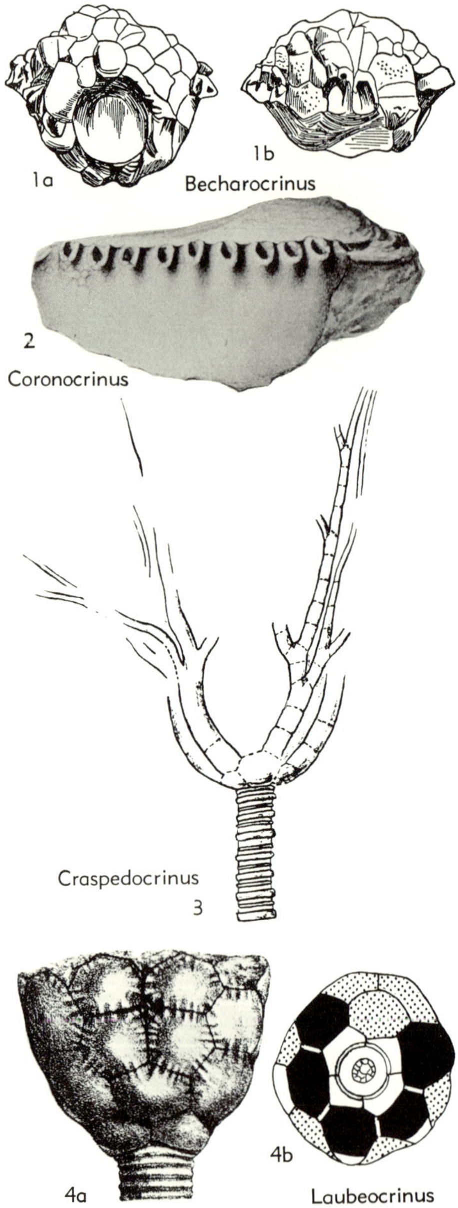

FIG. 318. Order, Suborder, Superfamily, and Family Uncertain (p. *T516-T518*).

?radials separated all around; interradial areas many-plated; arms pinnulate, uniserial, isotomously branching; brachials slightly higher than wide; stem transversely round, heteromorphic; axial canal quinquelobate. [Doubtfully referred by DAHMER to Reteocrinidae; may be early post-larval growth stage of an unknown genus and species.] *L.Dev.*, Eu.(Ger.).——FIG. 318,*3*. **C. schmidti*, Rammelsberger Schichten; crown and part of stem, ×5 (Dahmer, 1921).

Helicocrinus CHAPMAN, 1903, p. 107 [**H. plumosus*; M]. Monocyclic; basals 3; primibrachs ? 3; tegmen fairly high, vaulted over with numerous polygonal plates; rami 2 per ray, simple, densely pinnulate; brachials apparently cuneate in zigzag. Stem pentagonal, with small cirri at intervals. *Sil.*, E.Australia. [Based on a single incompletely known specimen. Possibly a hapalocrinid.]

Katarocrinus WAAGEN & JAHN in JAEKEL, 1918, p. 32 (no species named). [Placed by JAEKEL, 1918, in family Spyridiocrinidae along with *Spyridiocrinus* and *Beyrichocrinus*.] *L.Dev.*, Eu.(Czech.).

Laubeocrinus WAAGEN & JAHN, 1899, p. 31 [**L. barrandei*; M]. Monocyclic; calyx bowl shaped, with 4 unequal basals; primanal in radial circlet, followed by 2 plates, which in turn support 2 other plates; single rather large interprimibrach preserved in each interray, resting on radials; stem round; distal part of cup, tegmen, and arms unknown. [Based on an incomplete calyx; according to BATHER (1900a), may be an actinocrinitid of uncertain position; placed by BASSLER (1938) and BASSLER & MOODEY (1943) among Actinocrinitidae.] *Sil.*, Eu.(Czech.).——FIG. 318,*4*. **L. barrandei; 4a,b,* lat. and basal views, ×2 (anals and interbrachials stippled) (Waagen & Jahn, 1899).

Mitrocrinus MILLER & GURLEY, 1894, p. 22 [**M. wetherbyi*; OD]. Based on six-rayed crinoid with low bowl-shaped calyx, 3 equal basals, 3 interprimibrachs between rays, small primanal in radial circlet followed by 2 plates. Tegmen high, orals distinct and anal tube excentric. Arms and stem not preserved. [According to BATHER (1900a), this genus may be founded on an abnormal periechocrinid or carpocrinid; it was made type genus of a monotypical family, Mitrocrinidae, by MILLER & GURLEY, 1894, p. 22.] *M.Ord.*, USA (Tenn.).——FIG. 319,*1*. **M. wetherbyi; 1a-c,* dorsal, ventral, and *CD*-interray views, ×1 (Miller & Gurley, 1894b).

Pachyocrinus BILLINGS, 1859, p. 22 [**P. crassibasalis*; M] [=*Pachycrinus* SCUDDER, 1882, Nom. Zool. suppl. list, p. 243 *(nom. van.)*]. Founded on single part of calyx comprising 4 small proximal (?basal) plates, followed by 5 thick and large (?radial) plates which alternate with them. [According to HUDSON (1907, p. 120), this insufficiently known genus may belong to Eucalyptocrinitidae.] *M.Ord.*, N.Am.(Can.).——FIG. 319,*2*. **P. crassibasalis*, Chazyan; holotype, ×2 (Hudson, 1907).

Patinocrinus JAEKEL, 1918, p. 36 (no species named). [Proposed for *Stelidiocrinus* WACHSMUTH & SPRINGER, ?date (*non* ANGELIN), by JAEKEL (1918) and defined as a patelliocrinid with depressed tripartite base, radials and primibrachs low and wide, several secundibrachs, single large interprimibrach in each interray, and 2 anal plates

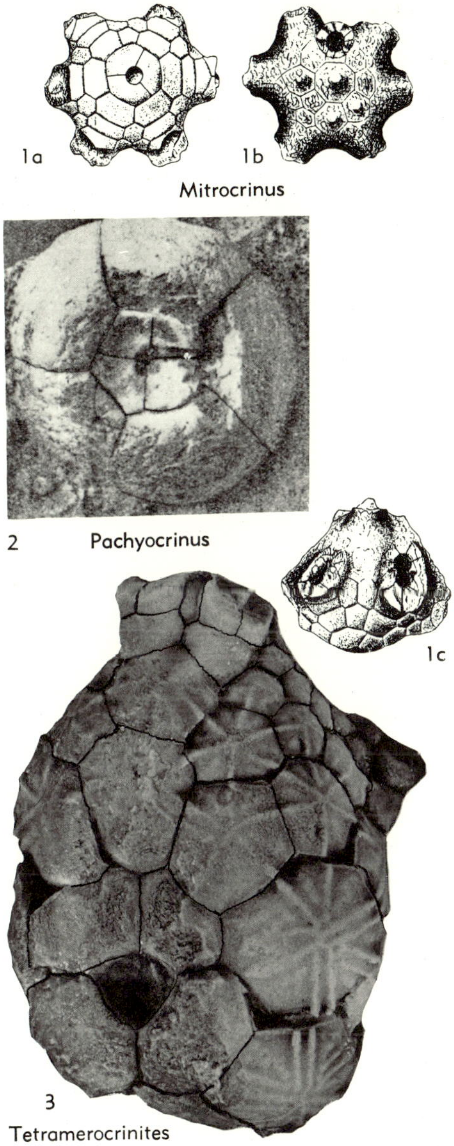

FIG. 319. Order, Suborder, Superfamily, and Family Uncertain (p. *T*518-*T*519).

located above level of radials.] ?Occurrence (unknown).

Phillipsocrinus M'COY, 1844, p. 183 [**P. caryocrinoides*; M]. Basals 4; radial circlet composed of 7 plates among which 5 support primibrachs, and others followed by 2 plates; single interprimibrach between primibrachs; ornamentation stellate. [Probably based on abnormal actinocrinitid. Type specimen apparently lost (GEORGES SEVASTOPULO, personal commun.).] *L.Carb.,* Eire.

Tetramerocrinites AUSTIN & AUSTIN, 1843, p. 203 [**T. formosus*; M] [=*Tetracrinites* AUSTIN & AUSTIN, 1842, p. 110 *(nom. nud.)*; *Tetramerocrinites* AUSTIN & AUSTIN, 1842, p. 110 *(nom. nud.)*; *Tetramerocrinus* HUXLEY & ETHERIDGE, 1865, p. 129 *(nom. subst. pro Tetramerocrinites)*]. Basals 4, 3 subequal and 4th smaller; radial circlet composed of 8 plates; one ray with single primibrach, which is axillary; other rays probably with 2 primibrachs; ornamentation stellate. [Genus apparently based on abnormal actinocrinitid.] *L. Carb.,* Eire.——FIG. 319,*3*. **T. formosus,* Tournais., oblique lat. view of calyx, lectotype, ×2 (Sevastopulo, n).

Trochocrinites PORTLOCK, 1843, p. 345 [**T. laevis*; M] [=*Trochocrinus* BATHER, 1900a, p. 168 *(nom. subst. pro Trochocrinites)*]. Camerate, probably monocyclic, with axial ridges. [Founded on single internal mold of basal part of the calyx; according to RAMSBOTTOM (1961, p. 29), it shows no diagnostic features and is probably not synonymous with *Periechocrinus,* as stated by WACHSMUTH & SPRINGER (1881).] *U.Ord.(Caradoc.),* Eire.

INADUNATA

By R. C. Moore, N. Gary Lane, H. L. Strimple, James Sprinkle, and R. O. Fay

[University of Kansas; University of Indiana; University of Iowa; University of Texas; Oklahoma Geological Survey]

Subclass INADUNATA Wachsmuth & Springer, 1885

[*nom. transl.* Moore & Laudon, 1943a, p. 21 (*ex* suborder Inadunata Wachsmuth & Springer, 1885, p. 228)] [=order Pentacrinoidea Jaekel, 1894, p. 114 (*partim*)] [Diagnosis by Georges Ubaghs]

Crinoidea with aboral cup composed of close-sutured plates; fixed brachials and interbrachials lacking (exception in few primitive forms); anal plates commonly present in aboral cup; mouth subtegminal; arms pinnulate or nonpinnulate. *L.Ord.-U.Perm.; M.Trias.*

Order DISPARIDA Moore & Laudon, 1943

[*nom. correct.* Moore, in Moore, Lalicker, & Fischer, 1952a, p. 613 (*pro* Disparata Moore & Laudon, 1943a, p. 24)] [=order Myelodactyloidea S. A. Miller, 1883a; branch Larviformia Wachsmuth & Springer, 1885, p. 305; suborder Monocyclica Bather, 1893a, p. 20; suborder Larvata Jaekel, 1894, p. 116; order Monocyclica Inadunata Bather, 1899b, p. 921; Turbata Jaekel, 1918, p. 82]

By R. C. Moore, N. Gary Lane, H. L. Strimple, and James Sprinkle

Monocyclic inadunates with weak to very prominent bilateral symmetry developed in different planes (*E-BC* in Homocrinacea, Calceocrinacea, Pisocrinacea, Allagecrinacea; *D-AB* in Heterocrinacea; *C-EA* in Myelodactylacea, Anomalocrinacea; and *A-CD* in Belemnocrinacea, Hybocrinacea). *L.Ord.-U.Perm.*

Most disparids are small and characterized by simplicity in structure of the crown and cup; many are microcrinoids, with diameter and height of the theca less than 2 mm. Chief plates are radials, which are main components of the cup and orals of the tegmen. The arms are slender and uniserial, branched at several heights above the cup or unbranched; pinnules are lacking.

A well-marked tendency toward development of bilateral symmetry is observed, oriented in different planes. These are respectively designated according to the ray and opposite interray in which they are placed, namely 1) homocrinoid (through the *E* ray), 2) heterocrinoid (through the *D* ray), 3) iocrinoid (through the *C* ray, and belemnocrinoid (through the *A* ray). The last is also called crinoid plane, since it is widely prevalent in cladid Inadunata, Flexibilia, Camerata, and Articulata. Genera having perfect pentamerous symmetry are bilaterally symmetrical with respect to planes passing through any ray and its opposite interray. Such symmetry is found in some crinoid groups, but not the Disparida.

A feature having general importance in the evolution, and hence classification of crinoids, is fusion of skeletal elements and another is change in structural relationships of morphological elements. Both of these result in simplicities derived from complexities. Among the disparid crinoids illustrations are offered by 1) reduction in number of basals from five to three to one, 2) lateral fusion of axillary brachials with each other and subjacent radials to produce the large multifaceted radials of Catillocrinidae, Anamesocrinidae, and Calceocrinidae (Fig. 320), 3) fusion of infer-and superradials to yield simple undivided radials as in Haplocrinitidae and Pisocrinidae derived from Heterocrinacea and Iocrinidae of unknown ancestry, and 4) disappearance of proximal brachials fixed in the cup of Homocrinidae and Heterocrinidae, evidently by change of their status from fixed to free brachials. Cladid inadunates evolved similarly in the direction of reducing the number and complexity of thecal and arm components. For example, the Aethocrinidae, near-oldest (L.Ord.) of all known cri-

(*See facing page.*)

Fig. 320. Diagram showing large multifaceted radials produced by lateral fusion of axillary brachials and subjacent radials (Moore, n).——*1*. Hypothetical primitive calceocrinid with separate axillary brachials in the *A* and *D* rays.——*2*. Advanced calceocrinid with axillaries fused to *A* and *D* radials. Higher arm branches typical of calceocrinids omitted for clarity.

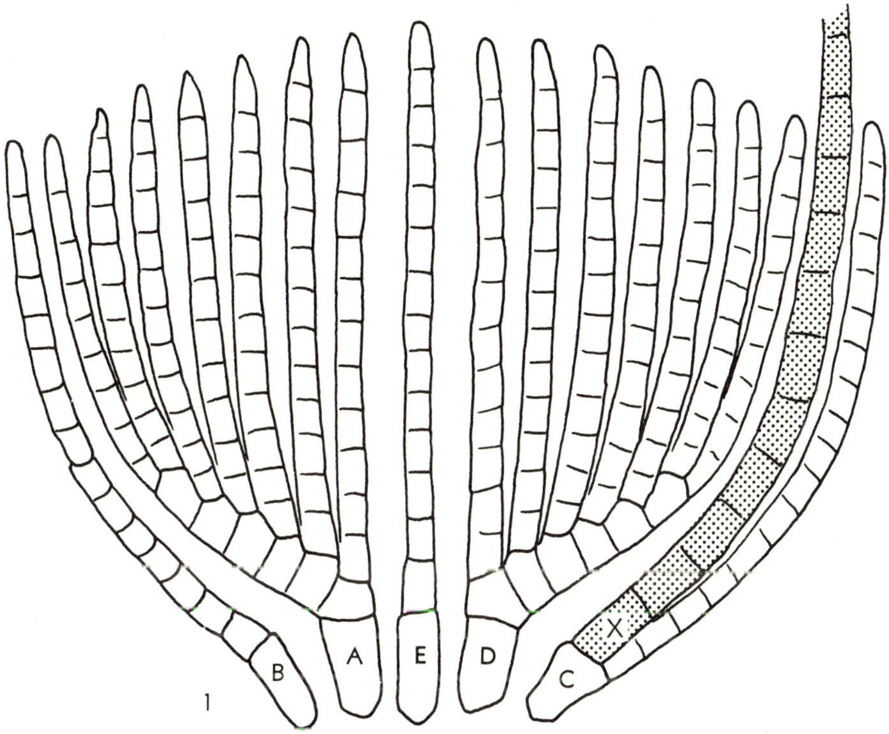

1

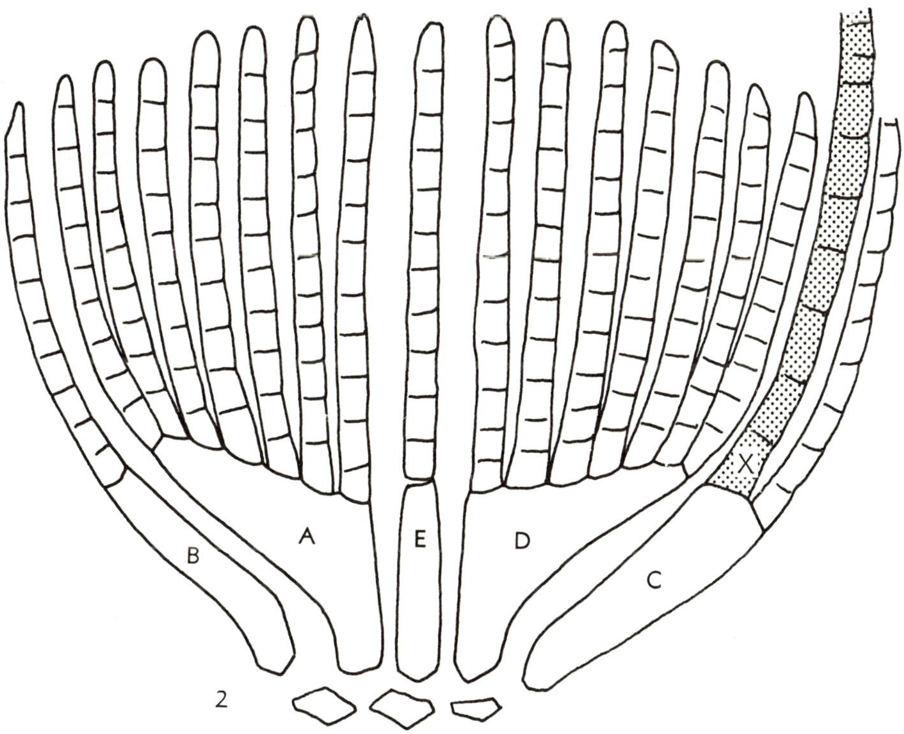

2

noids are followed soon (M.Ord.-U.Ord.) by much more simply organized genera.

Classification of the Disparida in the *Treatise* recognizes eight superfamilies defined mainly by shape of the cup and by the nature of bilateral symmetry. A large majority of genera have more or less steep-sided conical cup shapes; not very many are characterized by rounded bowl form and almost none are discoid. A basal concavity of the cup invariably is lacking.

Superfamily HOMOCRINACEA Kirk, 1914

[*nom. correct.* MOORE & LANE, in MOORE & STRIMPLE, 1973, p. 16 (*pro* Homocrinicae *nom. transl.* UBAGHS, 1953, p. 746, *ex* Homocrinidae KIRK, 1914, p. 479)] [Materials for this superfamily prepared by R. C. MOORE and N. GARY LANE]

Bilateral symmetry of crown and cup in *E-BC* plane; superradial and inferradial, present in *B, C,* and *E* rays, *A* and *D* rays with simple radials; radial articular facets as wide as radials, or composite and supporting several uniserial, atomous arms to a ray; anal *X* typically placed between *C* and *D* radials. Arms atomous, isotomously branched or with complex ramules. Anal sac tubular, with longitudinal row of enlarged plates posteriorly. *M.Ord.-U.Ord.; M.Sil.*

Family HOMOCRINIDAE Kirk, 1914

[Homocrinidae KIRK, 1914, p. 479]

Cup small, conical; *B, C,* and *E* radials compound; anal *X* incorporated in cup between *C* and *D* radials. *M.Ord.-U.Ord.; M.Sil.*

Homocrinus HALL, 1852, p. 185 [**H. parvus*; SD S. A. MILLER, 1889, p. 255 (*non* BATHER, 1893a)]. Cup small and cylindrical; basals elongate; *B, C,* and *E* rays with infer- and superradials; *A* and *D* rays with large, simple radials; anal *X* pentagonal, situated in notch between upper edges of *D* radial and *C* superradial. Anal sac not known. Arms long and slender, with first primibrach fixed in top of cup; atomous, or branching once on second primibrach, with widely spaced unbranched ramules on alternate sides of every third to sixth secundibrach. Column transversely round. [Species originally assigned to *Homocrinus* from the Lower Devonian of Maryland and from the Devonian of Germany are now placed in the Botryocrinidae.] *M.Sil.(Niagaran),* USA(N.Y.-Ind.-Ill.).——FIG. 321,*3*. **H. parvus,* N.Y.; *3a-d, E*-ray, *CD*-interray, *BC*-interray, and *A*-ray views of theca, ×8; *3e, A*-ray view of crown and stem, ×5 (Hall, 1852); *3f,* plate diagram of theca (radials and superradials black, inferradials horizontally ruled, anal *X* stippled) (Moore, 1962b).

Daedalocrinus ULRICH, 1925, p. 97 [**D. kirki*; OD]. Cup low cone shaped; plates arranged as in *Homocrinus*; anal *X* large, pentagonal, supporting longitudinal row of anal sac plates which taper in width distally, sac large, inflated and balloon shaped, composed of numerous small, irregular plates. Proximal brachials not fixed in cup; primibrachs *3* axillary in all but *C* ray, which has axillary primibrach *4*. Two stout rami to a ray, with long, robust, unbranched ramules on admedial side of every third or fourth secundibrach. Column transversely pentagonal proximally, becoming round distally. *M.Ord.(Trenton.),* Can. (Ont.).——FIG. 321,*2*. **D. kirki; 2a-c,* plate diagram of partial crown, side and transverse parts of column (Ulrich, 1925); *2d,* plate diagram of partial crown (patterns as in Fig. 321,*3f*) (Moore, 1962b).

Drymocrinus ULRICH, 1925, p. 96 [**Heterocrinus geniculatus* ULRICH, 1879, p. 16; OD]. Cup small, plates arranged as in *Homocrinus*. Anal *X* large, supporting median row of posterior sac plates. Anal sac narrow, elongate, or with several longitudinal rows of thin, quadrate plates. Primibrachs *1* as wide as radials proximally, tapering in width distally; primibrach *2* axillary in all rays, supporting 2 main rami in each ray. Long, slender, unbranched ramules given off on alternate sides of every second or third secundibrach. Ramule-bearing brachials large and wedge shaped, producing zigzag appearance of each ramus. Column transversely round, with numerous long, slender cirri directed proximally; cirri near cup concealing lower part of crown. *U.Ord.(Cincinnat.),* USA (Ohio-Ky.)-Can.(Ont.).——FIG. 321,*1*. **D. geniculatus* (ULRICH), Ohio; *1a,b,* plate diagram of partial crown and lat. view of stem, ×1 (Ulrich, 1925).

Ectenocrinus MILLER, 1889, p. 242 [**Heterocrinus simplex* HALL, 1847, p. 280; OD]. Cup cylindrical; basals short. Anal *X* small, pentagonal, barely in top of cup, supporting long, slender anal sac. Primibrachs normally free above radials, but first primibrachs may be fixed in cup; primibrach *2* axillary in all rays. Two stout rami to a ray, composed of wedge-shaped brachials that support long, slender ramules on alternate sides of every other, or every third or second, brachial. Stem transversely round, columnals composed of trimeres. *M.Ord.(Trenton.)-U.Ord.(Cincinnat.),* USA (Ky.-Mich.-Ohio-Iowa-Minn.-N.Y.)-Can. (Ont.).——FIG. 322,*1a-e*. **E. simplex* (HALL), U.Ord., Ohio; *1a,b,* post. and *?AB*-interray views of crown and stem, ×2 (Moore, n; Univ. Chicago 599A, 599B); *1c,d,* plate diagram of partial crown and cross section of stem (Ulrich, 1925); *1e, BC*-interray view of partial crown and stem, ×2 (Meek, 1873).——FIG. 322,*1f,g*. *E. grandis* (MEEK),

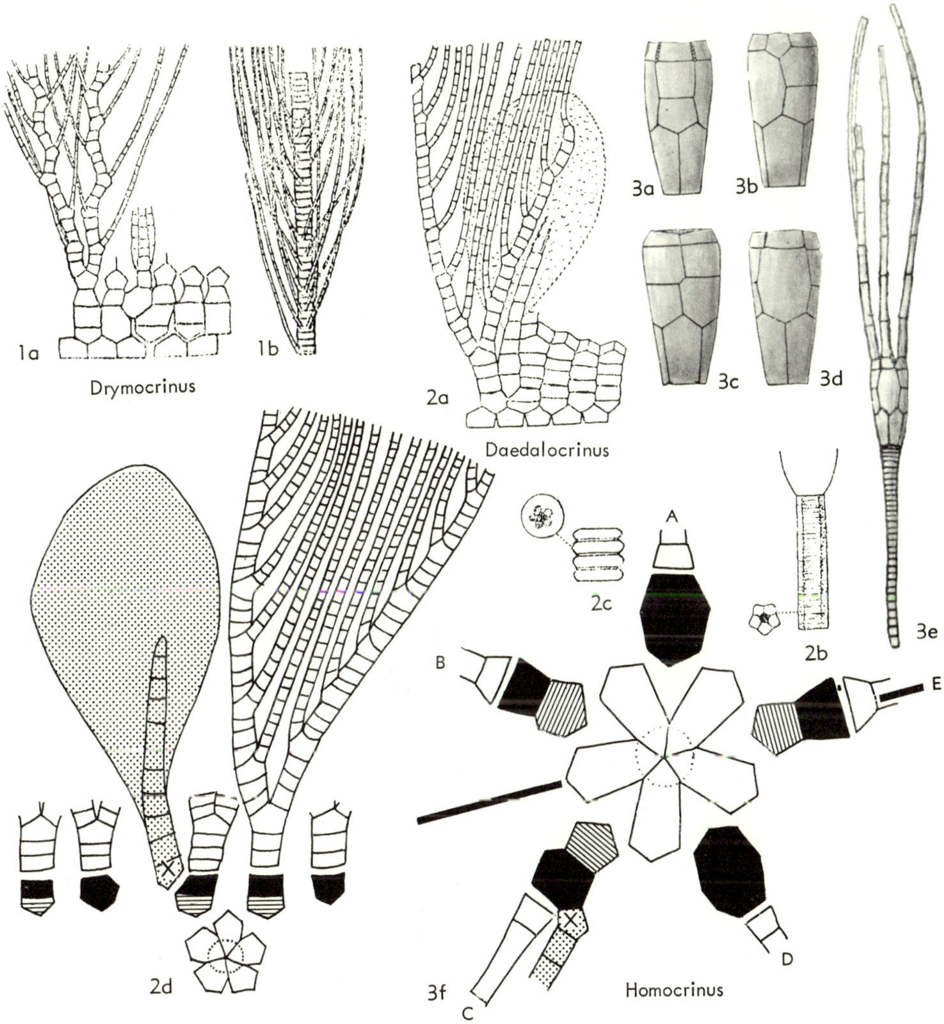

FIG. 321. Homocrinidae (p. T522).

U.Ord., Ohio; *1f,g*, *B*-ray view of crown and stem and lat. view of part of arm and ramules, ×1, enl.(Meek, 1873). [Also Fig. 324,*1*.]

Ibexocrinus LANE, 1970, p. 12 [**I. lepton*; OD]. Crown slender and elongate; cup barely wider than proximal columnals. Compound radials in *B, C,* and *E* rays; *D* and *A* radials simple; *C*-ray superradial axillary, supporting anal sac plates on left side and arm on right; primibrach *2* axillary in 2 rays; each arm with 2 main rami; a long, stout, unbranched ramule given off on alternate sides of every sixth to eighth secundibrach. Columnals composed of pentameres and proximally confluent with base of cup. [*Ibexocrinus,* one of the oldest known genera of the Homocrinidae, is most similar to the slightly younger *Ectenocrinus,* differing mainly in having pentameres rather than trimeres composing the columnals and in having more widely spaced ramules on the 2 branches of each arm.] *M.Ord.(Trenton.),* USA(Utah).——FIG. 323,*1*. **I. lepton; 1a, A*-ray view of crown, ×3.3; *1b,c, CD*-interray diagram of cup and *A*-ray diagram of crown (Lane, 1970a).

Sygcaulocrinus ULRICH, 1925, p. 98 [**S. typus*; OD]. Cup steep sided, small, continuous below with much enlarged proximal columnals, which taper in width distally. Anal *X* minute, diamond shaped. Primibrachs *2* axillary. Two rami to a ray, with ramules on alternate sides of every third or sixth brachial. Column transversely round. *U.Ord.(Cincinnat.),* USA(Iowa).——FIG. 323,*2*. **S. typus; 2a,b,* plate diagram and *D*-ray view of

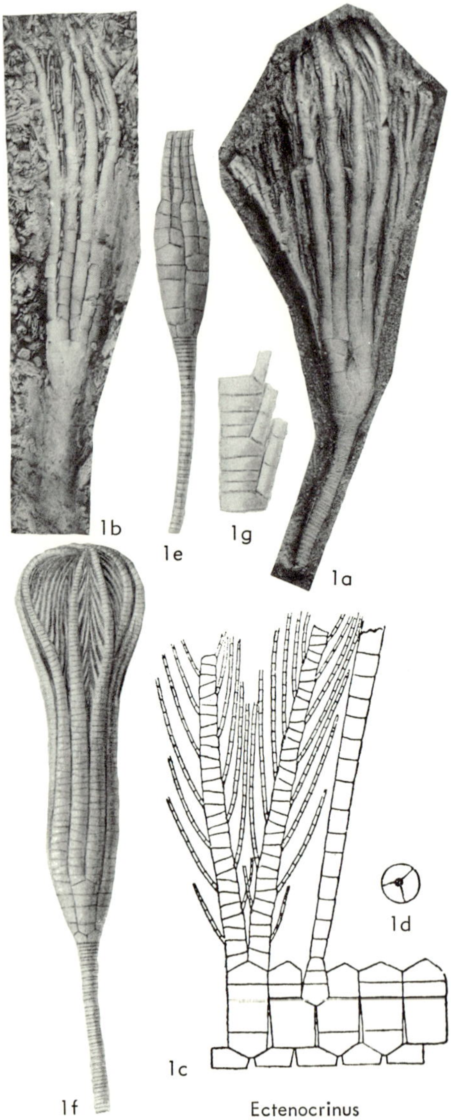

FIG. 322. Homocrinidae (p. T522-T523).

partial crown (patterns as in Fig. 321,*3f*) (Moore, 1962b).

Superfamily CALCEOCRINACEA Meek & Worthen, 1869

[*nom. transl.* MOORE & LANE, in MOORE & STRIMPLE, 1973, p. 16 (*ex* Calceocrinidae MEEK & WORTHEN, 1869a, p. 73)] [Materials for this superfamily prepared by R. C. MOORE and N. GARY LANE]

"Bent crown" monocyclic crinoids with mostly almost perfect bilateral symmetry developed (like Homocrinidae) in the *E*-ray *BC*-interray plane. Cup commonly quadrangular in outline and flattened on the *A-D-E* side and moveably hinged with plates on opposite side to which the stem is attached. Main axils composed of joined proximal axillary plates of contiguous successive arms borne by the large undivided *A* and *D* radials are distinguishing morphological features, as are ramule-bearing axil arms joined to the main axil. Anal *X* associated with *C* ray supports tubular anal sac. Stem transversely round, sharply bent near attachment to cup. *M.Ord.-L.Miss.; L.Perm.*

Family CALCEOCRINIDAE Meek & Worthen, 1869

[Calceocrinidae MEEK & WORTHEN, 1869a, p. 73] [=Chirocrinidae ANGELIN, 1878, p. 22; Cheirocrinidae ZITTEL, 1879, p. 343; Cremacrinidae ULRICH, 1886, p. 105; Castocrinidae JAEKEL, 1918, p. 88; Calceocrinites JAEKEL, 1918, p. 86 (suborder) (*partim*)]

Crown pendent or recumbent on stem, with imperfect to perfect bilateral symmetry developed in the plane of the *E* ray and *BC* interray. Cup with basal circlet subtriangular in outline consisting of four, three, or one plates which bears all or part of stem impression at apex of triangle and is muscularly hinged along base of triangle with radial plates: median *(E)* radial compound, with inferradial forming middle part of all hinge contact with basals and superradial connected broadly to narrowly with inferradial or disconnected by intervention of adjoined edges of large lateral *(A, D)* radials; *B* and *C* radials compound, in some genera with infer- and superradial elements of each distinct, but in others with superradials fused together and in one genus *(Senariocrinus)* with inferradials fused in accompaniment with unfused superradials; lateral radials *(A, D)* large and undivided, oppositely placed with respect to the plane of bilateral symmetry. Median (*E* ray) arm unbranched or isotomously divided; lateral arms multiple, characterized by compound proximal axillary brachials that form main-axil series and more or less numerous axil-arms with heterotomously developed unbranched ramules produced alternately from abanal and adanal sides of the axil-arms (except in *Chiropinna*). Anal tube composed of stout,

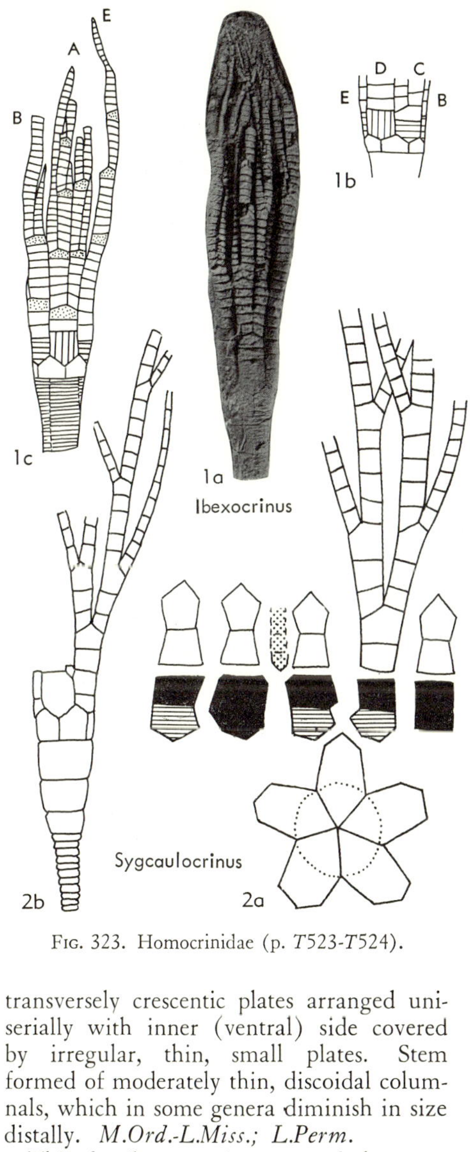

FIG. 323. Homocrinidae (p. *T*523-*T*524).

transversely crescentic plates arranged uniserially with inner (ventral) side covered by irregular, thin, small plates. Stem formed of moderately thin, discoidal columnals, which in some genera diminish in size distally. *M.Ord.-L.Miss.; L.Perm.*

This family comprises one of the most interesting groups of crinoids in peculiarities of their structure developed as adaptation to a particular mode of life. Most remarkable is their attainment of complete bilateral symmetry in a plane other than that coinciding with the anteroposterior axis of the theca (Fig. 324). Also, successive steps in evolution of the symmetry are clearly recognizable and its derivation from an ancestral homocrinid stock cannot be doubted (Fig. 321).

The heterotomous pattern of arm branching in the Calceocrinidae corresponds to that seen in various homocrinid genera (e.g., *Ectenocrinus, Drymocrinus*), but in most calceocrinids it is peculiarly accentuated in the *A* and *D* rays by development of so-called main-axils next above the radials and a series of axil-arms which give off unbranched ramules in a constant manner (Fig. 325) (MOORE, 1962a).

The structure of arms borne by the large lateral radials, mirror images of each other, is very remarkable and has importance in generic classification. First, is a linking together of the most proximal plates of the laterally successive arms to form so-called **main axils.** These are composed of axillary brachials at the base of all arms, with addition of a quadrangular nonaxillary plate at the base of arms next to the *E* ray. These plates are primibrachs and arms above them are termed **primaxil arms.** Arms next farther from the *E* arm begin with an axillary secundibrach and are called **secundibrach arms.** Following this plan farther from *E* we have **tertaxils** as main-axil components and **tertaxil arms, quartaxils,** and **quartaxil** arms, and so on.

Arms of each axil arm are divisible in **rami,** centrally placed main parts that are wider than bordering lateral parts, and possess successive axillary brachials separated by one or two nonaxillary brachials from preceding and following axils. Slender unbranched armlets are called **ramules.** These diverge from the rami first on the side nearest to the *E* ray and then on the opposite side. Successive axils and ramules in each axil arm are designated by Greek letters, alpha (α), beta (β), gamma (γ), delta (δ), and so on. The single ramule attached to the main axil farthest from *E* is known as omega (ω) ramule. The features are illustrated in Figure 325.

Calceocrinus HALL, 1852, p. 352 [**Cheirocrinus chrysalis* HALL, 1860b, p. 123; SD WACHSMUTH & SPRINGER, 1886, p. 205(281)] [=*Cheirocrinus* HALL, 1860, p. 122 (type, *C. chrysalis*; SD BASSLER & MOODEY, 1943, p. 468) (obj.) (*non* EICHWALD, 1856, p. 123); *Eucheirocrinus* MEEK & WORTHEN, 1869a, p. 73 (type, *C. chrysalis* HALL, 1860, p. 123; SD SPRINGER, 1926, p. 111) (obj.); *Proclivocrinus* RINGUEBERG, 1889, p. 396 (type, *Calceocrinus radiculus* RINGUEBERG, 1882, p. 120; OD) (=*C. chrysalis* HALL, 1860) (obj.);

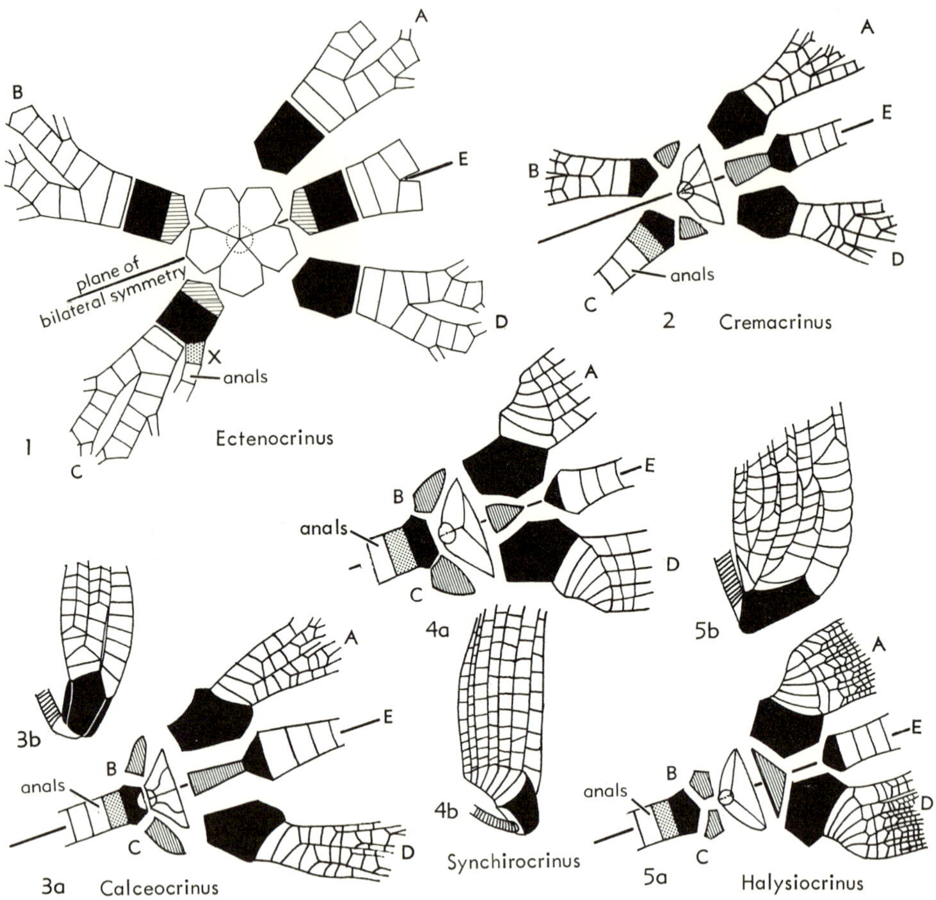

FIG. 324. Diagrams illustrating bilateral symmetry in the *E-BC* plane (homocrinid type) (mod. from Moore, 1962a).——*1. Ectenocrinus* (Ord.), typical homocrinid representative of inferred ancestral stock from which Calceocrinidae evolved, showing 5 subequal basals, 3 compound radials, and 2 undivided radials; anal series armlike succession of plates borne by *C* superradial.——*2. Cremacrinus* (Ord.-Sil.), bilateral symmetry well marked but imperfect; anal series not associated with remnant arm of *C* ray.——*3. Calceocrinus* (Ord.-Sil.), bilateral symmetry advanced by loss of arm in *B* ray.——*4. Synchirocrinus* (Sil.-Dev.), basals reduced to 3, main axils of *A* and *D* rays prominent.——*5. Halysiocrinus* (Dev.-Miss.), *B* and *C* superradials migrated to adjoin one another.

Euchirocrinus BATHER, 1893a, p. 65 (type, *C. chrysalis* HALL, 1860; OD) (obj.); *non Calceocrinus* AUCTT. (=*Chirocrinus* ANGELIN, 1878, p. 22)]. Crown with complete but not strikingly evident bilateral symmetry; median arm (*E* ray) generally branched isotomously but may be unbranched; 2 laterally placed arm systems (*A, D* rays) on opposite sides of crown composed of weakly developed main-axils, which commonly include nonaxillary plates alternating with successive axillaries that may have subequal distal facets, axil-arms and ramules few but developed in normal manner. Cup with four basals, each of which forms part of stem impression, suture between 2 small median basals *(DE, EA)* tending to be obscure; oppositely placed large undivided radials of *A* and *D* rays, superradial of *E* ray mostly connected narrowly with inferradial, which widens proximally, inferradials of *B* and *C* rays small and spear shaped, well separated from each other by basals and fused superradial situated obliquely above them, this superradial forming subanal plate that supports anal *X* plate and anal tube adjoined by stem in median plane of bilateral symmetry (MOORE, 1962a). *M.Ord.-M.Sil.,* N.Am.-Eu.(Eng.-Sweden).——FIG. 326,*5*. **C. chrysalis*

(HALL), M.Sil.(Niagaran), USA(N.Y.); *5a,b,* lower part of crown from *A-* and *E*-ray sides, ×2; *5c,* basal circlet showing 4 basals, stem impression above, hinge for articulation with radials below, ×2; *5d, A*-ray view of crown with part of attached stem (holotype of type species of *Proclivocrinus* RINGUEBERG), ×2 (all Springer, 1926a). [Also Fig. 324,*3*.]

[HALL's original description and illustrations of *Calceocrinus* were unaccompanied by a specific name, and therefore when RINGUEBERG (1889, p. 403) redescribed and figured a fragmentary fossil reported to be HALL's original specimen of *Calceocrinus,* naming it *C. halli,* various authors, including BASSLER (1938, p. 56), BASSLER & MOODEY (1943, p. 345), and RAMSBOTTOM (1952, p. 34), judged that the failure by HALL to publish a specific name for this single known representative of the genus has been rectified by RINGUEBERG. Accordingly, they cited *C. halli* RINGUEBERG as the type species of *Calceocrinus.* RINGUEBERG, however, did not consider the very incomplete type specimen named *C. halli* adequate to serve for delineating characters of the genus and he therefore published description and illustrations of a crown which he named *C. typus.* RINGUEBERG (1889, p. 402) explicitly designated *C. typus* as the type species of *Calceocrinus* and this action was accepted by BATHER (1893a, p. 65), SPRINGER (1926a, p. 115), and others as valid. ¶ Actually, neither *Calceocrinus halli* nor *C. typus* is eligible under the Rules to be the type species of the genus, for this must be either 1) the first species with valid "indication" assigned definitely to *Calceocrinus* by a subsequent author, if only one such species was so assigned, or 2) a species selected from a group of definitely referred species, if two or more were so referred at the same time, this selection being made either by the author who first assigned them to the genus or, if no selection then was made, by the first subsequent author who made a choice from the assigned group (all other species being excluded from consideration). Search of the literature shows no species were unequivocally placed in *Calceocrinus* until 1877 when S. A. MILLER (p. 72-73) listed 12 validly established species (including *Cheirocrinus chrysalis* HALL, 1860) as belonging to this genus. The type species must be one of this group, not some later-named form, but MILLER failed to make a choice. WACHSMUTH & SPRINGER (1886, p. 205 [281]) were first to do this when they explicitly designated *Cheirocrinus chrysalis* as the type species.]

Anulocrinus RAMSBOTTOM, 1961, p. 8 [**A. thraivensis*; OD]. Crown pendent, subparallel to stem, like *Cremacrinus* in having 4 basals and 4 arm-bearing rays, *E* arm unbranched or divided isotomously on about the sixth primibrach and fifth secundibrachs, lateral rays *(A, B, D)* with arms that branch isotomously or nearly so within main-axil series but heterotomously in all axil arms; ramules few to numerous, all arranged precisely in conformance with the normal calceocrinid pattern of heterotomy; *E* superradial wider than high, narrowly touching slender inferradial, which is slightly sinuous sided and appreciably taller than wide; large undivided lateral radials of *A* and *D* rays, forming most of opposite sides of cup; *B* superradial small, subquadrangular, *B* and *C* inferradials spear shaped, in type species *C* superradial (termed subanal) seemingly wide, low, nearly median plate at proximal extremity of anal tube, visible part of which consists of large, stout plates with height of each about equal to width. Stem moderately slender, composed of discoid to barrel-shaped columnals, circular transversely, thinnest in proximal region near cup and thickening distally until height may appreciably exceed diameter (MOORE, 1962a; RAMSBOTTOM, 1961).

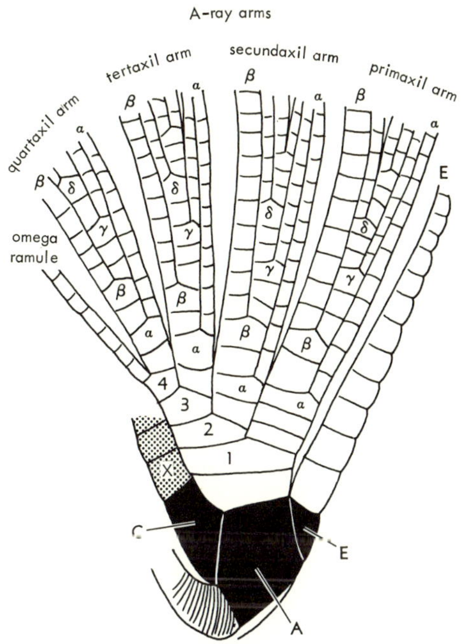

FIG. 325. Diagram of *A*-ray side of *Chirocrinus* illustrating differentiation of main-axil and axil-arms borne by it; abanal brachial and ramules shaded (mod. from Moore, 1962a).

U.Ord.-U.Sil., Eu.-N.Am.——FIG. 327,*2a-d.* **A. thraivensis,* U.Ord.(Ashgill), Scot.(Girvan); *2a, E*-ray view of paratype, ×2.8; *2b, A*-ray view of holotype, stout *E* arm at right, ×3.5; *2c,d, D*-ray view of arms, and post. view of partial crown, ×2.8 (Ramsbottom, 1961).——FIG. 327,*2e,f. A. simplex* (SPRINGER), M.Sil.(Beech River), USA (Tenn.); *2e,* view of holotype from *A*-ray side, *B* ray at left and *E* ray at right, ×2.4; *2f,* same from opposite (*D* ray) side, *E* ray at left, ×2.4 (Springer, 1926a).

Chirocrinus ANGELIN, 1878, p. 22 [**C. gotlandicus*; M] [=*Calceocrinus* AUCTT. (*non* HALL, 1852) (*non Cheirocrinus* EICHWALD, 1856, *nec* HALL, 1852, *nec* SALTER, 1873)]. Crown with perfect bilateral symmetry; median *(E)* ray undivided, small inferradial generally separated from superradial; lateral rays *(A, D)* with large radials succeeded by 2 primibrachs, primaxil with unequal distal facets, narrower one on abanal side supporting axil-arm and broader one bearing secundaxil of main-axil series, which, though definite, is not strongly distinguished or appreciably curved adanally; each axil-arm bearing moderately stout unbranched ramules that rise from axillaries separated by 1 or 2 quadrangular brachials, ramules of all series subequal in size. Cup showing mod-

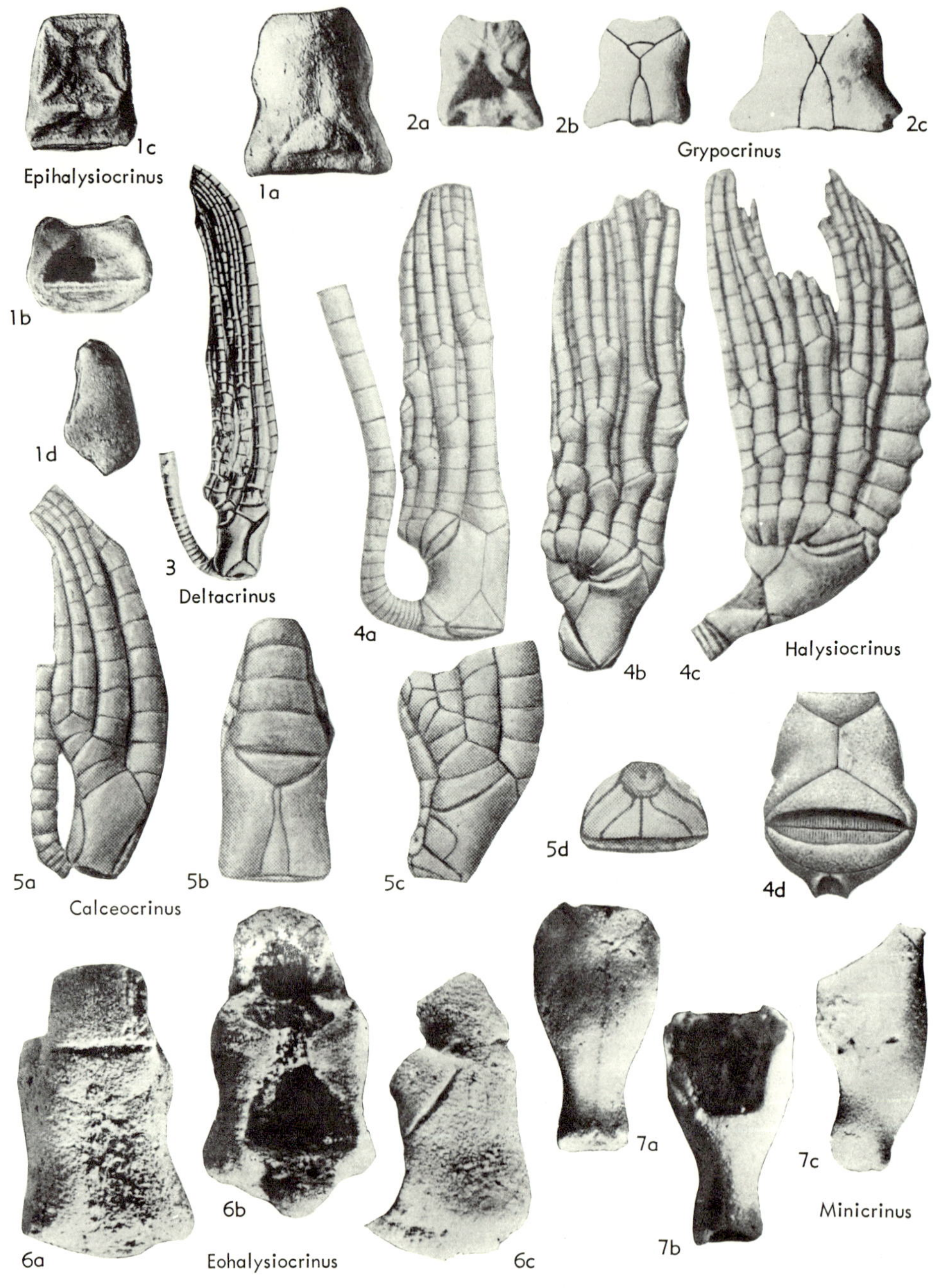

FIG. 326. Calceocrinidae (p. *T*525-*T*527, *T*530, *T*532).

erately large spear-shaped inferradials of *B* and *C* rays well separated by basals and fused superradial beneath anal *X* plate; basals reduced to 3 by fusion of *DE* and *EA* plates, which form low, triangular basal that generally does not extend into stem facet (MOORE, 1962a). *L.Sil.*, N.Am.; *M.Sil.*, Eu.——FIG. 327,*4*; 328. *C. fletcheri* (SALTER), M.Sil.(Wenlock.), Eng.(Dud-

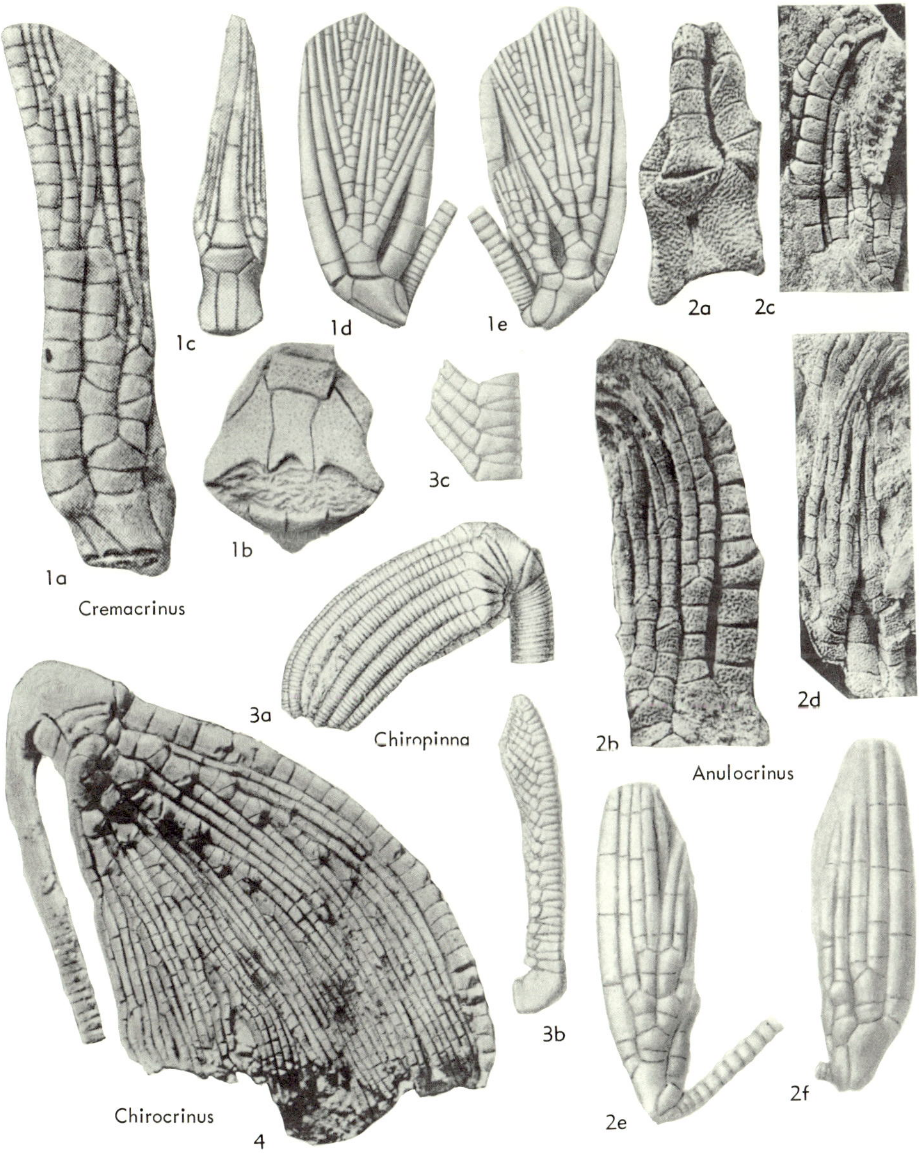

FIG. 327. Calceocrinidae (p. *T*527-*T*530).

ley); 327,*4*, *D*-ray view of nearly complete crown (holotype) with part of attached stem, ×1 (Ramsbottom, 1952); 328, drawing of same showing separate axil arms (Moore, n). [Also Fig. 325.]

Chiropinna MOORE, 1962, p. 28 [**Calceocrinus pinnulatus* BATHER, 1893a, p. 96; OD]. Crown compactly subcylindrical, cup of normal type, surmounted by pinnulate axil-arms in lateral rays and pinnulate dichotomous branches of median *(E)* ray, main-axils well developed (MOORE, 1962a). [This crinoid is not only a unique sort of calceocrinid but it stands alone among all disparid crinoids in having pinnulate arm structure (but interpretable also as extraordinarily numerous and regular ramules borne by brachials not distinguished as axils). Compared with all other calceocrinids, with which this genus unquestionably belongs, the unbranched offshoots of brachials on alternating sides are not different from ramules, except for the fact that every brachial plate of

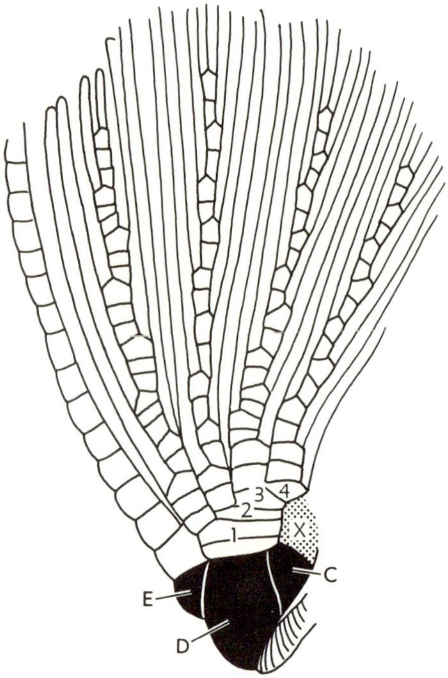

FIG. 328. Drawing of *Chirocrinus* showing separate axial arms *1-4* (anal plates stippled) (Moore, n).

each axil-arm is an axillary, giving off a branchlet (pinnule) to left or right. Thus the axil-arm structure of *Chiropinna* differs in no way from the pinnulate arms of such cladid inadunate genera as *Decadocrinus,* and numerous others. Whatever its significance may be, the brachials of *Chiropinna* give rise to first pinnules on the adanal side, rather than the abanal side of the *E* arm as characterizes the ramules in other genera of calceocrinids. The median *(E)* ray of *Chiropinna* is unique among calceocrinids in showing a dichotomous division into two arms that are unbranched (except for abundant pinnules) above the single axillary primibrach which bears them. The first pinnules of the median arms are given off from the adanal side of the second brachials above the axillary primibrach.] *U.Sil.(Wenlock.),* Sweden(Gotl.).——FIG. 327,*3.* **C. pinnulata* (BATHER); *3a, A*-ray view of holotype showing unusually thick stem, well-developed main-axil and parallel axil-arms without visible pinnules, *E* arm at right edge of view, ×1.3; *3b,* side view of single axil-arm of *A* ray showing pinnules borne by successive brachials, ×2; *3c,* detail of axil-arm with pinnules, ×4 (Bather, 1893a).

Cremacrinus ULRICH, 1886, p. 107 [**C. punctatus;* OD] [=*Castocrinus* RINGUEBERG, 1889, p. 389 (type, *Calceocrinus furcillatus* W. R. BILLINGS, 1887, p. 51; OD); *Castrocrinus* WACHSMUTH & SPRINGER, 1891, p. 387 *(nom. null.)*]. Bilateral symmetry of crown imperfectly developed, calyx asymmetrical, arm-bearing rays 4, stem not in plane of anal tube. Median *(E)* ray typically unbranched but in some species bifurcating distally, quadrilateral inferradial rather broadly in contact with superradial. Lateral arms 3, those of *A* and *B* rays on one side and of *D* ray alone on opposite side, each arm with ill-defined but definite main-axil series consisting of 2 primibrachs (axillary with subequal distal facets) and 1 or 2 secundibrachs; axil-arms few, generally 2 borne by each main-axil, ramules numerous, strictly conforming to calceocrinid pattern of heterotomy. Anal tube supported by *C* superradial, *B* superradial supporting arm that generally is distinctly smaller than that borne by adjoining *A* ray; *B* and *C* inferradials spear shaped, separated from one another by basals and superradials. Basals 4, all partly covered by stem impression (MOORE, 1962a). *M.Ord.-U.Sil.,* N.Am.——FIG. 327,*1a,b.* **C. punctatus,* M.Ord.(Black River), Minn.; *1a,b, D*-ray view of crown and *E*-ray view of cup, ×2 (Springer, 1926a).——FIG. 327,*1c-e.* *C. tubuliferus* SPRINGER, M.Sil.(Beech River), USA (Tenn.); *1c, E*-ray view of crown (paralectotype); *1d,e,* lat. views of crown (lectotype) from *D* and *AB* sides, robust unbranched *E* arm at left and right margins, anal tube shown in *1d* between *D* ray and stem; all ×2 (Springer, 1926a). [Also Fig. 324,*2.*]

Cunctocrinus KESLING & SIGLER, 1969, p. 340 [**C. fortunatus;* OD]. Basals 3; *A* and *D* radials laterally fused with *B* and *C* ray inferradials; *E*-ray infer- and superradials widely separated by admedial edge of *A* and *D* radials. Main axils *5* or *6,* bearing heterotomously branched axil-arms with 2 to 4 alphabrachs, and at least two higher divisions. *M.Dev.,* USA(Ohio).— FIG. 329,*1.* **C. fortunatus; 1a-d,* post., *A*-ray, *D*-ray, and ant. views of crown, ×5.2 (Kesling & Sigler, 1969).

Deltacrinus ULRICH, 1886, p. 100 [**Cheirocrinus clarus* HALL, 1862, p. 116; OD]. Crown slender and elongate, with perfect bilateral symmetry, median *(E)* ray undivided, lateral rays with very ill-defined main-axils and relatively few axil-arms which lack evident differentiation of alpha-, beta-, and gamma-ramules; inferradial of median ray triangular (delta-shaped) and widely separated from superradial by adjoined lateral radials; basals 3 closely resembling those of *Chirocrinus* (MOORE, 1962). [This genus is distinguished by relatively advanced characters of the cup combined with primitive features of the main-axils and slender axil-arms.] *L.Sil.-U.Sil.,* USA(Mo.-Ind.-N.Y.); *M.Dev.,* USA(N.Y.).——FIG. 326,*3.* **D. clarus* (HALL), M.Dev.(Hamilton), USA(N.Y.); holotype showing unbranched *E* ray toward right and

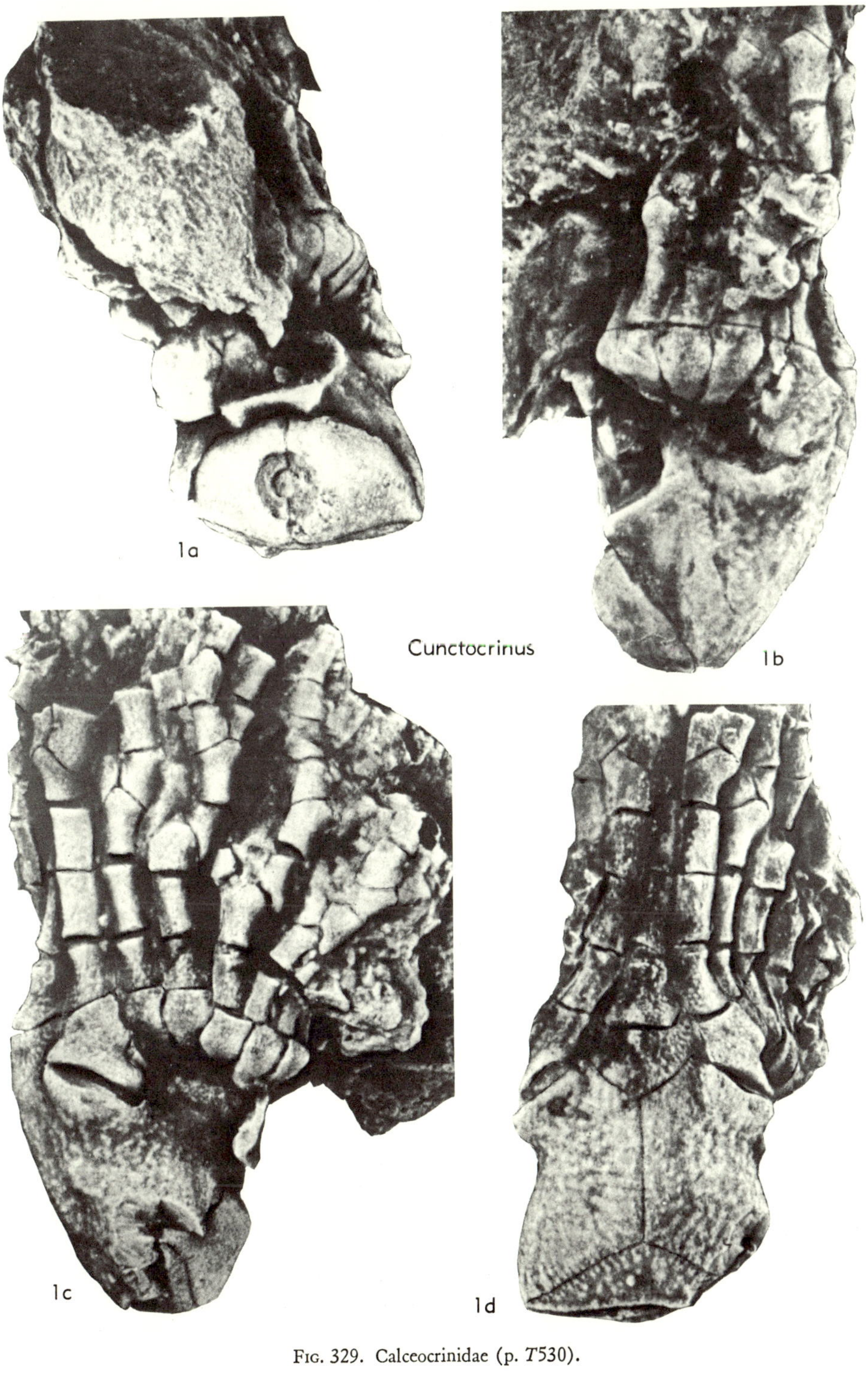

Fig. 329. Calceocrinidae (p. *T*530).

axil-arms of *A* ray at left, ×1.3 (Goldring, 1923).

Eohalysiocrinus Prokop, 1970, p. 97 [**E. convexus*; OD]. Cup small, elongate; basals 3, unfused; *B* and *C* inferradials small, in lateral contact, overlapped by large elongate *A* and *D* radials, bounded distally by fused, quadrangular *B* and *C* superradials; *E* ray inferradial small, triangular, widely separated from larger, triangular *E* superradial. Arms unknown. Stem slender, composed of minute, cylindrical columnals. *L. Dev.-M.Dev.*, Eu.(Boh.).——Fig. 326,*6*. **E. convexus; 6a-c, E*-ray, *BC*-interray, and lat. views of theca and first primibrach, ×8 (Prokop, 1970).

Epihalysiocrinus Arendt, 1965, p. 90 [**Halysiocrinus? tuberculatus* Yakovlev, 1927b, p. 188; OD]. Basals fused into a single plate; *E*-ray infer- and superradial widely separated by the *A* and *D* radials; small *B* and *C* inferradials present in cup. Arms and anal tube unknown. *L.Perm. (Artinsk.)*, Eu.(USSR).——Fig. 326,*1*. **E. tuberculatus* (Yakovlev); *1a, E* ray, ×2; *1b*, basal, ×3; *1c,d*, int. and lat. views of cup, ×2.3 (Arendt, 1965).

Espanocrinus Webster, 1976, p. 681 [**E. lemonei*; OD]. Crown slender, elongate, bilaterally symmetrical. Cup like that of *Halysiocrinus*. *E*-ray arm unbranched, plates convex; 3 unbranched lateral axil arms in *A* and *D* rays, main axils *I* and *III* axillary in *A* and *D* rays. Anal sac short, strongly protuberant. Column round, tapering distally. *L.Dev.(Ems.)*, Eu.(Spain).

Grypocrinus Strimple, 1963, p. 59 [**G. genuinus*; OD]. Basals unknown, probably 3; *E* ray infer- and superradial narrowly separated by *A* and *D* radials; superradial small, triangular, enclosed by large crescentic primibrach in contact with distal corners of *A* and *D* radials. Arms unknown. *U.Sil.*, USA(Okla.).——Fig. 326,*2*. **G. genuinus*; *2a-c*, int. and ext. *E*-ray views of cup, ×0.7 (Strimple, 1963a).

Halysiocrinus Ulrich, 1886, p. 110 [**Cheirocrinus dactylus* Hall, 1860, p. 123; OD]. Crown slender to moderately full in girth, perfectly symmetrical, median (*E* ray) with infer- and superradial plates well separated by adjoined lateral radials between them and generally with dichotomously branched medial and distal parts of median ray, lateral rays with large radials that support curved main-axils and numerous axil-arms which bear large but not dominant beta-ramules, as in *Synchirocrinus*; cup composed of large lateral radials that adjoin one another between infer- and superradial of median radial, inferradials of *B*- and *C*-rays that adjoin above stem impression and beneath fused superradial of these rays, and 3-plate basal circlet in which median element along hinge is low, arcuate plate (Moore, 1962a). *M.Dev.-L. Miss.(Osag.)*, N.Am.(Iowa-Ind.-Ky.-Tenn.).——Fig. 326,*4a,b*. **H. dactylus* (Hall), L.Miss. (Osag.), USA(Iowa); *A*- and *E*-ray views of crown, ×1.6 (Springer, 1926a).——Fig. 326,*4c*. *H. nodosus* (Hall), L.Miss.(Osag.), USA(Ind.); *A*-ray view of crown, ×1 (Springer, 1926a).——Fig. 326,*4d*. *H. perplexus* (Shumard), L.Miss. (Osag.), USA(Ky.); exterior of cup showing radials and basals fully extended from hinge, ×1.3 (Springer, 1926a). [Also Fig. 324,*5*.]

[Bather's (1893, p. 63) designation of *Calceocrinus ventricosus* (=*Cheirocrinus ventricosus* Hall, 1860) as type species of *Halysiocrinus* is invalid, whatever may be one's judgment as to the status of this species as a probable synonym of *Cheirocrinus dactylus* as generally held, or as an independent species.]

Minicrinus Prokop, 1970, p. 109 [**M. inflatus*; OD]. Theca vase shaped, narrow and elongate in basal area, broad and inflated in radial area; *A* and *D* radials large, composing most of theca, in lateral contact with anal plate and separating small fused *B* and *C* inferradials from basal plates. Arms and stem unknown. *L.Dev.-M.Dev.*, Eu. (Boh.).——Fig. 326,*7*. **M. inflatus; 7a-c, E*-ray, *BC*-interray, and lat. views of theca, ×12 (Prokop, 1970).

Senariocrinus Schmidt, 1934, p. 16 [**S. maucheri*; M]. Bilaterally symmetrical crown distinguished by single dichotomous division of *A* and *D* arms, neither of which bears ramules, and by long unbranched *E* arm; stout series of U-shaped anal plates supports fine-plated sac with anal pyramid at its tip. Cup plates doubly hinged, normal calceocrinid hinge between single triangular basal and *E* inferradial being supplemented by another hinge located between fused *B*+*C* inferradial and *B* superradial, which with oblique suture underlies *C* superradial, with anal series rising from latter. Slender cylindrical stem attached to extremity of basal hinge, composed of thin columnals and tapering distally to point (Schmidt, 1934; Moore, 1962a). [Inferred to have been free-swimming.] *L.Dev.(Ems.)*, Eu.(W.Ger.).——Fig. 330,*1*. **S. maucheri*, Hunsrück Sh., near Bundenbach; *1a*, view of holotype from *D*-ray side, *E* arm at right and elongate anal sac left, ×0.85 (Schmidt, 1934); *1b,c*, tip of anal sac, *D*-ray view of crown, ×0.85, enl. (Moore, 1962a).

Synchirocrinus Jaekel, 1918, p. 86 [**S. anglicus* (=*Calceocrinus nitidus* Bather, 1893a, p. 91); OD] [=*Cheirocrinus* Salter, in Murchison, 1859, p. 535 *(nom. nud.)*; *Cheirocrinus* Salter, 1873, p. 118 (type, *C. serialis*; SD Moore, 1962a, p. 26 (*non* Eichwald, 1856, *nec* Hall, 1860); *Pendulocrinus* Salter, 1859, p. 535 *(nom. nud.)*]. Crown with perfect bilateral symmetry, resembling *Chirocrinus* except in having strongly marked main-axil series in lateral rays and axil-arms that in side view of most specimens seem to consist of parallel rows of subequal brachials, whereas in fact they are chiefly composed of exceptionally robust beta-ramules (see Fig. 12) (Moore, 1962a). *L.Sil.-U.Sil.(Wenlock.)*, N.Am.(Ohio-N.Y.-Tenn.)-Eu.(Eng.-Sweden); *L.Dev.*, N.Am. (Iowa). [One of the most distinctive calceocrinid genera, differing from *Deltacrinus* in its well-marked, curved main-axils and much more numer-

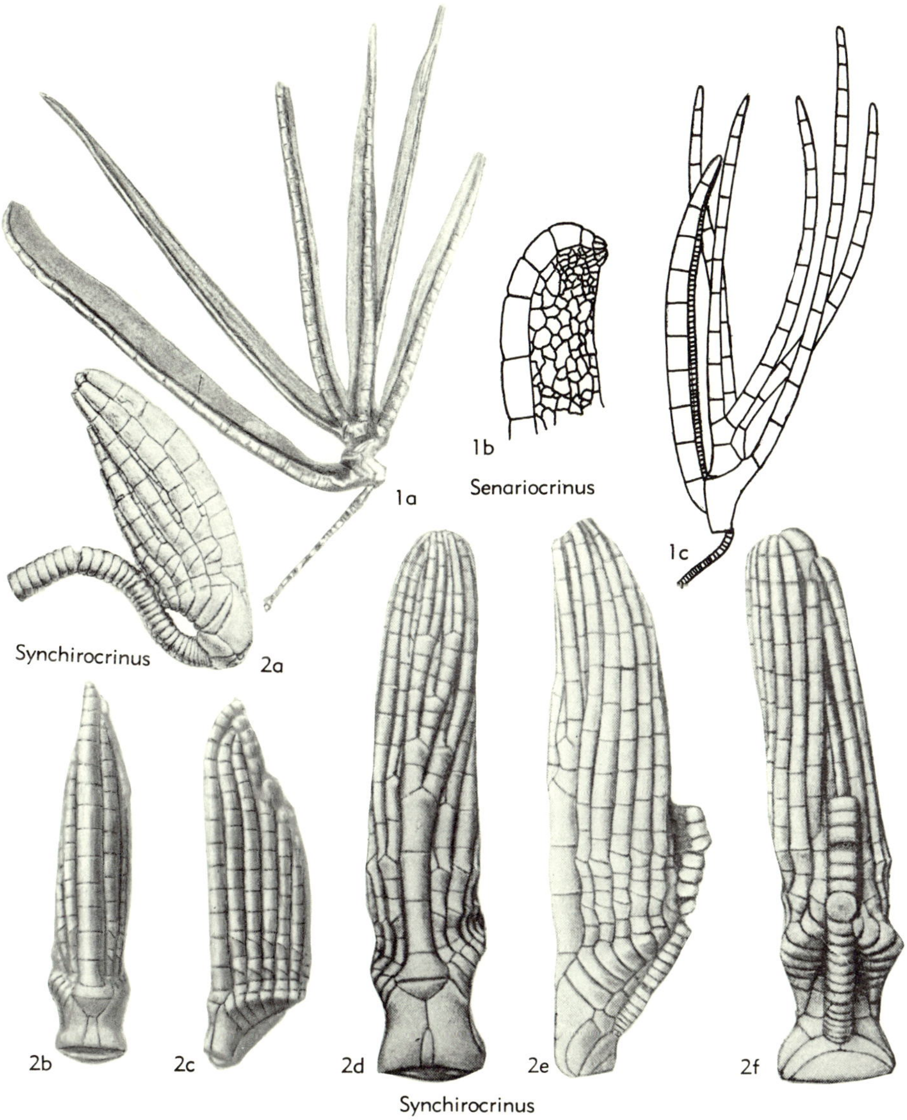

FIG. 330. Calceocrinidae (p. *T*532-*T*533).

ous axil-arms. It most closely resembles *Halysiocrinus,* but differs especially in its more robust beta-ramules and largely concealed axil-arms.]——FIG. 330,*2a*. *S. tucanus* (BATHER), U.Sil.(Wenlock.), Sweden(Gotl.); *D*-ray view of crown with part of attached stem, showing stout beta-ramules of this ray and, at right edge, unbranched *E* arm, ×2 (Bather, 1893a).——FIG. 330,*2b,c*. *S. foerstei* (SPRINGER), M.Sil.(Beech River F.), USA(Tenn.); *2b,c*, *E*- and *D*-ray views of lectotype, latter showing only beta-ramules visible in upper part of crown, ×1.3 (Springer, 1926a).——FIG. 330,*2d-f*. *S. bifurcatus* (SPRINGER), M.Sil.(Beech River F.), USA(Tenn.); *2d-f,* holotype viewed from *E, D,* and posterior sides, showing well-developed main-axil and parallel robust beta-ramules which conceal axil-arms, ×2 (Springer, 1926a). [Also FIG. 324,*4*; 331,*1*.]

Superfamily PISOCRINACEA Angelin, 1878

[*nom. transl.* MOORE & LANE, in MOORE & STRIMPLE, 1973, p. 16 (*ex* Pisocrinidae ANGELIN, 1878, p. 20)] [Materials for this superfamily prepared by R. C. MOORE and N. GARY LANE]

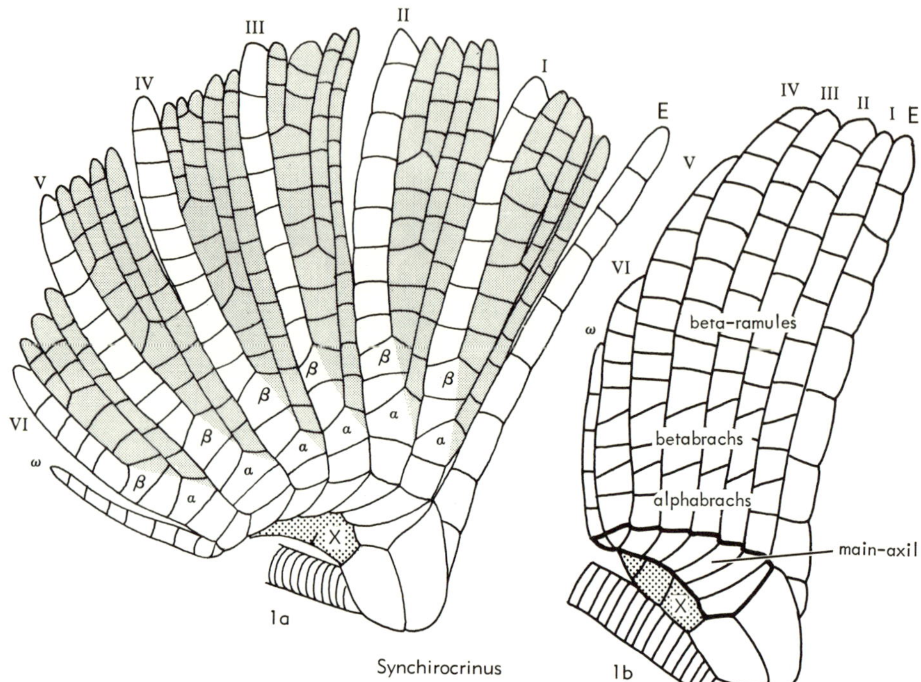

FIG. 331. Diagrams showing many-branched structure of *A* ray in *Synchirocrinus* (Sil.-Dev.) (anal series stippled) (Moore, 1962a).——*1a.* Axil-arms drawn in spread-out positions to reveal hidden parts of axil-arms and ramules (shaded).——*1b.* Arm branches above main-axil in normal position, visible parts being mostly betabrachs and beta-ramules.

Crown slender, with erect rounded unbranched arms above conical to globose cup which displays homocrinid type of bilateral symmetry through plane of *E* ray and *BC* interray. Basals three to five, low and inconspicuous; *A* and *D* radials large, undivided, those of other rays compound, radial articular facets narrow, leaving prominent interradial prongs. In Allagecrinidae, Catillocrinidae, and Anamesocrinidae some large multifaceted radials with threadlike unbranched arms. Small stem transversely round. *M.Sil.(Wenlock., Niagaran)-U.Dev. (Chemung.).*

Family PISOCRINIDAE Angelin, 1878

[Pisocrinidae ANGELIN, 1878, p. 20] [=Pisocrinidées DE LORIOL, 1882-84, p. 46; Triacrinidae JAEKEL, 1918, p. 89; Cicerocrinidae JAEKEL, 1918, p. 89]

Small crinoids with globose or conical cup. Basals three or five; three small radials, which correspond to the *B, C,* and *E* compound radials of Homocrinidae, and two large radials, inferradial of the *B* ray hypertrophied and displaced obliquely to right, forming with the large *A* and *D* radials most of the cup. Tegmen poorly known, anal opening above cup. Arms atomous and nonpinnulate or with two main ramule-bearing rami to a ray. *M.Sil.-U.Dev.*

Pisocrinus DE KONINCK, 1858, p. 104 [**P. pilula*; M] [=*Triacrinus* RINGUEBERG, 1884, p. 144 (type, *T. pyriformis*), *non* MÜNSTER, 1839]. Cup small, globose or rarely conical with flat base or basal concavity; basals 5, unequal in size, *AE* and *BC* basals smaller than other 3 basals and with truncated rather than acute distal edge. Radials unequal in size, *C* and *E* radials small, triangular and not in contact with basals; *B* ray with small triangular superradial and large inferradial which is shifted obliquely to left and situated directly above *BC* basal; *D* and *A* radials are large, simple, in contact with basals, and together with the *B* inferradial comprise most of the theca. Anal *X* small, situated above cup and in contact with upper corners of *C* and

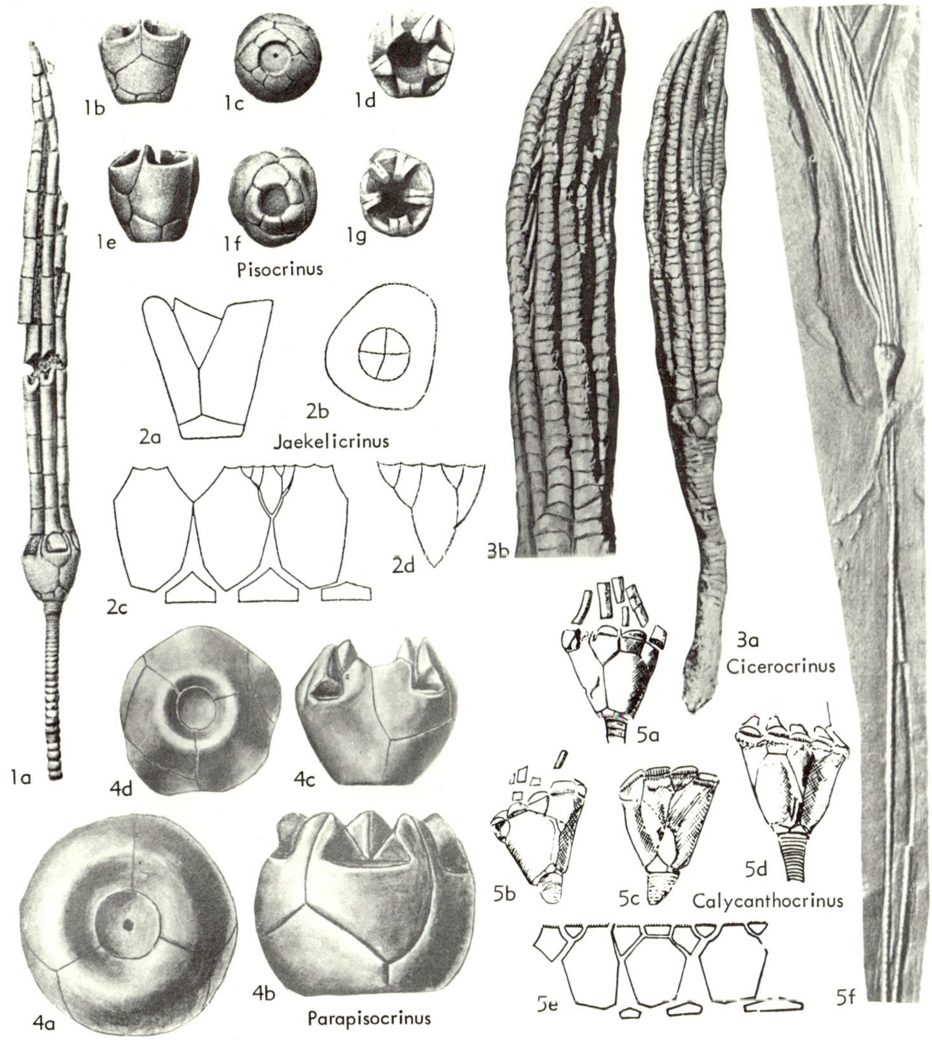

FIG. 332. Pisocrinidae (p. *T*534-*T*536).

D radials. Arm facets deeply notched into upper surfaces of radials; articular surfaces with fine radial ridges and grooves, or a transverse ridge; facets bounded laterally and internally by raised outer edges of the radials. First primibrach short, remainder of brachials slender and elongate; arms atomous and nonpinnulate. Anal sac narrow and elongate, closely resembling an arm, triangular or crescentic in cross section, supported directly by anal *X* and confined to posterior part of the tegmen. Tegmen arched by 5 oral plates which interlock medially and are in sutural contact with the radial processes. *M.Sil.(Niagaran)-L.Dev.,* Eu. (Gotl.-Boh.-Eng.-USSR)-USA (N.Y.-Tenn.-Ill.-Mo.-Ind.)-Australia.——FIG. 332,*1*. **P. pilula; 1a-d,* U.Sil., Gotl., lat, view of crown and stem, lat., dorsal, and ventral views of cup, ×2; *1e-g,* U.Sil., Eng.; lat., dorsal, and ventral views of cup, ×2 (Springer, 1926a).

Calycanthocrinus FOLLMAN, 1887, p. 115 [**C. decadactylus*; M]. Cup cone shaped; basals three, equal and triangular in shape. *A* and *D* radials large, each supporting a single atomous arm; *B* inferradial large, supporting small *C* radial and *B* superradial on upper left and right shoulders. Additional arm-supporting ray plates in the cup (pararadials) are present directly above the *B* inferradial, between *C* radial and *B* superradial, between the small *E* radial and large *D* radial, between the *E* radial and large *A* radial, and be-

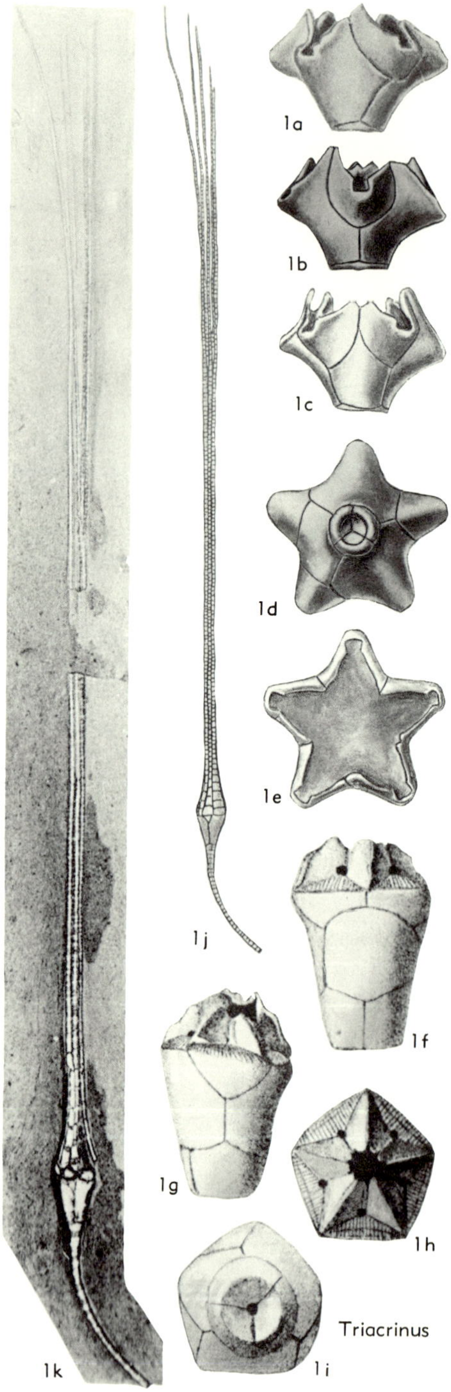

FIG. 333. Pisocrinidae (p. T536-T537).

tween the *B* superradial and *A* radial. Each of these 4 extra plates as well as the 5 radials bears a long, slender, atomous arm so that nine arms are present in the crown. Anal *X* a small plate at the upper right corner of *D* radial; anal sac not known. *L.Dev.-M.Dev.*, Eu.(Ger.).——FIG. 332,*5*. **C. decadactylus*, L.Dev., Ger.; *5a-d*, *C*-ray, *AE*-interray, *B*-ray, and *E*-ray views of cup and proximal brachials, ×1; *5e*, plate diagram of cup, ×1; *5f*, lat. view of crown and stem, ×0.5 (Schmidt, 1934).

Cicerocrinus SOLLAS, 1900, p. 267 [**C. elegans* SOLLAS, 1900 (May), p. 27; M] [=*Lagarocrinus* JAEKEL, 1901 (Sept.), p. 480 (type, *L. osiliensis*)]. Cup high and conical, with upright elongate basals. Arrangement of cup plates like that of *Pisocrinus*. Anal *X* small, on upper right shoulder of *D* radial. Arms branch on second primibrach in all rays, each arm with 2 main rami or with an additional division high above cup. Secundibrachs bear stout unbranched ramules on alternate sides of every second brachial. *U.Sil.*, Eu.(Eng.-Sweden-USSR).——FIG. 332,*3*. **C. elegans*, Eng.; *3a*, lat. view of crown and stem, ×1.5; *3b*, lat. view of arms, ×2 (Sollas, 1900).

Jaekelicrinus YAKOVLEV, 1949, p. 435 [**J. bashkiricus*; OD]. Cup small, cone shaped; basals 3 or 4; *A* and *D* radials large, each with 2 small arm facets; *B* inferradial large, with single arm facet; several small pararadials interposed between upper angles of *A* radial and *B* inferradial, each with single arm facet; 10 arm facets in all; arms and stem unknown. *U.Dev.*, Eu.(USSR).——FIG. 332,*2*. **J. bashkiricus*; *2a-d*, diagrams of lat. view of cup, basal view, plate diagram of cup, ×2, and enlargement of pararadials (Yakovlev, 1949b).

Parapisocrinus MU, 1954, p. 331 [**Pisocrinus ollula* ANGELIN, 1878, p. 21; OD] [=*Ollulocrinus* BOUŠKA, 1956, p. 28 (obj.)]. Like *Pisocrinus* except basal circlet small, circular in outline and confined to bottom of basal concavity. Basals 5, equal, and completely covered by proximal columnal. Recurved proximal edges of the *D* and *A* radials and *B* inferradial form lateral walls of basal concavity; interradial prongs prominent. *U.Sil.-L.Dev.*, Eu.(Gotl.-Czech.)-USA(Tenn.-Ill.-Mo.)-Australia.——FIG. 332,*4a,b*. **P. ollula* (ANGELIN), U.Sil., Boh.; basal and *B*-ray view of cup, ×5 (Bouška, 1956).——FIG. 332,*4c,d*. *P. quinquelobus* (BATHER), U.Sil., Boh.; *CD*-interray and basal views of cup, ×5 (Bouška, 1956).

Triacrinus MÜNSTER, 1839, p. 3 [**T. pyriformis*; SD WACHSMUTH & SPRINGER, 1886, p. 175] [=*Trichocrinus* MÜLLER, 1856, p. 354 (type, *T. altus*); *non Triacrinus* RINGUEBERG, 1884, p. 144 (=*Pisocrinus* DE KONINCK, 1858, p. 104)]. Like *Pisocrinus* except there are 3, rather than 5, basals in the cup. Cup commonly pentalobate in oral or aboral view. *U.Sil.-U.Dev.*, Eu.(Ger.-Gotl.-Boh.-USSR).——FIG. 333,*1a-e*. *T. regnelli* BOUŠKA, U.Sil., Boh.; *1a-e*, *CD*-interray, *E*-ray, *BC*-interray, basal, and ventral views of cup, ×5.6 (Bouška, 1956).——FIG. 333,*1f-i*. *T. altus* (MÜLLER), M.Dev., Ger.; *1f-i*, *BC*-interray, *E*-ray, ventral, and basal views of cup, ×1.7 (Schultze, 1867).——FIG. 333,*1j,k*. *T. elongatus* (FOLLMAN), L.Dev., Ger.; *1j-k*, lat. views of crown and stem, ×0.75, ×1.12 (Schmidt, 1934).

Superfamily ALLAGECRINACEA Carpenter & Etheridge, 1881

[*nom. transl.* MOORE & STRIMPLE, 1973, p. 16 (*ex* Allagecrinidae CARPENTER & ETHERIDGE, 1881, p. 292)] [Materials for this superfamily prepared by R. C. MOORE and H. L. STRIMPLE]

Crown slender, moderately tall, mostly with many rope- or threadlike unbranched uniserial arms, rarely with fewer than five arms or armless as adults. Cup conical or bowl shaped; basals five or fused to one; radials upflaring, nearly equal or very unequal, with *A* and *D* much larger than others, large radials with multiple arm facets. Anal tube cylindrical, tall, composed of superposed brachial-like plates. Stem transversely circular. Bilateral symmetry in plane through *E* ray and *BC* interray remarkably developed in several genera. *Ord.; ?Sil.; Dev.-Perm.*

Family ALLAGECRINIDAE Carpenter & Etheridge, 1881

[Allagecrinidae CARPENTER & ETHERIDGE, 1881, p. 292] [=Allagecrininae MOORE, 1940b, p. 111]

Microscopic to relatively small monocyclic crinoids with slender, tall crown having few to very many or closely appressed arms directed straight upward, rarely with fewer than five arms or armless as adults. Cup ranging from narrowly conical (among microcrinoids) to medium high or low bowl shaped, truncate at summit. Transverse outline normally circular to slightly elliptical but may be pentalobate or stelliform; horizontal base largely or entirely covered by proximal columnal; basals visible or invisible from side, consisting of five diminutive plates or fused to 3 or single moderately large solid one; radials five, sloping steeply upward, or longitudinally rounded to bulbous, planate radial articular facets gently hollow to rather deeply excavate, facets on different radials ranging from single ones to as many as 34, all relatively narrow and elongate in direction toward interior of cup, devoid of transverse ridge or distinct muscle areas. Dome-shaped circlet of orals in microcrinoid adults, as well as in immature stages of larger forms and in diminutive adults. Anal *X* plate present in most genera but absent in some. Followed by linear series of medium-sized extraordinarily thick plates resembling brachials, all together forming tall, erect, anal tube which may extend well above distal extremities of arms. Unbranched cylindrical arms well rounded externally and grooved internally, lacking pinnules. Elongate brachials with smooth truncate articular facets, oriented with planes normal to arm axis. Stem transversely circular or slightly elliptical, composed of homeomorphic or moderately heteromorphic columnals, stem tapering distally. *U.Dev.-U.Perm.*

The most distinctive morphological features of crinoids included in the Allagecrinidae are 1) occurrence of multiple unbranched arms on one or more radials of most genera (larviform adults in such genera as *Kallimorphocrinus, Desmacriocrinus,* and *Trophocrinus* excluded), and 2) distinct to strongly accentuated development of homocrinoid bilateral symmetry. Commonly, the *C* radial is developed as an aniradial but in some allagecrinine forms it is not.

Best development of multiple arms is seen in the Catillocrinidae, all genera of which show this feature. Maximum development of arms is found in the *A* and *D* rays (Tables 6, 7). In *Catillocrinus* as many as 34 arm facets are found on the *A* radial and 24 on the *D* radial, whereas the *B, C,* and *E* radials bear only single arm facets, thus indicating a total of 61 arms on the cup as a whole (see Table 6). The cup of *Eucatillocrinus* has up to 50 arms, with 22 on the *A* radial and 17 on the *D* radial (see Table 6). *Isocatillocrinus* is unique in that multiple arm facets occur on all radials.

Homocrinoid bilateral symmetry is recognized but weakly developed in most genera of the Allagecrinidae (lacking in *Desmacriocrinus, Kallimorphocrinus,* and *Trophocrinus*) (see Table 6). In the Catillocrinidae, on the other hand, homocrinoid bilateral symmetry is very distinct to strongly accentuated (Fig. 334).

Allagecrinus CARPENTER & ETHERIDGE, 1881, p. 282 [**A. austinii*; OD] [=*Hybochilocrinus* J. M. WELLER, 1930a, p. 21 (type, *Allagecrinus ameri-*

TABLE 6. *Arms and Basal Plates of Allagecrinidae and Catillocrinidae.*
[Explanation: A-E, CARPENTER designations of radials; BB, number of basals; Ref., reference; Total, total number of arms in adult; *, type species.]

	D	E	A	B	C	Total	BB	Ref.
ALLAGECRINIDAE								
*Allagecrinus austinii	2	1	6	2	1	12	3	1,14
Allagecrinus americanus	2	1	3	2	1	9	1	6a,b
*Desmacriocrinus weldenensis	?	1	?	1	1	3+	5	9
*Isoallagecrinus bassleri	5	1	4	3	1	14	3	8
Isoallagecrinus pecki	2	1	1	2	1	7	1	4
*Kallimorphocrinus astrus	1	1	1	1	1	5	1	13a
Kallimorphocrinus odiosus	0	0	0	0	0	0	1	13b
*Metallagecrinus quinquebrachiatus	1	1	1	1	1	5	1	11b
Metallagecrinus dux	2	1	5	3	1	12	1	12
Metallagecrinus multibrachiatus	3	1	3	4	1	12	1	16
*Stereobrachicrinus pustullosus	?	?	?	?	?	?	?	3
*Thaminocrinus biplex	4	1	5	1	2-5	13-16+	3	15
*Trophocrinus tumidus	0	1	0	0	0	1	3	2
Trophocrinus corpulentus	1	l	0	0	1	3	1	7a
Trophocrinus exsertus	1	1	1	1	1	5	1	7b
*Wrightocrinus jakovlevi	4	1	3	1	1	10	1	11a
CATILLOCRINIDAE								
*Catillocrinus tennesseeae	24	1	34	1	1	61	3	5
*Allocatillocrinus carpenteri	7	1	8	5	1	22	3	5
*Eucatillocrinus bradleyi	17	1	22	6	4	50	3	5
*Isocatillocrinus indicus	7	9	8	6	4	34	1	5
*Metacatillocrinus bulbosus	9	1	8	1	1	20	1	5
*Mycocrinus boletus	6	1	7	1	1	16	3	5
*Neocatillocrinus incissus	7	1	9	6	1	24	1	5
*Paracatillocrinus granulatus	8	1	10	1	1	21	3	5
*Xenocatillocrinus wrighti	8	1	10	5	4	28	?3	5

REFERENCES: 1, Carpenter & Etheridge, 1881, pl. 15, fig. 1b.——2, Kirk, 1930, p. 211.——3, Mather, 1915, p. 108.——4, Moore, 1940b, p. 125.——5, Moore, 1962b, p. 27.——6, Peck, 1935 (a, p. 766; b, p. 768).——7, Peck, 1936 (a, p. 285; b, p. 287).——8, Strimple, 1959, p. 116.——9, Strimple, 1966c, p. 109.——10, Strimple & Koenig, 1956, p. 1233.——11, Wanner, 1929b (a, p. 15; b, p. 24).——12, Wanner, 1930a, p. 23.——13, Weller, 1930a (a, p. 17; b, p. 28).——14, Wright, 1933, p. 203.——15, Wright, 1941, p. 294.——16, Yakovlev, 1927b, p. 191.

canus ROWLEY, 1895, p. 219; OD)]. Cup truncate bowl shaped or with straight sides flaring upward (greatest diameter in adults 1.5 to 6 mm.); basals 3 or fused to single plate; radials nearly equal to distinctly unequal in width, with largest ones in *A* and *D* rays, single arm facets on *C* and *E* radials, others multifaceted, with observed maximum of 6 on *A* radial; left shoulder on *C* radial with curved or straight bevel for reception of anal *X* plate (Fig. 334,*3*). Circlet of orals forming dome above cup in immature stages but generally lost in adults. Subcylindrical arms nearly equal in diameter, moderate in height, composed of elongate brachials. Stem composed of circular low columnals narrowing distally. *U.Dev.-L.Miss.(Kinderhook.-Osag.),* USA(Mo.-N. Mex.); *L.Carb.(Visean),* Eu.(Scot.).——FIG. 335, *7a-d*. **A. austinii*, L.Carb.(Visean), Scot.; *7a,b*, *E*- and *B*-ray views of specimen with attached proximal primibrachs and part of tapering stem, ×10; *7c,d*, *C*-ray and dorsal views (*A* ray below) of cup, ×10 (all Wright, 1951-54).——FIG. 335,*7e,f*. *A. americanus* (ROWLEY), U.Dev.(Louisiana Ls.), USA(Mo.); *C*-ray and ventral views of cup, ×20 (J. M. Weller, 1930).

Desmacriocrinus STRIMPLE, 1966, p. 109 [**Kallimorphocrinus weldenensis* STRIMPLE & KOENIG, 1956, p. 1234; OD]. Similar to *Kallimorphocrinus* but distinguished by 5 small basals and prominent dome of oral plates, articular facets for arms known only on *B, C,* and *E* radials; no anal notch in posterior interray. *U.Dev.(Louisiana Ls.),* USA (Mo.); *L.Miss.(Welden F.),* USA(Okla.).——FIG. 335,*1*. **D. weldenensis* (STRIMPLE & KOENIG), L.Miss.(Welden F.), USA(Okla.); *1a-c*, ventral, *CD*-interray, and *A*-ray views of holotype, ×17; *1d-f*, ventral, *CD*-interray, and dorsal views of paratype, ×17 (Strimple & Koenig, 1956).

Isoallagecrinus Strimple, 1966, p. 105 [**Allagecrinus bassleri* Strimple, 1938, p. 1; OD]. Similar to *Allagecrinus* but distinguished by its strongly depressed bowl-shaped cup and markedly convex radials, which may be distinctly tumid, in immature stages giving cups stelliform shape; basals 3, forming low, flat disc. Arms moderately tall, slender and threadlike, except for single robust ones borne by *C* and *E* radials, maximum observed number of arms 14 (see Table 6); where known, sequence in appearance of first arms is on radials *C, E, B, D,* and *A* (see Table 6). *M. Penn.(Atokan)-L.Perm(Wolfcamp.),* USA(Texas-Okla.-Kans.-Mo.-Ill.).——Fig. 335,*3*. **I. bassleri* (Strimple), U.Penn.(Missour., Wann F.), USA (Okla.); *3a, D*-ray view of crown, with smaller arms borne by this radial and robust ones in *C* (at right) and *E* (at left) radials, ×4; *3b,c,* ventral (*A* ray above) and dorsal (*A* ray below) views of typical cup, ×4; *3d,* dorsal view of small immature stage showing stelliform shape of cup, ×4 (all Strimple, 1938).

Kallimorphocrinus J. M. Weller, 1930, p. 15 [**K. astrus typicus* (=*K. astrus astrus*), p. 17; OD] [=*Aidemocrinus* J. M. Weller, 1930, p. 28 (type, *A. odiosus*; OD); *Callimorphocrinus* Peck, 1936, p. 287 *(nom. van.) (obj.)*]. Cup truncate conical, mostly taller than wide, pentamerously symmetrical with evenly expanded to nearly vertical sides; basals 3 or fused to solid disc; radials each with single crescentic arm facet (except *K. elongatus* Wright, which may have total of 8 arm facets) or armless (*K. odiosus* (Weller), Table 6), left shoulder of *C* radial not beveled for support of anal plate; summit with circlet of moderately prominent orals, that of *CD* interray larger than others and bearing small hydropore (Fig. 334,*1*). Only primibrachs known as part of arms. Stem composed of low, circular columnals, tapering slightly in distal direction, generally not preserved. [Ontogenetic studies indicate that various allagecrinid and catillocrinid genera (e.g., *Allagecrinus, Allocatillocrinus*) have an initial larval stage which is similar to or identical with *Kallimorphocrinus* and on this basis the genus is interpreted as a primitive member of the family. Distinct asymmetry of the cup and development of multiple arm facets on some of the radials characterize ontogeny subsequent to the *Kallimorphocrinus* stage.] *L.Miss.(Kinderhook.-Osag.),* USA(Okla.-Mo.-N.Mex.-Nev.); *L. Carb.(Visean),* Eu.(Scot.); *L.Penn.(Morrow.),* USA(Ind.-Ill.-Mo.).——Fig. 335,*4*. **K. astrus,* L.Penn., USA(Ind.); *4a,b,* lat. and dorsal views of holotype, ×20; *4c,* ventral view of another specimen, ×20 (all J. M. Weller, 1930).

Metallagecrinus Strimple, 1966, p. 109 [**Allagecrinus quinquebrachiatus* Wanner, 1929b, p. 24; OD]. Cup evenly expanded upward, truncate conical; basals 3 or solidly fused, low to medium in height and visible from side; radials 5, subequal, each normally with single arm facet; circlet of 5 orals moderately elevated, posterior oral separating *BC* and *DE* orals; like *Kallimorphocrinus* in lacking anal notch. *L.Perm.(Artinsk.)-U.Perm. (Basleo beds),* Eu.(USSR)-Indon.(Timor).——Fig. 335,*8*. **M. quinquebrachiatus* (Wanner), U.Perm., Timor(Basleo); *8a,b, A*-ray and dorsal views of holotype, ×5; *8c,* ventral view of holotype (*A* ray directed upward), ×7.5; *8d,e, B*-ray and ventral views of another specimen, ×5; *8f, CD*-interray views of another cup, ×5 (all Wanner, 1929b).

Table 7. *Sequence of Appearance of First Arms During Ontogeny of Allagecrinid Species.*

[Explanation: A-E, Carpenter designations of radials; numbers 1-5 indicate sequence in appearance of arms.]

	D	E	A	B	C
Allagecrinus austinii	4	2	5	3	1
Allagecrinus americanus	4	2	5	3	1
Isoallagecrinus pecki	4	2	5	3	1
Kallimorphocrinus angulatus	3	2	5	4	1

?Stereobrachicrinus Mather, 1915, p. 108 [**S. pustullosus*; OD]. Cup unknown; genus based on dissociated solidly fused uniserial arms with nearly straight sides, well-rounded outer surface, and narrowly grooved ambulacrum on inner side. *L.Penn.(Morrow.),* USA(Okla.-Ark.).——Fig. 335,*6*. **S. pustullosus,* Brentwood Ls. mbr., Bloyd Sh., NW.Ark.(Washington Co.); *6a,b,* side and outer surface of typical arm, ×1.5 (Moore & Plummer, 1938).

Thaminocrinus Strimple & Watkins, 1969, p. 217 [**Allagecrinus biplex* Wright, 1933, p. 206; OD]. Similar to *Wrightocrinus* but much smaller, in having more numerous arms in *A* and *C* rays (see Table 6), anal series borne by left shoulder of *C* radial, and in lacking preserved oral circlet in adults. *L.Carb.(Visean),* Eu.(Scot.).——Fig. 336,*1*. **T. biplex* (Wright); *1a-d, C*-ray, *E*-ray, ventral, and *B*-ray views of typical specimens, ×12 (Wright, 1941).

Trophocrinus Kirk, 1930, p. 210 [**T. tumidus*; OD]. Calyx tall conical to pyriform, morphologically similar to *Kallimorphocrinus* except for hypertrophic extensions of summit parts of some radials so as to form pouchlike structure or structures; basals 3, unequal, with azygous small plate in *CD* interray; radials 5, *D* and *E* largest, single arm facet on *E* radial in type species but additional facets (none multiple) on other radials in different species; oral circlet evenly domelike, with posterior oral distinctly largest. [Kirk's inter-

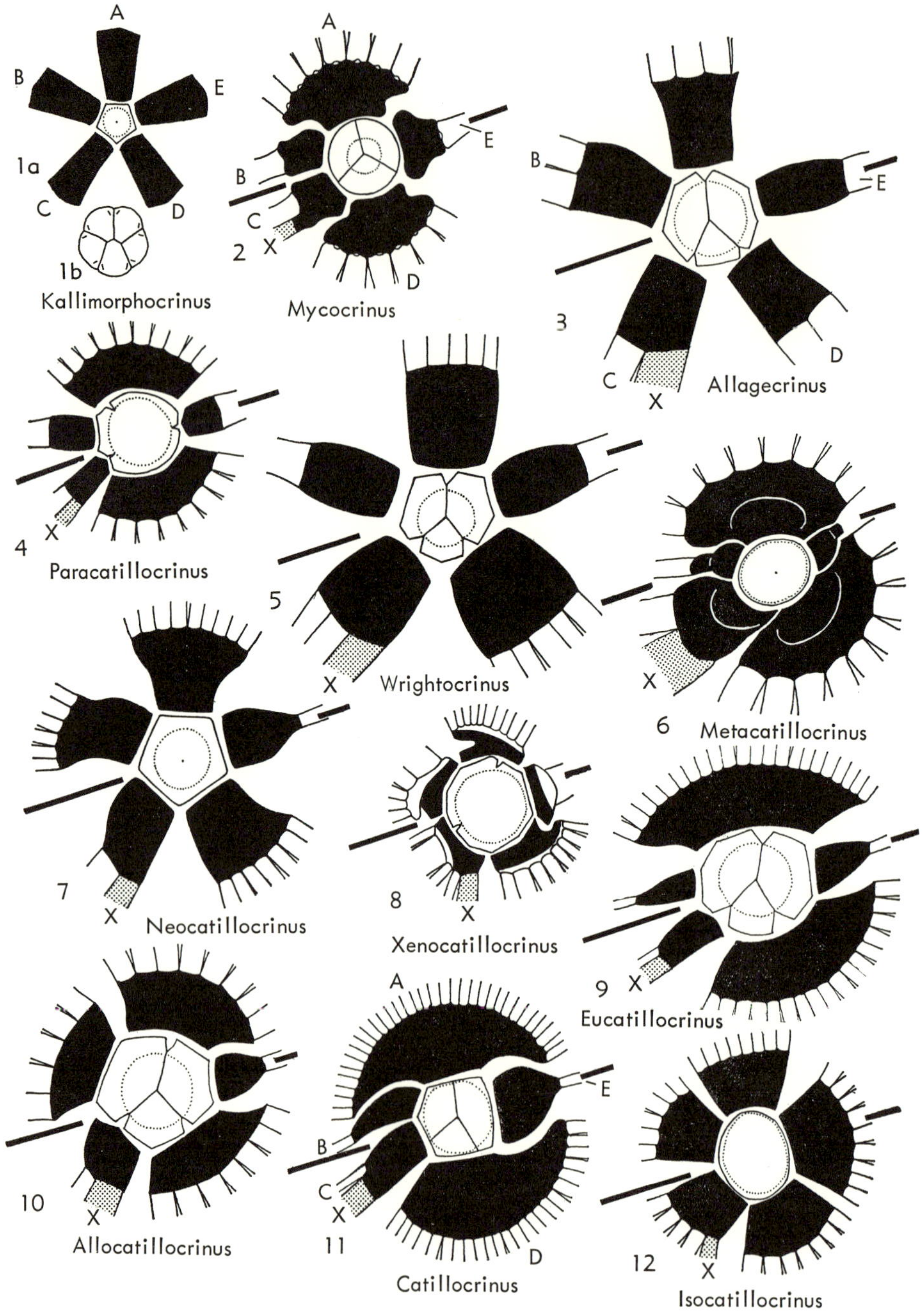

FIG. 334. Diagrammatic dorsal side views (not to scale) showing structure of allagecrinid cups with proximal parts of attached arms (radials black) and anitaxis (stippled) borne by *C* radius [*A* ray directed upward in all figures] (Moore, 1962a).

pretation of the marginal extensions borne by some radials as development of a type of brood pouch is discounted, especially since 2 such pouches appear in *T. bicornis* (STRIMPLE &

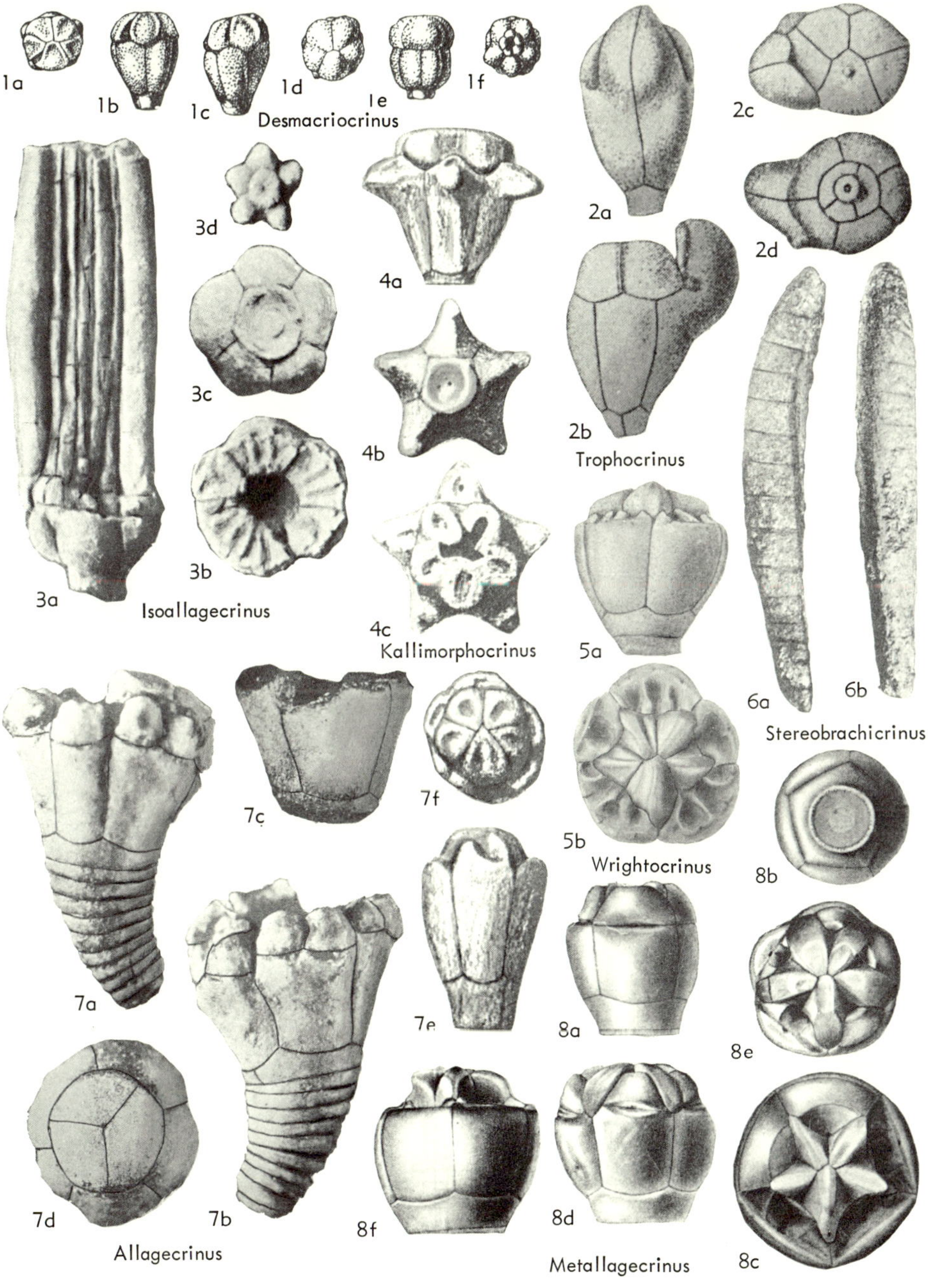

FIG. 335. Allagecrinidae (p. *T*537-*T*542).

KOENIG, 1956, p. 1238).] *L.Miss.(Kinderhook.-Osag.),* USA(Okla.-Mo.).——FIG. 335,*2*. **T. tumidus,* Kinderhook (Sycamore Ls.), USA(S. Okla.); *2a,b, DE*-interray and *A*-ray views of holotype, ×25; *2c,d,* ventral (posterior oral down) and dorsal (*C* radial down) views of holotype, ×25 (Kirk, 1930).

Wrightocrinus MOORE, 1940, p. 92 [**Allagecrinus jakovlevi* WANNER, 1929b, p. 15; OD]. Calyx small (height 1.5 to 4.5 mm.), pyriform or low

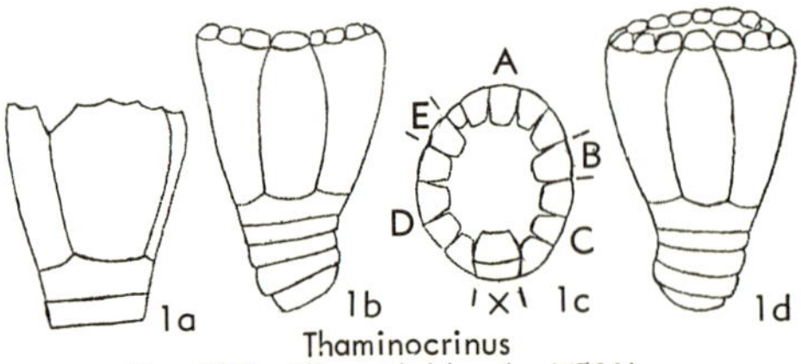

FIG. 336. Allagecrinidae (p. T539).

conical; basals 3 in some specimens but commonly fused to solid moderately high circlet visible from side; radials nearly equal in width but *B* and *E* radials narrower than others and with single arm facet instead of multiple facets (Fig. 334,*5*; see Table 6); *CD* interray with or without small notch for reception of anal plate; orals persistent in adults as moderately elevated dome, posterior oral distinctly larger than others. Arms known only from quadrangular dissociated primibrachs preserved with some specimens. Stem composed of low, round columnals, tapering somewhat distally. *U.Perm.(Basleo beds),* Indon.(Timor).——FIG. 335,*5*. **W. jakovlevi* (WANNER), Basleo; *5a,b, CD*-interray and ventral views of holotype, ×5 (Wanner, 1929b).

Family CATILLOCRINIDAE Wachsmuth & Springer, 1886

[Catillocrinidae WACHSMUTH & SPRINGER, 1886, p. 118(194)] [=Catillocrininae MOORE, 1940b, p. 112]

Cup low truncate conical or bowl shaped to discoid, characterized by strong asymmetry produced by enlargement of two radials *(A, D)* which bear numerous unbranched slender arms, accompanied by marked bilateral symmetry with respect to homocrinoid plane *(E-BC). L.Ord.; M.Dev.; L.Miss.-M.Penn.; U.Perm.*

Catillocrinus SHUMARD, 1865 (Oct.), p. 357 [catalogue publ. in parts (Aug., 1865, to Feb., 1866) and ultimately in complete volume issued in 1868] [**C. tennesseeae*; M] [=*Catillocrinites* TROOST, 1850a, p. 419 *(nom. van.); Nematocrinus* MEEK & WORTHEN, 1866b (July), p. 251 (type, *Synbathocrinus wachsmuthi* MEEK & WORTHEN, 1866b, p. 251; M); *Catillicrinus* S. A. MILLER, 1889, p. 231 *(nom. null.)*]. Crown relatively tall, subcylindrical, with multitudinous threadlike arms closely appressed to one another and rising vertically. Cup low truncate cone or bowl shaped, with flat discoid basal circlet normally composed of 3 plates not visible or only barely visible from side; radials exceptionally dissimilar in size, with many-armed *A* and *D* radials occupying at least 0.8 of summit periphery of cup, *B, C* and *E* radials with single arm facets (Fig. 334,*11*); *C* radial with raised process on left shoulder for support of large anal tube, which is composed of much thickened, massive, armlike plates, crescentic in cross section, anal tube extended to distal extremities of arms or above. Transversely circular stem composed of wide low columnals next to cup, tapering distally. *L.Miss.(Osag.),* USA(Ky.-Tenn.-Ind.-Iowa).——FIG. 337,*2a-c*. **C. tennesseeae,* New Providence F., USA(Ky.); *2a-c, C*-ray, ventral (*C* radial downward, *A* and *D* radials at left and right with extremely narrow groovelike arm facets), and dorsal (*E* radial downward) views of holotype cup, all ×1 (Moore & Laudon, 1943a).——FIG. 337,*2d,e*. *C. wachsmuthi* (MEEK & WORTHEN), up. Burlington Ls., USA(SE.Iowa); *2d, B*-ray view of well-preserved crown, ×1.5 (Wachsmuth & Springer, 1886); *2e, BC*-interray view of another crown, ×1.5 (Springer, 1923a).

Acolocrinus KESLING & PAUL, 1971, p. 228 [**A. hydraulicus*; OD]. Cup truncate conical below rounded midheight and vertical sided above; basal circlet pentagonal in outline, composed of 2 large plates *(CDE, EAB)* and a small one *(BC)*; 5 compound radials, consisting of subquadrangular inferradials and somewhat smaller superradials with pronglike upward projections on right sides of summits, several slender unbranched arms at left of prongs, their concave round articulation facets confluent with shallow ambulacral grooves which converge inward; small trapezoidal anal *X* between *C* and *D* superradials; outer surface of basals and radials deeply reticulate, that of anal *X* finely granulose. Stem transversely round, small, attached to cup in steep-sided basal concavity. *L.Ord.,* USA(Va.-Tenn.).——FIG. 338,*2*. **A. hydraulicus,* Benbolt F., Ottosee Gr., Russell and Scott Counties, SW.Va., and Hawkins County, SE.Tenn.; *2a-c,* dorsal, *D*-ray, and *E*-ray views of holotype, ×4 (Kesling & Paul, 1971); *2d,* plate diagram showing proximal parts of arms (Moore, n, based on Kesling & Paul, 1971, fig. 3).

[*Acolocrinus* is like *Allagecrinus* and at least five catillocrinid genera (*Catillocrinus, Wrightocrinus, Eucatillocrinus, Allocatillocrinus,* and *Paracatillocrinus*) in having a tripartite basal circlet with 1 small and 2 large plates, although only in *Paracatillocrinus* is the small basal located in the *BC* interray. This resemblance is not considered to be especially significant, however. Likewise, the presence of a small anal *X* supported by the *C* or *D* radial in the *CD* interray in *Acolocrinus* and these genera is not diagnostic. It is the numerous, simple, small, unbranched arms joined to summits of the radials that provides basis for placement of *Acolocrinus* in the Catillocrinidae, among which the division of radials into lower and upper parts by a transverse suture and prongs on right corners of superradial summits distinguishes this genus from all others. The biradial construction is undoubtedly a primitive trait. Segregation of the arms on left parts of radials links *Acolocrinus* with *Agostocrinus*.]

Agostocrinus KESLING & PAUL, 1971, p. 222 [**A. xenus*; OD]. Cup rotund bowl shaped, truncate with 3 equal basals, 5 moderately tall radials with prominent leftward extensions at summit consisting of semifused arm main-axils similar to those found in several calceocrinid genera, separated on posterior side by large quadrangular anal *X* plate which rests on *CD* basal. Orals 5, large posterior one clearly or indistinctly divided longi-

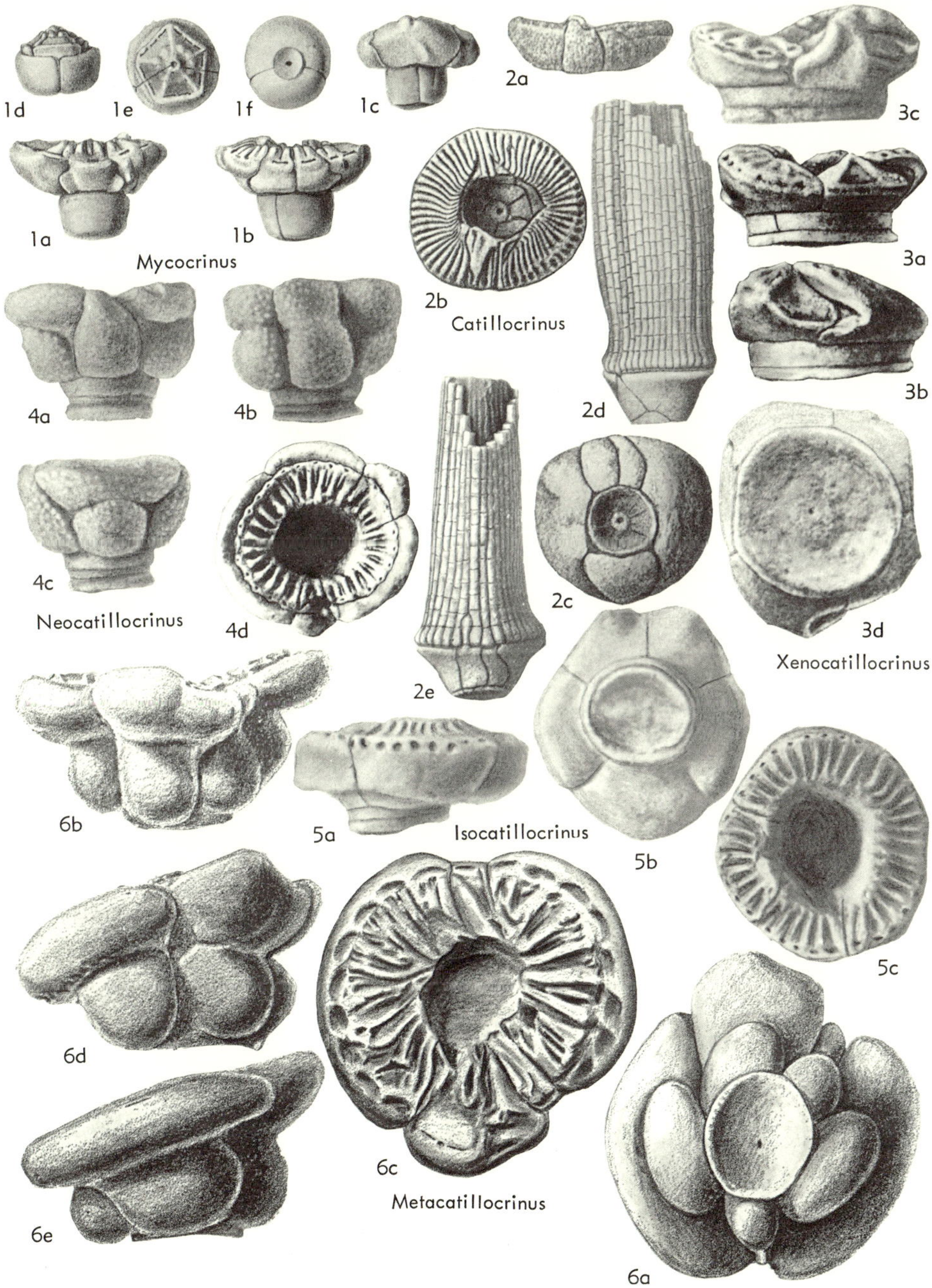

FIG. 337. Catillocrinidae (p. *T*542, *T*545-*T*547).

tudinally into halves and containing large, round anal opening subequal in size to centrally located circular mouth and provided with cylindrical lining which reaches downward in cup halfway to its base. Arms very slender, unbranched, 10 to 16 on each radial (50 to 72 in all), with somewhat elevated round facets on outer edges of radials formed by incorporated arm main-axil ele-

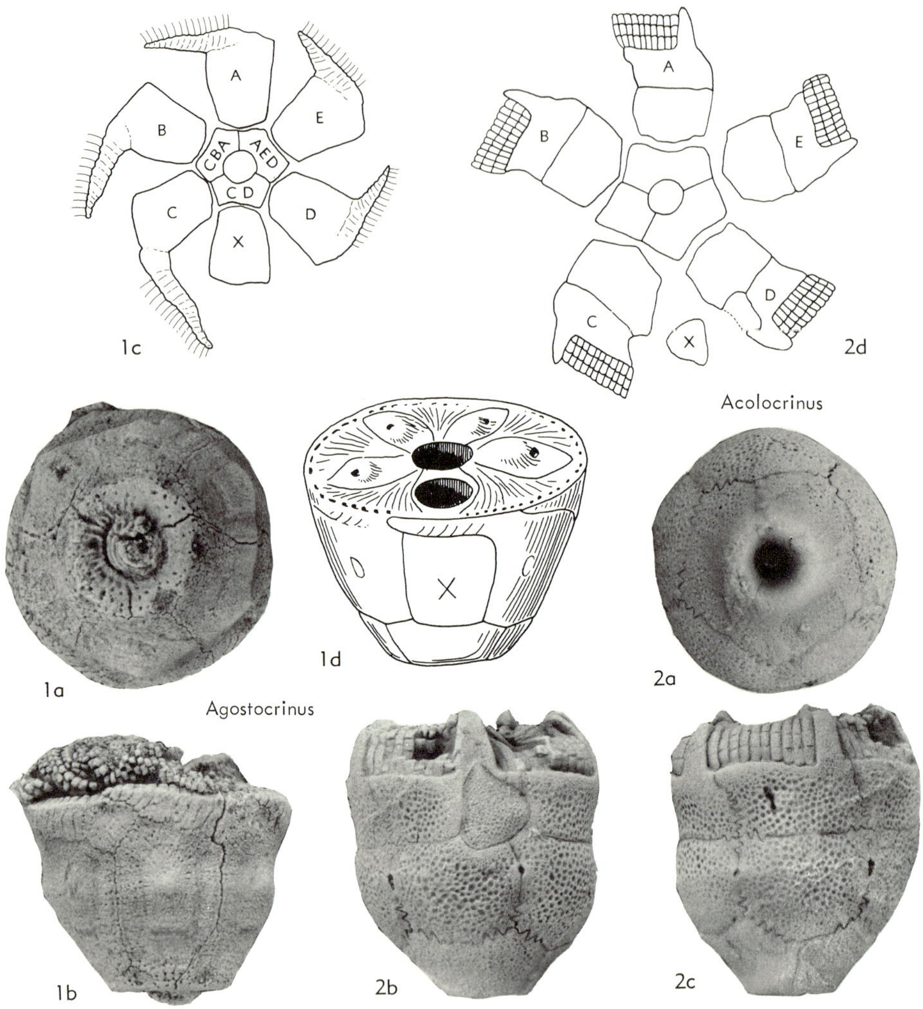

FIG. 338. Catillocrinidae (p. T542-T544).

ments, shallow ambulacral grooves converging toward mouth. Stem transversely circular. *L.Ord.*, USA(Va.).——FIG. 338,*1*. **A. xenus*, Benbolt F. (Ottosee Gr.), Russell and Scott Counties, SW. Va.; *1a,b*, dorsal and *C*-ray views of holotype, ×4 (Kesling & Paul, 1971); *1c*, plate diagram showing leftward extensions of radials formed by fused to semifused arm main axils and attached tiny arms (Moore, n, based on Kesling & Paul, 1971, fig. 1); *1d*, oblique post. diagramm. view of cup showing radials with leftward extensions and on summit mouth and anal apertures, arm facets and ambulacral grooves (Moore, n).

[This genus has exceptionally interesting morphological features that contribute to studies of disparid evolution. Relationship to calceocrinids is indicated by the presence of arm main-axils as components of summit parts of the radials in manner demonstrating the origin of catillocrinid radials, solidly fused plates each of which support multiple slender unbranched arms—as many as 30 or more borne by the *A* and *D* radials of *Catillocrinus*. The three equal basals of *Agostocrinus* correspond exactly to the basal circlet of *Mycocrinus*, a catillocrinid, although anal *X* of the latter genus is carried by the *C* radial instead of by the *CD* basal as in *Agostocrinus*. The plane of bilateral symmetry observed in *Agostocrinus* differs in orientation from that found in all catillocrinids and calceocrinids (*E-BC*, homocrinoidal plane), for it runs through midlines of the *A* ray and *CD* interray (crinoidal plane). Also, distinctive characters of *Agostocrinus* are the uniform leftward extensions of arm main-axil portions of the radials and nature of the large, round anal aperture in the posterior oral (or paired orals). A tall anal tube found in various catillocrinids is lacking in *Agostocrinus*, but in spite of this and other considerations the genus defined by KESLING & PAUL is assigned here to the Catillocrinidae.]

Allocatillocrinus WANNER, 1937, p. 109 [**Allagecrinus carpenteri* WACHSMUTH, 1882, p. 40; OD]. Crown tall and slender. Cup generally similar to

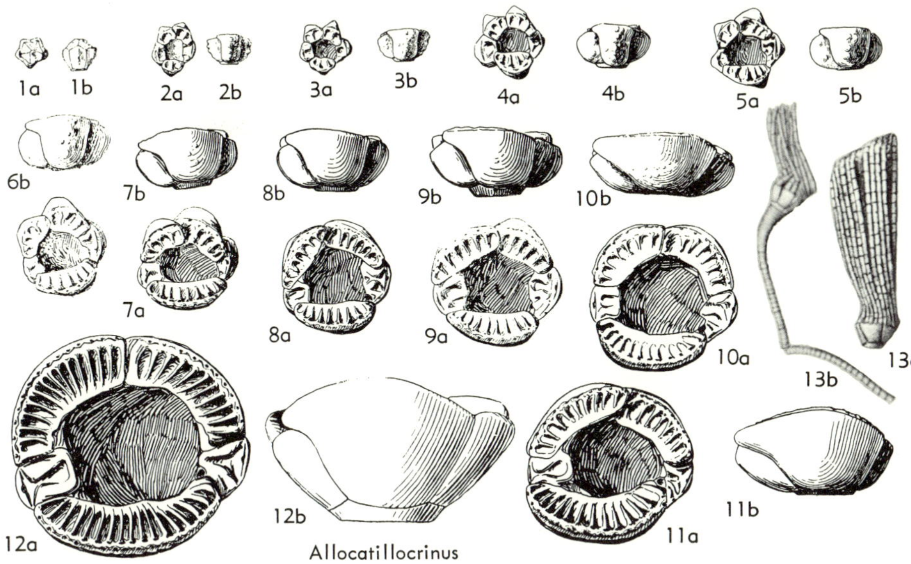

FIG. 339. Catillocrinidae (p. T544-T545).

Catillocrinus except for presence of 3 large multifaceted radials *(A, B, D)* and beveled or even left shoulder of *C* radial for contact with anal *X* plate (Fig. 334,*10*); armlike anal tube with diameter 2 or 3 times that of arms, very elongate, commonly rising well above tips of arms. *U.Miss. (Chester.),* USA(Ill.-Ind.-Ala.-Tenn.); *L.Carb. (Visean),* Eu.(Scot.); *L.Penn.(Morrow.),* USA (Okla.).——FIG. 339,*1-12*. *A. rotundus* (MOORE), L.Penn.(Morrow., Brentwood Ls.), USA(Okla.); *1-12,* ventral and *D*-ray views of ontogenetic series of cups (*1, Kallimorphocrinus* stage; *2,3, Allagecrinus* stage; *4-12,* immature to fully grown adult *Allocatillocrinus* stage), all ×3 (Moore, 1940b).——FIG. 339,*13*. **A. carpenteri* (WACHSMUTH), U.Miss.(Chester.); *13a, B*-ray view of crown from Monroe Co., Illinois, ×1.5 (Moore & Laudon, 1943a); *13b, C*-ray view of crown with attached stem from Tennessee, ×0.7 (Wachsmuth & Springer, 1897).

Eucatillocrinus SPRINGER, 1923, p. 19 [**Catillocrinus bradleyi* MEEK & WORTHEN, 1868, p. 343; OD]. Crown tall and slender. Cup truncate conical, medium in height; basal circlet composed of 3 plates or fused to solid piece with upflaring sides plainly visible in side view of cup; radials generally similar to *Catillocrinus,* with large, multifaceted *A* and *D* rays and others single armed (Fig. 334,*9*); *C* ray with even summit, its left part supporting first 2 plates of extremely long anal tube, which extends well above tips of threadlike arms. Long stem composed of low, round columnals, tapering distally near cup, but approximately uniform in diameter below this zone. *L.Miss.(Osag.),* USA (Ind.); *M.Penn. (?Atokan),* USA(Alaska).——FIG. 340,*2*. **E. bradleyi* (MEEK & WORTHEN), Miss.(Keokuk), Crawfordsville, Ind.; *C*-ray view of incomplete crown with attached long stem and anal tube, ×1 (Springer, 1923a).

Isocatillocrinus WANNER, 1937, p. 109 [**I. indicus*; OD]. Cup very low, discoid, with greatest diameter about 14.5 mm.; basal disc solidly fused, barely visible from side; radials nearly equal in size and all multifaceted (Fig. 334,*12*); anal series unknown. *U.Perm.,* Indon.(Timor, Basleo).——FIG. 337,*5*. **I. indicus; 5a-c, B*-ray, dorsal (*E* ray below), and ventral (*E* ray above) views of holotype, all ×2 (Wanner, 1937).

Metacatillocrinus MOORE & STRIMPLE, 1942, p. 79 [**M. bulbosus*; OD]. Cup low, very asymmetrical, distinguished by strongly bulbous nature of radial-plate sides, all of which are divided by transverse constriction into upper flangelike part and laterally less extended subspherical or egg-shaped lower part; multifaceted *A* and *D* radials much enlarged, forming 0.8 of cup periphery, other radials single armed (Fig. 334,*6*), left part of distal surface of *C* radial flattened for attachment of anal tube; body cavity much restricted by lengthened and thickened radials; basals fused into single elliptical disc, which is very low and not visible or barely visible from side. Arms and stem unknown. *M.Penn.(Desmoines.),* USA(Okla.).——FIG. 337,*6*. **M. bulbosus,* Altamont Ls., Tulsa Co., Okla.; *6a-e,* dorsal (*C* ray upward), *C*-ray, ventral (*E* ray upward), *CD*-interray, and *D*-ray views of holotype, distinct inclination of summit plane to base of cup illustrated in *6d* and *6e,* all ×5 (Moore & Strimple, 1942).

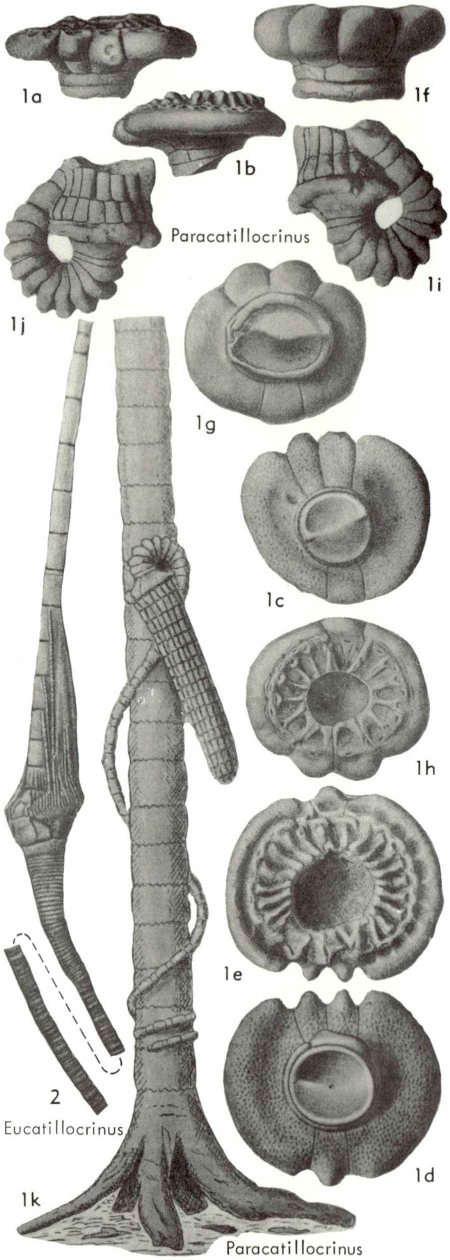

FIG. 340. Catillocrinidae (p. T545-T546).

Mycocrinus SCHULTZE, 1867, p. 110 [**M. boletus*; M]. Cup with lower half consisting of massive, vertical-sided basal circlet composed of 2 or 3 plates and upper half of broadly flaring radial circlet similar to *Catillocrinus* in having wide multifaceted *A* and *D* radials, others bearing single arms (Fig. 334,2; see Table 6), left shoulder of *C* radial elevated for support of anal plate. Stem impression moderately small, circular. *M. Dev.*, Eu.(Ger.).——FIG. 337,*1*. **M. boletus*; *1a-c,* complete cup from *E*-ray, *BC*-interray, *A*-ray sides, ×2; *1d-f,* basal circlet from *A*-ray, ventral, and dorsal sides, ×2 (all Schultze, 1867).

Neocatillocrinus WANNER, 1937, p. 117 [**N. incissus*; OD]. Cup moderately small (greatest width about 9.5 mm.), bowl shaped, with low, fused, vertical-sided circlet, visible from side, surmounted by strongly bulbous radials, in all except *E* ray divided into upper and lower halves by transverse constriction as in *Metacatillocrinus*, lower parts rather evenly bulbous and upper parts flaring sideward; arm distribution on radials similar to that of *Allocatillocrinus* (Fig. 334,*7*), left shoulder of *C* radial with attachment for anal plate. *U.Perm.(Basleo beds)*, Indon.(Timor, Basleo).——FIG. 337,*4*. **N. incissus; 4a-d, E*-ray, *C*-ray, *A*-ray, and ventral (*E* ray downward) views of holotype, ×3 (Wanner, 1937).

Paracatillocrinus WANNER, 1916, p. 6 [**P. granulatus*; OD]. Crown moderately tall, cylindrical, with slender, closely appressed vertical arms, articular facets of brachials transversely aligned. Cup turreted-discoid, with vertical-sided basal circlet composed of 3 or 2 plates or single fused plate, visible from side; strongly convex radials flare outward, distribution of arms in pattern of *Catillocrinus* (see Table 6) but facets wider and much fewer (Fig. 334,*4*). Stem transversely elliptical next to cup and joined to it somewhat obliquely, proximal part strongly curved and narrowing downward to merge with part of stem composed of moderately tall cylindrical columnals. *U.Perm.(Basleo beds)*, Indon.(Timor, Basleo).——FIG. 340,*1a-e, i-k*. **P. granulatus; 1a-c,* CD-interray, *A*-ray, and dorsal (*CD* interray upward) views of holotype, ×2 (Wanner, 1916a); *1d,e,* dorsal (*B* ray downward) and ventral (*B* ray upward) views of paratype, ×2 (Wanner, 1916a); *1i,j,* complete crown with attached proximal part of stem from *A*- and *D*-ray sides, ×2 (Wanner, 1930a); *1k,* complete specimen (reconstr.) showing presumed mode of attachment and pendent attitude of crown, ×1 (Wanner, 1930a).——FIG. 340,*1f-h*. *P. ellipticus* WANNER, *BC*-interray, dorsal (*BC* interray upward), and ventral (*BC* interray downward) views of cup with attached proximal columnal, ×2 (Wanner, 1924).

Xenocatillocrinus WANNER, 1937, p. 119 [**X. wrighti*; OD]. Cup very low, turreted-discoid (greatest width about 10.5 mm.), with vertical-sided basal circlet of 2 or 3 plates surmounted by very uneven, differently inclined radials, with all except *E* ray multifaceted (Fig. 334,*8*; Table 2), facet for attachment of anal plate on left shoulder of *C* radial. Stem facet elliptical, nearly featureless. *U.Perm.(Basleo beds)*, Indon.(Timor,

Basleo).——Fig. 337,*3*. **X. wrighti; 3a-d, EA-, AB-, CD*-interray views and dorsal side (*A* ray downward) of holotype cup, ×3 (Wanner, 1937).

Family ANAMESOCRINIDAE Goldring, 1923

[Anamesocrinidae Goldring, 1923, p. 323]

Cup small, truncated below. Radials and superradials with multiple facets, each bearing slender unbranched arms. *M.Dev.*

Anamesocrinus Goldring, 1923, p. 323 [**A. lutheri*; OD]. Cup small, conical; 5 low basals; large *(A, D)* radials and half as large superradials *(B, C, E)* each supporting 5 atomous arms; anal *X* small, triangular, wedged between upper angles of *C* superradial and *D* radial, supporting several small, quadrangular tube plates. Stem transversely round, tapering in width distally. *M.Dev.*, USA(N.Y.).——Fig. 341,*3*. **A. lutheri; 3a-c, E*-ray, *E*-ray, and *CD*-interray views of 2 crowns and stem, ×3 (Goldring, 1923); *3d*, plate diagram of cup and arm bases (radials and superradials black, inferradials horizontally ruled, anal *X* stippled), ×7 (Moore, 1962b).

Family HAPLOCRINITIDAE Bassler, 1938

[*nom. correct.* Bassler, 1938, p. 16 (*pro* Haplocrinidae Roemer, 1855, p. 229)] [=Aplocrinidae d'Orbigny, 1852, p. 19]

Cup globose to conical, with five basals; *B, C,* and *E* radials compound; anal *X* lacking. Arms atomous. *?Sil., Dev., ?L. Carb.*

Haplocrinites Steininger, 1837, p. 232 [**H. sphaeroideus* (=*Eugeniacrinites mespiliformis* Goldfuss, 1831, p. 213) (subj.); M] [=*Asterocrinus* Münster, 1839, p. 4 (type, *A. murchisoni*); *Asterocrinites* Roemer, 1851, p. 390; *Haplocrinus* Roemer, 1855, p. 63 *(nom. van.); Aplocrinus* d'Orbigny, 1852, p. 63 *(nom. van.)*]. Theca small, bipyramidal; basals 5, low; *A* and *D* radials simple, others compound, superradials larger than inferradials; no anal plates in cup. Tegmen composed of 5 large triangular oral plates, posterior one with median pore. Radial articular facets narrowly rounded, in deep notches at radial summits; grooves which housed proximal parts of arms extending from radial facets along lateral edges of orals to summit of theca. Arms atomous. Column unknown. *?Sil., Dev., ?L.Carb.*, Eu. (Ger.-Eng.)-USA(Tenn.-N.Y.).——Fig. 341,*2a,b*. **H. mespiliformis* (Goldfuss), M.Dev., Ger.; lat. and ventral views of theca, ×8 (Wachsmuth & Springer, 1886).——Fig. 341,*2c-f*. *H. clio* (Hall), M.Dev., N.Y.; dorsal, ventral, *C*-, and *D*-ray views of theca, ×4 (Goldring, 1923).

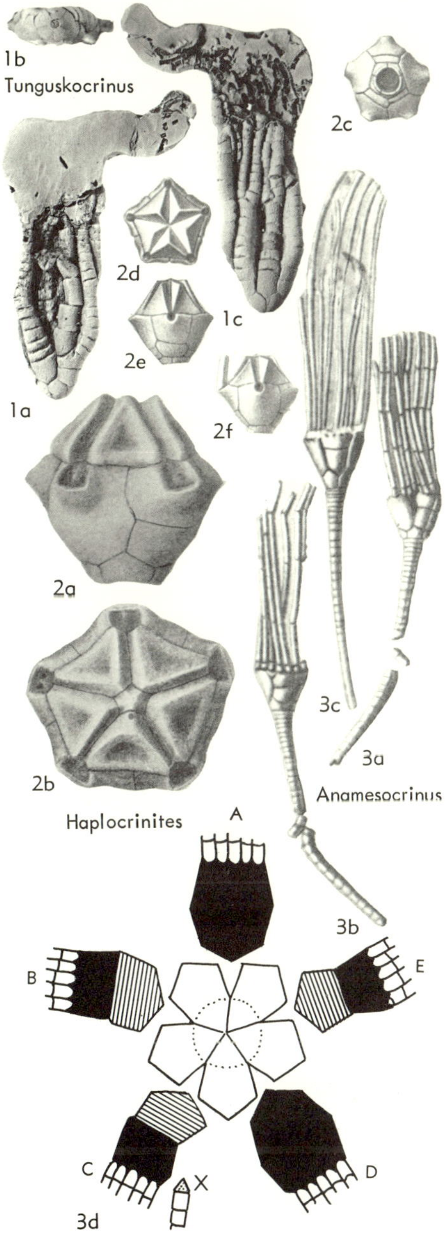

Fig. 341. Anamesocrinidae *(3)*; Haplocrinitidae *(2)*; Tunguskocrinidae *(1)* (p. *T*547-*T*548).

Family TUNGUSKOCRINIDAE Arendt, 1963

[Tunguskocrinidae Arendt, 1963, p. 132]

Cup with homocrinid (*E*-ray) bilateral symmetry. Arms branching once isoto-

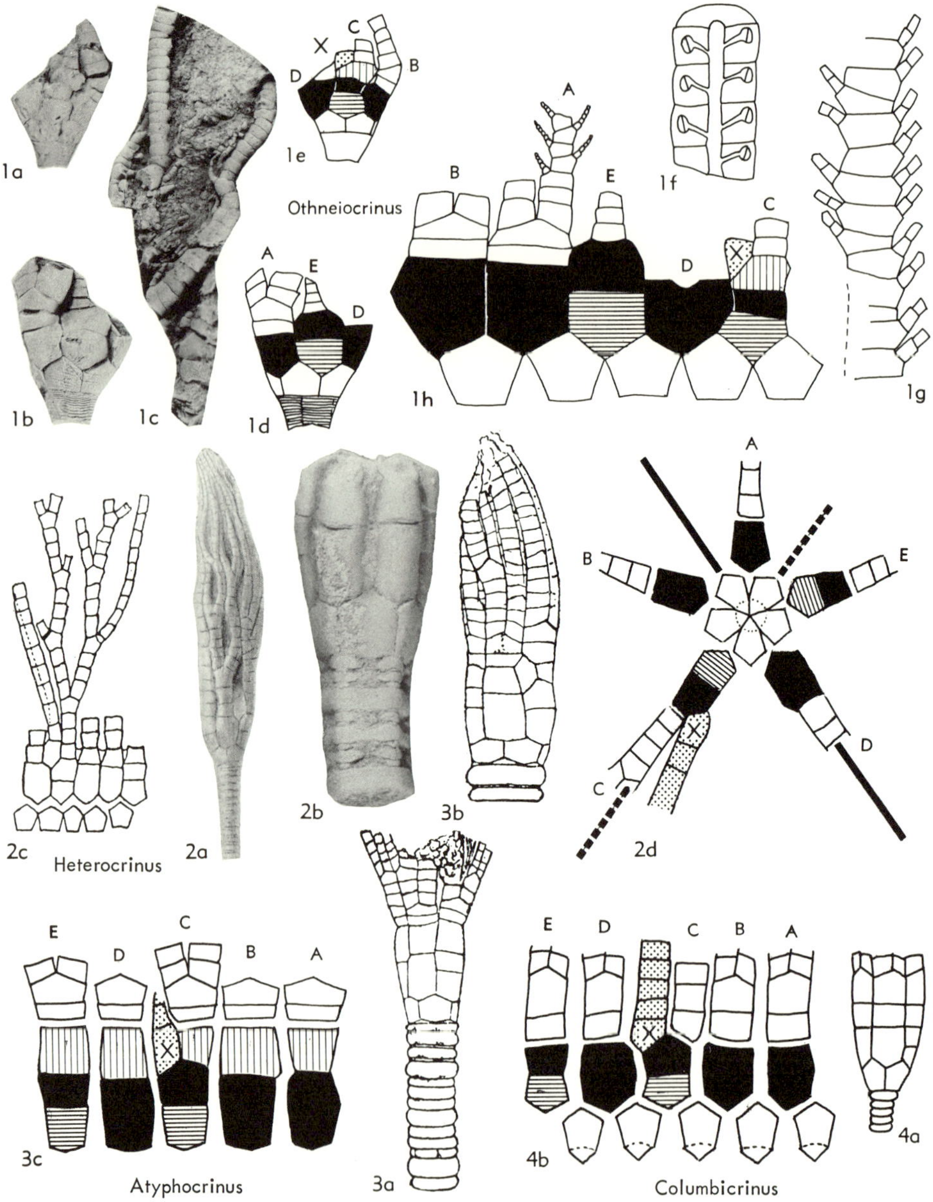

FIG. 342. Heterocrinidae (p. T549).

mously on primibrachs 7 or 8. Column minute, probably not functional. *Ord.*

Tunguskocrinus ARENDT, 1963, p. 132 [**T. ivanovae*; OD]. Cup conical; basals spear shaped and upflaring; *B, C,* and *E* radials divided into infer- and superradials, radials of *A* and *D* rays simple; anal *X* large, on upper left shoulder of *C* superradial, supporting longitudinal series of large tube plates which taper in width distally. Arms stout, rounded, branching once isotomously on primibrachs *7* in all but *E* ray, which has 8 primibrachs. Stem facet on basals quite small and round. Column probably not functional for support of crown. *Ord.,* Eu.(USSR).——FIG. 341,*1*. **T. ivanovae; 1a-c,* post., dorsal, and ant. views of crown, ×1.3 (Arendt, 1963).

Superfamily HETEROCRINACEA Zittel, 1879

[*nom. correct.* MOORE & LANE, in MOORE & STRIMPLE, 1973, p. 17, *pro* Heterocrinicae UBAGHS, 1953, p. 747 (*nom. transl. ex* Heterocrinidae ZITTEL, 1879, p. 343)] [=Heterocrinites JAEKEL, 1918, p. 84 (suborder) (*partim*)] [Materials for this superfamily prepared by R. C. MOORE and N. GARY LANE]

Compound radials in *C* and *E* rays only, radials of other rays simple; anal *X* between *C* superradial and *D* radial. *M.Ord.(Mohawk.)-U.Ord.(Cincinnat.).*

Family HETEROCRINIDAE Zittel, 1879

[Heterocrinidae ZITTEL, 1879, p. 343] [=Heterocrinidées DE LORIOL, 1882-84, p. 51]

Cup small, steeply conical; anal *X* supporting longitudinal row of enlarged tube plates. Anal tube elongate, may be spirally coiled. Proximal one to three primibrachs may be fixed in top part of cup, brachials occupying full width of radials. Arms uniserial, nonpinnulate, branching isotomously several times. *M.Ord.(Mohawk.)-U.Ord. (Cincinnat.).*

Heterocrinus HALL, 1847, p. 278 [**H. heterodactylus*; SD WACHSMUTH & SPRINGER, 1880, p. 70] [=*Stenocrinus* WACHSMUTH & SPRINGER, 1886, p. 13 (obj.)]. Cup small and slender; basals large and upright; *C* and *E* radials compound, others simple; anal *X* small, on upper left sloping shoulder of *C* superradial. Anal sac slender, armlike. Arms slender, rounded, branching on primibrachs *4* and with 2 or 3 higher equal divisions. Column transversely pentagonal. *M. Ord. (Mohawk.)-U. Ord. (Cincinnat.),* Can. (Ont.)-USA(Ohio-Ky.-N.Y.).——FIG. 342,*2*. **H. heterodactulus* HALL, U.Ord., Ohio; *2a, D*-ray view of crown, ×2 (Meek, 1873); *2b, AB*-interray view of cup, ×4 (Moore, n; Univ. Chicago 3136); *2c,d,* plate diagrams of cup and parts of crown, *2d* showing primary plane of bilateral symmetry—full heavy line through *D* ray—and secondary plane—broken line through *C* ray (Ulrich, 1925; Moore, 1962b).

Othneiocrinus LANE, herein, *nom. subst. pro Atopocrinus* LANE, 1970, p. 14 (*non* CLARK, 1912) [**A. priscus*; OD]. Cup low conical; basals 5; inferand superradials in *C* and *E* rays only. Primibrach *1* axillary in *C* ray, supporting anal *X* and row of sac plates on left and an arm on right. Anal sac unknown. Primibrachs *2* axillary in *A* and *B* rays, arms of *E* and *D* rays unknown. Each arm with 2 main rami, brachials with small unbranched ?ramules on one or both sides of each secundibrach. Columnals composed of pentameres. *M.Ord.(Mohawk.),* USA(Utah).——FIG. 342,*1*. **A. priscus; 1a,b, C*-ray and *E*-ray views of cup, ×1.3, ×1.7; *1c, A*-ray view of arms, ×1.7; *1d,e, E*-ray and *C*-ray diagrams of cup; *1f-h,* oral and aboral surfaces of brachials, and plate diagram of cup (simple radials and superradials black, inferradials horizontally ruled, *C*-ray primaxil vertically ruled, anal *X* stippled) (Lane, 1970a).

Atyphocrinus ULRICH, 1925, p. 85 [**Heterocrinus (Atyphocrinus) corryvillensis* ULRICH, 1925, p. 85; M]. Cup small, slightly wider than column. Plates of cup arranged as in *Heterocrinus,* 3 primibrachs to a ray, with first primibrachs fixed in cup. Arms branching heterotomously on every fourth or fifth brachial. *U.Ord.(Cincinnat.),* USA (Ohio).——FIG. 342,*3*. **A. corryvillensis* ULRICH; *3a, D*-ray view of crown, ×1.5 (Ulrich, 1925); *3b,* ant. view of crown, ×2 (Ulrich, 1925); *3c,* plate diagram of cup (inferradials horizontally ruled, superradials black, fixed primibrachs vertically ruled, anal plates stippled) (Moore 1962b).

Columbicrinus ULRICH, 1925, p. 91 [**C. crassus*; M]. Cup conical; basals long hexagonal; *C* and *E* radials compound; proximal 2 brachials not fixed in cup and as wide as radials; second primibrachs axillary in all but *C* ray, which has 3 primibrachs preserved above right shoulder of *C* superradial; anal *X* supporting longitudinal series of 4 large anal plates. Distal parts of arms and anal sac not known. *M.Ord.(Mohawk.),* USA (Tenn.).——FIG. 342,*4*. **C. crassus; 4a,b, A*-ray view of cup and plate diagram of cup and proximal arms (symbols as in Fig. 342,*3c*) (Moore, 1962b).

Dystactocrinus ULRICH, 1925, p. 85 [**Heterocrinus (Homocrinus) constrictus* HALL, 1871, p. 210; M]. Plates of cup arranged as in *Heterocrinus*; anal *X* large, pentagonal, supporting elongate series of armlike anal sac plates. Anal sac elongate and tapering, composed of small, loosely united plates except for posterior anals. First primibrachs fixed in cup; second primibrachs axillary in all rays; brachials thin, rounded; arms heterotomous, with 2 main rami to each ray; large elongate armlets branching twice or more on alternate sides of every third or fourth secundibrach. *U.Ord.(Cincinnat.),* USA(Ohio).——FIG. 343,*3*. *D. crassus* ULRICH; *3a,* plate diagram (symbols as in Fig. 342,*3c*) (Moore, 1962b, after Ulrich, 1925); *3b-d,* cross sec. of cup, column, and anal sac, ×1 (Ulrich, 1925); *3e,f,* post. and *B*-ray views of crown, ×1.5 (Ulrich, 1925).

Isotomocrinus ULRICH, 1925, p. 86 [**I. typus*; M]. Cup as in *Heterocrinus.* Anal sac elongate, with longitudinal series of posterior plates which are quadrangular proximally and become narrow and elongate distally; remainder of sac composed of small, loosely united plates. Proximal tip of anal *X* in contact with upper corner of *C* inferradial. Arms long and slender, free above radials, primibrachs *5* axillary, branching isotomously several times. *M.Ord.(Mohawk.),* Can.(Ont.).——FIG. 343,*2*. **I. typus; 2a,b,* plate diagrams of crown

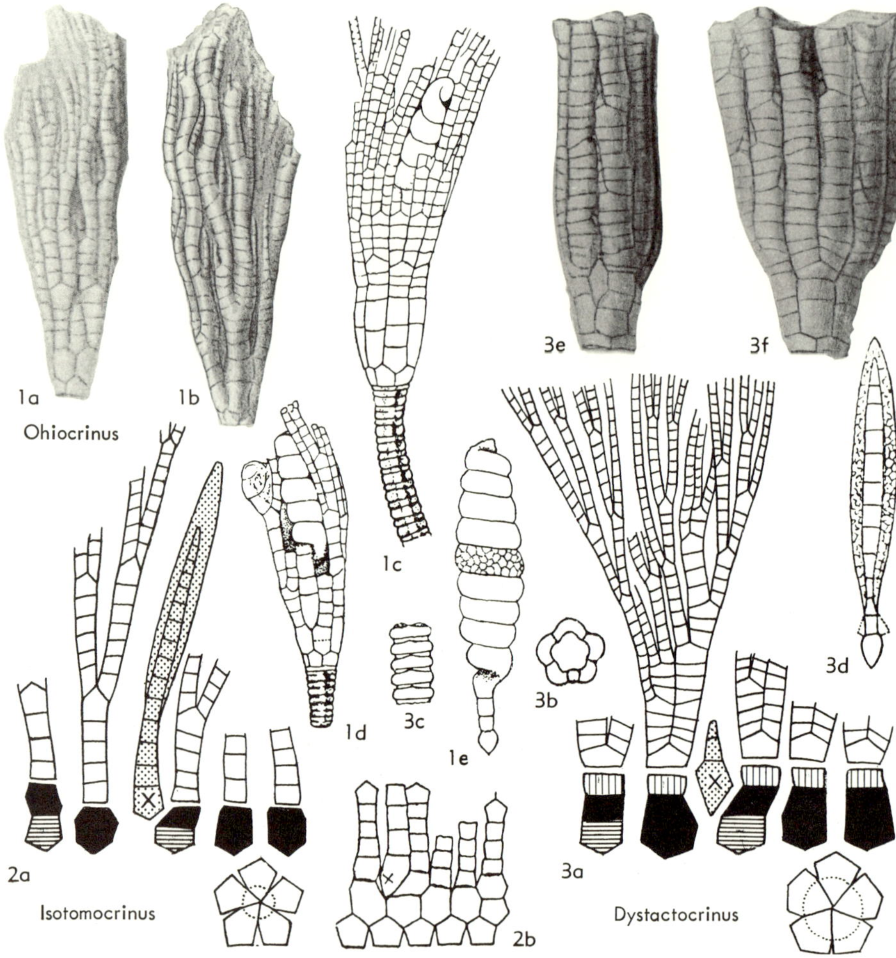

FIG. 343. Heterocrinidae (p. T549-T550).

and anal sac, ×1 (Symbols in *2a* as in Fig. 342,*3c*) (Moore, 1962b, Ulrich, 1925).

Ohiocrinus WACHSMUTH & SPRINGER, 1886, p. 208 (132) [**Heterocrinus laxus* HALL, 1871, p. 211; OD]. Cup as in *Heterocrinus.* First 2 primibrachs of each ray fixed in cup. Anal sac conspicuous, spirally coiled, composed of small rounded, imbricating plates. Arms heterotomous, branching first on primibrachs *4,* with 2 main rami to each ray; slender unbranched ramules diverge from alternate sides of every fourth secundibrach. Stem transversely subpentagonal. *M.Ord.(Mohawk.)-U. Ord. (Cincinnat.),* USA (Ky.-Ind.-Ohio).——FIG. 343,*1a,b.* **O. laxus* (HALL), U.Ord., Ohio; *1a,b,* post. and *AB*-interray views of crown, ×2, ×1.5 (Meek, 1873).——FIG. 343,*1c-e.* *O. brauni* ULRICH, U.Ord., Ind.; *1c-e, E*-ray and post. views of crown, and sketch of anal sac, ×3.6 (Ulrich, 1925).

Superfamily MYELODACTYLACEA S. A. Miller, 1883

[*nom. transl.* MOORE & LANE, in MOORE & STRIMPLE, 1973, p. 17 (*ex* Myelodactylidae S. A. MILLER, 1883a, p. 278)] [=superfamily Iocrinicae UBAGHS, 1953, p. 744 (*nom. transl. ex* Iocrinidae MOORE & LAUDON, 1943, p. 29)] [*non* order Myelodactyloidea S. A. MILLER, 1883a, p. 278] [Materials for this superfamily prepared by R. C. MOORE and N. GARY LANE]

Crown small and elongate, cup conical, monocyclic, with compound radials present in all five rays or in *C* ray only; anal *X* supported on left side of *C* superradial or directly above it. Bilateral symmetry developed in crown through *C-AE* plane. Arms uniserial, isotomous or heterotomous.

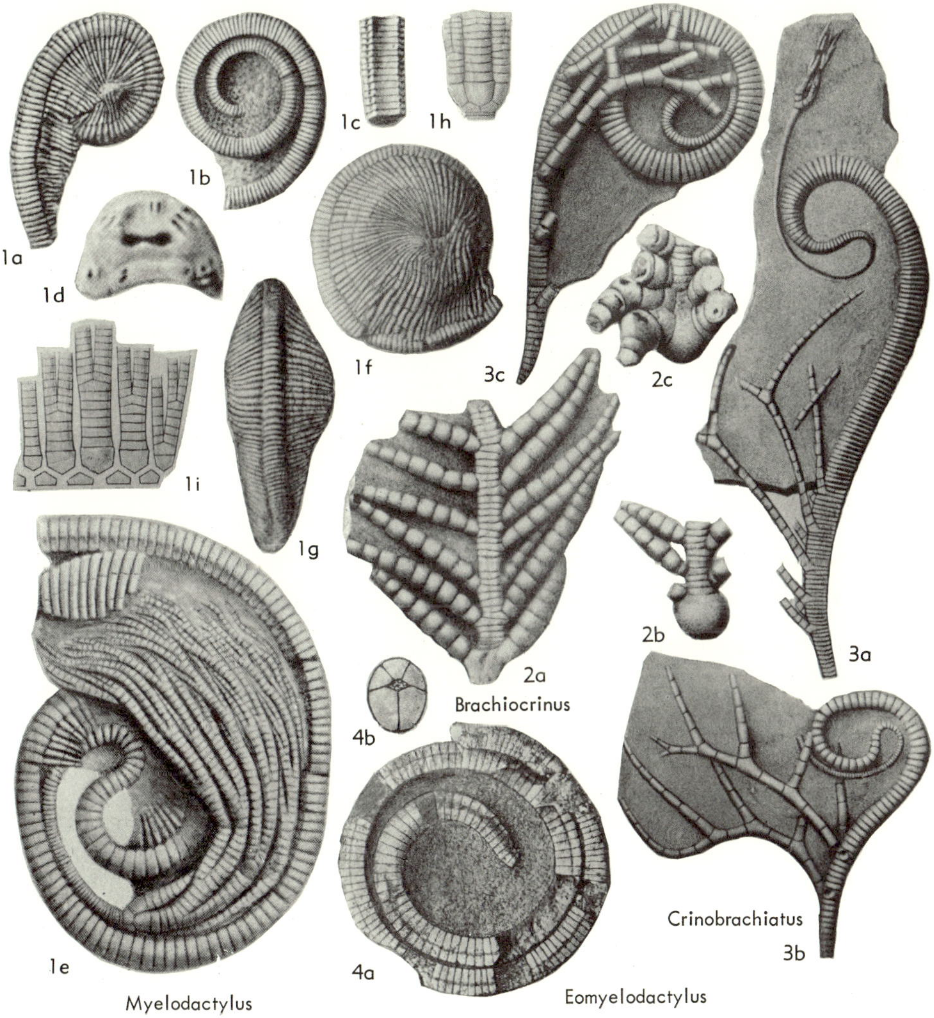

FIG. 344. Myelodactylidae (p. *T*551-*T*552).

Column transversely pentagonal, round, or bilaterally symmetrical subcrescentic and disposed in doubly recurved coil. *L.Ord.-L.Dev., ?U.Dev.*

Family MYELODACTYLIDAE S. A. Miller, 1883

[Myelodactylidae S. A. MILLER, 1883a, p. 278]

Stem composed of bilateral columnals arranged in a doubly recurved coil; cirri specialized for concealment of elongate slender crown; compound radial in *C* ray only, superradial bearing series of anal sac plates on left side and arm on right. Arms slender, branching isotomously or heterotomously. *L.Sil.-L.Dev.(Helderberg.), ?U. Dev.(Chemung.).*

Myelodactylus HALL, 1852, p. 191 [**M. convolutus*; SD SPRINGER, 1926b, p. 8]. Cup small, with 5 basals and compound radials in *C* ray only. Superradial of *C* ray in line with first primibrachs of other rays, bearing anal *X* on its left side and an arm on right. Arms slender, branching heterotomously. Proximal part of stem composed of thin circular columnals without cirri; distal part spirally coiled, composed of crescentic, bilaterally symmetrical columnals with paired cirri and shallow concavity between cirral sockets facing toward inner side of stem coil, long slender distal cirri disposed on each side of coil parallel to one another so that proximal parts of stem and crown are concealed. Crown lies against inner curve of coiled stem in reversed position, with tips of arms pointed backward in direction of distal part of stem. *M.Sil.(Niagaran)-L.Dev.(Helderberg.), ?U.Dev.(Chemung.)*; Eu.(Gotl.-Eng.-?Ger.)-USA (N.Y.-Tenn.-Ind.-W.Va.-Ill.).——FIG. 344,*1a-d*. **M. convolutus*, U.Sil., N.Y.; *1a,b*, lat. views of coiled stem and cirri, ×1; *1c*, inward side of distal columnals showing concave inner surface and cirral sockets along edges, ×1.5; *1d*, articular surface of distal columnal, ×4 (Springer, 1926b). ——FIG. 344,*1e-i*. *M. keyserensis* SPRINGER, L. Dev., W.Va.; *1e*, lat. view of stem and crown, ×1.5; *1f,g*, views of stem and cirri from side and in plane of coil, ×1; *1h,i*, *E*-ray view of partial crown and plate diagram, ×1.5 (Springer, 1926b). [=*Ophiocrinus* CHARLESWORTH, 1865 (*non* SALTER, 1856).]

Brachiocrinus HALL, 1858, p. 278 [**B. nodosarius* HALL, 1859b, p. 118; M]. Crown and proximal part of stem not known. Column with very stout, moderately short cirri given off from opposite sides of stem; cirri thickening in diameter away from stem and then tapering toward outer end, each pair borne by 2 successive columnals, which are separated by 2 to 5 noncirriferous columnals; distal extremity of stem a rounded to spheroidal, smooth, bulbous enlargement. *L.Dev.(Helderberg.)*, USA(N.Y.).——FIG. 344,*2*. **B. nodosarius*; *2a-c*, stems with paired cirri and bulbous distal extremities, ×1 (Springer, 1926b).

Crinobrachiatus MOORE, 1962, p. 43 [**Myelodactylus brachiatus* HALL, 1852, p. 232; OD]. Crown poorly known, apparently like *Iocrinus* in structure. Stem exceptionally large distally and threadlike proximally, bent into open S-shaped curve which does not enclose crown; cirri few, robust, and some very long, confined to part of stem on distal side of S-shaped coil and somewhat widely spaced, cirri composed of elongate cylindrical cirrals and branching in 2 or 3 places, directed obliquely upward; stem tapers to a point at distal extremity. *M.Sil.(Rochester)*, USA(N.Y.).——FIG. 344,*3*. **C. brachiatus* (HALL); *3a*, lat. view of stem and crown, ×1.5; *3b,c*, lat. views of distal part of stem, ×2, ×1.5 (Springer, 1926b).

Eomyelodactylus FOERSTE, 1919 [**Myelodactylus (Eomyelodactylus) rotundatus* FOERSTE, 1919, p. 19; OD]. Crown unknown. Stem closely coiled, without reverse curvature in proximal region, composed of moderately stout columnals circular to elliptical in outline, not concave on inner side, each columnal divided into pentameres of unequal size, those on inner side of stem being largest, relatively large lens-shaped lumen excentrically placed slightly outward from center. *L.Sil.(Brassfield Ls.)*, USA(Ohio).——FIG. 344,*4*. **E. rotundatus* FOERSTE; *4a,b*, lat. view of column and articular surface of columnal showing pentameres and lens-shaped lumen, ×1.5, ×4 (Foerste, 1919).

Herpetocrinus SALTER, 1873, p. 118 [**H. fletcheri*; M]. Cup with 4 basals and 4 radials. Anal sac larger in diameter than arms, and very long. Arms 4, small, heterotomous, branching repeatedly; cirrals short, rounded and beadlike. *U.Sil.(Wenlock.)*, Eu.(Eng.-Sweden)-USA(Tenn.-Ind.).——FIG. 345,*1*. **H. fletcheri*, Eng.; *1a,b*, lat. views of crown and stem, ×1, ×1.5; *1c,d*, *C*-ray and *?E*-ray views of cup, ×6; *1e*, plate diagram of cup and proximal brachials, ×1 (Springer, 1926b); *1f-k*, same species from Gotland, *1f,g*, anal tube plates from right and left sides, ×6; *1h,i*, lat. and transv. views of distal columnals and cirri, ×3; *1j*, lat. view of crown and column, ×1.5; *1k*, lat. view of partial crown and proximal stem, ×3 (Bather, 1893a).

Family IOCRINIDAE Moore & Laudon, 1943

[Iocrinidae MOORE & LAUDON, 1943a, p. 29]

Compound radial in *C* ray only; superradial supports anal *X* on left side and free arm on right. Arms with many isotomous branches. Column transversely pentagonal to pentastellate, composed of alternating thick and thin columnals. *?L.Ord., M.Ord.-U.Ord.*

Iocrinus HALL, 1866, p. 5 [**Heterocrinus (Iocrinus) polyxo* HALL, 1866, p. 5 (=*Heterocrinus subcrassus* MEEK & WORTHEN, 1865b, p. 148 [subj.], see MOORE, 1962b, p. 40); M]. Cup conical, basals 5; radials large, supporting several primibrachs in each ray. Infer- and superradial in *C* ray only; superradial axillary, at level of first primibrachs, supporting arm on right side and anal sac plates on left. Anal sac tall, with row of large anal plates on posterior side and thin, horizontally plicate plates laterally. Arms branching isotomously up to 8 times. Column transversely pentalobate. *?L.Ord.(Arenig.), M.Ord.(Llanvirn.)-U.Ord.*, Eu.(Wales-Eng.), USA (Ill.-Ohio-N.Y.)-Can.(Ont.).——FIG. 346,*2a,b*. **I. subcrassus* (MEEK & WORTHEN), U.Ord., Ohio; lat. views of crown, ×1 (Hall, 1872).——FIG. 346,*2c-f*. *I. crassus* (MEEK & WORTHEN), U.Ord., Ohio; *2c,d*, *A*- and *D*-ray views of cup, ×1.5 (Hall, 1872); *2e,f*, *A*-ray and *CD*-interray views of crown, ×1.5 (Meek, 1873).

Caleidocrinus WAAGEN & JAHN, 1899, p. 106 [**C. multiramus*; SD BASSLER, 1938, p. 56] [=*Caleidocrinus* WAAGEN & JAHN, 1892, p. 399 *(nom. nud.)*; *Kaleidocrinus* JAEKEL, 1902, p. 1075 (obj.) *(nom. van.)*]. Cup small, low, plates arranged as in *Iocrinus*. Small, rounded interbrachial plates present between proximal parts of arms above radials. Arms isotomous, branching repeatedly. Stem transversely round. *M.Ord.(Llanvirn.)*, Eu.

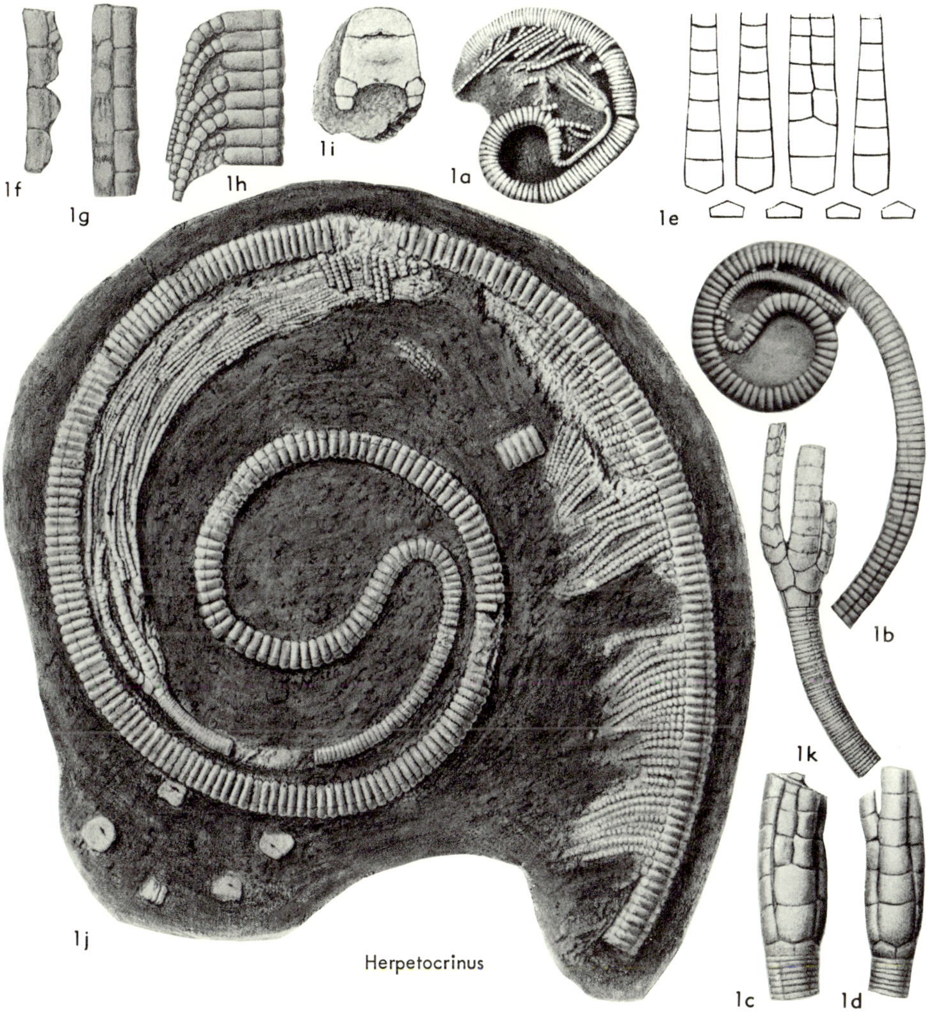

FIG. 345. Myelodactylidae (p. T552).

(Scot.-Boh.).——FIG. 346,*1a,b*. **C. multiramus*, M.Ord., Boh.; lat. and *CD*-interray views of crown, ×1, ×3 (Springer, 1920).——FIG. 346,*1c*. *C. turquidulus* RAMSBOTTOM, M.Ord., Scot.; lat. view of crown, ×2 (Ramsbottom, 1961).

Family EUSTENOCRINIDAE Ulrich, 1925

[Eustenocrinidae ULRICH, 1925, p. 99]

Compound radials in all five rays. Anal *X* supported directly by *C*-ray superradial, or by left side of first fixed primibrach in *C* ray. *L.Ord.-M.Ord.*

Eustenocrinus ULRICH, 1925, p. 99 [**E. springeri*; OD]. Crown barely wider than stem; basals 5, infer- and superradials present in all 5 rays; *C* superradial supports anal *X* and row of anal sac plates directly above, without an arm. First primibrachs fixed in cup. Arms branching isotomously on sixth to tenth primibrach. *M.Ord. (Trenton.),* Can.(Ont.).——FIG. 347,*1*. **E. springeri; 1a,* plate diagram of crown (superradials black, inferradials horizontally ruled, fixed brachials vertically ruled, anal plates stippled) (Moore, 1962b); *1b,c, AE*-interray and *C*-ray views of crown, ×5 (Lane, n; Springer Coll., U.S. Natl. Museum).

Peniculocrinus MOORE, 1962, p. 33 [**Heterocrinus? milleri* WETHERBY, 1880b, p. 153; OD]. Crown

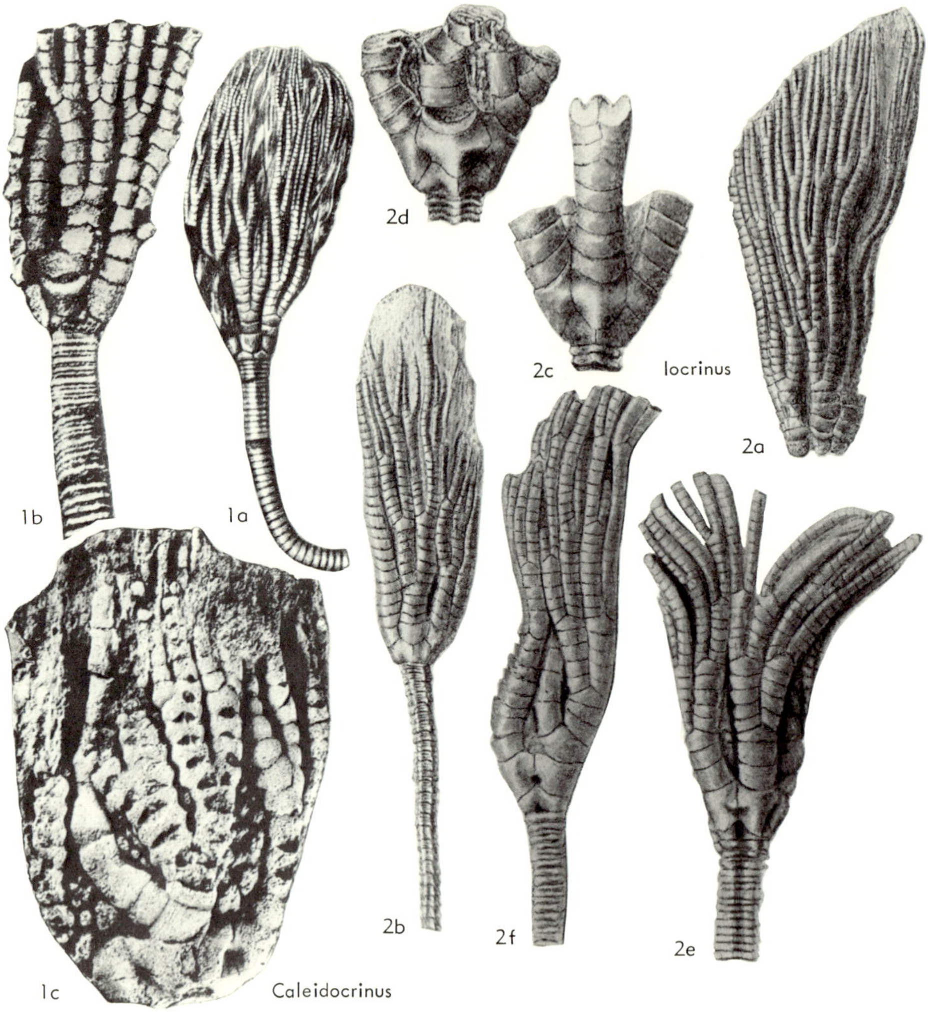

FIG. 346. Iocrinidae (p. T552-T553).

tall and very slender; basals 5; infer- and superradials present in all rays; first fixed primibrach of *C* ray supporting anal *X* and series of anal sac plates on left side and arm on right. First 2 primibrachs fixed in top of cup in each ray. Arms rounded, branching on sixth or seventh primibrachs, and with at least 2 higher isotomous branches. Stem subpentagonal in section. *M.Ord. (Trenton.)*, USA(Ky.).——FIG. 347,2. **P. milleri* (WETHERBY); *2a, AE*-interray view of crown, ×2.25 (Wetherby, 1880b); *2b, AE*-interray view of crown, ×3.75; *2c,d,* post. and ant. views of proximal part of crown and stem, ×6 (Moore, 1962b).

Ramseyocrinus BATES, 1968, p. 406 [**Dendrocrinus cambriensis* HICKS, 1873, p. 50; OD]. Cup composed of ?3 or ?4 basals; infer- and superradials present in each ray. Those of *C* ray lower than in other rays and supporting series of anal plates directly above. Primibrachs 9 or more, proximal brachials not fixed in cup. Arms strongly rounded, branching isotomously 3 or more times. Stem transversely wide, quadrilobate, columnals irregular in height. *L.Ord.(Tremadoc.)*, Eu. (Wales).——FIG. 348,2. **R. cambriensis* (HICKS); *2a,b,* ant. and post. views of crown and stem, ×5 (Bates, 1968).

Ristnacrinus ÖPIK, 1934, p. 3 [**R. marinus*; OD]. Cup high conical, truncate below; basals fused with proximal columnal or absent, not visible; large quadrate inferradials present in each ray supporting smaller superradials above, that of *C* ray next below small anal *X* and succeeding anal sac plates on left side and large primibrach on right. Primibrachs *1* axillary in *B, D,* and *E* rays, primibrachs *2* axillary in *A* and *C* rays; brachials not fixed in cup. Arms strongly rounded, branching at least twice isotomously. Stem transversely round, composed of thin and thick columnals, latter with bifascial articular surfaces. *M.Ord. (Caradoc.),* Eu.(Est.).——Fig. 348,*1*. **R. marinus; 1a, E*-ray view of crown and stem, ×2; *1b, C*-ray view of cup and proximal arms, ×2.9 (Öpik, 1934).

Superfamily ANOMALOCRINACEA Wachsmuth & Springer, 1886

[*nom. transl.* Moore & Lane, in Moore & Strimple, 1973, p. 17 (*ex* Anomalocrinidae Wachsmuth & Springer, 1886, p. 135[211])] [Materials for this superfamily prepared by R. C. Moore and N. Gary Lane]

Cup globose, with convex base, upflared basals clearly visible from side; radial articular facets narrow, circular and small. Tegminal plates firmly united, with modified ambulacrals and interambulacrals. Arms heterotomous, with ramule developed on each brachial, arranged in series on one side of branch. *M.Ord.-U.Ord.*

Family ANOMALOCRINIDAE Wachsmuth & Springer, 1886

[Anomalocrinidae Wachsmuth & Springer, 1886, p. 135 (211)]

Characters of superfamily. *M.Ord.-U. Ord.*

Anomalocrinus Meek & Worthen, 1865, p. 148 [**Heterocrinus? (Anomalocrinus) incurvus* Meek & Worthen, 1865b, p. 148; M] [=*Ataxiacrinus* Lyon, 1869, p. 464 (type, *A. caponiformis*); *Ataxocrinus* Bather in Lankester, 1900a, p. 146, *nom. van.*]. Cup large, globose; basals 5, small, subequal, pentagonal in outline; or with extra small quadrangular plate inserted in *D*-ray circlet. Radials large, *A* and *B* radials simple, *C* and *E* rays with infer- and superradials; *D* radial divided into 2 equal plates by a vertical suture; radial articular facets deeply inserted in plates of *D* and *A* radials; others on central raised area of plates and facing adorally; large quadrangular anal plate (?anal *X*) situated between upper shoulders of *C* superradial and *D* radial; arms branching several times on second, third, or fourth primibrachs, branches above axillaries may be unequal in size; brachials large, rounded, slender unbranched ramules present at irregular intervals in distal parts of arms. *U.Ord.(Cincinnat.),* USA(Ohio).——Fig. 349,*1*. **A. incurvus* (Meek & Worthen); *1a,b,* post. and ant. views

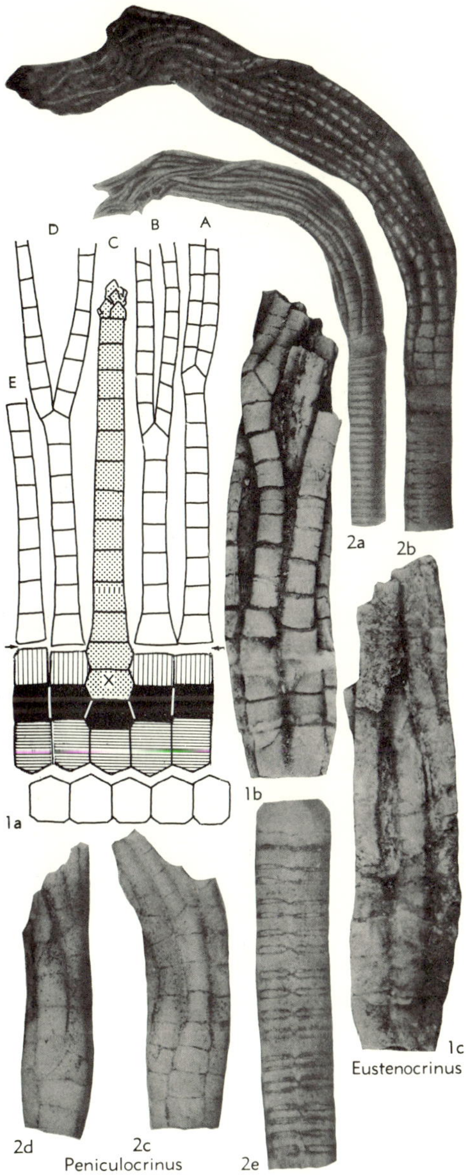

Fig. 347. Eustenocrinidae (p. *T*553-*T*554).

of crown, ×1; *1c,* articular surface of columnal, ×2 (Meek, 1873).

Geraocrinus Ulrich, 1925, p. 92 [**G. sculptus*; M]. Cup small, low; basals 5, subequal; *C* and *E* radials compound, others simple; large anal *X* on upper shoulders of posterior radials, supporting stout anal sac composed of 4 longitudinal rows of large convex plates. Arms long, stout, 2 to a ray, each brachial with simple, long ramule on

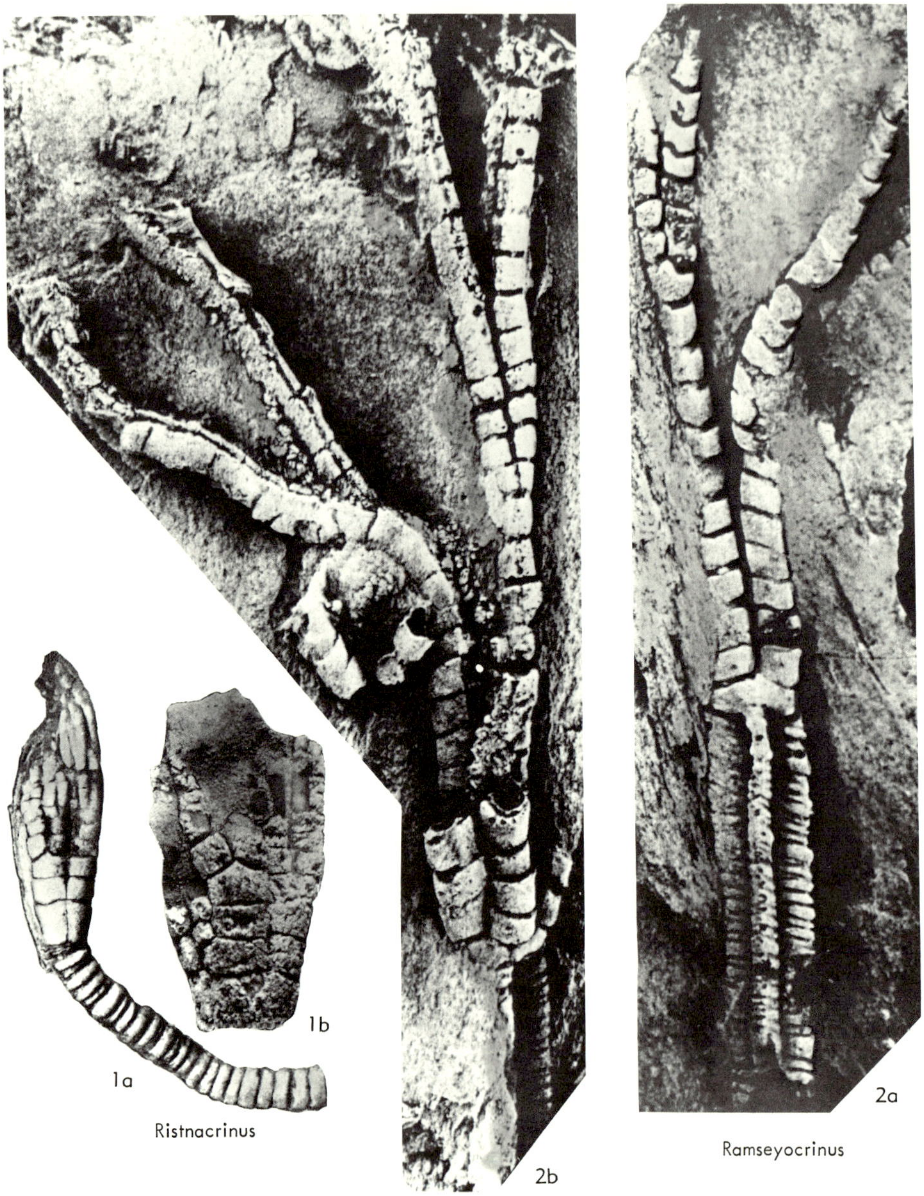

FIG. 348. Eustenocrinidae (p. T554-T555).

its admedial side; primibrachs nearly as wide as radials, 2 in all except anterior ray, which has only one primibrach. *M.Ord.(Trenton.)*, USA (Tenn.).——FIG. 349,2. **G. sculptus*; *2a*, post. view of partial crown, ×1; *2b,c*, plate diagram and arms of one ray with ramules given off on inward sides, ×2; *2d,e*, aboral surface of brachials and ramules and articular surface, ×4 (Ulrich, 1925).

Glaucocrinus PARKS & ALCOCK, 1912, p. 43 [**G. falconeri*; M]. Cup globose, basals 5, equal and rather large; *D* and *B* rays with small infer- and superradials, commonly lacking articular facet and arm, other radials simple, large, with narrow rounded facets; anal *X* a small, triangular plate on upper left shoulder of *C* radial. Arms rounded, stout, dividing on first primibrach, and with 2 or 3 higher equal divisions. *M.Ord.(Trenton.)*,

N.Am.(Can., Ont.).——FIG. 349,*3*. **G. falconeri*; post. view of crown, ×1 (Parks & Alcock, 1912).

Superfamily BELEMNOCRINACEA S. A. Miller, 1883

[*nom. transl.* MOORE & LANE, in MOORE & STRIMPLE, 1973, p. 17 (*ex* Belemnocrinidae S. A. MILLER, 1883a, p. 276)] [Materials for this superfamily prepared by R. C. MOORE and N. GARY LANE]

Crown moderately tall, cylindrical. Cup with bilateral symmetry defined by plane through *A* ray and *CD* interray. Cup steep sided conical, five or three strongly upflared large basals; radials three to five, generally with narrow horseshoe-shaped articular facets sloping outward-downward or occupying full width of plate summit; single anal plate (*X* or radianal) in cup. Arms uniserial, branching isotomously on primibrachs 4 or unbranched. Stem transversely round. *M.Sil.-U.Carb.; U.Perm.*

Family BELEMNOCRINIDAE S. A. Miller, 1883

[Belemnocrinidae S. A. MILLER, 1883a, p. 276] [=Belemnocrinidae WACHSMUTH & SPRINGER, 1886, p. 193]

All five radials simple; basals highly elongate and rather large, making up most of cup. Anal sac large and with pores along plate sutures. Arms with two main branches, bearing branched or unbranched ramules. *L.Miss.(Osag.).*

Belemnocrinus WHITE, 1862, p. 14 [**B. typus*; M] [*non Belemnocrinus* MUNIER-CHALMAS, 1876, p. 105, =*Belocrinus* (blastoid)]. Cup small, narrow and tall; basals 5, elongate, conspicuously larger than radials, which are small and quadrangular. Radianal or anal *X* directly above *CD* basal in line with and equal in size to radials, supporting large anal sac composed of hexagonal plates with transverse slitlike pores across plate sutures. Arms long slender rounded, bifurcating once on fourth or fifth primibrachs, each arm with ramules on alternate sides of every second or third secundibrach. *L.Miss.,* USA(Iowa-Mo.).——FIG. 350,*1*. **B. typus,* U.Burlington, Iowa; lat. view of crown and proximal part of column, ×1 (Wachsmuth & Springer, 1885).

Whiteocrinus JAEKEL, 1918, p. 58 [**Belemnocrinus florifer* WACHSMUTH & SPRINGER, 1877, p. 256; M]. Like *Belemnocrinus* except basal-radial sutures strongly constricted and radials outflaring. Arms arranged as in *Belemnocrinus* but unbranched. Column strongly pentagonal in section

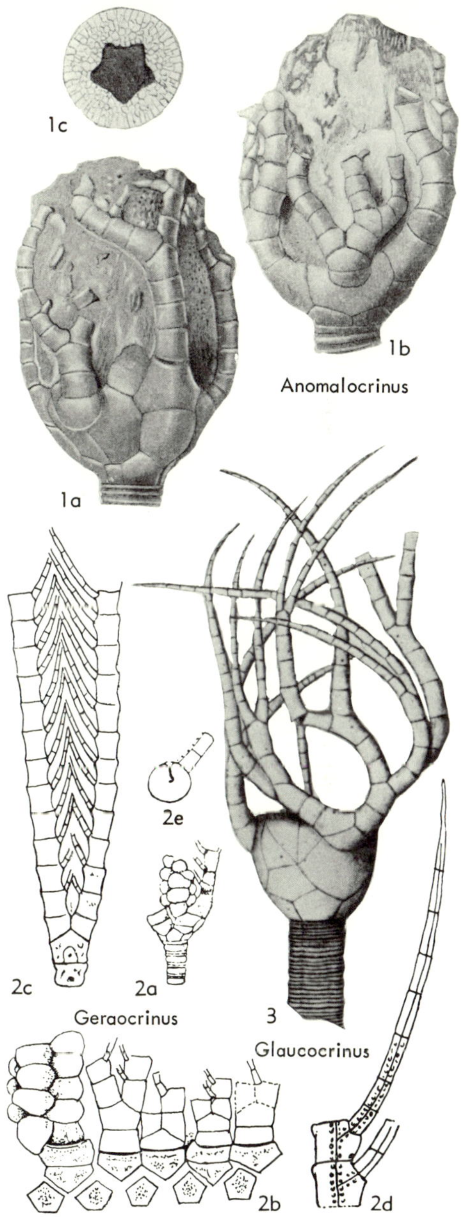

FIG. 349. Anomalocrinidae (p. T555-T557).

with cirriferous nodals. *L.Miss.,* USA(Iowa).——FIG. 350,*3*. **W. florifer* (WACHSMUTH & SPRINGER); lat. view of crown, ×1 (Wachsmuth & Springer, 1885).

Family HOLYNOCRINIDAE Bouška, 1948

[Holynocrinidae BOUŠKA, 1948, p. 521]

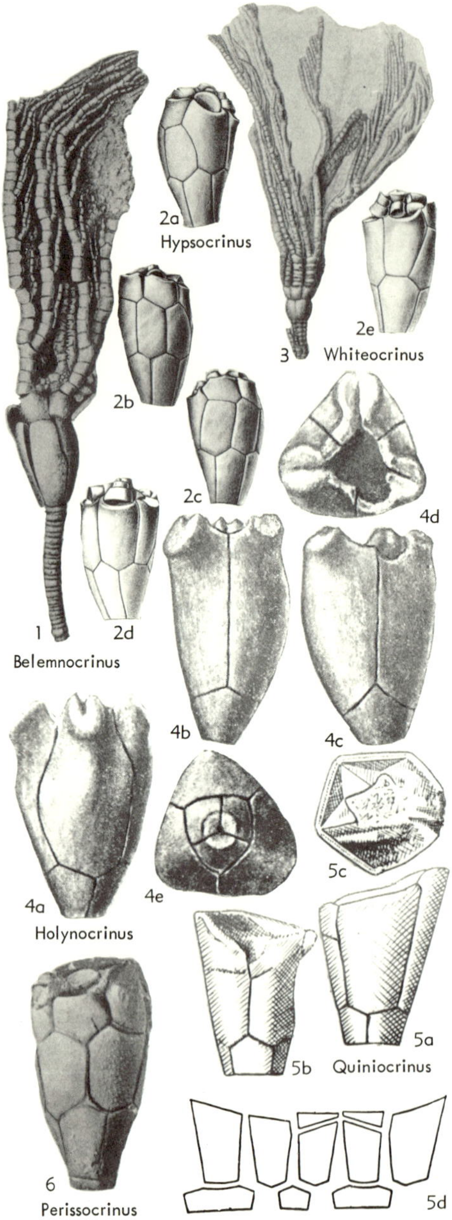

FIG. 350. Belemnocrinidae *(1,3)*; Holynocrinidae *(4)*; Perissocrinidae *(2,5,6)* (p. *T557-T558*).

Cup trimerous; *A, C,* and *D* radials only ones in cup. *M.Dev.*

Holynocrinus BOUŠKA, 1948, p. 521 [**H. moorei*; OD]. Cup small, trilobate in cross section; 3 equal basals; 3 large subequal radials *(A, C, D)* in cup; other radials, radianal, and anal *X* plates lacking. Upper left shoulder of *C* radial notched by proximal edge of anal opening. Arm facets small, narrowly rounded, horseshoe-shaped, with deep ambulacral notches. *M.Dev.*, Eu.(Boh.).——FIG. 350,*4*. **H. moorei; 4a-e, A*-ray, *AC*- and *CD*-interrays, ventral and dorsal views of theca, ×20 (Bouška, 1948).

Family PERISSOCRINIDAE Strimple, 1963

[Perissocrinidae STRIMPLE, 1963a, p. 24]

Cup conical, with three or five basals, and two, three, or all five rays with infer- and superradials. Anal *X* either in cup or above it. *M.Dev.*

Perissocrinus GOLDRING, 1936, p. 20 [**P. papillatus*; OD]. Cup cylindrical, basals large and elongate; all rays with infer- and superradials; anal *X* large, pentagonal, separating *C* and *D* superradials and in contact below with *C* and *D* inferradials. Radial facets shallow, concave, filling most of distal surface of each radial. Three nonaxillary primibrachs known in 2 rays, other arms unknown. *M.Dev.*, USA(N.Y.).——FIG. 350,*6*. **P. papillatus; CD*-interray view of theca, ×2 (Goldring, 1923).

Hypsocrinus SPRINGER & SLOCUM, 1906, p. 267 [**H. fieldi*; OD]. Cup narrow, conical; basals 5, large and elongate, one *(?DE)* basal larger than others, truncate distally and directly below *D* radial; anterior superradial large and expanded, so that *A* inferradial is shifted to left, partly below *B* superradial; *B* and *C* superradials bounded below by *A, B,* and *C* inferradials, *E* radial simple. Anal *X* not in cup. Two or 3 proximal brachials only known parts of arms. *M.Dev.*, USA(N.Y.).——FIG. 350,*2*. **H. fieldi; 2a-e, AE*-interray, *C-, B-, E-,* and *D*-ray views of theca, ×1.5 (Springer & Slocum, 1906).

Quiniocrinus SCHMIDT, 1942, p. 33 [**Q. erectus*; OD]. Cup small, conical; 3 unequal basals, small one in *CD* interray; *A, D,* and *E* rays with large simple radials; *B* and *C* rays with large inferradials and small, wedge-shaped superradials; radial articular facets as wide as radials. Arms and anal sac unknown. *M.Dev.*, Eu.(Ger.).——FIG. 350,*5*. **Q. erectus; 5a-d, A*-ray, *BC*-interray, and ventral views of theca, ×2, and plate diagram of theca (Schmidt, 1942).

Family PYGMAEOCRINIDAE Strimple, 1963

[Pygmaeocrinidae STRIMPLE, 1963a, p. 36]

Theca small, with perfect pentameral symmetry, or with three unequal basals; no anal plates in cup. First primibrachs set in deep notch at top of each radial plate. *U.Sil.-M.Dev.*

Pygmaeocrinus BOUŠKA, 1947, p. 2 [**P. kettneri*;

OD]. Theca small, with perfect pentameral symmetry; bipyramidal in lateral view and pentastellate as seen from above or below; 5 small basals; 5 radials laterally protruding and keeled medially. Arms composed of only 2 primibrachs, first one small, quadrangular, and set into notch in radial, second primibrachs large, spear shaped, set into notches in top of cup and completely covering ventral surface of theca. *U.Sil.-M.Dev.*, Eu.(Boh.).——FIG. 351,*1*. **P. kettneri*, L.Dev.; *1a-c*, lat., dorsal, and ventral views of crown (large keeled second primibrachs shown in *1a,c*), ×15; *1d-f*, ventral view of cup, int. and lat. views of arm, ×10 (Bouška, 1947).

Storthingocrinus SCHULTZE, 1867, p. 181 [**Platycrinus fritillus* MÜLLER, 1855, p. 80; OD]. Cup small; basals 3, unequal, small one in *DE* interray; radials 5, quadrangular, with deeply inset narrow articular facets. *L.Dev.-M.Dev.*, Eu.(Ger.-Eng.-Spain)-India.——FIG. 351,*2*. **S. fritillus* (MÜLLER), M.Dev., Ger.; *2a-d*, *C*-ray, *D*-ray, ventral, and dorsal views of cup, ×2 (Schultze, 1867).

Family SYNBATHOCRINIDAE S. A. Miller, 1889

[Synbathocrinidae S. A. MILLER, 1889, p. 215] [=Symbathocrinidae BATHER, 1899b, p. 921 (*nom. van.*); suborder Symbathocrinites JAEKEL, 1918, p. 89 (*partim*)]

Crown elongate and slender. Cup small, broadly truncate conical or bowl-shaped; basals three or five, small one in *AE* interray; radials simple and subequal, or with traces of transverse division into infer- and superradials in *B*, *C*, and *E* rays; radial articular facets wide, with transverse ridge and may extend inward almost to longitudinal axis of cup; lateral edges of facets may be produced distally into radial processes which abut sides of first primibrachs. Anal *X* either notching upper corners of *C* and *D* radials or out of cup. Anal sac slender, elongate, confined to posterior edge of tegmen, closely resembling an arm externally. Tegmen composed of five oral plates, *CD* oral slightly larger than others. Arms atomous, first primibrachs fully as wide as radials. *M.Sil.-U.Carb.; U.Perm.*

Synbathocrinus PHILLIPS, 1836, p. 206 [**S. conicus*; M] [=*Synbathocrinites* AUSTIN & AUSTIN, 1842, p. 109 (*nom. van.*); *Donacicrinites* TROOST, 1849, p. 420 (*nom. nud.*); *Symbathocrinus* D'ORBIGNY, 1852, p. 157 (*nom. van.*); *Donacicrinus* BATHER in LANKESTER, 1900a, p. 202 (*nom. nud.*)]. Cup low conical or bowl shaped; basals low, 3, unequal, with small one in *AE* interray; radials large, radial articular facets as wide as radials, with transverse ridge and ligament pit; anal *X* slender and elongate, with proximal part notching upper corners of *C* and *D* radials, extending distally to top of adjacent first primibrachs. Anal sac elongate and narrow, armlike, confined to posterior edge of tegmen. Tegmen composed of 5 large oral plates, *CD* oral slightly larger than others and interlocked at its apex with tips of *AB* and *AE* orals. Arms long and slender, atomous. *M.Dev.-U.Penn.*, cosmop.——FIG. 352,*1a*. **S. conicus*, L.Carb., Eng.; lat. view of cup, ×2 (Wright, 1951-54).——FIG. 352,*1b*. *S. anglicus* WRIGHT, L.Carb., Eng.; lat. view of crown, ×1 (Wright, 1946).——FIG. 352,*1c,d*. *S. robustus* WACHSMUTH & SPRINGER, L.Miss.(Osag., Burlington Ls.), SE.Iowa; dorsal and ventral views of cup, ×1 (Moore & Laudon, 1943a).——FIG. 352,*1e-p*. *S. texasensis* MOORE & EWERS, L.Miss. (Osag., Chappel Ls.), Mason Co., Texas; *1e,f*, *CD*-

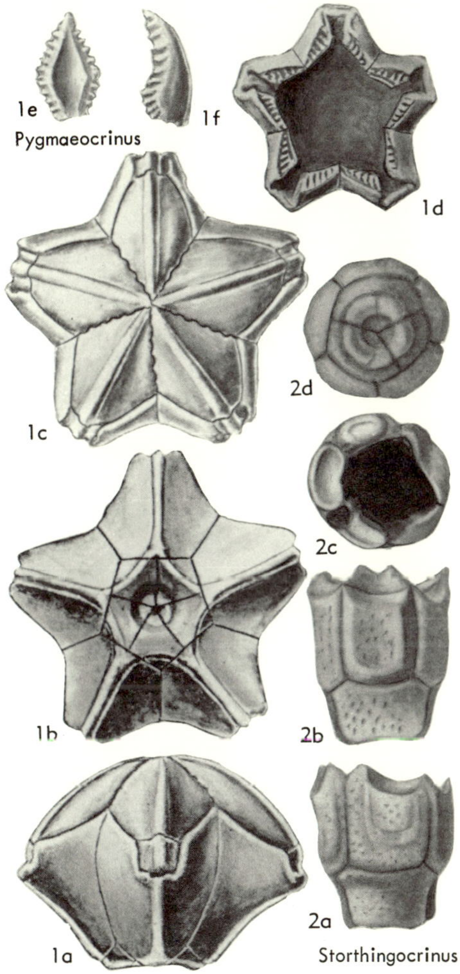

FIG. 351. Pygmaeocrinidae (p. T558-T559).

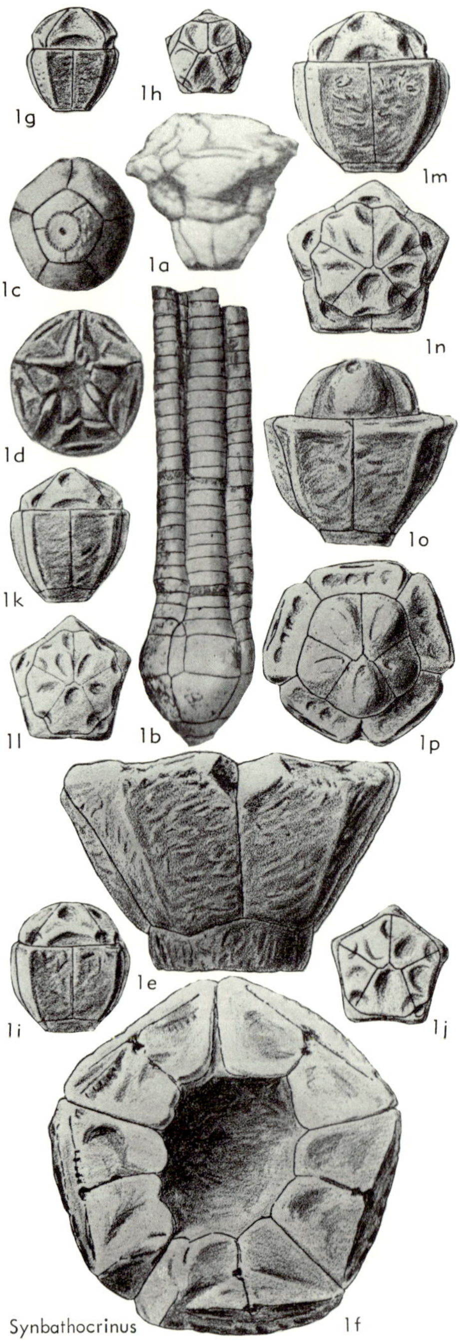

FIG. 352. Synbathocrinidae (p. T559-T560).

interray and ventral views of adult cup, ×10; *1g,h, 1i,j, 1k,l, 1m,n, 1o,p, CD*-interray and ventral views of successively more mature juvenile cup, ×10 (Moore & Ewers, 1942).

Abyssocrinus STRIMPLE, 1963, p. 29 [**Synbathocrinus antiquus* STRIMPLE, 1952c, p. 76; OD]. Like *Theloreus* except transverse sutures separating infer- and superradials of *B, C,* and *E* rays not present consistently. Body cavity quite small and restricted ventrally by medially extended parts of radial articular facets. *C* and *D* radials notched by anal *X. M.Sil.(Niagaran),* USA(Okla.).——FIG. 353,*3*. **A. antiquus* (STRIMPLE); *3a-c, CD*-interray, *A*-ray, and ventral views of theca, ×2 (Strimple, 1952c).

Phimocrinus SCHULTZE, 1866, p. 29 [**P. laevis*; SD WACHSMUTH & SPRINGER, 1886, p. 170]. Cup high conical, truncate below; basals 5, simple; notch for anal *X* plate visible in side view of cup; posterior edges of *C* and *D* radial articular facets excavated at sides of anal tube, facets horizontal or inclined outward-upward, not visible in side view. *L.Dev.-M.Dev.,* Eu.(Ger.)-USA (Tenn.).——FIG. 353,*1*. **P. laevis,* M.Dev., Ger.; *1a-c,* lat., ventral, and dorsal views of theca, ×2 (Schultze, 1867); *1d,* plate diagram of cup (radials black, *A* ray upward) (Moore, n).

Ramacrinus PROKOP, 1969, p. 375 [**R. multiformis*; OD]. Basals 5, equal, high; *A* and *D* radials undivided, *B, C,* and *E* radials divided transversely by convex sutures into tall inferradials and short broad superradials. Arms unknown. Stem composed of low, cylindrical columnals. *L.Dev.-M.Dev.,* Eu.(Boh.).——FIG. 354,*1*. **R. multiformis,* L.Dev., Boh.; *1a,b,* post., and oral views of theca, ×7.5 (Prokop, 1970).

Stylocrinus SANDBERGER & SANDBERGER, 1856, p. 399 [**S. scaber* (attributed to *Platycrinites scaber* GOLDFUSS [*nom. nud.*] by SANDBERGER & SANDBERGER); M]. Cup low; basals 5, or rarely 3, with small basal in *DE* interray; posterior radials not notched by anal *X*; radial articular facets extending inward horizontally even with tops of radial plates, without distally directed radial processes. *U.Sil.-M.Dev.,* Eu.(Ger.)-USA(N.Y.-Okla.).——FIG. 353,*2*. **S. scaber,* M.Dev., Ger.; *2a-c,* lat. views of cup and crown and dorsal view, ×3, ×6, ×12 (Sandberger & Sandberger, 1856).

Taidocrinus TOLMACHEV, 1924, p. 289 [**T. poljenowi*; M]. Like *Synbathocrinus* except that anal *X* is completely above cup and does not notch upper corner of either *C* or *D* radials. *U.Carb.; U.Perm.,* Eu.(USSR, Kuzbas)-Indon.(Timor).——FIG. 353,*5*. *T. inflatus* (WANNER), U.Perm., Timor(Noil Tonini); *5a-c,* ant., dorsal, and ventral views of theca (*CD* interray downward in *5b,c*), ×2 (Wanner, 1916a).

Theloreus MOORE, 1962b, p. 44 [**Phimocrinus jouberti* OEHLERT, 1882, p. 353; OD]. Cup tall and conical; basals 5, large, and distally upright; *B, C,* and *E* radials divided by transverse suture into large inferradial and short small superradial; upper corner of *D* radial and *C* superradial notched for reception of anal *X. M.Dev.,* Eu.(Ger.).——

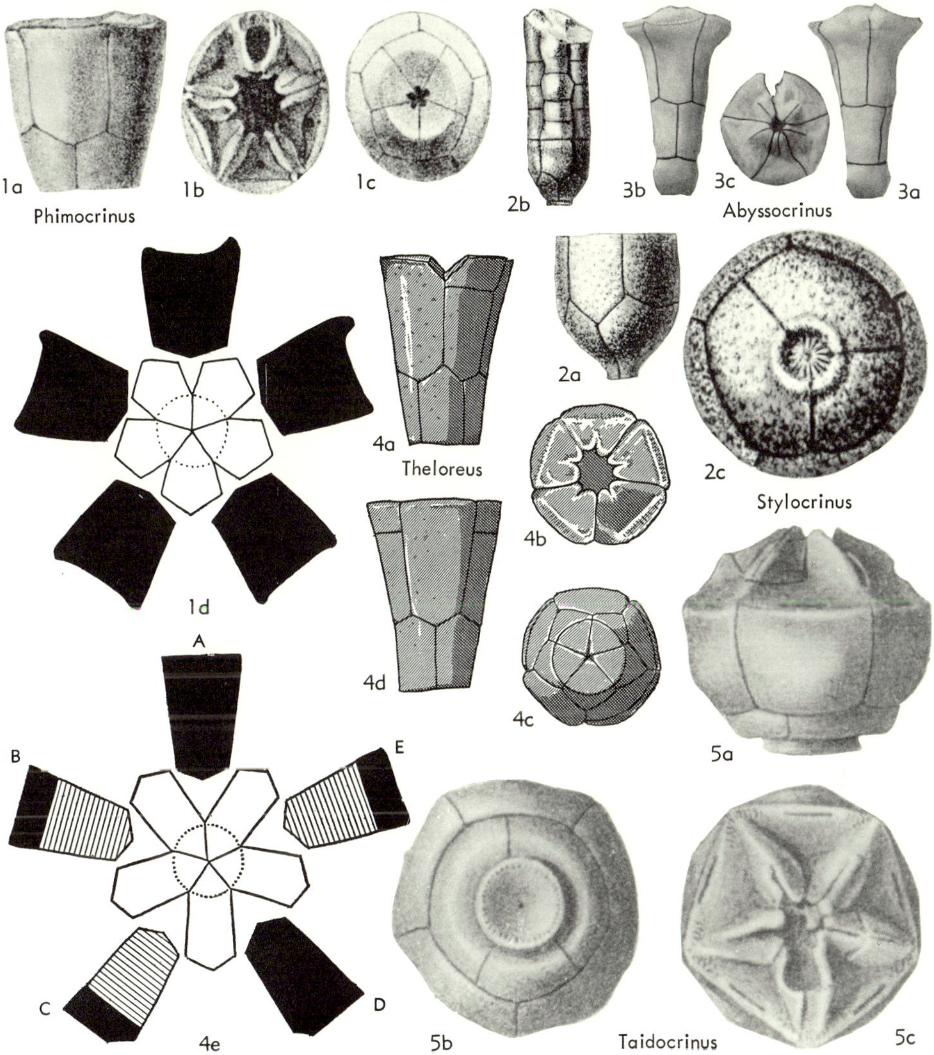

FIG. 353. Synbathocrinidae (p. *T*560-*T*561).

FIG. 353,*4*. **T. jouberti* (OEHLERT); *4a-e, CD*-interray, ventral, dorsal, and *A*-ray views of cup, plate diagram (radials and superradials black, inferradials horizontally ruled), ×2.5 (Moore, 1962b).

Family ZOPHOCRINIDAE S. A. Miller, 1892

[Zophocrinidae S. A. MILLER, 1892b, p. 642] [=Tiaracrinidae BATHER in LANKESTER, 1900a, p. 57]

Cup small, globose; basals three or five; radials four or five; transverse epispiral canals may be present in radials; anal plates unknown; tegmen composed of five oral plates. Arms much reduced or absent, brachials interposed between oral plates. *U. Sil.-M.Dev.*

Zophocrinus S. A. MILLER, 1891, p. 32 [**Z. howardi*; M]. Cup small and high, pear shaped or obconical; basals 3, large, *CD* basal larger than others; radials 4, elongate, radial directly above *CD* basal larger than others and may represent laterally fused *C* and *D* radials. Anal plates unknown. Tegmen composed of 5 spear-shaped plates, *CD* oral slightly larger than others and interlocking with tips of *AB* and *AE* orals. Five groups of 3 small ?fixed brachial plates disposed around periphery of tegmen directly above each

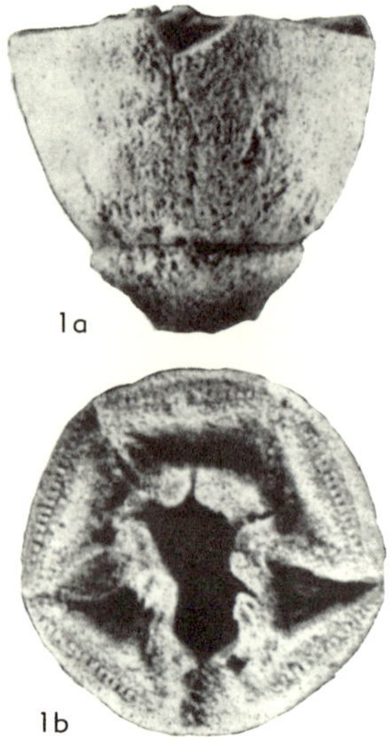

FIG. 354. Synbathocrinidae (p. T560).

interoral suture and in contact with distal edges of radials. No ambulacral or other opening in theca known; center plates of each group bearing a small ?arm facet at its admedial end and bounded by curved ends of 2 lateral plates of each group. Arms and stem unknown. *U.Sil.*, USA(Ill.-Ind.-Okla.-Tenn.).——FIG. 355,*1*. **Z. howardi; 1a,b* (Ind.), post. and ant. views of theca, ×4 (Springer, 1926b); *1c-e* (Tenn.), dorsal, lat., and ventral views, ×3 (Miller, 1892b).

Parazophocrinus STRIMPLE, 1963, p. 61 [**P. callosus*; OD]. Like *Zophocrinus* except that 5 basal and 5 radial plates are present. *U.Sil.*, USA (Okla.).——FIG. 356,*1*. **P. callosus; 1a,b,* ventral and dorsal views, ×1.5; *1c,* dorsal view of another specimen, ×1.5 (Strimple, n).

Tiaracrinus SCHULTZE, 1867, p. 226 [**T. quadrifrons*; M] [=*Staurosoma* BARRANDE, 1887, p. 21 (obj.) (type, *S. rarum* BARRANDE, 1887, =*T. quadrifrons*)]. Cup small, globular, composed of 3 small, low basals which form basal concavity of cup; radials 4, large, with 4 sets of transverse epispiral ridges and intervening grooves; epispires increase in width distally, most proximal one just above basals having single opening above intraradial suture; epispiral canals extending under each ridge through skeletal material of radial from one external opening to another at extremities of each ridge; 8 to 25 canals present in each epispire. Tegmen composed of 5 large interlocking orals. ?Fixed brachials numerous, in 5 sets, and in contact with radials admedially, each set composed of 11 or 12 small thin plates and 1 or 2 small triangular plates admedially. Arm facets obscure or absent. Arms and column unknown. *L.Dev.-M.Dev.*, Eu.(Ger.-France-Boh.).——FIG. 355,*2*. **T. quadrifrons*; M.Dev., Ger.; *2a,b, CD*-interray and dorsal views, ×2; *2c,d,* dorsal and ventral views, ×3 (Springer, 1926b).

Superfamily PERITTOCRINACEA Ubaghs, 1971

[*nom. transl.* SPRINKLE & LANE, herein (*ex* Perittocrinina UBAGHS, 1971a, p. 329)] [Material for this superfamily prepared by JAMES SPRINKLE and N. GARY LANE]

Disparid crinoids characterized by accessory plates between the basals and radials; anal *X* well developed on side of cup; only four basals; one genus has endothecal pore slits between cup and accessory plates, whereas the other presently known genus has enlarged proxistele composed of tetrameres. *Up.L.Ord. or low.M.Ord.*

Although classified by UBAGHS (1971a) as a suborder of the order Hybocrinida, perittocrinids do not seem to be closely related to hybocrinids. Both perittocrinid genera have numerous triangular accessory plates

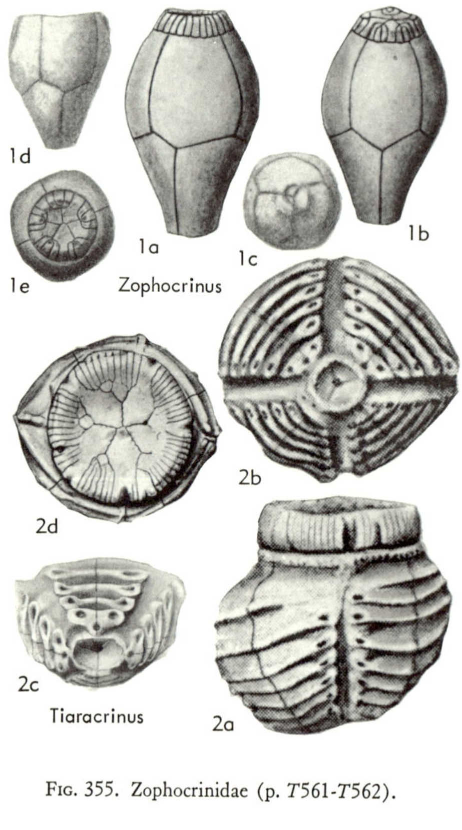

FIG. 355. Zophocrinidae (p. T561-T562).

between the larger basal and radial plates and only four basals; in addition, *Perittocrinus* has endothecal pore slits between most of the cup plates (not present in *Tetracionocrinus,* which has a proximally enlarged stem with primitive four-piece columnals). The stem is unknown in *Perittocrinus.* None of these features are found in hybocrinids, which have a standardized pattern of cup plates without accessory ones or pore slits and a slender stem composed of normal one-piece columnals. Perittocrinids are here classified as a separate superfamily of the Disparida because they do not seem to be closely related to any other Ordovician disparids.

Family PERITTOCRINIDAE Abel, 1920

[Perittocrinidae ABEL, 1920, p. 264]

Cup globose, composed of four basals, five large radials, radianal, anal *X,* and numerous small to medium-sized accessory plates between primary cup plates; radial articular facets narrow and suboval; endothecal pore slits (respiratory folds) or hollow ridges crossing most cup plate sutures at right angles. *Up.L.Ord. or low.M.Ord.*

Perittocrinus JAEKEL, 1902, p. 1093 [**Porocrinus radiatus* BEYRICH, 1879, p. 61; SD BASSLER, 1938, p. 150]. Perittocrinids with endothecal pore slits crossing sutures between major cup plates and smaller triangular accessory ones. Stem, arms, and tegmen unknown. *Up.L.Ord. or low.M.Ord.,* Eu.(NW.USSR).——FIG. 357,*2a-c.* **P. radiatus* (BEYRICH); *2a,* plate diagram; *2b,c,* lat. post. and dorsal views of cup, ×2 (Ubaghs, 1971a).

Tetracionocrinus UBAGHS, 1971, p. 333 [**Perittocrinus transitor* JAEKEL, 1918, p. 49]. Perittocrinids with raised hollow ridges extending across sutures between all cup plates, including accessory plates. Enlarged proximal columnals composed of tetrameres. Arms and tegmen unknown. *Up.L. Ord. or low.M.Ord.,* Eu.(NW.USSR).——FIG. 357,*1a-c.* **T. transitor* (JAEKEL); *1a,* plate diagram; *1b,c,* ant. and post. views of cup, ×5 (Ubaghs, 1971a).

Superfamily UNCERTAIN

Family PARADOXOCRINIDAE Moore & Laudon, 1943

[Paradoxocrinidae MOORE & LAUDON, 1943a, p. 102] [Materials for this family prepared by N. GARY LANE]

Diagnosis same as for genus. *U.Perm.*

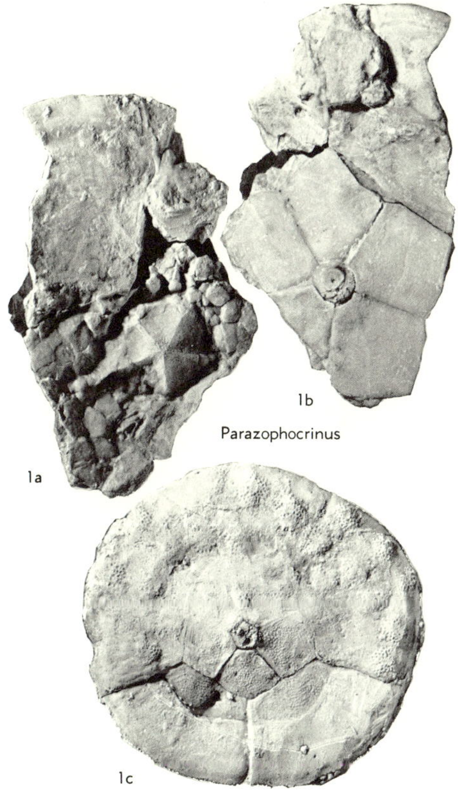

FIG. 356. Zophocrinidae (p. *T*562).

Paradoxocrinus WANNER, 1937, p. 205 [**P. patella;* OD]. Cup bowl shaped, basally concave, constricted distally; basals 3, subequal in size; radials 5; basals and radials surrounded and mostly covered by 5 to 13 bulbous accessory plates; 5 triangular orals alternate with radials; anal vent opening below posterior oral and between *C* and *D* radials. Arms unknown. Stem round. *U.Perm.,* Indon.(Timor).——FIG. 358,*1.* **P. patella,* Basleo; *1a,b,* lat. and ventral views of theca with orals in place, ×1.5 (Marez Oyens, 1940b); *1c-f,* lat., dorsal, central, and dorsal views of cup, ×2 (Wanner, 1937).

Family UNCERTAIN

Vosekocrinus JAEKEL, 1918, p. 56 [**V. granulatus;* M]. Cup low, bowl shaped; basals and radials large; radial facets rounded, somewhat narrower than radials; anal plates unknown. Arms isotomous, branching on second primibrachs and with at least two higher divisions. Stem pentagonal. [Poorly known, traditionally placed in the Dendrocrinidae of the cladid inadunates, but JAEKEL (1918) states clearly that infrabasal plates are lacking and he places the genus in the family Vosekocrinidae with *Caleidocrinus* WAAGEN &

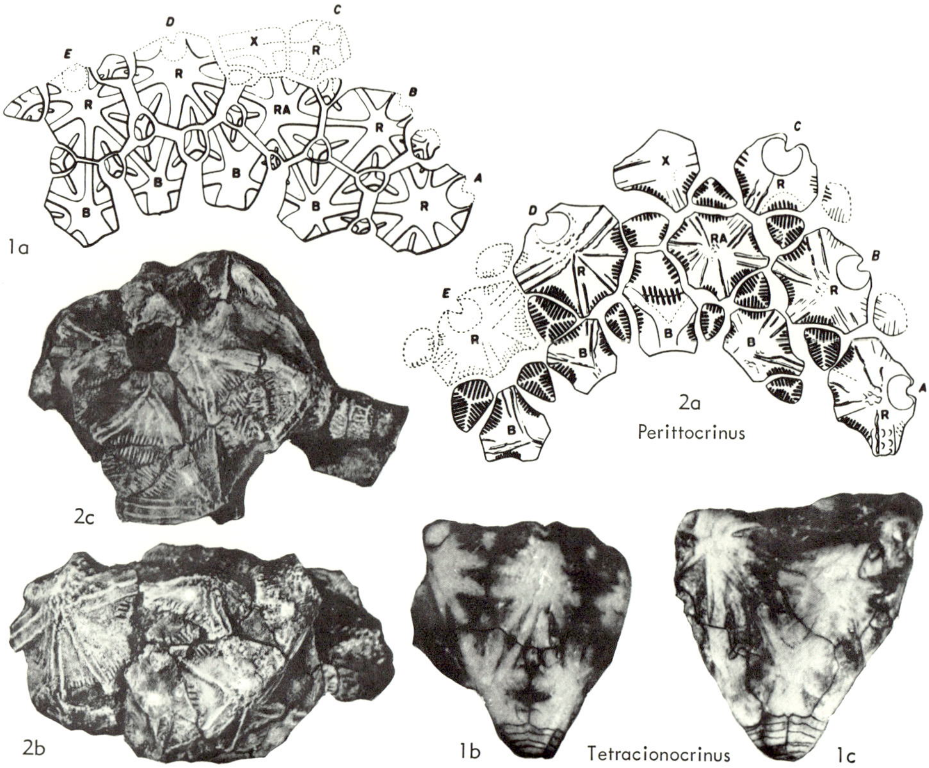

FIG. 357. Perittocrinidae (p. T563).

JAHN, a genus that has been accepted as a monocyclic disparid inadunate.] *L.Ord.*, Eu.(Boh.).——FIG. 359,*1*. **V. granulatus*; diagram, lat. view of cup, ×2 (Jaekel, 1918). [LANE]

HYBOCRINIDA

By JAMES SPRINKLE and RAYMOND C. MOORE

INTRODUCTION

The Hybocrinida are a small group of early Paleozoic monocyclic inadunate crinoids that show a tendency toward reduction or loss of the arms and replacement by recumbent ambulacral grooves. Only seven genera and about 27 species are known, ranging from the Early to Late Ordovician. At present, these genera are grouped into four families in the order Hybocrinida, based on the number and type of appendages present. Although the food-gathering appendages have been drastically modified in several advanced genera, the thecal plating and development of the water vascular system has remained very stable in nearly all hybocrinids, enabling them to be differentiated easily from other early echinoderm groups with similar morphology and recumbent ambulacra.

THECA

The theca of most hybocrinid genera consists of a large cup-shaped to globular cup with a reduced tegmen. No anal tube or sac is present in any of the known genera. The plating is monocyclic and only a single compound radial is present in the *C* ray. The lateral profile of the theca usually shows a slight to moderate asymmetry with a bulged *CD* interray, unequal-sized basals, and in some species a slightly offset stem facet.

All hybocrinids except *Cornucrinus* have standardized monocyclic arrangement of the cup plates with only slight variation from genus to genus; five basals, five radials with the *C* radial generally much smaller than the others and supported by a large radianal below, anal *X* and several accessory anals in the posterior interray, and five small orals around the mouth on the summit (Fig. 360). The specialized genus *Cornucrinus* has two of its basals fused, no anal *X*, the anal opening displaced down to the side of the cup, and either very small orals on the tegmen or none at all (*see* Fig. 366,*1a*).

The only plating feature to show any systematic variation is the size and position of the anal *X* with respect to the *C* radial and radianal. In *Revalocrinus,* the most primitive hybocrinid, the anal *X* (missing) must have been quite small and confined to the tegmen obliquely above the *C* radial (*see* Fig. 364,*3*); it apparently did not reach the radianal below. In slightly more advanced genera *(Baerocrinus, Hoplocrinus),* the anal *X,* still relatively small, is mostly confined to the domed tegmen but does reach the elongate radianal (Fig. 364,*2a*). In relatively advanced genera such as *Hybocrinus, Hybocystites,* and *Tripatocrinus,* the anal *X* has moved down to become part of the cup above the radianal and has increased to about the same size as the *C* radial, which is located directly beside it (Fig. 360).

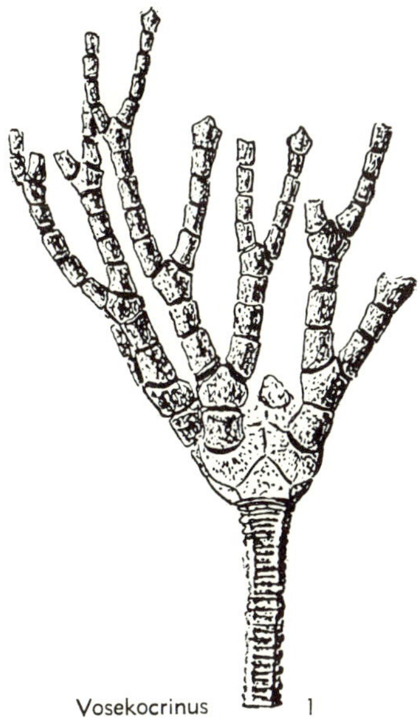

FIG. 359. Family Uncertain (p. *T*563-*T*564).

Hybocrinids show two unequally-developed planes of bilateral symmetry in the organization of the theca, in addition to the basic pentameral symmetry. Because only one radial is compound (*C* radial plus radianal in the *C* ray), hybocrinids should have an eustenocrinoidal plane *(C-AE)* of bilateral symmetry, but in most hybocrinid genera this is far outweighed by a crinoidal plane of bilateral symmetry passing through *A-CD* (Fig. 360). The sizes and shapes of all three circlets of cup plates, location of the hydropore and anal opening, comparison of the combined *C* radial plus radianal with the larger *D* radial, and type of food-gathering appendage in each ray all correspond with this latter symmetry plane. The crinoidal plane is best developed in genera either with recumbent ambulacra or no food-gathering appendages in some rays, because the ray dominance in these structures conforms strongly to this bilateral symmetry plane.

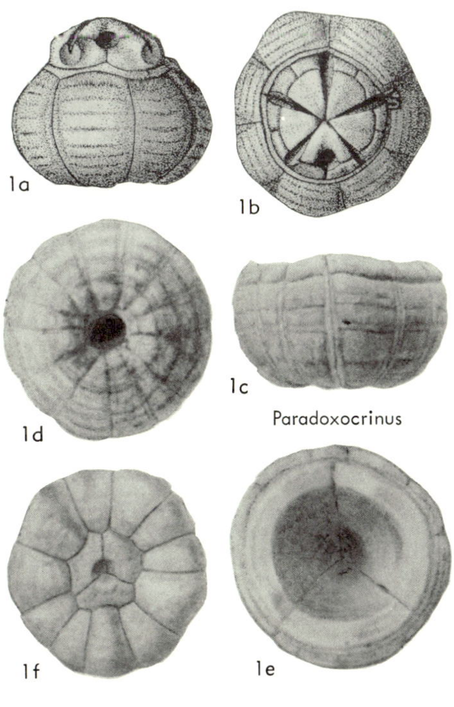

FIG. 358. Paradoxocrinidae (p. *T*563).

Most hybocrinids have fine to relatively

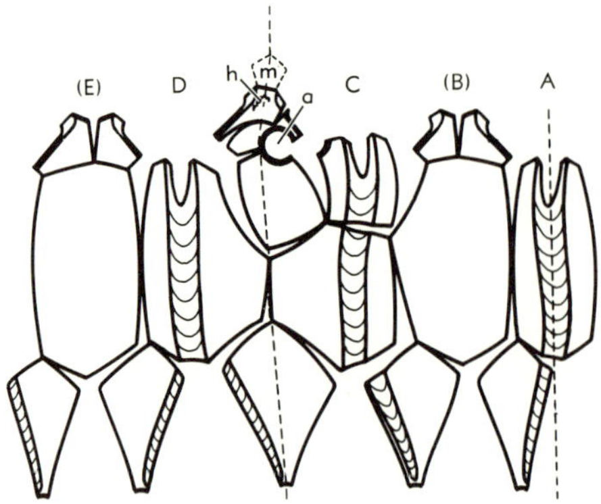

FIG. 360. Plate diagram of *Tripatocrinus* (Hybocrinacea, Cornucrinidae) (Sprinkle, 1973b, mod.). [Explanation: *a*, anus; *h*, hydropore; *m*, mouth.]

coarse pustulose or ridgelike ornament on the exterior of the cup plates. Concentric growth lines are also present on the interior of a few plates of *Tripatocrinus* (SPRINKLE, 1973b), implying that these plates have increased in size by overlayering primary plate growth, as is known to occur in several other fossil crinoids.

FOOD-GATHERING APPENDAGES

Hybocrinids show great variability in their type of food-gathering structures. These can be developed 1) as normal erect arms attached to the radial articular facets, 2) as modified and shortened arms with a recurved ambulacral groove, 3) as recumbent ambulacral grooves extending down over the surface of the cup plates, or 4) lack of any food-gathering structure at all in a ray. Normal arms, where present, are relatively short, uniserial, unbranched (atomous), and nonpinnulate. In the few complete specimens known, they range up to about three times the cup length (*see* Fig. 364,*1c*). An ambulacral furrow or food groove protected by a set of tiny ambulacral cover plates extends up the adoral side of each arm. These arms in hybocrinids seem to represent the simplest (and perhaps ancestral) type of arm found in all crinoids.

Modified arms are known only in the genus *Hybocystites* where they occur in the *A, C,* and *D* rays. These are very short, curved arms (less than the cup height) and contain only two to six brachials. The ambulacral groove is recurved, running up the adoral side, over the rounded distal brachial, and down the aboral side, in some individuals even reaching the cup plates below (*see* Fig. 365,*1d*). These unusual arms in *Hybocystites* are thought to represent an intermediate stage in the change from an erect arm to a recumbent ambulacrum (Fig. 361).

Recumbent ambulacral grooves are found in three hybocrinid genera, but are practically unknown in other crinoids. In *Hybocystites* or *Tripatocrinus,* these U-shaped grooves extend from the central mouth down over the cup plates or along the sutures between them, in some specimens reaching the stem facet (Fig. 360). In *Cornucrinus,* these ambulacral grooves extend out over the humped or pronged radials and continue back along part of the dorsal side. No flooring (or side) plates occur in these recumbent grooves, probably because no other plated appendages were mounted on them. Like the food grooves on erect or modified arms, the recumbent ambulacral grooves were protected by one or two sets of minute cover plates. Plates of the cup beneath each of these ambulacral grooves are usually thickened to support the deeply embayed groove.

The *B* and *E* rays of three hybocrinid genera have no food-gathering structures at all. The radial and oral plates in these rays appear to be normally developed, but in contrast to the other three rays (*A, C,* and *D*) which bear either arms *(Baerocrinus)* or ambulacral grooves *(Tripatocrinus, Cornucrinus),* no appendages are present here. This apparently represents a terminal stage in the reduction of an erect arm or a continuation of the transition from an erect arm to a recumbent ambulacrum (Fig. 361).

Hybocrinids with modified arms or recumbent ambulacra have a large elliptical coelomic notch (SPRINKLE, 1973b) extending through the bottom of the ambulacrum into the cup interior at the oral-radial suture or radial articular facet. This opening is also present, although smaller, in hybocrinid genera with erect arms, but is absent from rays where no food-gathering appendage occurs. Extensions from the major coelomic cavities and from the water-vascular, hemal,

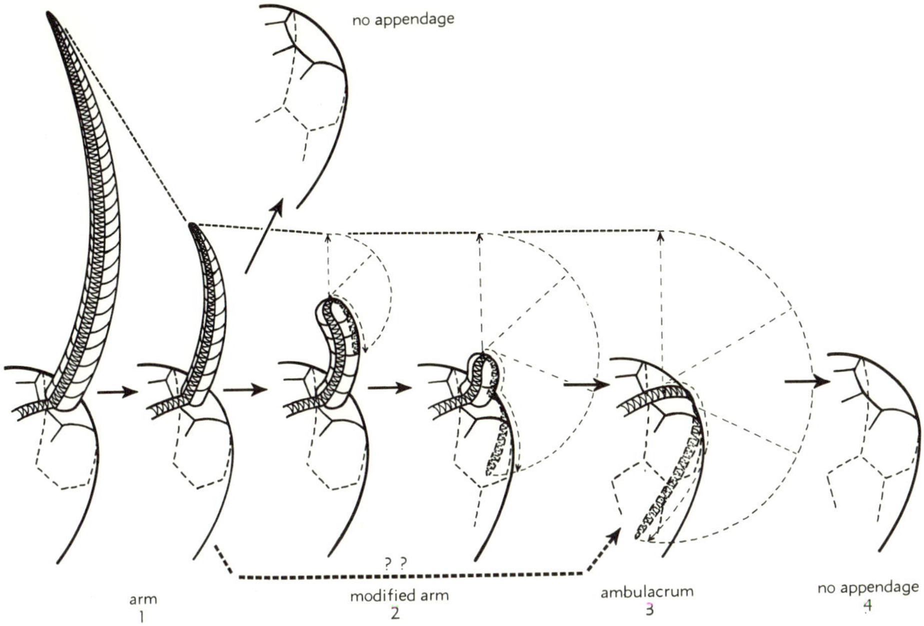

FIG. 361. Inferred modification of food-gathering appendages during hybocrinid evolution. Normal erect arms (state 1) are either directly lost (short side branch at top), or shortened to modified arms with recurved ambulacral grooves (state 2), then to recumbent ambulacral grooves (state 3), and finally lost (state 4, right) (Sprinkle, n).

perihemal, and nervous systems probably extended out through each coelomic notch to the ambulacral groove or arm, where feeding, respiration, sensory perception, and other important functions took place. The coelomic notches in *Tripatocrinus* show apparent allometric growth when compared to the size of the surrounding cup plates (SPRINKLE, 1973b), implying that their area served the internal volume of the cup. The presence of coelomic notches indicates that a water-vascular canal and tube feet were almost certainly present in the recumbent ambulacra of these hybocrinids, as would be expected in such crinoids. However, recumbent ambulacral areas in similar-appearing blastozoan echinoderms lack coelomic notches, implying a possible major difference in the development of the water-vascular system.

In hybocrinids retaining five arms, no ray dominance is seen because the arms are equally developed in all rays. However, in more specialized hybocrinids, which have developed modified arms, recumbent ambulacra, or no appendages in some rays, it is invariably the equally developed *A, C,* and *D* rays that have retained the more ancestral state, whereas rays *B* and *E* show an equal but greater change or reduction in food-gathering structures. Ray dominance in hybocrinids can therefore be indicated by the "equation": $(A,C,D) \geqq (B,E)$ where $A=C=D$ and $B=E$.

Dominance of one or more rays over the others is observed in several other crinoid groups, such as Calceocrinidae where *A* and *D* are the dominant rays and retain the two sets of modified arms.

STEM

All hybocrinids have a normal stem of unknown length composed of disclike, one-piece, circular columnals. Near the cup the stem either has a small diameter *(Tripatocrinus)* or is slightly larger than its more distal portions *(Hoplocrinus, Hybocystites)*. In most specimens of *Hybocystites,* the proxistele shows a strong curvature toward the bulged posterior *(CD)* side of the cup (*see* Fig. 365,*1e*), perhaps implying that the theca may have been held either in a tilted position or upright just above the bottom with a distally recumbent stem. STRIMPLE (1975) has proposed that asymmetrical hybocrinids were bottom living and had a short trailing (or buried) stem. No terminal attachment structures have been found on hybocrinid stems.

HYBOCRINID EVOLUTION

Hybocrinids apparently evolved from an ancestral inadunate crinoid with a relatively large cup, five short simple arms, reduced monocyclic arrangement of the cup plates with a divided radial in only one ray *(C)*, and a normal columnal-bearing stem. No Late Cambrian or Early Ordovician crinoid ancestor with these features is known at present.

Once they had evolved, some hybocrinid genera (three out of seven) retained the primitive five-armed condition; these apparently represent the most conservative group of hybocrinids. However, four other genera have diverged from this arrangement either by losing two of the arms (*Baerocrinus, see* Fig. 365,*2a*) or by reducing and modifying all of the appendages to varying degrees (Fig. 361). *Hybocystites* (see Fig. 365,*1c,d*) has the most unusual arrangement with three modified arms and two ambulacral grooves. If we assume that *Hybocystites* represents an intermediate stage of arm reduction, we can construct a chart showing possible steps in this arm reduction and possible loss (Fig. 361).

These advanced genera apparently reduced the relatively short atomous arms found in conservative hybocrinids, but kept the food-gathering ambulacral groove nearly constant in total length by running it over the top of the rounded distal brachial and down the ventral side, in some crowns even reaching the cup plates (Fig. 361). From this stage, shown by the *A*-, *C*-, and *D*-ray arms of *Hybocystites,* it would have been only a short additional step to do away altogether with the reduced remnant of the arm and run about the same length of ambulacral groove down over the cup surface or out onto long radial prongs. The genera with no appendage at all in the *B* and *E* rays could have achieved this state either by reducing an arm to zero with no modification (short side branch, Fig. 361), or by continuing the arm-to-ambulacrum reduction and finally eliminating the ambulacral groove also. The arm-reduction sequence appears to be paedomorphic in nature because growth of the arm by budding off new brachials at the distal tip stops at an earlier and earlier stage during this transition (Fig. 361).

Based on this inferred evolutionary sequence, an appendage-state "matrix" and evolutionary chart can be set up by plotting the appendage type in rays *A, C,* and *D* against the appendage type in rays *B* and *E* (1 = arm; 2 = modified arm; 3 = ambulacral groove; 4 = no appendage) and connecting by arrows all possible appendage reduction steps (Fig. 362,*A,B*). Conservative hybocrinids with five normal arms (*Revalocrinus, Hoplocrinus,* and *Hybocrinus*) fit into box [1, 1] of this appendage-state chart and matrix (Fig. 362,*B*). *Baerocrinus,* with 3 arms, belongs in box [1, 4]; *Hybocystites,* with 3 modified arms and 2 ambulacral grooves, fits into box [2, 3]; and *Tripatocrinus* and *Cornucrinus,* with 3 ambulacral grooves, belong in box [3, 4] (Fig. 362*A,B*). Attention may be called to the fact that appendage states [2, 2], [3, 3], and [4, 4] are all unknown in fossil hybocrinids. This suggests that once arm reduction in the hybocrinids began, the *A, C,* and *D* rays became dominant over rays *B* and *E,* and these equal-reduction states may never have been reached. Therefore, a simplified evolutionary matrix and chart can be set up by eliminating these unlikely positions (Fig. 362,*C,D*). Three intermediates still remain unknown in this inferred phylogeny.

A normal hybocrinid ancestor with five equal arms could have given rise to *Baerocrinus* either by direct loss of the *B* and *E* arms or by reduction to modified arms, ambulacra, and then complete loss (Fig. 362,*D*). Also, *Hybocystites* could have evolved from the second of these unknown intermediates by reduction of the normal *A, C,* and *D* arms to modified arms. *Tripatocrinus* and *Cornucrinus* could have descended from either *Baerocrinus* (less likely) or *Hybocystites* (more likely) by means of an unknown intermediate with

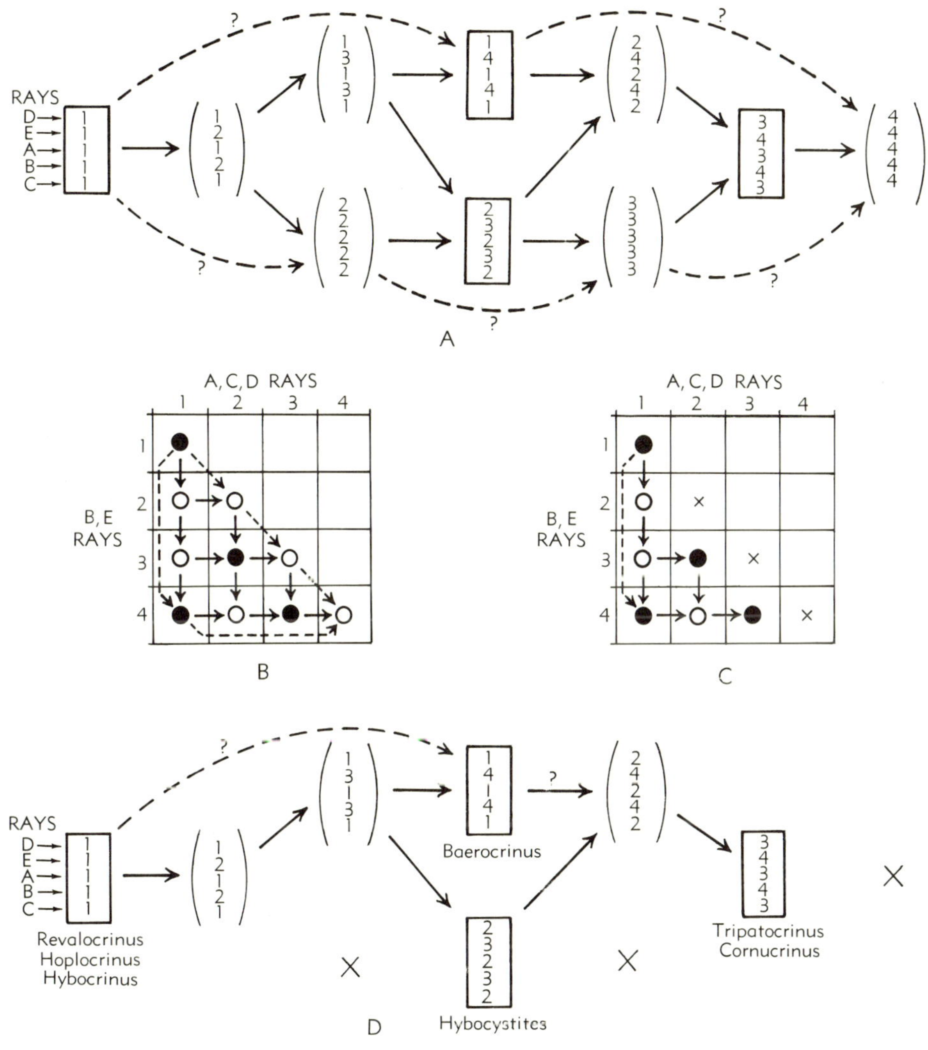

FIG. 362. Diagrams *(A, D)* and "matrices" *(B, C)* showing possible appendage states and pathways for hybocrinid evolution either with no assumptions *(A-B)* or assuming ray dominance [rays *A, C, D* > *B, E* for *B, E* > 1] *(C-D)*. Note that even in the simplified phylogeny *(C-D)*, three intermediates ([1, 2], [1, 3], and [2, 3]) are still missing from the fossil record of hybocrinids (Sprinkle, n). [Explanation of symbols: *1*, arm; *2*, modified arm; *3*, ambulacrum; *4*, no appendage; ● and □, morphology represented by genera in fossil record; O and ( ), morphology of unknown intermediates; *X*, improbably morphology because of appendage dominance.]

only three modified arms [2, 4] (Fig. 362,*C,D*). This sequence of appendage reductions apparently ended here because reduction to no appendages (box [4, 4]) was apparently impossible for crinoids as large as hybocrinids; however, some microcrinoids have apparently reached this final stage.

This modification and reduction of arms to ambulacral grooves in advanced hybocrinids such as *Tripatocrinus* and *Cornucrinus* has produced a plate arrangement in the cup very similar to that found in many blastozoan echinoderms (eocrinoids, rhombiferans, parablastoids, and blastoids)

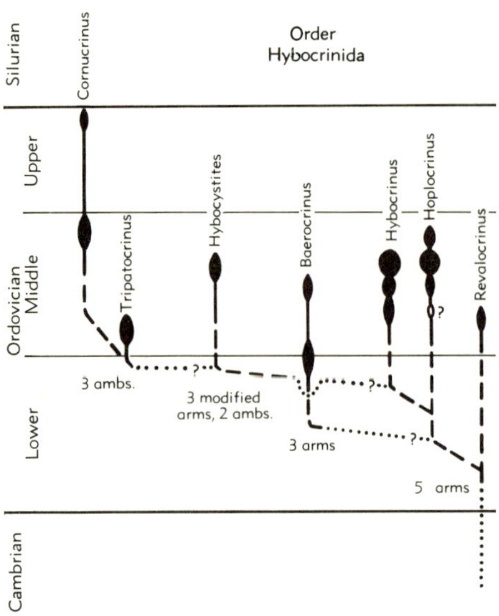

FIG. 363. Stratigraphic distribution of hybocrinids (Sprinkle, n).

(SPRINKLE, 1973a). These convergent hybocrinids have adopted a blastozoan "design" and perhaps way of life with a large globular theca, recumbent ambulacra, reduced cup plates, and strongly developed pentameral symmetry. However, they have made these changes while retaining a variety of features conclusively showing their crinoid affinities and ancestry—normal hybocrinid plating, unplated ambulacral grooves bearing a coelomic notch, overlayering plate growth, and inferred presence of ambulacral tube feet. The observed differences in ambulacral development in these two convergent groups may indicate several possible differences in soft part morphology (SPRINKLE, 1973b). Unlike blastozoans which have plated brachioles mounted on their recumbent ambulacra, hybocrinids apparently had only soft tube feet which were hydraulically controlled and needed no special supporting side or flooring plates. Flooring plates might also have interfered with the overlayering type of thecal growth found in these crinoids (SPRINKLE, 1973a).

Hybocrinids range from Early to Late Ordovician (Fig. 363), when they apparently became extinct without leaving any descendants.

Order HYBOCRINIDA Jaekel, 1918

[*nom. transl. et correct.* MOORE in MOORE, LALICKER & FISCHER, 1952, p. 613 (*ex* suborder Hybocrinites JAEKEL, 1918, p. 90)]

Monocyclic crinoids having moderately large ovoid or trigonal to pentagonal conical theca with slight to well-defined asymmetry and dominance of bilateral symmetry in crinoidal plane *(A-CD)* over eustenocrinoidal plane *(C-EA)*. Five relatively short, uniserial, atomous, and nonpinnulate arms which are erect in conservative genera, but recumbent or reduced to ambulacral grooves on cup plates in others and eliminated entirely from *B* and *E* rays in most specialized forms. Thecal plating simple (five basals and radials, one or more anals, five orals and small additional plates on tegmen); radials undivided except in *C* ray, no anal tube or sac, and tegmen considerably reduced. Small-diameter stem with circular columnals. *L.Ord.-U.Ord.*

Family HYBOCRINIDAE Zittel, 1879

[Hybocrinidae ZITTEL, 1879, p. 350]

Hybocrinids having five relatively short uniserial arms, one in each ray. *M.Ord.*

Hybocrinus BILLINGS, 1857, p. 274 [**H. conicus*; SD WACHSMUTH & SPRINGER, 1880, p. 74] [=*Indianocrinus* MILLER & GURLEY, 1895b, p. 83 (type, *I. punctatus*)]. Cup moderately asymmetrical with normally arranged plates including medium-sized *C* radial and anal *X* on side of theca above radianal, several very small accessory anals present around multiplated periproct above anal *X*; posterior oral containing raised hydropore; radial articular facets relatively large and circular; arms relatively short; ambulacral cover plates in multiple series on peristome, one series aborally. Most species have fine granular ornament on cup plates. Stem slightly enlarged proximally. *M.Ord.*, Can.(Ont.-Que.), USA(N.Y.-Ky.-Va.-Tenn.-Okla.).——FIG. 364,*1a-f*. **H. conicus*, Ont.; *1a,b*, plate diagram and reconstruction of specimen with appendages restored (*h*, hydropore; *m*, mouth) (Sprinkle, n); *1c,d*, side views of 2 nearly complete specimens, showing relatively short uniserial arms, ×1 (Billings, 1859); *1e,f*, side and oral views of well-preserved theca showing multiplated anal pyramid, hydropore, and ambulacral cover plates, ×1 (Springer, 1911b).

Hoplocrinus GREWINGK, 1867, p. 7 [**Apiocrinus dipentas* LEUCHTENBERG, 1843, p. 17; M]. Cup slightly asymmetrical; radianal reaching level of radial summits, obliquely supporting medium-sized *C* radial and smaller anal *X* which is confined to rim of cup; plates with fine-coarse pus-

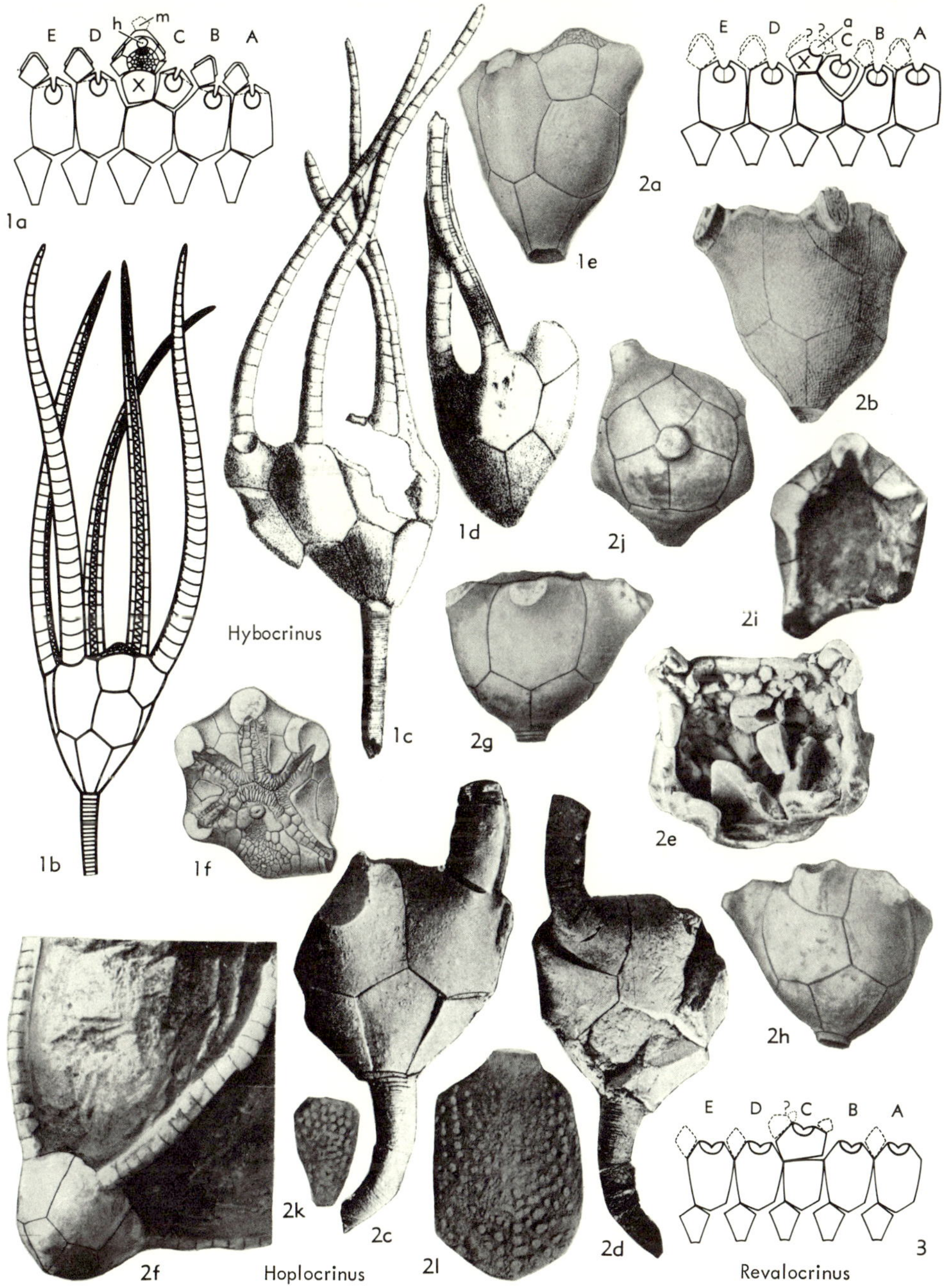

FIG. 364. Hybocrinidae (p. *T570-T572*).

tulose ornament; tegminal features poorly known. Arms relatively short, brachials U-shaped, with relatively large facets occupying about one-half of radial width. Stem slightly enlarged proximally. *M.Ord.*, Eu. (NW. USSR-Sweden)-?USA (Utah).——FIG. 364,*2a,b*. **H. dipentas* (LEUCHTENBERG), USSR; *2a*, plate diagram (*a*, anus) (Sprinkle, n); *2b*, complete specimen showing thecal shape, ?×1 (Grewingk, 1867).——FIG. 364,*2c,d*. *H. laevis* MYANNIL, USSR (Est.); side

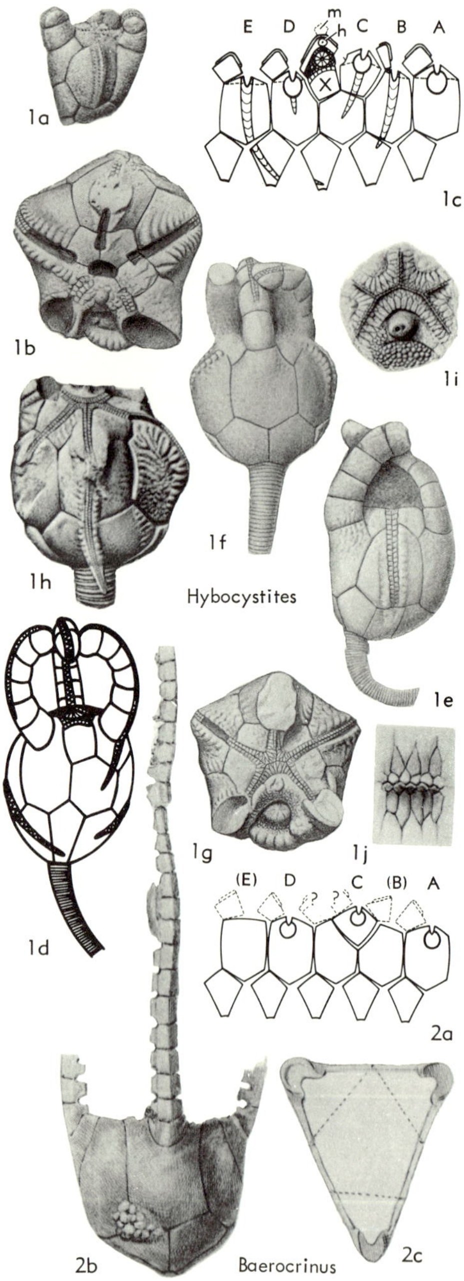

FIG. 365. Baerocrinidae *(2)*; Hybocystitidae *(1)* (p. *T*572-*T*574).

views of specimen with slightly enlarged proximal stem and attached arm, ×1.1 (Myannil, 1959). ——FIG. 364,*2e*. *H. estonus* ÖPIK, USSR(Est.); specimen showing details of crushed summit, incl. anal *X*, U-shaped brachials, and scattered ambulacral cover plates, ×1.2 (Öpik, 1935).——FIG. 364,*2f*. *H. tallinnensis* ÖPIK, USSR(Est.); specimen showing relatively short incomplete arms, ×1 (Öpik, 1935).——FIG. 364,*2g-j*. *H. dalecarlicus* REGNÉLL, Sweden; lat., lat., ventral, dorsal views of well-preserved cup, ×2 (Regnéll, 1948a). ——FIG. 364,*2k,l*. *?H.* sp. LANE, USA(Utah); *2k,l*, isolated basal and radial plates which may belong to this genus, ×1.3 (Lane, 1970a).

Revalocrinus JAEKEL, 1918, p. 90 [**R. costatus*; M]. Cup composed of 5 small equal-sized basals, 5 large elongate radials with relatively large *C* radial directly above pentagonal radianal, small anal *X* obliquely above *C* radial on summit and not reaching radianal, 5 relatively large ?orals extending from side of cup to summit of tegmen. Arms uniserial with wide massive brachials, radial articular facets occupying over half of radial width; tegmen structures, ornament, and stem poorly known. *M.Ord.*, Eu.(NW.USSR).——FIG. 364,*3*. **R. costatus*, plate diagram of only described specimen, (Sprinkle, n).

Family BAEROCRINIDAE Jaekel, 1918

[Baerocrinidae JAEKEL, 1918, p. 90]

Hybocrinids having only three arms *(A, C,* and *D* rays), with no appendages in *B* and *E* rays. *L.Ord.-M.Ord.*

Baerocrinus VOLBORTH, 1864, p. 65 [**B. ungerni*; M]. Cup triangular in ventral view, with one arm at each corner *(A, C,* and *D* rays); plate arrangement normal, radianal and *C* radial large, anal *X* small and confined to summit, orals unknown. Arms 3, composed of uniserial brachials quadrangular in side view, with large notches ventrally for insertion of muscles. Stem attachment scar relatively small. *L.Ord.-M.Ord.*, Eu. (NW.USSR).——FIG. 365,*2a-c*. **B. ungeri; 2a,* plate diagram (Sprinkle, n); *2b,c,* lat. and ventral views showing shape of cup and development of arms, ×2 (Grewingk, 1867).

Family HYBOCYSTITIDAE Jaekel, 1918

[*nom. correct.* REGNÉLL, 1948a, p. 2 *(pro* Hybocystidae JAEKEL, 1918, p. 90)]

Hybocrinids having three short arms with long recurved ambulacral grooves in *A, C,* and *D* rays, in addition to long ambulacra running downward to base of cup in *B* and *E* rays. *M.Ord.*

Hybocystites WETHERBY, 1880b, p. 150 [**H. problematicus*; M] [=*Hybocystis* BATHER, 1890a, p. 324 *(nom. van.)*]. Hybocrinids having normally arranged plates showing slight to moderate asymmetry; *C* radial and anal *X* on side of theca above radianal; basals asymmetrical, with stem attachment slightly shifted forward; anal *X* distal, edges of radials, and areas alongside ambulacral grooves ornamented by raised ridges. Three short,

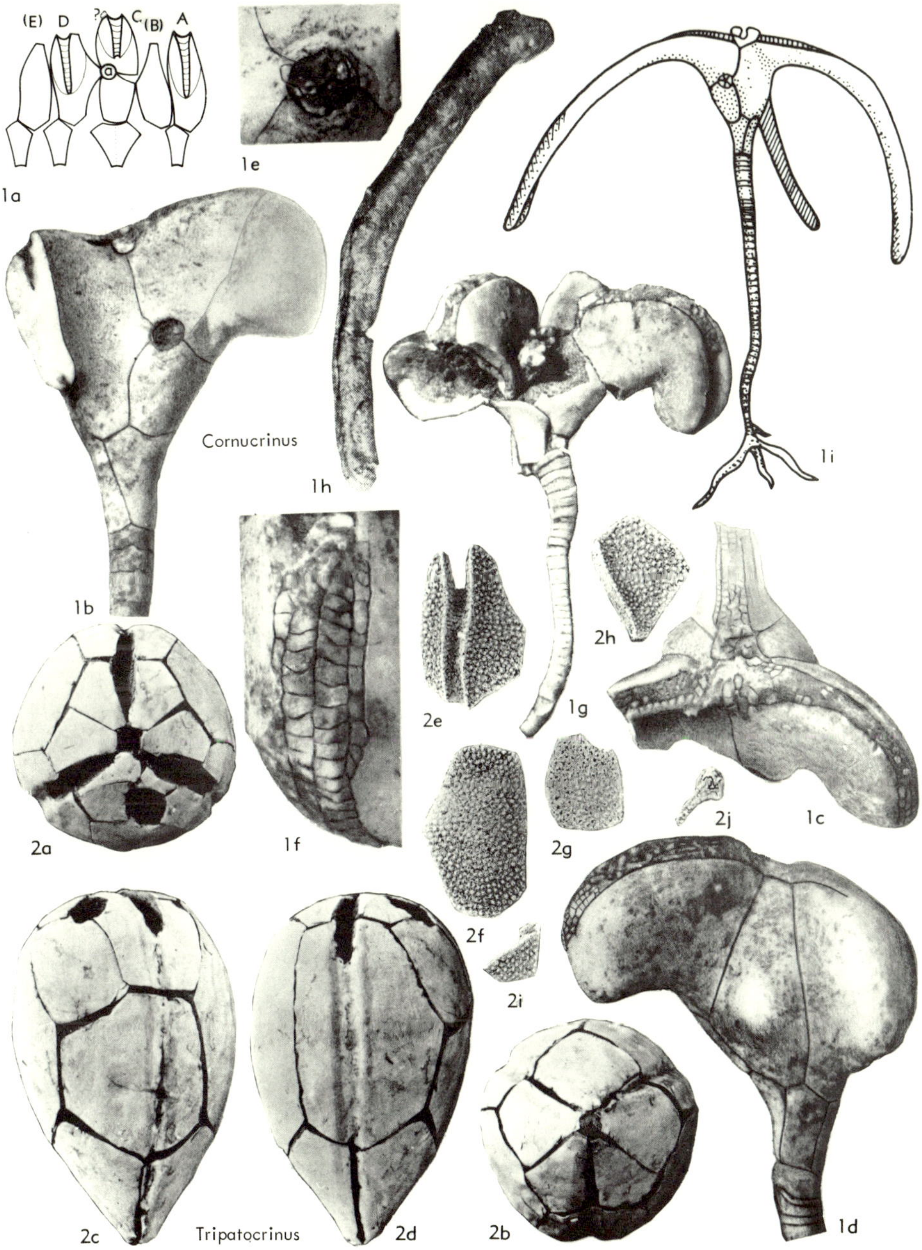

FIG. 366. Cornucrinidae (p. *T*574).

modified arms present in *A, C,* and *D* rays, each containing 2 to 6 uniserial brachials and rounded at distal tip, ventral side bears covered ambulacral groove which extends over tip and down dorsal side, in some cases reaching radial plates or even basals; similar grooves in armless *B* and *E* rays located on radials and commonly extended diagonally across basals; all ambulacral grooves apparently provided with elongate coelomic notches between radials and orals near central mouth; 3 sets of ambulacral cover plates on peristome, but seemingly only single set present aborally.

Tegmen with round anal pyramid formed by 6 to 9 plates set in integument of tiny periproctal plates above anal *X*; small hydropore located between anus and mouth. Curved stem slightly enlarged proximally. *M.Ord.*, Can.(Ont.)-USA (Ky.).——FIG. 365,*1a,b*. **H. problematicus*, Ky.; *1a*, small specimen showing arms with only 2 brachials and short ambulacra; *1b*, ventral view of larger specimen showing coelomic notches through ambulacra, orals, and ornamented radials, ×1.3 (Springer, 1911b).——FIG. 365,*1c-j*. *H. eldonensis* (PARKS), Ont.; *1c,d*, plate diagram and specimen with restored appendages (Sprinkle, n); *1e,f*, lat. views (*E* and *A* rays) of 2 complete specimens showing arms and stems, ×2; *1g*, ventral view showing peristomial and ambulacral cover plates in place, anal pyramid, and hydropore, ×1.5; *1h*, ant. view of abnormal specimen with ambulacrum from aborted or damaged *A*-ray arm extending down over cup onto proximal portion of stem, ×1.3; *1i*, peristomial cover plates and hydropore, ×2.3; *1j*, 3 rows of peristomial cover plates, ×8 (*1e-j*, Springer, 1911b).

Family CORNUCRINIDAE Regnéll, 1948

[Cornucrinidae REGNÉLL, 1948a, p. 8]

Hybocrinids having three recumbent ambulacra running down over thecal plates in *A*, *C*, and *D* rays, with no appendages in *B* and *E* rays. *M.Ord.-U.Ord.*

Cornucrinus REGNÉLL, 1948, p. 11 [**C. mirus*; M]. Cup with 3 widely flared recumbent ambulacra running down over strongly humped or elongated radials in *A*, *C*, and *D* rays; only 4 basals (*CD* and *BC* fused); no anal *X*; anal opening displaced down side of theca and bounded by *C* and *D* radials and radianal; orals absent or covered with wide adoral ambulacra. *M.Ord.-U.Ord.*, Eu.(Sweden).——FIG. 366,*1a-g*. **C. mirus*, M.Ord.; *1a*, plate diagram (Sprinkle, n); *1b-d*, post., ventral, and right lat. view of complete specimen, ×1.5; *1e,f*, periproct and ambulacral cover plates, ×2.5, ×4; *1g*, post. view of another specimen with long stem segment attached, ×1.3 (*1b-g*, Regnéll, 1948a).——FIG. 366,*1h,i*. *C. longicornis* REGNÉLL, U.Ord.; *1h*, separate radial, ×1; *1i*, (reconstr.) complete specimen in life position, ×1 (both Regnéll, 1972).

Tripatocrinus SPRINKLE, 1973b, p. 872 [**T. pustulatus*; M]. Cup. ovoid and slightly asymmetrical, with 3 long ambulacra running down over thecal plates in *A*, *C*, and *D* rays; plate arrangement normal, *C* radial and anal *X* equally developed above radianal on side of theca; 2 small accessory anals above anal *X*; elongate coelomic notch present through bottom of each ambulacrum between radial and orals; slitlike hydropore opening through posterior oral; small-diameter stem attached to basals. Known only from disarticulated plates. *M.Ord.*, USA(Nev.-Calif.).——FIG. 360; 366,*2*. **T. pustulatus*, Nev.; 360, plate diagram; 366,*2a-d*, ventral, dorsal, *C*-, and *D*-ray views of cup (reconstr. model), ×1; 366,*2e-j*, ext. views of separate plates (*D* radial, *B* radial, anal *X*, *DC* basal, *EA* oral, *CD* oral), showing pustulose ornament and diagnostic plate shapes, ×1 (all Sprinkle, 1973b).

Order CORONATA Jaekel, 1918[1]

[Coronata JAEKEL, 1918, p. 109]

By R. O. FAY

Crinozoans with three circlets of regularly arranged thecal plates, without pores and infolds of any type, with five biserially arranged arms. Basal circlet composed of three plates, the azygous basal located in anterior right *(AB)* interradial position; with five radials and five large interradial plates on ventral part of theca; radials and interradial plates prolonged ventrally into high coronal processes. Small primaxil plate attached rigidly to adorally constricted end of each radial plate, adoral part of which rests internally on adjacent interradial plates. Each primaxil surmounted by two plates, and these by two more, and so on for at least nine sets of plates (Fig. 367). Bottom pair of plates each appearing to give rise to a biserial arm, second pair so arranged that plate on left supports a bottom plate of second armlet, and plate on right supports top plate of second arm on that side, bottom plate receiving support from first plate below. Thus, from this point upward, armlets are arranged alternately and are not paired. Each armlet coiled planispirally adorally, with three or four revolutions to each armlet and single food groove on adoral surface. Food grooves on left side lead to a food groove on left side of main plates of arm on that side, and food grooves on right side lead to a food groove on right side of main plates. Thus, primaxils have two food grooves on their adoral surface. Arms appear to have flexible sutures above primaxils. Food grooves extend adorally along margins of interradial plates, meeting along suture between interradial plates about one-

[1] The author, R. O. FAY, would prefer to raise the order Coronata to the level of subclass because of incorporation of the lower brachial circlet into the top of the cup, the presence of a coronal process, the unique sausage-shaped cover plates, and the rare biserial arrangement of the arms.

third of distance from mouth, forming single food groove in each ambulacrum that leads to mouth. Two large, sausage-shaped cover plates above food grooves on each ambulacrum abutting against primaxil and slightly overlapping adoral portion of primaxil. Ambulacra floored by adjacent interradial plates which meet along midline of each ambulacrum. Oral opening covered by five large oral plates which are interradial in position, each abutting against interradial and adjacent cover plates. Single large anal opening on aboral side of anal interradial plate at its junction with adjacent radial limbs, with four subtriangular cover plates. *M.Ord.-U.Ord.; M.Sil.-U.Sil.*

Of five genera referred to this order, only one, *Stephanocrinus,* is fairly well understood. The characters of the order are therefore taken mainly from descriptions of this genus. It is obvious that the five genera assigned to the Coronata are closely related, and that they belong to the Crinoidea. It is difficult to place them, however, within known groups of the class. The genera are distinguished mainly by their external form and stratigraphic and geographic occurrence.

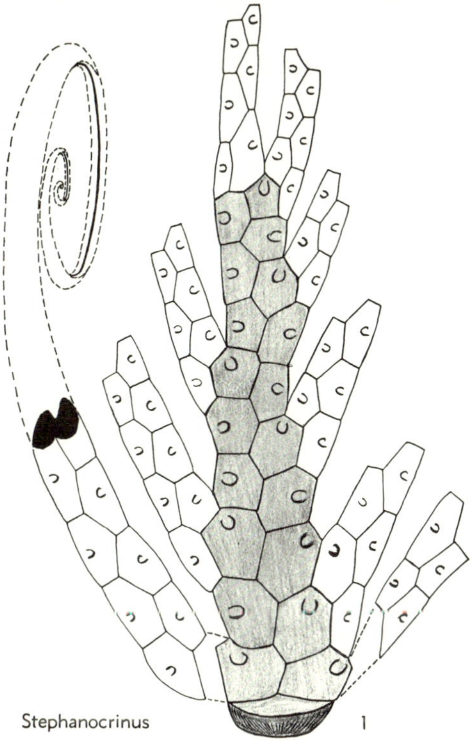

FIG. 367. *Stephanocrinus angulatus,* schematic diagram of aboral side of the anterior *(A)* arm, ×31 (Fay, 1962a).

Family STEPHANOCRINIDAE Wachsmuth & Springer, 1886

[Stephanocrinidae WACHSMUTH & SPRINGER, 1886, p. 282] [=Stephanoblastidae JAEKEL, 1918, p. 110]

Characters of order. *M.Ord.-U.Ord.; M. Sil.-U.Sil.*

Stephanocrinus CONRAD, 1842, p. 278 [**S. angulatus*; M]. Conical theca with angular ridges on dorsal side, and high-pointed coronal processes, with sharp adoral surfaces and ridged lateral surfaces; ornamentation consists of fine ridges, in the form of pore-rhombs. [Small rounded forms referred to this genus may belong to another genus.] *M.Sil.,* N.Am., *?M.Sil.,* Eu.(Eng.).——FIG. 367; 368,*1*. **S. angulatus,* USA(N.Y.); 367, *A* ambulacrum, ×31; 368,*1a,* oral view, ×5; 368,*1b, D*-ray view, ×4; 368,*1c,* anal view, ×15; 368,*1d,* oral view with *A*-ray arms in place, ×4.9; 368,*1e,* side view of *A* amb., showing aboral face of arms, ×20 (Fay, 1962a).——FIG. 369,*1a-d*. *S. gemmiformis* HALL, Waldron Sh., Tenn.; *1a,* oral view showing "pores" in weathered coronal processes, anal opening toward bottom, ×14 (Fay, 1960); *1b-d,* side view, ×8; right ant. *(AB)* interambulacrum showing deeply weathered coronal process with divided coelomic cavity appearing like 2 pores, ×24; *"C"* ambulacrum showing cover plates and primaxil in place, ×18 (Fay, n; Illinois Dept. Geology).——FIG. 369,*1e*. *S. osgoodensis* S. A. MILLER, Niagaran, Ind.; *A* ambulacrum showing sutures between interambulacral plates and radial plate, ×22 (Fay, n).

Mespilocystites BARRANDE, 1887, p. 162 [**M. bohemicus*; M] [=*Mespilocystis* BATHER, 1889c, p. 269 (obj.)]. Theca cup shaped, almost as wide as high, with coronal processes as long as theca, and apparently with interradial plates visible in side view, with strong ornamentation of thecal plates similar to pore-rhombs. *M.Ord.*$(D_2\text{-}D_4)$ *(Llandeil.-Caradoc.),* Eu.(Czech.).——FIG. 370,*1*. **M. bohemicus; 1a-c,* oral, side, and aboral views, all ×10 (Fay, 1962b); *1d,* plate layout of *B* and *C* radials, ×10 (Fay, n; Univ. Oklahoma Dept. Geology 2380A, 2380B, 2380C).

Paracystis SJÖBERG, 1915, p. 173 [**P. ostrogothica*; M]. Cup-shaped theca with low rounded coronal processes and large, wide basal plates. Base concave adjacent to stem. Ornamentation consists of strongly developed ridges similar to pore-rhombs. *M.Ord.,* Eu.(Sweden).——FIG. 368,*2*. ***P. ostrogothicus,*** Chasmops Ls., Östergötland; *2a-e, A, B-C, E,* aboral, and oral views, all ×5 (Regnéll, 1945).

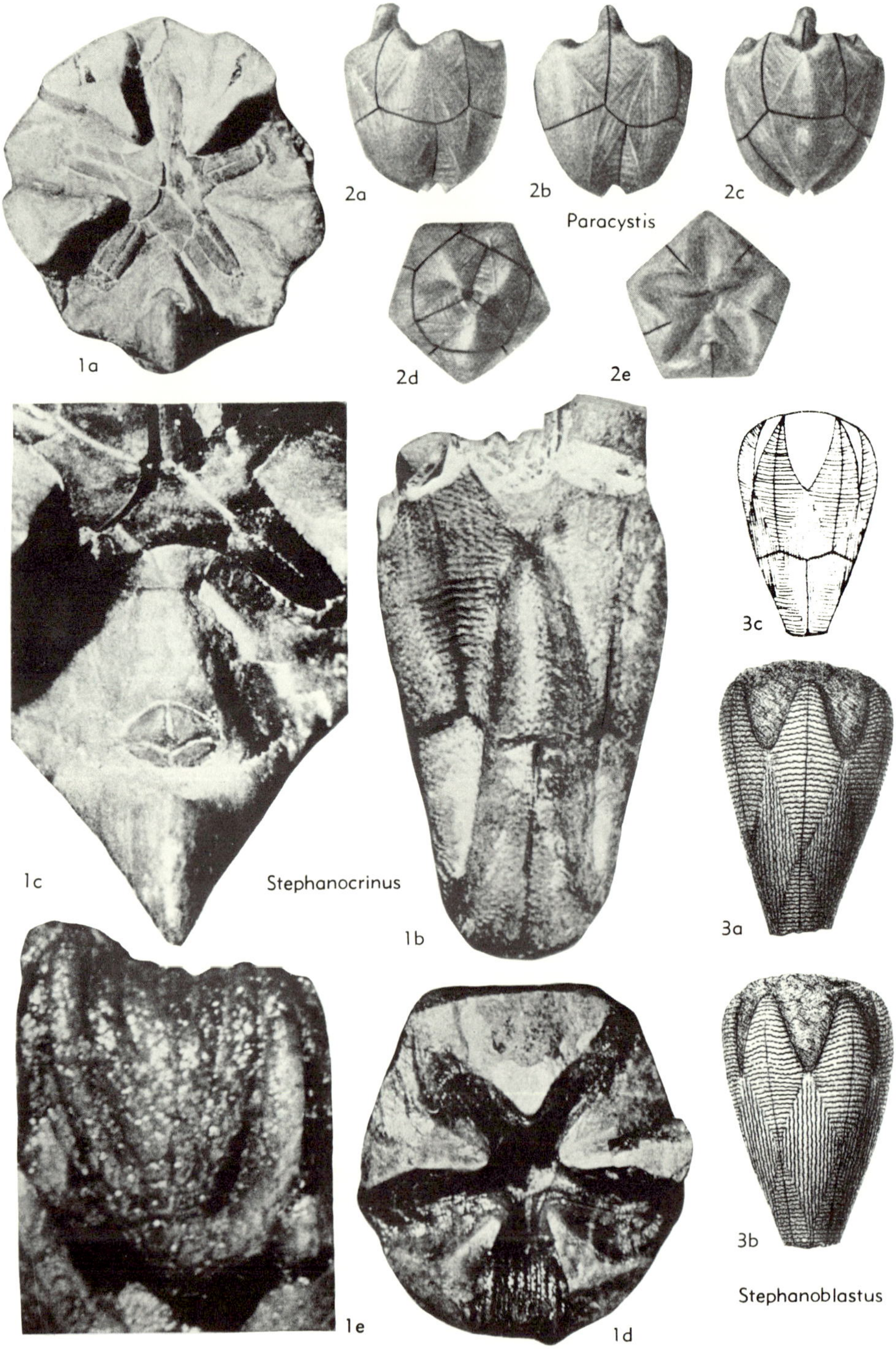

FIG. 368. Stephanocrinidae (p. T575, T577).

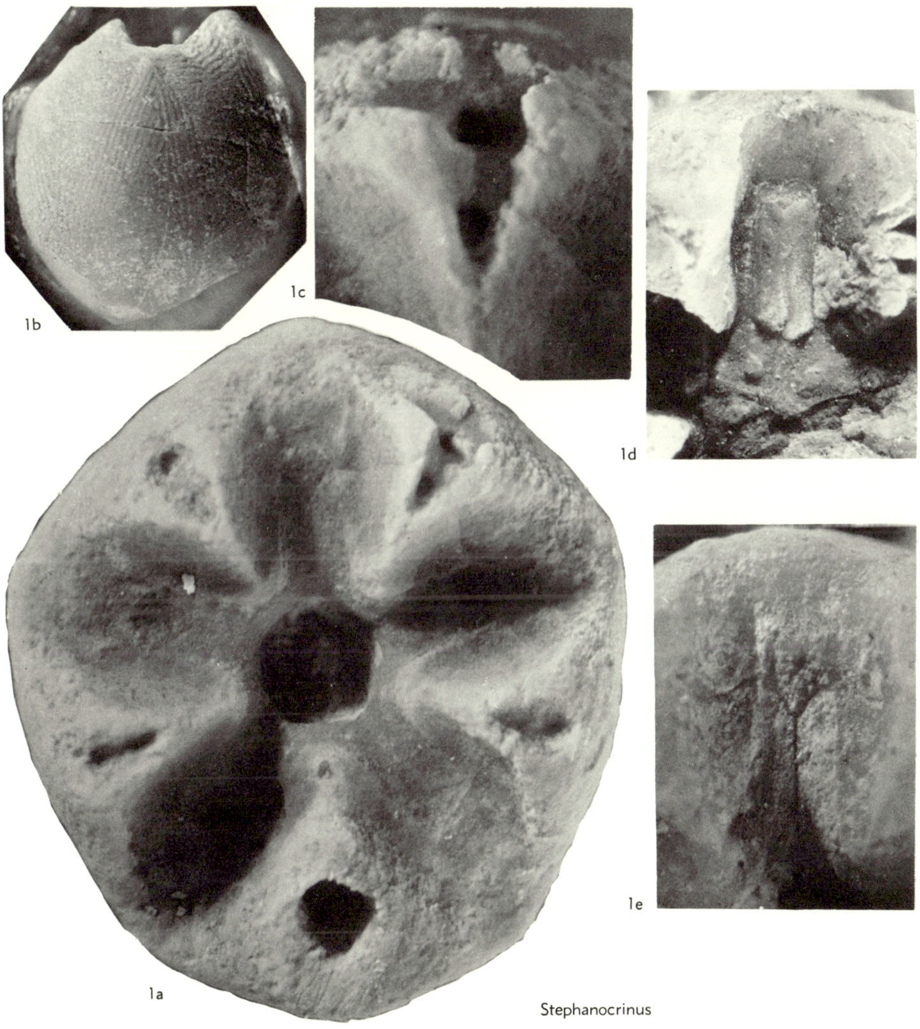

FIG. 369. Stephanocrinidae (p. *T*575).

Stephanoblastus JAEKEL, 1918, p. 110 [**Rhombifera? mira* BARRANDE, 1887, p. 80; M]. Club-shaped theca with rounded top and conical sides, and broadly triangular base; coronal processes are high and wide, with sharply rounded ventral terminations; ornamentation consists of strongly developed ridges resembling pore-rhombs. *U.Sil.* *(E_2)(Ludlov.),* Eu.(Czech.).——FIG. 368,*3*. **S. mirus* (BARRANDE), near Lodĕnice; *3a,b,* side views, ×5 (Barrande, 1887); *3c,* side view, ×3.2 (Fay, n, after Jaekel, 1918).

Tormoblastus JAEKEL, 1927, p. 1 [**T. bodae*; M]. Cup-shaped theca with high, rounded coronal processes as long or longer than theca, which is as wide as high, with strongly developed ridges on basals and radial bodies; a few short indented grooves occur on the radial limbs; interradial plates not visible in side view. *U.Ord.,* Eu.(Sweden).——FIG. 370,*2*. **T. bodae,* Leptaena Ls., Boda, Dalarne; *2a-e,* oral, *(A-B),* aboral, anal, and side coronal process views, all ×10 (Jaekel, 1927a).

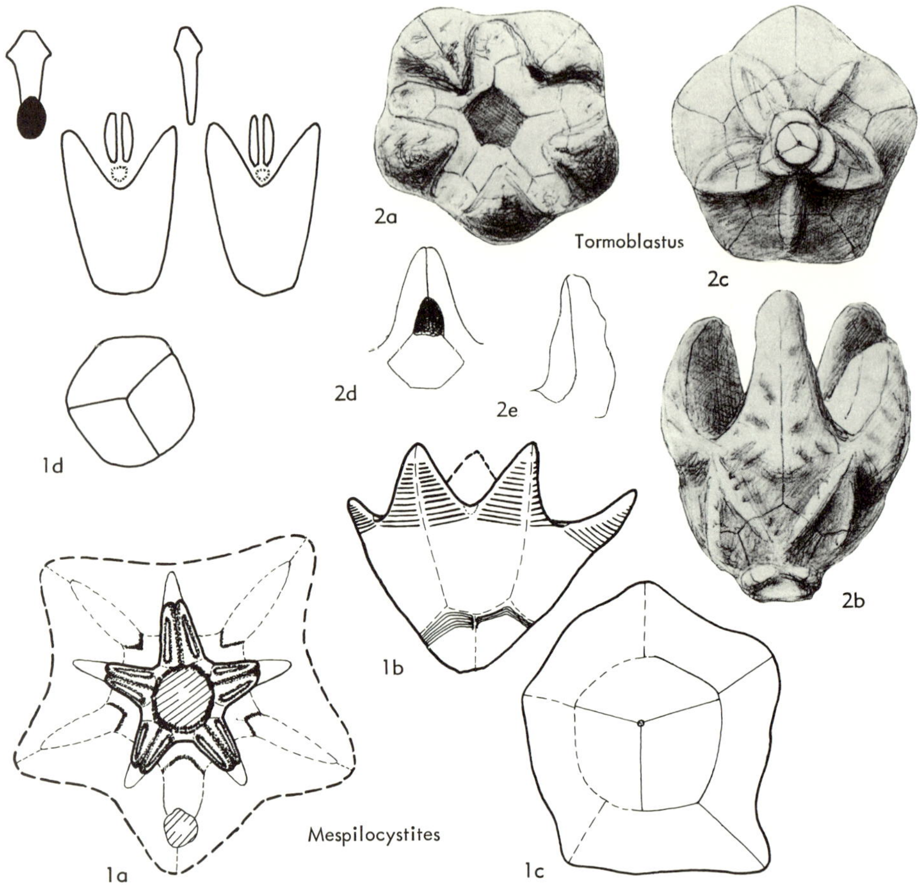

FIG. 370. Stephanocrinidae (p. *T*575, *T*577).

Order CLADIDA Moore & Laudon, 1943

[*nom. correct.* MOORE, in MOORE, LALICKER, & FISCHER, 1952, p. 613 (*pro* Cladoidea MOORE & LAUDON, 1943a, p, 32)] [=Fistulata WACHSMUTH & SPRINGER, 1885, p. 305 (83); Dicyclica BATHER, 1893a, p. 100; Inadunata Dicyclica BATHER, 1899b, p. 921]

By R. C. MOORE, N. GARY LANE, and H. L. STRIMPLE

Dicyclic inadunate crinoids having two circlets of plates below radials, mostly three, two, or one anal plates in cup but a few lacking any. Anal sac generally prominent. Arms branched or unbranched. Stem mostly circular in section. *L.Ord.(Tremadoc.)-U.Perm.; M.Trias.*

[KNAPP (1969, p. 351) split off from the Inadunata a new order Declinida having the following characteristics: Dicyclic cup bowl shaped with flat or concave base, infrabasals downflaring to horizontal, basals large or small, anal plate and radianal present but may decrease in number, arms simple or multibranched, brachial plates uniserial or biserial. Considered mainly significant is structure of lower plates of cup, with evolution toward strongly downflared infrabasals and proximal parts of basals. U.Miss.-L.Perm. This order is here judged unacceptable as a natural taxonomic division of inadunate crinoids.—R. C. MOORE.]

Suborder CYATHOCRININA Bather, 1899

[*nom. correct.* MOORE, in MOORE, LALICKER, & FISCHER, 1952,

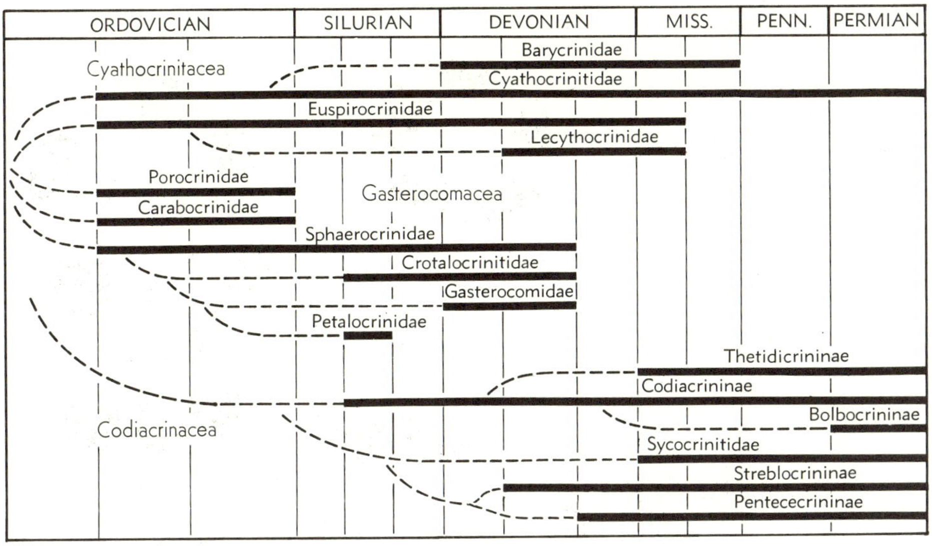

FIG. 371. Stratigraphic distribution and inferred phylogeny of families of Cyathocrinina (Moore, n).

p. 613 (*pro* suborder Cyathocrinoidea BATHER, 1899b, p. 922)] [Materials for this suborder prepared by N. GARY LANE and R. C. MOORE]

Dicyclic inadunates with bowl- to globe-shaped theca; anal plates in cup five to none, commonly anal *X* only present; radial facets narrower than width of radials, smooth or rarely with obscure transverse ridge, axial canal commonly present. Tegmen low or with short, generally nonporous, anal sac; anal opening flush with tegmen or at distal end of sac. Orals prominent, posterior one imperforate, a madreporite or with hydropore. Arms narrowly rounded, uniserial, nonpinnulate; atomous or branching isotomously or rarely heterotomously, commonly with stout ambulacral covering plates. Stem transversely round, with large axial canal or small and associated with three or four accessory peripheral canals. *M.Ord.-U.Perm.*

At least 65 genera considered to be valid here and distributed among three superfamilies are assigned to the Cyathocrinina. Stratigraphic distribution and inferred phylogeny of the family divisions are shown graphically in Figure 371.

Superfamily CYATHOCRINITACEA Bassler, 1938

[*nom. transl.* LANE, 1967b, p. 9 (*ex* family Cyathocrinitidae BASSLER, 1938, p. 20; *nom. correct. pro* Cyathocrinidae ROEMER, 1854, p. 227)]

Theca bowl to cone shaped; anal *X* large, radianal small or absent or with one to three additional anal plates in cup (Fig. 372,*1*). Tegmen produced into a short, stout or slender, elongate anal sac that is generally imperforate and has terminal anal opening. Arms typically isotomous with few to many branches, or rarely heterotomous. *M.Ord.-U.Miss.; U.Perm.*

Family CYATHOCRINITIDAE Bassler, 1938

[*nom. correct.* BASSLER, 1938, p. 20 (*pro* Cyathocrinidae ROEMER, 1854, p. 227)]

Cup bowl shaped, infrabasals three or five; single anal plate in cup. Arms branch isotomously several times, first on axillary primibrachs *1* to *7*. Anal sac short, slender to massive, imperforate, confined to posterior area of tegmen. Posterior oral may be a madreporite. *?M.Ord.; M.Sil.-L.Miss.; U.Perm.*

Cyathocrinites MILLER, 1821, p. 85 [**C. planus*; SD WACHSMUTH & SPRINGER, 1880, p. 81 (*non C. tuberculatus* MILLER, 1821; SD ROEMER, 1851, p. 365, type of *Temnocrinus* SPRINGER, 1902); ICZN pend.] [=*Cyathocrinus* AGASSIZ, 1836, p. 197 (*nom. van.*); *Saccosomopsis* MEEK & WORTHEN, 1870, p. 25]. Cup low bowl shaped to globose, with flat or gently convex base; radials

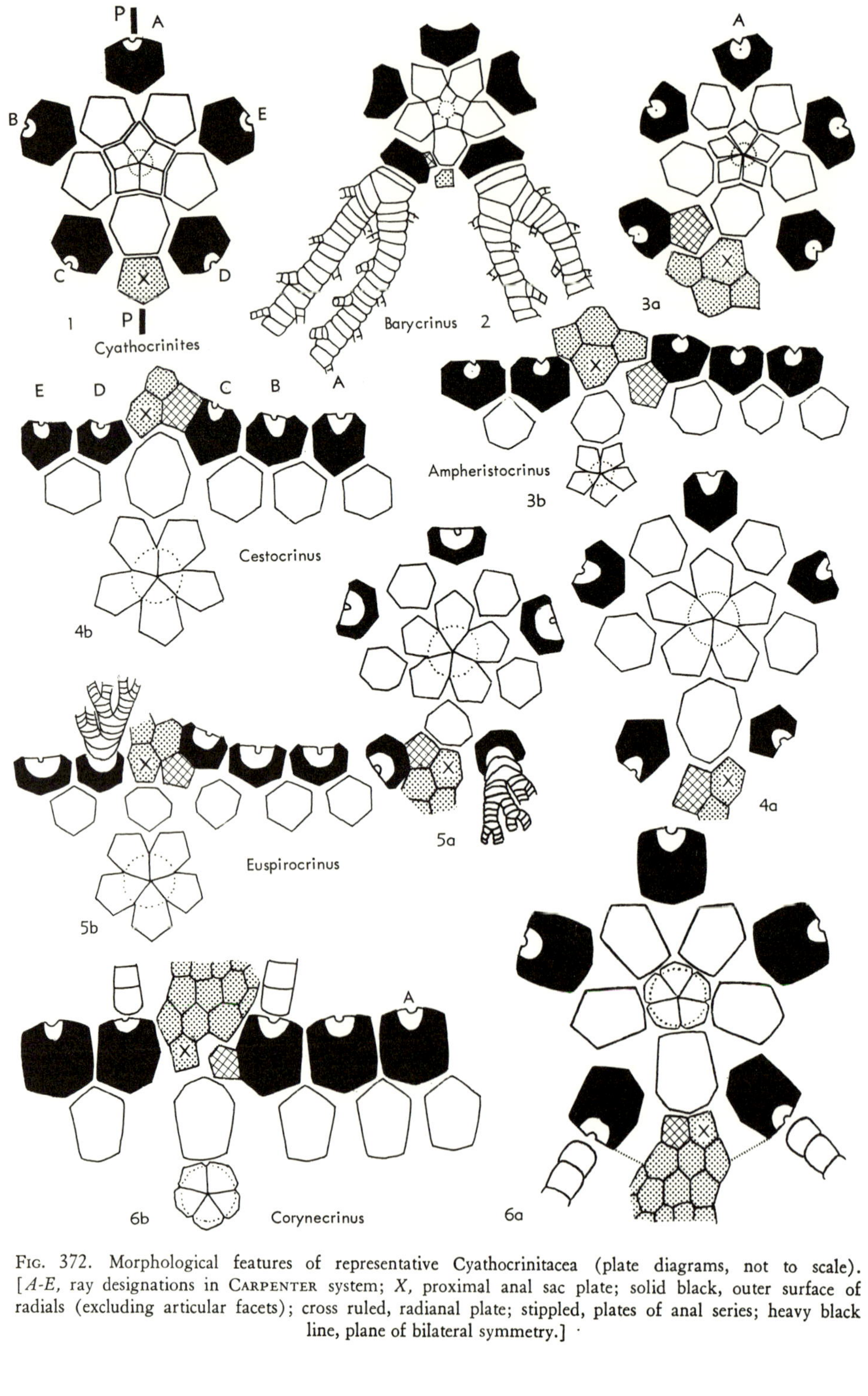

FIG. 372. Morphological features of representative Cyathocrinitacea (plate diagrams, not to scale). [*A-E*, ray designations in CARPENTER system; *X*, proximal anal sac plate; solid black, outer surface of radials (excluding articular facets); cross ruled, radianal plate; stippled, plates of anal series; heavy black line, plane of bilateral symmetry.]

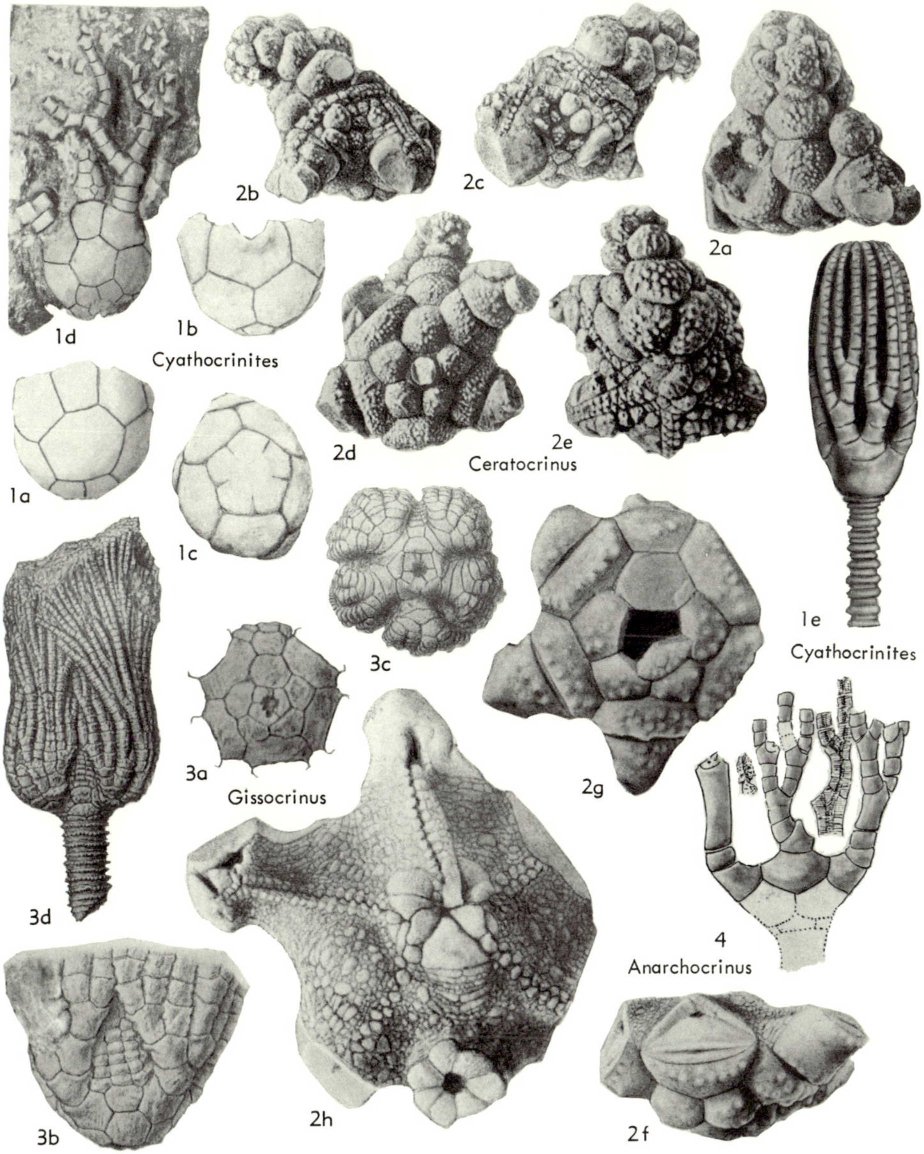

FIG. 373. Cyathocrinitidae (p. *T*579, *T*581-*T*582).

quadrangular, with narrow, horseshoe-shaped articular facets; large anal *X* alone in cup, directly above *CD* basal (Fig. 372,*1*). Tegmen stout, with 5 large orals, posterior one a madreporite; anal sac narrow, cylindrical, confined to posterior area of tegmen. Arms narrow, branching isotomously twice or more, first on axillary primibrachs 1 to 7; ambulacral covering plates large, stout, commonly preserved. Stem round or pentagonal in section. *M.Sil.(Wenlock.)-L.Miss.(Osag.), ?U.Perm.*, cosmop.——FIG. 372,*1*; 373,*1a-d*. **C. planus*, L. Carb.(Tournais.), Eu.(Eng.); 372,*1*, plate diagram of cup (radials black, anal *X* stippled) (Moore, 1962b); 373,*1a-c*, post., ant., and dorsal views of cup, ×1; 373,*1d*, post. view of crown, ×1 (Wright, 1951-54).——FIG. 373,*1e*. *C. milleri* (M'COY), L.Carb.(Tournais.), Eu.(Ire.); lat. view of crown, ×2 (M'Coy, 1862). [Revision of M'COY's species of Irish crinoids by G. SEVASTOPULO has resulted in transfer of *Atocrinus* M'COY in GRIFFITH, 1844, from *Cyathocrinites* to *Platycrinites* (G. SEVASTOPULO, pers. comm.)]

?**Anarchocrinus** JAEKEL, 1918, p. 51 [**A. rossicus*; OD]. Cup wide, low bowl shaped, base unknown; cup plates smooth; radials large, with articular facets narrowly rounded and directed gently outward-downward. Uniserial arms branching on primibrachs *1, 2*, or *3*; primaxils short to extremely high; arms branch isotomously again on secundibrach *3* and tertibrach *3* in only known ray. Anal tube with large sculptured plates. [Poorly known.] *M.Ord.(Caradoc.)*, Eu.(Est.). ——FIG. 373,*4*. **A. rossicus*; lat. view of crown (part. reconstr., ×0.7) (Jaekel, 1918).

Ceratocrinus WANNER, 1937, p. 177 [**C. exornatus*; OD]. Cup low, infrabasals mostly covered by stem, cup plates highly nodose; radial articular facets almost as wide as radials, declivate or vertically directed, with proximal parts of arms subhorizontal; single anal plate in cup. Arms stout, primibrachs *1* axillary. Tegmen composed of numerous small ambulacral and interambulacral plates; anal sac short, massive, confined to posterior side of tegmen. *U.Perm.*, Indon.(Timor, Basleo). ——FIG. 373,*2a-e*. **C. exornatus; 2a-e*, post., *BC*- and *DE*-interray, dorsal and ventral views of theca and proximal arms, ×1 (Wanner, 1937). ——FIG. 373,*2f-h*. *C. gracilis* WANNER; *2f,g*, *B*-ray and dorsal views, ×2; *2h*, ventral view of theca, ×3 (Wanner, 1937).

Gissocrinus ANGELIN, 1878, p. 10 [**G. typus* BATHER, 1893a, p. 155 (*nom. subst. pro G. arthriticus* ANGELIN, 1878, *non* PHILLIPS in MURCHISON, 1839, p. 674, =*Actinocrinites? arthriticus*) (=*G. punctuosus* ANGELIN, 1878, p. 10, *partim; G. nudus* ANGELIN, 1878, p. 10; *G. tubulatus* ANGELIN, 1878, p. 11); SD BATHER, 1893a, p. 155]. Cup bowl shaped, 3 or 5 infrabasals; radials low, wide, with articular facets almost as wide as radials. Single anal plate in cup, directly above *CD* basal. Arms branching isotomously several times, primibrach *1* axillary; brachials narrow, spindle shaped. Anal sac stout, composed of longitudinal rows of heavy plates. *M.Sil.-L.Dev.*, Eu.(Boh.-Ger.-Eng.-Sweden-USSR), USA(Tenn.-Ind.-Ky.-Mo.-Okla.).——FIG. 373,*3a,b*. **G. typus* BATHER, U.Sil., Gotland; *3a,b*, dorsal view of cup and post. view of crown, ×2 (Bather, 1893a). ——FIG. 373,*3c,d*. *G. campanula* BATHER, U.Sil., Gotland; *3c,d*, dorsal and post. views of crown, ×1.5 (Bather, 1893a).

[The problem of defining the type species of *Gissocrinus* is an involved one, as reviewed at length by BATHER (1893a), who properly introduced a substitute name for *G. arthriticus* ANGELIN, although incorrect in stating that this was ANGELIN's choice as type species of the genus. ANGELIN simply placed this species first, without indicating it or any other described form as the type. BASSLER's (1938, p. 103) subsequent designation of *G. arthriticus* ANGELIN as type species was erroneous and moreover was superseded by the action of BATHER (1893a).]

Family BARYCRINIDAE Jaekel, 1918

[Barycrinidae JAEKEL, 1918, p. 58]

Crown large, cup broad bowl shaped; radials with relatively wide articular facets which are semicircular in outline, concave, and strongly declivate, with weak median cross-ridges extending laterally from deep V-shaped notch on inner side of facet, thus providing trifascial articulation with primibrachs *1*. Anal *X* large, quadrangular, directly above *CD* basal; small quadrangular radianal (rarely absent) at left below *C* radial. Arms 10 to 16, long, strongly rounded, with narrow ambulacral groove, branching isotomously on primibrachs *2* or *3* and bearing nonpinnulate ramules given off on alternate sides from every second brachial. Anal sac short, stout; anal opening terminal. Column transversely round, columnals divided into pentameres or not. *L.Dev.(Ems.); L.Miss.(Osag.)-U.Miss.(Meramec.)*.

Barycrinus WACHSMUTH & WORTHEN, 1868c, p. 338 [**Cyathocrinus spurius* HALL, 1858b, p. 625; OD] [=*Pottsicrinus* JILLSON, 1960, p. 37 (type, *P. quinstilus*; OD)]. Crown large; commonly with anal *X* and radianal in cup, rarely with anal *X* only (Fig. 372,*2*). Arms long, robust, primibrachs *2* axillary, followed by isotomous or unbranched half-rays. Every second brachial generally a subaxil, supporting branched armlets or unbranched ramules. Anal sac low, with large plates, anal opening terminal. Circular columnals composed of discrete pentameres. *L.Miss.(Osag.)-U.Miss.(Meramec.)*, N.Am.(C.USA).——FIG. 372, *2*; 374,*1a,b*. **B. spurius* (HALL), Osag., USA(Ia.); 372,*2*, plate diagram (radials black, radianal cross ruled, anal *X* stippled) (Moore, n); 374,*1a,b*, post. and dorsal (*C* ray up) views, ×1 (Hall, 1858b).——FIG. 374,*1c,d*. *B.* sp., Osag.(Keokuk), Iowa; *1c,d*, radials, showing large, rounded articular facets, ×1 (Springer, 1911c).

Pellecrinus KIRK, 1929, p. 15 [**Cyathocrinites hexadactylus* LYON & CASSEDAY, 1860, p. 74; M] [=*Vasocrinus* LYON *auctt.* prior to 1929 (*non Vasocrinus* LYON, 1857, *emend.* KIRK, 1929a)]. Crown large. Cup low bowl shaped, commonly with anal *X* and radianal, rarely with anal *X* only. Arms long, slender, dividing isotomously on primibrachs *3*, regularly heterotomous, with armlets alternately on opposite sides of each second brachial, which is subaxil; armlets similarly bearing ramules alternately on opposite sides of each second brachial. Anal sac conical, with longitudinal series of large plates above anal *X*. Columnals not divided into pentameres. *L.Miss.(Osag.)*, N.Am.(C.USA).——FIG. 374,*3*. **P. hexadactylus* (LYON & CASSEDAY), USA(Ind.); *3a,b*, *D*-ray view of crown and part of arm with armlets and ramules on one side, ×1 (Hall, 1872).

Situlacrinus BREIMER, 1962, p. 153 [**S. costatus*;

OD]. Cup high conical; primibrachs joined to tegmen by small, interradial plates; posterior oral madreporitic. *L.Dev.(low. Ems.)*, Spain.——FIG. 374,*2*. **S. costatus*, LaVid Sh., León Prov.; *2a,c*, post., ventral, and dorsal views of cup, ×4 (Breimer, 1962).

Family EUSPIROCRINIDAE Bather, 1890

[Euspirocrinidae BATHER, May, 1890b, p. 346] [=Ampheristocrinidae S. A. MILLER, 1889, p. 346 (Dec., 1890); Vasocrinidae JAEKEL, 1918, p. 61; Parisocrinidae JAEKEL, 1918, p. 51]

Cup cone or bowl shaped; infrabasals three to five, visible in side view; radial articular facets vertically or subhorizontally directed, with axial canal; three anal plates in cup, anal *X* large. Arms slender, branching isotomously several times, ambulacral grooves with stout covering plates; primibrachs *2* to *4* axillary. Anal sac present or not, porous or nonporous. Orals five, large, posterior one a madreporite. Stem transversely round, wide. *M.Ord.(Mohawk.)-L.Carb.(Tournais.)*.

Euspirocrinus ANGELIN, 1878, p. 24 [**E. spiralis*; M]. Cup low, wide bowl shaped; 5 infrabasals; radianal and anal *X* in cup. Anal sac short, slender, composed of few large plates. Tegmen of 5 large orals and numerous small interambulacrals and ambulacrals. Arms branching isotomously and curved spirally inward; primibrachs *2* axillary. Stem wide. *M.Ord.(Mohawk.)-U.Sil.*, Can. (Ont.)-Eu. (Sweden)-USA (Ohio).——FIG. 372,*5*; 375,*3*. **E. spiralis*, U.Sil., Gotland; 372, *5a,b*, plate diagrams of cup and *D*-ray arms (radials black, radianal cross ruled, anal *X* and other anals stippled) (Moore, n; 1962b); 375,*3a*, lat. view of crown and stem, ×1.5; 375,*3b,c*, post. view of crown and ant. view of theca, ×1.5; 375,*3d*, ventral view of tegmen, ×1.5 (Bather, 1893a).

Ampheristocrinus HALL, 1879, p. 11 [**A. typus*; OD]. Cup cone shaped, plates with strong ridges; infrabasals 3 or 5, visible in side view, *C* infrabasal smaller than others. Anal *X* large, supporting 3 tube plates above, radianal pentagonal, below and left of *C* radial. Arms slender, primibrachs *3* axillary, branching isotomously 3 or more times. Anal sac long, slender, projecting above arms. Stem transversely round, stout, composed of short, alternating nodals and internodals. *M.Sil. (Niag.)*, USA(Tenn.-Ind.-Ill.).——FIG. 372,*3*; 375,*6*. **A. typus*, USA (Ind.); 372,*3a,b*, plate diagrams of cup (radials black, radianal cross ruled, anal *X* and other anals stippled) (Moore, n; 1962b); 375,*6*, lat. view of crown, ×1 (Springer, 1926a).

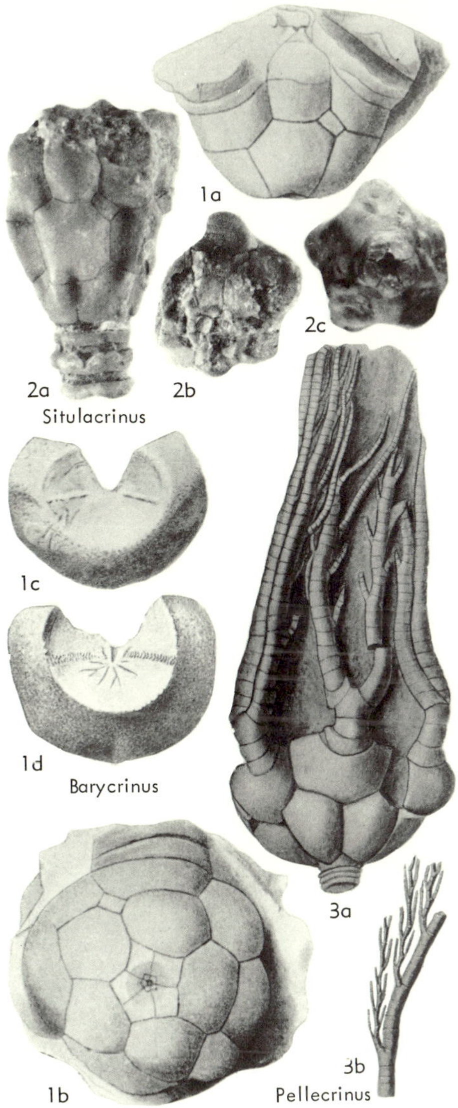

FIG. 374. Barycrinidae (p. *T*582-*T*583).

Caelocrinus XU, 1962, p. 46 [**C. stellifer*; OD]. Cup low, bowl shaped, with moderate basal concavity, basals and radials ornamented with strong ridges that radiate from the center of each of the 5 basals. Infrabasals 3, the small one in the *C*-ray position. Basals large, pentagonal; radials wider than high, and apparently with wide radial facets. Radianal narrow, quadrangular, between *C* radial and *CD* basal. Anal *X* large, separating 2 posterior radials. Stem, arms and tegmen unknown. *L.Sil.*, China(Szechwan Prov.).

?**Closterocrinus** HALL, 1852, p. 179 [**C. elongatus*;

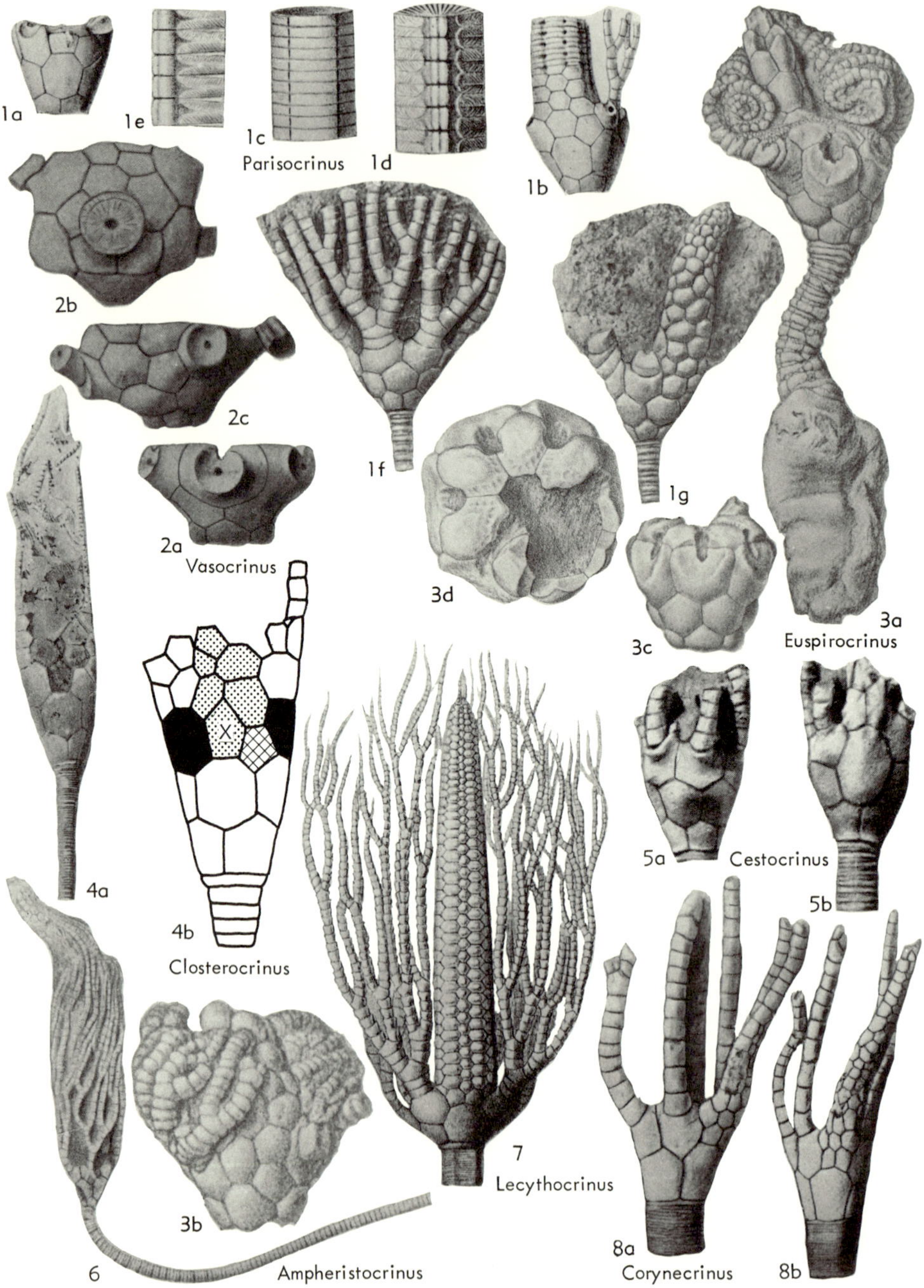

FIG. 375. Euspirocrinidae *(1-4,6)*; Lecythocrinidae *(5,7,8)* (p. *T*583-*T*585).

OD]. Cup narrow, high, steeply conical, with wide flat base; infrabasals 3 or 5, high, visible in side view; radials ?5, with subhorizontal articular facets. Anal *X*, radianal, and right tube plate in cup. Arms branching on primibrachs ?*3*. Anal sac apparently present. Stem transversely round, stout, expanded proximally just below cup. *M.Sil. (Niagaran)*, USA(N.Y.).——FIG. 375,*4*. **C. elon-*

gatus; 4a, post. view of crown, ×1 (Hall, 1852); *4b,* plate analysis of post. side of cup (radials black, radianal cross ruled, anal *X* and other anals stippled) (Hall, 1852, mod.).

Parisocrinus WACHSMUTH & SPRINGER, 1880, p. 109 [**Poteriocrinites perplexus* MEEK & WORTHEN, 1869c, p. 138; OD]. Cup bowl to cone shaped, as wide as high; infrabasals large, visible in side view; radials pentagonal, with narrow horseshoe-shaped articular facets; anal *X,* radianal, and right tube plate in cup. Arms narrowly rounded, branching isotomously several times, primibrachs *3* or *4* axillary. Anal sac large and porous or short and without pores, anal opening terminal. Madreporite present on anterior side of tegmen at base of anal sac. [Devonian species have separate axial canal in radials.] *M.Sil.(Niagaran)-L. Carb.(Tournais.),* Eu.(Ger.-Eng.-Belg.-Ire.-USSR)-USA(Ind.-N.Y.-Ill.-Iowa).——FIG. 375,*1a,b. P. siluricus* SPRINGER, U.Sil., USA(Ind.); *1f,g,* ant. and *D*-ray views of crown, ×1.5 (Springer, 1926a).——FIG. 375,*1c-g. P. curtus* (MÜLLER), M.Dev.(Givet.), Ger.; *1a,b,* post. views of cup and partial crown, ×1; *1c-e,* lat. view and transv. sections of stem, *1c,d,* ×2; *1e,* ×3 (Schultze, 1867).

Vasocrinus LYON, 1857, p. 485 [**V. valens*; OD] [*non Vasocrinus* LYON, pre-1929 *auctt.* (=*Pellecrinus* KIRK, 1929a, p. 82)]. Cup low cone shaped, plates thin and smooth or with sharp ridges radiating from basals and crossing adjacent radials; base wide, 5 low infrabasals visible in side view; radials wide, low, with large vertical horseshoe-shaped articular facets with large axial canal and obscure transverse ridge at either side; 3 anal plates in cup. Arms massive, directed horizontally or declivate at low angle; ambulacral grooves covered by stout, high cover plates, 3 to a brachial; primibrachs *1* and *2* (all known) not axillary. Tegmen low, convex, composed of numerous small plates and 5 large orals, posterior one a madreporite. *Up.L.Dev.-M.Dev.,* Eu.(Ger.-Spain)-USA(Ky.).——FIG. 375,*2. *V. valens,* M.Dev., USA(Ky.); *2a-c,* lat., basal, and post. views of cup, ×1 (Kirk, 1929a).

?**Zygotocrinus** KIRK, 1943 [**Z. fragilis*; OD]. *L. Miss.(Osag., Burlington Ls.),* USA(Iowa).

Family LECYTHOCRINIDAE Kirk, 1934

[Lecythocrinidae KIRK, 1934, p. 4]

Cup small, bowl or cone shaped; two anal plates directly above *CD* basal, none to three additional anal plates in cup. Anal sac narrow and tall, with well-marked vertical rows of polygonal plates. Arms branching isotomously three times or more, many slender arms in upper part of crown. Stem large, transversely round, with large central axial canal and four peripheral canals. *M.Dev.(Givet.); L.Miss.(Osag.).*

Lecythocrinus MÜLLER, 1858, p. 196 [**L. eifelianus*; M]. Cup low bowl shaped; infrabasals small, may be undivided; *CD* basal large, followed above by 2 small, equal anal plates. Primibrachs *3* axillary, with 5 higher isotomous divisions. Anal tube long, with strong plates in longitudinal rows. Stem transversely subquadrangular, with large axial canal and 4 peripheral canals. *M.Dev.(Givet.),* Eu.(Ger.)——FIG. 375,*7. L. briareus* (SCHULTZE); post. view of crown, ×1 (Schultze, 1867).

Cestocrinus KIRK, 1940, p. 221 [**C. striatus*; OD]. Cup cone shaped; large infrabasals visible in side view; *CD* basal projecting upward between *C* and *D* radials, supporting 2 large anal plates (radianal at right, anal *X* at left). Arms slender, subcylindrical, 3, 4, or more than 6 primibrachs to a ray, remainder of arms unknown. *L.Miss.(Osag.),* USA(Ind.).——FIG. 372,*4*; 375,*5. *C. striatus*; 372,*4a,b,* plate diagrams of cup (radials black, radianal cross ruled, anal *X* and higher anal plates stippled) (Moore, n; 1962b); 375,*5a,b,* ant. and post. views of incomplete crown, ×1 (Kirk, 1940a).

Corynecrinus KIRK, 1934, p. 1 [**C. romingeri*; OD]. Cup cone shaped; infrabasals low, wide, barely visible in side view; *CD* basal large, supporting subequal radianal at right and anal *X* at left, followed by additional anal plates in cup. Arms slender, brachials with stout covering plates and axial canal; primibrachs *9* or *10* axillary. Anal tube composed of 5 longitudinal series of plates. Stem large, composed of thin, round columnals with large obscurely quadripartite axial canal. *M.Dev.,* USA(Ind.).——FIG. 372,*6*; 375,*8. *C. romingeri*; 372,*6a,b,* plate diagrams of cup (radials black, radianal cross ruled, anal *X* and higher anal plates stippled) (Moore, n; 1962b); 375,*8a,b, D*-ray and post. views of incompl. crown, ×1.15 (Kirk, 1934).

Tetrapleurocrinus WANNER, 1942 [**T. eifelensis*; OD]. *M.Dev.,* Eu.(Ger.).

Superfamily GASTEROCOMACEA Roemer, 1854

[*nom. transl.* MOORE & LANE in MOORE & STRIMPLE, 1973, p. 18 (*ex* Gasterocomidae ROEMER, 1854, p. 299)] [=superfamily Palaeocrinacea BATHER, 1899b (*nom. transl.* LANE, 1967b, p. 9, *ex* Palaeocrinidae BATHER, 1899b, p. 922); superfamily Hypocrinacea WANNER, 1916a (*nom. transl.* ARENDT, 1970, p. 94, *ex* Hypocrinidae WANNER, 1916a, p. 88) *(partim)*]

Theca bowl shaped; radianal and anal *X* present, or one or both may be absent; infer- and superradianal present primitively. Tegmen low, anal opening through tegmen or side of cup; posterior oral with hydropore or developed as madreporite. Arms atomous or isotomous; brachials fused in each ray or laterally united in specialized families. *M.Ord.(Black River.)-U.Ord.; M.Sil.-M.Dev.(Givet.).*

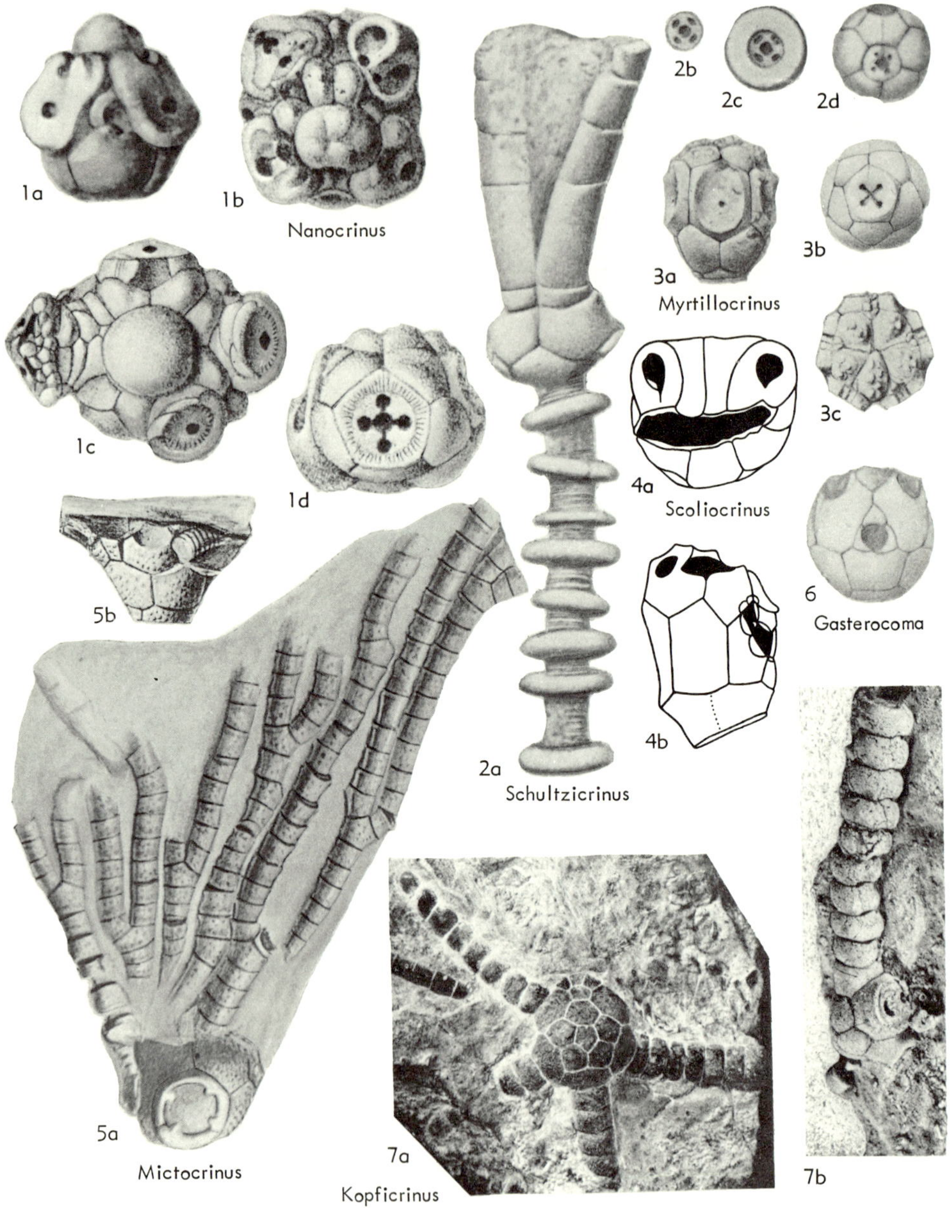

FIG. 376. Gasterocomidae (p. *T*587-*T*588).

Family GASTEROCOMIDAE Roemer, 1854

[Gasterocomidae ROEMER, 1854, p. 299] [=Gastrocomidae JAEKEL, 1895, p. 73; Schultzicrinidae JAEKEL, 1918, p. 82]

Theca conical bowl shaped to globose; infrabasals five, three, or one; radial articular facets narrow horseshoe shaped; with axial canal; radianal lacking, anal *X* present or absent, anal opening generally through side of cup below posterior radials. No anal sac. Moderately thick uniserial arms spread sideward, branching isotomously. Transversely round to quadrangular stem with axial and three or four peripheral canals. *L.Dev.-M.Dev.*

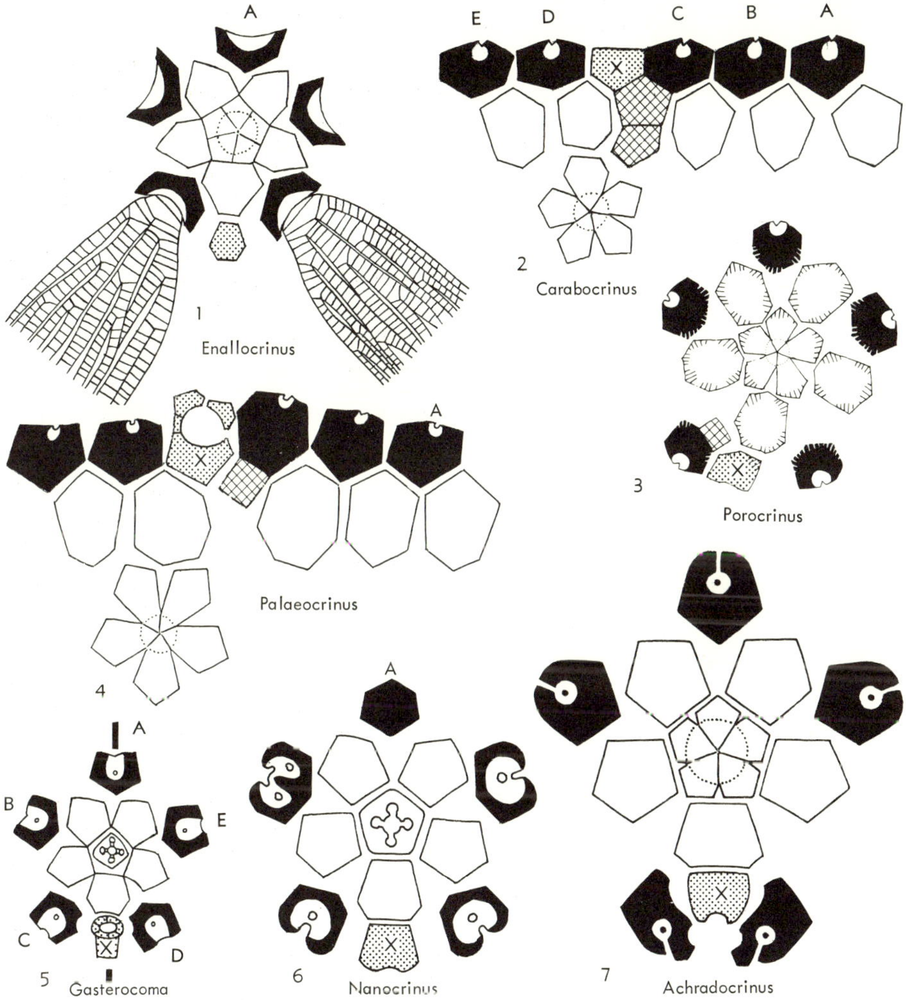

FIG. 377. Morphological features of representative Gasterocrinacea (plate diagrams, not to scale). [Letters and patterns as in Fig. 372.]

Gasterocoma GOLDFUSS, 1839, p. 350 [**G. antiqua*; M] [=*Ceramocrinus* MÜLLER, 1855, p. 83; *Epactocrinus* MÜLLER, 1855, p. 84; *Gastrocoma* BEYRICH, 1871, p. 50 *(nom. van.)*]. Theca globular; infrabasals low, fused, barely visible in side view; arm facets horseshoe shaped, very steeply declivate, with large axial canal; quadrangular anal *X* directly above anal opening and between *C* and *D* radials, or irregular in disposition. Arms, tegmen, and column unknown. *M.Dev.(Givet.)*, Eu.(Ger.).——FIG. 376,*6*; 377,*5*. **G. antiqua*; 376,*6*, *CD* interray view of cup, ×2 (Springer, 1911c); 377,*5*, plate diagram of cup (radials black, anal *X* stippled) (Moore, 1962b).

Kopficrinus GOLDRING, 1954, p. 11 [**K. pustuliferus*; OD]. Cup low, bowl shaped, flaring slightly ventrally; three unequal infrabasals, small one anterior; basals large; radials with flat, steeply inclined, horseshoe-shaped facets; one anal plate above enlarged posterior basal, followed by three small anal plates; tegmen unknown; arms heavy rounded, consisting of at least 13 primibrachs; stem round, small. *L.Dev.*, USA(N.Y.).——FIG. 376,*7*. **K. pustuliferus; 7a,b,* basal view of cup and partial arms; lat. *B*-ray view of cup and arms, ×1.3 (Goldring, 1954).

Mictocrinus GOLDRING, 1923, p. 362 [**M. robustus*; OD]. Cup low cone shaped, plates with coarse nodes; infrabasals 3, low, visible in side view; *CD* basal supporting small, quadrangular anal *X*; anal opening above anal *X* and between *C* and *D* radials. Arms robust, strongly rounded, branching isotomously on primibrachs *5* or *6*, and with 2 higher bifurcations. Column transversely round, large, with large axial canal and 4 peripheral canals. *M.Dev.*, USA(N.Y.).——FIG. 376,*5*. **M. robustus*; *5a,b*, dorsal view of crown and *C*-ray view of cup, ×1 (Goldring, 1923).

Myrtillocrinus SANDBERGER & SANDBERGER, 1856, p. 388 [**M. elongatus*; M] [=*Tripleurocrinus* WOOD, 1904, p. 56 (type, *T. levis*)]. Cup higher than wide, infrabasals fused, anal plate absent; arm facets large, horseshoe shaped, vertically directed and occupying most of outer surface of radials, with large axial canal; tegmen low, composed of 5 large orals; anal opening through tegmen. Column with 4 small peripheral canals joined to small central axial canal. *M.Dev.*, Eu.(Ger.)-USA (N.Y.).——FIG. 376,*3*. *M. americanus* HALL, N.Y.; *3a,b*, lat. and dorsal views of theca, ×2 (Springer, 1911c); *3c*, ventral view of theca, ×1 (Springer, 1926a).

Nanocrinus MÜLLER, 1856, p. 355 [**N. paradoxus*; M]. Theca small, low, infrabasals fused; basals 5; faceted radials 4, facet not developed on *A*, radial articular facets large, steeply declivate, with large axial canal; anal *X* directly above *CD* basal, anal opening above anal *X*, laterally directed; tegmen composed of several small plates and large, central bulbous *CD* oral. *M.Dev.(Givet.)*, Eu.(Ger.).——FIG. 376,*1*; 377,*6*. **N. paradoxus*; 376,*1a-d*, *CD*-interray, ventral views of 2 specimens, and dorsal view of cup, ×3 (Schultze, 1867); 377,*6*, plate diagram of cup (radials black, anal *X* stippled) (Moore, n).

Schultzicrinus SPRINGER, 1911, p. 132 [**S. typus*; OD]. Cup like *Arachnocrinus*. Arms unbranched, primibrachs *1* short, subsequent primibrachs high; brachials and radials with axial canal. Stem with 4 peripheral canals and thin nodals about twice as wide as internodals. *M.Dev.*, USA(N.Y.).——FIG. 376,*2*. **S. typus*; *2a*, lat. view of crown and stem, ×2; *2b,c*, articular surfaces of nodal and internodal, ×1.3, ×2; *2d*, dorsal view of cup, ×1.3 (Springer, 1911a).

Scoliocrinus JAEKEL, 1895, p. 76 [**S. eremita*; OD]. Cup asymmetrical, high cone shaped; infrabasals fused, basals 5, unequal in size, *AE* basal largest; radials unequal, large *A* and *E* radials with arm facets, other radials small, without facets; arm facets large, horizontally directed, with large axial canal; anal *X* small, quadrangular, directly below large anal opening and in line with basals; anal opening laterally directed, in side of cup, bounded above by small *B*, *C*, and *D* radials; tegmen unknown, space at top of cup lozenge shaped. *M.Dev.(Givet.)*, Eu.(Ger.).——FIG. 376, *4*. **S. eremita*; *2a,b*, ventral and *DE*-interray views of cup, ×1 (Jaekel, 1895).

Family SPHAEROCRINIDAE Jaekel, 1895

[Sphaerocrinidae JAEKEL, 1895, p. 64] [=Palaeocrinidae BATHER, 1899b, p. 922]

Theca globose or conical; angles of cup plates with radially directed pore slits in one Ordovician genus (*Palaeocrinus*); infrabasals five, high; radials five, with narrow articular facets; large anal *X* and radianal present; anal opening in side of cup or through low, flat tegmen; posterior oral larger than others and a madreporite where known; arms branching isotomously or unbranched in one genus (*Palaeocrinus*) where known. *M.Ord.*; *M.Sil.*; *M.Dev.*

Sphaerocrinus ROEMER, 1851, p. 369 [**Cyathocrinites geometricus* GOLDFUSS, 1831, p. 189; OD] [=*Sphaerocrinites* STEININGER, 1853, p. 38 (*nom. van.*)]. Theca globose; infrabasals 5 or rarely 3; radials pentagonal, articular facets semicircular, with large axial canals; radianal and anal *X* large, in normal position; anal opening large, on side of cup below summit of radials, surrounded by anal *X* below and 3 to 5 small accessory anal plates. Tegmen low, flat, composed of 4 small orals and large madreporitic posterior oral; central tegminal area composed of numerous small irregular plates. Arms and stem unknown. *M.Dev. (Givet.)*, Eu.(Ger.).——FIG. 378,*1a-c*. *S. pauci-sculptus* WANNER; *A*-ray, *CD*-interray, and ventral views of theca, ×1.15 (Wanner, 1942a).——FIG. 378,*1d-f*. **S. geometricus* (GOLDFUSS); *1d*, *AB*-interray view of typical specimen from Kerpen; *1e,f*, dorsal and ventral views of smooth-surfaced theca, from same locality, all ×1 (Schultze, 1867).

Palaeocrinus BILLINGS, 1859, p. 24 [**P. striatus*; SD MILLER, 1889, p. 267]. Theca small, cone shaped, with prominent hollow ridges radiating from center of each basal to adjacent infrabasals and radials; infrabasals 5, high, visible in side view; basals large; radials low, wider than high with narrow, rounded, subhorizontal articular facets; radianal large, quadrangular, below and left of *C* radial; anal *X* large, pentagonal, with concave upper surface forming proximal edge of anal opening. Tegmen composed of 10 plates, 4 orals above interradial sutures, 3 plates bordering anal opening, and large central madreporitic *CD* oral. [Other species assigned to this genus have an anal sac and probably do not belong in this family.] *M.Ord.(Chazy.-Mohawk.)*, USA(Minn.-Ky.)-Can.——FIG. 377,*4*; 378,*2*. **P. striatus*, Chazy., Que.; 377,*4*, plate diagram of cup (radials black, radianal cross ruled, anal *X* and higher anals stippled) (Moore, 1962); 378,*2a,b*, *AB*-interray and ventral views of theca, ×2 (Lane, n; Springer Coll., U.S. Natl. Museum).

Thalamocrinus MILLER & GURLEY, 1895b, p. 82 [**T. ovatus*; OD]. Cup small, cylindrical; infrabasals large, making up one-third to one-half of cup side; basals higher than wide, pentagonal; radials small, decreasing in width distally, incurved at top, each with narrow rounded subhorizontal articular facet; radianal small, rectangular, anal *X* smaller than radials, directly above *CD* basal. *M.Sil.(Niagaran)*, USA(Tenn.).——FIG. 378,*3a-c*. **T. ovatus*; *CD*-interray, ventral, and dorsal views of cup, ×1 (Springer, 1926a).——FIG. 378,*3d*. *T. cylindricus* MILLER & GURLEY; *CD*-interray view of cup, ×1 (Springer, 1926a).

Family POROCRINIDAE Miller & Gurley, 1894

[Porocrinidae MILLER & GURLEY, 1894b, p 24]

Theca globose or conical, anal *X* and radianal present; cup plates with one or more folds and ridges at plate corners. *M.Ord.(Mohawk.)-U.Ord.(Richmond.)*.

Porocrinus BILLINGS, 1857, p. 279 [**P. conicus*; M]. Theca small, steeply conical, base narrow, truncate; infrabasals 5, high; basals hexagonal; radials higher than wide, small circular articular facets declivate; radianal small, rectangular, anal *X* almost as large as radials, directly above *CD* basal and between posterior radials; cup-plate angles bear goniospire areas, with slits directed toward points where 3 plates come together, slit areas rounded triangular in outline, present between all infrabasals and basals, basals and radials, radianal, and at upper corners of adjacent radials, but lacking from upper juncture of anal *X* and adjacent radials. Arms slender, rounded, unbranched to height of primibrach *14*. *M.Ord. (Mohawk.)-U.Ord.(Richmond.)*, USA(Ky.-Ill.-Iowa-Minn.-Wis.)-Can. (Ont.-Baffin Island)-Eu. (Scot.).——FIG. 377,*3*; 379,*2*. **P. conicus*, Mohawk., Ont.; 377,*3*, plate diagram of cup (radials black, radianal cross ruled, anal *X* stippled) (Moore, n); 379,*2a*, lat. view of crown, ×3.3 (Lane, n; Springer Coll., U.S. Natl. Museum); 379,*2b,c*, post. and ventral views of cup, ×7 (Lane, n); 379,*2d,e*, *E*-ray view of crown, ×2.7, and plate diagram (Kesling & Paul, 1968).

Triboloporus KESLING & PAUL, 1968, p. 13 [**T. cryptoplicatus*; OD]. Like *Porocrinus* except goniospires single, consisting of median ridge with deep trough on each side of plate corner. *M.Ord. (Mohawk.)*, USA(Pa.-Va.).——FIG. 379,*1*. **T. cryptoplicatus*, Va.; *1a-d*, ventral, oblique *CD*-interray, *C*-ray, and *A*-ray views, ×2.7; *1e*, thecal diagram; *1f*, *CD*-interray view, ×5.3 (Kesling & Paul, 1968).

Family CARABOCRINIDAE Bather, 1899

[Carabocrinidae BATHER, 1899b, p. 922]

Theca egg shaped; infer- and superradianal and anal *X* present in cup. *M.Ord.-U.Ord.*

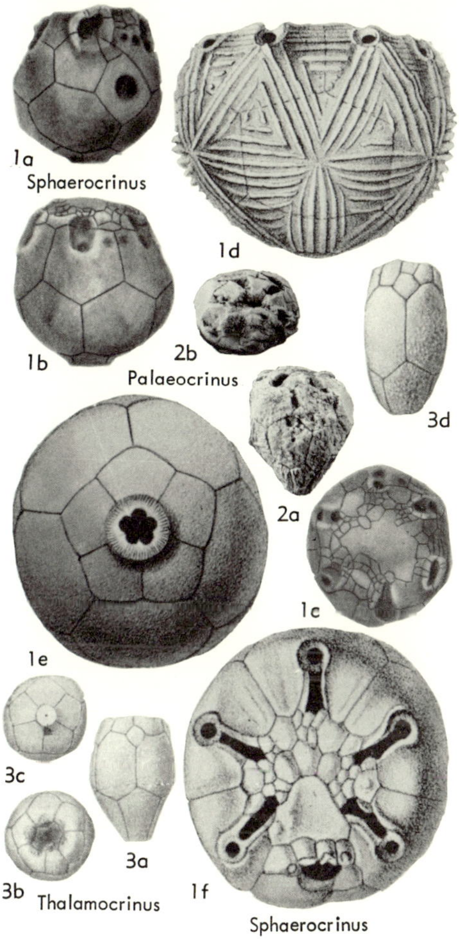

FIG. 378. Sphaerocrinidae (p. *T*588-*T*589).

Carabocrinus BILLINGS, 1857, p. 276 [**C. radiatus*; M] [=*Strophocrinus* SARDESON, 1899, p. 264 (type, *S. dicyclicus*)]. Cup egg shaped, base wide, sides gently convex; infrabasals large, visible in side view; radials wider than high, inclined inward distally toward narrow tegminal area; arm facets small, round, vertically inclined; infer- and superradianal in cup, inferradianal in contact with 2 infrabasals, *CD* and *BC* basals, superradianal above inferradianal, below and left of *C* radial; anal *X* large, above and right of *CD* basal, between *C* and *D* radials; arms narrowly rounded, short, branching on primibrachs *2* or *3*. *M.Ord. (Chazy.)-U.Ord.(Richmond.)*, USA(Minn.-Iowa-Ky.-N.Y.-Okla.-Wis.)-Can. (Ont.)-Eu. (Est.).——FIG. 377,*2*. *Carabocrinus* BILLINGS; plate diagram of cup (radials black, infer- and superradianal cross ruled, anal *X* stippled) (Moore, 1962b). [Restudy of SARDESON's type specimens by J.

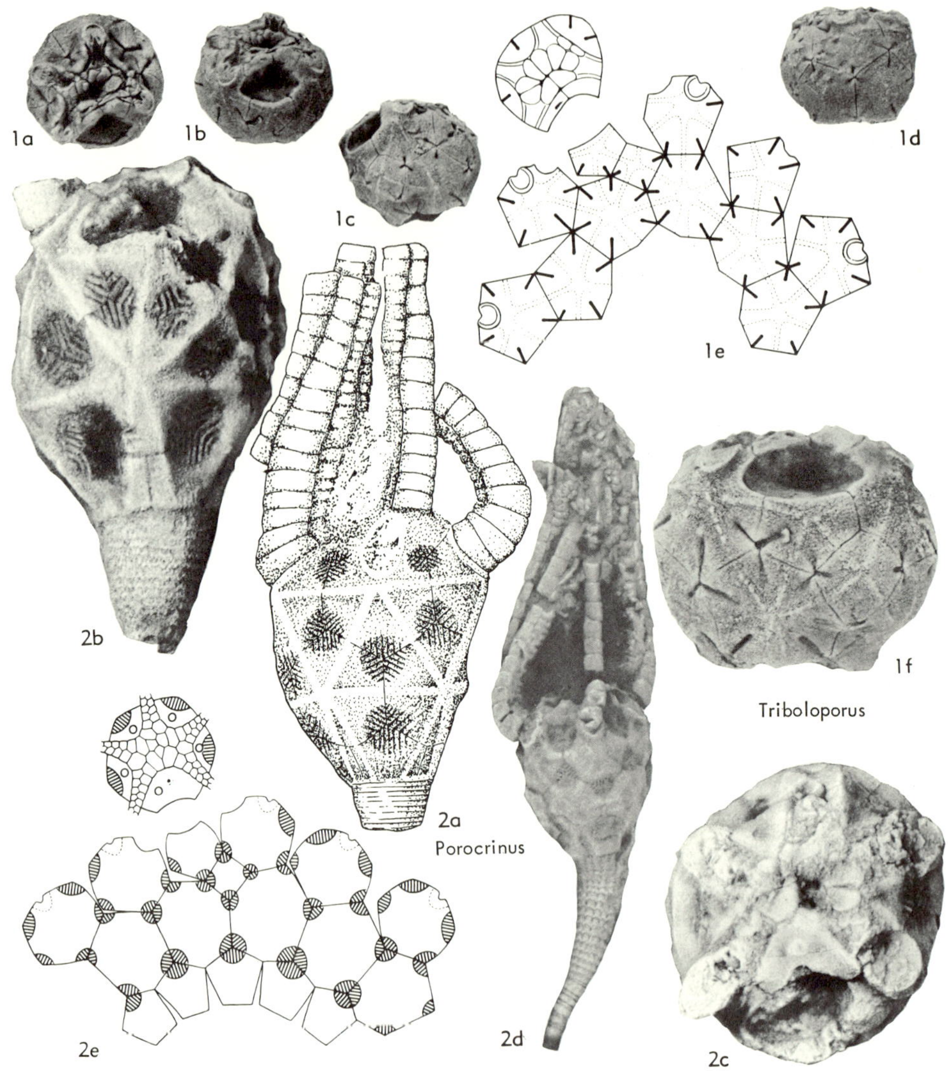

FIG. 379. Porocrinidae (p. T589).

BROWER, results as yet unpublished, reveals that *Strophocrinus* is a junior subjective synonym of *Carabocrinus*.—J. BROWER (pers. comm., Aug., 1968).]

Family CROTALOCRINITIDAE Bassler, 1938

[*nom. correct*. BASSLER, 1938, p. 18 (*pro* Crotalocrinidae ANGELIN, 1878, p. 26)] [=Enallocrinidae ANGELIN, 1878, p. 25]

Cup low, with wide flat base; anal *X* in cup or absent; radials support narrow primaxil, secundaxil *1*, and in some genera, tertaxil *1*; axial canal in radials and brachials; tegmen low, flat, composed of numerous plates; madreporitic posterior oral and other orals recognized in one genus; anal opening directly through tegmen. Arms composed of numerous, exceptionally thick brachials laterally joined together proximally or completely joined within a ray, or both within and between rays, forming large, laterally directly fanlike appendages; ventral side of arms with stout ambulacral covering plates; arms enrolled orally at lateral edges. Stem transversely circular, wide, with large, circular lumen; stout and short, with radicular cirri in some genera. *M.Sil.-M.Dev.(Givet.).*

Crotalocrinites Austin & Austin, 1843, p. 198 [**Cyathocrinites rugosus* Miller, 1821, p. 89; OD] [=*Crotalocrinites* Austin & Austin, 1842, p. 109 *(nom. nud.); Crotalocrinus* Morris, 1843, p. 50 *(nom. van.); Anthocrinus* Müller, 1854, p. 192 (type, *A. loveni*)]. Cup low bowl shaped, base wide and flat; plates strongly ornamented with radiating ridges or rows of nodes; infrabasals 5, visible in side view; *CD* basal truncate above, in contact with anal *X* situated between *C* and *D* radials; radials pentagonal, incurved and narrow distally, upper surface occupied by multiple articular facets; axial canal present in radials and brachials; radianal absent. Arms with minute axillary primibrachs *1*, confined to center of radial; secundibrachs *1* and tertibrachs *1* in contact with radial; arms composed of numerous, isotomously branching brachials laterally united by articular processes at sides of each brachial, which in each ray are united into large flexible fan-shaped appendage or completely joined crown with edges enrolled adorally; individual brachials elongate, slender, smooth and rounded dorsally, with deep ambulacral groove and alternating covering plates ventrally. Tegmen composed of numerous small plates, 5 orals, posterior one a madreporite; anal opening above anal *X* flush with tegmen surface or at end of short anal tube. *U.Sil.*, Eu.(Boh.-Eng.-Sweden-USSR)-USA(Wis.-Iowa-Ill.).——Fig. 380,*1a-c*. **C. rugosus* (Miller), Eng.; *1a-c, CD*-interray view of theca, radial and proximal brachials, and ventral side of brachials, ×2 (Springer, 1926a).——Fig. 380,*1d*. *C. pulcher* (Hisinger), Gotl.; lat. view of crown, ×0.7 (Springer, 1926a).——Fig. 380,*1e*. *C.* sp., Gotl.; ventral view of theca, ×1 (Wachsmuth & Springer, 1889b).

Achradocrinus Schultze, 1867, p. 213 [**A. ventrosus*; M]. Cup bowl shaped; infrabasals 5, visible in side view; anal *X* quadrangular, directly above truncate *CD* basal; *C* and *D* radials concave on posterior sides, surrounding anal opening directly above anal *X*; radial articular facets narrow, vertically directed, with axial canals. Arms unknown. Stem transversely circular, with circular lumen, no peripheral canals. *M.Sil.*, USA(Ill.), *M.Dev.*, Eu.(Ger.).——Fig. 377,*7*; 380,*2*. **A. ventrosus*, M.Dev., Ger.; 377,*7*, plate diagram of cup (radials black, anal *X* stippled) (Moore, n); 380,*2a-c*, *CD*-interray, ventral, and dorsal views of cup, ×1.5 (Schultze, 1867).

Arachnocrinus Meek & Worthen, 1866, p. 177 [**Cyathocrinus bulbosus* Hall, 1862, p. 123; OD]. Cup small, globose; infrabasals fused; anal opening directly above *CD* basal, bounded above by lower edges of *C* and *D* radials; anal *X* absent. Arms robust, branching isotomously on primibrachs *3* to *12*, and with 4 or 5 higher bifurcations; brachials strongly rounded, axillaries large and bulbous; axial canal in brachials and radials. Stem round or subquadrate in section, with 4 peripheral canals. *M.Dev.*, USA(N.Y.-Ky.)-Can.(Ont.).——Fig. 381,*2a-d*. **A. bulbosus* (Hall), N.Y.; *2a,b*, lat. and ventral views of crown, ×2; *2c,d*, *CD*-interray and dorsal views

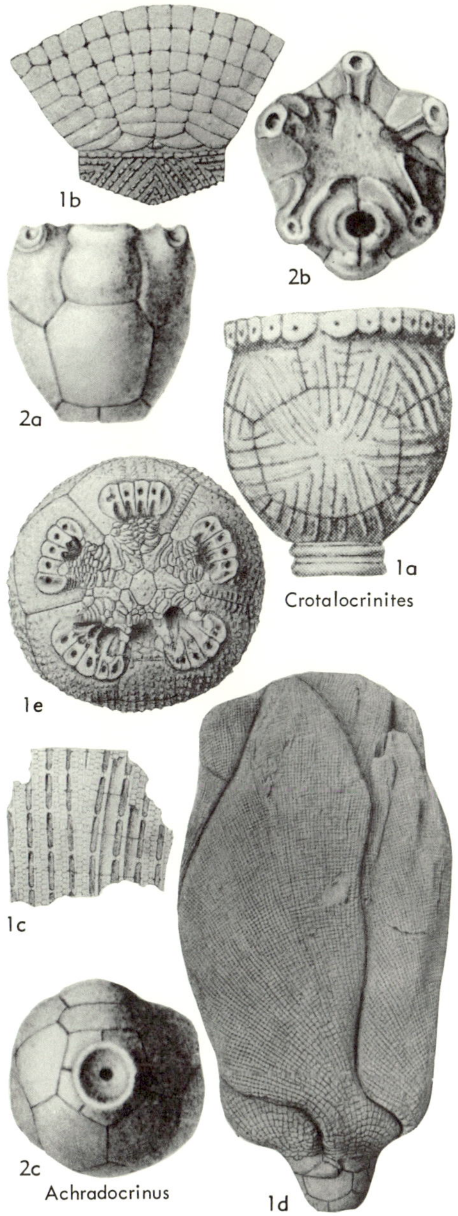

Fig. 380. Crotalocrinitidae (p. T591).

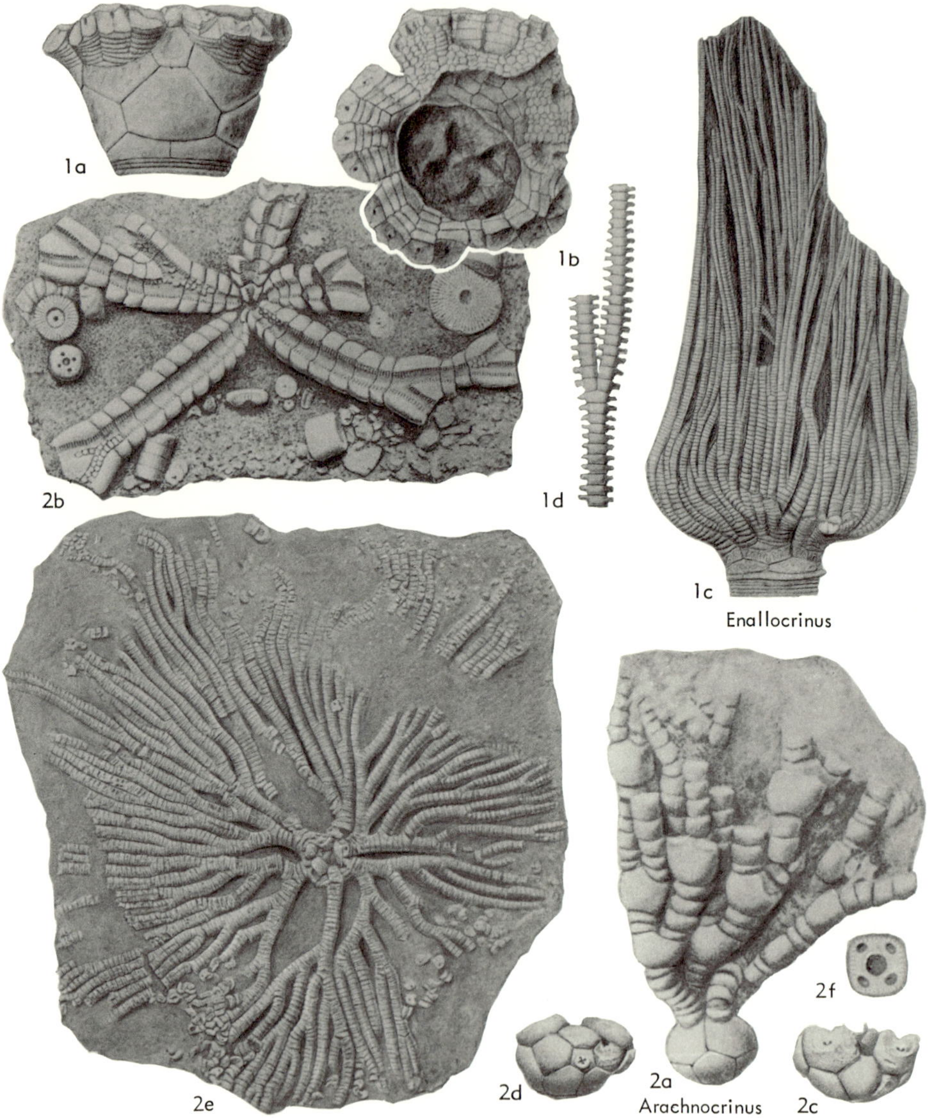

FIG. 381. Crotalocrinitidae (p. *T*591-*T*592).

of cup, ×2 (Springer, 1911c).——FIG. 381,*2e,f*. *A. extensus* WACHSMUTH & SPRINGER; *2e*, dorsal view of crown, ×0.5; *2f*, articular surface of columnal, ×2 (Springer, 1911c).

Enallocrinus D'ORBIGNY, 1850, p. 46 [**Apiocrinites scriptus* HISINGER, 1828, p. 217; M]. Like *Crotalocrinites* except brachials laterally joined only in proximal parts of each arm. *U.Sil.*, Eu.(Eng.-Sweden).——FIG. 377,*1*; 381,*1*. **E. scriptus* (HISINGER), Wenlock., Gotland; 377,*1*, plate diagram of cup and arms of *C* and *D* rays (radials black, anal *X* stippled) (Moore, n); 381,*1a,b*, *CD* interray and ventral views of theca and proximal brachials, ×1; 381,*1c*, lat. view of crown, ×0.7; 381,*1d*, dorsal view of brachials, ×1.5 (Wachsmuth & Springer, 1889b).

Parapernerocrinus YAKOVLEV, 1949, p. 17 [**P. sibiricus*; M]. Cup composed of numerous, small, irregularly arranged plates; brachials thick, closely appressed and laterally joined together, with axial canal and angular ambulacral groove. Arms branching isotomously several times, on primibrachs *5* or *6*, and on about secundibrach *12*. Stem large, composed of thin, wide, circular col-

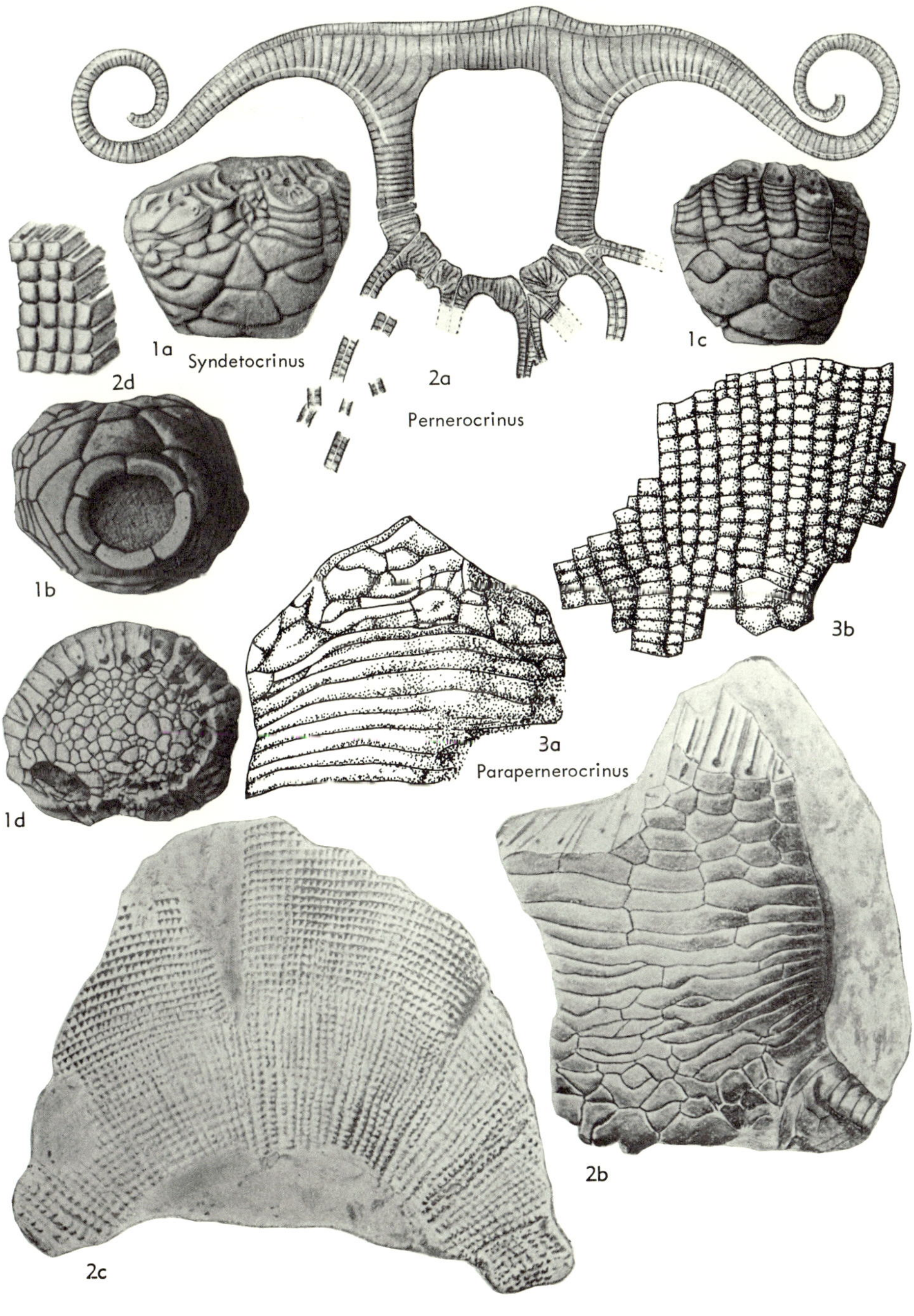

FIG. 382. Crotalocrinitidae (p. *T592-T594*).

umnals with wide circular lumen about four-fifths width of stem. *L.Dev.*, USSR(C.Urals).——FIG. 382,3. **P. sibiricus; 3a,b,* lat. view of fragmentary proximal columnals and cup, and dorsal view of arm fragment, ×1.3 (Lane, n; mod. from Arendt & Gekker in Orlov, 1965).

Pernerocrinus BOUŠKA, 1946, p. 18 [**P. paradoxus*; OD]. Crown large, mushroom shaped; stem, cup,

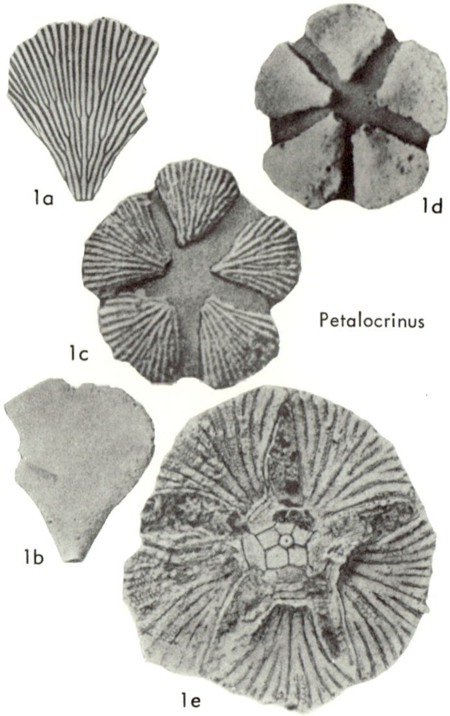

FIG. 383. Petalocrinidae (p. T594).

and proximal part of arms obscured by thick layers of stereom; brachials numerous, high and narrow, completely in contact laterally around crown, forming wide expanse orally enrolled at edges; brachials with prominent axial canal and ambulacral groove concealed by imbricated covering plates. Tegmen flat, composed of numerous small polygonal plates; anal opening at summit of low protuberance of tegmen. Numerous stout radicular cirri penetrate stereom distally. *L.Dev.*, Eu.(Boh.).——FIG. 382,*2*. **P. paradoxus*; *2a,* vert. sec. through crown, ×0.3; *2b,d,* lat. view of partial crown and oblique view of brachials, ×1; *2c,* dorsal side of arms, ×0.5 (Bouška, 1946).

Syndetocrinus KIRK, 1933, p. 345 [**S. dartae*; OD]. Cup low, with wide flat base and straight or gently convex sides; infrabasals low, mostly covered by wide proximal columnal; radials wide, low, separated on posterior side by pentagonal anal *X;* CD basal truncate above for contact with anal *X* or separated from it by edges of *C* and *D* radials; axial canal present in radials and brachials. Tegmen low, composed of numerous small irregular plates; orals not certainly identified. Arms with axillary primibrachs *1* occupying one-half to full width of radial; where primaxil is narrow, secundibrach *1* in contact with lateral edge of radial; secundibrachs *2* and tertibrachs *1* and tertibrachs *2* or *4* axillary; arms laterally in contact and firmly joined together at sides. Circular columnals wide, low, with large circular lumen. *U.Sil.,* Can.(Que.)-Eu.(Boh.-USSR).——FIG. 382,*1*. **S. dartae,* Que.; *1a-d, CD*-interray, dorsal, *BC*-interray, and ventral views of partial crown, ×1 (Kirk, 1933).

Family PETALOCRINIDAE Weller & Davidson, 1896

[Petalocrinidae WELLER & DAVIDSON, 1896, p. 167]

Cup small, radianal and anal *X* lacking; brachials of each ray laterally fused into large, rigid, fan-shaped plates bearing covered ambulacral grooves on ventral side. *M.Sil.(Niagaran, Wenlock.).*

Petalocrinus WELLER & DAVIDSON, 1896, p. 167 [**P. mirabilis*; OD]. Cup small, bowl shaped; infrabasals hidden under proximal columnal; radials large, with large, vertically directed articular facets and large axial canal; brachials of each ray completely fused into single, large, fan-shaped arm plate bearing covered ambulacral grooves on ventral side; arm plates smooth on dorsal side, articulated to radials by single, short, primibrach. *M.Sil.(Niagaran, Wenlock.),* USA(Iowa-Ind.-Ohio)-Eu.(Sweden).——FIG. 383,*1a,b*. *P. inferior* BATHER, Iowa; *1a,b,* ventral and dorsal views of arm plate, ×1 (Springer, 1926a).——FIG. 383, *1c-e*. **P. mirabilis,* Iowa; *1c,d,* ventral and dorsal views of arm plates, artificially mounted, ×1; *1e,* dorsal view of crown with arm-fans broken away, ×2 (Springer, 1926a).

Superfamily CODIACRINACEA Bather, 1890

[*nom. transl.* LANE, 1967, p. 10 (*ex* subfamily Codiacrinites BATHER, 1890b, p. 385)] [=superfam. Hypocrinacea WANNER, 1916a (*nom. transl.* ARENDT, 1970a, p. 94, *ex* Hypocrinidae WANNER, 1916a, p. 88) *(partim)*]

Theca small, typically globose or cone shaped; radials five, three, one, or absent; infrabasals five, three or fused; radianal absent, anal *X* present only primitively or secondarily; tegmen low, composed of five orals; posterior oral large, separating *BC* and *DE* orals, commonly with hydropore, or rarely and secondarily developed as madreporite (Fig. 384; see Fig. 388). Arms uniserial, isotomous, rarely preserved; absent or internal to orals in genera which lack radials or radial arm facets. *M.Sil.; L.Dev.-U.Perm.*

Family CODIACRINIDAE Bather, 1890

[*nom. transl. et correct.* JAEKEL, 1895, p. 63 (*ex* subfamily Codiacrinites BATHER, 1890b, p. 385)] [=Embryocrinidae WANNER, 1916a, p. 130; Hypocrinidae WANNER, 1916a, p. 88]

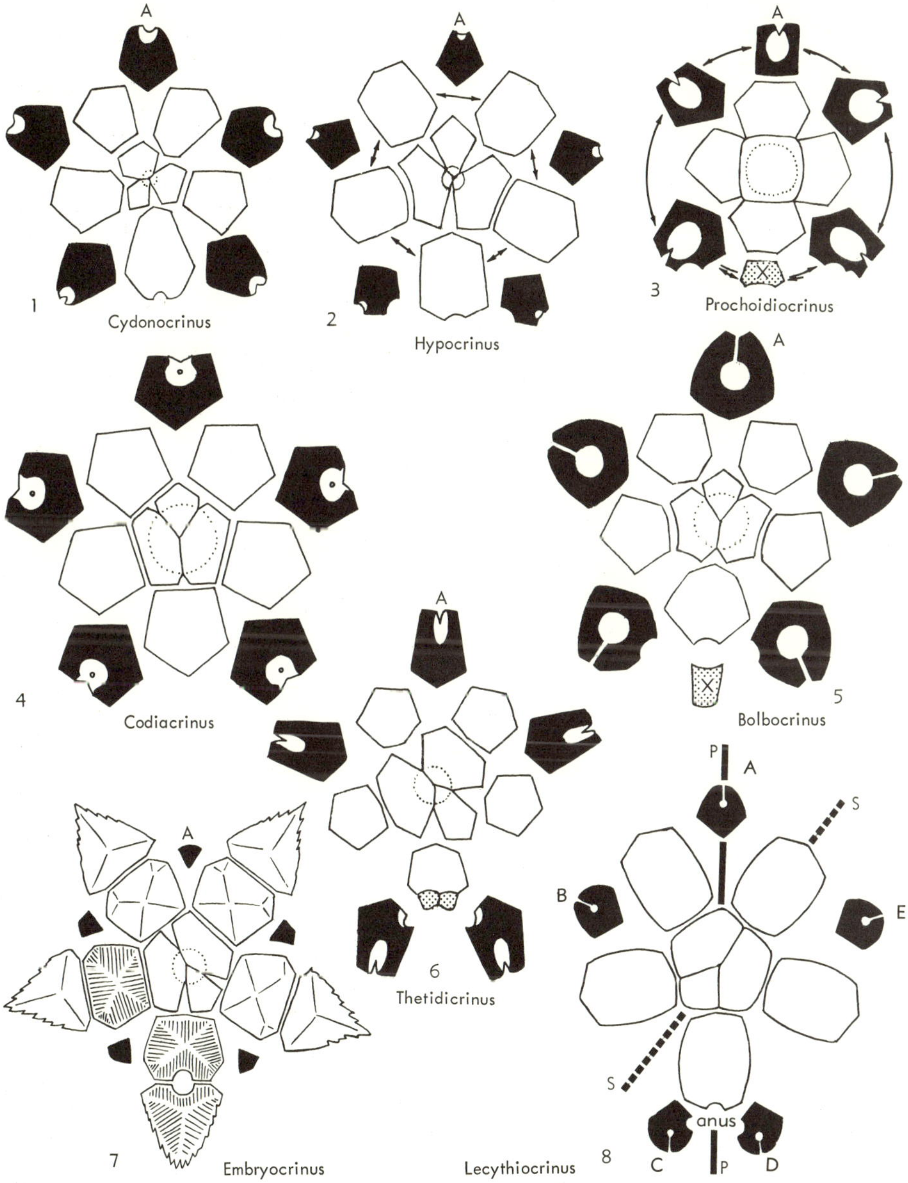

FIG. 384. Morphological features of representative Codiacrinacea (Codiacrinidae) (plate diagrams, not to scale). [Letters and patterns as in Fig. 372; broken heavy black line in Fig. 384,*8* marks secondary plane of bilateral symmetry.]

Theca with five facet-bearing radials, or rarely with five secondarily reduced, non-facet-bearing radials of equal size; anal *X* absent, or secondarily developed above or below opening in advanced genera; anal opening between radials and basals, or radials and orals, posterior oral a madreporite or with hydropore. *M.Sil.; L.Dev.-M.Dev.; L.Miss.-U.Perm.*

Subfamily CODIACRININAE Bather, 1890

[*nom. correct.* LANE, 1967, p. 11 (*pro* subfamily Codiacrinites BATHER, 1890b, p. 385)] [=Hypocrininae WANNER, 1929b, p. 38; section Hypocrinites WANNER, 1929b, p. 38] [incl. Cranocrininae ARENDT, 1970a, p. 140; Amphipsalidocrininae ARENDT, 1970a, p. 192]

Theca without anal plate; posterior oral not a madreporite. *M.Sil.; L.Dev.-M.Dev.; L.Miss.-U.Perm.*

Codiacrinus SCHULTZE, 1867, p. 143 [**C. granulatus*; OD]. Cup high, conical; 3 or ?5 infrabasals visible in side view; 5 large basals and 5 large radials with small articular facets; no anal plate in cup. Arms branching dichotomously on primibrachs *3,* with at least 2 higher divisions. Stem round. *L.Dev.-M.Dev.,* Eu.(Ger.-Spain).——FIG. 384,*4*; 385,*7*. **C. granulatus,* M.Dev., Ger.; 384,*4,* plate diagram of cup (radials black) (Moore, n); 385, *7a-c,* lat., dorsal, and ventral views of cup, ×2 (Schultze, 1867).

Abrachiocrinus WANNER, 1920, p. 29 [**Sycocrinites clausus* AUSTIN & AUSTIN, 1843, p. 206; OD]. Theca pear shaped; composed of 4 circlets (3 infrabasals, and 5 basals, radials, and orals); radial articular facets and anal plate lacking; anal opening in side of cup, bounded by *CD* basal and *C* and *D* radials; orals alternating with radials, forming swollen vault over radial circlet. Stem round. *L.Carb.(Tournais.),* Eu.(Eng.), *U.Perm.,* Indon.(Timor, Basleo).——FIG. 385,*1*. **A. clausus* (AUSTIN & AUSTIN), L.Carb., Eng.; *1a-e,* ant., *CD* interray, *B*-ray, ventral, and dorsal views of theca, ×3; *1f,g,* diminutive juvenile theca and columnar articular views, ×3 (all Bather, 1914a).

Amphipsalidocrinus J. M. WELLER, 1930, p. 33 [**A. scissurus*; OD]. Theca small, mushroom shaped; cup low cone shaped; infrabasals absent, atrophied or hidden; basals 3, equal, supporting 5 small radials with large articular facets; anal opening small, bordered below by projecting lip, bounded by *CD* basal, *C* and *D* radials, and *CD* oral, situated closer to *C* than to *D* radial facet; orals large, protruding, each with conspicuous elongate depression above radial facet that may have partly enclosed short, atomous arm. *L.Miss.-L.Perm.,* USA(Mo.-Ind.).——FIG. 385,*3*. **A. scissurus,* L.Penn., Ind.; *3a,b, CD*-interray and dorsal views of theca, ×30 (Peck & Connelly, 1951).

Asymmetrocrinus WANNER, 1937, p. 130 [**A. poteriocrinoides*; OD]. Cup cone shaped, asymmetrical, composed of 3 circlets of 5 plates each; infrabasals visible in side view; *CD* basal separates *C* and *D* radials; *E* radial and radial facet twice as large as *B* and *C* radials and facets, and projecting ventrally; facets semicircular, obliquely inclined. Stem round. *U.Perm.,* Indon.(Timor, Basleo).——FIG. 385,*8*; 386,*4*. **A. poteriocrinoides;* 385,*8a-d, CD*-interray, ventral, *B*-ray, and *A*-ray views of cup, ×3 (Wanner, 1937); 386,*4a-c,* dorsal, post., and ventral views of holotype (radials black), ×3 (Arendt, 1970a).

Cranocrinus WANNER, 1929, p. 41 [**C. timoricus*; OD]. Theca composed of 4 circlets; infrabasals fused into single plate visible in side view; basals and radials more or less asymmetrical; *CD* basal elongate, 6-sided; *C* and *D* radials separated by anal opening, all others in lateral contact; radial articular facets projecting, horseshoe shaped, narrow, steeply declivate, unequal in size; anal opening in side of cup, bounded by *CD* basal, *C* and *D* radials, and *CD* oral; orals 5, in normal interradial position, forming helmet-shaped vault over cup, hydropore on *CD* oral, each oral centrally concave. *L.Perm.(Artinsk.),* USSR(NC.Sib.); *U. Perm.(Basleo beds),* Indon.(Timor).——FIG. 385,*5*. **C. timoricus,* U.Perm., Timor(Basleo); *5a-c, CD*-interray, ventral, and dorsal views of theca, ×6 (Wanner, 1929b).——FIG. 386,*3*. *C. praestans* ARENDT, L.Perm., Kolva R., NC.Sib.; *3a-e, EA-, DE-, CD-, BC-,* and *AB* interray views (radials black, anus oblique ruled) of theca showing size variation of radial articular facets and hydropore on *CD* oral, ×10 (Arendt, 1970a).

Cydonocrinus BATHER, 1913, p. 391 [**C. parvulus*; OD]. Cup small, globular; 3 infrabasals visible in side view, small infrabasal in *C* ray; basals large, *CD* basal truncate above for anal opening between *C* and *D* radials; posterior side of cup bulging and distal edge of *CD* basal inflated around ventrally directed anal opening; radials 5, smaller than basals, incurved distally; arm facets semicircular, raised above level of surrounding radial, horizontally directed. *L.Carb.(Visean); U.Perm.,* Eu.(Eng.)-Indon.(Timor, Basleo).——FIG. 384,*1*; 385,*4*. **C. parvulus,* L.Carb., Eng.; 384,*1,* plate diagram of cup (radials black) (Moore, n); 385,*4a-d, CD*-interray, ventral, dorsal, and *B*-ray (with hypothetical reconstruction of anal sac and arms) views of cup, ×3 (Bather, 1913d).

Embryocrinus WANNER, 1916, p. 130 [**E. hanieli*; OD] [*non Embryocrinus* HUDSON, 1918, p. 162 (=*?Dendrocrinus* HALL, 1852, p. 193)]. Theca small, pear shaped; 4 circlets of massive plates forming closed capsule with anal opening between *CD* basal and oral; 3 infrabasals, small one in *D* ray; radials small, trigonal, not forming continuous circlet around cup, isolated between projecting distal edges of basals; orals forming dome-shaped arch with broad base resting on basals and radials. *U.Perm.,* Indon.(Timor, Basleo).——FIG. 384,*7*; 385,*6*. **E. hanieli*; 384,*7,* plate diagram of theca (radials black) (Moore, n); 385,*6a-c,* post., dorsal, and ventral views of theca, ×3 (Wanner, 1916a).

Hydroporocrinus ARENDT, 1970, p. 146 [**H. obliquus*; OD]. Theca low, transversely pentagonal in outline, cup conical and moderately elevated oral dome quinquestellate owing to strong median depressions and raised lateral margins of orals; infrabasals solidly fused together and similarly joined to *DE* basal; 5 basals, *CD* and *DE* larger than others and asymmetrical; radials with side-

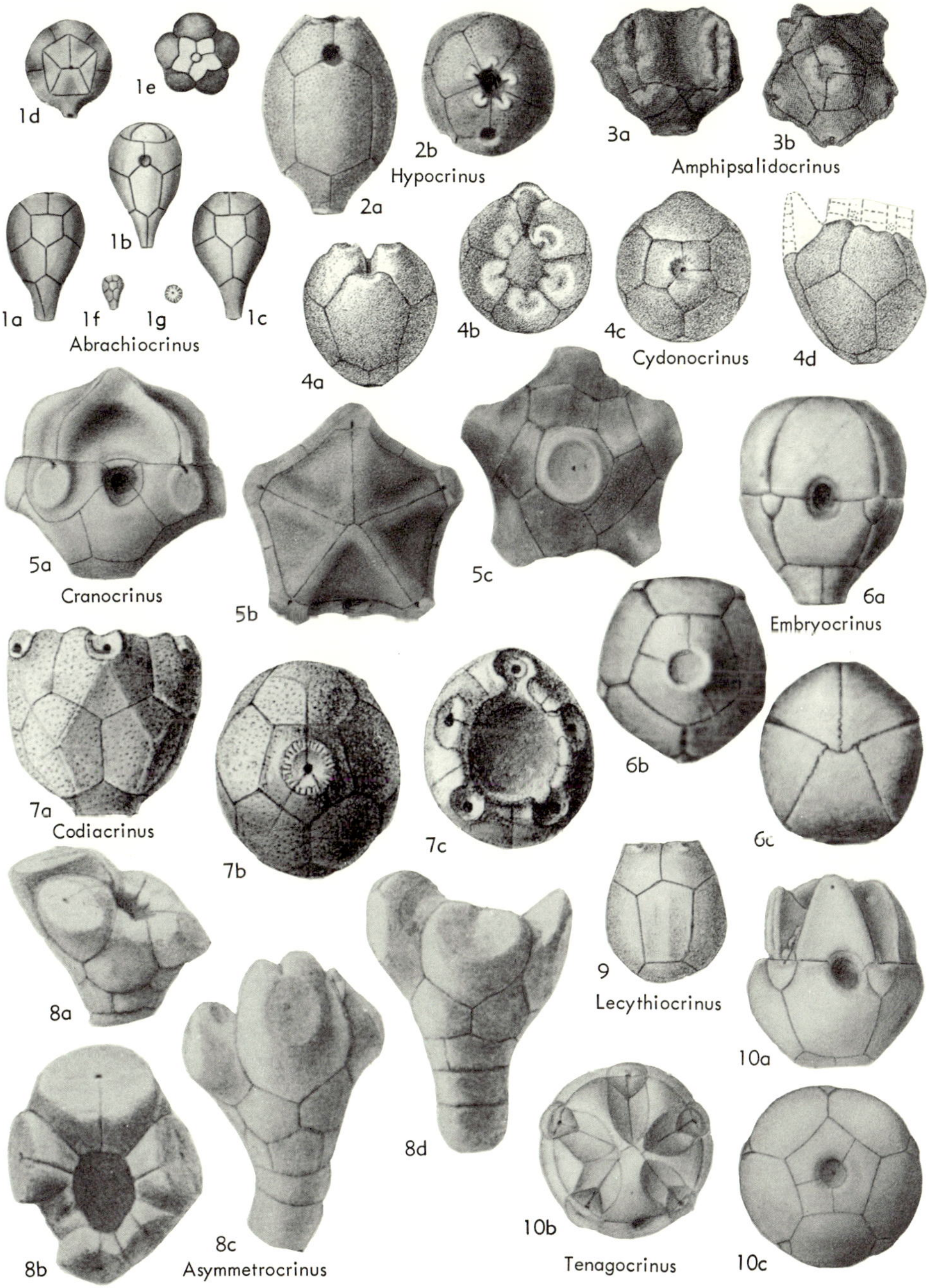

FIG. 385. Codiacrinidae (Codiacrininae) (p. *T596*, *T598-T599*).

ward projections which bear subvertical circular articular facets slightly notched at upper margins. *CD* oral distinctly wider than others and bearing large hydropore on prominent tubercle; smaller hydropores and tubercles on other orals, except *DE* which lacks them. *L.Perm.(up.Artinsk.)*, USSR(Kolva River, NC.Sib.).——FIG. 386,2. **H. obliquus; 2a-d,* dorsal, post., *DE* interray, and ventral views of holotype, ×5.3; *2e-i,* diagram. views of theca from *EA-*, *DE-*, *CD-*, *BC-*, and

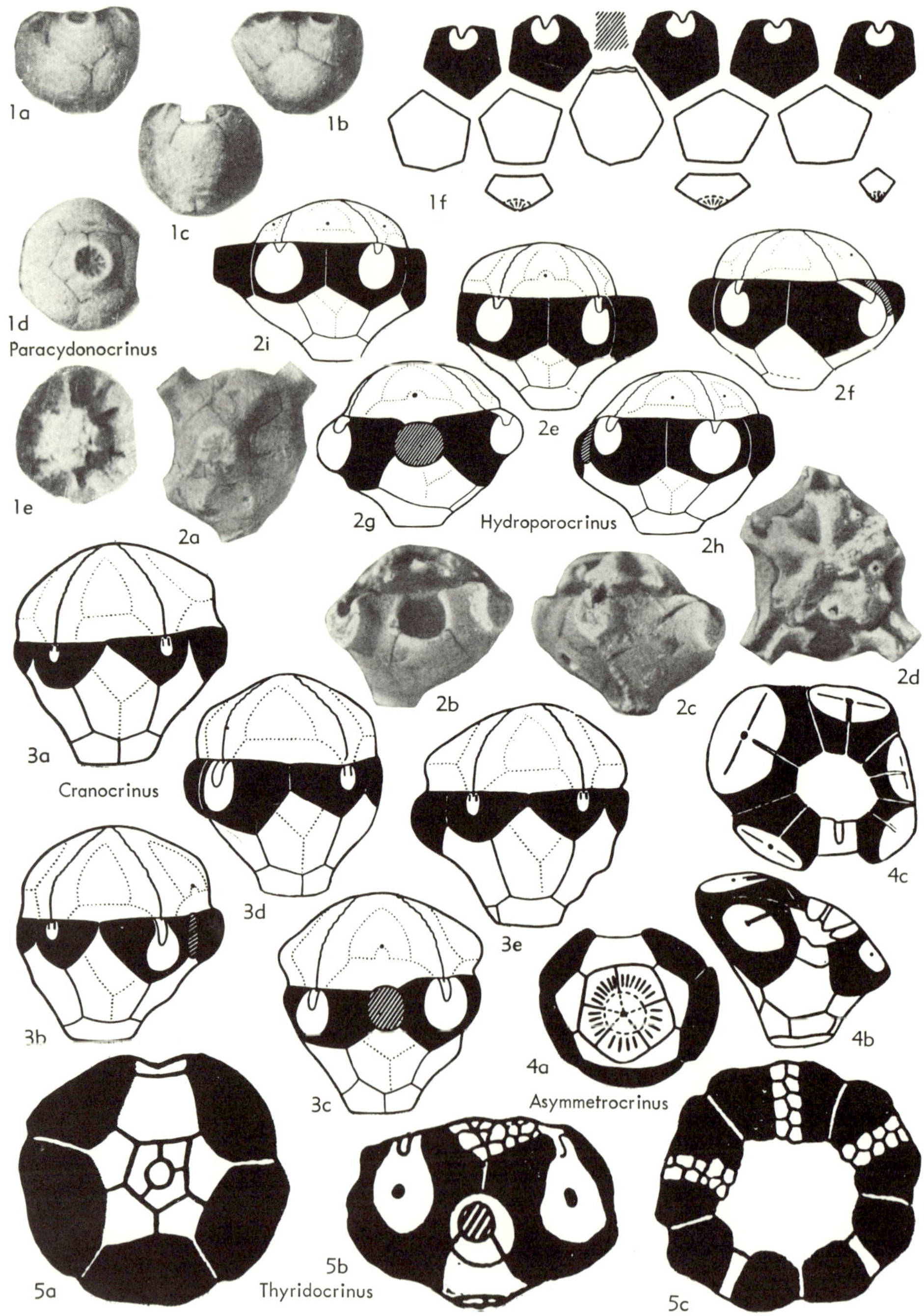

FIG. 386. Codiacrinidae (Codiacrininae) (p. T596-T599).

AB-interray sides (radials black, anus oblique ruled), ×6 (Arendt, 1970a).

Hypocrinus BEYRICH, 1862, p. 537 [**H. schneideri*; M]. Cup small, globose; infrabasals high, 3, small one in anterior position, restricted in width proximally just above small stem facet; basals high, convex, circlet forming widest part of theca; radials small, shield shaped, distally restricted, inclined into center of distal cup border, in contact around cup; arm facets small, horizontally directed, with narrow ambulacral groove; opening at top of cup narrow; anal opening between *C* and *D* radials, notching top of *CD* basal and sides of radials. *U.Perm.*, Indon.(Timor, Basleo).——FIG. 384,*2*; 385,*2*. **H. schneideri*; 384,*2*, plate diagram of cup (radials black) (Moore, n); 385,*2a,b*, *CD*-interray, and ventral views of cup, ×2 (Wanner, 1916a).

Lecythiocrinus WHITE, 1880, p. 256 [**L. olliculaeformis*; OD] [=*Menocrinus* MILLER, 1889, p. 262]. Cup globose; 3 or 5 low infrabasals, visible in side view; basals high, bulging distally below radials, which are small, low, distally restricted and inclined toward center of cup; articular facets small, horizontally directed; anal opening large, bounded above and at sides by *C* and *D* radials, and proximally by bulging upper edge of *CD* basal. *L.Penn.-U.Penn.*, USA(Ind.-Ill.-Okla.-Kans.).——FIG. 384,*8*; 385,*9*. **L. olliculaeformis*, U. Penn., Kans.; 384,*8*, plate diagram of cup (radials black, *P-P* primary and *S-S* secondary planes of bilateral symmetry) (Moore, 1962); 385,*9*, lat. view of cup, ×2 (White, 1880b).

Paracydonocrinus ARENDT, 1970, p. 106 [**P. diviensis*; OD]. Cup very diminutive, globose; small infrabasal in *A* ray, rather than *C* ray as in *Cydonocrinus*, *CD* basal distally truncate, distinctly larger than others; anus between posterior radials. Stem small, transversely circular. *L.Perm. (up.Artinsk.)*, USSR(Sib., Krasnoufimsk).——FIG. 386,*1*. **P. diviensis*; *1a-e*, *A*-ray, *AB*- and *CD*-interray, dorsal, and ventral (*A* ray downward) views of holotype cup, ×2; *1f*, plate diagram of cup (radials black), ×2.5 (all Arendt, 1970a).

Tenagocrinus WANNER, 1929, p. 48 [**Embryocrinus sulcatus* WANNER, 1916a, p. 141; OD]. Theca small; 3 infrabasals, small one in *A*-ray, barely visible in side view; basals large, in lateral contact; radials 5, small, 3-sided, not in contact, with stunted articular facets for arms; orals large, in normal interradial position, forming high vault, with deep, recessed furrow above each arm facet for reception of small unbranched arm; large anal opening in side of theca, shifted to right of *CD* midline, bounded by *CD* basal and oral, and *C* radial. *U.Perm.*, Indon.(Timor).——FIG. 385,*10*. **T. sulcatus* (WANNER); *10a-c*, *CD*-interray, ventral, and dorsal views of theca, ×5 (Wanner, 1929b).

Thyridocrinus KIRK, 1944, p. 388 [**Lecythiocrinus? problematicus* SPRINGER, 1926a, p. 133; OD] [=*Prolecythiocrinus* ARENDT, 1970a, p. 117 (obj.)]. Differs from *Lecythiocrinus* in low spheroidal form of theca instead of tall pear shape, as well as in proportionally much smaller basals and larger radials which extend from summit almost to basal plane; large subvertical radial articular facets horseshoe shaped, with central round canal opening; *A*, *B*, and *E* rays with biserially arranged cover plates traversing ambulacra on distal part of radials leading to tegmen. *M.Sil.(Niagaran)*, USA.——FIG. 386,*5*. **T. problematicus* (SPRINGER), Laurel Ls., USA(Ind.); *5a-c*, dorsal, post., and ventral views of theca (radials black), ×4 (Arendt, 1970a).

Subfamily BOLBOCRININAE Wanner, 1930

[*nom. transl.* LANE, 1967, p. 11 (*ex* section Bolbocrinites WANNER, 1930a, p. 4)]

Anal opening in side of cup, bounded above or below by secondarily developed anal plate; without madreporite. *L.Perm.-U.Perm.*

Bolbocrinus WANNER, 1916, p. 115 [**B. hieroglyphicus*; OD]. Cup bowl shaped; 3 infrabasals low but visible in side view, with wide stem facet; small pentagonal infrabasal in *A* or *C* ray; basals large, regularly pentagonal; radials almost as large as basals, sharply constricted distally; *C* and *D* radials not in lateral contact, separated by anal opening and quadrangular anal plate which does not project above radial circlet; articular facets large, horseshoe shaped, with long deep ambulacral canals; anal opening large, rounded, bounded by *CD* basal, *C* and *D* radials, and anal plate. *L.Perm.-U.Perm.*, Indon.(Timor, Basleo)-USSR.——FIG. 384,*5*; 387,*3a-c*. **B. hieroglyphicus*, U.Perm., Timor; 384,*5*, plate diagram of cup (radials black, anal *X* stippled) (Moore, n); 387, *3a-c*, *CD*-interray, dorsal, and ventral views of cup, ×1.5 (Wanner, 1916a).——FIG. 387,*3d,e*. *B. rex* WANNER, U.Perm., Timor; *3d,e*, *CD*-interray and ventral views of cup, ×0.7 (Wanner, 1916a).

Nereocrinus WANNER, 1924, p. 127 [**N. antiquus*; OD] [=*Oceanocrinus* WANNER, 1924, p. 131 (type, *O. granulatus*)]. Like *Bolbocrinus* except anal plate directly above *CD* basal and anal opening not in cup. *L.Perm.-U.Perm.*, Indon.(Timor, Basleo)-USSR(Krasnoufimsk).——FIG. 387,*4*. **N. antiquus*; *4a,b*, post. and ventral views of cup, ×2 (Wanner, 1924).

Subfamily THETIDICRININAE Wanner, 1930

[*nom. transl. et correct.* LANE, 1967, p. 11 (*ex* section Thetidicrinites WANNER, 1930a, p. 3)]

Codiacrinidae with one or two anal plates in cup and madreporitic posterior oral. *L.Carb.(Tournais.)*; *U.Perm.*

Thetidicrinus WANNER, 1916, p. 111 [**T. piriformis*; OD]. Theca high, cone shaped; 3 infrabasals in funnel-shaped circlet; small plate in *D* ray; *CD* basal protruding distally, with concave upper border supporting 2 equal anal plates; radials large, constricted distally, in lateral contact; articular facets large, elongate, horseshoe shaped,

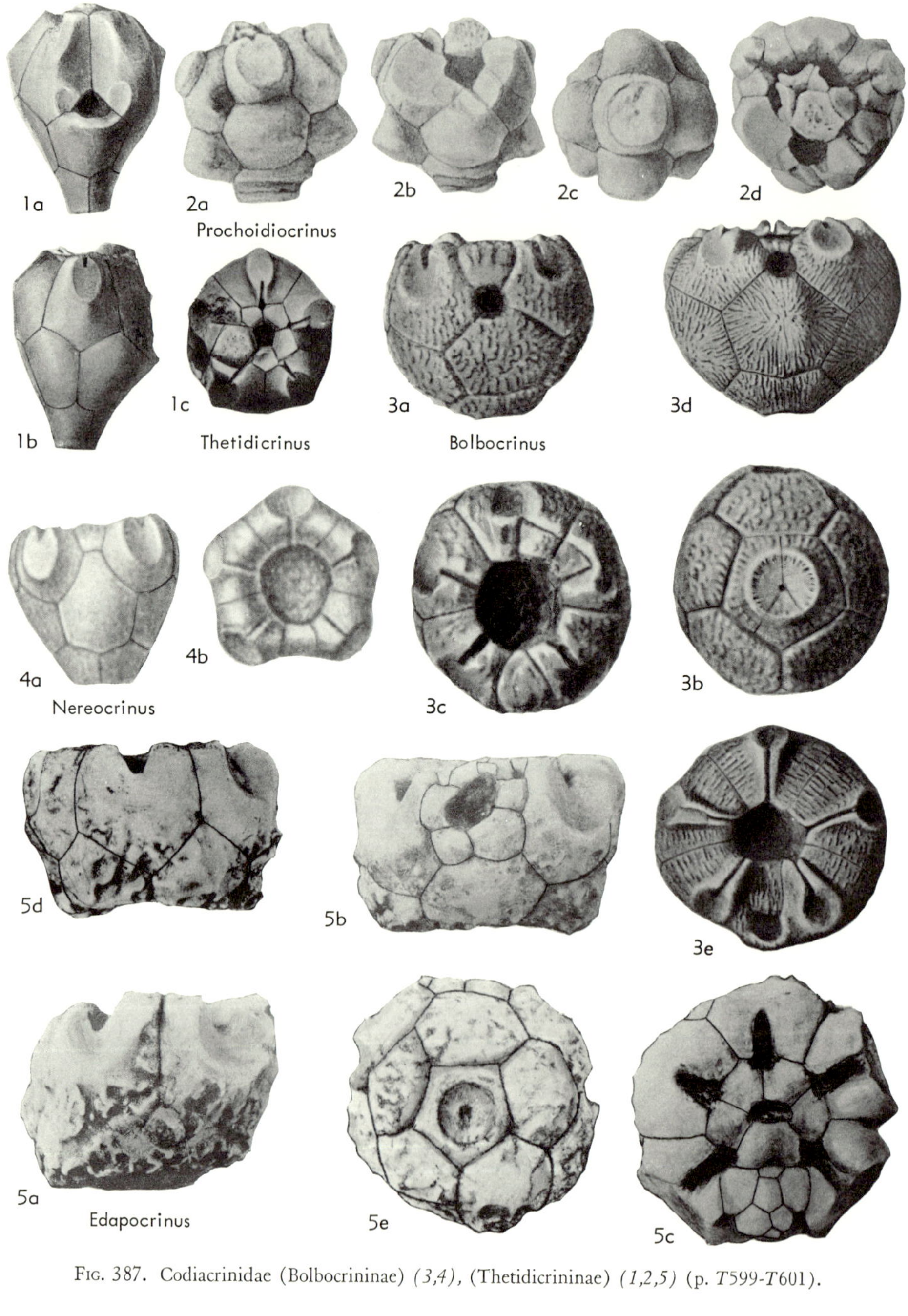

FIG. 387. Codiacrinidae (Bolbocrininae) *(3,4)*, (Thetidicrininae) *(1,2,5)* (p. *T*599-*T*601).

inclined outward, with deeply incised ambulacral canals; anal opening directed obliquely upward, bounded by *C* and *D* radials above, and externally by 2 equal anal plates; small pentagonal orals in lateral contact; *CD* oral larger and madreporite. *U.Perm.*, Indon.(Timor, Basleo).——FIG. 384,*6*; 387,*1*. **T. piriformis*; 384,*6*, plate diagram of cup (radials black, anals stippled) (Moore, n); 387, *1a-c*, *CD*-interray, *E*-ray, and ventral views of theca, ×1.5 (Wanner, 1916a).

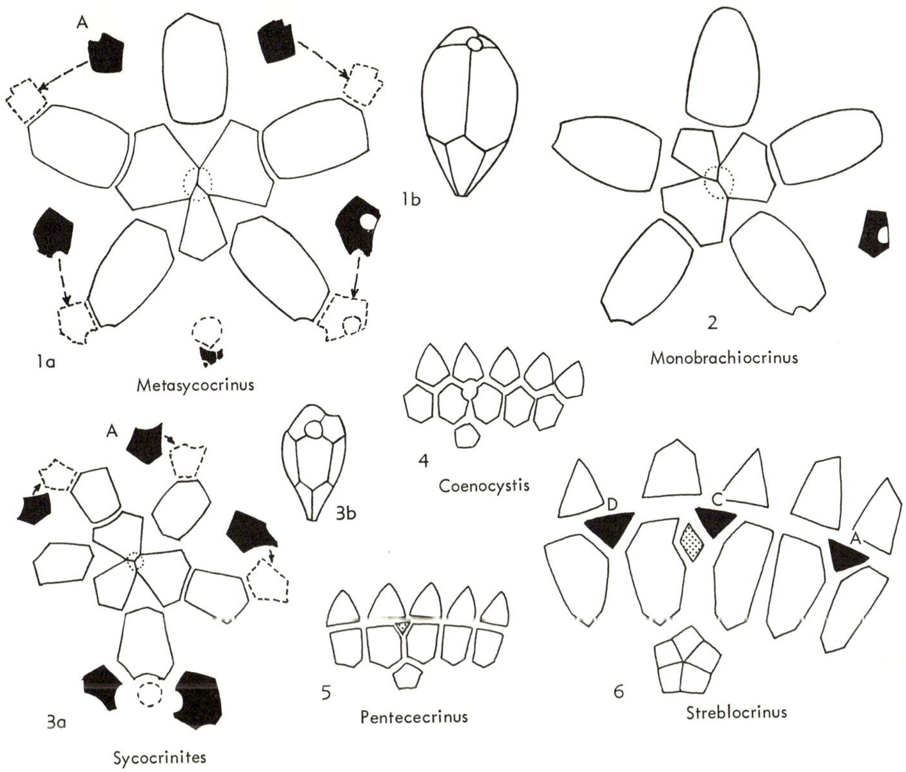

FIG. 388. Morphological features of representative Codiacrinacea (Sycocrinitidae, Streblocrinidae) (plate diagrams, not to scale). [Letters and patterns as in Fig. 372.]

Edapocrinus WRIGHT, 1935, p. 195 [**E. rugosus*; OD] [=*Edaphocrinus* WANNER, 1937 *(nom. van.)*]. Cup bowl shaped, base wide, flat; infrabasals fused, not visible in side view; radials 5, with large, horseshoe shaped, vertically directed facets and deep ambulacral notches; anal opening in side of cup between *C* and *D* radials, directly above truncate *CD* basal, surrounded by 7 small accessory anal plates; anal *X* and radianal not definitely identified; tegmen narrow, flat, with 5 orals, posterior one larger and a madreporite, tegmen with central, oval opening. *L.Carb.(Tournais.),* Eu.(Eng.).——FIG. 387,*5*. **E. rugosus; 5a-e, AB*-interray, *CD*-interray, ventral, *A*-ray, and dorsal views of theca, ×1.5 (Wright, 1951-54).

Prochoidiocrinus WANNER, 1937, p. 127 [**P. nodosus*; OD]. Theca bowl shaped, infrabasals fused, barely visible in side view; 4 nodose basals, *AB* and *AE* basals fused below *A* radial, radials 5, with large, vertically directed articular facets occupying most of outer plate surfaces; anal opening in side of cup, between *C* and *D* radials, bounded below by small quadrangular anal *X* directly above *CD* basal; orals 5, small except for large posterior oral developed as madreporite and bounding ventral edge of anal opening. *U.Perm.,* Indon.(Timor, Basleo).——FIG. 384,*3*; 387,*2*. **P. nodosus*; 384,*3,* plate diagram of cup (radials black, anal *X* stippled) (Moore, n); 387,*2a-d, A*-ray, *CD*-interray, dorsal, and ventral views of theca, ×1.5 (Wanner, 1937).

Family SYCOCRINITIDAE Lane, 1967

[Sycocrinitidae LANE, 1967, p. 12]

Theca with one arm-bearing radial; other radials present or absent, commonly unequal and reduced in size; infrabasals three; anal *X* plate absent; anal opening bounded by basals and radials, or by radials. Orals small, commonly not preserved. *L.Carb. (Tournais.); L.Perm.-U.Perm.*

Sycocrinites AUSTIN & AUSTIN, 1842, p. 111 [**S. anapeptamenus*; SD BATHER, 1914a, p. 253] [=*Sycocrinus* MORRIS, 1843, p. 54 *(nom. van.)*]. Cup high globular with narrow base; infrabasals 3, high, small plate in *C* ray; 5 radials, *D* radial

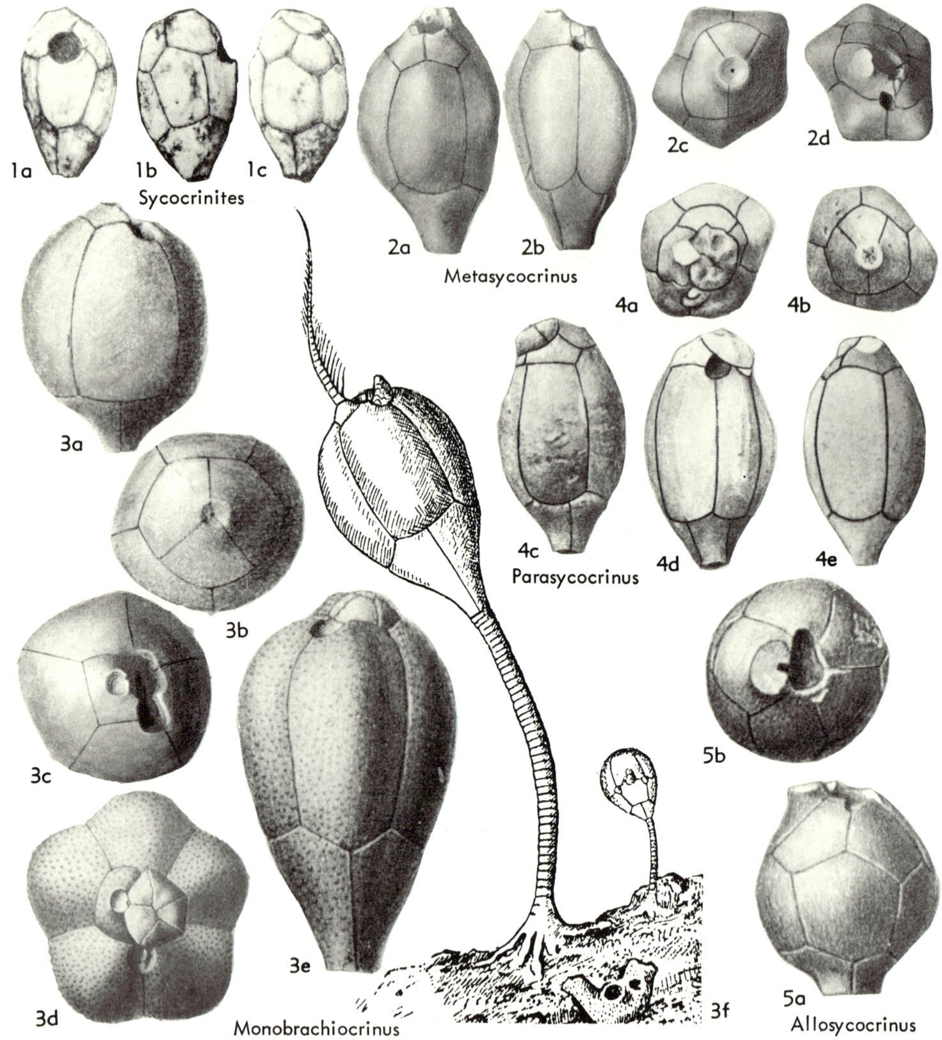

FIG. 389. Sycocrinitidae (p. T601-T603).

large, higher than others and with large horseshoe-shaped articular facet; other radials small, without facets; anal opening large, bounded by concave edges of *CD* basal and *C* and *D* radials. *L.Carb.(Tournais.)*, Eu.(Eng.).——FIG. 388,*3*; 389,*1*. **S. anapeptamenus*; 388,*3a,b*, plate diagram of cup (radials black) and post. view of theca (Moore, n); 389,*1a-c*, *CD*-interray, *E*-ray, and *A*-ray views of cup, ×2 (Wright, 1951-54).

Allosycocrinus WANNER, 1924, p. 141 [**A. pusillus*; OD]. Like *Sycocrinites* except anal opening forms notch in distal edge of one of nonarm-bearing radials, and is strongly variable in position; tegmen a vault of 5 small oral plates. *U.Perm.*, Indon.(Timor, Basleo).——FIG. 389,*5*. **A. pusillus*; *5a,b*, *CD*-interray, and ventral views of cup, ×3 (Wanner, 1924).

Metasycocrinus WANNER, 1920, p. 25 [**Hypocrinus pyriformis* ROTHPLETZ, 1892, p. 75; OD]. Cup globular, base attenuate; 3 high narrow infrabasals, small one in *C* ray; basals large, inflated; *D* radial largest and highest in circlet; radials inclined inward toward axis of cup, arm facet on *D* radial only; anal opening on upper right shoulder of *CD* basal, bounded by *CD* and *BC* basals, *B* and *D* radials and by minute plate above that may be *C* radial. *U.Perm.*, Indon.(Timor, Basleo).——FIG. 388,*1*; 389,*2*. **M. pyriformis* (ROTHPLETZ); 388,*1a,b*, plate diagram of cup (radials black) and post. outline of cup (Moore, n); 389,*2a-d*, *AE*- and *CD*-interray, dorsal, and ventral views of cup, ×0.7 (Wanner, 1916a).

Monobrachiocrinus WANNER, 1916, p. 103 [**M. ficiformis*; OD]. Cup inverted pear shaped; infrabasals 3, forming conspicuous part of cup, small infrabasal in *A* ray; basals 5, higher than wide; *D* radial enlarged, rounding off top of *DE* basal and limited laterally by *CD* and *AE* basals, other radials absent; anal opening bounded by *CD* and *BC* basals, *D* radial, and 2 orals; orals 5, small, *CD* one largest. *L.Perm.-U.Perm.*, Indon.(Timor, Basleo), Eu.(USSR, Krasnoufimsk-Sicily).——FIG. 388,*2*; 389,*3c-e*. **M. ficiformis*, Basleo beds, Timor; 388,*2*, plate diagram of cup (*D* radial black) (Moore, n); 389,*3c-e*, dorsal, ventral, and *CD*-interray views of cup, ×1.5 (Wanner, 1916a).——FIG. 389,*3a,b,f*. *M. granulatus* WANNER, Basleo beds, Timor; *3a,b*, *BC*-interray and dorsal views of theca, ×1.5 (Wanner, 1924); *3f*, reconstr. (with *Embryocrinus hanieli*), ×1 (Wanner, 1920).

Parasycocrinus MAREZ OYENS, 1940, p. 321 [**P. fastigate-pileatus*; OD]. Cup high, cone shaped; infrabasals 3, unequal, small one in *C* ray; basals large, high; radials 4, *C* radial absent, *A* and *B* radials reduced, *D* radial only with arm facet; anal opening large, circular, bounded by *CD* and *BC* basals below and *D* radial above. *L.Perm.*, Indon.(Timor, Tae Wei)——FIG. 389,*4*. **P. fastigatepileatus; 4a-e*, ventral, dorsal, *E*-ray, post., and *D*-ray views of cup, ×2 (Marez Oyens, 1940b).

Family STREBLOCRINIDAE Lane, 1967

[Streblocrinidae LANE, 1967, p. 12]

Theca typically composed of infrabasals, basals, and orals; abrachiate; radials and anal plate absent in all but most primitive genera; posterior oral commonly with hydropore and separating *BC* and *DE* orals; orals primitively above basals in interradial position, and anal opening on lower right side of posterior oral and upper right shoulder of *CD* basal. *M.Dev.-U.Perm.*

Subfamily STREBLOCRININAE Lane, 1967

[Streblocrininae LANE, 1967, p. 12] [incl. Pilidiocrininae ARENDT, 1970a, p. 184]

Anal opening shifted to left of *A-CD* axis and orals uniformly rotated clockwise to radial position, alternating with basals. *M. Dev.-L.Penn.; Perm.*

Streblocrinus KOENIG & MEYER, 1965, p. 393 [**S. brachiatus*; OD]. Theca small, composed of 4 circlets of plates; infrabasals 5, low; basals large, elongate; 3 small spoutlike radials in *A*, *C*, and *D* rays, other radials absent; single anal plate may have been present in cup (only known specimen is damaged in *CD* area); orals 5, large, directly above basals in interradial position; *CD* oral with hydropore. *M.Dev.*, USA(N.Y.).——FIG. 388,*6*; 390,*5*. **S. brachiatus*; 388,*6*, plate diagram of cup (radials black, anal plate stippled) (Moore, n); 390,*5a-g*, ventral, *CD*-interray, *BC*-interray, *AB*-interray, dorsal, *AE*-interray, and *DE*-interray views of theca, ×50 (Koenig & Meyer, 1965).

Atremacrinus WANNER, 1929, p. 82 [**A. calyculus*; OD]. Theca small, bud shaped; infrabasals and basals apparently fused, radials and anal opening absent; tegmen dome-shaped vault of 5 orals; stem thick, round. *U.Perm.*, Indon.(Timor, Basleo).——FIG. 390,*6*. **A. calyculus; 6a,b*, lat. and ventral views of theca, ×3 (Wanner, 1929b).

Coenocystis GIRTY, 1908, p. 109 [**C. richardsoni*; OD] [=*Coencystis* PECK, 1936 *(nom. null.)*]. Theca small, pear shaped, composed of infrabasals, basals, and orals; infrabasals visible in side view, completely fused on most specimens; basals 5, in lateral contact, radials lacking; anal opening in midline of *CD* oral, bounded by oral and 2 basals; orals 5, forming cap-shaped vault, radial in position, alternating with basals; *CD* oral with hydropore. *L.Miss.(Osag.)*, USA(Mo.); *L.Perm.-U.Perm.*, USA(Texas)-Indon.(Timor, Basleo).——FIG. 388,*4*; 390,2. **C. richardsoni*, U.Perm.(Guadalup.), Texas; 388,*4*, plate diagram of theca (Moore, n); 390,*2a-d*, ventral and lat. views of large cup, *CD*-interray views of 2 small thecae, ×4 (Girty, 1908).

Dichostreblocrinus J. M. WELLER, 1930, p. 36 [**D. scrobiculus*; OD]. Cup cone shaped, longitudinally twisted slightly in counter-clockwise direction; infrabasals low, fused; basals elongate, supporting 5 large orals, radials absent; anal opening bordered below by upper right edge of *CD* basal and on right by *BC* basal; orals shifted about 9° to left of normal interradial position; *CD* oral larger than others, with minute, central hydropore. *L.Miss.(Osag.)*, USA(N.Mex.-Mo.); *L.Penn.*, USA (Ind.); *U.Perm.*, Indon.(Timor, Basleo).——FIG. 390,*4*. **D. scrobiculus*, L.Penn., Ind.; *4a-c*, *CD*-interray, *E*-ray, and ventral views of theca, ×1.5 (Weller, 1930a).

Hemistreptacron YAKOVLEV, 1926, p. 52 [**H. abrachiatum*; M]. Theca small, composed of infrabasals, basals, and orals; infrabasals 3 or fused, visible in side view; basals large, in lateral contact; radials lacking; orals 5, forming cap-shaped vault above cup, shifted to left 12° from interradial position; anal opening large, between *CD* oral and basal, not in midline of these plates, shifted somewhat to left side of basal and to right side of oral; *CD* oral with hydropore. *L.Perm.*, Eu.(USSR, Krasnoufimsk)-*U.Perm.* Indon.(Timor, Basleo).——FIG. 390,*1*. **H. abrachiatum*, L.Perm., USSR; *1a-d*, post., ant., dorsal, and ventral views of theca, ×5 (Yakovlev & Ivanov, 1956).

Pilidiocrinus WANNER, 1937, p. 138 [**P. permicus*; OD]. Theca adnate on other echinoderms, without column; infrabasals lacking or fused with 5 basals which are proximally adherent to substrate; orals 5 equal plates alternating with basals.

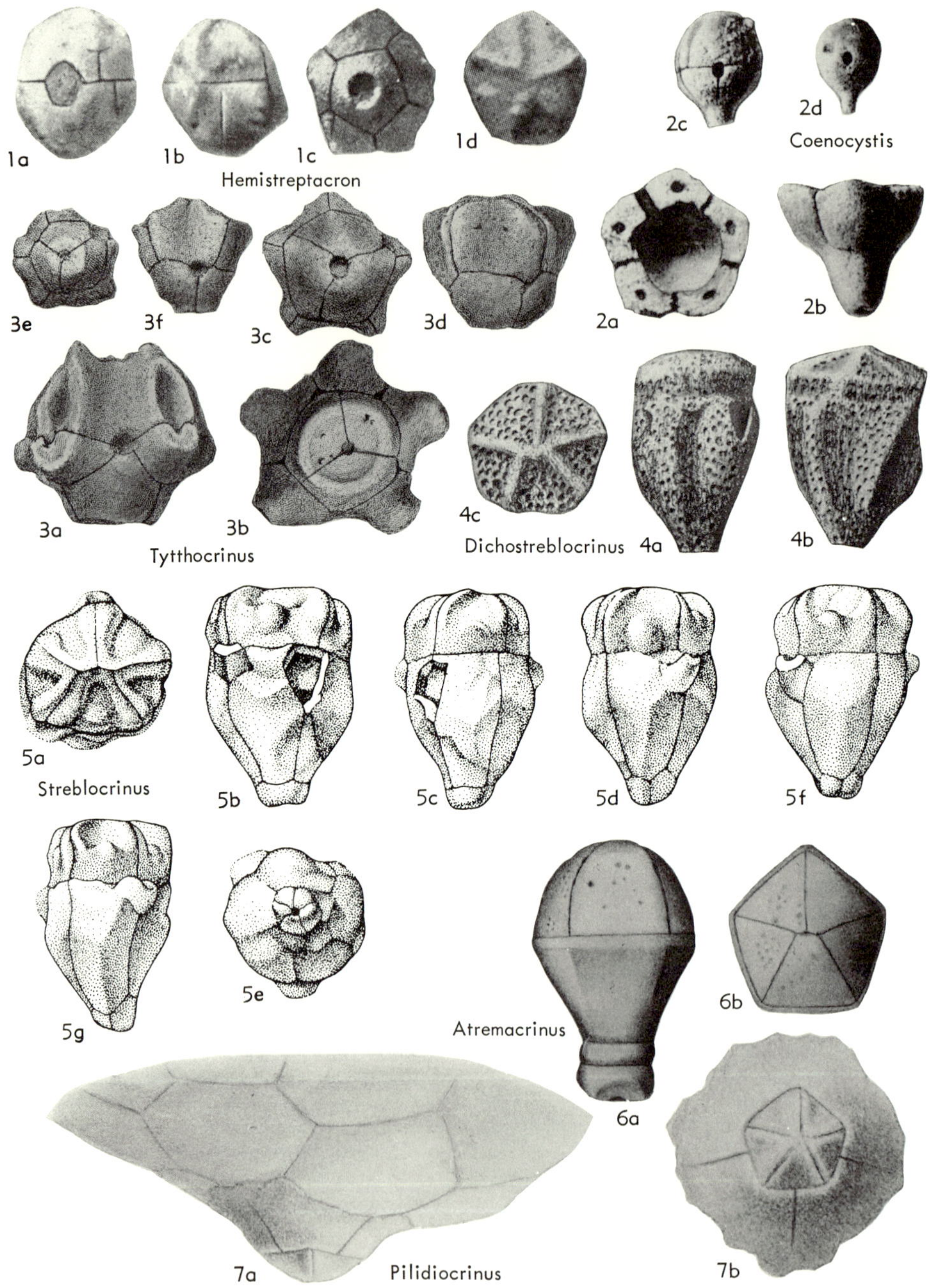

FIG. 390. Streblocrinidae (Streblocrininae) (p. T603-T605).

U.Perm., Indon.(Timor, Basleo).——FIG. 390,*1*. **P. permicus; 7a,b,* lat. and ventral views of theca, ×4 (Wanner, 1937).

Tytthocrinus J. M. WELLER, 1930, p. 35 [**T. comptus*; OD] [=*Octocrinus* PECK, 1936, p. 291 (type, *O. inconsuetus*)]. Theca cone shaped; small specimens with 3 infrabasals and basals, and 5 orals; large specimens with 3 basals and 5 orals, infrabasals fused or atrophied; orals large, centrally depressed, *CD* oral larger than others; anal opening marked by notch at distal end of interbasal suture just below right side of *CD* oral. *M.Dev.-L. Penn.,* USA (Ind.-Mo.-N.Mex.-N.Y.).——Fig. 390,*3a-d*. *T. inconsuetus* (Peck), L.Miss., Mo.; *3a,b,* post., and dorsal views of large individual; *3c,d,* dorsal and post. views of small individual, ×30 (Peck & Connelly, 1951).——Fig. 390,*3e,f*. **T. comptus,* L.Miss., Mo.; *3e,f,* dorsal and post. views of theca, ×30 (Peck & Connelly, 1951).

Subfamily PENTECECRININAE Lane, 1967

[Pentececrininae Lane, 1967, p. 13] [incl. Acariaiocrininae Arendt, 1970a, p. 169; Lageniocrininae Arendt, 1970a, p. 158]

Anal opening shifted secondarily to *A-CD* axis, and orals rotated to radial position by left-lateral extension of *CD* oral. *?U. Dev., L.Miss.-L.Penn.; L.Perm.-U.Perm.*

Pentececrinus Koenig & Niewoehner, 1959, p. 464 [**P. parvus*; OD]. Theca small, globe shaped; 5 infrabasals, basals, and orals; radials lacking; single anal plate in notch on upper right shoulder of *CD* basal and below lower right edge of *CD* oral; orals large, in interradial position directly above basals, each with single vertical protuberance. *?U.Dev.* or *?L.Miss.,* USA(Mo.).——Fig. 388,*5*; 391,*1*. **P. parvus,* Louisiana F.; 388,*5*, plate diagram of theca, anal plate stippled (Moore, n); 391,*1a-g*, *CD*-interray, *C-, B-, E-, D*-ray, ventral, and dorsal views of theca, ×22 (Koenig & Niewoehner, 1959).

Acariaiocrinus Wanner, 1924, p. 141 [**A. clavulus*; OD] [=*Streptostomocrinus* Yakovlev, 1927, p. 186 (type, *S. caryophylloides*)]. Theca small, flask or club shaped; infrabasals 3, visible in side view; basals 5; radials lacking; orals 5, large, in radial position, alternating with basals; anal opening in notch between *CD* basal below and *C-* and *D*-ray orals above. *L.Perm.,* Eu.(USSR, Krasnoufimsk)-*U.Perm.,* Indon.(Timor, Basleo).——Fig. 391,*5*. **A. clavulus,* U.Perm., Timor; *5a-c, A*-ray, *CD*-interray, and ventral views of theca, ×8.7 (Wanner, 1924).

Clistocrinus Kirk, 1937, p. 106 [**C. pyriformis*; OD] [=*Clithrocrinus* Kirk, 1937b *(nom. van.); non Cleistocrinus* Springer, 1920]. Theca small, globe shaped; infrabasals 3, unequal, small one in *A* ray; basals 5, large; anal opening large, situated between upper right shoulder of *CD* basal and upper left shoulder of *BC* basal, bounded above by lower left edge of *C* oral; orals shifted to radial position by enlargement of left side of originally *CD* oral, which occupies a *D*-ray position; *D*-ray oral with central hydropore. *L.Penn.,* USA(Alaska).——Fig. 391,*2*. **C. pyriformis; 2a-d, CD*-interray, *A*-ray, ventral, and dorsal views of theca, ×4.5 (Kirk, 1937a).

Lageniocrinus deKoninck, in deKoninck & LeHon, 1854, p. 187 [**L. seminulum*; M]. Theca small; cup cone shaped; infrabasals 3, small one in *A*

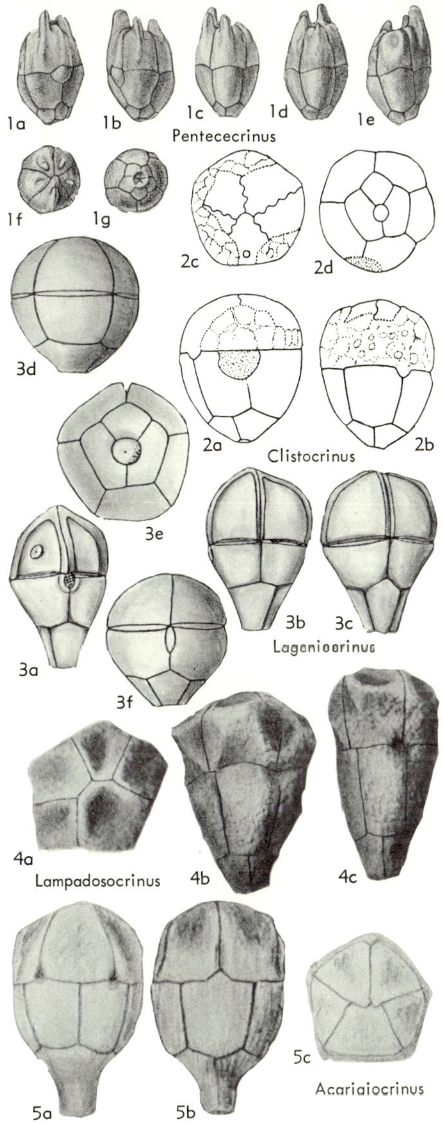

Fig. 391. Streblocrinidae (Pentececrininae) (p. *T*605-*T*606).

ray; basals 5, anal opening between upper right and upper left corners of *CD* and *BC* basals; orals 5, large, each plate with raised borders next to interoral sutures, in interradial position directly above basals; *CD* oral with hydropore. [One Permian species referred to this genus by Wanner (1920) was judged by Kirk (1940e) to be an immature *Embryocrinus*.] *L.Carb.(Tournais., ?Visean),* Eu.(Belg.-Eng.); *L.Penn.,* USA(Ind.).——Fig. 391,*3a-c*. **L. seminulum,* L.Carb.(Visean), Belg.; *3a-c, CD*-interray, *BC*-interray, and *AB*-interray views of theca, ×2.6 (Kirk, 1940e).

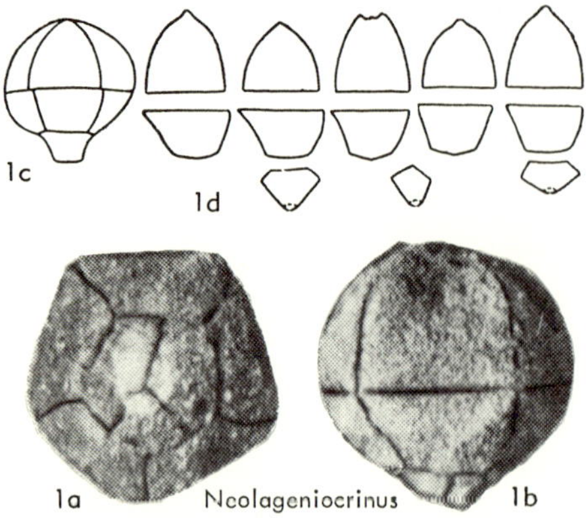

FIG. 392. Streblocrinidae (Pentececrininae) (p. T606).

——FIG. 391,*3d-f*. *L. jacksoni* (AUSTIN & AUSTIN), L.Carb.(Tournais.), Eng.; *3d-f*, lat., dorsal, and *CD*-interray views of theca, ×5 (Kirk, 1940e).

Lampadosocrinus STRIMPLE & KOENIG, 1956, p. 1244 [**Dichostreblocrinus minutus* PECK, 1936, p. 289; OD]. Theca small, cone shaped; infrabasals 5, high; basals short, each with prominent central node; anal opening between upper right shoulder of *CD* basal and upper left shoulder of *BC* basal; orals high, flattened ventrally, each oral centrally depressed with raised borders; orals nearly radial in position except *CD* oral, which is only slightly to left of *CD* basal; posterior oral with central hydropore. *L.Miss.*, USA(Mo.-N. Mex.-Okla.).——FIG. 391,*4*. **L. minutus* (PECK), Mo.; *4a-c*, ventral and post. views of theca, post. view of another theca, ×30 (Peck, 1936).

Neolageniocrinus ARENDT, 1970a, p. 162 [**N. shichanensis*; OD]. Closely similar to *Lageniocrinus* but infrabasals may be fused together, anal opening in theca is absent, and posterior oral commonly lacks hydropore. *L.Penn.*, USA; *L.Perm.(Artinsk.)*, NC.Sib.; *U.Perm.(Basleo beds)*, Indon.(Timor).——FIG. 392,*1*. **N. shichanensis*, L.Perm.(Artinsk.), Krasnoufimsk, NC.Sib.; *1a,b*, dorsal and post. views of holotype, ×5; *1c,d*, *BC* side of paratype, ×6.5, and plate diagram of holotype, ×3 (Arendt, 1970a).

Superfamily and Family UNCERTAIN

Elicrinus PROKOP, 1973, p. 221 [**E. procerus*; OD]. Cup high, conical; infrabasals 3, 2 large and one small; 5 high basals and radials; no anal plate in cup; radial facet narrow, rounded, horseshoe shaped, with distinct fulcral ridge. Tegmen, arms, and stem unknown. *L.Dev.(Prag.)*, Eu.(Boh.).——FIG. 393,*2*. **E. procerus*; *2a,b*, lat. views of cup, ×71 (Prokop, 1973).

Parastephanocrinus SPRINGER, 1926, p. 139 [**P. typus*; OD]. Dicyclic small cup with basal concavity. Five small infrabasals form bottom of basal invagination; 5 large basals form sides of concavity and most of cup height, posterior basal elongated and extending to the anal opening; radials small in exterior but extending inward toward a small oral center, ambulacral grooves prominent and bounded by projecting processes, articular facets small, directed outward. *M.Sil.(Niagaran)*, USA(Tenn.).——FIG. 393,*1*. **P. typus*; *1a-c*, cup from post., summit, and base, ×3; *1d*, detail of radials and food grooves, ×3 (Springer, 1926a).

Suborder DENDROCRININA Bather, 1899

[*nom. correct.* MOORE, in MOORE, LALICKER, & FISCHER, 1952, p. 613 (*pro* suborder Dendrocrinoidea BATHER, 1899b, p. 178)] [=suborder Dendrocrinites JAEKEL, 1918, p. 52] [Materials for this suborder prepared by R. C. MOORE, N. GARY LANE, and H. L. STRIMPLE]

Crown small to moderately large, mostly tall, cylindrical to ovoid or pear shaped, uncommonly explanate, with widely separated horizontal or pendent arms. Cup prevailingly conical but bowl shaped or discoid in many genera, base convex, flat, or concave; infrabasals five, uncommonly three or fused as a single solid circlet, mostly visible from side but not visible in flat- and concave-based forms; radials with narrowly rounded articular facets occupying small part of radial summits or wide, with gently curved external margins, plane of facets sloping outward-downward, essentially horizontal, or exceptionally inclined outward-upward, in terms of inner to outer facet margins; anal plates in cup three, two, one, or none visible from side of cup, anal sac tall, cylindrical, inflated, or not evident, many with prominent plicate plates. Arms atomous, branched isotomously or heterotomously, few or very many, brachials arranged uniserially or biserially, pinnulate or lacking pinnules. Stem pentagonal or mostly circular. *L.Ord.-L.Miss.*

The suborder Dendrocrinina includes lower and middle Paleozoic crinoids that gave rise on one hand to the Flexibilia in the Ordovician and to the more advanced cladid inadunates, the Poteriocrinina in the Devonian, on the other. Although dendrocrinines are present in the Ordovician and Silurian, they are neither abundant nor diverse. The few genera that have been

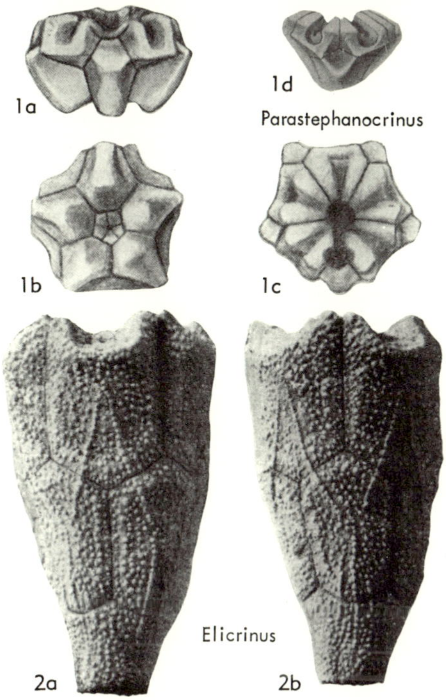

FIG. 393. Cyathocrinina Superfamily and Family Uncertain (p. *T*606).

described, however, are quite important to knowledge of the main evolutionary pathways that developed within this group. During the Devonian the Dendrocrinina diversified rapidly and many genera are recognized currently, most of them placed in the families Dendrocrinidae and Botryocrinidae. Crinoids in both of these families exhibit a tendency to evolve from types that have isotomously branched arms to ones with strongly heterotomous arms. These latter reached the verge of pinnulation in the Dendrocrinidae and true pinnules developed in Late Devonian members of the Poteriocrinina.

Shape of the cup observed in crinoids of the Inadunata include 1) cylindroconical, 2) moderately steep-sided conical with narrowly to broadly truncate base equal to or wider than the stem impression, 3) deep to shallow bowl shaped with longitudinally curved sides and gently convex, flat, or slightly concave base, all classed as crateriform, 4) basally invaginated, with wide to narrow and moderate to very deep central concavity, and 5) discoid (patelliform). The Dendrocrinina are distinguished by cups of the first two types, which appear to be the most primitive of cup shapes.

The radial articular facets are divisible into three types: 1) small circular or semicircular, horseshoe-shaped facets much narrower than summit width of the radials and hence called **angustary** (narrow), 2) broad outwardly shallow crescentic facets occupying much but not all of the radial summits, called **peneplenary** (almost full width), and 3) facets fully equal in width to the radial summits, termed **plenary** (fully wide). Crinoids with angustary arm facets exceed others in the Dendrocrinina and are judged to be the most primitive type. They characterize all or nearly all Cyathocrinina. Finally, the attitude of articular facetal planes on radials of inadunate crinoids has taxonomic importance, for in different genera it is found to range from facing outward vertically to sloping steeply or gently outward-downward, subhorizontal, and uncommonly outward-upward. Descriptive terms for types of facetal attitudes are 1) **declivate,** sloping outward-downward; 2) **planate,** subhorizontal, and 3) **sursumate,** sloping outward upward.

Superfamily DENDROCRINACEA Wachsmuth & Springer, 1886

[*nom. transl.* MOORE & LANE in MOORE & STRIMPLE, 1973, p. 19 (*ex* Dendrocrinidae WACHSMUTH & SPRINGER, 1886)] [Materials for this superfamily prepared by R. C. MOORE and N. GARY LANE]

Crown tall, mostly somewhat narrowly cylindrical. Cup conical, with very steep to moderately sloped, longitudinally straight sides; infrabasals five, distal parts upflared, readily visible in side views of cup; radial articular facets narrow, with strongly rounded to horseshoe-shaped outer margins, interradial notches commonly prominent, surface of facets curved or plane, sloping outward-downward; two or three anal plates in cup, anal sac mostly cylindrical, very tall, and formed partly of plicate plates. Arms well rounded externally, slender, branching isotomously and heterotomously, mostly numerous, brachials uniserial, in some genera laterally bearing ramules. Stem transversely circular or less commonly pentagonal. *M.Ord.-U.Dev.*

Family DENDROCRINIDAE Wachsmuth & Springer, 1886

[*nom. transl.* BATHER, 1890b (May), p. 383 (*ex* section Dendrocrinites WACHSMUTH & SPRINGER, 1886, p. 115 (191))] [=Dendrocrinidae S. A. MILLER, 1890 (Dec.), see MOORE, 1962b, p. 37; Bactrocrinidae JAEKEL, 1918, p. 55; Esthonocrinidae JAEKEL, 1918, p. 53]

Cup steeply conical; infrabasals strongly upflared, visible from side; radianal directly below or obliquely left below *C* radial, anal *X* in line with radials and above *CD* basal. Radial arm facets narrow, horseshoe shaped, smooth or with fine radial culmina. Arms rounded, slender, branching repeatedly, with strong tendency toward heterotomy. *M.Ord.-U.Dev.*

Chief primitive features of the Dendrocrinidae are the steeply conical form of the cup, structure of its posterior side, and the smooth, outwardly directed radial facets. The rounded facets are horseshoe shaped and narrow. The arms are strongly rounded and branch isotomously several times, representing an advanced evolution of nonpinnulate ray structure. An archaic feature of *Dendrocrinus* is the occurrence of a radianal directly below the *C* radial (Fig. 394,*5*). The very tall anal sac, formed of a number of vertical rows of thin plicate plates, is a distinctive attribute of the family that is very unlike the sacs of merocrinids and cupulocrinids but closely similar to those of botryocrinid genera and others, which are inferred to belong in the same genetic line as the Dendrocrinidae.

Key to Genera of Dendrocrinidae

A. Cup tall and narrow cylindroconical ....
- I. Infrabasals very tall; radial articular facets narrow, horseshoe shaped; round slender arms branching on primibrachs *4-5*; stem transversely round .. *Bactrocrinites*
- II. Infrabasals moderately tall; radial articular facets wider than horseshoe shaped; arms branching on primibrachs *3* or *4*; stem transversely pentagonal .. .. *Atractocrinus*

B. Cup moderately wide conical ....
- I. Infrabasals 3; anals in cup 2 .... *?Alsopocrinus*
- II. Infrabasals 5; anals in cup 3 or 4.
  - a. Radianal directly below *C* radial, anal sac very tall cylindrical with plicate plates; arms branching first on primibrachs *5-7* .... *Dendrocrinus*
  - b. Radianal obliquely left below *C* radial; arms many (60 or more), branching first on primibrachs 3 .... *Grenprisia*
  - c. Radianal obliquely left below *C* radial; arms ?few, branching on primibrachs 3 .... *Esthonocrinus*
- III. Cup very minute, plates ill-known, extremely slender long arms, numerous (approximately 60) well separated, branching first on primibrachs *6-9* .... *Parisangulocrinus*

Dendrocrinus HALL, 1852, p. 193 [**D. longidactylus*; M]. Crown elongate, slender. Cup high conical, with large radianal directly below *C* radial; anal *X* large, between posterior radials and supporting series of large anal plates at posterior base of anal sac, which is very elongate, tapering to a point distally, composed of several longitudinal rows of thin, plicate, laterally interlocking plates. Arms rounded, slender, branching several times isotomously; primibrachs *5* to *7* axillary. Radial articular facets narrow, rounded, with outward-downward slope and wide interradial notches. Stem circular transversely, tapering distally. *M.Ord.-U.Sil.*, N.Am.(Ont.-N.Y.-Ky.-Ill.-Ind.-Ohio-Wis.-Iowa)-Eu.(Eng.).——FIG. 395,*1a-d*. **D. longidactylus*, U.Sil., N.Y.; *1a*, post. view of cup, tall anal sac, and proximal parts of arms and stem, ×1; *1b*, fragment of anal sac, ×2; *1c*, post. view of theca and slender uniserial arms, ×1 (Hall, 1852); *1d*, plate diagram (radials black, radianal cross ruled, anal plates stippled) (Moore, n).——FIG. 395,*1e,f*. *D. caduceus* (HALL, 1866), U.Ord., Ohio; *1e,f*, ?ant. and lat. views of crown, ×2 (Moore, n; Univ. Chicago 2302A, 2302B).

?**Alsopocrinus** TANSEY, 1924, p. 181 [**A. anna*; M]. Cup small, truncated below; infrabasals 3, small azygous one in *C* ray, barely visible in side view, with wide concavity for proximal columnal; basals large and hexagonal; radials with straight articular facets that bear transverse ridge; radianal below and to left of *C* radial, anal *X* in line with radials. Tegmen, arms and column unknown. [Poorly known and possibly not a dendrocrinid.] *L.Dev.*, USA(Mo.).——FIG. 395,*2*. **A. anna; 2a-c*, lat., post., and dorsal views of cup, with attached pleurotomariid gastropod, ×2 (Tansey, 1924).

Atractocrinus KIRK, 1948, p. 701 [**A. concinnus*; OD]. Cup narrow, elongate; infrabasals nearly equal in size to large basals, both elongate; radials small, *A* radial smaller than others and with reduced articular facet, other articular facets larger, rounded, sloping outward-downward; radianal, anal *X* and additional anal plate in cup. Arms stout, branching first on third or fourth primibrach. Stem pentagonal. *M.Dev.-U.Dev.*, USA (Mich.-Iowa), Can.(Ont.).——FIG. 396,*1*. **A. concinnus*; M.Dev., Ont.; *1a-d*, ant., post., ventral and dorsal views of cup, ×1.5 (Kirk, 1948).

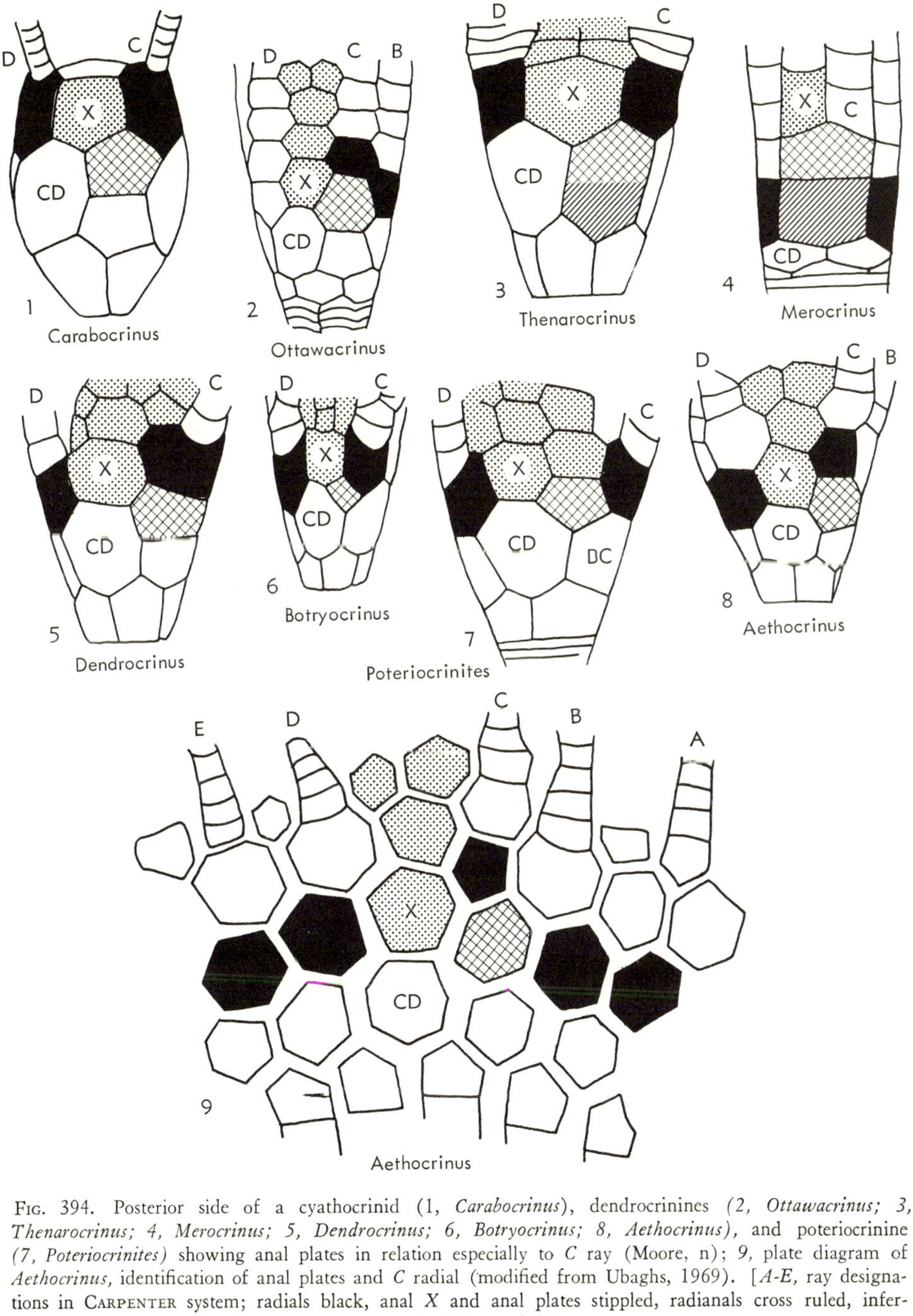

FIG. 394. Posterior side of a cyathocrinid (1, *Carabocrinus*), dendrocrinines *(2, Ottawacrinus; 3, Thenarocrinus; 4, Merocrinus; 5, Dendrocrinus; 6, Botryocrinus; 8, Aethocrinus)*, and poteriocrinine *(7, Poteriocrinites)* showing anal plates in relation especially to *C* ray (Moore, n); *9*, plate diagram of *Aethocrinus*, identification of anal plates and *C* radial (modified from Ubaghs, 1969). [*A-E*, ray designations in CARPENTER system; radials black, anal *X* and anal plates stippled, radianals cross ruled, inferradianals oblique ruled.]

Bactrocrinites SCHNUR, 1849, p. 22 [**Poteriocrinus fusiformis* ROEMER, 1844, p. 61; M] [=*Bactrocrinus* QUENSTEDT, 1875, p. 533 *(nom. van.)*]. Theca narrow and very tall; infrabasals and basals

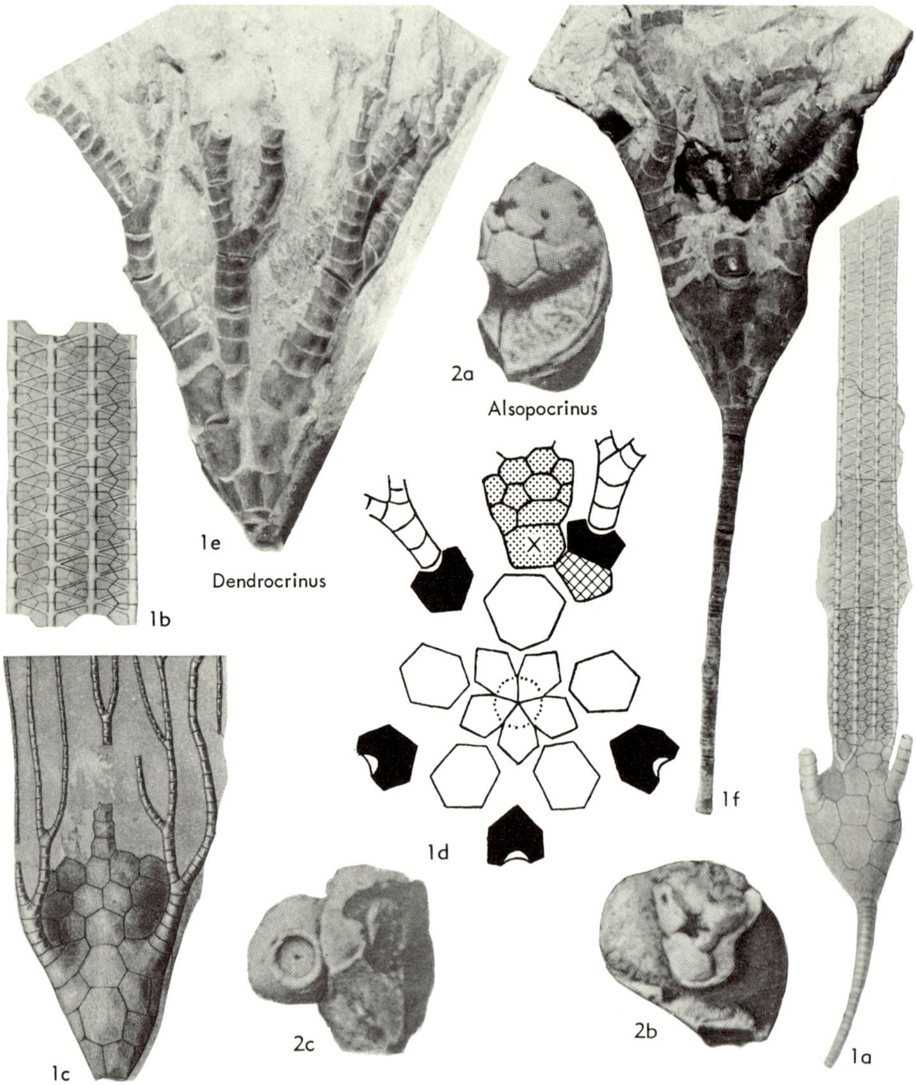

FIG. 395. Dendrocrinidae (p. *T*608).

large and elongate; radials small, quadrangular, radial articular facets sloping obliquely outward-downward; small radianal and anal *X* in cup; tegmen with 4 large interambulacral plates. Anal sac narrow and elongate, base confined to posterior part of tegmen. Arms long, slender, branching 3 or 4 times isotomously, first on primibrachs 4 or 5. *U.Sil.-M.Dev.*, Eu.(Ger.)-Asia(India)-USA (Iowa-N.Y.-Okla.).——FIG. 396,*2a,b*. *B. oklahomaensis* STRIMPLE, U.Sil., Okla.; *2a,b*, *BC* and post. views of theca, ×1 (Strimple, 1952c).——FIG. 396,*2c,d*. *B. reimanni* GOLDRING, M.Dev., N.Y.; ant. and post. views of theca, ×2 (Goldring, 1954).——FIG. 396,*2e,f*. **B. fusiformis* (ROEMER), M.Dev., Ger.; post. and ventral views of theca, ×2 (Schultze, 1867).

Esthonocrinus JAEKEL, 1918, p. 53 [**E. laevior*; OD]. Theca large, relatively broad, with depressions at each cup plate angle; infrabasals upright, basals slightly larger than radials; articular facets rounded and narrow, interradial notches prominent, radianal large, below and to left of *C* radial; anal *X* above radianal and almost out of cup. Base of anal sac composed of longitudinal row of large posterior anal plates and small lateral plates. Arms narrow, rounded, branching on

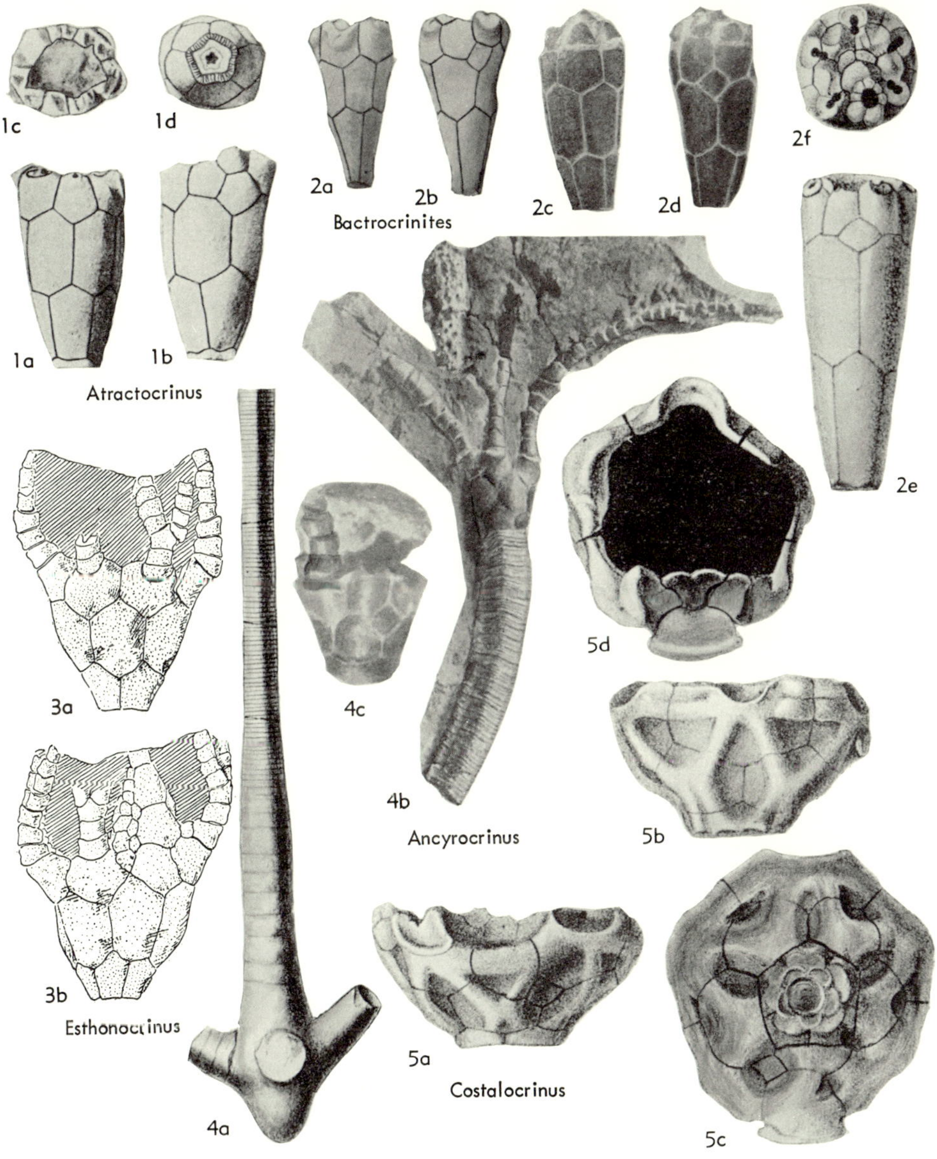

FIG. 396. Dendrocrinidae *(1-3)*; Botryocrinidae *(4,5)* (p. *T*608-*T*611, *T*614-*T*615).

primibrachs 3. [Poorly known.] *M.Ord.*, Eu. (Est.).——FIG. 396,*3*. **E. laevior; 3a,b, AB-* and *CD*-interray views of cup and proximal parts of arms, ×0.7 (Jaekel, 1918).

Grenprisia MOORE, 1962, p. 38 [**Ottawacrinus billingsi* SPRINGER, 1911b, p. 40; OD] [=*Ottawacrinus* SPRINGER, 1911b (*non* W. R. BILLINGS, 1887) (obj.)]. Cup high; infrabasals and basals large, radials equal to or smaller than basals; large radianal in line with basals, supporting anal *X* and an anal sac plate above; small interbrachial plates above interradial notches between proximal brachs. Anal sac very large, cylindrical, composed of numerous small stellate plates. Arms uniserial, strongly heterotomous, branching first on primibrachs *3*, ramules that have 1 or 2 branches given off on alternate sides of main rami. Stem transversely subpentagonal. *M.Ord.*, Can.(Ont.).——FIG. 397,*2a-g*. **G. billingsi* (SPRINGER); *2a,b, A-* and *D*-ray views of crown,

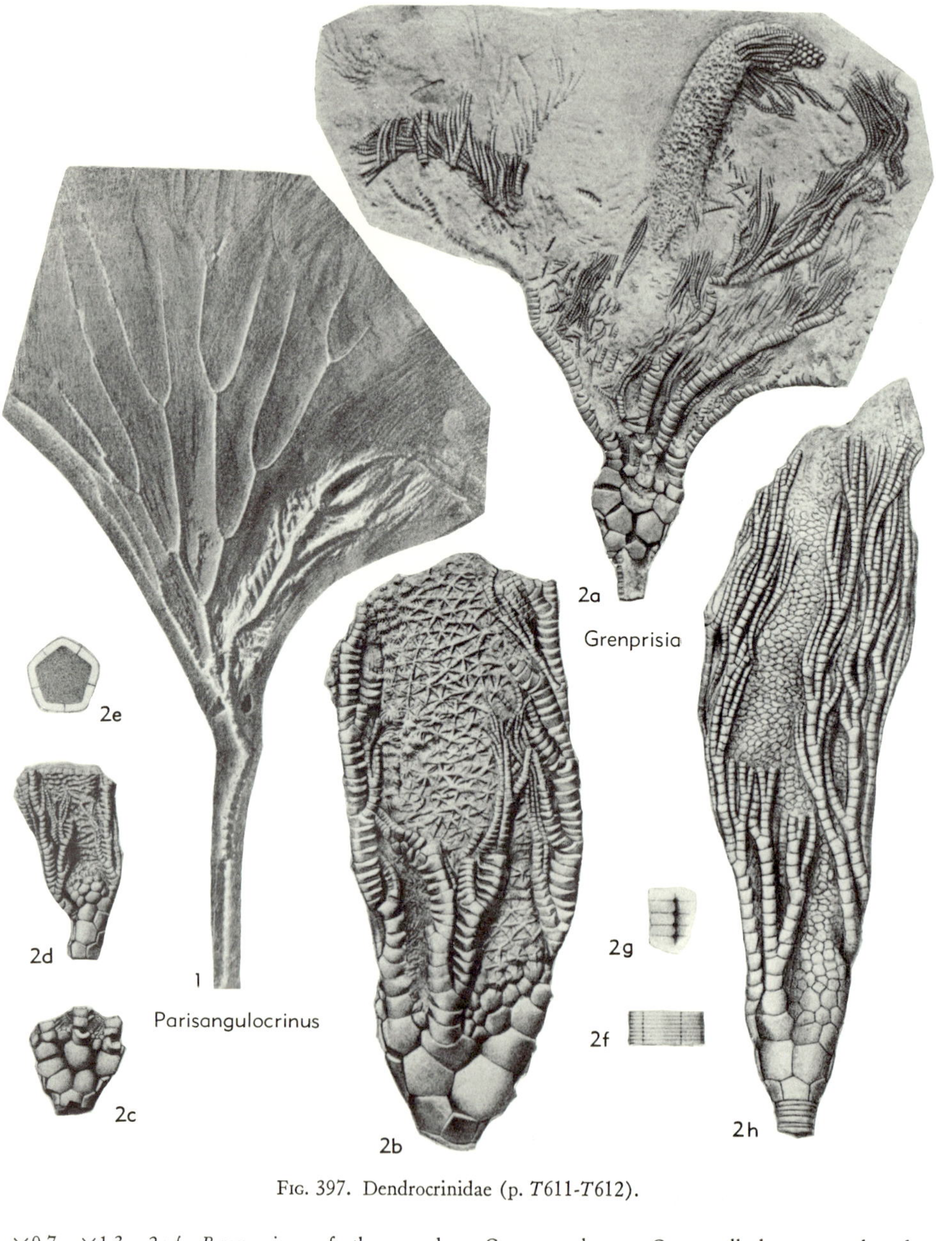

FIG. 397. Dendrocrinidae (p. *T*611-*T*612).

×0.7, ×1.3; *2c-f, B*-ray view of theca and proximal brachs, post. view of crown, transv. and long. views of columnals, ×0.7; *2g,* view of suture between columnal pentameres, ×2.7 (Springer, 1911b).——FIG. 397,*2h. G. springeri* MOORE; post. view of crown, ×2 (Springer, 1911b).

Parisangulocrinus SCHMIDT, 1934, p. 59 [**Poteriocrinus zeaeformis* SCHULTZE, 1866, p. 43; OD]. Crown explanate. Cup small, low, cone shaped; 3 anal plates in cup; radial articular facets narrow and rounded. Anal sac large, elongate, composed of several rows of thin, plicate plates. Arms slender, rounded, branching isotomously several times. Stem transversely round. *L.Dev.,* Eu. (Ger.).——FIG. 397,*1. P. furcaxialis* SCHMIDT; lat. view of crown and stem, ×1 (Schmidt, 1934).

Family BOTRYOCRINIDAE Wachsmuth & Springer, 1886

[*nom. transl.* BATHER, 1899b, p. 921 (*ex* section Botryocrinites WACHSMUTH & SPRINGER, 1886, p. 191)] [=Gothocrinidae JAEKEL, 1918, p. 59; Pandoracrinidae JAEKEL, 1918, p. 56; Rhadinocrinidae JAEKEL, 1918, p. 58]

Radial articular facets typically much narrower than radials, sloping outward-downward (declivate); anal plates in cup generally three but may be only two, radianal small and quadrangular, or large, pentagonal, or absent; anal *X* situated between posterior radials. Anal sac large, with plicate plates, distally recurved, with anal opening directed anteriorly in some genera. Arms uniserial, typically heterotomous, with two main rami, but may be isotomous with numerous branches, or with single ramus; armlets on alternate sides of brachials, branched or unbranched; ramules developed in advanced forms. Stem mostly round transversely and noncirriferous, but may be pentagonal and cirriferous. *?M.Ord.; M.Sil.-U.Dev.*

[WACHSMUTH & SPRINGER (1886, p. 191) clearly used the name Botryocrinites to discriminate a suprageneric assemblage of crinoids and not as synonymous with *Botryocrinus* ANGELIN (1878, p. 24)].

Members of this family are progressive early and middle Paleozoic Dendrocrinina, which are interpreted as close relatives of the Dendrocrinidae but not derived from them. The botryocrinid arm facets resemble those of the Cyathocrinina, but the nature of the tegmen and thin-walled plicate anal sac are characters of the Dendrocrinina. In *Botryocrinus* and several other members of the family a quadrangular radianal occurs obliquely below the *C* radial at left (Fig. 394,*6*). Also, the arms divide heterotomously, as well as isotomously. Some crinoids placed in the family show a tendency toward disappearance of the radianal plate, which is an evolutionary specialization.

Key to Genera of Botryocrinidae

A. Cup steep-sided conical, radial articular facets narrow horseshoe shaped; 2 anal plates in cup (small radianal, large anal *X*) ....
  I. Arms 20 or more, branching first on primibrachs *4*; anal sac tall composed of small, polygonal plates; stem transversely quadrangular, terminating distally in massive grapnel-shaped holdfast .... *Ancyrocrinus*
  II. Arms 10 or more, branching first on primibrachs *4-5*; anal sac tall, stem transversely round .... *Gothocrinus*

B. Cup wide conical; radianal small or lacking, anal sac slender, tall ....
  I. Cup very diminutive, partly ill known; infrabasals 5, except *Parabotryocrinus*; 1-4 anal plates in cup, anal sac tall; arms 10-40 or more, branching first on primibrachs *2-8*, commonly bearing ramules ....
    a. Infrabasals 3; anal plates in cup 2; arms ?10 .... *Parabotryocrinus*
    b. Large anal *X*; arms 10, branching on primibrachs *2*, ramules abundant .... *Imitatocrinus*
    c. Anal plates in cup 3, anal sac extremely slender and tall; arms ?10, branching on primibrachs ?2-3 .... *Schmidtocrinus*
    d. Anal plates in cup 3, anal sac cylindrical and very tall; arms 10, branching on primibrachs *4*, widely spaced armlets branched .... *Rhadinocrinus*
    e. Anal plates in cup 4, anal sac large, cylindrical; arms 30-40, branching first on primibrachs *8* or higher; stem transversely pentagonal, bearing cirri .... *Gastrocrinus*
  II. Cup medium sized; anal plates in cup 1-4, anal sac slender, cylindrical; arms 5-40 or more ....
    a. Radianal small, quadrangular, anal *X* large; arms 10-40 or more.
      1. Two anal plates in cup above anal *X*, anal sac tall, cylindrical, with plicate plates; arms 30-40 or more, branching first on primibrachs *2* .... *Botryocrinus*
      2. No anal plates in cup above anal *X*; anal sac unknown; arms 10 or more, branching on primibrachs *3-4* .... *Jahnocrinus*
      3. No anal plates in cup above small anal *X*; anal sac and arms unknown; cup plates strongly fluted .... *Costalocrinus*
    b. Radianal absent, anal *X* small; arms 5 .... *Sigambrocrinus*
    c. Cup anals unknown, anal sac cylindrical, composed of short and wide hexagonal plates; arms 10 or more, first branching on primibrachs *2-4* .... *Pandoracrinus*

Botryocrinus ANGELIN, 1878, p. 24 [**B. ramosissimus*; SM BATHER, 1893a, p. 117] [=*Sicyocrinus* ANGELIN, 1878, p. 23 (type, *S. cucurbitaceus*)]. Cup conical, plates smooth or ornamented with radial ridges, spines or longitudinal folds; infrabasals 5, upflared; radianal small, quadrangular;

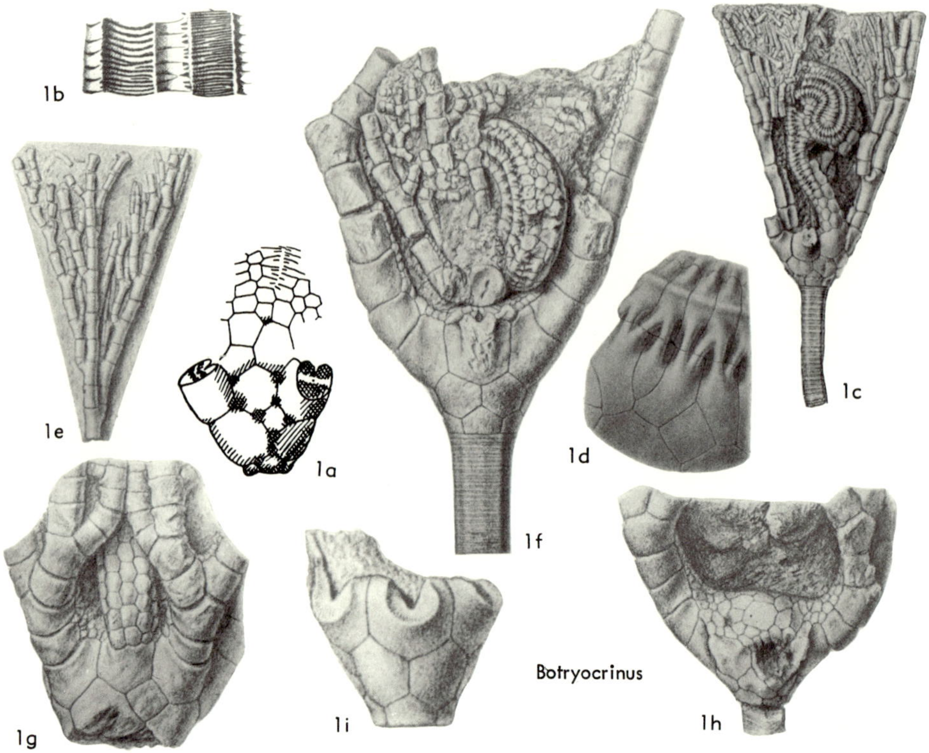

FIG. 398. Botryocrinidae (p. T613-T614).

anal *X* large, situated between *BC* and *CD* basals and supporting 2 or 3 small sac plates. Anal sac large, composed of numerous rows of small hexagonal plates, with anal opening at distal tip or recurved with anal opening directed anteriorly and lateral sac plates sharply plicate. Arms stout, with 2 or 4 main branches, primibrachs 2 commonly axillary; stout covering plates preserved over ambulacral grooves; ramules on alternate sides of main branches. Stem transversely round or subpentagonal, pentameres commonly developed. *M.Sil.-M.Dev.*, Eu.(Sweden-Ger.-Eng.), USA (N.Y.-Tenn.-Mich.-Iowa-Ill.-Ind.), Can.(Ont.).——FIG. 398,*1a,b*. *B. schultzei* HAARMANN, L.Dev., Ger.; *1a,* diagram of post. part of cup, ×1; *1b,* view of anal sac plates, ×2 (Schmidt, 1942).——FIG. 398,*1c-e*. *B. cucurbitaceus* (ANGELIN), M.Sil., Gotl.; *1c, D*-ray view of crown showing recurved slender anal sac, and proximal part of stem, ×1.5; *1d,* part of anal sac, ×10; *1e,* arm, ×1.5 (Bather, 1893a).——FIG. 398,*1f-i*. **B. ramosissimus*, U. Sil., Gotl.; *1f,g, A*-ray view of crown and stem, post. view of theca and proximal parts of arms and anal sac, ×1.5; *1h,* lat. view of cup with arm removed, showing tegmen, ×3; *1i, A*-ray view of cup, ×1.5 (Bather, 1893a).

Ancyrocrinus HALL, 1862, p. 118 [**A. bulbosus*; SD S. A. MILLER, 1889, p. 223]. Cup cone shaped, broadly truncate below; radianal small, quadrangular; anal *X* large, between posterior radials; median parts of radial plates protruding, radial articular facets directed obliquely outward-downward, narrow and rounded. Anal tube large, elongate, composed of numerous small plates. Arms branching on primibrachs *4,* with 2 main rami, and ramules on opposite sides of alternate secundibrachs. Stem massive, quadrangular in cross section proximally; columnals thin proximally becoming fused distally and merging into a large, grapnel-like terminal holdfast bluntly rounded distally, with 4 or 5 blunt or sharply pointed projections directed obliquely back toward crown; lumen quadrilobate or rarely pentalobate; articular facets quadrangular in outline. *L.Dev.-M.Dev.*, N.Am.(N.Y.-Penn.-Ont.-Ind.-Ohio)-Eu. (France).——FIG. 396,*4*. **A. bulbosus*, M.Dev., N.Y.; *4a,* lat. view of distal part of stem and holdfast, ×1; *4b, B*-ray view of crown and stem, ×1; *4c,* post. view of cup and proximal brachs, ×2 (Goldring, 1923, 1942).

Costalocrinus JAEKEL, 1918, p. 60 [**Poteriocrinus dilatatus* SCHULTZE, 1867, p. 161; M]. Cup low bowl shaped; plates thin, ornamented with strong ridges radiating from center of each radial and basal; radianal small, quadrangular, situated in depression between ridges; anal *X* large, bearing part of prominent transverse ridge between posterior radials; radial articular facets about one-half as wide as radials, thin and arcuate; posterior facets obliquely inclined toward anterior side of cup by large, medially projecting edges of primanal and posterolateral edges of radials, which together form large concave surface projecting admedially and presumably providing support for an anal sac. Arms, anal sac, and stem not known. *M.Dev.*, Eu.(Ger.).——FIG. 396,*5a-d*. **C. dilatatus* (SCHULTZE); post., ant., dorsal, and ventral views of cup (*CD* interray directed downward), ×1.5 (Schultze, 1867).

Gastrocrinus JAEKEL, 1895, p. 60 [**Poteriocrinus patulus* MÜLLER, 1859, p. 192; M]. Cup small, low, poorly known; plates stellate. Anal sac very large, cylindrical, composed of several vertical rows of convex interlocking hexagonal plates which bear transverse median ridges and small folds perpendicular to sutures between plates of adjacent rows; anal opening terminal. Arms slender, short, branching first high above cup, with 2 or 3 additional branches and widely spaced branching armlets in distal parts of arms. Stem transversely pentagonal, with cirriferous nodals at intervals along its entire length; cirri long, slender, and directly obliquely back toward cup. *L.Dev.*, Eu.(Ger.-France).——FIG. 399,*1*. *G. giganteus* SCHMIDT, Ger.; lat. view of crown and stem, ×0.5 (Schmidt, 1934).

Gothocrinus BATHER, 1893, p. 114 [**G. gracilis*; OD]. Cup steeply conical; plates smooth; basals 5; small pentagonal radianal below and to left of *C* radial; arm facets well rounded externally, much narrower than radials; anal *X* large, situated between posterior radials, supporting 2 large anal plates which are succeeded by several rows of smaller hexagonal plates of anal sac. Arms uniserial, composed of large, quadrangular brachials; primibrachs *4* or *5* axillary; each ray with 2 main rami; long, slender, unbranched ramules given off on alternate sides of each arm. *M.Sil.*, Eu.(Sweden).——FIG. 399,*2*. **G. gracilis*, Gotl.; *2a*, crown with attached stem, ×1.5; *2b*, theca and proximal brachials, ×3 (Bather, 1893a).

Imitatocrinus SCHMIDT, 1934, p. 104 [**Cyathocrinus gracilior* ROEMER, 1863, p. 149; OD]. Cup small, low conical, with smooth plates; radianal apparently not developed, anal *X* directly above *CD* basal; radial articular facets about two-thirds as wide as radials. Anal sac not known. Arms uniserial, long, slender, with 2 main rami in each ray, primibrachs *2* axillary; long slender ramules present on secundibrachs. Stem transversely round, slender. *L.Dev.*, Eu.(Ger.).——FIG. 399,*4*. **I. gracilior* (ROEMER); lat. view of crown and stem, ×1 (Schmidt, 1934).

Jahnocrinus JAEKEL, 1918, p. 54 [**J. minutus*; M]. Cup low bowl shaped; infrabasals small, low; basals with prominent longitudinal median folds; radial articular facets one-half as wide as radials; radianal small, quadrangular. Anal sac unknown. Arms slender, first primibrachs short, succeeding 2 to 3 primibrachs elongate, primibrachs *3* or *4* axillary; distal parts of arms not known. Stem transversely pentagonal (poorly known). *M.Dev.*, Eu.(Boh.).——FIG. 399,*3*. **J. minutus*; *3a-c*, *A*- and *D*-ray views of partial crown and stem, plate diagram of cup, ×4 (Jaekel, 1918).

?**Pandoracrinus** JAEKEL, 1918, p. 57 [**P. pinnulatus*; OD]. Cup low, broad, with smooth plates; infrabasals low or possibly entirely covered by stem; radial articular facets small and narrow; radianal and anal *X* unknown. Anal sac apparently large, composed of wide, low, hexagonal plates. Arms branching at least twice isotomously; primibrachs *2* to *4* axillary; brachials with ?2 ramules on each side possibly joined ?laterally. Stem transversely round. *M.Ord.*, Eu.(Boh.-Eng.).——FIG. 400,*4*. **P. pinnulatus*, Boh.; lat. view of incomplete crown and proximal part of stem, ×2 (Jaekel, 1918).

[The type species of this genus is poorly known and the holotype of **P. pinnulatus* may be a monocyclic inadunate referable to the Disparida. JAEKEL (1918) placed together within his suborder Merocrinites five genera, two of which are now accepted as monocyclic inadunates (*Iocrinus*, *Caleidocrinus*), one clearly has infrabasals (*Merocrinus*), and two others (*Pandoracrinus*, *Vosekocrinus*), all of which JAEKEL illustrated as lacking infrabasals, but nevertheless assigned by him to the dicyclic Botryocrinidae. JAEKEL believed that these apparently monocyclic crinoids had secondarily lost their infrabasal plates, and therefore should be classified as dicyclic inadunates. This interpretation is unacceptable. Until the type specimens of these forms or duplicates of them are re-studied their systematic position must remain in doubt.]

Parabotryocrinus YAKOVLEV, 1941, p. 327 [**P. tschudovensis*; OD]. Cup low, bowl shaped; infrabasals small, 3, barely visible in side view; radials larger than basals; radianal and anal *X* present in cup. Arms with slender ramules, mode of branching unknown. Stem round in section. *U.Dev.*, Eu.(USSR).——FIG. 400,*3*. **P. tschudovensis*; *3a*, lat. view of cup, ×1; *3b*, brachials and ramules, ×3 (Yakovlev, 1941b).

Rhadinocrinus JAEKEL, 1895, p. 87 [**R. rhenanus*; OD]. Cup small, cone shaped, plates smooth; infrabasals upflared; radial articular facets strongly curved at outer margins, two-thirds as wide as radials; 3 anal plates in cup, radianal small, quadrangular, separated from anal *X* by right tube plate in line with posterior radials. Anal sac very long and slender, composed of few rows of small hexagonal plates, or distally recurved and with plicate plates. Arms long, slender, branching on primibrachs *4*, with 2 main rami bearing short and slender armlets, densely branched and widely spaced, or quite long, slender and unbranched, or with single branch and more closely spaced. *L.Dev.-M.Dev.*, Eu.(Ger.).——FIG. 400,*1a*. *R. nanus* (ROEMER), L.Dev.; lat. view of crown and stem, ×0.7 (Schmidt, 1934).——FIG. 400,*1b*. **R. rhenanus*, L.Dev.; post. view of crown and prox-

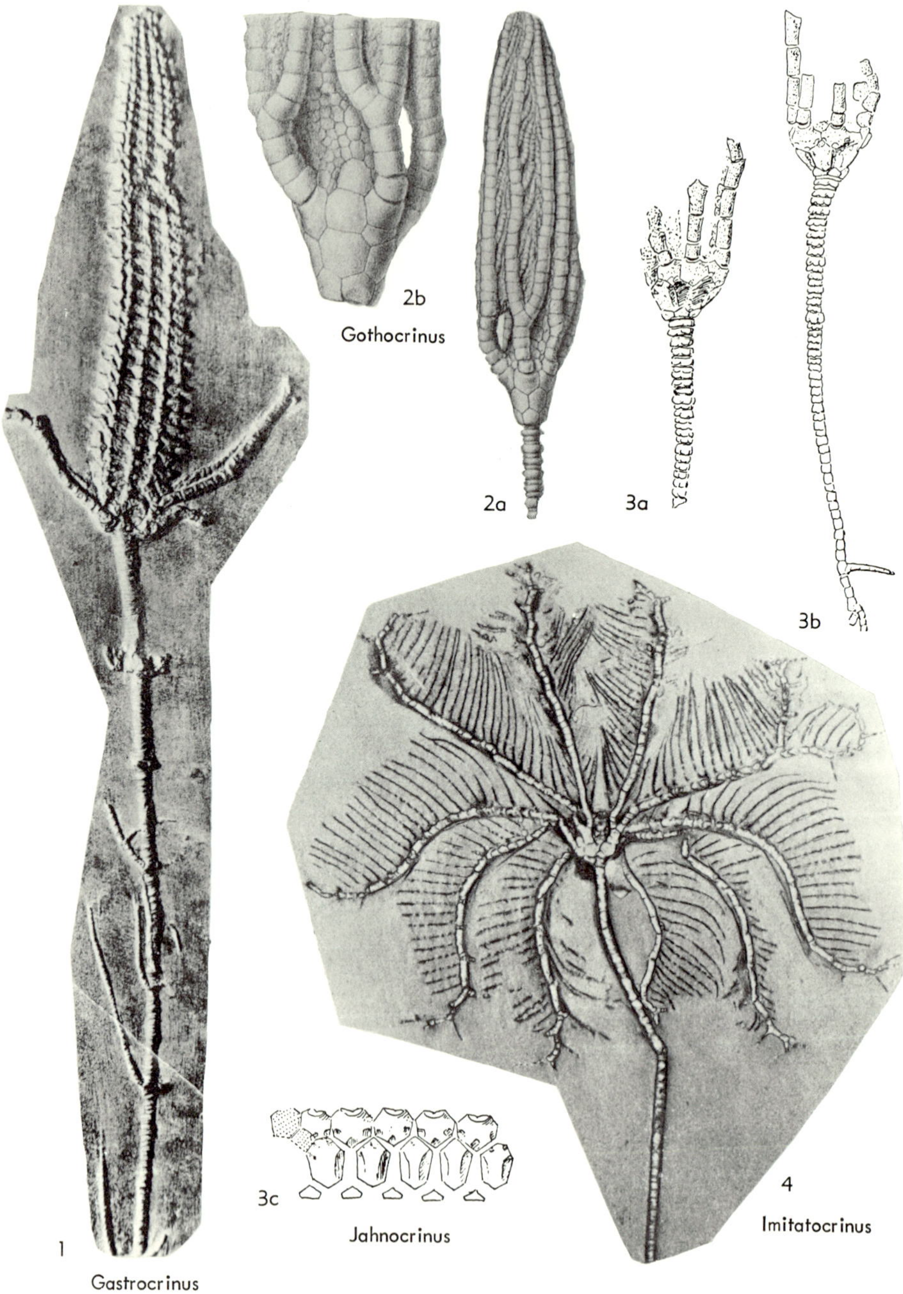

FIG. 399. Botryocrinidae (p. T615).

imal stem, ×1 (Jaekel, 1895).

Schmidtocrinus HAARMANN, 1921, p. 41 [**Rhenocrinus winterfeldi* SCHMIDT, 1906, p. 544; M]. Cup small, conical, arrangement of plates not known. Anal sac extremely slender and tall. Arms long and delicate, branching once just above cup and again in distal region; brachials elongate, arranged in zigzag pattern, with long slender

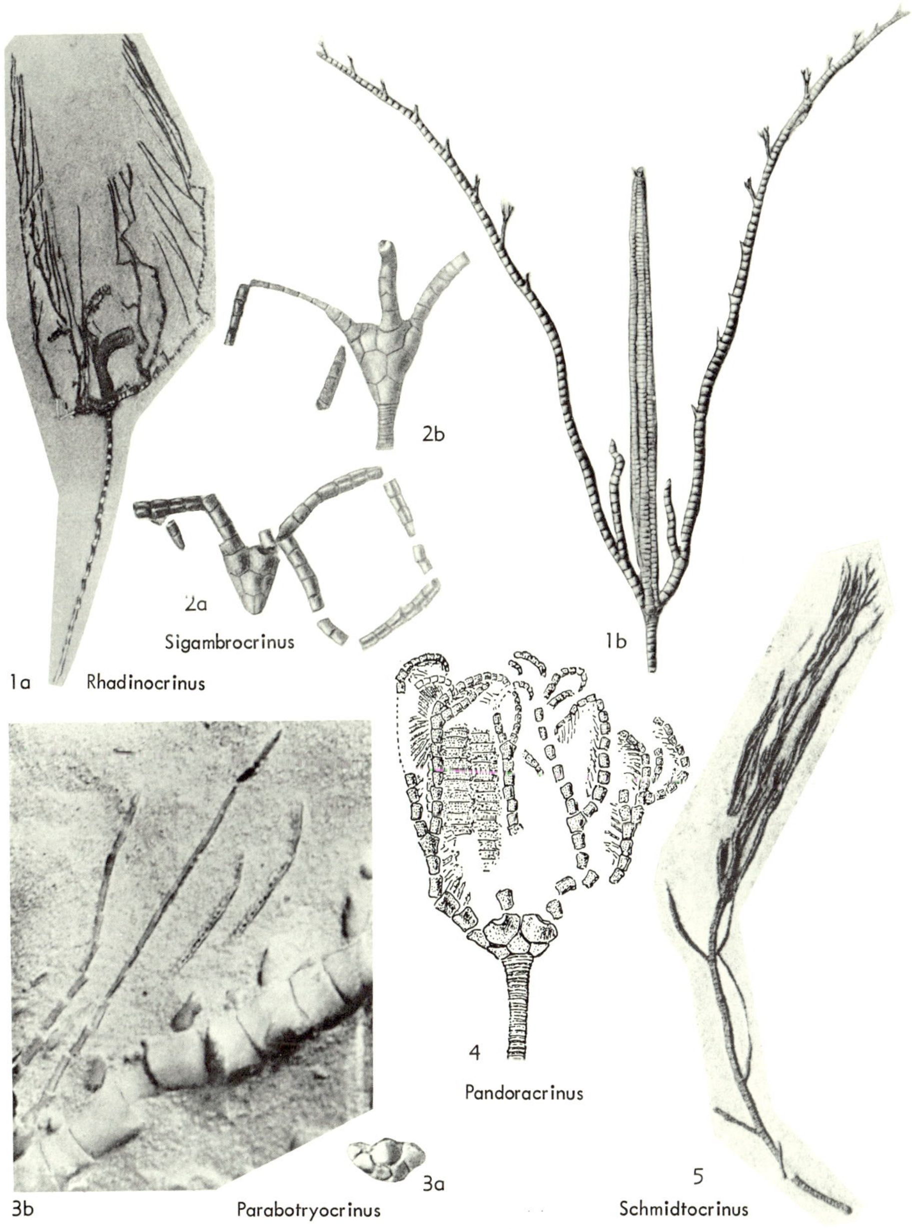

FIG. 400. Botryocrinidae (p. *T*615-*T*618).

ramules on opposite sides of every second brachial. Stem pentagonal in section, with widely spaced cirri directed obliquely up toward crown. *M.Dev.*, Eu.(Ger.).——FIG. 400,*5*. **S. winterfeldi* (SCHMIDT); lat. view of crown and stem, ×2 (Haarmann, 1921).

[Although the type species of this genus has been credited to JAEKEL by SCHMIDT (1906), HAARMANN (1921) pointed out that SCHMIDT did not ascribe authorship of this species to JAEKEL, although he did cite JAEKEL as the author of *Rhenocrinus* and of *R. ramosissimus* on the same page where *R. winterfeldi* is described. HAARMANN recorded SCHMIDT as the author of *R. winterfeldi*, and his view is accepted here.]

Sigambrocrinus SCHMIDT, 1942, p. 178 [**S. laevis*; M]. Cup high, cone shaped, plates smooth; radianal absent from cup, anal *X* small, situated between posterior radials. Anal sac unknown. Arms uniserial, long and relatively stout, unbranched, with ramules; ambulacral grooves with small covering plates. Stem transversely round. *L.Dev.*, Eu.(Ger.).——FIG. 400,*2a,b*. **S. laevis*; ant. and post. views of cup and proximal part of arms, ×1 (Schmidt, 1942).

Superfamily MASTIGOCRINACEA Jaekel, 1918

[*nom. transl.* MOORE & LANE in MOORE & STRIMPLE, 1973, p. 19 (*ex* Mastigocrinidae JAEKEL, 1918, p. 55)] [Materials for this superfamily prepared by R. C. MOORE and N. GARY LANE]

Crown and cup similar to those of Dendrocrinacea except for articular radial facets, which are wide (peneplenary) but less than full width of radials, and mostly have gently curved outer margins, interradial notches present. *L.Ord.-M.Ord.; M.Sil.-L.Miss.*

Family AETHOCRINIDAE Ubaghs, 1969

[Aethocrinidae UBAGHS, 1969, p. 2]

Crown moderately tall and slender. Cup steep sided conical, summit width 0.7 of height; infrabasals five, with distal portion subvertical; basals and radials large hexagonal plates (except heptagonal *CD* basal); one or two primibrachs of different rays included in cup; all plates of cup bearing rounded ridges which cross interplate sutures at right angles; articular facets at base of free arms wide but peneplenary (less than width of highest cup plate in each ray; *C* ray with super- and inferradianals directly below *C* radial; *CD* interray containing six anal plates, lowermost one interpreted by UBAGHS as primanal *X*) (Fig. 394,*8,9*). Anal sac low inverted cone, sharp pointed at summit, formed by vertical rows of very narrow polygonal plates. Arms rounded, very slender branching isotomously on approximately eighth primibrach above cup and again on secundibrachs *13-16*. Stem pentalobate in section, heteromorphic, columnals divided by sutures into pentameres with lobate expansions radial in position aligned with infrabasals. *L.Ord. (Tremadoc.-Arenig.).*

The Aethocrinidae are assigned to the Mastigocrinacea because the comparatively wide proximal facets of fixed primibrachs *2* and succeeding free primibrachs are distinctly narrower than fixed primibrachs *1* and radial plates.

One of the oldest of all crinoids known (Lower Ordovician, Tremadoc or Arenig) is *Aethocrinus,* which has a steep straight-sided conical cup, and it must be close to the root stock of other cladid inadunates and possibly disparids also. Five anal plates and lowermost primibrachs are fixed in the cup (Fig. 394,*8,9*) and articular facets for attachment of the free arms are on fixed brachials rather than summit edges of radials. The free arms are uniserial, transversely round, and branched isotomously first on free primibrachs *8*. The relatively large stem is transversely quinquelobate, with wide axial canal, and columnals divided into pentameres. All of these characters denote primitiveness in stage of evolution.

Aethocrinus UBAGHS, 1969, p. 2 [**A. moorei*; M]. Characters of family. *L.Ord.(Tremadoc.-Arenig.),* Eu.(Montagne Noire, S.France).——FIG. 401; 402, *1a-d*. **A. moorei*; 401, plate diagram of cup with part of arms and stem, enl. (radials black, anals vertically ruled, and interradial plates stippled); 402,*1a,* post. view of crown and stem (reconstr.), ×1; 402,*1b,* view of uncrushed paratype with part of attached stem, ×1.7; 402,*1c,d,* opposite sides of crushed holotype, ×1.7 (all Ubaghs, 1969).

Family THENAROCRINIDAE Jaekel, 1918

[Thenarocrinidae JAEKEL, 1918, p. 55]

Crown tall, expanding in width upward, many-armed. Cup steeply straight sided, conical, truncate at base; infrabasals strongly upflared outside of stem impression; radial articular facets peneplenary, with gently curved outer margins, declivate; large radianal obliquely at left below *C* radial, supporting on its left shoulder large anal *X* plate, third anal barely extending below *C* radial summit (Fig. 394,*3*). Very prominent large anal sac composed of vertical rows of plicate plates, sac pointed at top. Nonpinnulate uniserial arms branching several times, first on primibrachs *1*. Stem wide, formed of thin circular columnals, noncirriferous. *M.Sil.*

Thenarocrinus BATHER, 1890, p. 334 [**T. callipygus*; OD]. Characters of family. *M.Sil.,* Gotl.-Eu.(Eng.).——FIG. 403,*1a-c*. **T. callipygus,* Gotl.; *1a,b,* ant. and post. views of crown, ×0.7 (Bather, 1891a); *1c,* plate diagram of cup (radials black, radianal cross ruled, anal *X* and higher anal plates stippled) (Moore, n).

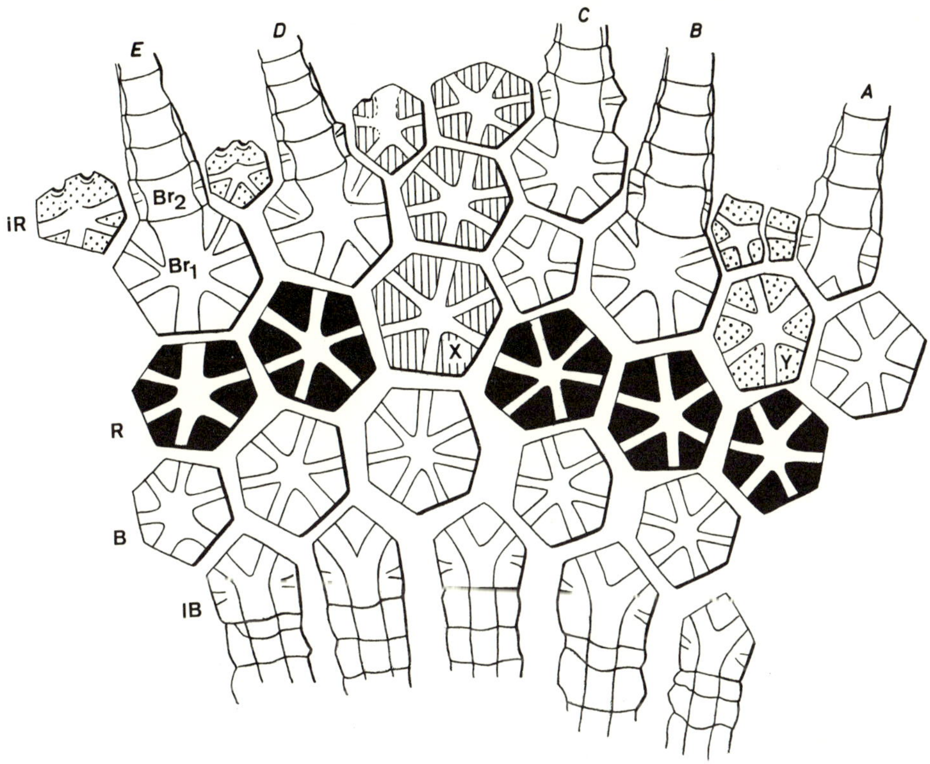

FIG. 401. *Aethocrinus moorei* UBAGHS, 1969, from Lower Ordovician, France; plate diagram of cup with part of arms and stem, enl. (Ubaghs, 1969). [*A-E,* ray designations in CARPENTER system; *B,* basal; *Br₁, Br₂,* primibrachs *1,2; IB,* infrabasal; *iR,* interradial plate; *X,* primanal; *Y,* supplementary interradial; radials black, anal plates vertically ruled, interradial plates stippled.]

Family MASTIGOCRINIDAE Jaekel, 1918

[Mastigocrinidae JAEKEL, 1918, p. 55]

Cup conical with height approximately equal to width; infrabasals upflared, clearly visible in side view of cup; radial articular facets wide but somewhat narrower than radial summits, sloping outward-downward (declivate); anal plates in cup, one or mostly two. Anal sac mostly tall; cylindrical. Arms uniserial, slender, branching isotomously on primaxils and isotomously or heterotomously higher in crown. Stem circular or pentagonal in section. *M.Ord.; M.Sil.-L.Miss.*

The Mastigocrinidae are mostly diminutive crinoids with few upflared, uniserial, slender arms and conical cups. Probably they constitute an offshoot from the poteriocrinitids or botryocrinids.

Key to Genera of Mastigocrinidae

A. Cup very steep sided, narrow conical ....
- I. Infrabasals and basals tall, narrow; anal sac tall, slender, composed of small polygonal or plicate plates, arms 10 ....
  - a. Two anal plates in cup, radianal like basals, touching infrabasals; plates fluted, with pits at angles; arms branching isotomously on primibrachs *4* or *5*; anal sac tall, with plicate plates, stem transversely pentagonal .... *Atelestocrinus*
  - b. Three anal plates in cup; arms 10, branching isotomously on primibrachs *4* or *5*; stem transversely pentagonal .... *Quantoxocrinus*
  - c. Four anal plates in cup; arms branching isotomously on primibrachs *4* and secundibrachs *4*; stem transversely round .... *Antihomocrinus*
  - d. Three anal plates in cup; arms 10; extra plates between radials and

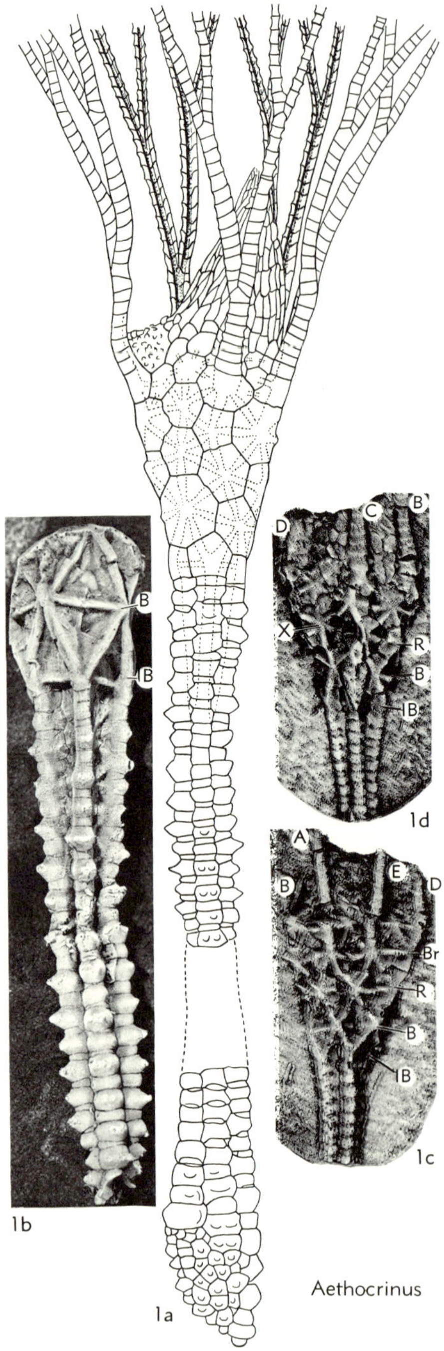

FIG. 402. Aethocrinidae (p. T618).

basals .... *Belanskicrinus*

II. Infrabasals low; 2 anal plates in cup (small quadrangular radianal, large anal *X*); stem transversely pentagonal ..

a. Anal sac tall, with plicate plates; arms 10, branching isotomously on primibrachs *3-5* .... *Dictenocrinus*

b. Anal sac tall, composed of small polygonal plates; arms 60 or more, branching first isotomously on primibrachs *4* and heterotomously higher in crown .... *Lasiocrinus*

c. Basals tall and slender, radials small, pits at plate angles; arms many, branching first isotomously on primibrachs *3* or *4* and heterotomously higher in crown .... *Cradeocrinus*

III. Infrabasals low; only anal *X* plate in cup; stem transversely pentagonal .... *Kalpidocrinus*

B. Cup conical, with height subequal to width ....

I. Cup with fluted plates, pits at their angles; 2 anal plates in cup (small quadrangular radianal, large anal *X*) ..

a. Arms approximately 90, branching first isotomously on primibrachs *3* and heterotomously higher in crown; anal sac tall and slender .... *Goniocrinus*

b. Arms at least 10, branching isotomously on primibrachs *6-8*; anal sac short, distally pointed, composed of polygonal plates .... *Nassoviocrinus*

II. Cup smooth plated; anals in cup 1 or 2 ....

a. Arms numerous, branching isotomously on primibrachs *4* and heterotomously higher; anal plates in cup 2, anal sac with laterally directed summit spines .... *Bathericrinus*

b. Arms numerous, branching isotomously first on primibrachs *10-12* and 5 or 6 times higher in crown; anal plate in cup 1 (radianal absent, anal *X* large), tall anal sac composed of small polygonal plates with posterior "backbone" of large convex plates .... *Mastigocrinus*

c. Arms branching first isotomously on primibrachs *2* or *3* and heterotomously higher in crown; anal plates in cup 2 .... *Iteacrinus*

d. Arms 20-40, branching isotomously first on primibrachs *4* or *5* and higher in crown; anal sac low spheroidal, composed of stellate polygonal plates .... *Follicrinus*

e. Main arms unbranched, bearing very numerous short slender ramules on opposite sides; anal sac tall, slender; stem tapering distally to prehensile point .... *Eifelocrinus*

f. Arms 40 or more, first branching

isotomously on primibrachs *4* or *5*; single anal plate in cup; anal sac composed of small polygonal plates .... *Polycrinus*

C. Cup medium sized; anal plates in cup 1-4, anal sac slender, cylindrical; arms 5-40 or more. Radianal lacking, anal *X* large; slender anal sac spirally coiled; arms 10 or more, branching on primibrachs *1-4* .... *Streptocrinus*

Mastigocrinus Bather, 1892, p. 194 [**M. loreus*; OD]. Cup conical; infrabasals large, upflaring; radial articular facets peneplenary, slightly narrower than radials; radianal absent, anal *X* large, situated between posterior radials; anal sac large, elongate, composed of wide, low plates, each with transverse ridge that bifurcates at lateral edges of sac plates, in confluence with bifurcate ridges of sac plates of adjacent rows. Arms uniserial, slender, long, bifurcating on primibrachs *10* to *12*, with 5 or 6 higher, isotomous divisions; ramules not developed. Stem round in section, columnals composed of pentameres. *M.Sil.*, Eu.(Eng.).——Fig. 404,*3*. **M. loreus*, Wenlock.; *3a*, post. view of crown and proximal part of stem, ×0.7; *3b,c*, side view of columnal and proximal part of arms and anal sac, ×5; *3d*, post. view of theca and proximal part of arms and stem, ×1.5; *3e*, anal sac plates, ×10; *3f*, diagram of stem divided into pentameres, ×1.5 (Bather, 1892b).

Antihomocrinus Schmidt, 1934, p. 79 [**Homocrinus tenuis* Bather, 1893a, p. 105; OD]. Cup steeply conical and elongate; small, oblique radianal and larger anal *X* in cup; radial articular facets peneplenary, less than equal to full width of plates, leaving rounded interradial notches. Anal sac large and elongate. Arms slender, rounded, uniserial, branching isotomously on primibrachs *4* and secundibrachs *4*. *M.Sil.-L.Dev.*, Eu.(Ger.-Sweden).——Fig. 404,*1*. **A. tenuis* (Bather), M.Sil., Gotl.; *1a,b*, post. view of crown and part of attached stem, ant. view of theca, ×1 (Bather, 1893a).

Atelestocrinus Wachsmuth & Springer, 1886, p. 221 [**A. delicatus*; SD S. A. Miller, 1889, p. 226] [=*Attelesocrinus* Wachsmuth & Springer, 1886, p. 223 *(nom. null.)*; *Atelestrocrinus* Bassler & Moodey, 1943, p. 314 *(nom. null.)*]. Cup inverted bell shaped; infrabasals and basals large, elongate; radials small, radial articular facets peneplenary, rounded; *A* radial narrow, armless; radianal large, elongate, and similar in size and shape to basals; anal *X* elongate, constricted medially between posterior radials. Anal sac not known. Arms uniserial, branching on primibrachs *2* to *5*, ramules on opposite sides of every second secundibrach. *L.Miss.*, USA(Iowa).——Fig. 404,*2*. *A. robustus*; post. view of partial crown and stem, ×1.5 (Wachsmuth & Springer, 1886).

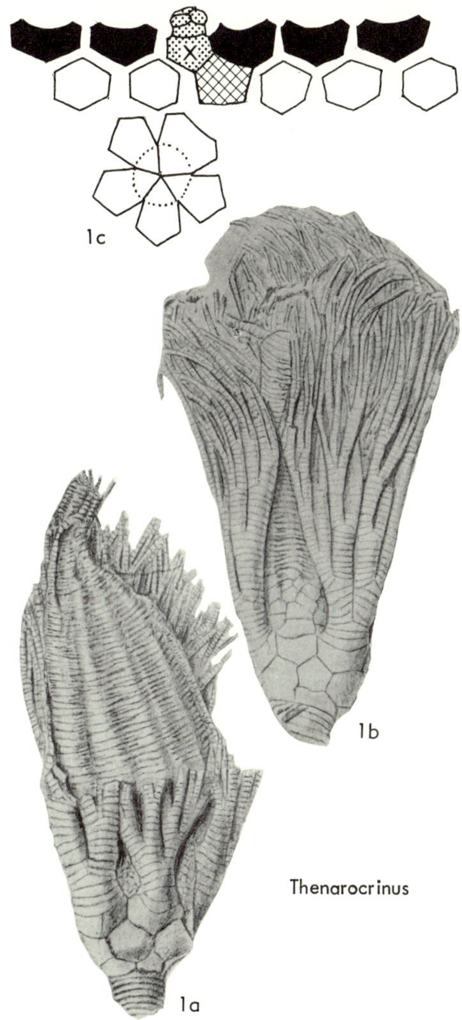

Fig. 403. Thenarocrinidae (p. *T*618).

Bathericrinus Jaekel, 1918, p. 58 [**Botryocrinus ramosus* Bather, 1891a, p. 394; M]. Cup small, low, plates smooth or basals with median longitudinal fold; radial articular facets almost as wide as radials, interradial notches present; radianal small, quadrangular, anal *X* large and projecting above adjacent radials. Anal sac large, elongate, plates strongly plicate, distal extremity with ring of large spinose plates. Arms uniserial, branching isotomously on primibrachs *4*, each arm with 2 main rami; large ramules on opposite sides of every second secundibrach. *M.Sil.-L.Dev.*, Eu.(Eng.-Ger.).——Fig. 404,*4a,b*. **B. ramosus* (Bather), M.Sil., Eng.; *4a,b*, post. and *B*-ray views of crown, ×0.7 (Bather, 1891a).——Fig. 404,*4c*. *B. semipinnulatus* Schmidt, L.Dev., Ger.;

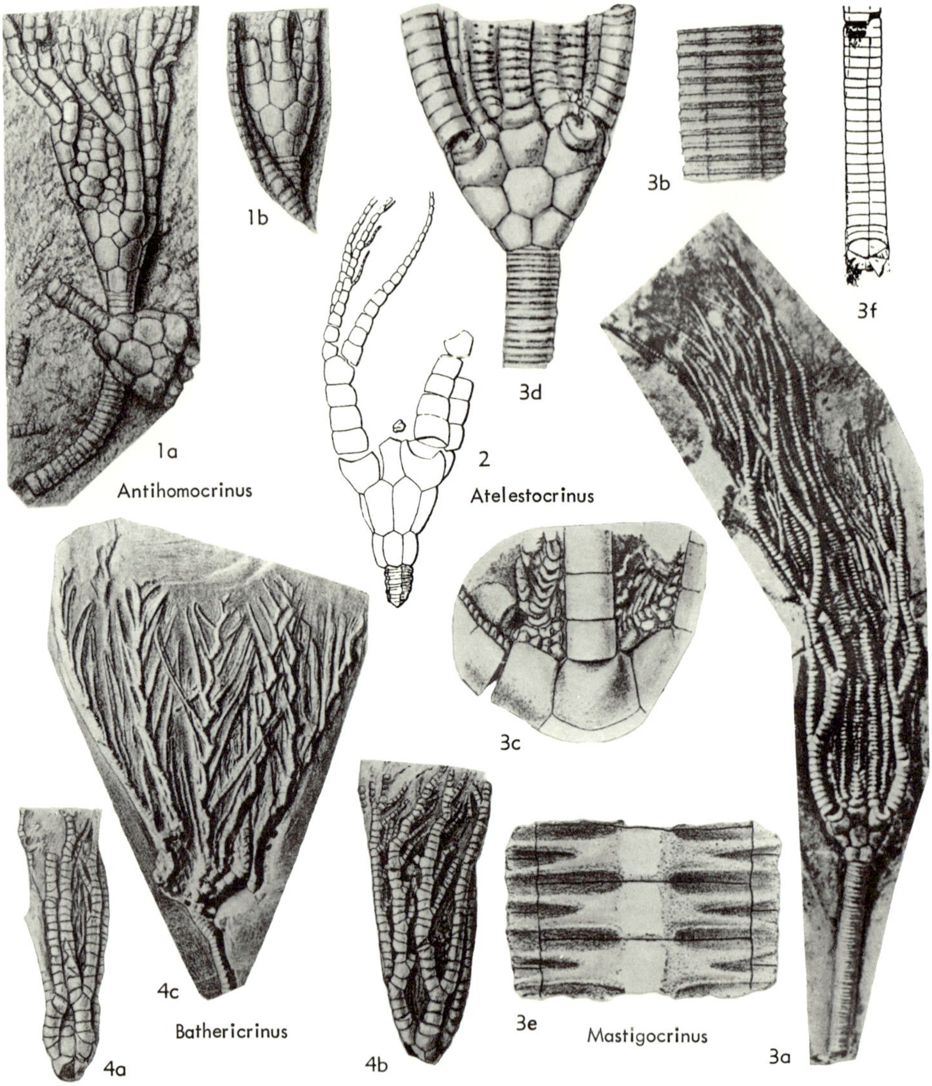

FIG. 404. Mastigocrinidae (p. T621-T622).

lat. view of crown, ×1 (Schmidt, 1934).

Belanskicrinus STRIMPLE & LEVORSON, in STRIMPLE *et al.*, 1969, p. 18 [**Bactrocrinus westoni* BELANSKI, 1928, p. 177; OD]. Cup straight sided, steeply conical, circlet of partially fused upflared infrabasals tending to break away from basals, extra plates common between basals and radials; 3 anals in cup, anal sac exceptionally tall and slender. Arms 10, uniserial, branching isotomously on primibrachs *4* or *5*. Stem transversely round, tapering distally from cup. *M.Dev.*, USA(Iowa).——FIG. 405,*1*. **B. westoni* (BELANSKI), Shellrock F. (Mason City Member), near Rockport, Floyd Co., Iowa; *1a*, *AB*-interray view of crown, ×1.7; *1b*, *D*-ray view of hypotype crown, ×1.8; *1c*, lat. view of crown showing anal sac, ×1.8 (Strimple & Levorson, 1969).

Cradeocrinus GOLDRING, 1923, p. 347 [**C. elongatus*; OD]. Crown tall and slender. Cup narrowly conical; infrabasals small, upright; basals large and elongate; radials small; deep depressions at junctures of radial-basal sutures; radial articular facets almost as wide as radials, each supporting 3 or 4 large, rounded primibrachs; each ray with 2 main rami above primaxil, with long slender ramules given off alternately on opposite sides of every third or fourth secundibrach. *U.Dev.-L.Miss.*, N.Am.(N.Y.-Mont.-Alta.).——FIG. 406,*2*. **C.*

elongatus, U.Dev., N.Y.; *2a,b,* lat. views of crown and part of attached stem, ×1.5 (Goldring, 1923).

Dictenocrinus JAEKEL, 1918, p. 59 [**Botryocrinus decadactylus* BATHER, 1891b, p. 395 (=*Cyathocrinus decadactylus* SALTER, 1873); SD BASSLER, 1938, p. 87]. Cup high, steeply conical; plates smooth or with inconspicuous longitudinal folds on basals; radial articular facets slightly narrower than radials. Anal sac elongate, narrow, composed of several rows of wide, low, strongly plicate plates, recurved at distal tip, with anal opening just below sac summit on anterior side, large spinose plates may cap summit. Arms uniserial, branching on primibrachs *3* to *5,* with 2 main rami or with one or two higher branches; brachials bearing ramules; stout covering plates on arms and ramules. Stem subpentagonal in section, columnals simple or composed of pentameres. *U.Sil.-L.Dev.,* Eu.(Ger.-Eng.).——FIG. 406,*4*. *D. cyathiformis* (HAARMANN), L.Dev., Ger.; post. view of crown and stem, ×1 (Haarmann, 1921).

Eifelocrinus WANNER, 1916, p. 200 [**Ptilocrinus dohmi* WANNER, 1916, p. 344a; OD] [=*Ptilocrinus* WANNER, 1916a, p. 343 (obj.), *non* CLARK, 1907]. Cup steeply conical, plates convex; radial articular facets occupying nearly full width of radials. Anal sac slender, about one-third as long as arms, confined to posterior interray of tegmen, composed of several rows of hexagonal plicate plates, distal extremity may bifurcate; tegmen composed of large interradial plates, small median orals, and interlocking ambulacral plates in each radial area. Arms uniserial, long and slender, one main ramus to each ray; heterotomous armlets alternate sides of every second brachial; armlets composed of long, stout main branch and side ramules given off distally from opposite sides of every second plate. Stem subpentagonal in section proximally, becoming round distally and tapering to a prehensile point. *L.Dev.,* Eu.(Ger.).——FIG. 406,*5*. **E. dohmi* (WANNER); post. and ant. views of 2 reconstructions of crown and stem, ×0.7 (Wanner, 1916a).

Follicrinus SCHMIDT, 1934, p. 74 [**Taxocrinus arebei* FOLLMAN, 1887, p. 118; OD]. Cup small, conical, as wide as high; radial articular facets peneplenary; radianal and anal *X* in cup. Anal sac greatly inflated, balloon shaped, composed of numerous, small stellate plates. Arms slender, uniserial, spreading, with 3 or 4 isotomous branches. *L.Dev.,* Eu.(Ger.).——FIG. 406,*6*. **F. arebei* (FOLLMAN); *6a,* lat. view of crown and attached part of stem, ×0.5; *6b,* sketch of anal sac and disarticulated arm fragments, ×1 (Schmidt, 1934).

Goniocrinus MILLER & GURLEY, 1890, p. 32 [**G. sculptilis*; OD] [=*Goniacrinus* SPRINGER, 1913, p. 218 *(nom. null.)*]. Cup small, plates convex with depressions at corners; infrabasals low; small oblique radianal and anal *X* in cup, supporting elongate anal sac which has posterior longitudinal row of larger anal plates that resemble brachials, radial articular facets curved, nearly as wide as radials, small interradial notches distinct; primibrach *3* axillary, supporting 2 main rami in each ray, with long, stout ramules alternating on opposite sides of every third or fourth secundibrach proximally and every second secundibrach distally. Stem transversely pentagonal, with proximally directed cirri on nodals just below cup. *L.Miss.,* N.Am.(Ind.-Iowa-Can.).——FIG. 406,*1*. **G. sculptilis,* Iowa; *1a,b,* lat. views of crown, ×2 (Miller & Gurley, 1890).

Iteacrinus GOLDRING, 1923, p. 344 [**I. flagellum*; OD]. Cup small, low, with highly convex plates; radianal large, anal *X* partly out of cup; radial articular facets almost as wide as radials, supporting 2 or 3 primibrachs in each ray. Anal sac slender,

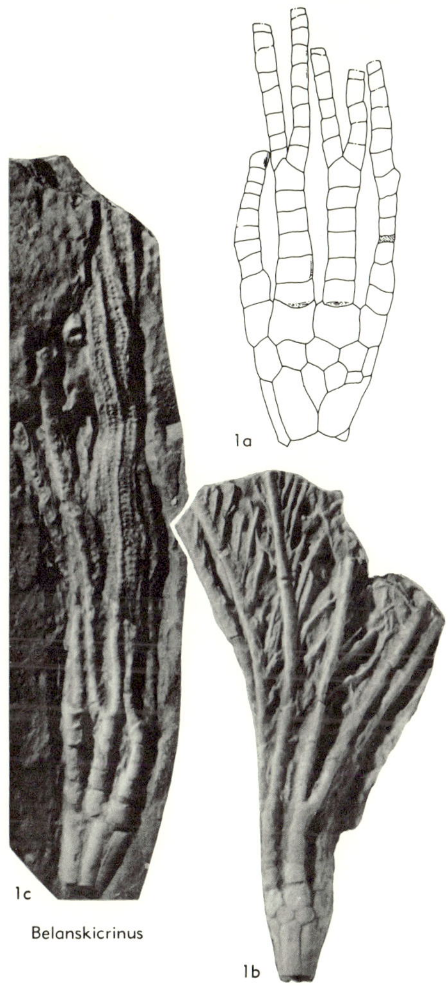

FIG. 405. Mastigocrinidae (p. *T*623).

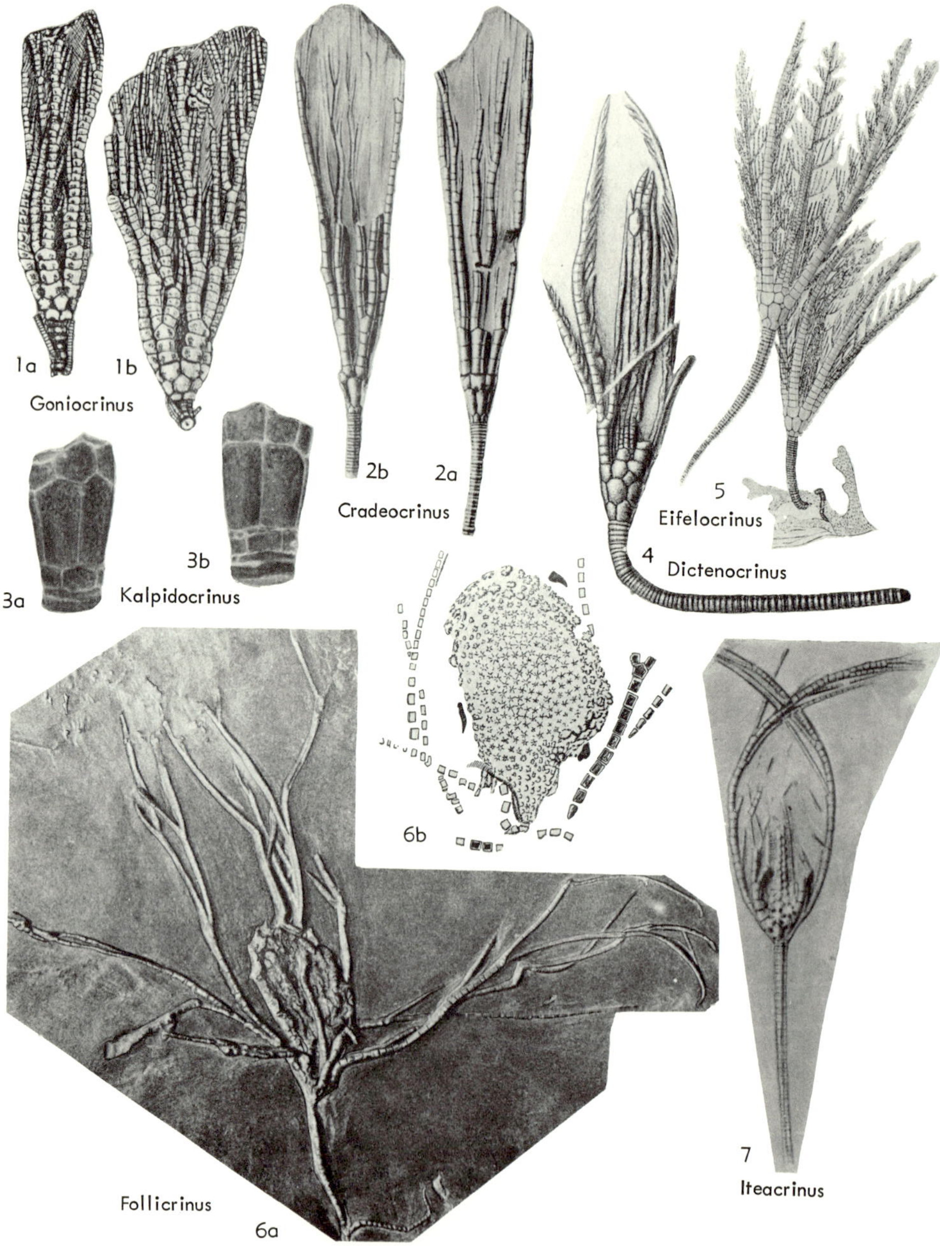

FIG. 406. Mastigocrinidae (p. *T622-T625*).

elongate, composed of several longitudinal rows of delicate plates. Arms uniserial, slender, rounded, with two main isotomous rami in each ray heterotomously supporting long, slender ramules on opposite sides of every second or third brachial. Stem transversely pentagonal. *U.Dev.*, USA(N.Y.).——FIG. 406,7. **I. flagellum*; post. view of crown and part of stem, ×1 (Goldring, 1923).

Kalpidocrinus GOLDRING, 1954, p. 17 [**K. eriensis*; OD]. Cup cylindrical, tall; infrabasals low, but steeply upflared; basals very large and elongate, radials small, one anal plate (anal *X* or radianal) in cup directly above *CD* basal; arm facets wide, peneplenary. Stem transversely subpentagonal. *M.Dev.*, USA(N.Y.).——FIG. 406,*3*. **K. eriensis*; *3a,b*, *A*-ray and post. views of cup and proximal columnals, ×4 (Goldring, 1954).

Lasiocrinus KIRK, 1914, p. 482 [**Homocrinus scoparius* HALL, 1859b, p. 102; OD]. Cup conical, wide at base; infrabasals elongate, steeply upflared; radials longer than wide, with straight linear facets occupying nearly full width of plates, deep interradial notches present; radianal small, quadrangular; anal *X* large, in line with posterior radials. Anal sac long, composed of several longitudinal rows of small hexagonal plates. Arms uniserial, branching isotomously on fourth primibrachs and heterotomously above, with long, stout ramules alternately on opposite sides of every fourth to ninth brachial. *L.Dev.-L.Miss.*, Eu.(Ger.)-USA (Iowa-N.Y.).——FIG. 407,*3*. **L. scoparius* (HALL), L.Dev., N.Y.; *3a,b*, ant. and post. views of crown with short attached part of stem, ×2 (Kirk, 1914).

Nassoviocrinus JAEKEL, 1918, p. 54 [**?Heterocrinus pachydactylus* SANDBERGER & SANDBERGER, 1855, p. 402; M]. Cup high, conical, plates with elongate, longitudinal ridges; radianal small, quadrangular; anal *X* large; radial articular facets slightly narrower than radials. Anal sac short, pyramidal, composed of few large plates proximally, sac plates diminishing in size distally. Arms elongate, slender, brachials quadrangular; primibrachs *6* to *8* axillary; distal parts of arms or ramules not known. Stem transversely pentagonal; columnals composed of pentameres. *L.Dev.-U.Dev.*, Eu.(Ger.)-USA(Iowa).——FIG. 407,*1*. **N. pachydactylus* (SANDBERGER & SANDBERGER), L.Dev., Ger.; post. view of partial crown and stem, ×0.7 (Jaekel, 1918).

Polycrinus JAEKEL, 1918, p. 55 [**P. ramulatus*; OD]. Cup low, broad, truncate below; infrabasals barely visible in side view; radials large, radial articular facets nearly as wide as radials; radianal absent from cup; anal *X* between posterior radials. Anal sac large, with posterior longitudinal row of large, quadrangular plates, anterior side composed of small plates. Arms uniserial, robust, branching isotomously on primibrachs *4* to *6*, with 2 main rami and widely spaced, long, slender ramules on alternate sides of rami. Stem round, wide proximally, tapering distally. *M.Ord.*, Eu. (Boh.).——FIG. 407,*2a*. **P. ramulatus*; *D*-ray view of partial crown and stem, ×0.7 (Jaekel, 1918).——FIG. 407,*2b*. *P. kosoviensis* JAEKEL; *DE*-interray view of crown and stem, ×0.7 (Jaekel, 1918).

Quantoxocrinus WEBBY, 1965, p. 12 [**O. ussheri*; OD]. Crown small, slender. Cup conical, infrabasals large and steeply upflared; radial articular facets curved, nearly equal to radials in width, small interradial notches; 3 anal plates in cup. Anal sac large, erect or recurved, composed of

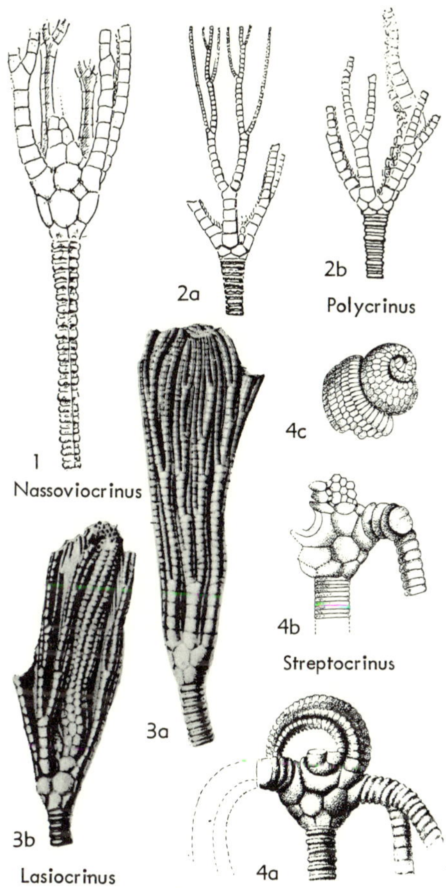

FIG. 407. Mastigocrinidae (p. *T*625-*T*626).

several vertical rows of hexagonal, plicate plates. Arms uniserial, isotomous with 2 main rami, bearing ramules, primibrachs *4* or *5* axillary. Stem transversely pentagonal. *M.Dev.(Givet.)*, Eu. (Eng.); *U.Dev.*, USA(Iowa).——FIG. 408,*1*. **O. ussheri*; *1a-c*, post., *C*- and *B*-ray views of crown, ×2.8 (Webby, 1965).

Streptocrinus WACHSMUTH & SPRINGER, 1886, p. 224 (148) [**Ophiocrinus crotalurus* ANGELIN, 1878, p. 24; M] [=*Ophiocrinus* ANGELIN, 1878, p. 24 (obj.) (*non* SALTER, 1856, *nec* SEMPER, 1868)]. Cup low bowl shaped; infrabasals small; radianal absent; radial articular facets large, concave, directed abmedially and occupying most of outer surface of radials. Anal sac recurved and spirally coiled, tapering distally, composed of regular rows of alternating hexagonal plates. Arms recumbent and directed downward proximally; primibrachs wide and thin, primibrachs *1* to *4* axillary, 2 main rami to an arm, apparently with ramules; ambulacral groove with cover-

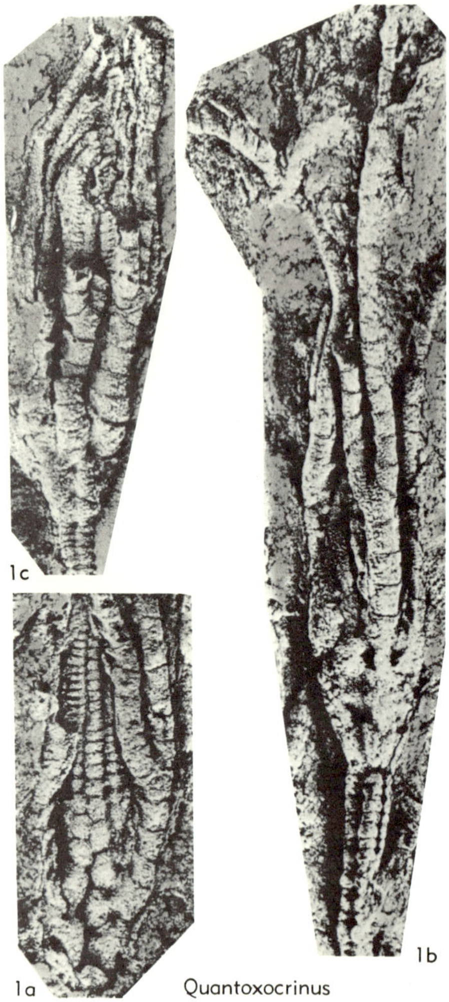

FIG. 408. Mastigocrinidae (p. *T*625).

ing plates. Stem transversely round. *U.Sil.,* Eu. (Sweden).——FIG. 407,*4*. **S. crotalurus* (ANGELIN), Gotl.; *4a-c, D*-ray view of theca showing part of arms, stem, and coiled anal sac, post. view of partial theca and arm, spiral anal sac, all ×1 (Angelin, 1878).

Superfamily MEROCRINACEA S. A. Miller, 1890

[*nom. transl.* MOORE & LANE in MOORE & STRIMPLE, 1973, p. 19 (*ex* Merocrinidae S. A. MILLER, 1890, p. 352)] [Materials for this superfamily prepared by R. C. MOORE and N. GARY LANE]

Crown tall, slender cylindrical or moderately expanded upward. Cup straight-sided narrowly to broadly conical; five upflared infrabasals clearly visible from side; basals and radials large, arm facets plenary, occupying full width of radials, anals mostly three in cup. Anal sac prominent. Arms well rounded externally, branched isotomously at or near base and isotomously or heterotomously higher in crown, mostly bearing ramules. Stem circular in section or rarely pentagonal. *?L.Ord., M.Ord.-U. Ord.; U.Sil.-L.Dev.*

Family MEROCRINIDAE S. A. Miller, 1890

[Merocrinidae S. A. MILLER, 1890, p. 352] [=Merocrinidae AUSTIN & AUSTIN, 1843, p. 205, *pro* Merocrinoidea AUSTIN & AUSTIN, 1842, p. 110 (invalid, no type-genus named)]

Radial articular facets wide. Infer- and axillary superradial present in *C* ray, supporting anal *X* and longitudinal series of anal plates on left side and branched arm on right. *M.Ord.-U.Ord.*

The Merocrinidae, composed so far as now known of the single genus *Merocrinus,* differ from most other dendrocrinines in having an inferradianal and superradianal (Fig. 394,*4*). This character indicates a primitive evolutionary stage. The cup is perfectly symmetrical, inasmuch as the inferradianal corresponds exactly in size and shape to the four adjacent radial plates. The superradianal and *X* plates are not incorporated in the cup, but there is no indication that they have been displaced upward from an original position within the cup. As in *Iocrinus,* the radials and inferradianal of *Merocrinus* may be interpreted to represent the most archaic of all structural plans among inadunate crinoids, in which there is no differentiation of compound radials (consisting of unfused primitive radials, otherwise known as inferradials, and primitive first brachials, otherwise called superradials) and "simple" radials (consisting of fused lower elements of the rays). If this diagnosis is correct, it is pertinent to call attention to the cyclic course of normal evolutionary changes in the cup of inadunate crinoids, that is, from initial or near-initial perfect pentamerous symmetry to bilateral symmetry, and finally back to perfect pentamerous symmetry.

The radial articular facets of the Merocrinidae are wide (plenary), which suggests origin of the group in an unknown stock that surely stands far apart from the early Cyathocrinina. The characters of the cup, arms, and strikingly armlike anal tube most closely resemble those of some families of Disparida, but this similarity is not known to have significance as to genesis of *Merocrinus*. The repeated isotomous branching of the arms indicates an early evolutionary stage, but not the most primitive one, in development of this part of the organism. No descendants of the Merocrinidae are known.

Merocrinus WALCOTT, 1884, p. 208 [**M. typus*; OD]. Characters of family. Cup small, short, extending laterally only short distance beyond uppermost columnal; infrabasals wide, low, visible in side view; basals short. Arms long, bifurcating isotomously several times, 6 or more primibrachs to a ray. Cup plate in *C* ray, which is identical to normal radials of other rays, is homologous to inferradianal and next higher plate with form and position of axillary; first primibrach bears a normal arm on its right shoulder and strong longitudinal series of anal tube plates on its left shoulder, is equivalent to a superradianal. This latter plate has been termed anibrachial. The presence of two radianals in the *C* ray has been long recognized as an archaic structure correlated with a primitive evolutionary character. Stem transversely round, stout, composed of quite thin columnals. *M.Ord.-U.Ord.*, N.Am.(N.Y.-Ky.-Ohio)-Eu.(Eng.).——FIG. 409,*3a*. **M. typus*, M. Ord., N.Y.; *CD*-interray view of crown and part of stem, ×1 (Moore & Laudon, 1943a).——FIG. 409,*3b*. Plate diagram of cup (radials black, *C* inferradial horizontally ruled; anal *X* and sac plates stippled; *PS*, plane of bilateral symmetry) (Moore, 1962b).

Family CUPULOCRINIDAE Moore & Laudon, 1943

[Cupulocrinidae MOORE & LAUDON, 1943a, p. 52]

Infra- and superradial present in *C* ray only; anal *X* in line with radials, supporting longitudinal row of large anal sac plates; radial articular facets fully as wide as radials. *M.Ord.-U.Ord.*

The Cupulocrinidae constitute an independent line of Ordovician dicyclic inadunates that seem to be related closely to the Flexibilia. They do not give rise to known descendants. The posterior side of the cup is primitive in having the radianal directly below the right posterior *(C)* radial; there are two other anals in the cup. The cupulocrinids differ from most other Ordovician cladids in having radial articular facets that are fully equal to the greatest width of the radial plates. The anal sac differs also from that of associated inadunates.

Cupulocrinus D'ORBIGNY, 1850, p. 23 [**Scyphocrinus heterocostalis* HALL, 1847, p. 85; M] [=*Scyphocrinus* HALL, 1847 (*non* ROEMER, 1849) (obj.)]. Cup conical; infrabasals 5; radianal directly in line below *C* radial. Anal sac composed of small, loosely united plates except for longitudinal row of large plates on posterior side. Brachials wide, low; arms branch several times isotomously, third of fourth primibrachs axillary, no pinnules. *M.Ord.-U.Ord.*, N.Am.(Ont.-N.Y.-Ky.-Ohio)-Eng.——FIG. 409,*2a,b*. *C. jewetti* (BILLINGS, 1857), M.Ord., Ont.; *2a,b*, post. views of crown, ×1 (Springer, 1911b).——FIG. 409,*2c*. *C. kentuckiensis* SPRINGER, M.Ord., Ky.; post. view of crown, ×1 (Springer, 1911b).——FIG. 409,*2d*. Plate diagram of *Cupulocrinus* (radials black, radianal cross ruled, anal *X* and sac plates stippled) (Moore, n).

Family ONTARIOCRINIDAE Jaekel, 1918

[Ontariocrinidae JAEKEL, 1918, p. 83] [Materials for this family prepared by N. GARY LANE]

Poorly known and may be unrecognizable or incorrectly assigned. Diagnosis same as for *Ontariocrinus*. *M.Ord.*

Ontariocrinus JAEKEL, 1918, p. 83 [**O. deviatus*; OD]. Cup wide, bowl shaped, plates thin; infrabasal circlet low, wide, with 4 plates; basals apparently 4; radials 5, *C* radial in contact with infrabasals; radianal directly above infrabasals, anal *X* above radianal and between posterior radials; radial articular facets narrow, rounded; proximal brachials uniserial, rounded; cup plates with prominent radiating ridges. Stem wide, columnals thin. [Known only from JAEKEL's (1918) brief description and text-figures. Similar crinoids have not been described from the well-known Kirkfield locality from which JAEKEL's specimen is reputed to have been collected. Inquiry at the University of Greifswald indicates that the holotype is not present in their collections.] *M.Ord.*, Can.(Ont.).

Family OTTAWACRINIDAE Moore & Laudon, 1943

[Ottawacrinidae MOORE & LAUDON, 1943a, p. 55] [Materials for this family prepared by N. GARY LANE]

Cup cylindroconical, radianal directly below *C* radial, anal *X* above truncate *CD* basal and at left above radianal, radials followed by fixed primibrachs in each ray forming part of cup (two above *C* radial, three above others). Stem quinquepartite. *?L.Ord., M.Ord., ?U.Ord.*

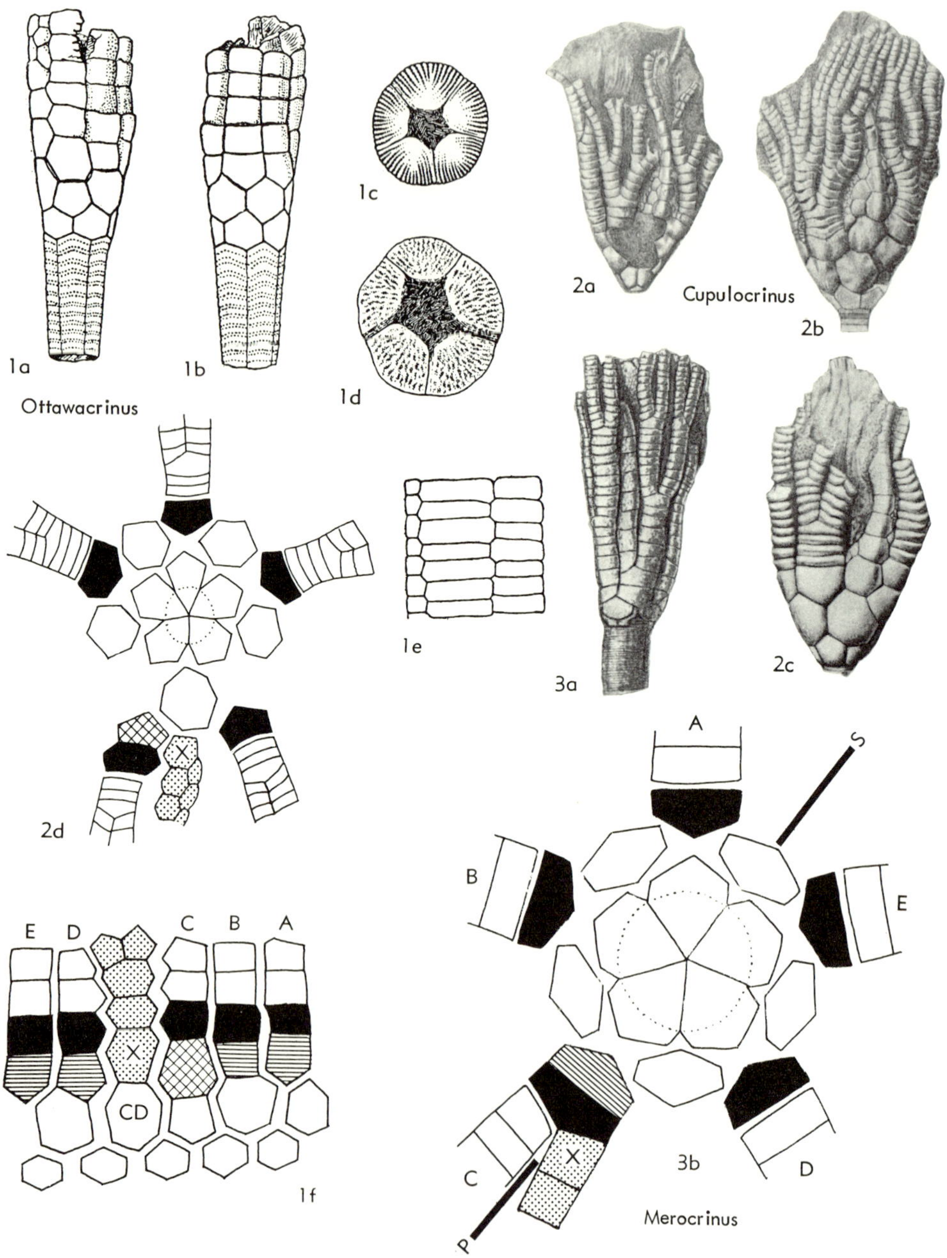

FIG. 409. Merocrinidae *(3)*; Cupulocrinidae *(2)*; Ottawacrinidae *(1)* (p. *T*627-*T*629).

Primitive features of *Ottawacrinus* are its nearly vertical-sided narrow cup, the presence of a radianal directly below the *C* radial, and presence of fixed brachials in the cup. The anal *X* and higher anal plates are directly above the distally truncate *CD* basal.

Ottawacrinus W. R. BILLINGS, 1887, p. 49 [**O. typus*; M]. Characters of family, articular facets on superradials wide, straight, and sutures beneath first primibrachs slightly gaping; large anal *X* directly above truncate *CD* basal supporting longitudinal row of hexagonal anal plates, most proximal 2 of which are within cup. Arms with several primibrachs, distal parts of arms and ramules unknown. Stem subcircular in transverse section, each columnal composed of pentameres; lumen large, pentagonal. *M.Ord.,* Can.(Ont.); *?Ord.,* S.Am.(Arg.).——Fig. 394,*2*; 409,*1*. **O. typus,* M.Ord., Ont.; 394,*2,* post. side of cup showing anals (inferradianal oblique ruled, superradianal cross ruled, anal *X* and higher anals stippled) (Moore, n); 409,*1a,b,* post., ant. views of cup, ×3 (Bather, 1913b); 409,*1c-e,* articular faces of proximal and distal columnals, and side view of columnals, ×5 (Bather, 1913b); 409,*1f,* plate diagram (radials black, inferradials horizontally ruled, radianal cross ruled, anal *X* and other anals stippled) (Moore, n).

Family METABOLOCRINIDAE Jaekel, 1918

[Metabolocrinidae Jaekel, 1918, p. 83] [Materials for this family prepared by N. Gary Lane]

Infra- and superradial present in *C* ray; *B*- and *D*-ray radials simple, other radials not known. Arms with two main rami, ramules present on admedial side of every second secundibrach. *M.Ord.; U.Sil.-L.Dev.*

Metabolocrinus Jaekel, 1902, p. 1100 [**M. rossicus*; M]. Crown oval, compact. Cup low, broad, bowl shaped, base wide and flat; infrabasals low, barely visible in side view, mostly covered by wide stem; basals large; radials pentagonal, inferradial in *C* ray equivalent to primitive radianal, superradial axillary, simple radials in *B* and *D* rays, other radials not known; large, quadrangular anal plate directly above *CD* basal and between posterior radials, supporting longitudinal row of large, wide, low anal sac plates. Arms stout, each with 2 main rami, primibrachs *1* or *2* axillary; long, stout ramules on admedial side of arms in each ray borne by successive second secundibrach. Stem transversely round. *M.Ord.,* Eu.(Est.).——Fig. 410,*3*. **M. rossicus; C*-ray view of diagram. reconstr. of crown, ×1.5 (Jaekel, 1918).

[In the plate diagram of the cup of this genus Jaekel (1918, p. 85), reconstructed the anterior side of the cup with compound radials in the *E* ray, which would indicate that the symmetry of the cup was comparable to that of the Heterocrinacea, which are monocyclic inadunate crinoids. The illustration of the crown of *Metabolocrinus* by Jaekel shows clearly infrabasal plates below the basals, but his reconstruction of the cup cannot be accepted in this respect until the type or other authentic specimens have been re-studied.]

Cyliocrinus Jaekel, 1918, p. 60 [**Melocrinus? rigidus* Angelin, 1878, p. 20; OD]. Cup small, low, with wide truncate base; radial articular facets as wide as radials; radianal small, quadrangular. Anal sac ?recurved, lateral plates strongly plicate, base of sac situated in concavity formed by admedial part of anal *X* and posterior sides of *C* and *D* radials. Arms uniserial, massive, brachials large and quadrangular; primibrachs *3* axillary; 2 main rami to ray, each with large ramules alternately on opposite sides of each brach proximally, distally each brach bears 2 ramules, one on each side; ambulacral grooves of rami and ramules with stout cover plates. Stem transversely subpentagonal, columnals composed of pentameres. *U.Sil.,* Eu.(Swed.).——Fig. 410, *2a*. **C. rigidus* (Angelin), Gotl.; post. view of crown, ×1.5 (Angelin, 1878).——Fig. 410,*2b,c*. *C. scolopendra* (Bather), Gotl.; ventral surface of arm showing cover plates of ambulacra, base

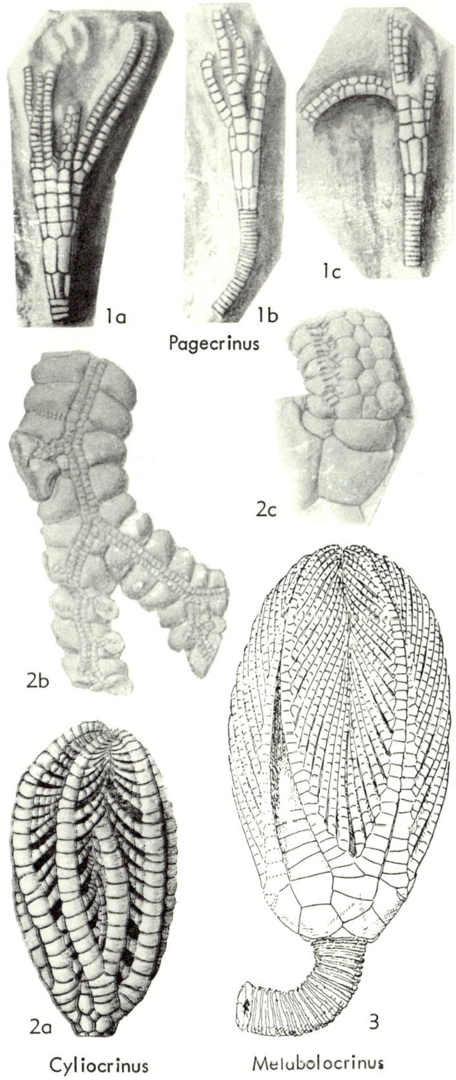

Fig. 410. Metabolocrinidae (p. *T629-T630*).

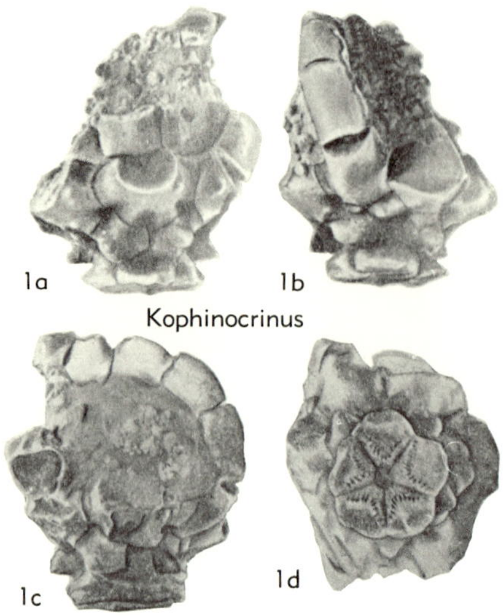

FIG. 411. Superfamily and Family Uncertain (p. T630).

of anal sac and adjacent cup, ×3 (Bather, 1893a).

Pagecrinus KIRK, 1929, p. 1 [**P. gracilis*; OD]. Crown narrow and elongate. Cup high, cylindrical; infrabasals tall, basals very large and elongate; radials small with wide linear straight articular facets; fully equal to radials in width; single anal plate (anal *X* or radianal) directly above *CD* basal and in line with radials, supporting longitudinal row of large anal sac plates which resemble brachials. Anal sac stout, elongate, composed of 6 or 7 rows of hexagonal plates. Arms uniserial, rounded, primibrachs as wide as radials, primibrachs *3* or *4* axillary; 2 main rami to each ray; long, stout ramules given off alternately on opposite sides of every second secundibrach. *L.Dev.*, USA(Ind.).——FIG. 410,*1*. **P. gracilis; 1a-c, D*-ray, *B*-ray, and lat. views of crown and proximal part of stem, ×2 (Kirk, 1929).

Superfamily and Family UNCERTAIN

Kophinocrinus GOLDRING, 1954, p. 36 [**K. spiniferus*; M]. Cup truncate conical, height approximately 0.5 of summit width, plates with prominent rounded ridges and furrows crossing sutures at right angles; infrabasal circlet swollen, expanded laterally; radials largest plates in cup, with outward-downward semicircular articular facets on middle part of summits, notches between upper corners of radials; 2 anal plates in cup, a very small quadrangular radianal and large anal *X*, latter supporting vertical series of large long narrow plates, bordered at sides by numerous small spinose plates. Arms unknown except for quadrangular first primibrachs, which are wider than high. Stem unknown but impression of proximal columnal indicates large diameter and pentalobate outline with exceptionally large stellate lumen. *M.Dev.*, USA(N.Y.).——FIG. 411,*1*. **K. spiniferus*, Hamilton Gr.; *1a-d, A*-ray, *CD*-interray, *D*-ray and dorsal views, ×3 (Goldring, 1954). [LANE]

Suborder POTERIOCRININA Jaekel, 1918

[*nom. correct.* MOORE, in MOORE, LALICKER, & FISCHER, 1952, p. 614 (*pro* suborder Poteriocrinites JAEKEL, 1918, p. 60)] [=Poteriocrinitina UBAGHS, 1953, p. 752] [Materials for this suborder prepared by R. C. MOORE, H. L. STRIMPLE, and N. GARY LANE]

Crown much varied in shape, size, number of arms, and mode of branching, but arms prevailingly pinnulate and biserial arrangement of brachials common. Cup high conical, gently convex-, flat-, or concave-based bowl shaped (craticuliform), or extremely low, discoid (patelliform). Radial articular facets narrow (angustary) to wide (peneplenary, plenary), with facetal surfaces declivate, planate, or rarely sursumate; transverse ridge, ligament pits and fossae, and muscle fields generally prominent. Four, three, two, one, or no anal plates in cup. Anal sacs virtually duplicate kinds found in the Dendrocrinina. Stem mostly round transversely. *L.Dev.-U.Perm.; M.Trias.*

The combined characters of the Poteriocrinina mark them as distinctly more advanced in evolution than the Cyathocrinina and Dendrocrinina. As a whole, they exhibit remarkable diversity, exceeding the other suborders in this regard and in number of recognized taxa at all levels. Inferred phylogeny of poteriocrinine families is indicated diagrammatically in Fig. 412.

Superfamily POTERIOCRINITACEA Austin & Austin, 1842

[*nom. transl.* MOORE & STRIMPLE, 1973 (*ex* Poteriocrinitidae BASSLER, 1938, p. 20, *nom. correct. pro* fam. Poteriocrinidae AUSTIN & AUSTIN, 1843, p. 195)] [Materials for this superfamily prepared by R. C. MOORE and H. L. STRIMPLE]

Crown tall, formed of steeply conical cup and many-branched uniserial arms; radial articular facets narrow, with strongly rounded outer margins; three anal plates in cup. Anal sac tall, commonly reaching

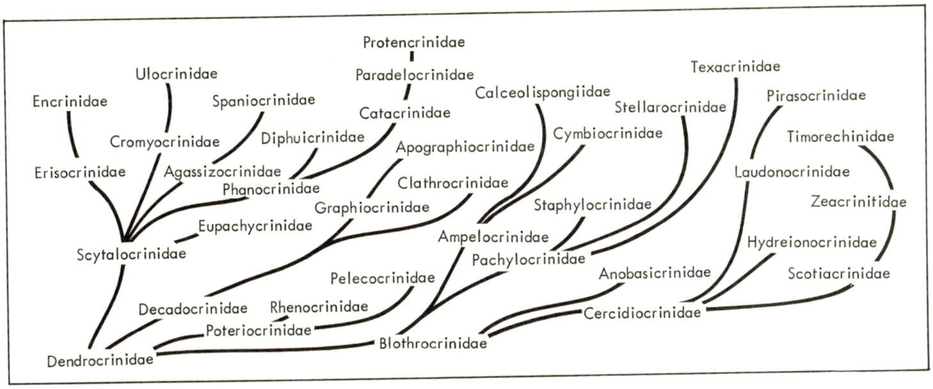

FIG. 412. Inferred phylogeny of suprageneric divisions of Poteriocrinina.

to or above tips of arms, composed of small polygonal plates. Relatively large stem round transversely. *L.Dev.-L.Penn., ?M. Penn.-?U.Perm.*

Family POTERIOCRINITIDAE Austin & Austin, 1842

[*nom. correct.* BASSLER, 1938, p. 20; *pro* fam. Poteriocrinoidea AUSTIN & AUSTIN, 1842, p. 108 *(nom. imperf.)* (=Poteriocrinidae AUSTIN & AUSTIN, 1843, p. 195)] [=Poteriocrininae, Poteriocrinidae JAEKEL, 1918, p. 61; Rhabdocrinidae WRIGHT, 1944, p. 266]

Crown large. Cup conical, with three or four anal plates in cup; radial articular facets narrow or peneplenary, all strongly rounded, mostly with well-developed transverse ridge; anal sac large, elongate, commonly with spinose summit plates, and may have pores or slits at plate boundaries. Arms uniserial, branching once or several times isotomously; pinnulate. *L.Dev.-L. Penn., ?M.Penn.-?U.Perm.*

[Although this family name was credited to ROEMER (1851) by BASSLER (1938), the first usage of the genus *Poteriocrinites* (=*Poteriocrinus*) as a type of a family-group name was by AUSTIN & AUSTIN (1842), who classified this group as the second family of their order Cionacineti. In 1843 they corrected the family name to Poteriocrinidae and the name was used in this form in their monograph (1846, p. 68). The family name was also used by BRONN, GÖPPERT, & VON MEYER (1848, p. 173) and credited to the AUSTINS, prior to citation of the family by ROEMER.]

This assemblage is now somewhat narrowly defined. According to common earlier classification, the family was so loosely and broadly expanded that virtually all dicyclic inadunates having a transverse ridge on the radial articular facets were included in it. The family seems to have no significant differences from the Botryocrinidae except in the beginning of signs of muscular articulation on the radial articular facets and in the higher development of pinnulation of the arms. The facets are narrow, rather strongly outward-downward sloping, and marked by a fairly distinct transverse ridge. Characters of the anal sac and the arrangement of anal plates of the cup suggest that the poteriocrinitids may have been derived from the Dendrocrinidae, rather than the Botryocrinidae. The oldest known representative of the Poteriocrinitidae occurs in Lower Devonian rocks of Europe; the type species of *Poteriocrinites* and several others occur in Lower Carboniferous deposits. Whether or not identifications of several Pennsylvanian and some Permian cups as belonging to *Poteriocrinites* are valid is not known, but the occurrence of undoubted representatives of the family in post-Mississippian rocks is not established.

Key to Genera of Poteriocrinitidae

A. Cup conical with truncate base; anal plates in cup 3-4, anal sac mostly large, elongate cylindrical ....
 I. Radial articular facets narrow; horseshoe shaped ....
  a. Arms 10, branching first on primibrachs 2; stem pentagonal, cirriferous .... *Denariocrinus*
  b. Arms 30 or more, branching first on primibrachs *1-14*; tall anal sac with laterally directed spines at summit .... *Poteriocrinites*
  c. Arms 30-40, branching first on primibrachs *1-8*; anal sac with plicate plates ....
   1. Arms approximately 30, branching first on primibrachs *4*, anal sac

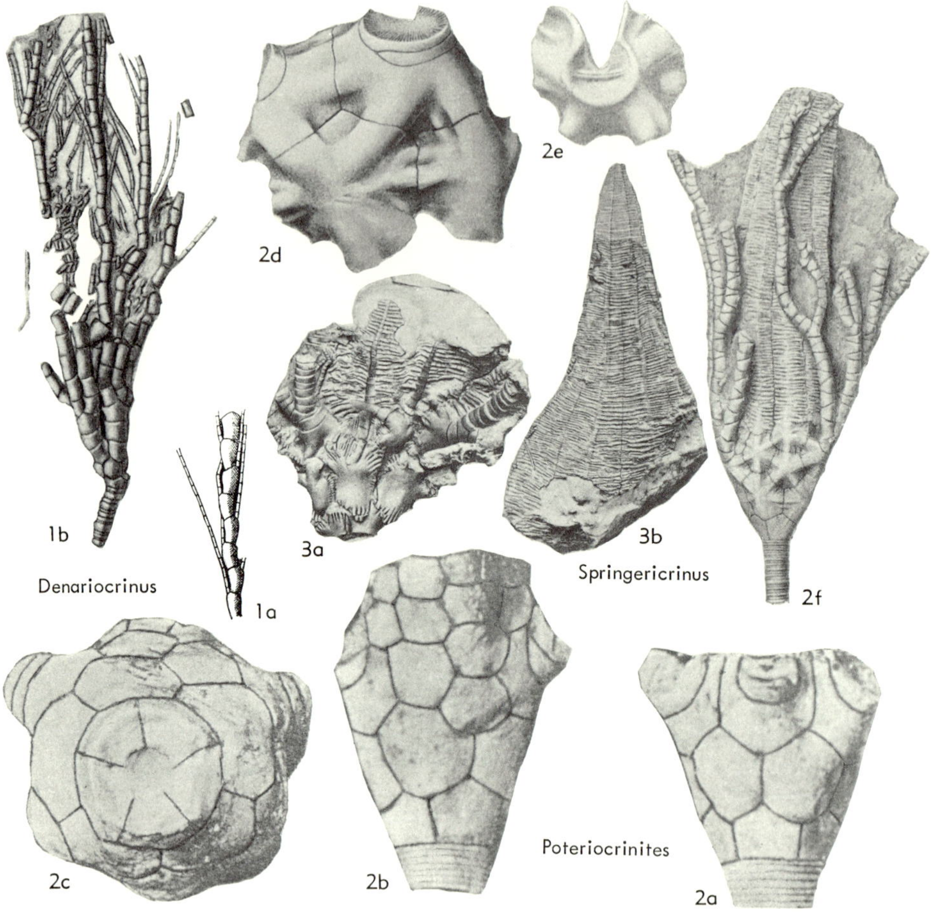

FIG. 413. Poteriocrinitidae (p. T632-T633).

with spines at summit .... *Propoteriocrinus*

2. Arms 40 or more, branching first on primibrachs *1-8*; anal sac very broad at base, terminating distally in point without summit spines; composed of plicate plates; cup plates fluted .... *Springericrinus*

II. Radial articular facets concave, broadly rounded externally, lacking transverse ridge, crenulated along outer margin; many arms, branching first on primibrachs *13* or above .... *Rhabdocrinus*

Poteriocrinites MILLER, 1821, p. 14 [**P. crassus*; M] [=*Poteriocrinus* AGASSIZ, 1836, p. 197]. Crown elongate, subcylindrical, or expanding uniformly upward. Cup conical, surface of plates smoothly rounded or with broad flutes normal to sutures; infrabasals 5, basals 5; radials wider than high, *C* radial slightly smaller than others, and elevated above their level, articular facets narrow, horseshoe shaped, with weak transverse ridge; radianal, anal *X* and 1 or 2 other anal plates in cup. Anal sac large, cylindrical or club shaped in some, commonly with long, spinose hexagonal plates at summit, pierced at sutures by pores or slits that penetrate margins of plates; anal opening rarely observed, but in all known, situated laterally and low on side of sac. Arms uniserial, long, branching isotomously several times; primibrachs *1* to *14* axillary; pinnulate. Stem transversely round or subpentagonal, composed of thin columnals with small, round lumen. *L.Dev.-U.Miss.(L.Carb.), ?U.Perm.,* USA-Eu. (Eng.-Ger.-Belg.-Ire.-France-Scot.)-Australia.——FIG. 413,*2a-c*. **P. crassus,* L.Carb., Eng.; ant., post., and dorsal views of cup, ×1.5 (Wright, 1950-54).——FIG. 413,*2d,e*. *P. macropleurus* (HALL), L.Miss., Iowa; basal view of incomplete cup and radial plate, ×1

(Springer, 1911c).——Fig. 413,*2f*. *P. doris* (Hall), L.Miss., Iowa; post. view of crown, ×0.7 (Springer, 1911c).

Denariocrinus Schmidt, 1942, p. 163 [**D. ferula*; M]. Crown narrow and elongate. Cup steeply conical; infrabasals very low; basals large and elongate; radials high and narrow; 4 anal plates in cup, posterior interray wide, radianal and anal *X* large, supporting 2 tube plates. Anal sac not known. Arms uniserial, branching isotomously on primibrachs *2*, with 10 main rami; brachials elongate, with inclined sutures that give slightly zigzag appearance to arms; pinnules long and very slender. Stem transversely subpentagonal, cirriferous. *M.Dev.*, Eu.(Ger.).——Fig. 413,*1*. **D. ferula; 1a,* distal brachials and pinnules, ×2; *1b,* ant. view of crown, ×1 (Schmidt, 1942).

Propoteriocrinus Schmidt, 1942, p. 110 [**P. follmanni*; OD] [=*Propoteriocrinus* Schmidt, 1934, p. 109 (*nom. nud.,* no description or figure of species)]. Crown tall, cylindrical. Cup small, bowl shaped, infrabasals low, basals relatively large; 3 anal plates in cup. Anal sac very large, elongate and cylindrical, composed of several rows of hexagonal, plicate plates, summit capped by plates with short, blunt spines. Arms uniserial, slender; primibrachs *4* axillary, with 2 higher isotomous branches in each half-ray; pinnules stout, directed obliquely upward, producing comblike appearance of arms. Stem transversely pentagonal. *L.Dev.*, Eu.(Ger.).——Fig. 414,*1a,b*. **P. follmanni; A*-ray and post. views of 2 partial crowns, one with anal sac preserved, ×1.5 (Schmidt, 1941).——Fig. 414,*1c*. *P. scopae* Schmidt; lat. view of crown, ×0.7 (Schmidt, 1934).

Rhabdocrinus Wright, 1944, p. 266 [**Poteriocrinus scotocarbonarius* Wright, 1937, p. 397; OD]. Cup cone shaped, with wide base and convex plates, depressions at plate angles; infrabasals large, upflaring, convex, widely truncate above large proximal columnal; basals large, convex; radials small, wider than high, with arcuate distal border, curved articular facets sloping strongly outward-downward; 3 large anal plates in cup. Anal sac not known. Primibrachs 13 or more, higher parts of arms not known; brachials very thin, rounded, proximal 3-5 brachials fixed into theca by small interradial plates in all interrays; articular surfaces of brachials with radiating fine crenellae next to outer edges. *L.Carb.(Tournais.)-U.Carb.(Namur.),* Eu.(Scot.-Eng.-Ire.-Belg.-USSR).——Fig. 414,*2*. **R. scotocarbonarius* (Wright), L.Carb., Scot.; *2a,b,* post. and ant. views of cup and proximal brachials, ×1; *2c-e,* ventral, lower, and int. views of radial plate, enl. (Wright, 1944).

Springericrinus Jaekel, 1918, p. 62 [**Poteriocrinus magniventrus* Springer, 1911c, p. 155; OD]. Crown large and explanate. Cup large, cone shaped, plates with distinctive radiating folds, subsidiary ridges and troughs disposed normal to sutures; rounded radial articular facets narrow, with weak transverse ridges, sloping outward-downward as in *Poteriocrinites*; 3 anal plates in cup. Anal sac very large, broad at base and tapering to point distally, anal opening apparently low on side of sac, plates wide, low, each with sharply elevated, rounded ridges and adjacent grooves; base of sac composed of thick, small ossicles. Arms uniserial, branching isotomously on primibrachs *1-3,* with 2 or 3 higher bifurcations; brachials short, wedge shaped, pinnules short and stout. Stem composed of alternating thick and thin round columnals; lumen very large. *L.Miss.,* USA(Ind.-Iowa).——Fig. 413,*3*. **S. magniventrus* (Springer), Ind.; *3a,b,* post. view of theca and base of anal sac, and isolated anal sac, ×0.5 (Springer, 1911c).

Superfamily RHENOCRINACEA Jaekel, 1918

[*nom. transl.* Moore in Moore & Strimple, 1973, p. 19 (*ex* Rhenocrinidae Jaekel, 1918, p. 59)] [Materials for this superfamily prepared by R. C. Moore, H. L. Strimple, and N. Gary Lane]

Similar to Poteriocrinitacea, differing in having peneplenary radial articular facets, mostly with shallowly curved outer facet margins, interradial notches at rim of cup narrow to moderately wide and distinct. *L.Dev.-U.Miss.; M.Penn.*

Family RHENOCRINIDAE Jaekel, 1918

[Rhenocrinidae Jaekel, 1918, p. 59] [=Glossocrinidae Goldring, 1923, p. 389]

Cup low cone shaped; differing from Poteriocrinitidae in shallowly curved outer margin of its radial articular facets; three anal plates in cup. Anal sac large, elongate, mostly with posterior longitudinal row of large sac plates bordered laterally by small, low plicate plates. Arms slender, uniserial, pinnulate, with one or several isotomous branches, or, rarely, atomous; two, three, or numerous primibrachs present in a ray. Stem transversely pentagonal or rarely round. *L.Dev.-U.Miss.; M.Penn.(Atokan).*

Crinoids of this group are differentiated from the Poteriocrinitidae only by wider radial facets and some details of arm structure. Several rhenocrinid genera have a prominent row of plates extending upward on the posterior side of the anal sac. The family seems to be a branch of the poteriocrinitid stock.

Key to Genera of Rhenocrinidae

A. Cup plates smooth, not fluted; anal sac

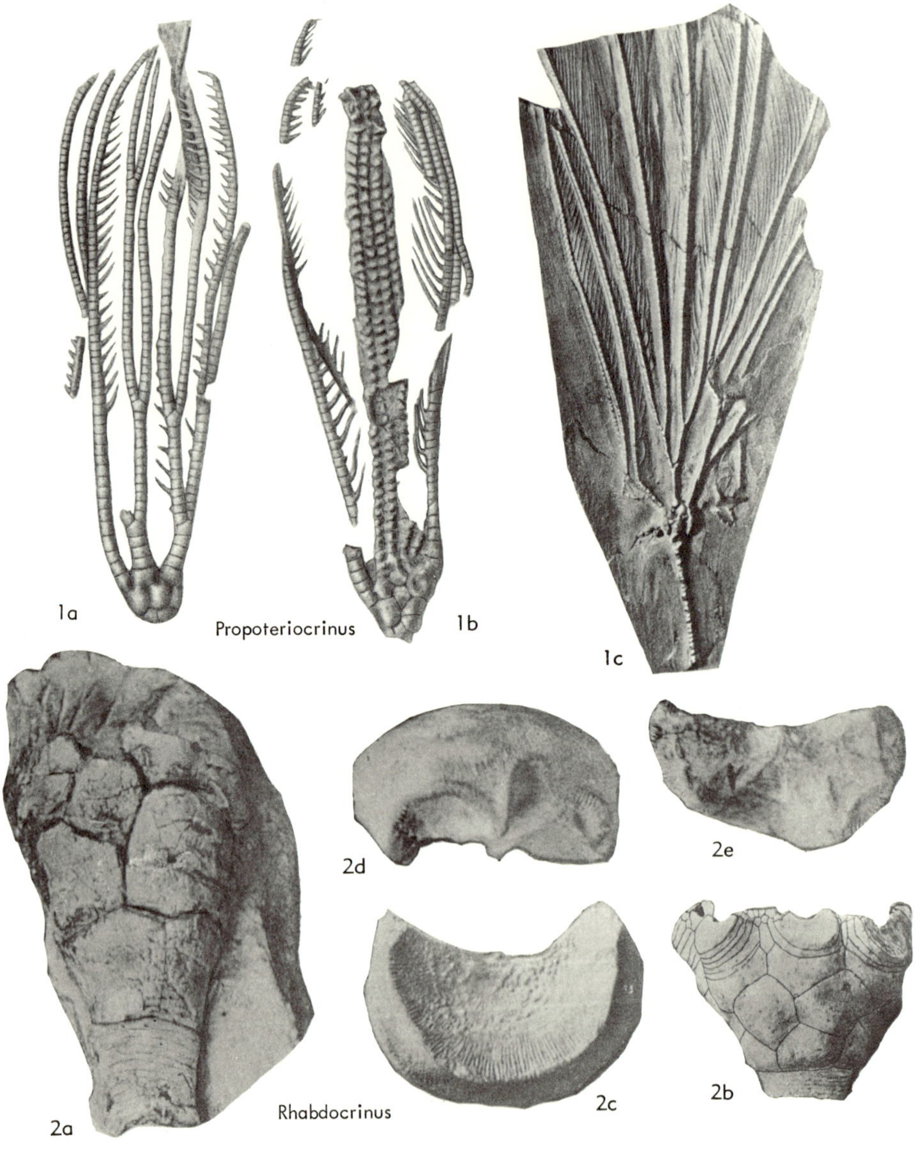

FIG. 414. Poteriocrinitidae (p. T633).

lacking prominent "backbone" of large polygonal plates on posterior side; stem transversely round ....

I. Arms few, uniserial, pinnulate; radial articular facets moderately wide, well curved outwardly ....
   a. Arms 20 or more, branching isotomously first on primibrach *4* or *5*, very tall anal sac composed of vertical rows of plicate plates .... *Araeocrinus*
   b. Arms 20 or more, each half ray endotomous, anal sac unusually broad, tall, composed of large nodose plates with radiating ridges .... *Cydrocrinus*
   c. Arms 7-10, branching on primibrachs *6-36*, some arms atomous .... *Rhenocrinus*
   d. Arms 10, without endotomous smaller arms, anal sac composed of longitudinal small hexagonal plates

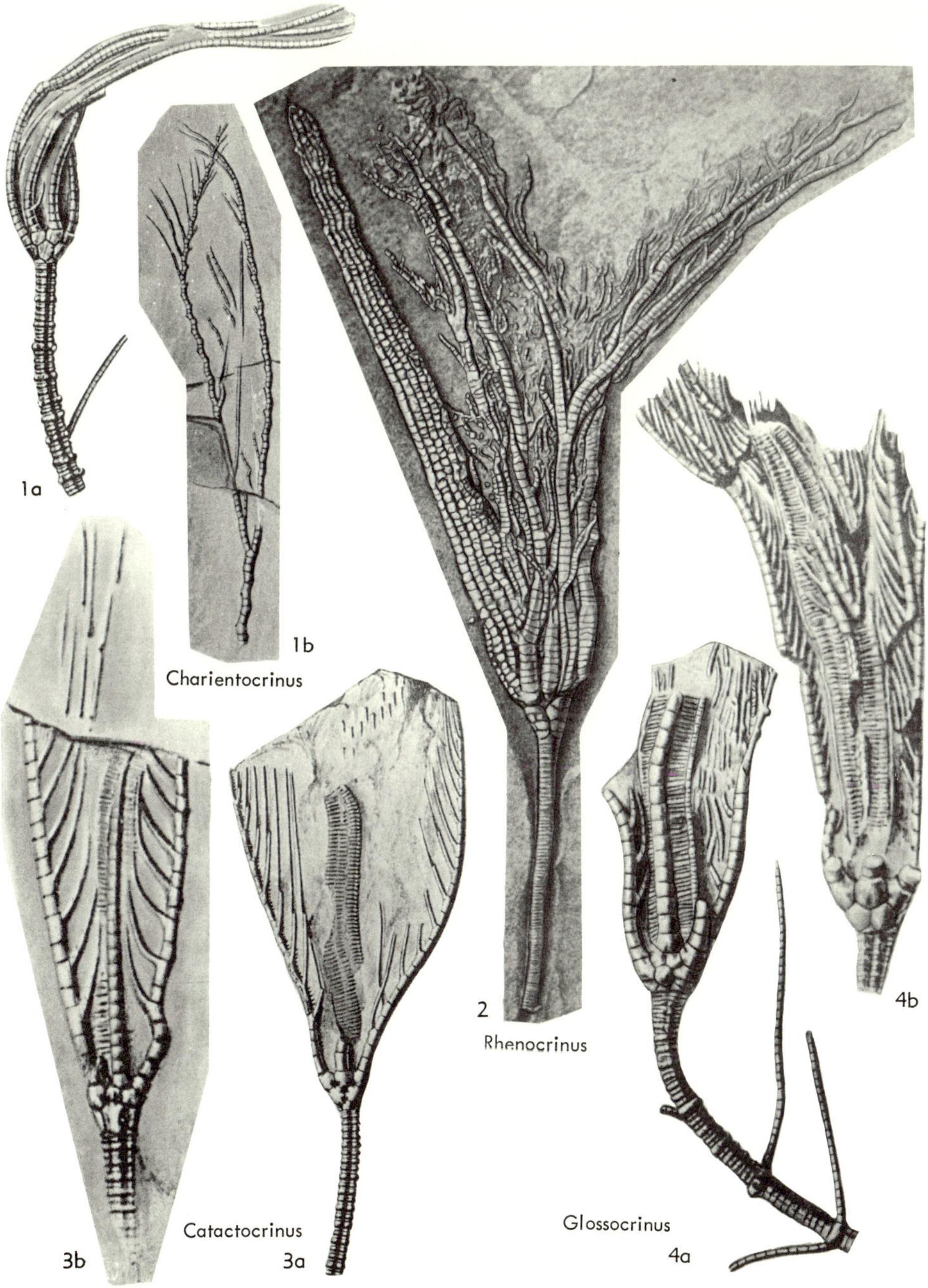

FIG. 415. Rhenocrinidae (p. T636-T637).

.... *Maragnicrinus*

II. Arms 80-90, branching first isotomously on primibrach *3*, and then heterotomously ....

In endotomous pattern; anal sac plates plicate .... *Hallocrinus*

B. Cup plates fluted, with pits at plate angles; anal sac with prominent longitudinal row of polygonal plates on posterior side; stem transversely pentagonal or pentalobate ....

I. Arms 5; stem pentalobate ...... *Catactocrinus*

II. Arms 5-10; stem cirriferous ....

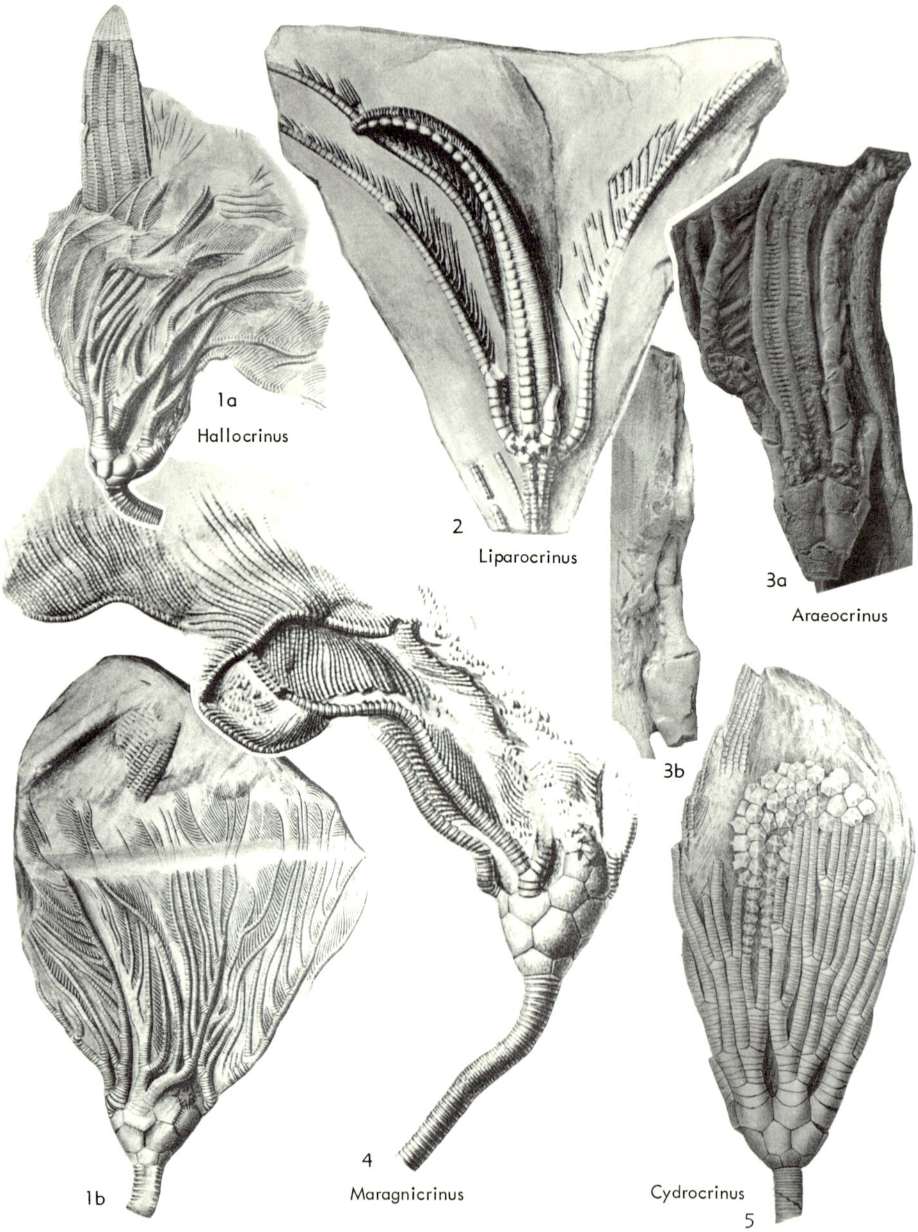

FIG. 416. Rhenocrinidae (p. T637-T638).

a. Arms branching on primibrachs *19-20*; anal sac with rows of small polygonal plates; stem pentagonal .... *Charientocrinus*

b. Arms branching on primibrachs *2-6*; anal sac with rows of plicate plates; stem pentagonal .... *Glossocrinus*

c. Arms branching on primibrachs *7* or higher; anal sac with rows of plicate plates; stem pentalobate .... *Liparocrinus*

Rhenocrinus JAEKEL in SCHMIDT, 1906, p. 544 [**R. ramosissimus*; M]. Cup small, low, broadly truncate proximally, plates smooth; cup anal plates supporting large, elongate, cylindrical sac composed of several vertical rows of small, hexagonal plates. Arms long and slender, unbranched or with single isotomous division high above cup; long stout ramules on alternate sides of every second brachial. Stem transversely round. *L.Dev.*, Eu.(Ger.).——FIG. 415,*2*. **R. ramosissimus; B*-ray view of crown and part of stem, ×1 (Schmidt, 1934).

Araeocrinus STRIMPLE & WATKINS, 1969, p. 196 [**A. bassus*; OD]. Cup high cone shaped; infrabasals rising evenly from columnar attachment area; radial articular facets not quite filling width of plates. Arms uniserial, 20 or more, branching first isotomously on primibrachs *4-5*, cuneate brachials, syzygial pairs common, large pinnules. Tall anal sac composed of thin plicate plates. Column transversely round. *M.Penn.(Atokan)*, USA (Texas).——FIG. 416,*3*. **A. bassus*, Marble Falls F., San Saba Co., Texas; *3a,b*, holotype crown from *D*-ray and post. sides, ×1.5 (Strimple & Watkins, 1969).

Catactocrinus GOLDRING, 1923, p. 405 [**C. leptodactylus*; OD]. Crown slender, delicate. Cup small, cone shaped; plates with radial ridges; anal sac like that of *Glossocrinus*. Arms atomous, brachials elongate, pinnules very long and robust. Stem pentagonal proximally, becoming round distally, without cirri in proximal parts. *U.Dev.*, USA(N.Y.).——FIG. 415,*3*. **C. leptodactylus; 3a,b*, ant. and post. views of incomplete crowns with attached stems, ×2 (Goldring, 1923).

Charientocrinus GOLDRING, 1923, p. 402 [**C. ithacensis*; OD]. Crown long and slender. Cup low, broad; vertical row of large anal sac plates as in *Glossocrinus*. Arms very long, with numerous primibrachs, first isotomous division occurring on primibrachs *17-20*, with single higher division rarely in some rays; brachials quadrangular proximally, becoming wedge shaped distally, with long pinnule on every second enlarged brachial, producing zigzag appearance of arms; pinnules long, slender, composed of long, quadrangular ossicles. Stem pentagonal, with cirriferous nodals each of which bears 2 cirri, with successive pairs apparently borne by different faces of nodals next above or below, cirri directed obliquely upward. *M.Dev.-U.Dev.*, USA(N.Y.).——FIG. 415,*1*. **C. ithacensis*, U.Dev.; *1a,b*, lat. view of crown with proximal part of stem, and distal parts of arms, ×1 (Goldring, 1923).

Cydrocrinus KIRK, 1940, p. 323 [**Poteriocrinus coxanus* WORTHEN, 1882, p. 4; OD]. Crown of medium height, compact, expanding gradually to about three-fourths height, then contracting. Cup broadly turbinate, plates relatively thin, unornamented; infrabasals large, making up appreciable amount of cup; basals large. Radianal large, pentagonal, not entering deeply between posterior and right posterior basals; anal *X* large, extending well above level of radials, right tube plate small. Anal sac unusually broad and stout, three-fourths to three-fifths height of crown, composed of large nodose plates with radiating ridges, but not spinose. Arms parendotomous (i.e., each half-ray endotomous); 2 primibrachs in all rays (variation possible in anterior ray, as apparently in type specimen of *C. concinnus*, the only example known; 2 or 3 divisions in each half ray, giving a great number of closely crowded rami. Brachials wedge shaped to cuneate, pinnules long, moderately stout; arm facets plenary, slightly crescentic, sutures not gaping. Stem transversely circular, lumen pentalobate. [Most nearly resembles *Blothrocrinus* but differs in relatively short, compact crown and numerous rami.] *L.Miss.*, USA(Iowa-Ind.-Ill.).——FIG. 416,*5*. **C. coxanus* (WORTHEN), Keokuk, Ill.; ant. view of crown, ×0.5 (Worthen, 1883).

Glossocrinus GOLDRING, 1923, p. 389 [**G. naplesensis*; OD]. Crown long and slender. Cup low cone shaped, plates convex and with radiating ridges; right tube anal plate supporting vertical row of large convex quadrangular anal sac plates; other plates of elongate sac with 4 or 5 high thin folds, giving sac a plicate appearance. Arms uniserial, stout and long, brachials quadrangular or slightly wedge shaped in side view; primibrachs *2* to *6* axillary, one higher isotomous division; except axillaries all brachials bearing long pinnules. Stem pentagonal in section, with long, widely spaced, upwardly directed cirri along its length. *U.Dev.*, USA(N.Y.).——FIG. 415,*4*. **G. naplesensis; 4a,b*, post. and ant. views of crown and stem, ×1.4 (Goldring, 1923).

Hallocrinus GOLDRING, 1923, p. 374 [**Cyathocrinus ornatissimus* HALL, 1843, p. 247; OD]. Cup low cone shaped, plates with depressions at angles; infrabasals low; radials low and broad. Anal sac large, composed of several longitudinal rows of low, broad, transversely plicate plates. Arms uniserial, with low broad brachials; 3 primibrachs in each ray, first 2 primibrachs may be constricted so that lateral edges of primaxil are in contact with radial; arms endotomously branched, with main outer branch and 7 or 8 inner branches of each half-ray pinnulate. *L.Dev.-L.Carb.*, Eu. (Ger.)-USA(N.Y.).——FIG. 416,*1*. **H. ornatissimus* (HALL), U.Dev., N.Y.; *1a,b, D*-ray and *CD*-interray views of crown, ×0.6 (Goldring, 1923).

Liparocrinus GOLDRING, 1923, p. 397 [**L. batheri*; OD]. Cup low, broad, with deep depressions at plate angles; radials small, fluted; 3 anal plates in cup, right tube plate large, radial-like in appearance and supporting longitudinal row of large quadrangular convex sac plates bordered laterally by thin, plicate plates; anal sac large, tapering to point distally. Arms long, stout, brachials quadrangular in side view and convex; one isotomous division in each ray high above cup; pinnules stout and long. Stem strongly pentalobate proximally, with stout cirri directly obliquely upward. *U.Dev.* *(Chemung.),* USA(N.Y.).——FIG. 416,2. **L.* *batheri; CD*-interray view of crown and proximal part of stem, ×0.93 (Goldring, 1923).

Maragnicrinus WHITFIELD, 1905, p. 17 [**M. portlandicus*; M]. Cup moderately high bowl shaped; plates with fine sculpture; infrabasals high; radials with widely rounded peneplenary articular facets; 3 anal plates in cup, anal *X* and right tube plate supporting 3 large sac plates followed by several longitudinal rows of hexagonal plates bearing radiating ridges. Arms uniserial, long, slender, branching on primibrachs *3,* secundibrachs low, outwardly wedge shaped, pinnulate. Stem transversely round. *U.Dev.(Portage Gr.),* USA(N.Y.). ——FIG. 416,4. **M. portlandicus; D*-ray view of crown and partial stem, ×0.93 (Whitfield, 1905).

Family PROCTOTHYLACOCRINIDAE Kier, 1952

[Proctothylacocrinidae KIER, 1952, p. 71]

Cup small, highly ornamented; radials with wide, deep articular facets not extending full widths of the plates; three anal plates in cup; anal sac large, cylindrical, plates highly plicate. Arms branching isotomously several times; nonpinnulate. *M.Dev.*

Proctothylacocrinus KIER, 1952, p. 71 [**P. longus*; OD]. Cup steeply conical and broadly truncate at base; cup plates with strong radiating ridges; infrabasals low, spear shaped, projecting distally between basals, circlet only slightly wider than proximal columnal; basals pentagonal, separated by deep indentations, bridged across sutures by ridges; radials pentagonal, slightly wider than high, and smaller than basals; radiating ridges connecting to all adjacent plates except first primibrachs; radial articular facets wide, deep, subovate, not extending full width of plates; leaving small interradial notches; 3 anal plates in cup, radianal slightly smaller than radials, pentagonal; anal *X* and 2 to 4 succeeding sac plates larger, more inflated than remaining anals. Anal sac large, plates low, broad, bearing prominent transverse ridges. Arms branching isotomously several times, composed of quadrangular brachials, nonpinnulate; primibrachs *3* to *5* axillary. Stem composed of thin round columnals. *M.Dev.,* USA(Ohio).——FIG. 417,*1.* **P. longus; 1a,b, CD*-interray views of incomplete crown with very tall anal sac and attached stem, ×0.87, ×1.6; *1c,* ant. of crown with stem, ×0.87 (Kier, 1952).

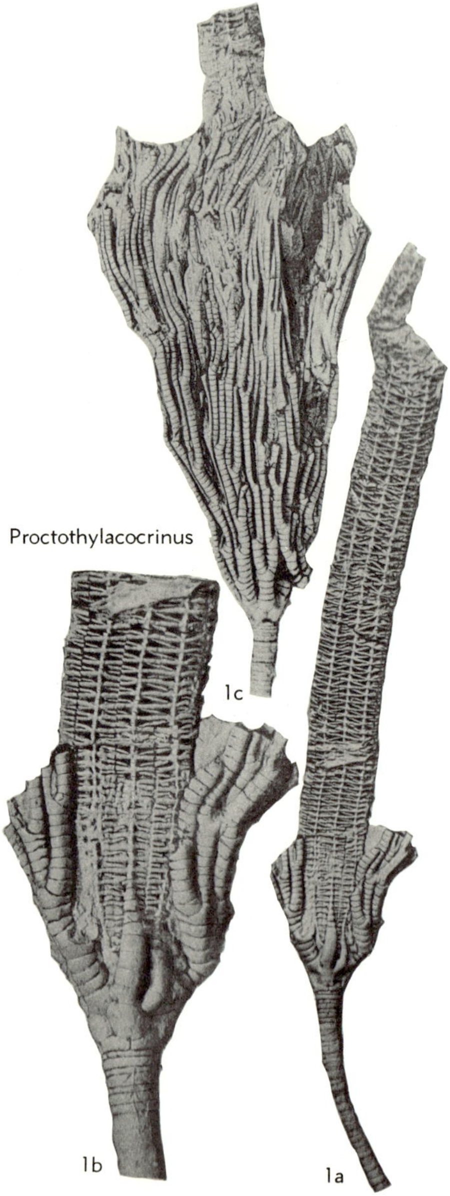

FIG. 417. Proctothylacocrinidae (p. *T*638).

Superfamily SCYTALOCRINACEA Moore & Laudon, 1943

[*nom. transl.* MOORE & STRIMPLE, 1973, p. 19 (*ex* Scytalocrinidae MOORE & LAUDON, 1943a, p. 59)] [incl. Hydriocrininae JAEKEL, 1918, p. 63] [Materials for this superfamily prepared by R. C. MOORE and H. L. STRIMPLE]

Crown mostly tall, with few to very many erect uniserial arms. Cup conical, with infrabasals visible in side view; radial articular facets plenary. Arms branching isotomously. Stem transversely round. *M. Dev.-U.Perm.*

Family SCYTALOCRINIDAE Moore & Laudon, 1943

[Scytalocrinidae MOORE & LAUDON, 1943a, p. 59] [=Scytalecrinidae BATHER, 1899b, p. 922]

Crown slender, tall. Cup conical; infrabasals five, visible from side; radial articular facets wide, bearing transverse ridge and ligament pits; three anal plates in cup. Anal sac tall, slender. Arms uniserial (exceptionally biserial), branching on primibrachs *2* in early forms and usually on primibrachs *1* in later ones but may remain unbranched; pinnulate. *M.Dev.-U.Perm.*

Several families thought to have been derived from dendrocrinid stock show development along lines that differ from those of the blothrocrinids and their presumed allies. The oldest of this group is the Scytalocrinidae, which appear in the Middle Devonian and range to Upper Permian. The cup of the scytalocrinids is only a little changed from primitive form and structure. The longitudinally straight sides slope steeply, and the infrabasal plates are visible from the side, but not uncommonly the base is truncate. Three anal plates occur on the posterior side of the cup. The arms are distinctly advanced in evolution, although they may be interpreted as a development toward simplified structure. A single isotomous division occurs in each ray immediately above the lowermost brachial; the uniserially arranged arm segments bear pinnules. This arm plan does not differ from that of a moderately advanced mastigocrinid, such as *Lasiocrinus,* except that the isotomous division is closer to the cup, and the branchlets, or pinnules, are much more numerous, being borne by every brachial. The blothrocrinids and most allied genera retain the early dendrocrinid habit of several isotomous arm divisions; the scytalocrinids suppress this tendency. Evidence is not now at hand to prove that many-branched crinoids, such as the blothrocrinids, were not evolved from five- or ten-armed types, like the scytalocrinids, or vice versa, but the phylogeny of the families seems to be interpreted best as represented diagrammatically in Figure 412. The tall anal sac of the Scytalocrinidae is of dendrocrinid type; its many small plates seem to be less firmly united than in other families so far considered.

Key to Genera of Scytalocrinidae

A. Arms 20 or more ....
- I. Branching isotomously on primibrach *1,* secundibrachs *4* and higher ....
  - a. Cup cone shaped, anal sac reflexed tube of stout rugose plates .. *Haeretocrinus*
  - b. Cup bowl shaped with upflared infrabasals, anal sac unknown, 3 anal plates advanced, anal *X* and right tube plate with confluent distal surfaces .... *Tundracrinus*
- II. Branching on primibrachs *1* or *2* and secundibrachs *7-11,* anal sac thin-plated recurved tube .... *Sostronocrinus*

B. Nine or ten uniserial arms (except *Hydriocrinus*) ....
- I. High, steep-sided cup, prominent infrabasals ....
  - a. Brachials somewhat elongated ....
    - 1. Arms divided on primibrachs *1* or *2*; anal *X* and right tube plate with common distal articulating facets .... *Melbacrinus*
    - 2. Arms divided on primibrachs *2* or *3* .... *Logocrinus*
    - 3. Arms divided on primibrachs *2* ..
      - a). Impressed sutures, depressions at plate angles, transversely pentagonal stem .. *Corematocrinus*
      - b). Smooth cup, sutures not impressed .... *Bridgerocrinus*
      - c). Advanced anal plates, radianal in CD position ...... *Prininocrinus*
  - b. Brachials short ....
    - 1. *A* ray undivided commonly or invariably ....
      - a). Cup low truncate conical, arms relatively stout .. *Histocrinus*
      - b). Cup high conical, arms slender .... *Hypselocrinus*
    - 2. All rays divided on primibrachs *1* (except *Phacelocrinus* which may have 1 or 2 primibrachs) ....
      - a). Pentagonal stem ....
        - 1). Three anal plates in cup with normal arrangement, 10 arms .... *Phacelocrinus*
        - 2). Anal *X* and right tube plate with confluent distal articulating facets, arms may branch twice .... *Hydriocrinus*
      - b). Round stem ....
        - 1). Narrow arms ....
          - aa. Three anal plates (normal) .... (part) *Scytalocrinus*
          - bb. One anal plate .... *Atrapocrinus*
        - 2). Wide arms ...... *Morrowcrinus*

3. All rays divided on primibrachs *2*
   a). Brachials of moderate width .... *Pegocrinus*
   b). Wide brachials ...... *Bollandocrinus*

II. Medium cone-shaped cup, infrabasals not prominent ....
All rays divided on primibrach *1*, brachials short .... (typical) *Scytalocrinus*

C. Four to six arms ....
   I. Uniserial arms ....
      a. High cone-shaped cup, prominent infrabasals ....
         1. Four arms (*B* radial armless) .... *Linobrachiocrinus*
         2. Five or six arms .... (typical) *Ophiurocrinus*, (part) *Anemetocrinus*, *Gilmocrinus*
      b. Low cone-shaped cup, infrabasals not prominent .... (part) *Ophiuriocrinus*
   II. Five incipiently biserial or equibiserial arms .... (part) *Anemetocrinus*

D. Arms unknown ....
   I. High cone-shaped cup, prominent infrabasals, round stem, single anal plate .... *Roemerocrinus*

Scytalocrinus WACHSMUTH & SPRINGER, 1880, p. 116 [**Scaphiocrinus robustus* HALL, 1861a, p. 315; SD S. A. MILLER, 1889, p. 281] [=*Dactylocrinus* SLADEN, 1878, p. 245 (*non* QUENSTEDT, 1876) (type, *D. loreus*); *Scytalecrinus* BATHER in LANKESTER, 1900a, p. 180 (obj.)]. General form of crown very slender, almost cylindrical. Calyx obconical or bell shaped; infrabasals well developed, bent upward; radials and primibrachs of nearly same form; articular facets between them, straight, occupying entire width of plates; radianal obliquely on *BC* and *CD* basals, anal *X* may not touch *CD* basal. Arms long, composed of externally rectangular or slightly cuneiform joints, unbranched above primibrachs, pinnules of moderate size; anterior ray may bear single unbranched arm. Stem transversely round or obtusely pentagonal. *U.Dev.-U.Penn.*, USA.——FIG. 418,*5a-c*. **S. robustus* (HALL), L. Miss.(Keokuk), Ind.; *A*-ray, *CD*-interray, and dorsal views, ×1.5 (Hall, 1872). ——FIG. 418,*5d,e*. *S. validus* WACHSMUTH & SPRINGER, L.Miss.(Keokuk), Ind.; *5d*, ant. view with exposed sac, anal opening at midheight; *5e*, another specimen showing single arm in ant. ray; both ×1 (Springer, 1926a). [=*Pininocrinus* GOLDRING, 1938.]

Anemetocrinus WRIGHT, 1938, p. 339 [**A. biserialis*; OD]. Cup conical, low or high; anal area normal with radianal, anal *X*, and right tube plate within limits of cup. Arms 5, simple unbranched, wholly or partly biserial, brachials wedge shaped to interlocking. *L.Carb.*, Eu.(Scot.). ——FIG. 419,*2a,b*. **A. biserialis*, Low. Ls. Gr.; Fife; *2a*, dorsal view with arms, ×0.7; *2b*, dorsal view, ×1.5 (Wright, 1950-54).——FIG. 419,*2c*. *A. ardrossensis* WRIGHT, Calciferous Ss. Ser., Fife; post. view, ×1.9 (Wright, 1950-54).

Atrapocrinus STRIMPLE, 1975, p. 5 [**A. mutatus*; OD]. Crown cylinder shaped; cup low cone shaped with flat or mildly convex base; large basals with proximal tips curved into basal plane, large radials, one narrow anal plate. Arms 10, uniserial, brachials long, narrow, primibrachs *1* axillary. *M.Penn.*, USA(Okla.).——FIG. 420,*1*. **A. mutatus*, Atokan; holotype, *1a-c*, crown from *E* ray, *CD* interray, and *B* ray, ×2.5 (Strimple, 1975b).

Bollandocrinus WRIGHT, 1951, p. 62 [**Poteriocrinus conicus* PHILLIPS, 1836, p. 205; OD]. Cup moderately large, turbinate, plates thick; height of cup approximately equal to width across radials; infrabasals high, steeply upturned distally; basals and radials large; articular facets horizontal, full width of radials, with strong transverse ridge and deep ligamental furrow; anal area normal, with radianal and anal *X* adjoining *CD* basal. Anal sac unknown. Arms 10, stout, recti-uniserial, pinnulate, branching on short primibrachs *1*. Stem transversely round, heteromorphic, lumen pentalobate. *L.Carb.(Tournais.-Visean)*, G.Brit. ——FIG. 418,*4a*. **B. conicus* (PHILLIPS), Mtn. Ls., Eng. (Lancash.); post. view, ×0.7 (Wright, 1950-54).——FIG. 418,*4b,c*. *B. erectus* WRIGHT, Eng.(Yorks.); ant., post. views of holotype, ×1 (Wright, 1950-54).

Bridgerocrinus LAUDON & SEVERSON, 1953 [**B. fairyensis*; OD]. Cup steeply conical; infrabasals upturned distally, confluent with higher circlets of cup; articular radial facets occupying entire upper surface of plates; radianal pentagonal, resting on *BC* and *CD* basals, anal *X* on truncated upper surface of *CD* basal. Anal sac long and slender, composed of small hexagonal plates in parallel rows. Arms uniserial, long, slender, bifurcating once on primibrachs *1*, or *2*, or *3*, one long slender pinnule on each brachial. *U.Dev.-L.Miss.(Osag.)*, USA.——FIG. 418,*1*. **B. fairyensis*, L.Miss.(Lodgepole Ls.), Mont.; *1a*, post. view of paratype; *1b,c*, lat. views of holotype and paratype, all ×1.5 (Laudon & Severson, 1953).

Corematocrinus GOLDRING, 1923, p. 434 [**C. plumosus*; OD]. Crown moderately tall, expanding upward. Cup rather low truncate conical, incipient ridges and furrows at plate borders; infrabasals small, pentagonal, with distal parts visible from side; basals large, hexagonal, except for heptagonal *BC* basal; radial articular facets occupying full width of these plates; anals in cup as in *Scytalocrinus*. Anal sac tall and slender, composed of hexagonal plates arranged in vertical rows. Arms 10, uniserial, not appressed, branching isotomously once on primibrachs *2*, pinnules long and moderately stout. Stem strongly pentagonal transversely. *U.Dev.*, USA.——FIG. 418,*6*. **C. plumosus*, low. Chemung F.(Portage Ss.), N.Y.; *C*-ray view, ×1.5 (Goldring, 1923).

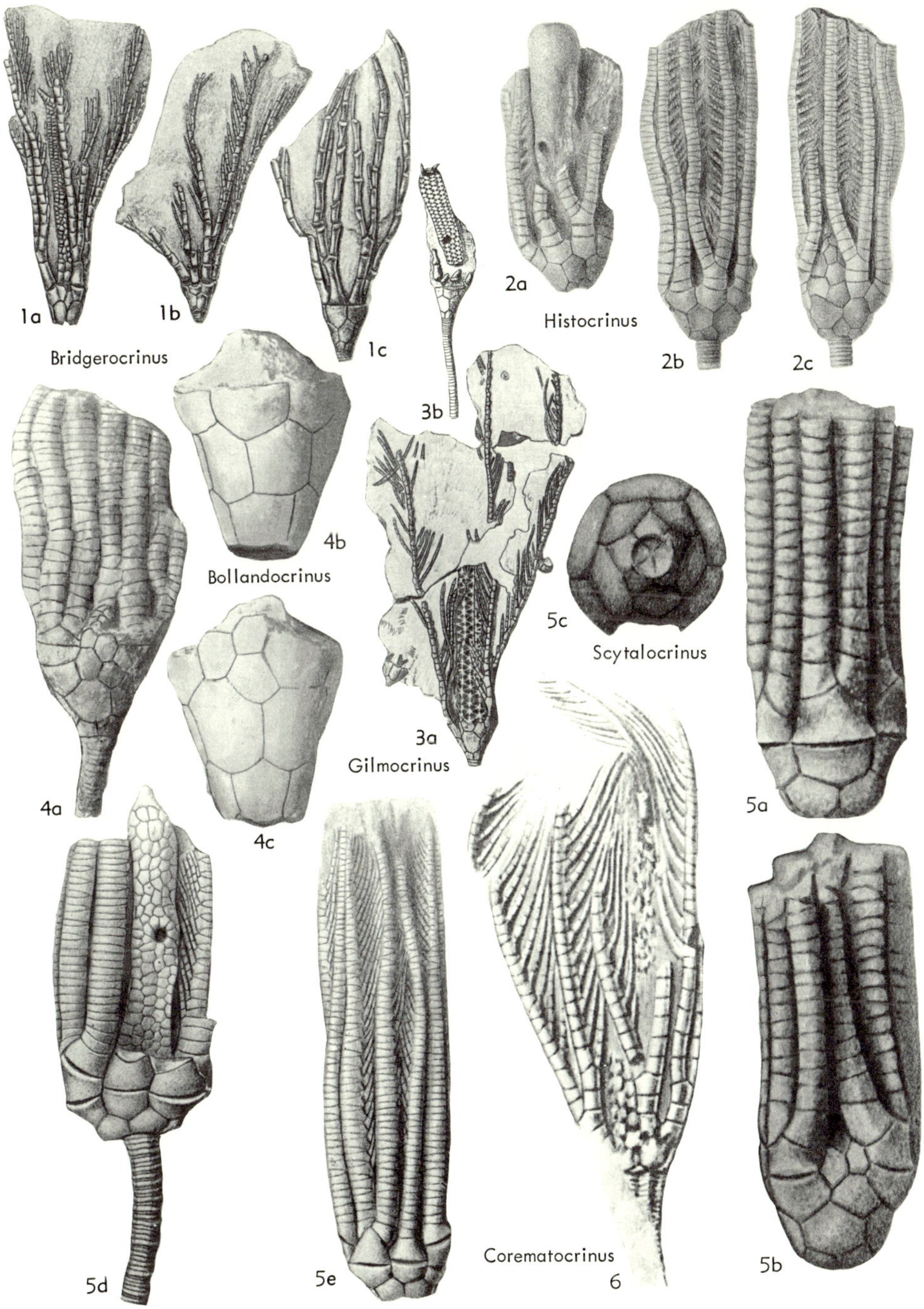

FIG. 418. Scytalocrinidae (p. *T*640-*T*643).

Gilmocrinus LAUDON, 1933, p. 70 [**G. iowensis*; OD]. Cup steeply conical, basals hexagonal except for *BC* and *CD* plates, which are heptagonal; radianal in contact laterally on one side with

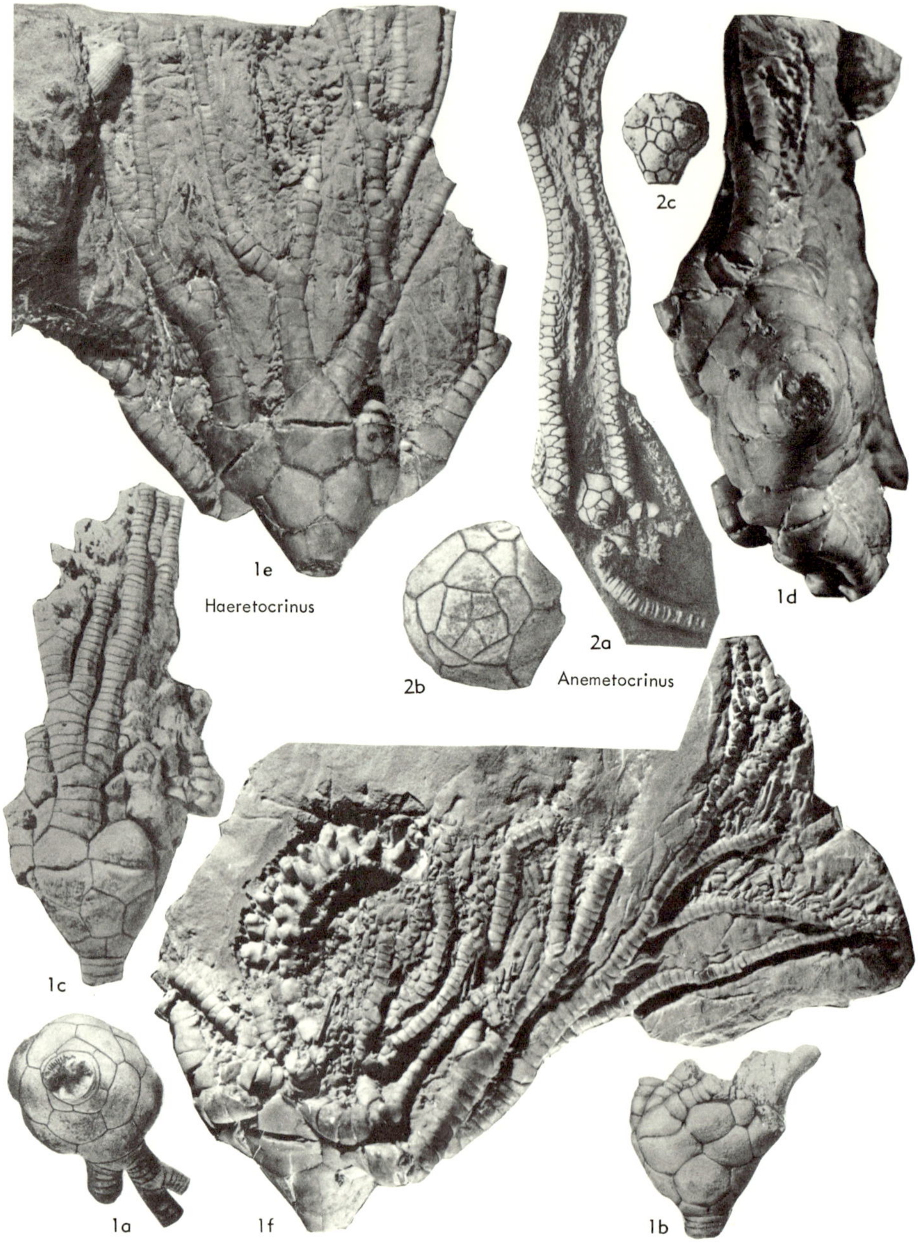

FIG. 419. Scytalocrinidae (p. T640, T643).

radial and basal, on other with anal *X* and *CD* basal; anal sac tall, cylindrical, plates in vertical rows, ornamented by raised radiating ridges, summit of sac carrying several spines, anus low on anterior side, much as in *Culmicrinus.* Arms slender, exceptionally long in adult specimens and comparatively short in immature ones, latter characterized by very long first brachials, arms uniserial, cuneiform brachials bearing stout pinnules; arms usually single but one of them may branch once, apparently with no regularity as to ray. *L.Miss.(Kinderhook.), Miss.,* USA(Mont.-Iowa); *L.Carb.,* G.Brit.(Scot., Isle of Man).——FIG. 418,*3.* **G. iowensis,* L.Miss.(Gilmore City Ls.),

Iowa; *3a,* holotype from post. side, ×0.7; *3b,* paratype, ant. view, ×1 (Laudon, 1933).

Haeretocrinus Moore & Plummer, 1940, p. 109 [**H. missouriensis*; OD]. Cup moderately high, cone shaped, infrabasals prominent, upflared confluently with higher circlets of cup; radials with wide articular facets; 3 anals in cup. Anal sac large recurved, tube composed of rugose plates with anal opening at midheight on anterior side. Arms robust, uniserial, first branching isotomously on primibrachs 1, then on secundibrachs *4* and at higher points, pattern exotomous. Stem large, transversely circular. *M.Penn.(Desmoines.)-U. Penn.(Virgil.),* USA(Ill.-Mo.-Okla.-Kans.-Texas).——Fig. 419,*1a,b.* **H. missouriensis,* Missour. (Lane Sh.), Mo.; holotype from base and post. side, ×1 (Moore & Plummer, 1940).——Fig. 419,*1c. H. turbinatus* Strimple, Missour.(Wann F.), Okla.; paratype crown from left *(DE)* side, ×1 (Strimple, 1952e).——Fig. 419,*1d-f. H. wagneri* Strimple & Moore, Missour.(LaSalle Ls.), Ill.; hypotype viewed from dorsal side (*CD* interray directed downward to right), *D* ray, and *AB* interray showing large, recurved anal sac, ×0.93 (Strimple & Moore, 1971a). [=*Haerteocrinus* Strimple, 1952e *(nom. null.)*.]

Histocrinus Kirk, 1940, p. 327 [**Poteriocrinus (Scytalocrinus) grandis* Wachsmuth & Springer, 1880, p. 118 (=*P. coreyi* Worthen in Worthen & Meek, 1875, p. 516); OD]. Crown compact, high, cylindrical but arms not appressed. Cup subturbinate, plates smooth, or faintly plicated; infrabasals well exposed in side view; basals medium sized; radials with notches between them, articular facets slightly crescentic, extending full width of radial summits, sutures not gaping; large radianal resting subequally on *BC* and *CD* basals, anal *X* large, extending well above radial summits, right tube plate smaller and almost entirely above level of radials. Anal sac subcylindrical, expanding slightly upward, reflexed so as to carry tip of sac and anal opening well down on anterior side, posterior side with 5 vertical series of plates; in distal recurved portion sac plates are tumid to subspinose. Arms moderately stout, composed of cuneate brachials, 2 primibrachs in all rays except anterior, which may be atomous; pinnules moderately stout, standing out stiffly from arms. Stem transversely circular, heteromorphic, made up of prominent series of nodals and internodals. *L.Miss.(Osag.),* USA(Miss. Valley-Ind.).——Fig. 418,*2.* **H. coreyi* (Worthen), L.Miss.(Keokuk), Ind.; *2a-c,* ant., left ant. (*E*-ray), post. views, ×0.7 (Worthen in Worthen & Meek, 1875).

Hydriocrinus Trautschold, 1867, p. 16 [**H. pusillus*; OD]. Crown moderately tall, inverted pear shaped. Cup high conical, with 5 prominent, upflared infrabasals; articular facets of radials wide, sutures gaping; 3 anal plates with anal *X* in broad contact with *CD* basal and having an outer ligament pit for articulation with first tube plate, right tube plate on radianal, also with upper articulating facet, distal edges of anal *X* and right tube plate not extending appreciably above radial summits and forming a confluent plane. Anal sac small, cylindrical, composed of rows of diminutive plates. Arms slender with well-rounded exteriors, elongate cuneate brachials with long side pinnule bearing; primibrachs *1* elongate, midsection slightly constricted, axillary; a second dichotomy may appear high in arms. Stem transversely pentagonal or subpentagonal at cup. [*Hydriocrinus* is thought to have evolved from *Phacelocrinus* and older forms assigned to the genus are probably not congeneric.] *M.Penn.(Desmoines.)-*

Fig. 420. Scytalocrinidae (p. *T*640).

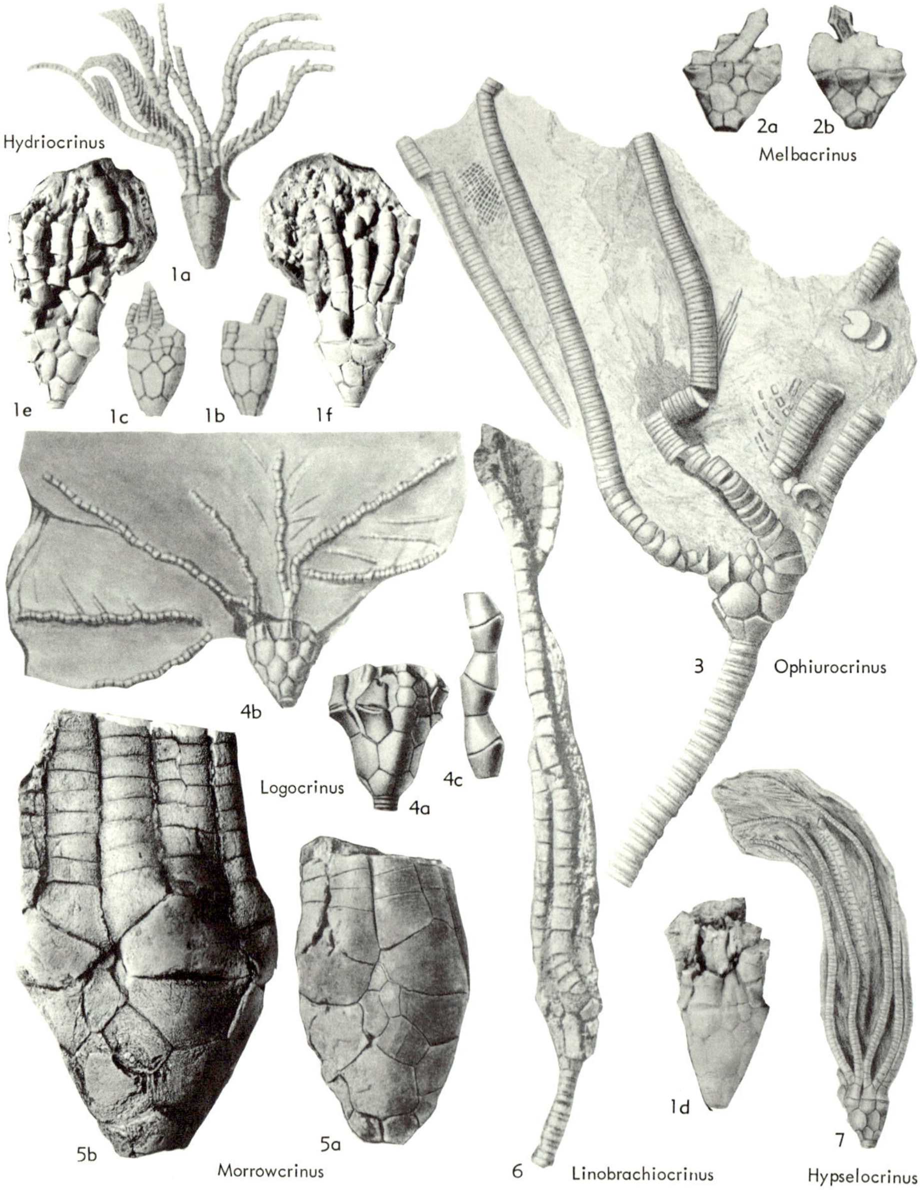

FIG. 421. Scytalocrinidae (p. *T643-T645*).

U.Penn.(Missour.), USA; *U.Carb.*, USSR.——FIG. 421,*1a-d*. **H. pusillus*, U.Carb.(Moscov.), Myachkovo, USSR; *1a-c*, ant. view of crown, ant. and post. views of cup, ×0.93 (Trautschold, 1867); *1d*, topotype showing anal tube, ×0.93 (Strimple, 1971d).——FIG. 421,*1e,f*. *H. turbinatus* STRIMPLE, U.Penn.(Missour., Wann F.), Okla.; holotype crown from post. and ant. sides, ×0.93 (Strimple, 1971d).

Hypselocrinus KIRK, 1940, p. 325 [**Poteriocrinus hoveyi* WORTHEN in WORTHEN & MEEK, 1875, p. 516; OD]. Crown very tall and slender. Cup high, narrowly turbinate; infrabasals long, forming lower 0.25 of cup; basals and radials large, articular radial facets not gaping; radianal high on upper shoulders of *BC* and *CD* basals, anal *X* large, right tube plate approximately half within cup. Anal sac long, slender, 6 or 7 times height of cup, composed of 10 or more vertical series of hexagonal plates. Arms very long, slender, typically not dividing above primibrachs *1* and may be atomous in *A* ray, brachials cuneate, pinnules long, slender. Stem circular in section. [Differs from *Scytalocrinus* in taller, steeply conical cup with long upturned infrabasals, slender arms, and tall anal sac.] *L.Miss.(Osag.)*, USA(Ind.-Iowa). ——FIG. 421,*7*. **H. hoveyi* (WORTHEN), Keokuk (Borden Gr.), Ind.; lat. view, ×0.47 (Worthen in Worthen & Meek, 1875).

Linobrachiocrinus GOLDRING, 1939, p. 354, *pro Linocrinus* GOLDRING, 1938b (Aug.), p. 19 (*non* KIRK, 1938[April]) [**Linocrinus kindlei* GOLDRING, 1938b, p. 21; OD]. Crown extremely tall and slender, arms almost appressed. Cup elongate obconical; infrabasals 5, comparatively small and inconspicuous, but visible from side; basals very large; radials considerably smaller than basals, with linear articular facets occupying full width of radial; *B* radial crowded by *C* radial, considerably reduced in size and apparently armless; radianal, anal *X*, and right tube plate within cup, *X* in line with radials. Arms unbranched throughout length, 4 in number, each brachial bearing 2 pinnules. Stem stout and transversely round. [Very similar to *Cradeocrinus* in character of cup, *Linobrachiocrinus* is distinguished by its 4 unbranched arms and irregularity shown by *B* and *C* radials.] *U.Dev.*, Can.(Mackenzie Basin, NW. Terr.).——FIG. 421,*6*. **L. kindlei* (GOLDRING); *C*-ray view, ×1 (Goldring, 1938b).

Logocrinus GOLDRING, 1923, p. 437 [**L. geniculatus*; OD]. Crown expanding upward, with widely separated arms. Cup steeply conical or bell shaped; infrabasals pentagonal, wider at top than at base, large, with length half that of basals; basals hexagonal except for heptagonal *BC* and *CD* basals; radials pentagonal, 2 posterior ones smaller than others, flaring out slightly near tops, articular facets straight, occupying entire upper face of radials; radianal pentagonal, fairly large, about same in size as infrabasals; anal *X* hexagonal, smaller than posterior radials, slightly larger than radianal, its upper edge even with or slightly above summits of radials. Arms very slender, well rounded, with zigzag succession of cuneate brachials; 2 primibrachs in *D* ray and 3 in others (or uncommonly with 2 primibrachs in *B* and *D* rays); stout pinnules borne by some brachials but pattern undetermined. Only lower plates of anal tube exposed; large, hexagonal. Stem heteromorphic, proximal columnals slightly subpentagonal but circular below. *M.Dev.*, USA (N.Y.).——FIG. 421,*4*. **L. geniculatus*, Hamilton Gr.; *4a*, *D*-ray view, ×2; *4b*, *A*-ray, ×1.4; *4c*, brachials, × 4.7 (Goldring, 1923).

Melbacrinus STRIMPLE, 1939, p. 17 [**M. americanus*; OD]. Cup evenly expanded, cone shaped; infrabasals flared evenly upward from columnal attachment; articular facets of radials straight, filling their full width; 3 anal plates in *CD* interray, anal *X* and *RX* with confluent distal faces even with cup summit. Primibrachs compound with division apparent in some rays. Column transversely round. *U.Penn.*, USA(Okla.).——FIG. 421,*2*. **M. americanus*, Missour.(Wann F.), near Bartlesville, Okla.; *2a,b*, holotype from post. and ant. side, ×2 (Strimple, 1939a).

Morrowcrinus MOORE & PLUMMER, 1938, p. 242 [**M. fosteri*; OD]. Cup conical, height approximately equal to width; infrabasals 5, upflared, clearly visible from side; basals moderately large (abnormal in *BC* interray of holotype of *M. fosteri* by occurrence of horizontal suture dividing plate into lower and upper parts but normal in *M. defendens*); radials 5, moderately large, with nearly straight articular facets occupying full width of these plates; 3 anal plates in cup comprising relatively large radianal obliquely at right above *CD* basal, followed above by anal *X* and right tube plate, which project slightly above radials. Well-developed anal sac indicated by cross section at height of secundibrachs *6* or *7*. Appressed arms at least 10 above first axillary primibrachs which are slightly larger than radials, arms uniserial, composed of large quadrangular segments; pinnules not observed. Stem transversely round, with strongly pentalobate axial canal. *L.Penn. (Morrow.)*, USA(Okla.-Ark.-Utah).——FIG. 421, *5a*. **M. fosteri*, Ark.; post. view, ×1.5 (Moore & Plummer, 1938).——FIG. 421,*5b*. *M. defendens* (WASHBURN), Morrow.(Oquirrh F.), Provo Canyon, Utah; post. view of holotype, ×1.4 (Strimple, n; Brigham Young Univ. 1496).

Ophiurocrinus JAEKEL, 1918, p. 62 [**Poteriocrinus originarius* TRAUTSCHOLD, 1867, p. 2; M]. Cup conical with upflared large infrabasals visible from side; basals and radials large, latter with linear straight articular facets equal to full width of these plates, large radianal and anal *X* on *CD* basal, only lower tip of small right tube plate in cup. Anal sac unknown. Five very robust uniserial long arms almost circular in section, resembling stout homeomorphic stems, brachials very short. Stem circular in section, distinctly heteromorphic, with alternating thick and thin columnals. *U.Carb.*, USSR-G.Brit.-Eire; *U.Miss.*, USA. ——FIG. 421,*3*. **O. originarius* (TRAUTSCHOLD), Moscov., USSR; *CD*-interray view, ×0.7 (Trautschold, 1967).

Pegocrinus KIRK, 1940, p. 331 [**Poteriocrinus bijugus* TRAUTSCHOLD, 1867, p. 14; OD]. Crown high, with long cylindrical well-separated arms. Cup broadly turbinate, thick plated; infrabasals small, partially anchylosed into firm circlet, clearly visible in side view, sutures distinct; basals variable in size and shape, pointed at top and barely

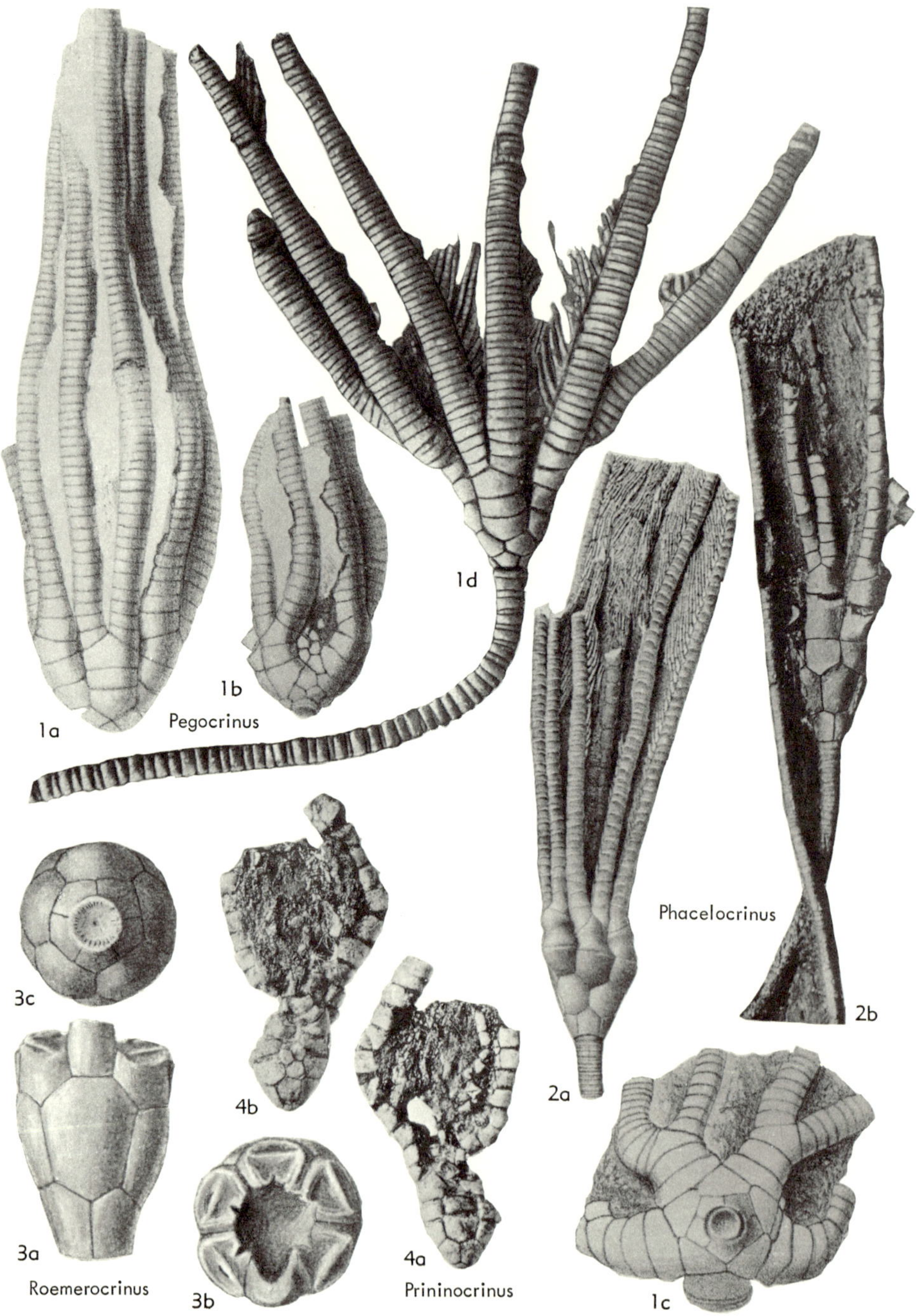

FIG. 422. Scytalocrinidae (p. T645-T647).

meeting laterally or on very short faces; *CD* basal high, narrow, supporting radianal on its long, upper right shoulder, barely meeting anal *X* above. Radials large, with articular facets full width of radial, apposed edges of radials and primibrachs along strongly crenulate sutures, not gaping, facets with prominent fulcral ridge, deep ligament pit, and large, well-defined pair of muscular fossae. Radianal small, elongate, narrow; *X* small, high narrow; right tube plate small. Anal sac poorly known (one specimen shows partially dissociated plates in approximately their original position indicating sac to be short, erect, and fairly stout in relation to cup, with small thin plates forming fragile structure). Arms uniserial, robust, long, with 2 undivided branches to each ray; 2 or 3 primibrachs variable in number without respect to rays (usually 2); brachials low, nearly cylindrical, quadrangular as seen from side, each bearing long, stout pinnule. Stem transversely circular, with stellate lumen. *U.Carb. (Moscov.)*, USSR.——FIG. 422,*1*. **P. bijugus* (TRAUTSCHOLD); *1a-c*, Moscow, *A*-ray, *CD*-interray, and dorsal views of 3 specimens, ×1 (Trautschold, 1867); *1d*, Myachkovo, ant. view of crown and stem, ×1 (Yakovlev & Ivanov, 1956).

Phacelocrinus KIRK, 1940, p. 329 [**Poteriocrinus wetherbyi* S. A. MILLER, 1879a, p. 36; OD]. Crown high, subcylindrical or spreading somewhat upward. Cup steeply conical; infrabasals high; basals medium sized; radials large, articular facets straight, occupying full width of radials, sutures gaping; radianal reaching well down between *BC* and *CD* basals; anal *X* extending above level of radials, right tube plate mostly out of cup. Anal sac cylindrical, with maximum height about 4 times that of cup, composed of vertical rows of hexagonal plates. Arms uniserial, very long and slender, primibrachs 2 in each ray or fusing to single primaxil constricted medially, 2 undivided arms to each ray typically, but irregular branching may occur in some Chester forms; brachials quadrangular or slightly cuneate, with long slender pinnules. Stem pentagonal in section. *L.Carb.*, Eu.; *U.Miss.(Chester.)*, USA(Miss.Valley-Ala.-Ky.-Ill.).——FIG. 422,*2a*. **P. wetherbyi* S. A. MILLER, U.Miss.(Kaskaskia Gr.), Ky.; ant. view of crown, ×1 (S. A. Miller, 1879a).——FIG. 422,*2b*. *P. rostratus* (AUSTIN & AUSTIN), L.Carb., Eire; lat. view, ×1.5 (Wright, 1950-54).

Prininocrinus GOLDRING, 1938, p. 18 [**P. robustus*; OD]. Cup cone shaped with small upflared infrabasals; basals elongated; radials moderate sized, articular facet occupying full width of radial; 3 anal plates in advanced arrangement (radianal in *CD* position with equidimensional anal *X* and right tube plate above). Anal sac unknown. Arms relatively large, uniserial, pinnulate, branching once on primibrachs 2. Stem apparently round. [*Prininocrinus* was assigned by GOLDRING to the family Poteriocrinidae (*nom. null.* =Poteriocrinitidae); its wide radial articular facets are atypical of this family, however, and assignment is made here to the Scytalocrinidae.] *U.Dev.*, N.Am.(Can.).

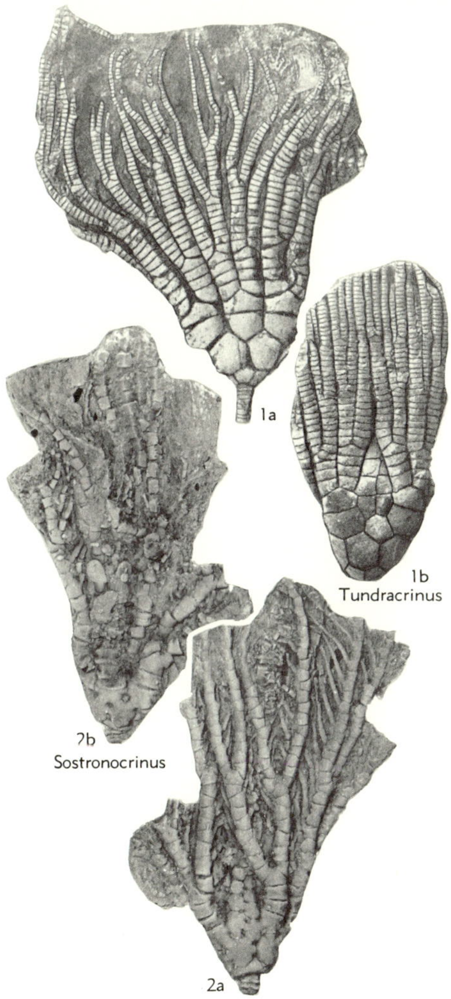

FIG. 423. Scytalocrinidae (p. *T*647-*T*648).

——FIG. 422,*4*. **P. robustus*; *4a,b*, holotype from ant. and post. sides, ×1.5 (Goldring, 1938b).

Roemerocrinus WANNER, 1916, p. 205 [**R. gracilis*; OD]. Characters of family; with one anal plate in cup. *L.Perm.-U.Perm.*, Indon.(Timor, Basleo)-Eu.(USSR, Krasnoufimsk).——FIG. 422,*3*. **R. gracilis*, Indon.(Timor); *3a-c*, post., ventral (*CD* interray downward), and dorsal views of cup, ×3 (Wanner, 1916).

Sostronocrinus STRIMPLE & MCGINNIS, 1969, p. 21 [**S. superbus*; OD]. Crown elongate, expanded, uniserial, pinnule-bearing arms not apposed. Cup truncate cone shaped, angles of plates impressed, distal ends of infrabasals visible in side view of cup; composed of 5 infrabasals, 5 basals, 5 radials and 3 anal plates in normal (primitive) arrangement. Primibrachs *2* or *3* and secundibrachs *7-11* axillary. Anal sac long, reflexed, plates small, thin, with radiating ridges. Column transversely round. *L.Miss.(Kinderhook.)*, USA(Iowa).——FIG. 423,*2*. **S. superbus*, Iowa; *2a,b*, holotype from *CD* interray and *A* ray, ×1.6 (Strimple & McGinnis, 1969).

Tundracrinus YAKOVLEV, 1928, p. 1 [**T. polaris*; OD]. Cup slightly rotund but essentially bowl shaped; infrabasals upflared; radial articular facets plenary; anal plates advanced, with radianal in direct *CD* position, followed by equidimensional anal *X* and right tube plate above, forming confluent horizontal facet. Arms well rounded externally, uniserial, branching isotomously on primibrachs *1*, secundibrachs *4* and at higher points. *L.Perm.*, USSR(NE.Eu.Russia).——FIG. 423,*1*. **T. polaris*; *1a,b*, ant. and post. views of crown, ×0.75 (Yakovlev, 1928b).

Family BLOTHROCRINIDAE Moore & Laudon, 1943

[Blothrocrinidae MOORE & LAUDON, 1943a, p. 55]

Crown tall, cylindrical or expanded upward. Cup conical, five or three upflared infrabasals readily visible from side, radial articular facets plenary, bearing transverse ridge and ligament pits; three anals in normal arrangement. Anal sac tall, cylindrical. Arms uniserial, pinnulate, branching two or more times. Stem long, transversely circular. *L.Miss.-L.Perm.*

The Blothrocrinidae constitute a main branch identified tentatively as derived from the Dendrocrinidae or Ottawacrinidae. They represent the central or at least the most generalized line of evolutionary development. The blothrocrinids resemble dendrocrinids in all characters except that the arms are advanced toward numerous isotomous divisions and an abundance of well-developed pinnules.

Key to Genera of Blothrocrinidae

A. Cone-shaped cup, with 5 prominent upflared infrabasals; 3 anal plates in cup ....
- I. First arms branching on primibrachs *1* in all rays ....
  - a. Arms slender, second branching on secundibrachs *1* or rarely on *2* unbranched above .... *Ulrichicrinus*
  - b. Arms slender, second branching on secundibrachs *1*, additional divisions above .... *Nebraskacrinus*
  - c. Arms very stout, second bifurcation on secundibrachs *7* or *8*, unbranched above .... *Woodocrinus*
- II. First arm branching on primibrachs *2*, brachials equiuniserial ....
  - a. Second arm branching on secundibrachs *6-13*, arms stout .... *Stinocrinus*
  - b. Second arm branching on secundibrachs *6* or *7*, arms slender .... *Blothrocrinus*
  - c. Second arm branching on secundibrachs *6-10*, arms stout .... *Fifeocrinus*
  - d. Second arm branching normally on secundibrachs *2* .... *Moscovicrinus*
- III. First branching of arms above primibrachs *2*, second above secundibrachs *2* ....
  - a. Radial articular facets wide, no interradial notches .... *Culmicrinus*
  - b. Prominent notches between radial articular facets .... *Carcinocrinus*

B. Same as A but with only 3 infrabasals, first branching on primibrachs *1* or *2* in all rays .... *Elibatocrinus*

Blothrocrinus KIRK, 1940, p. 321 [**Poteriocrinus jesupi* WHITFIELD, 1881, p. 7; OD]. Crown elongated, expanded. Cup steeply conical; 5 upflared infrabasals prominent; 3 anal plates in normal (primitive) arrangement. Arms numerous (50-60), uniserial, well-rounded exteriors, first branching on primibrachs *2* with several irregular isotomous bifurcations above. Stem large, transversely round, homeomorphic, composed of thin columnals. *L.Miss.(Kinderhook.-Osag.)*, USA; *L.Carb.(Tournais.)*, Eng.-Afr.(Morocco)-USSR(Kazakh.).——FIG. 424,*2*. **B. jesupi* (WHITFIELD), Miss. (Osag., Burlington Ls.), Iowa; *2a*, post. view of crown, ×0.5 (Whitfield, 1881); *2b*, plate diagram of cup (radials black, radianal cross ruled, other anal plates stippled) (Moore, 1962b).

Carcinocrinus LAUDON, 1941, p. 390 [**C. stevensi*; OD] [=*Carinocrinus* STRIMPLE, 1953b, p. 201 (*nom. null.*)]. Crown large, relatively slender arms extending about a massive, cylindrical anal tube. Cup sharply conical, infrabasals upflared, prominent; distal portions of radials separated by large notches, but articular facets equal in width to radial summits; 3 anal plates in normal (primitive) arrangement. Anal sac cylindrical, as wide as cup, exceptionally tall, with rounded summit well above arm tips, composed of series of hexagonal plates, with large snoutlike projection at about midheight of crown on anterior side. Arms uniserial, brachials cuneate above first bifurcation on primibrachs *5* (only *B* and *C* rays known), second isotomous branching in 3 half-rays and a third in one outer ray. Stem transversely round, slightly tapered, columnals thin. *U.Miss.(Chester.)*, USA.——FIG. 424,*3*. **C. stevensi*, NE.Okla.; *BC* side of holotype crown, ×0.7 (Laudon, 1941).

Culmicrinus JAEKEL, 1918, p. 62 [**Poteriocrinus*

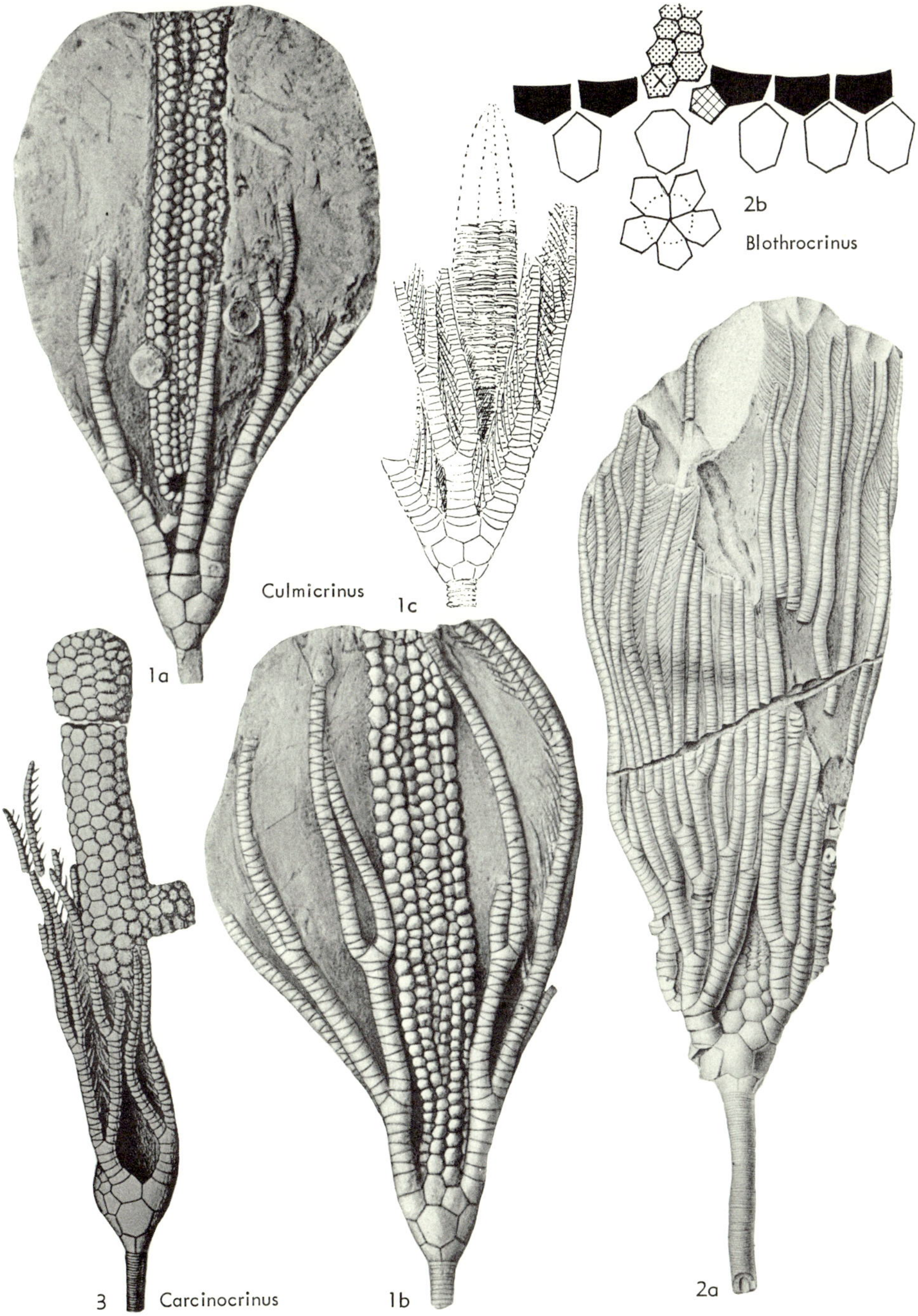

FIG. 424. Blothrocrinidae (p. *T648-T650*).

regularis MEYER, 1858, p. 119; OD]. Cup tall, cone shaped; infrabasals prominent; basals moderately long; 3 anal plates in normal (primitive) arrangement. Anal sac long, slender, recurved tube with anal opening on anterior side, sac plates small, in numerous vertical series. Arms uniserial,

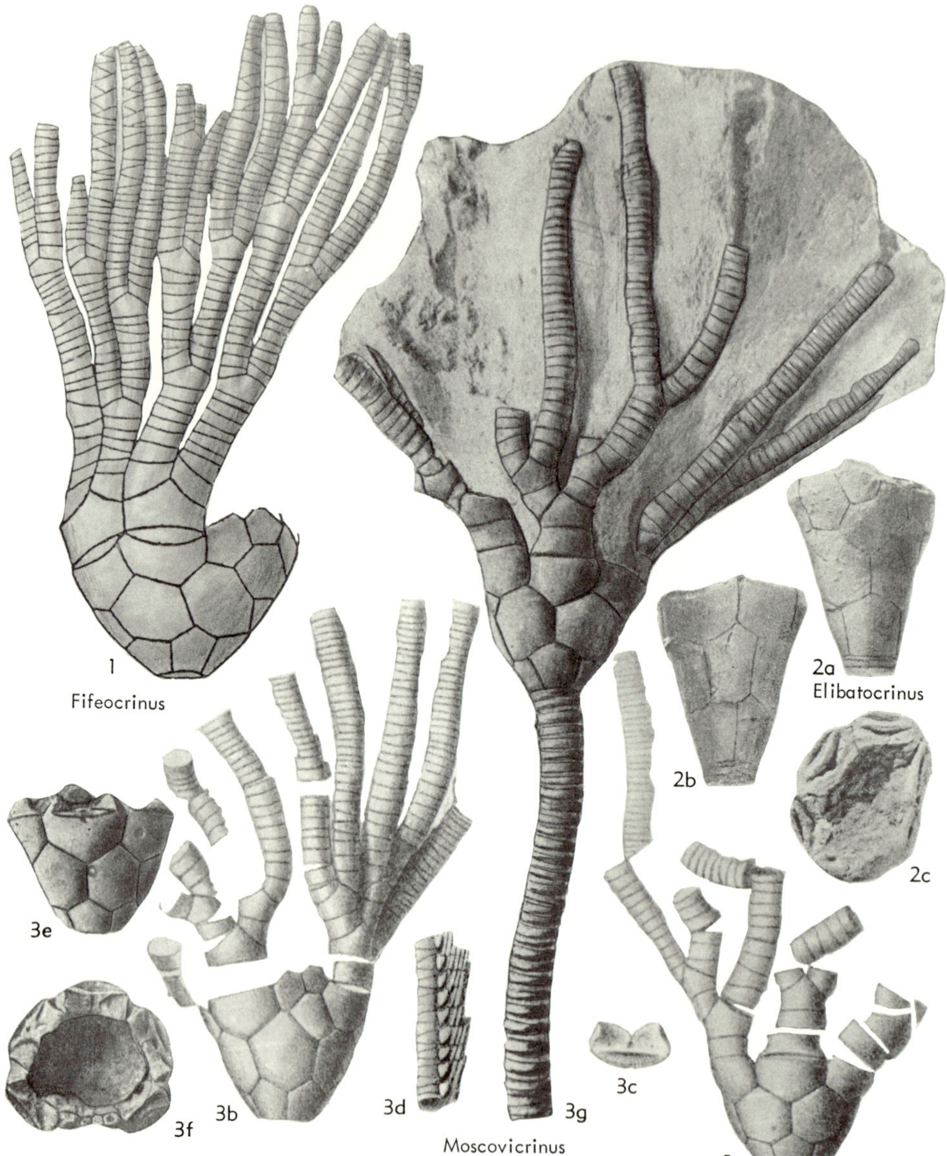

FIG. 425. Blothrocrinidae (p. T650-T651).

exteriors well rounded, first branching above primibrachs 2, usually *A*-ray branching much higher than in other 4 rays, second isotomous division at about midheight of crown. Stem transversely round (exception, *C. thomasi* LAUDON, which may have subpentagonal proximal columnals). *L.Miss.(Kinderhook.)-U.Miss.(Chester.)*, USA; *L.Carb.*, Eu.——FIG. 424,*1a,b*. *C. elegans* (SPRINGER), Chester.(Glen Dean), Ky.; ant. and post. views of crown, ×1 (Springer, 1926b).——FIG. 424,*1c*. **C. regularis* (MEYER), L.Carb. (Visean), Ger.; crown from ant. side, ×1.5 (Jaekel, 1918).

Elibatocrinus MOORE, 1940, p. 36 [**E. leptocalyx*; OD]. Crown moderately tall, slender. Cup steep sided conical; infrabasals 3, elongate, externally visible parts beyond stem forming one-third height of cup, much thickened proximally but thin at distal margins, small azygous plate in position of *C* ray; radial articular facets equal to radials in width, with strong transverse ridge and ligament fossae slightly shorter; 3 anal plates in cup. Anal sac tall but shape unknown, small hexagonal sac plates with slitted margins. Arms rounded, slender and very long, 2 in each ray, slightly cuneate brachials uniserial, bearing long pinnules. Stem transversely circular, heteromorphic. *M.Penn. (Desmoines.)-U.Penn.(Virgil.),* USA(Kans.-Okla.-Texas).——Fig. 425,*2*. **E. leptocalyx,* Missour. (Stoner Ls.), Wilson Co., Kans.; *2a-c,* holotype from post. and ant. sides and ventral view of cup (*A* radial upward), ×2 (Moore, 1940a).

Fifeocrinus Wright, 1951, p. 41 [**Pachylocrinus tielensis* Wright, 1936, p. 399; OD]. Cup medium tall conical; infrabasals upflared, prominent; 3 anal plates in normal (primitive) arrangement. Arms equiuniserial, moderately stout, usually with 3 isotomous branchings, first on primibrachs 2, second on or about secundibrachs *6-10,* with further erratic branching high in crown. *L.Carb. (Visean),* Eu.(G.Brit.).——Fig. 425,*1*. **F. tielensis* (Wright), Lower Ls. Gr., Scot.; *D*-ray view of holotype, ×1 (Wright, 1950-54)

Moscovicrinus Jaekel, 1918, p. 62 [**Poteriocrinus multiplex* Trautschold, 1867, p. 6; OD]. Cup conical, upflared infrabasals prominent, radial articular facets plenary, but interradial notches present. Arms isotomous, branching 3 times in some rays, first bifurcation on primibrach *2* (or may be *3* in *A* ray), second usually on secundibrach *2* but may be on *4* to *7,* third usually on tertibrach *2* but may be higher. *U.Carb.(Moscov.),* USSR; *L.Perm.(Wolfcamp.),* USA-Indon.(Belitung).——Fig. 425,*3*. **M. multiplex* (Trautschold), Moscov., USSR(Moscow Bas.); *3a,b,* crown from ant. and post. sides; *3c,* summit view of radial; *3d,* part of arm and attached pinnules in side view; all ×1 (Trautschold, 1867); *3e,f,* cup from ant. side and ventral view; *3g,* ant. view of crown with part of attached column; all ×1 (Yakovlev & Ivanov, 1956).

Nebraskacrinus Moore, 1939, p. 198 [**N. tourteloti*; OD]. Cup conical with sides somewhat steeper in upper part than in lower; infrabasals readily visible in side view; *CD* interray slightly bulged, 3 large anal plates in normal (primitive) arrangement. Anal sac strong, tapering upward, composed of several columns of plates without intervening slits. Arms uniserial, primibrachs *1* large, wide, axillary, branching thereafter heterotomous, with secundibrachs *2* axillary in some rays. Stem transversely round, lumen pentastellate. *L.Perm.,* USA(Neb.).——Fig. 426,*2*. **N. tourteloti,* Big Blue Series (Winfield Ls.); *2a,b,* holotype from *AB*-interray and post. sides, ×1.4 (Moore, 1939c).

Stinocrinus Kirk, 1941, p. 84 [**S. granulosus*; OD]. Crown medium in height. Cup low bowl shaped; infrabasals barely visible in side view, radials dominant cup elements, 3 anal plates in normal (primitive) arrangement. Arms equiuniserial, first branching on primibrachs *2* in all rays except *A* ray, which may have as many as *9* primibrachs, second isotomous branching high in crown on secundibrachs *6-13* (except *A* ray), with third branching in some outer half-rays. *L.Miss. (Osag.),* USA(Ky.).——Fig. 426,*4*. **S. granulosus,* New Providence F.; *4a,b,* holotype from *D*-ray and *AB*-interray sides; ×1.4 (Kirk, 1941b).

Ulrichicrinus Springer, 1926, p. 75 [**U. oklahoma*; OD]. Crown tall. Cup conical, infrabasals readily visible from side; first primibrachs low, axillary, second bifurcation on secundibrachs *1* or *2* in some half-rays. *L.Miss.(Osag.)-U.Miss. (Chester.),* USA(Okla.-Ga.-Ind.); *L.Miss.(Osag.)-M.Penn.(Desmoines.),* USA(Okla.-Texas).——Fig. 426,*1*. **U. oklahoma,* L.Penn.(Morrow.), Okla.; *1a,b,* holotype from post. *D*-ray and *AB*-interray sides, ×0.62 (Springer, 1926b).

Woodocrinus de Koninck, 1854, p. 210 [**W. macrodactylus*; OD]. Crown widely expanded. Cup low cone shaped; 5 infrabasals upflared but not prominent, tendency toward short sutures between basals, 3 anal plates in normal (primitive) arrangement. Arms typically 20, uniserial, composed of exceptionally wide, very short brachials, first branching on primibrachs *1* in all rays, second branching on about secundibrachs *6-8,* no higher bifurcations, branching variable among specimens and species, including lack of second bifurcation in some rays. *U.Carb.(Namur.),* Eu. (Eng.).——Fig. 426,*3*. **W. macrodactylus,* Namur.(Red Beds above Main Ls.); *3a,* lectotype crown from post. (*C* ray) side, ×0.62 (Springer, 1926b); *3b,* slab bearing several crowns of this species, approximately ×0.2 (Wright, 1950). [=*Philocrinus* de Koninck, 1863.]

Family CERCIDOCRINIDAE Moore & Laudon, 1943

[Cercidocrinidae Moore & Laudon, 1943a, p. 55]

Cup small, low; three anal plates in cup. Anal sac recurved, slender or with greatly expanded distal extremity. Arms stout, uniserial, branching isotomously on first primibrachs and then endotomously several times; primibrachs *1* axillary in all rays. Stem transversely round. *L.Miss.*

Suppression of certain isotomous bifurcations of the blothrocrinid arms produces the endotomous structure that characterizes the pinnule-bearing arms of the Cercidocrinidae. Except in arm structure, and possibly greater inflation of the anal sac, the cercidocrinids do not differ from the blothrocrinids, and doubtless they were developed from this stock.

Cercidocrinus Kirk, 1938, p. 163 [**Poteriocrinus bursaeformis* White, 1862, p. 10; OD]. Cup cone

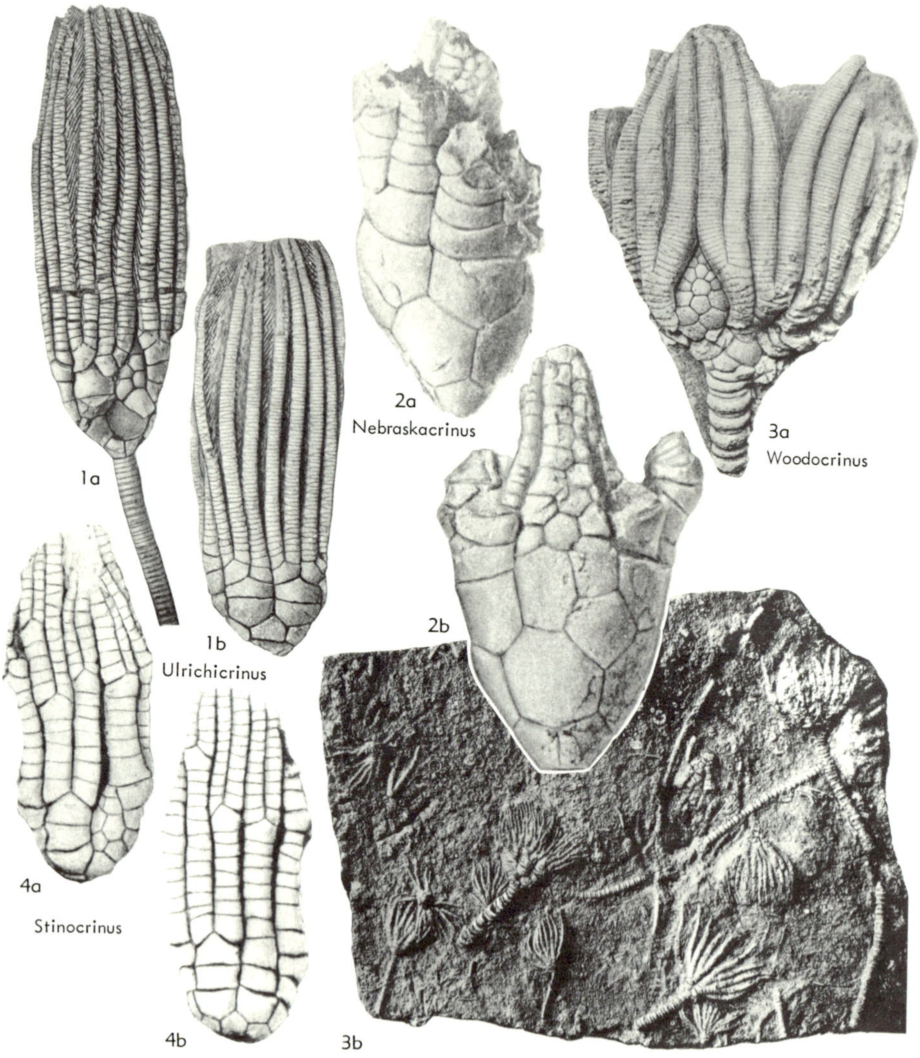

FIG. 426. Blothrocrinidae (p. *T*651).

shaped, plates smooth; infrabasals low, clearly visible in side view; radial articular facets fully as wide as radials. Arms stout, rounded, branching isotomously on first primibrachs and then endotomously several times; brachials low and broad, pinnulate. *L.Miss.*, USA(Iowa-Mo.).——FIG. 427,*3*. **C. bursaeformis* (WHITE), Burlington Ls.(Osag.), Iowa; post. view of crown, ×0.7 (Springer, 1926b).

Ascetocrinus KIRK, 1940, p. 324 [**Scaphiocrinus rusticellus* WHITE, 1863, p. 505; OD]. Crown elongate and slender. Cup small, cone shaped, with small pits at plate angles; infrabasals small, barely visible in side view; radials large, articular facets plenary; sutures with primibrachs gaping; right tube plate just below top of *C* radial. Anal sac slender, with 8 vertical rows of plates, recurved, anal opening about midway on anterior side, plates with pores at angles. Arms long, slender, branching endotomously 2 or 3 times above axillary first primibrachs; brachials high, wedge shaped, with marked lateral shoulders for support of long, slender pinnules. *L.Miss.*, USA (Iowa).——FIG. 427,2. **A. rusticellus* (WHITE), Burlington Ls. (Osag.); lat. view of crown and proximal stem, ×1 (Springer, 1926b).

Coeliocrinus WHITE, 1863, p. 498 [**Poteriocrinus dilatatus* HALL, 1861b, p. 6; SD S. A. MILLER, 1889, p. 233]. Cup low, broad bowl shaped; infrabasals barely visible in side view of cup;

radial articular facets as wide as radials. Anal sac large, strongly inflated distally, composed of numerous stout plates with strong radiating ridges or with spinose summit plates. Arms with up to 5 endotomous branches above large, axillary first primibrachs; brachials strongly wedge shaped, pinnulate. *L.Miss.*, USA(Iowa-Mo.-N.Mex.).——FIG. 427,*1a-d*. **C. dilatatus* (HALL), Burlington Ls. (Osag.), Iowa; *1a,b*, *D*-ray and *BC*-interray views of crown, ×1.5; *1c,d*, *D*-ray, and *BC*-interray views of another crown, ×1.5 (all Springer, 1926b).——FIG. 427,*1e,f*. *C. ventricosus* (HALL), Burlington Ls.(Osag.), Iowa; proximal and distal views of isolated inflated part of anal sac, ×1 (Springer, 1926b).

Family APHELECRINIDAE Strimple, 1967

[Aphelecrinidae STRIMPLE, 1967, p. 82]

Crown slender cylindrical, flared upward, ovoid. Cup moderately low cone shaped, height and width subequal; five small infrabasals upflared, rising directly from round columnar cicatrix; five relatively short basals; five radials and three anal plates in normal (primitive) arrangement. Arms uniserial, branching on primibrachs *1* in all rays except that in earlier forms first branching in *A* ray is above primibrach *2*, all rays with one or subsequent isotomous bifurcations, pinnules long and stout. Anal sac slender, moderately to strongly reflexed, composed of series of small, polygonal plates. Column round (exception, *Aphelecrinus mundus* KIRK, and *A. oweni* KIRK, reportedly with subpentagonal proximal columnals [KIRK, 1944b]), cirriferous in few forms. *L.Miss. (Kinderhook.)-U.Miss.(Chester.)*.

Key to Genera of Aphelecrinidae

Crown slender cylindrical or upflared, arms in most or all rays branching on primibrachs 1 ....

- I. Arms branching isotomously in all rays on primibrachs *1*, stem not cirriferous ..
  - a. Crown slender cylindrical, brachials externally cuneate .... *Cosmetocrinus*
  - b. Crown flared upward, brachials externally rectangular .... *Aphelecrinus*
- II. Arms branching in *A* ray above primibrach *1*, stem proximally very cirriferous .... *Paracosmetocrinus*

Aphelecrinus KIRK, 1944, p. 190 [**A. elegans*; OD] [=*Amphelecrinus* LAUDON & SEVERSON, 1953 *(nom. null.)*; *Ampelecrinus* LAUDON, PARKS, &

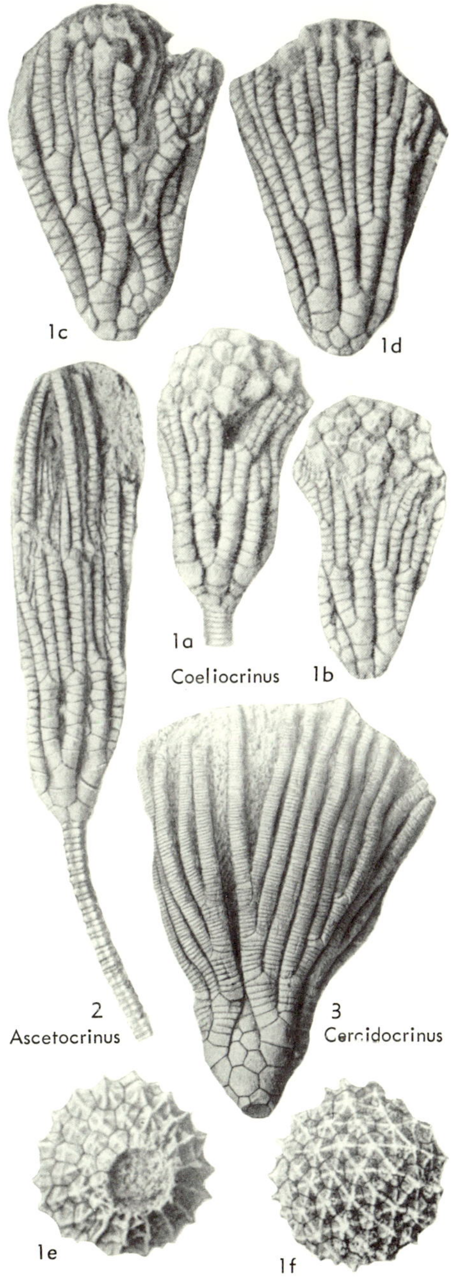

FIG. 427. Cercidocrinidae (p. *T*651-*T*653).

SPRENG, 1952 *(nom. null.)*]. Crown moderately high. Cup cone shaped, infrabasals rising evenly from columnar cicatrix; 3 anal plates in normal placement; arms branching on primibrachs *1* in all rays and once again in higher positions (at

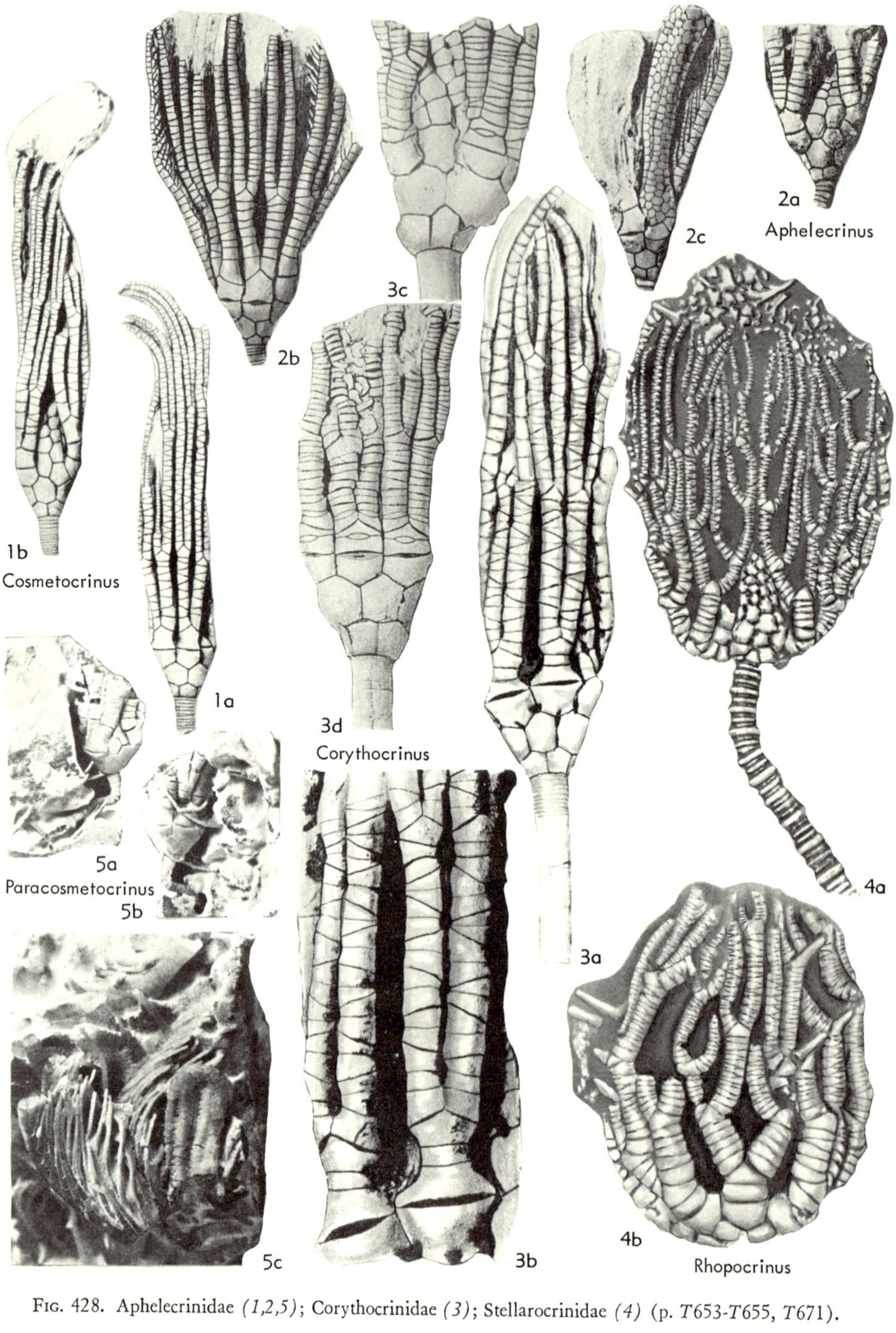

FIG. 428. Aphelecrinidae *(1,2,5)*; Corythocrinidae *(3)*; Stellarocrinidae *(4)* (p. *T*653-*T*655, *T*671).

about 0.3 to 0.5 height of arms), sporadic divisions above uncommon, brachials with slightly sloping articular facets in earlier species, becoming cuneate in later forms. Anal sac extending to about one-half or two-thirds height of arms, reflexed and composed of numerous vertical rows of small polygonal plates. Stem typically circular in section but in 2 species *(A. oweni, A. mundus)* proximal columnals are subpentagonal. *L.Miss.(Kinderhook.)-U.Miss.(Chester.),* USA; *L.Carb.,* Eu.(G.Brit.).——FIG. 428,2. **A. elegans,* Meramec.(Ste.Genevieve Ls.), USA(Ala.); *2a,b,* holotype crown from post. and ant. sides; *2c,* ant. view of paratype showing tall, cylindrical anal sac, ×1.5 (Kirk, 1944b).

Cosmetocrinus KIRK, 1941, p. 86 [**C. gracilis*; OD]. Crown tall, slender; 3 anals in cup; brachials rectangular to cuneate externally, arms branching isotomously on primibrachs *1* in all rays and once or twice at higher positions; sutures between radials and primibrachs *1* somewhat gaped. *L.Miss.(Osag.)-U.Miss.(Chester.),* USA (Ind.-Okla.).——FIG. 428,*1*. **C. gracilis,* Osag. (Keokuk, Borden Gr.), Ind.; *1a,b,* holotype from ant. and post. sides, ×0.7 (Kirk, 1941b).

Paracosmetocrinus STRIMPLE, 1967, p. 82 [**P. straki*; OD]. Like *Cosmetocrinus* except that first branching in *A* ray is above primibrach *1,* branching in all other rays on primibrachs *1*. Proximal columnals very cirriferous, with cirri extending upward around crown. *L.Miss.(Kinderhook.-Osag.),* USA(Mont.-Iowa-Ind.).——FIG. 428,*5*. **P. straki,* Kinderhook.(Wassonville Chert), Iowa; *5a,b,* holotype from post., ant. sides, ×1.3; *5c,* paratype showing termination of anal sac and arms, ×2.2 (Strimple, 1967).

Family CORYTHOCRINIDAE Strimple & Watkins, 1969

[Corythocrinidae STRIMPLE & WATKINS, 1969, p. 195]

Cup high, even sided, conical; five infrabasals with surfaces beyond stem impression abruptly turned upward so as to be confluent with higher cup plates, readily visible from side; radial articulating facets wide, filling distal face of these plates; one large anal plate (radianal) in cup in line with radials, obliquely in contact with *CD* basal. Anal sac tall, slender. Arms uniserial, pinnulate; usually branching isotomously two or more times, syzygial pairs of brachials present. Stem large, heteromorphic, composed of alternating thin and thick circular columnals with pronounced crenulations visible in side view. *L.Miss.(Osag.).*

Corythocrinus KIRK, 1946, p. 272 [**C. insculptus*; OD]. Crown subcylindrical, composed of conical cup surmounted by slender, well-rounded, closely spaced arms lacking observed pinnules. Cup with wide even base at stem impression and evenly steep-sloping sides formed by prominent circlets of 5 infrabasals, 5 moderately large basals, and 5 radials, interrupted on posterior side by large anal plate (radianal) resting mainly with oblique suture on *CD* basal and short oblique suture on *BC* basal, its upper part rising well above even line of radial summits; articular facets of radials with first primibrachs equal in width to radials, gaping, arms branched isotomously on primibrachs *3,* above which nearly all brachials are syzygially paired, with isotomous branching also once or twice in upper part of arms. Round homeomorphic stem composed of very thin columnals. *L.Miss.(Osag.),* USA(Ind.-Iowa).——FIG. 428,*3a,b*. **C. insculptus,* Borden Gr.(Keokuk), Ind.; *3a,* holotype crown from *D* ray side, ×1; *3b,* portion of *D* and *E* rays, ×2 (Kirk, 1946b).——FIG. 428, *3c,d*. *C. fragilis,* Borden Gr.(Keokuk), Ind.; post. and ant. views of holotype crown, ×0.7 (Kirk, 1946b).

Family SPANIOCRINIDAE Moore & Laudon, 1943

[Spaniocrinidae MOORE & LAUDON, 1943a, p. 60]

Cup low, cone or bowl shaped; infrabasals two to three, five, or ?lacking, may or may not be visible in side view of cup; radial articular facets occupying full distal surface of radials, horizontal or inclined outward-upward; one to three advanced anal plates in cup; when three, radianal in posterior position followed by *X* and right tube plate to left and right above, or anal plate lacking and cup with perfect pentameral symmetry. Arms five, single and unbranched, or branching on primibrachs *1* in all rays or in all but *A*-ray; brachials uniserial or biserial. Stem transversely round. *L.Miss.; L.Penn.-U.Perm.*

The Spaniocrinidae, a family of very doubtful lineage, consist chiefly of Permian crinoids having simple structure and perfect or nearly perfect pentamerous symmetry. The cup retains a primitive conical form, but anal plates commonly are reduced to one or none. The radial articular facets are very wide (plenary) and show highly developed muscular articulation of the arms, which in known genera are like those included in the Erisocrinacea. The Spaniocrinidae are not offshoots of the Erisocrinidae but may have come from slightly antecedent crinoids belonging to the same line.

Spaniocrinus WANNER, 1924, p. 292 [**S. validus*; OD]. Cup low conical, with perfect pentameral symmetry; infrabasals low, visible in side view; basals hexagonal; radials twice as wide as high and in contact all around cup; no anal plates or anal sac. Arms 5, simple and unbranched, uniserial, composed of massive, externally quadrangular brachials interlocked along their lateral

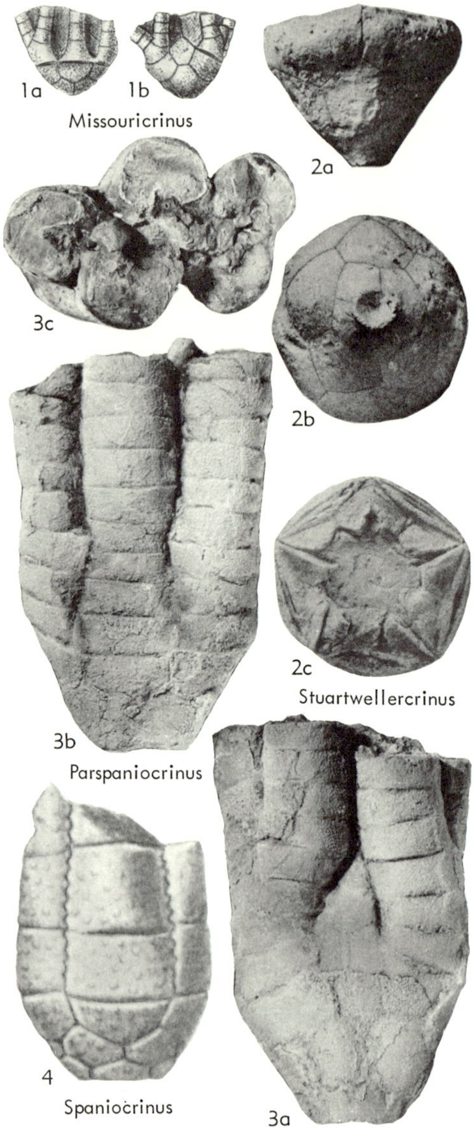

FIG. 429. Spaniocrinidae (p. T655-T656).

edges. *U.Perm.*, Indon.(Timor, Basleo).——FIG. 429,*4*. **S. validus*; lat. view of partial crown, ×2 (Wanner, 1924).

Missouricrinus S. A. MILLER in MILLER & GURLEY, 1891, p. 31 [**M. admonitus*; OD]. Cup bowl shaped; infrabasals ?lacking; basals large; radials pentagonal, with articular facets occupying full width of radials; single large anal plate directly above *CD* basal and between posterior radials. Arms rounded, primibrachs *1* axillary in all but *A* ray, which has at least 3 primibrachs; arms not known above secundibrach *3* in *A* ray. [May or may not have infrabasals and may be assigned incorrectly to Spaniocrinidae.] *L.Miss.*, USA(Mo.).——FIG. 429,*1*. **M. admonitus; 1a,b,* ant. and post. views of partial crown, ×1 (Miller in Miller & Gurley, 1891).

Parspaniocrinus STRIMPLE, 1971, p. 1040 [**P. beinerti*; OD]. Closely similar to *Spaniocrinus* but having 3 anal plates in cup, radianal obliquely across *CD* basal with 2 equidimensional anal plates above; anal *X* above *CD* basal. *L.Perm.*, Mex.(Coahuila).——FIG. 429,*3*. **P. beinerti,* Wolfcamp., Las Delicias; *3a-c,* post., ant., and ventral views of incomplete crown (holotype), ×1.3 (Strimple, 1971c).

Stuartwellercrinus MOORE & PLUMMER, 1938, p. 305 [**Cibolocrinus turbinatus* WELLER, 1909, p. 632; OD]. Cup small, steeply conical, base narrow; infrabasals 3, upflaring and visible in side view, small infrabasal generally in *A, C,* or *D* ray. Basals 5, *CD* basal typically larger than others and truncated beneath single anal plate; radials with broad articular facets projecting inward so as to constrict opening to visceral cavity; single anal present between posterior radials, or above cup. Arms not known. *L.Penn.(Morrow.)-L.Perm.(Wolfcamp.),* USA(Texas-Mo.)-Indon.(Timor).——FIG. 429,*2*. *S. symmetricus* (WELLER), L.Perm., Texas; *2a-c,* post., dorsal, and ventral views of cup, ×1.5 (Moore & Plummer, 1940).

Superfamily CUPRESSOCRINITACEA Roemer, 1854

[*nom. transl.* MOORE & LANE in MOORE & STRIMPLE, 1973, p. 20 (*ex* Cupressocrinitidae ROEMER, 1854, *nom. correct.* BASSLER, 1938, p. 16)] [Materials for this superfamily prepared by R. C. MOORE and N. GARY LANE]

Characters of family; differentiated as superfamily correlative with others of this rank among cladid inadunates because assignable to none of them. *L.Dev.-U.Dev.*

Family CUPRESSOCRINITIDAE Roemer, 1854

[*nom. correct.* BASSLER, 1938, p. 16 (*pro* Cupressocrinidae ROEMER, 1854, p. 227)]

Crown compact, arms five, massive, unbranched; first primibrachs short, wide; higher brachials three to many, large externally quadrangular, short or elongate, hyperpinnulate; stem quadrangular to round, with three or four peripheral canals. *L.Dev.-U.Dev.*

The Cupressocrinitidae, from Middle Devonian rocks of Germany and Spain, are set apart by specialized morphological characters and because of our entire uncertainty

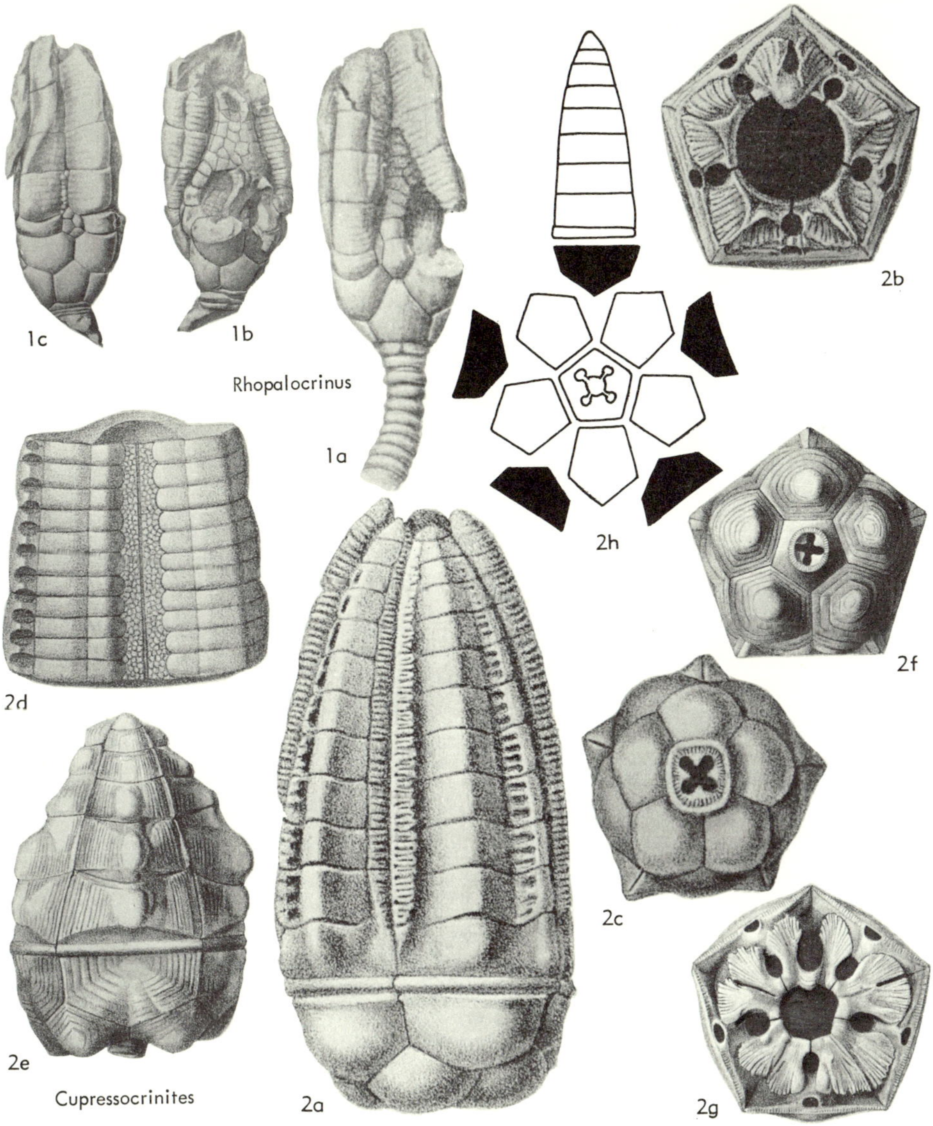

FIG. 430. Cupressocrinitidae (p. *T657-T658*).

as to their phylogenetic relations. Study of representative specimens and of discussions of these fossils by other workers has not resolved the questions. The cup is one of advanced evolutionary type in having perfect pentamerous symmetry and in characters of the base, which is flattened and mainly composed of solidly fused infrabasals. The radial articular facets are wide and straight and have decidedly special surface features. Five very peculiarly modified orals constitute a so-called "consolidating apparatus" at the summit of the cup, and surrounding them are the five unbranched arms, composed of broadly hollowed uniserially arranged brachials that bear multiple pinnules. The subquadrangular column has four peripheral canals and an axial canal.

Cupressocrinites GOLDFUSS, 1831, p. 212 [**C. crassus*; SD WACHSMUTH & SPRINGER, 1886, p. 105] [=*Cupressocrinus* AGASSIZ, 1836, p. 198

(nom. van.); *Cypellocrinites* STEININGER, 1849, p. 22 (type, *C. elegans*); *Cypressocrinites* STEININGER, 1849, p. 20; *Cypellocrinus* BATHER in LANKESTER, 1900a, p. 177; *Halocrinites* STEININGER, 1831, p. 23 (type, *H. schlotheimi*); *Halocrinus* D'ORBIGNY, 1852, p. 140; *Procupressocrinus* JAEKEL, 1918, p. 82 (type, *Cupressocrinites gracilis* GOLDFUSS, 1831, p. 213); *Aviadocrinus* ALMELA & REVILLA, 1950, p. 11 (type, *A. sampelayoi*)]. Cup cone to bowl shaped, symmetrical, without anal plates; infrabasals low, fused; radial facets wide (plenary) with large axial nerve canal. Tegmen with flattened orals enclosing ambulacral tracts from arms; posterior oral madreporitic; anal sac small. Five massive arms uniserial, with delicate pinnules covering deep ambulacral grooves. *L.Dev.-U.Dev.*, Eu.(Ger.-Spain-Belg.-Eng.).——FIG. 430,*2a-d*. **C. crassus*, M.Dev., Ger.; *2a-c*, lat. view of crown, ventral and dorsal views of cup, X1.5; *2d*, ventral view of brachials and pinnules, X3 (Schultze, 1867).——FIG. 430,*2e-g*. *C. abbreviatus* GOLDFUSS, M.Dev., Ger.; *2e*, lat. view of crown, X1; *2f*, dorsal view of cup, X1; *2g*, ventral view of cup and orals, X1 (Schultze, 1867).——FIG. 430,*2h*. Plate diagram for genus (radials black) (Moore, n).

Rhopalocrinus WACHSMUTH & SPRINGER, 1880, p. 57 [**Taxocrinus gracilis* SCHULTZE, 1866, p. 39 (*non* MEEK & WORTHEN, 1865a); OD]. Like *Cupressocrinites* except for single anal plate between posterior radials; anal sac wide based and massive; orals unknown. *M.Dev.*, Eu.(Ger.).——FIG. 430,*1*. **R. gracilis* (SCHULTZE); *1a*, post. view of crown, X1.5; *1b,c*, *C*-ray and *AE*-interray views of crown and proximal part of stem, X1.5 (Springer, 1920).

Superfamily MOLLOCRINACEA Wanner, 1916

[*nom. transl.* MOORE & STRIMPLE, 1973, p. 20 (*ex* Mollocrinidae WANNER, 1916a, p. 67)] [Materials for this superfamily prepared by R. C. MOORE and N. GARY LANE]

Cup bowl shaped to globose with convex or flattened base, exceptionally with small shallow basal concavity; infrabasals five, four, three or fused into single solid plate, generally but not universally visible from side; radial articular facets narrow, separated by interradial notches; one to three anal plates in cup. Anal sac low rounded. Stem transversely circular, slender to robust. *L.Penn.-U.Perm.*

Family MOLLOCRINIDAE Wanner, 1916

[Mollocrinidae WANNER, 1916a, p. 67]

Cup small, conical to globose; radial articular facets angustary (much narrower than radials), concave, sloping outward-downward; one to three anal plates in cup, but generally two. Arms and tegmen unknown. Stem round. *L.Penn.-U.Perm.*

Mollocrinus WANNER, 1916, p. 67 [**M. poculum*; OD]. Cup globose, infrabasals 3, low, azygous small one in *A* ray, visible from side; basals 5, large, as wide as high; radials pentagonal, smaller than basals, their articular facets with trace of transverse ridge and slitlike ligament groove; quadrangular radianal and anal *X* in cup. Arms and tegmen unknown. Stem transversely round, with large round central canal. *U.Perm.*, Indon. (Timor, Basleo).——FIG. 431,*3*. **M. poculum*; *3a,b*, post. and ventral (*CD* interray downward) views of cup, X1.5 (Wanner, 1916a).

Hemimollocrinus YAKOVLEV, 1930, p. 97 [**H. uralensis*; M]. Cup small, low conical; infrabasals 5, low, visible in side view; basals and radials about equal in size; radial articular facets narrower than radials, rounded, with well-developed transverse ridge; 3 anal plates in cup. *L.Perm. (Artinsk.)*, USSR(Krasnoufimsk).——FIG. 431,*1*. **H. uralensis*; *1a,b*, post. and ventral (*CD* interray downward) views of cup, X1.5 (Yakovlev, 1930a).

Strongylocrinus WANNER, 1916, p. 72 [**S. molengraffi*; OD]. Cup globose; infrabasals 3, low, but visible from side, azygous small one in *C* ray; basals large; radials smaller than basals, with narrow horseshoe-shaped articular facets facing directly upward; single anal plate in cup directly above *CD* basal and between posterior radials. *L.Penn.(Morrow.)-U.Perm.(Basleo beds)*, USA (Okla.)-Indon.(Timor).——FIG. 431,*2*. **S. molengraffi*, Perm., Timor; *2a-d*, post., ventral (*CD* interray downward), dorsal, and ant. views of cup, X2 (Wanner, 1916a).

Superfamily LOPHOCRINACEA Bather, 1899

[*nom. transl.* MOORE & STRIMPLE, 1973, p. 20 (*ex* Lophocrinidae BATHER, 1899b, p. 922)] [Materials for this superfamily prepared by R. C. MOORE, H. L. STRIMPLE, and N. GARY LANE]

Crown tall, expanding upward. Cup conical, with five upflared infrabasals readily visible from side; radial articular facets wide but less than full width of radials, leaving interradial notches at upper corners of these plates; three or single anal plates in cup. Anal sac tall, cylindrical or markedly narrowing upward, mainly composed of subhorizontally plicate plates. Arms five to highly branched, slender, uniserial. Stem transversely circular. *L. Miss.-U.Perm.*

Family LOPHOCRINIDAE Bather, 1899

[Lophocrinidae BATHER, 1899b, p. 922]

Cup small, conical single large anal plate in cup between posterior radials. Anal sac large, cylindrical. Arms slender, unbranched, with small, unbranched ramules or pinnules on alternate sides of successive brachials. *Up.L.Carb.*

Lophocrinus VON MEYER, 1858, p. 59 [**L. speciosus* (=*Poteriocrinus minutus* ROEMER, 1850b, p. 47); M] [=*Carduocrinus* VON KOENEN, 1895b, p. 210 (type, *C. jaekeli* VON KOENEN, =*P. minutus*)]. Cup small, conical, plates smooth; infrabasals and basals higher than wide; single large anal plate in cup between posterior radials. Anal sac large, cylindrical, composed of several longitudinal rows of small, convex plates. Arms slender, atomous, bearing long pinnules on alternating sides of successive brachials, which are low, wedge shaped. Stem transversely round, columnals alternating in width proximally, becoming higher and equal in width distally *Up.L.Carb.(Visean)*, Eu. (Ger.-Spain).——FIG. 432,6. **L. minutus* (ROEMER), Ger.; *6a*, lat. view of crown and column, ×0.7; *6b-f*, side views and articular surfaces of proximal and distal columnals, ×6 (Jaekel, 1895).

Family PELECOCRINIDAE Kirk, 1941

[Pelecocrinidae KIRK, 1941b, p. 82]

Crown tall, ovoid. Cup low, bowl shaped, with shallow basal concavity; infrabasals five, rarely four, barely visible or not visible in side view; radial articular facets peneplenary, narrower than radials, inclined slightly to strongly outward-downward, with prominent transverse ridge, interradial notches distinct; three anal plates in cup. Arms isotomous where known, pinnulate, with wedge-shaped brachials. *L.Miss.-U.Miss.; M.Penn.-U.Penn.; U.Perm.*

The Pelecocrinidae have rounded, relatively narrow radial articular facets that slope strongly outward-downward, as in the Poteriocrinitidae, but they show advancement in evolution of the shape of the cup and in structure of the arms. The cup is low bowl shaped, gently convex or slightly concave at the base. The arms have cuneate uniserial or biserial brachials that bear pinnules, and they branch isotomously twice or more.

Key to Genera of Pelecocrinidae

A. Base of cup flat or shallowly concave ....
 I. Infrabasals, basals, and radials 5 ....

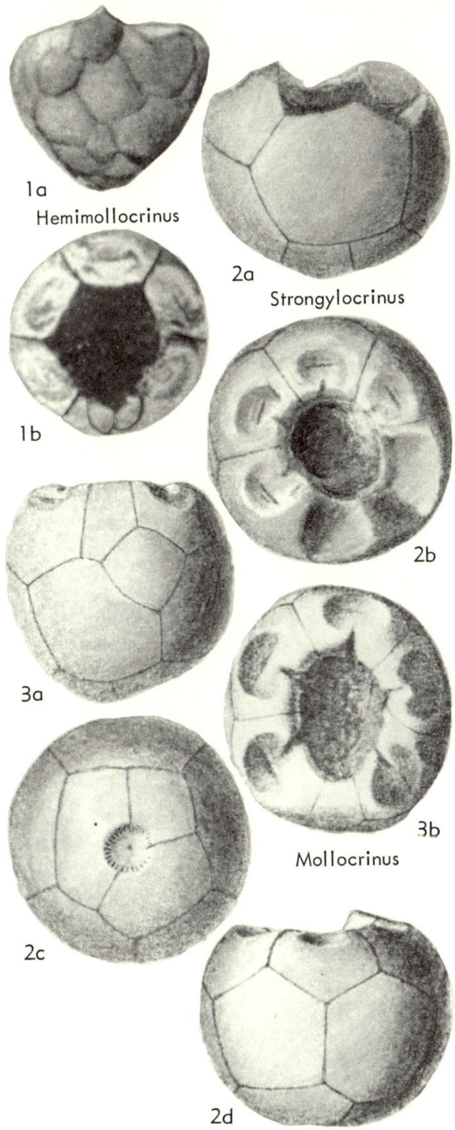

FIG. 431. Mollocrinidae (p. T659).

 a. Radial articular facets sloping gently outward-downward ....
  1. Stem impression narrow, round .. .. *Forthocrinus*
  2. Stem impression very wide, round .... *Tetrabrachiocrinus*
 b. Radial articular facets sloping steeply outward-downward, round stem, impression very wide .... *Malaiocrinus*
 II. Infrabasals, basals, and radials 4 .... *Depaocrinus*
B. Base of cup convex ....

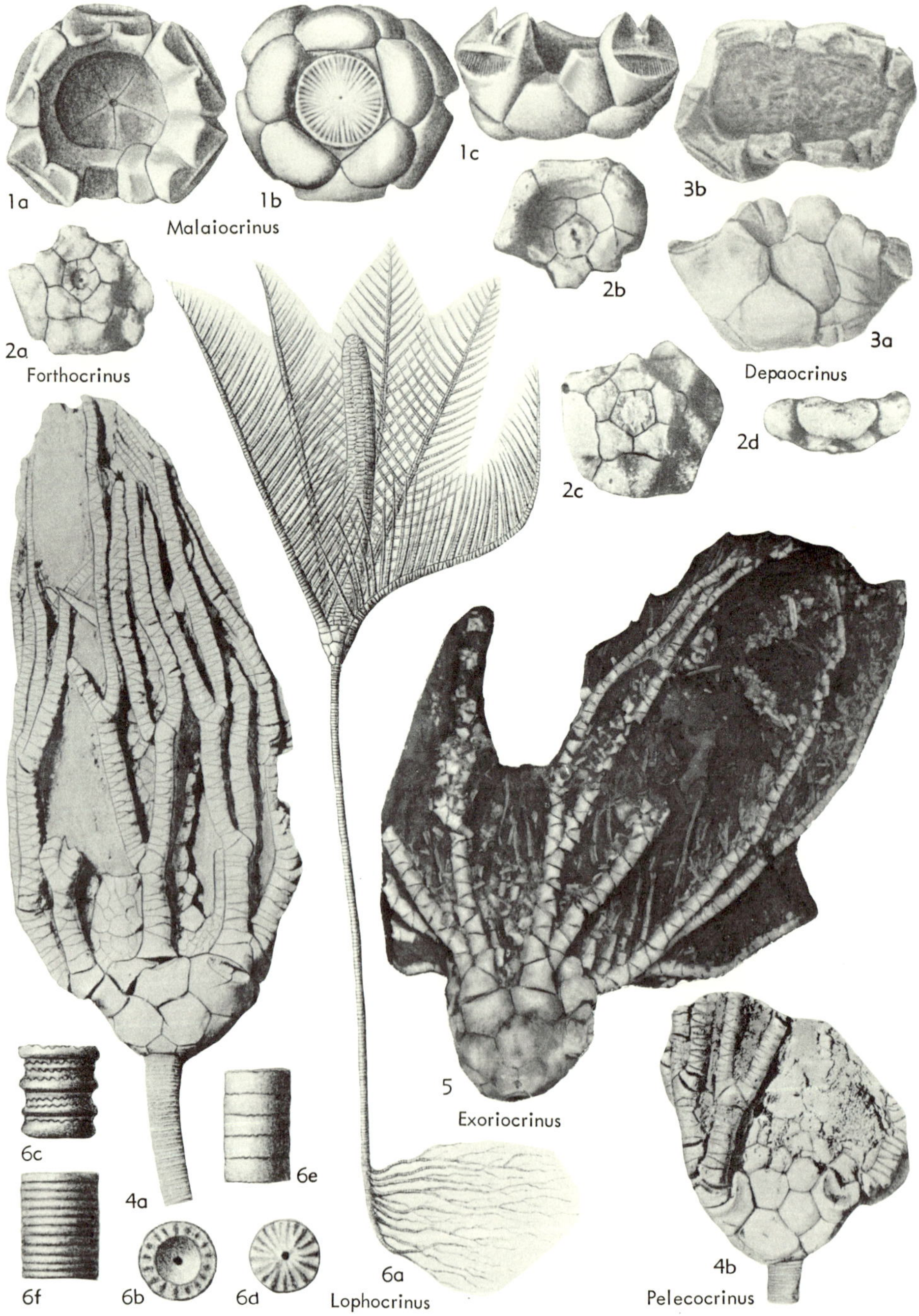

FIG. 432. Lophocrinidae *(6)*; Pelecocrinidae *(1-5)* (p. *T*659-*T*662).

I. Stem round .... *Pelecocrinus*
II. Stem pentagonal .... *Exoriocrinus*

Pelecocrinus KIRK, 1941, p. 82 [**P. insignis*; OD]. Crown large, expanding upward to about one-third of its height and then contracting in width; bases of arms widely spaced with divergent rami. Cup bowl shaped; infrabasals 5, low, visible in side view; basals large; radials large, pentagonal, with rounded peneplenary articular facets, sloping outward-downward, interradial notches shallow and wide; 3 anal plates in cup. Anal sac short, slender, composed of 10 or more vertical rows of irregular polygonal plates; summit with heavy, spinose plates. Arms branching isotomously 2 or 3 times; first primibrachs axillary in all but *A* ray, which has 10 to 12 primibrachs; brachials strongly wedge shaped, rounded, axillaries large, widely pentagonal to triangular and convex; pinnules long and stout. Stem transversely round. *L.Miss.*, USA (Iowa)-Can.(Alta.).——FIG. 432,*4*. **P. insignis*, Iowa; *4a,b*, *A*-ray view of crown and post. view of partial crown, ×0.7 (Kirk, 1941b).

Depaocrinus WANNER, 1937, p. 152 [**D. ottowi*; OD]. Cup bowl shaped, base flat to slightly concave, greatest width at top of radials; infrabasals 4, circlet quadrangular in outline, mostly covered by stem impression, not visible in side view; basals and radials 4; radial articular facets peneplenary, sloping outward-downward in side view and occupying about four-fifths of radial plate width; 3 anal plates in cup. Stem impression round. *U.Perm.*, Indon.(Timor, Basleo).——FIG. 432,*3*. **D. ottowi; 3a,b*, post. and ventral views of cup, ×1 (Wanner, 1937).

Exoriocrinus STRIMPLE & MOORE, 1971, p. 15 [**Poteriocrinus lasallensis* WORTHEN in WORTHEN & MEEK, 1875, p. 526; OD]. Cup moderately large, medium bowl shaped, infrabasals visible from side, 5 large basals; 5 large radials with notches between, subhorizontal plenary articular facets. Anal sac unknown. Arms 20, slender, rounded, not appressed, branching isotomously on primibrachs *1* and again higher in each ray, nonaxillary brachials strongly cuneate in external view; pinnules short and stout. Stem pentagonal in section proximally and circular distally, lumen pentalobate. *M.Penn.-U.Penn.*, USA(Okla.-Ill.).——FIG. 432,*5*. **E. lasallensis* (WORTHEN), U.Penn. (Missour., LaSalle Ls.), near Pontiac, Livingston Co., Ill.; *B*-ray view of hypotype crown, ×1.4 (Strimple & Moore, 1971a).

Forthocrinus WRIGHT, 1942, p. 274 [**F. lepidus*; OD]. Cup low, wide saucer shaped, with small shallow basal concavity; plates thin, internal cavity of cup large; plates ornamented with large nodes; infrabasals small, not visible in side view; basals convex and nodose; radials wide, with proximal tips in depressed areas between nodes; interradial sutures notched distally, articular facets slightly narrower than radials and inclined outward-downward; 3 anal plates in cup. *L.Carb.*, Eu.(Eng.).——FIG. 432,*2*. **F. lepidus; 2a,c*, dorsal views; *2b,d*, ventral and ant. views of cups, ×1 (Wright, 1954).

Malaiocrinus WANNER, 1924, p. 183 [**Zeacrinus? sundaicus* WANNER, 1916a, p. 147; OD]. Cup low, broad, plates convex; infrabasal circlet wide, flat, not visible in side view, mostly covered by stem impression; basals large, low, strongly convex; radials large, higher than wide, with articular facets large, rounded, steeply inclined outward-downward, with prominent transverse ridge; interradial notches prominent; 3 anal plates in cup. Arms and tegmen unknown. Stem impression round, with numerous fine culminae. *U.Perm.*, Indon.(Timor).——FIG. 432,*1*. **M. sundaicus* (WANNER), Basleo; *1a-c*, ventral, dorsal, and post. views of cup, ×2 (Wanner, 1924).

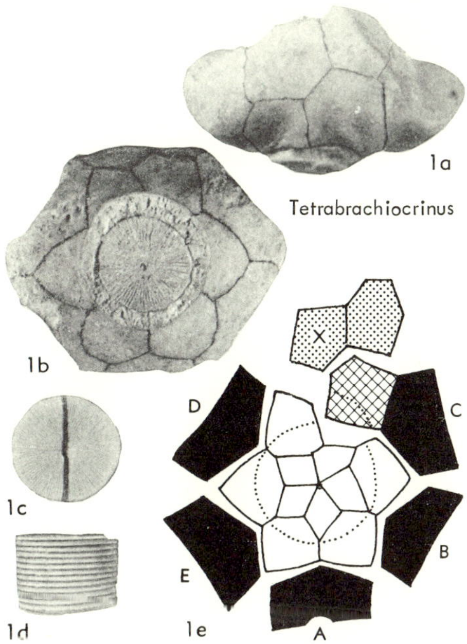

FIG. 433. Pelecocrinidae (p. *T*661).

Tetrabrachiocrinus YAKOVLEV, 1934, p. 276 [**T. fabianii*; OD] [=*Heterobrachiocrinus* YAKOVLEV, 1938, p. 249 (obj.)]. Cup large, hexagonal in outline in dorsal or ventral view, wider than high, base broad and flat; infrabasals small, covered by stem; proximal edges of basals covered by stem; radials large, unequal in size, *A* radial smaller than others and with a narrow facet, other facets large, rounded, steeply inclined; anal plates three, large, projecting above level of radials; arms unknown; stem large, round, with narrow lumen and wide, thin columnals. *U.Perm.*(Sicily).——FIG. 433,*1a-d*. **T. fabianii; 1a-d*, post. and dorsal views of cup, articular and lat. views of stem, ×2 (Yakovlev, 1934).——FIG. 433,*1e*. Plate diagram for genus, of cup, ant. ray up (radials black, anal plates stippled, radianal cross ruled) (Moore, n).

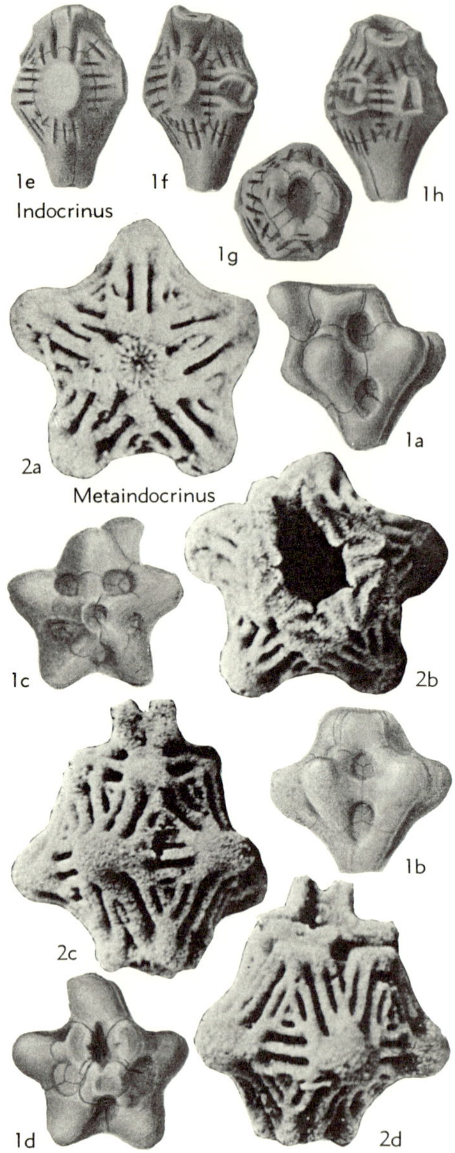

FIG. 434. Indocrinidae (p. *T662*).

[After YAKOVLEV published the name *Tetrabrachiocrinus* in 1934, he later (1938), in a short note, stated that this name was preoccupied by that of a Mesozoic crinoid, without giving author or reference for the earlier usage. Search of literature and correspondence with several authorities have failed to reveal the use of *Tetrabrachiocrinus* by any author prior to YAKOVLEV, 1934. YU. ARENDT believes that YAKOVLEV may have mistaken *Tetracrinus* for *Tetrabrachiocrinus*. For this reason YAKOVLEV's substitute name, *Heterobrachiocrinus*, is here rejected as a junior objective synonym of *Tetrabrachiocrinus*.—N. GARY LANE.]

Family INDOCRINIDAE Strimple, 1966

[Indocrinidae STRIMPLE, 1966b, p. 80]

Cup small, globose or biturbinate; infrabasals three, visible in side view; basals large, nodose or convex, cup plates mostly with epispires across plate sutures; radials small, restricted in width and outer surface incurved distally, *B* and *E* radials or only *B* radial armless; one or three anal plates in cup. *L.Perm.-U.Perm.*

Key to Genera of Indocrinidae

A. Cup biturbinate, pentalobate transversely, large protuberant basals; epispires crossing plate sutures ....

  I. *B* and *E* radials narrow, armless; single anal plate *(X)* in cup .... *Indocrinus*

  II. Only *B* radial armless; 3 anal plates in cup .... *Metaindocrinus*

B. Cup globose, constricted at top; no epispires; *B* and *E* radials armless; 3 anal plates in cup .... *Proindocrinus*

Indocrinus WANNER, 1916, p. 74 [**I. elegans*; OD]. Cup biturbinate in side view and pentalobate in basal view; infrabasals 3, high, circlet cone shaped, small infrabasal in *A* ray or rarely in *D* ray; basals large, protruding, strongly nodose, with deep depressions at plate angles which contain epispires transverse to sutures; radials small, restricted in width, unequal in size; *B* and *E* radials narrow and armless; other 3 radials larger and with narrow, rounded articular facets; single anal plate in cup, large, strongly convex and protruding, quadrangular, directly above *CD* basal. *U.Perm.*, Indon.(Timor).——FIG. 434,*1a-d*. **I. elegans*, Basleo; *BC*-interray, *A*-ray, dorsal, and ventral views of cup, ×1.7 (Wanner, 1916a).——FIG. 434,*1e-h*. *I. rimosus* WANNER, Basleo; post., *C*-ray, ventral, and *A*-ray views of cup, ×1.7 (Wanner, 1916a).

Metaindocrinus STRIMPLE, 1966, p. 81 [**M. cooperi*; OD]. Cup small, biturbinate, pentalobate in basal view; infrabasals 3, low, visible in side view; basals 5, large, with strong projecting central node; radiating epispires present across all cup plate sutures, epispires deeply depressed between prominent ridges; radials small, restricted in width, *B* radial armless, others with wide horizontal articular facets; large radianal above *CD* basal, succeeded by 2 small anal plates. Stem impression round. *U.Perm.*, USA(Texas).——FIG. 434,*2*. **M. cooperi*, Word F., Glass Mts., Texas; *2a-d*, dorsal, ventral, post., ant. views of cup, ×3.4 (Strimple, 1966b).

Proindocrinus YAKOVLEV, 1939, p. 832 [**Indocrinus piszowi* YAKOVLEV, 1926a; M]. Cup globose, plates smooth and without epispires; infrabasals large, visible in side view; basals large; radials small, restricted in width, circlet constricted distally, articular facets confined to *A, C,* and *D* radials; large radianal, small anal *X,* and right tube plate in cup. *L.Perm.(Artinsk.),* USSR (Krasnoufimsk).

Family LAUDONOCRINIDAE Moore & Strimple, 1973

[Laudonocrinidae MOORE & STRIMPLE, 1973, p. 20]

Cup bowl shaped, with flat or faintly convex base and subvertical sides near rim, arm facets peneplenary, interradial notches seen in dorsal and ventral views; two or three anals in cup; anal sac mushroomlike with horizontal girdling spines at summit; arms endotomous. *U.Miss.-L.Perm.*

Laudonocrinus MOORE & PLUMMER, 1940, p. 174 [**Hydreionocrinus subsinuatus* MILLER & GURLEY, 1894b, p. 40; OD] [=*Appalachiacrinus* BURKE, 1974, p. 5 (type, *A. erwini*; OD)]. Cup moderately low bowl shaped, characterized essentially by smooth contour of surface; base flat; summit of interradial sutures with well-defined notch and *CD* interray wide, concave at cup summit; infrabasals forming small subhorizontal regular pentagon, mostly covered by round stem impression; basals essentially equal but *CD* basal slightly longer and *BC* basal wider than others; basals flared upward and clearly visible from side; radials pentagonal, with width generally about twice length; peneplenary articular facets sloping outward-downward at moderate angle, transverse ridge and ligament areas clearly defined; anal plates 3, radianal larger than others, resting on *BC* and *CD* basals. *L.Penn.-U.Penn.(Missour.),* N.Am.(Midcontinent region).——FIG. 435,*2a*; 436,*5*. **L. subsinuatus* (MILLER & GURLEY), Missour.(Kansas City F.), Kansas City, Mo.; 435,*2a,* dorsal view of cup, ×2 (Moore & Plummer, 1940); 436,*5a,b,* diagram. long. secs. through midlines of *A* and *D* radials, ×2.25 (explanation on p. *T723*) (Moore, n).——FIG. 435,*2b-d*. *L. catillus* MOORE & PLUMMER, Millsap Lake Gr.(Desmoines.), Palo Pinto Co., Texas; dorsal, post., ventral views of cup, ×1.5 (Moore & Plummer, 1940).

Anchicrinus STRIMPLE & WATKINS, 1969, p. 206 [**A. toddi*; OD]. Cup low, bowl shaped, with broad basal plane and shallow concavity; infrabasals downflared except at distal tips, which are subhorizontal; sutures impressed, with dimplelike depressions at plate corners. Arms long slender uniserial, branching endotomously 5 times; axillaries spinose except primibrachs *1* which are tumid but seldom spinose. Tegmen terminates at summit of crown with small platform of few polygonal plates surrounded by 8 outwardly directed spinelike plates. *L.Penn.-M.Penn.,* USA (Okla.-Texas).——FIG. 437,*2*. **A. toddi,* M.Penn. (Millsap Lake F., Brannon Bridge Ls. Mbr.), near Brock, Parker Co., Texas; *2a,b,* holotype, lat. and post. views of crown, ×1.1; *2c,d,* paratype, dorsal,

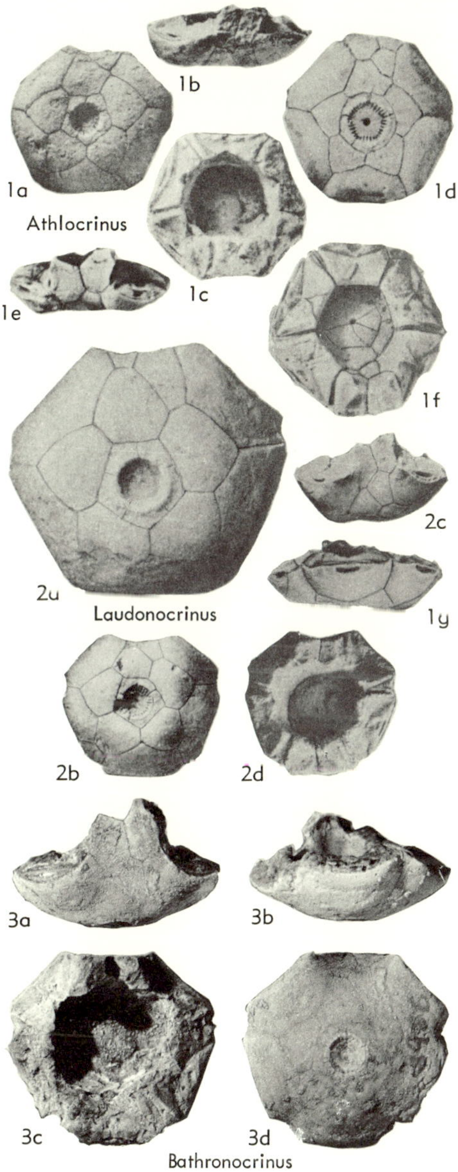

FIG. 435. Laudonocrinidae (p. *T663*, *T665-T667*).

ventral views, ×1.7 (Strimple & Watkins, 1969).

Athlocrinus MOORE & PLUMMER, 1940, p. 169 [**A. placidus*; OD]. Cup extremely shallow saucer shaped, with base almost perfectly flat, except for indentation of round stem impression; sides flaring gently upward to subvertical at summit plane; *CD* interray narrower than in *Laudonocrinus,* interradial notches small; sutures not at all impressed, and contour of surface of cup entirely smooth;

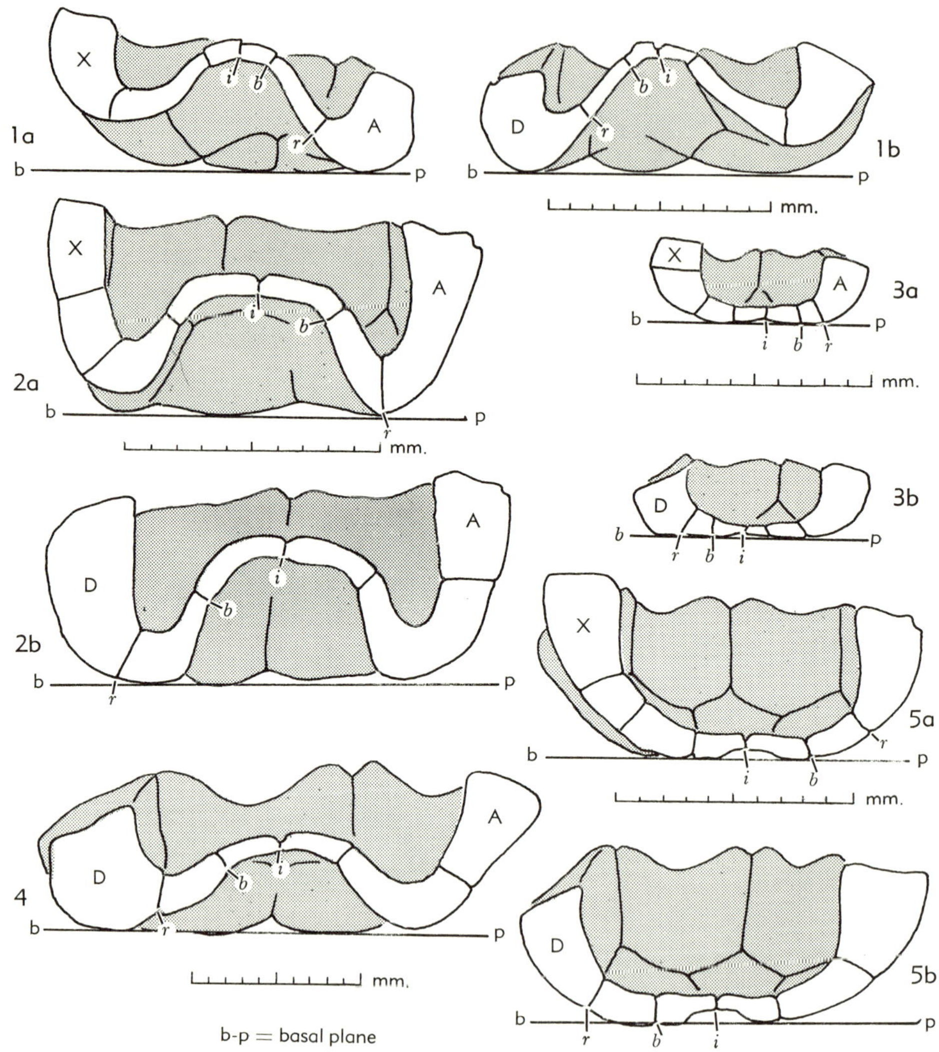

		A ray			D ray		
		i	*b*	*r*	*i*	*b*	*r*
1a,b.	Pirasocrinus	−	−	−	−	−	−
2a,b.	Perimestocrinus	−	−	0	−	−	0
3a,b.	Athlocrinus	−	0	0	−	0	0
4.	Aatocrinus				−	−	−
5a,b.	Laudonocrinus	−	0	+	−	0	+

FIG. 436. Cross sections of cups of Laudonocrinidae *(3,5)* and Pirasocrinidae *(1,2,4)*, portion of cup beyond the plane of the section shaded (Moore, n, after Moore & Plummer, 1940). [Explanation: *i*, proximal tip of infrabasal; *b*, proximal tip of basal; *r*, proximal tip of radial. Symbols: minus sign (−), infral (above basal plane within basal concavity of cup); zero (0), basiplanal (in or tangent to basal plane of cup); plus (+), supral (above basal plane on outer side of the cup).]

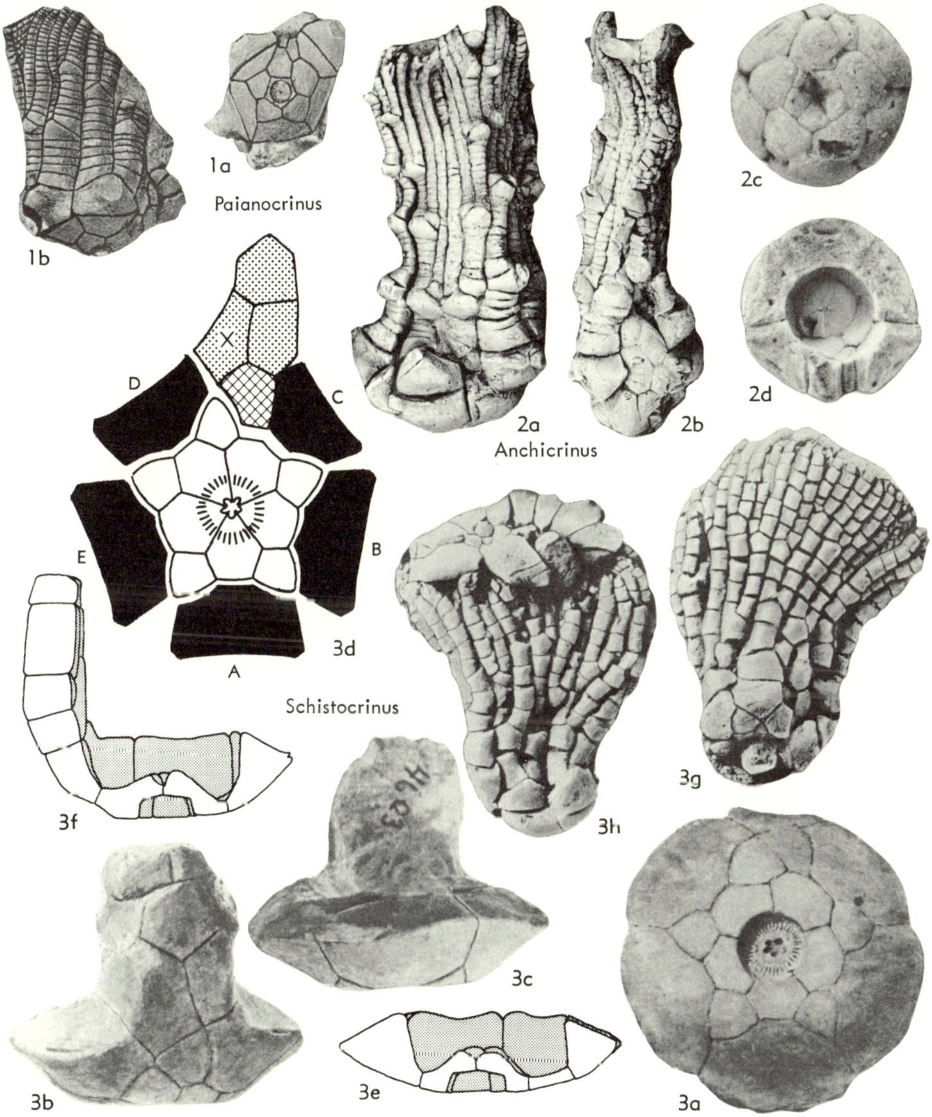

FIG. 437. Laudonocrinidae (p. *T*663, *T*667).

infrabasals forming regular pentagon covered largely by indented stem attachment, distal parts of infrabasals horizontal, not visible in side view; subequal basals, with horizontal proximal parts, distally upflared and visible in side view; radials with proximal tips in basal plane of cup (formula, 000), articular facets less than maximum width of radials, slope of facets gently outward-downward; anals 3, anal *X* resting on truncate distal tip of *CD* basal. *M.Penn.(Desmoines.)-U.Penn.(Missour.),* USA(Texas-Okla.-Kans.).——FIG. 435, *1a-c*; 436,*3*. **A. placidus,* U.Penn.(Plattsburg Ls.), Altoona, Kans.; 435,*1a-c,* dorsal, post., ventral views of holotype, ×2 (Moore & Plummer, 1940); 436,*3a,b,* diagram. long. secs. of cup through midline of *A* and *D* radials, ×2.25 (Moore, n).——FIG. 435,*1d-g*. *A. clypeiformis* MOORE & PLUMMER, U.Penn.(Merriman Ls., Graford F.), Palo Pinto Co., Texas; *1d-g,* dorsal, post., ventral, ant. views of cup, ×2 (Moore & Plummer, 1940).

Bathronocrinus STRIMPLE, 1962, p. 37 [**B. turioformis*; OD] [=*Hypermorphocrinus* ARENDT, 1968a, p. 99 (type, *H. magnospinosus*; OD)].

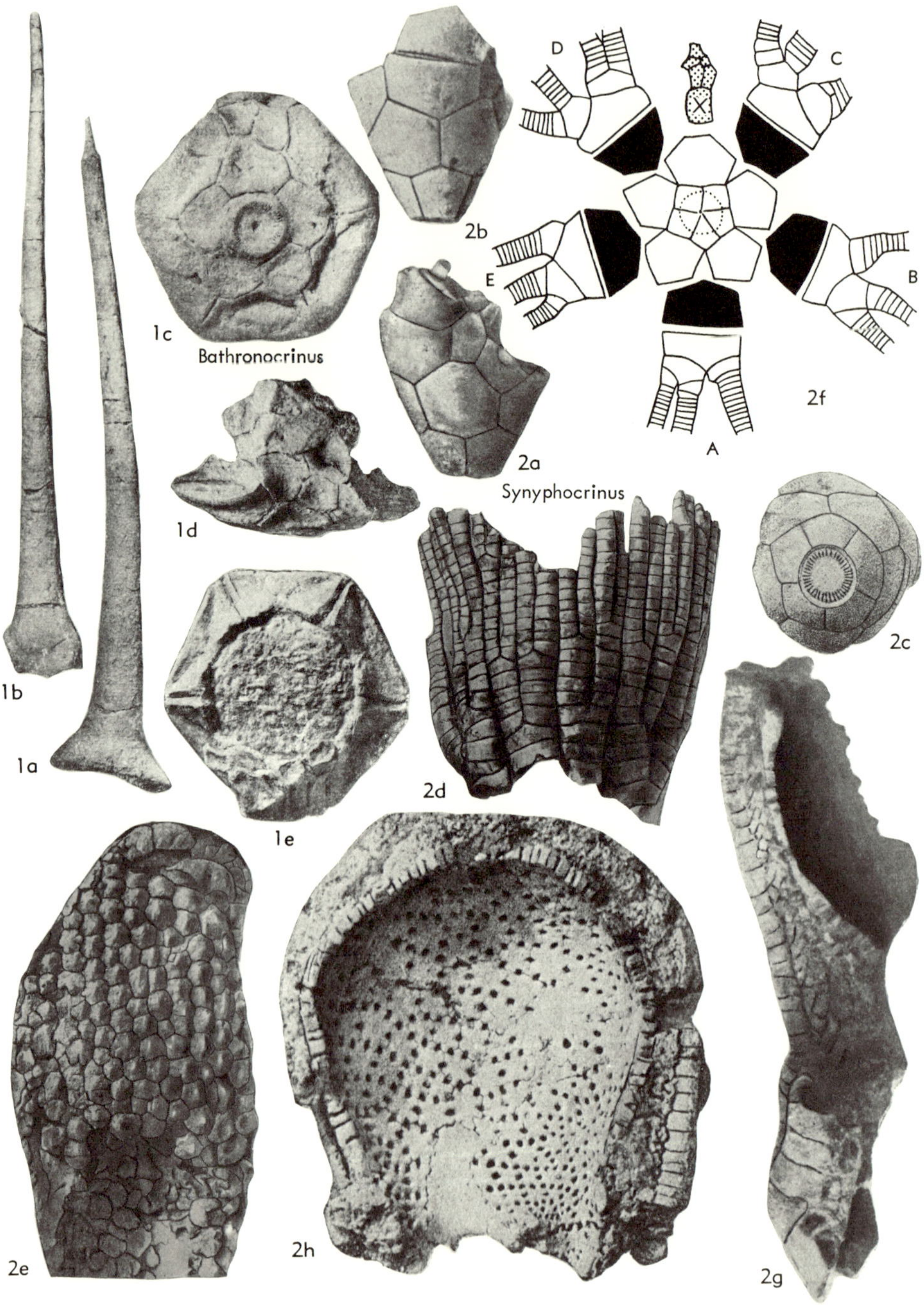

FIG. 438. Laudonocrinidae *(1)*; Anobasicrinidae *(2)* (p. *T*665-*T*667, *T*680).

Cup like *Paianocrinus* except for slight curvature of distal portions of radials and notched appearance between radials at summit. *Laudonocrinus* has a taller cup with proximal ends of basals flexed as though to form a basal concavity, distal ends of infrabasals upflared but barely visible in side view of cup. *M.Penn.*, USA(Okla.); *L.Perm.*, USSR(Krasnoufimsk, Urals.).——FIG. 435,*3*. **B*.

turioformis, M.Penn.(Oologah F.), Okla.(near Tulsa, Tulsa Co.); *3a-d,* post., ant., ventral, dorsal views, ×2.5 (Strimple, 1962a).——FIG. 438,*1*. *B. magnospinosus* (ARENDT), L.Perm., Sib.; *1a-e,* primibrach spine, anal sac spine, dorsal, post., ventral views of cup, ×1 (Arendt, 1968a).

Paianocrinus STRIMPLE, 1951, p. 669 [**P. durus*; OD]. Cup truncate cone shaped, infrabasals gently upflared and distal tips visible in side view of cup. Anal sac cylindrical, terminating in platform bordered by 6 outwardly directed spines with bases joined together. Arms uniserial, branching isotomously 3 times, axillaries nonspinose except for primibrachs *1,* which are extended as spines. Column round. *U.Miss.,* USA(Okla.-Ark.).——FIG. 437,*1*. **P. durus,* Pitkin F., near Braggs, Muskogee Co., Okla.; *1a,b,* holotype, dorsal, *D*-ray views, ×1.2 (Strimple, 1951d).

Schistocrinus MOORE & PLUMMER, 1940, p. 217 [**S. torquatus*; OD]. Cup low, shallow saucer shaped, base nearly flat except for sharply depressed, vertically walled, round stem impression; sides of cup gently and evenly upflared, posterior side with strongly elevated, firmly joined, thick lower anal plates; surface smooth, sutures distinct but not impressed; infrabasals not visible from side, except distal extremity of right posterior *(C)* infrabasal in posterior view, each infrabasal truncated distally for contact with radials; small, subquadrangular basals, shaped like arrowheads, not laterally in contact with other basals; large hexagonal radials, slightly wider than long, proximally touching infrabasals; radials with peneplenary articular facets and interradial notches; radianal on small right posterior *(C)* infrabasal supporting large anal *X* and right tube plate, which are mostly above summit line of radials but considered to be cup plates. Anal sac large, balloonlike. Arms branching isotomously on broad first primibrachs in all rays; succeeding brachials wide, short, externally rectangular segments or pentagonal axillary plates; division of branches in lower part of rays regularly isotomous. *M.Penn.(Desmoines.)-U.Penn.(Missour.),* USA(Midcontinent region).——FIG. 437,*3a-f*. **S. torquatus,* Missour.(Winterset Ls.), Kansas City, Mo.; *3a-c,* dorsal, post. and ant. views, ×2 (Moore & Plummer, 1940); *3d,* plate diagram of cup (radials black, radianal cross ruled, anals stippled), ×1.5 (Moore, n); *3e,f,* long. secs. of cup through midlines of *D* and *A* radials, ×1.5 (Moore & Plummer, 1940).——FIG. 437, *3g,h*. *S. parvus* MOORE & PLUMMER, Desmoines.(Millsap Lake F.), Parker Co., Texas; ant. and post. views of holotype crown, ×2 (Moore & Plummer, 1940).

Family STELLAROCRINIDAE Strimple, 1961

[Stellarocrinidae STRIMPLE, 1961d, p. 108]

Crown typically spreading sideward, low; arms well separated from neighbors. Cup medium to low bowl shaped, commonly with broad, shallow basal concavity; infrabasals subhorizontal; basals large, usually protruded about basal depression; radials wider than long, articular facets not as wide as radials, strongly declivate, typically leaving well-marked interradial notches; cup anal plates three in normal to advanced arrangement with single anal plate (radianal) on *CD* basal, followed above cup by two equidimensional plates. Anal sac cylindrical, tall, composed of moderately large protuberant hexagonal plates in alternating vertical series; terminating in *Stellarocrinus* and *Brabeocrinus* with a few thick, short spines, anus covered by small polygonal plates at summit of sac. Arms strongly cuneate-uniserial to mostly biserial, not apposed, branching on primibrachs *1* in geologically younger forms, primibrachs *2* in older ones, with one or more subsequent branchings in all rays; pinnules stout, attached to inner surfaces of brachials. Column usually round, but transversely pentastellate in some Chesterian and Morrowan species, round in Visean and later forms. *U.Miss.(Chester.)-L.Perm.*

Key to Genera of Stellarocrinidae

A. Arms narrow, formed of uniserially arranged cuneate brachials, 3 normal (primitive) anal plates in cup ....
- I. Strong plications on cup plates; primibrachs *2* axillary; stem pentastellate .... *Heliosocrinus*
- II. Moderate plications on cup plates, primibrachs 2 axillary, stem round .... *Rhopocrinus*
- III. Cup plates smooth, primibrachs *1,* axillary, stem round .... *Pedinocrinus*

B. Single anal plate in cup or one followed by two others, located between posterior radials; primibrachs *1* axillary; stem round ....
- I. Cup narrow box shaped, arms cuneate-uniserial to incipiently biserial .... *Brabeocrinus*
- II. Cup basally concave shallow bowl shaped, arms biserial ....
  - 1. Arms narrow, no plications on cup plates .... *Brychiocrinus*
  - 2. Arms relatively wide ....
    - a. Basals and radials strongly plicate .... *Stellarocrinus*
    - b. Cup plates smooth, not plicate .... *Celonocrinus*

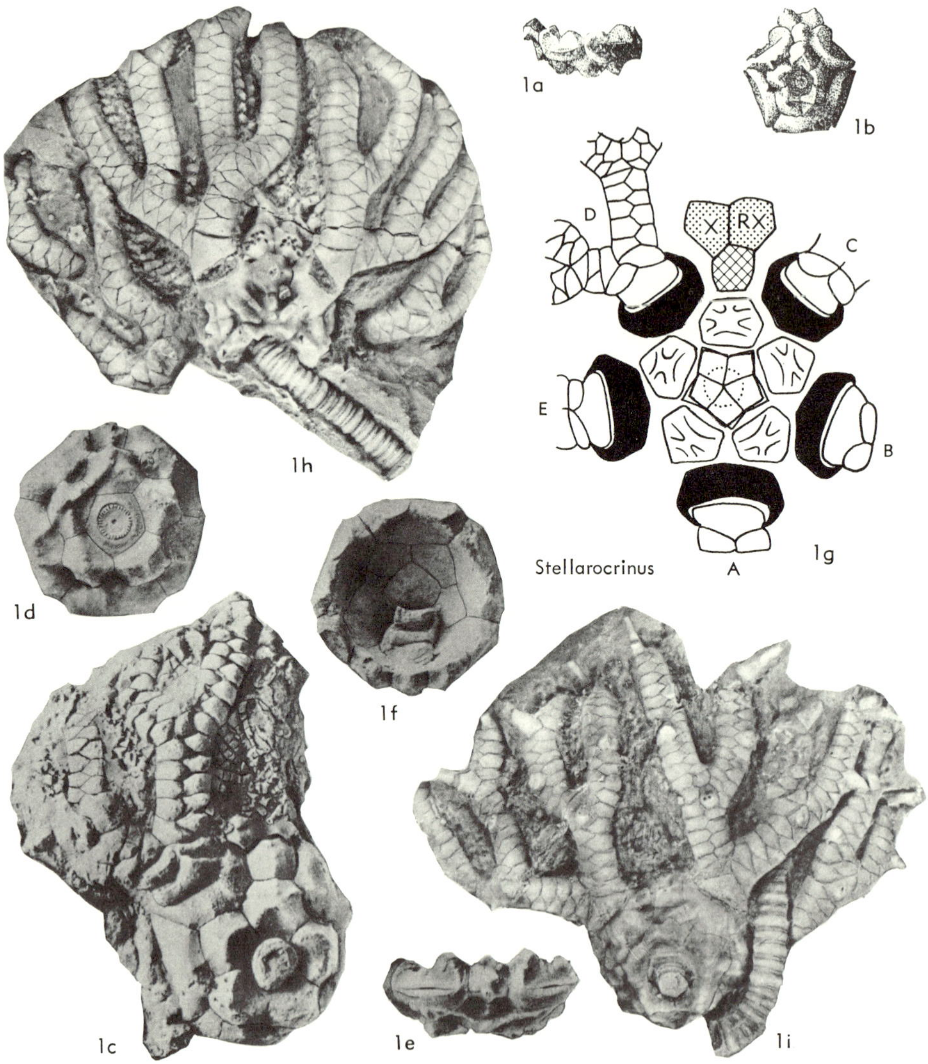

FIG. 439. Stellarocrinidae (p. T668-T669).

Stellarocrinus STRIMPLE, 1940(Mar.), p. 1 [*pro Whiteocrinus* STRIMPLE, 1939b, p. 4 (*non* JAEKEL, 1918, p. 58)] [**Cyathocrinus stillativus* WHITE, 1880b, p. 258; OD] [=*Apollocrinus* MOORE & PLUMMER, 1940 (May), p. 102 (type, *A. geometricus*)]. Arms equibiserial, wide; primibrachs *1* axillary, subsequent axillaries small, triangular in outline. Stem transversely round. *M.Penn.-L. Perm.*, USA(Texas-Ill.-Mo.-Kans.-Okla.).——FIG. 439,*1a,b*. **S. stillativus* (WHITE), U.Penn.(Virgil.), Kans.; *1a,b,* holotype viewed from *BC* interray, and dorsal view, ×1 (White, 1880b).——FIG. 439,*1c*. *S. florealis* (MOORE & PLUMMER), M.Penn.(Desmoines.), Texas; holotype, oblique view from below (*CD* interray upward), ×1.5 (Moore & Plummer, 1940).——FIG. 439,*1d-g*. *S. geometricus* (MOORE & PLUMMER), M.Penn.(Desmoines.), Texas; *1d-f,* holotype, dorsal, post., and ventral views, ×1.5 (Moore & Plummer, 1940); *1g,* plate diagram of cup and parts of arms (radials black, radianal cross ruled, anal *X* and right tube plate *RX* stippled), ×1.5 (Moore, n).——FIG. 439,*1h*. *S. bilineatus* STRIMPLE & MOORE, U.Penn.(Missour.), Ill.; post. view, ×1.2 (Strimple & Moore, 1971a).——FIG. 439,*1i*. *S. virgilensis* STRIMPLE, U.Penn.(Missour.), Okla.;

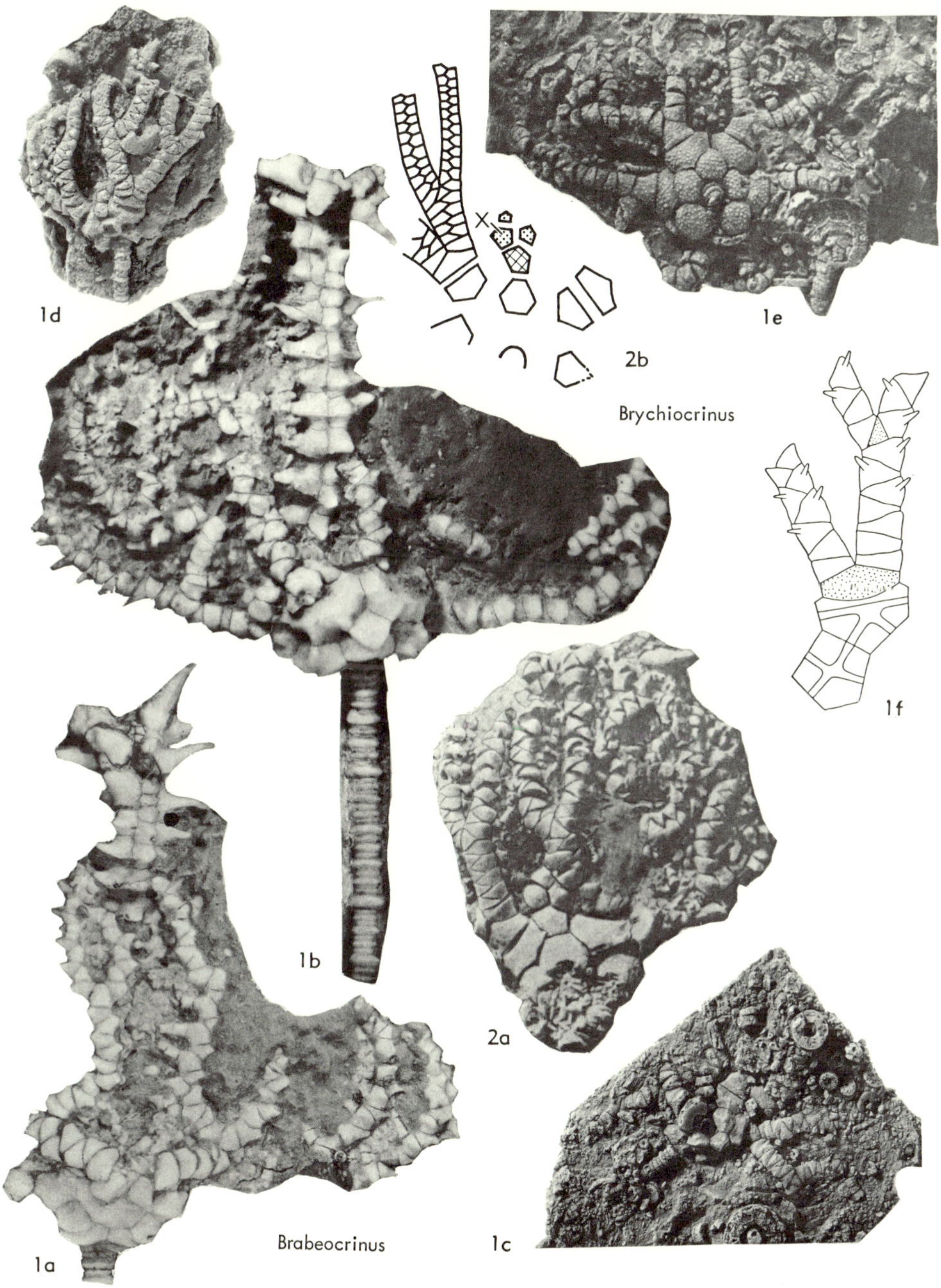

FIG. 440. Stellarocrinidae (p. *T669-T670*).

hypotype crown (*CD* interray upward), ×1.2 (Strimple & Moore, 1971a).

Brabeocrinus STRIMPLE & MOORE, 1971, p. 32 [**B. christinae*; OD]. Cup more boxlike than *Stellarocrinus* and arms uniserial (cuneate brachials) rather than biserial. *M.Penn.(Atokan)-L. Perm.(Wolfcamp.)*, USA.——FIG. 440,*1a,b*. **B. christinae,* U.Penn.(Missour.), Ill.; holotype viewed

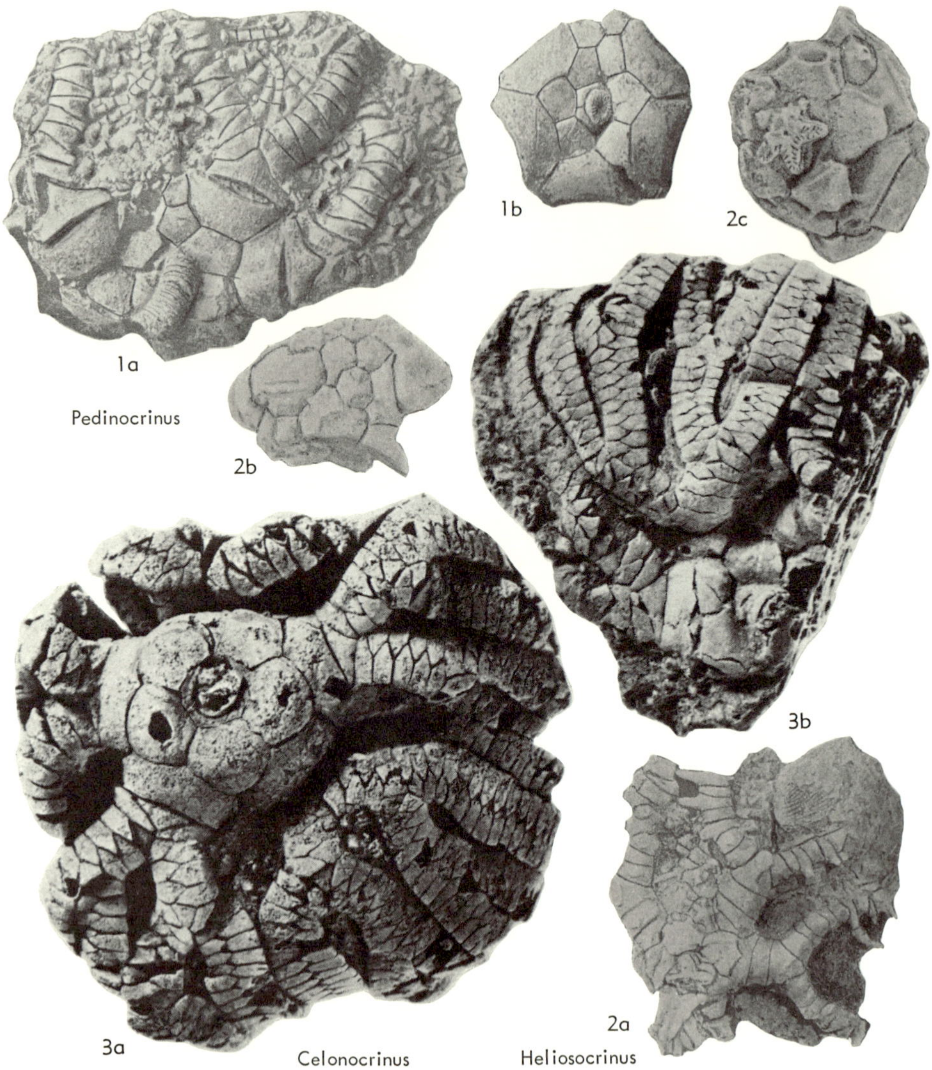

FIG. 441. Stellarocrinidae (p. T670-T671).

from *D* ray and opposite side (*AB* interray), both showing tall spinose anal sac, ×2 (Strimple & Moore, 1971a).——FIG. 440,*1c-f*. *B. cuneatus* (LANE & WEBSTER), L.Perm., Nev.; *1c,d*, dorsal (*CD* interray upward) and *C*-ray views of crown, ×0.7 (Lane & Webster, 1966); *1e*, dorsal (*AB* interray upward) view of crown, ×1 (Webster & Lane, 1967); *1f*, part of arm, ×2 (Webster & Lane, 1967).

Brychiocrinus MOORE & PLUMMER, 1940, p. 146 [**B. texanus*; OD]. Like *Stellarocrinus* except articular facets more nearly fill distal surface of radials and biserial arms are narrow. *M.Penn. (Desmoines.)*, USA.——FIG. 440,*2*. **B. texanus*, Texas; *2a*, holotype viewed from base (*CD* interray upward), ×1.5 (Moore & Plummer, 1940); *2b*, diagram of *D*-ray arms, ×1.5 (Moore, n).

Celonocrinus LANE & WEBSTER, 1966, p. 47 [**C. expansus*; OD]. Cup lower and broader than in typical *Stellarocrinus*, partially owing to lack of ridges and pronounced protrusions on plates; arms wider and externally flatter than in *Stellarocrinus*, with angulated edges. Column transversely round. *U.Penn.-L.Perm.*, USA(Nev.-Kans.-Texas).——FIG. 441,*3*. **C. expansus*, L.Perm., Nev.; *3a,b*, holotype viewed from base and side, ×1 (Lane & Webster, 1966).

Heliosocrinus STRIMPLE, 1951, p. 675 [**H. af-*

tonensis; OD]. Anal plates 3, in normal (primitive) arrangement; basals flattened in basal plane, extended as long subhorizontal spines, articular facets not filling distal face of radials. Arms uniserial (cuneate brachials), first branching on primibrachs *2*. Stem large, transversely pentastellate. *U.Miss.(Chester.)-L.Penn.(Morrow.),* USA (Okla.).——FIG. 441,*2*. **H. aftonensis,* U.Miss. (Chester.), Okla.; *2a,* post. view of holotype crown, ×1; *2b,c,* post. and dorsal views of paratype, ×1.5 (Strimple, 1951a).

Pedinocrinus WRIGHT, 1951, p. 77 [**Pachylocrinus clavatus* WRIGHT, 1937, p. 405; OD]. Crown explanate, short, recumbent well-separated arms spread out laterally. Cup discoid, base concave; infrabasals mostly concealed by stem, not visible from side; proximal part of basals in basal concavity; radials wide, with peneplenary articular facets and strong interradial notches; 3 anals in cup; surface of all plates smooth. Arms uniserial, widely divergent, and branching at least once above axillary primibrachs *1,* brachials strongly cuneate, approaching biserial arrangement; pinnules very stout, clublike. Stem transversely round. *L.Carb.(Visean),* Eu.(Scot.).——FIG. 441,*1*. **P. clavatus* (WRIGHT); *1a,b,* post. and dorsal (*CD* interray upward) views of holotype, ×1 (Wright, 1951-54).

Rhopocrinus KIRK, 1942, p. 152 [**R. spinosus*; OD]. Crown of medium height, compact. Cup low, broadly turbinate, plates smooth or with pits at angles of plates and low folds normal to edges, sharply defined ventral groovelike depression between adjacent radials in all known species; infrabasals small, but visible in lateral view; basals large; radials large, articular facets wide, extending full width of radial, linear to slightly crescentic, sutures not gaping; 3 large anals in cup. Anal sac formed by very many small nodose plates, with well-defined pits at their angles, sac wide and extending to 0.75 or more height of arms, distally bearing several stout spines. Arms uniserial, pinnulate, stout proximally, and slender distally, not appressed, primibrachs typically 2 in all rays, branching higher in crown parendotomous (i.e., above primaxils branches given off only to inside of each half-ray), axillaries tending to be spinose, brachials cuneate. Stem transversely circular, heteromorphic, with pentalobate lumen, nodals very prominent, bearing cirri to within 8 cm. of cup. *U.Miss.(Chester.),* USA(Ill.-Tenn.-Ala.-Ky.).——FIG. 428,*4*. **R. spinosus,* Ky.; *4a,* post. view of holotype, *4b,* ant. view of paratype, both ×1 (Kirk, 1942b).

Family PACHYLOCRINIDAE Kirk, 1942

[Pachylocrinidae KIRK, 1942b, p. 151]

Cup low, truncate bowl shaped, base slightly concave; infrabasals not visible from side; radial facets wide, peneplenary, bearing transverse ridge and ligament pits; three anals in cup. Anal sac slender, composed of vertical rows of small plates. Arms uniserial, branching isotomously twice or more, pinnulate. *L.Miss.-L.Perm.*

Advancement in evolution of the cup shape is seen in the truncate, concave base of crinoids belonging to the Pachylocrinidae. The arm structure closely resembles that of the blothrocrinids, and these families probably came from the same stock.

Pachylocrinus WACHSMUTH & SPRINGER, 1880, p. 115(338) [**Scaphiocrinus aequalis* HALL, 1861b, p. 8; OD]. Crown tall, ovoid. Cup low bowl shaped, with shallowly concave base; infrabasals not visible from side; basals notably larger than other cup plates; radials with peneplenary articular facets, interradial notches small but well defined; 3 anal plates in cup. Anal sac tall, subcylindrical, composed of irregularly polygonal small plates. Arms uniserial, branching isotomously on primibrachs *2* and additionally at 2 or more higher levels, making approximately 40 arms in upper part of crown, brachials mostly somewhat cuneate; pinnules long and slender. Stem transversely round or obtusely pentagonal in section. *L.Carb.-U.Carb.,* USSR(Voskresensk, Kalinin); *L.Miss.-U.Miss.,* central N.Am.——FIG. 442,*2*. **P. aequalis* (HALL), L.Miss.(Keokuk), Ind.; post. view of crown, *D* and *B* rays at left and right margins, ×1.5 (Springer, 1926b).

Plummericrinus MOORE & LAUDON, 1943, p. 46 [**Pachylocrinus mcguirei* MOORE, 1939c, p. 205; OD]. Crown slender. Cup truncate bowl shaped, base narrowly concave; 5 infrabasals not visible from side, concealed by stem; 5 basals, distal portions reaching about midheight of cup; 5 radials projecting outward so as to make prominent interradial notches, transverse ridges of articular facets not meeting neighbors at interradial sutures; 3 anal plates in cup; anal sac tubular, curved forward, side plates plicated. Arms uniserial, pinnulate, branching on primibrachs *1* and higher heterotomously. Stem round. *L.Penn.-L.Perm.,* USA(Okla.-Texas-Kans.).——FIG. 442,*1a-d*. **P. mcguirei* (MOORE), U.Penn., Virgil.(Brownsville Ls.), near Strohm, Okla.; *1a-c,* holotype cup from post., ventral, and dorsal (*CD* interray upward) sides, ×1.5; *1d,* paratype crown from *EA*-interray side, ×1.5 (Moore, 1939c).——FIG. 442,*1e*. *P. striatus* STRIMPLE, Missour.(Francis Sh.), Ada, Pontotoc Co., Okla.; hypotype crown from ant. side, showing cuneate brachials and long pinnules, ×1 (Strimple & Moore, 1971b).——FIG. 442,*1f,g*. *P. uddeni* (MOORE & PLUMMER), M.Penn., Desmoines.(Mineral Wells F.), near Mineral Wells, Palo Pinto Co., Texas; holotype from dorsal (*A* ray downward) and post. sides, ×1 (Moore & Plummer, 1940).

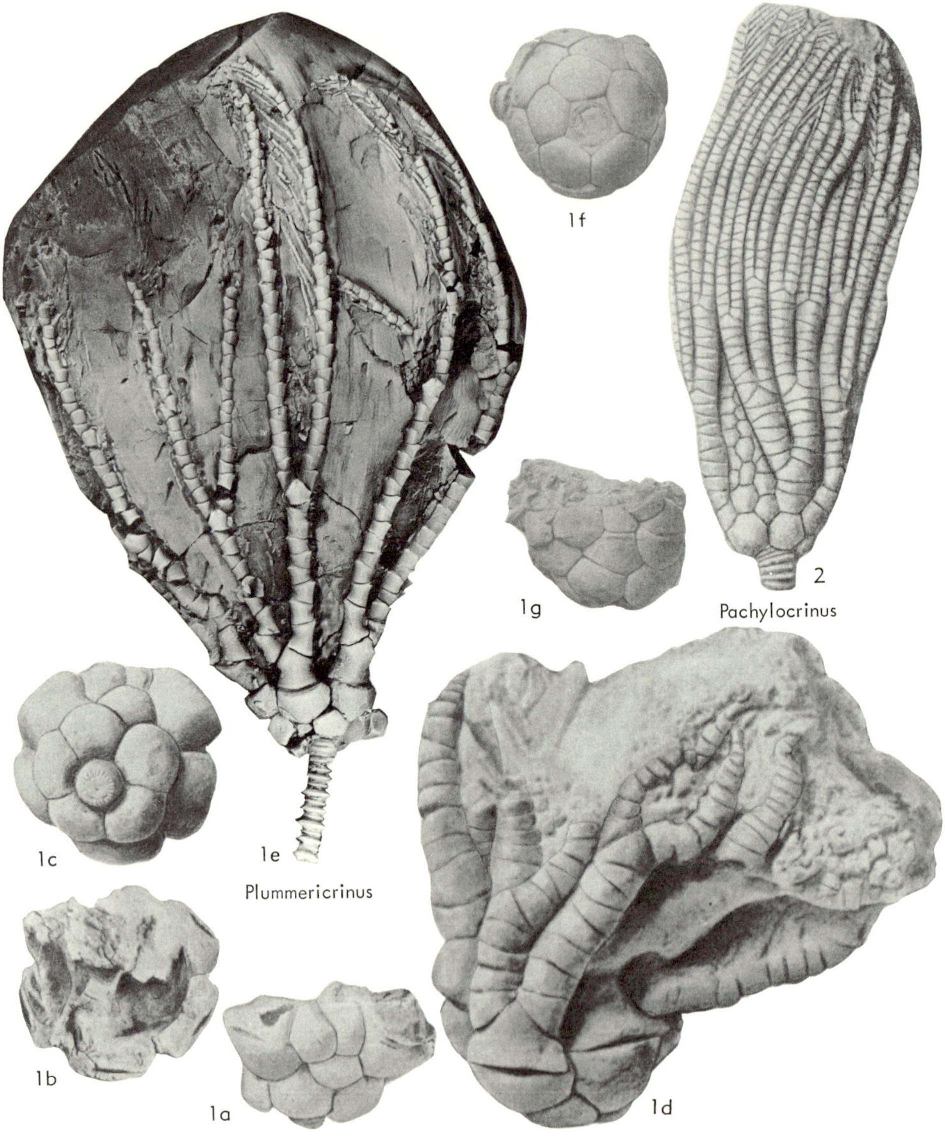

Fig. 442. Pachylocrinidae (p. T671).

Superfamily AGASSIZOCRINACEA S. A. Miller, 1889

[*nom. transl.* Moore & Strimple, 1973, p. 20 (*ex* Agassizocrinidae S. A. Miller, 1889, p. 214)] [Materials for this superfamily prepared by R. C. Moore, N. Gary Lane and H. L. Strimple]

Crown cylindrical to pear shaped. Cup bowl shaped with convexly round base; infrabasals five or fused into single solid plate, readily visible from side; radial articular facets wide, occupying full width of radial summits; anal plates in cup none to four, joined very tightly in some to few lower posterior sac plates so as to form in effect upward projection of cup margin. Anal sac variably weak to strong, cylindrical or mushroom shaped. Arms mostly numerous, isotomously branched on primaxils and heterotomously above. Stem transversely round or exceptionally pentagonal or pentalobate. *L.Miss.-U.Perm.*

Family BURSACRINIDAE Kirk, 1947

[Bursacrinidae KIRK, 1947, p. 287]

Crown compact; cup low, broadly conical; infrabasals small, upflared, visible from side; radial articular facets fully equal to radials in width; *X* only anal plate in cup. Anal sac short, rarely preserved. Arms stout, uniserial, closely appressed laterally, primibrachs *1* axillary; without higher branching or with several isotomous branches; pinnulate. *L.Miss.*

Bursacrinus MEEK & WORTHEN, 1861, p. 136 [**B. wachsmuthi*; M]. Cup with high basals and low broad radials; single anal plate in cup between posterior radials. Anal sac unknown. Arms broad and externally flattened, closely abutting laterally; first primibrachs axillary in all rays, with at least one higher division in some or all rays; brachials low, broad, wedge shaped, pinnulate. *L.Miss.*, USA(Iowa).——FIG. 443,*2*. **B. wachsmuthi*; *C*-ray view of partial crown, ×0.75 (Meek & Worthen, 1868b).

Lebetocrinus KIRK, 1940, p. 74 [**L. grandis*; OD]. Cup low, broad, plates thin; infrabasals and basals large, radials with plenary articular facets; single anal plate in cup directly above *CD* basal; anal *X* large, supporting longitudinal row of large sac plates. Anal sac short. Arms uniserial, stout, rounded, and closely appressed laterally, branching isotomously several times, pinnulate; primibrachs *1* rarely *2* axillary, narrow, and completely or partially enclosed by radial and first secundibrachs in some rays. Stem large, transversely round, with large lumen. *L.Miss.*, USA(Ind.).——FIG. 443,*3*. **L. grandis*; *3a,b*, *A*-ray and *CD*-interray views of crown and proximal part of stem, ×0.5 (Kirk, 1940d).

Nactocrinus KIRK, 1947, p. 288 [**N. nitidus*; OD]. Crown short and compact; cup cone-shaped, infrabasals barely visible in side view; radials and their articular facets as in *Bursacrinus*; anal *X* only in cup, supported by *CD* basal, or occupying a notch between the posterior radials; anal sac not known. Arms 10, uniserial, stout, closely appressed laterally; first primibrach axillary, large and broad; brachials large, externally rounded, pinnulate. *L. Miss.*, USA(Iowa).——FIG. 443,*1*. **N. nitidus*; *1a,b*, side and *CD*-interray views of crown, ×1.12 (Kirk, 1947).

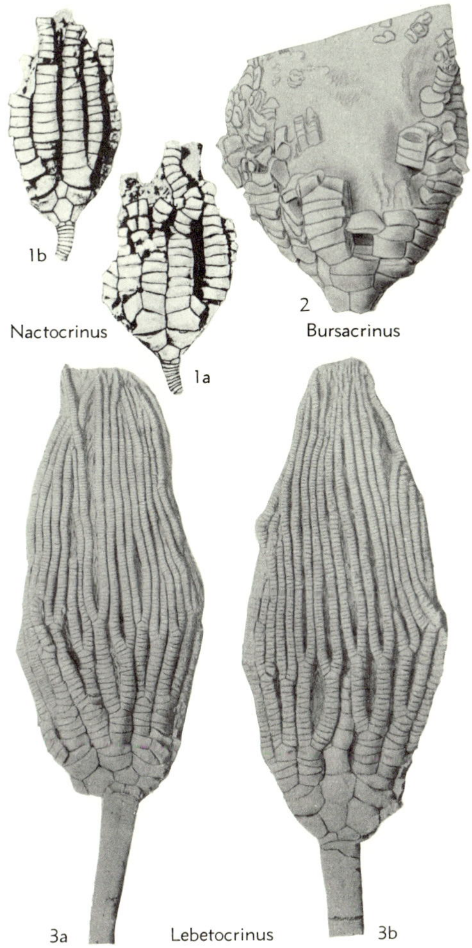

FIG. 443. Bursacrinidae (p. *T*673).

Family AMPELOCRINIDAE Kirk, 1942

[Ampelocrinidae KIRK, 1942c, p. 23] [incl. Ampelocrininae KIRK, 1942c (*nom. transl.* STRIMPLE & WATKINS, 1969, p. 185)]

Crown tall, subcylindrical or expanding. Cup low to medium bowl shaped or truncate conical, with upflared infrabasals, visible from side, or infrabasals in shallow basal concavity and not visible from side; radials large, articular facet wide, plenary, short, sloping outward-upward; one dominant anal plate in cup. Anal sac slender, may be recurved or straight. Arms usually ten but may branch twice or not at all; primibrachs *2-4* axillary except that in younger forms primibrachs *1* and *2* may fuse and become nonaxillary; long pinnules on opposite sides of alternate brachials unless syzygial pairs are present. Stem transversely round, subpentagonal, or decidedly pentagonal, with or without cirri. *L.Miss.-U.Perm.*

The ampelocrinids comprise a distinctive but not aberrant branch of the blothrocrinid stock which appeared in Early Mississippian time and persisted into the Permian. The cup is moderately advanced in evolution in that its base is very gently convex, flat, or slightly concave, and a single dominant anal plate occurs below the summit of the radials. Seemingly, this plate is the radianal, as indicated by obliquity of its contact with the *CD* basal in some specimens of *Ampelocrinus* and by its relation to overlying plates at the base of the anal sac. In other genera this anal rests squarely on the *CD* basal. Specialized characters of the family are its uncommon sursumate radial articular facets sloping outward-upward and the tall, slender but stout anal sac, formed of vertical rows of plates. Features of arm structure and development of cirri aid in differentiation of the family.

Key to Genera of Ampelocrinidae

A. Cup moderately tall, conical, infrabasals clearly visible from side, 1-2 anal plates in cup, 10 arms branching on primibrachs *2-4* ....

- I. Two anal plates, primibrachs *2* axillary .... *Proampelocrinus*
- II. One anal plate, primibrachs *2-4* axillary .... *Armenocrinus*

B. Cup bowl shaped with convex base, infrabasals visible from side, 3 anals in cup, arms branching isotomously on primibrachs *2* ....

- I. Brachials cuneate, pentagonal stem highly cirriferous .... *Ampelocrinus*
- II. Brachials externally rectangular, cirri not observed .... *Polusocrinus*

C. Arms in upper crown 10 to 30 or more, branching on primibrachs *2* ....

- I. Arms 10, branching above primaxils not observed, brachials cuneate, single anal plate in cup .... *Moundocrinus*
- II. Arms 14-16, branching above primaxil on secundibrach *1*. Three anal plates in advanced position (radianal dominant) .... *Arroyocrinus*
- III. Arms 20 or more; anal plate in cup 1
  - a. Arms branching above primaxil on secundibrachs *9-10*; highly cirriferous pentalobate stem .... *Chlidonocrinus*
  - b. Arms branching above primaxil on secundibrachs *6-8* .... *Sphenicocrinus*

D. Cup low, truncate cone shaped, arms branching on primibrachs *3* or *4* .... *Halogetocrinus*

Ampelocrinus Kirk, 1942, p. 23 [**A. bernhardinae*; OD]. Cup low or high turbinate; some species reported with syzygial pairs of brachials. Anal tube recurved. Column transversely subpentagonal or pentagonal. *U.Miss.(Chester.)-L.Penn.,* USA.——Fig. 444,*5*. **A. bernhardinae,* Chester.(Glen Dean F.), Ky.; side view of holotype crown, ×0.93 (Kirk, 1942c).

Armenocrinus Strimple & Horowitz, 1971, p. 23 [**A. watersi*; OD]. Like *Ampelocrinus* except that older species have more nonaxillary primibrachs and cup is proportionately larger and higher. [*Armenocrinus* is thought to be derived from a form like *Culmicrinus thomasi* Laudon (1933), from the Gilmore City Formation (Kinderhook.) and may represent the basic lineage of the Ampelocrinidae.] *L.Miss.(Osag.)-U.Miss.(Chester.),* USA(Ind.-Pa.-Ala.).——Fig. 445,*1a*. **A. watersi,* Geneviev.(Monteagle F.), Ala.; holotype from *CD* interray, ×1.93 (Strimple & Horowitz, 1971).——Fig. 445,*1b,c*. *A. neglectus* (Miller & Gurley), Osag.(Borden Gr.), Crawfordsville, Ind.; holotype from *A* ray and *CD* interray, ×1.93 (Strimple & Horowitz, 1971).——Fig. 445,*1d*. *A. collinsi* Strimple & Horowitz, Chester.(Greenbrier F.), Fayette Co., Pa.; holotype from *CD* interray, ×1.93 (Strimple & Horowitz, 1971).

Arroyocrinus Lane & Webster, 1966, p. 40 [**A. popenoei*; OD]. Infrabasals not visible from side, 3 anal plates advanced (radianal oblique, in dominant posterior position followed above by anal *X* and right tube plate). Arms uniserial, closely apposed, 14 to 16 in number, branching on primibrachs *1* and secundibrachs *1*. *L.Perm.,* USA (Nev.).——Fig. 444,*1*. **A. popenoei,* Wolfcamp.(Bird Spring F.); *1a,b,* holotype from *A* ray, paratype from *CD* interray, ×0.46, ×0.35 (Lane & Webster, 1966).

Chlidonocrinus Strimple & Watkins, 1969, p. 189 [**C. echinatus*; OD]. Cup bowl shaped, large subhorizontal infrabasals, pronounced plications passing between plates, single large anal plate (radianal), faceted for 2 plates above. Arms branching more than twice, some with syzygially paired brachials. Cirri abundant. *U.Miss.(Chester.)-U.Penn.(Missour.),* USA(Texas-Okla.).——Fig. 444,*3a*. **C. echinatus,* M.Penn.(Atokan), Texas; holotype viewed from ant. side, ×1.1 (Strimple & Watkins, 1969).——Fig. 444,*3b*. *C. ornatus* Strimple & Moore, Missour.(Francis Sh.), near Ada, Pontotoc Co., Okla.; holotype viewed obliquely from post. side, ×1.4 (Strimple & Moore, 1971b).

Halogetocrinus Strimple & Moore, 1971, p. 29 [**Aesiocrinus paucus* Strimple, 1951a, p. 22; OD]. Cup low, saucer shaped; subhorizontal infrabasals, not visible from side; basals usually small; radials large, width and length about equal, with articular facets filling their upper face; one large anal (radianal) in line with radials, obliquely above posterior *(CD)* basal, fol-

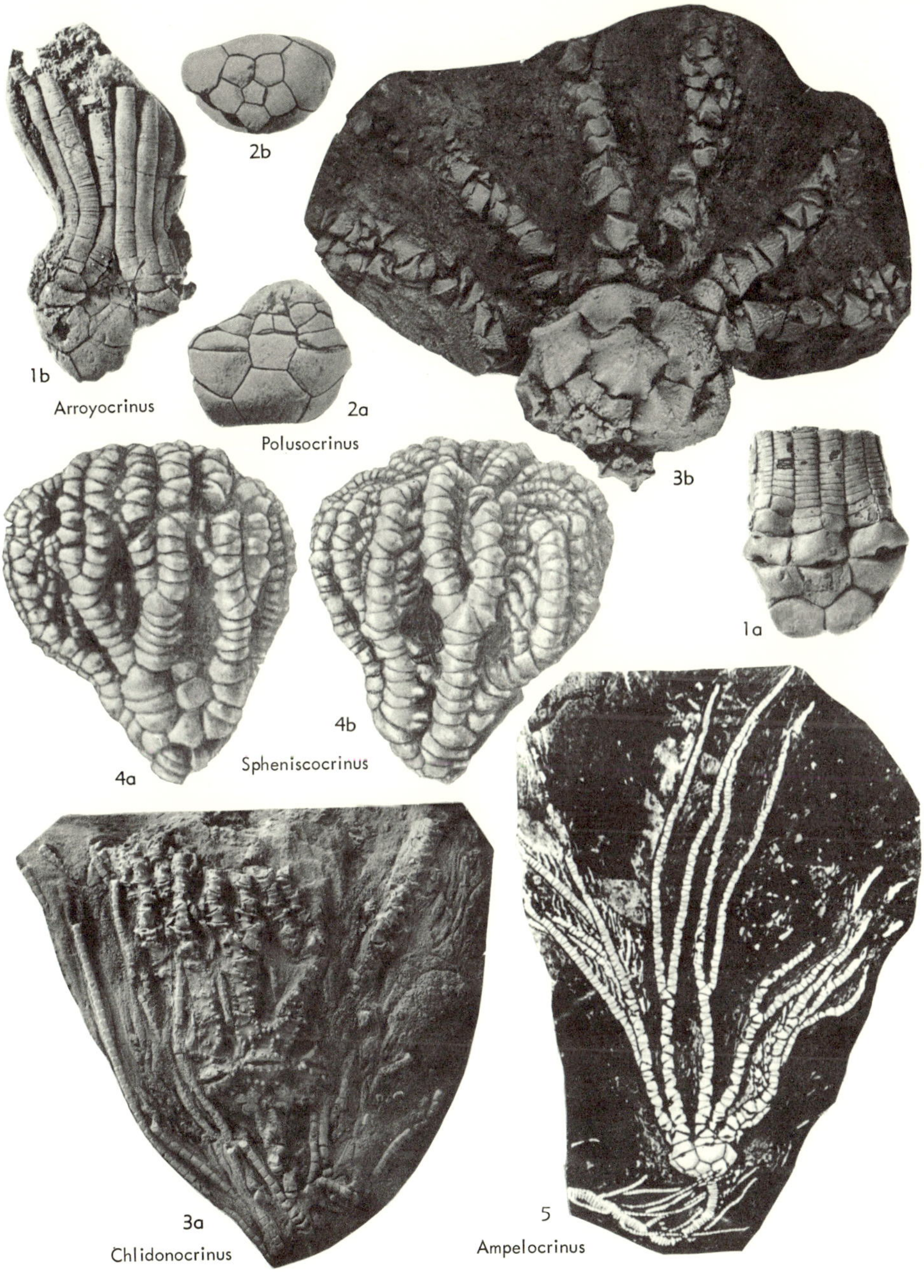

FIG. 444. Ampelocrinidae (p. *T674*, *T676*).

lowed evenly by 2 plates; brachials cuneiform above first branching, may be syzygial; armlets nonpinnulate, composed of rectilinear brachials, first branching on primibrachs *3* or *4*; stem very cirriferous. *U.Penn.(Missour.-Virgil.),* USA(Ill.-Okla.-Kans.-Neb.).——FIG. 446,*2a-c.* **H. paucus* (STRIMPLE), Missour., Okla.; *2a-c,* from Wann F., Bartlesville, Okla.; holotype cup from dorsal, post., and ventral sides, ×2.5 (Strimple, 1951a); *2d,* hypotype, from LaSalle Ls., near Pontiac,

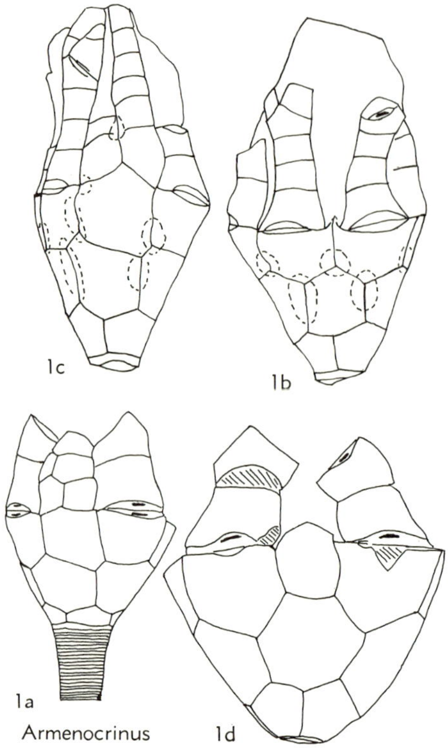

FIG. 445. Ampelocrinidae (p. *T674*).

Livingston Co., Ill.; crown from side, ×2.2 (Strimple & Moore, 1971a).——FIG. 446,*2e*. *H. tumidus* STRIMPLE & MOORE, Missour.(Francis Sh.), Ada, Pontotoc Co., Okla.; holotype crown from right post. side, ×2.7 (Strimple & Moore, 1971b).

Moundocrinus STRIMPLE, 1939, p. 9 [**M. osagensis*; OD]. Cup bowl shaped with subhorizontal base and nonimpressed sutures; infrabasals small, subhorizontal, or slightly downflared; basals wide, prominent; radials large, articular facets plenary but short; single anal plate extending only slightly about cup summit, faceted above for reception of one tube plate. Arms 10, very long, uniserial (mildly cuneate brachials), well-rounded exteriors, pinnules borne by alternate brachials on opposite sides, primibrachs *2* axillary unless fused with primibrachs *1*. Proximal columnals transversely pentagonal or subpentagonal. *U.Penn.(Missour.)*, USA(Okla.-Ill.).——FIG. 446,*3a-c*. **M. osagensis*, Wann F., near Bartlesville, Okla.; holotype from ventral, dorsal, and post. sides, ×1.7 (Strimple, n; National Museum of Natural History S 4306).——FIG. 446,*3d-f*. *M.* sp. cf. *M. osagensis* STRIMPLE & MOORE, Missour.(Bond F.), near Pontiac, Livingston Co., Ill.; holotype viewed from dorsal, ant., and ventral sides, ×2 (Strimple & Moore, 1971a).

Polusocrinus STRIMPLE, 1951, p. 24 [**P. avanti*; OD]. Cup deep, bowl shaped with erect lateral sides; infrabasals large, gently upflared; basals large; radials moderately large; single anal plate followed by two equidimensional tube plates. Arms uniserial; primibrachs *2* axillary. Proximal columnals pentagonal in outline becoming circular at short distance from cup. *U.Penn.*, USA.——FIG. 444,*2*. **P. avanti*, Missour.(Avant F.), Okla.; *2a,b*, holotype from post. and dorsal (*CD* interray upward), ×0.93 (Strimple, 1951a).

Proampelocrinus GUPTA & WEBSTER, 1974, p. 337 [**P. himalayaensis*; OD]. Crown slender, elongate; cup high, conical, with 2 anal plates in cup. Arms long, slender, 10, branching isotomously on second primibrach in all rays, brachials cuneate. Anal sac long, slender. Stem pentagonal, with nodals and two series of internodals. *L.Carb.*, India(Ladakh, near Zanskar).

Spheniscocrinus WANNER, 1937, p. 175 [**S. spinosus*; OD]. Dish-shaped cup; infrabasals entirely covered by stem; 1 anal plate in cup, resting on posterior basal. Anal sac unknown. Straight articular facets occupying whole width of radials. Arms not adjoining, branching isotomously on primibrachs *2* and secundibrachs *6* to *8*; pinnules stout, resembling ramules. *U.Perm.*, Timor.——FIG. 444,*4*. **S. spinosus*, Basleo; *4a,b*, *CD*-interray and *A*-ray views, ×1.4 (Wanner, 1937).

Family SUNDACRINIDAE Moore & Laudon, 1943

[Sundacrinidae MOORE & LAUDON, 1943a, p. 62]

Cup small, bowl to globe shaped, plates massive; infrabasals five or three, mostly low, and visible in side view or confined to slight basal concavity and not visible from side; basals large; radials typically unequal in size and shape, or may be small and subequal; *B* and *E* radials commonly small, narrow, and without articular facets or with single, small, incurved primibrach fused to their summit; other radials larger and arm bearing; one or three anal plates in cup. Stem transversely round. *L.Perm.-U.Perm.*

The Sundacrinidae, chiefly of Permian age, are plainly a very specialized group, inasmuch as they have two armless radials of much reduced size. Variation of the anal plates ranging from three to one is seen within a single species. The ancestry of this family is very uncertain.

Key to Genera of Sundacrinidae

A. Base of cup convex ....

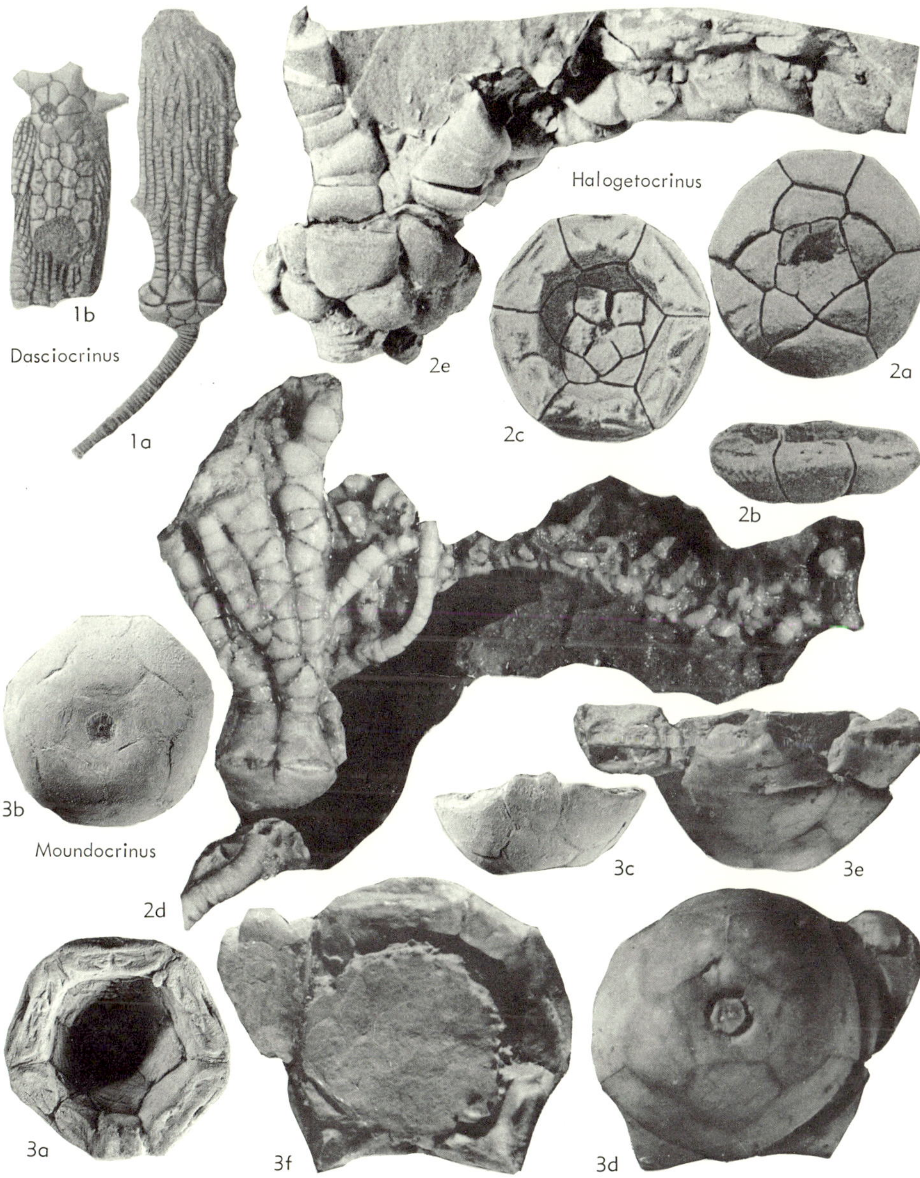

FIG. 446. Ampelocrinidae *(2,3)*; Pirasocrinidae *(1)* (p. *T674-T676, T726*).

I. Infrabasals 5 ....
  a. Small subequal articular facets on 5 radials; 3 anal plates in cup .... *Hemiindocrinus*
  b. Large subequal articular facets on 5 radials; anal *X* alone in cup .... *Laccocrinus*
  c. Large articular facets on *A, C,* and *D* radials, *B* and *E* radials armless; 1 or rarely 2 anal plates in cup .... *Sundacrinus*

II. Infrabasals 3 ....
  a. Articular facets on *A, C, D,* and *E* radials large (that on *D* radial twice size of others), *B* radial armless; single anal plate in cup .... *Parindocrinus*
  b. Articular facets on *A, C,* and *D* ra-

dials large, *B* and *E* radials armless; 3 anal plates in cup .... *Tribrachyocrinus*

III. Infrabasals 1, subequal radial articular facets, anal *X* alone in cup .... *Basleocrinus*

Sundacrinus Wanner, 1916, p. 219 [**S. granulatus*; OD]. Cup bowl shaped, base flat with central depressed stem impression, plates massive; infrabasals 5, low, distal parts visible from side; basals 4 or 5, unequal in shape, *BC* basal may be fused with *C* radial forming single plate; radials 5, those of *A, C,* and *D* rays large, low, twice as wide as high and bearing arm facets; *B* and *E* radials small, narrow, and without arm facets, all nearly as wide as radials, inclined outward-downward, with strong transverse ridge; single anal plate, or rarely 2 in cup, directly or obliquely above *CD* basal, or between upper edges of *CD* and *BC* basals; small, quadrangular radianal obliquely below *C* radial in some specimens. *U.Perm.*, Indon.(Timor).——Fig. 447,*3*. **S. granulatus,* Basleo; *3a-c,* post., dorsal, and ventral views of cup, ×2 (Wanner, 1916a).

Basleocrinus Wanner, 1916, p. 214 [**B. pocillum*; OD]. Cup bowl shaped; infrabasals low, fused into single plate visible from side; basals large; radials with wide articular facets bearing prominent ligament grooves and ridges; no interradial notches; single anal plate directly above *CD* basal and projecting well above radials. Stem round, narrow. *U.Perm.*, Indon.(Timor).——Fig. 447,*5*. **B. pocillum,* Basleo; *5a-c,* post., ventral (*CD* interray downward), and dorsal views of cup, ×3 (Wanner, 1916a).

Hemiindocrinus Yakovlev, 1926, p. 57 [**H. fredericksi*; M]. Cup small, globose, restricted in width distally; infrabasals 5, large, high, visible in side view; basals large, convex; radials small, narrow, equal in size and all with plenary articular facets; 3 anal plates in cup. *L.Perm.*, USSR (Urals-Krasnoufimsk).——Fig. 447,*1*. **H. fredericksi,* Krasnoufimsk; *1a,b,* lat. and ventral views of cup, ×1.5 (Yakovlev, 1926a).

Laccocrinus Wanner, 1949, p. 55 [**Basleocrinus scrobiculatus* Wanner, 1924, p. 267; OD]. Cup bowl shaped, with strongly convex plates and deep pits at plate angles, sutures impressed; infrabasals 5, small, low; radial articular facets wide (plenary), prominent ligament grooves and ridges, extended inward so as to restrict opening to visceral cavity; only anal *X* in cup, convex and projecting above radials. *U.Perm.*, Indon.(Timor).——Fig. 447,*4*. **L. scrobiculatus* (Wanner), Basleo; *4a-b,* post. and ventral (*CD* interray downward) views of cup, ×3 (Wanner, 1924).

Parindocrinus Wanner, 1937, p. 165 [**P. oyensi*; OD]. Cup pentalobate in basal view, asymmetrical, higher and protruding in *D* ray; infrabasals 3, low, broad, visible in side view, circlet truncate at base for wide, round stem impression; basals large, strongly convex and nodose, radiating ridges extending from nodes across sutures to adjoining cup plates; radials projecting outward, with wide large articular facet bearing prominent transverse ridge and ligament groove; *D* radial twice width of others; *C* and *A* radials normal in size, with wide arm facets; small narrow *E* radial with obscure articular facet; *B* radial small, without facet; single anal plate in cup between *C* and *D* radials. Arms unknown. *U.Perm.*, Indon.(Timor).——Fig. 447,*6*. **P. oyensi,* Basleo; *6a-c,* post., dorsal, and ventral views of cup, ×3 (Wanner, 1937).

Tribrachyocrinus M'Coy, 1847, p. 228 [**T. Clarkii*; OD] [=*Tribrachiocrinus* Wachsmuth & Springer, 1886, p. 250 *(nom. van.)*]. Cup globose, large, plates massive; infrabasals 3, unequal, small infrabasal in *A* ray; basals very large, higher than wide, irregular in outline; radials 5, irregular in size and shape, *B* and *E* radials smaller than others and without arm facets, other radials with wide articular facets; 3 large anal plates in cup. *L.Perm.*, Australia(New S.Wales-Tasmania).——Fig. 447,*7*. **T. clarkii,* New S.Wales; *7a,b,* ant. and post. views of cup, ×1 (M'Coy, 1847).

Family ANOBASICRINIDAE Strimple, 1961

[Anobasicrinidae Strimple, 1961d, p. 114]

Tall cone shaped to bowl shaped cup, flat or gently rounded at base; infrabasals subhorizontal to slightly upflared, mostly not visible in side view; three anal plates in cup. Large recurved anal sac may become bulbous. Arms 20 or more, long, slender, pinnulate, uniserial, with convex surfaces, primibrachs *1* axillary. Stem large, transversely round. *L.Penn.(Morrow.)-U.Perm.*

Key to Genera of Anobasicrinidae

A. Cup tall cone shaped to weakly convex bottomed bowl shaped with upward projection of anal plates in *CD* interray; peneplenary radial articular facets markedly declivate. Stem impression circular ....

I. Cup low bowl shaped ....

a. Base gently convex, infrabasals relatively large, very low, interradial notches small, anal sac balloon-like .... *Anobasicrinus*

b. Basal concavity but with tips of infrabasals flexed upward, anal sac large, looped .... *Terpnocrinus*

II. Tall cone-shaped cup, prominent upflared infrabasals, balloon-like anal sac .... *Synyphocrinus*

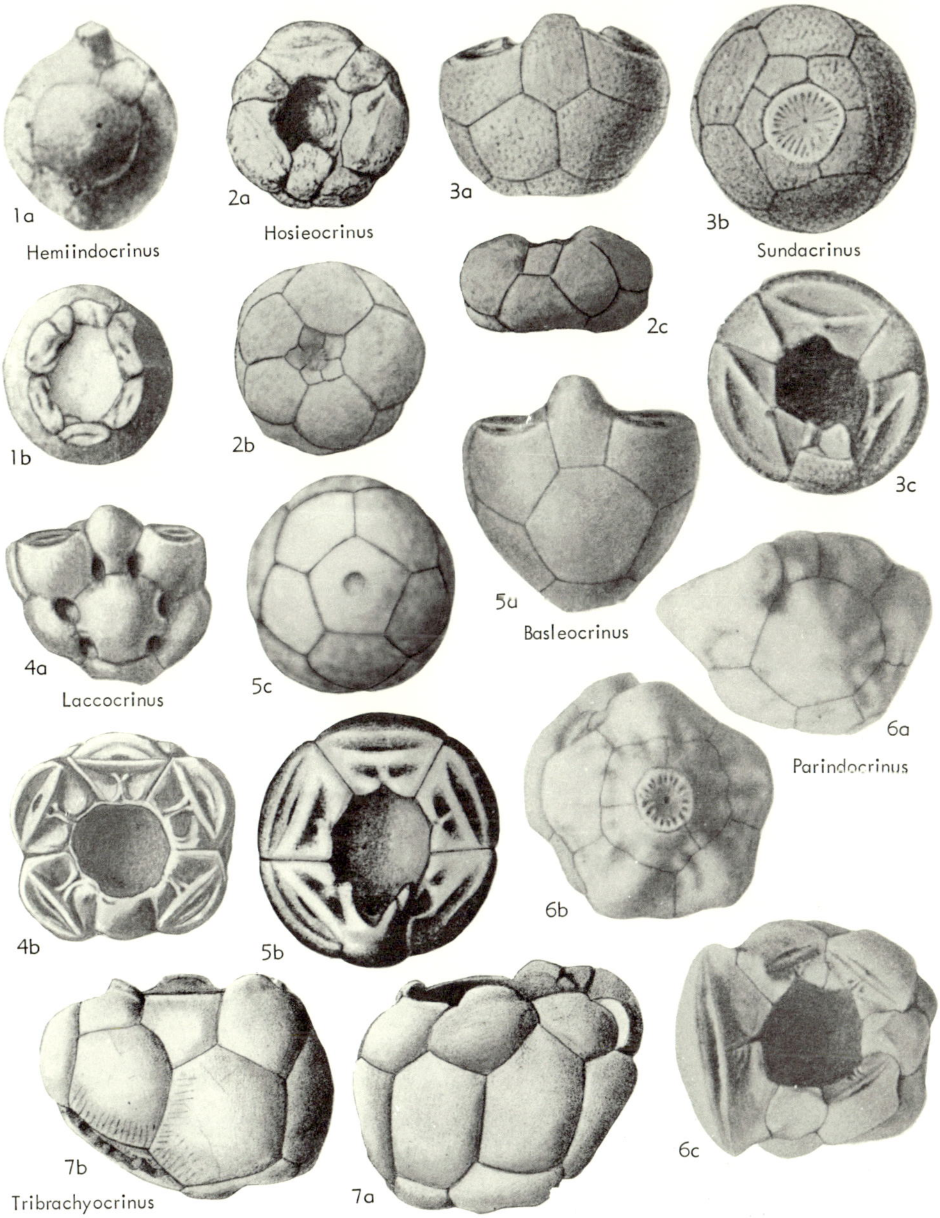

FIG. 447. Sundacrinidae *(1,3-7)*; Phanocrinidae *(2)* (p. *T*678, *T*692).

Anobasicrinus STRIMPLE, 1961, p. 115 [**A. bulbosus*; OD]. Crown expanding upward, with approximately 30 slender uniserial arms not apposed. Cup with evenly upflared longitudinally straight or curved sides and narrow flat to faintly convex or concave base; infrabasals generally not visible from side, radials large, with proximal tips well above basal plane of cup, their articular facets peneplenary, not quite equal in width to radial summits; 3 anal plates in cup. Anal sac low and broad balloon shaped distally, irregularly arranged rugose polygonal sac plates marginally slitted. Arms branching isotomously on primibrachs *1* and endotomously on secundibrachs *3* and higher, pinnulate. *L.Penn.(Morrow.)-U.Penn.(Missour.)*, USA (Okla.-Texas).——FIG. 448,*1a-c*. **A. bulbosus*,

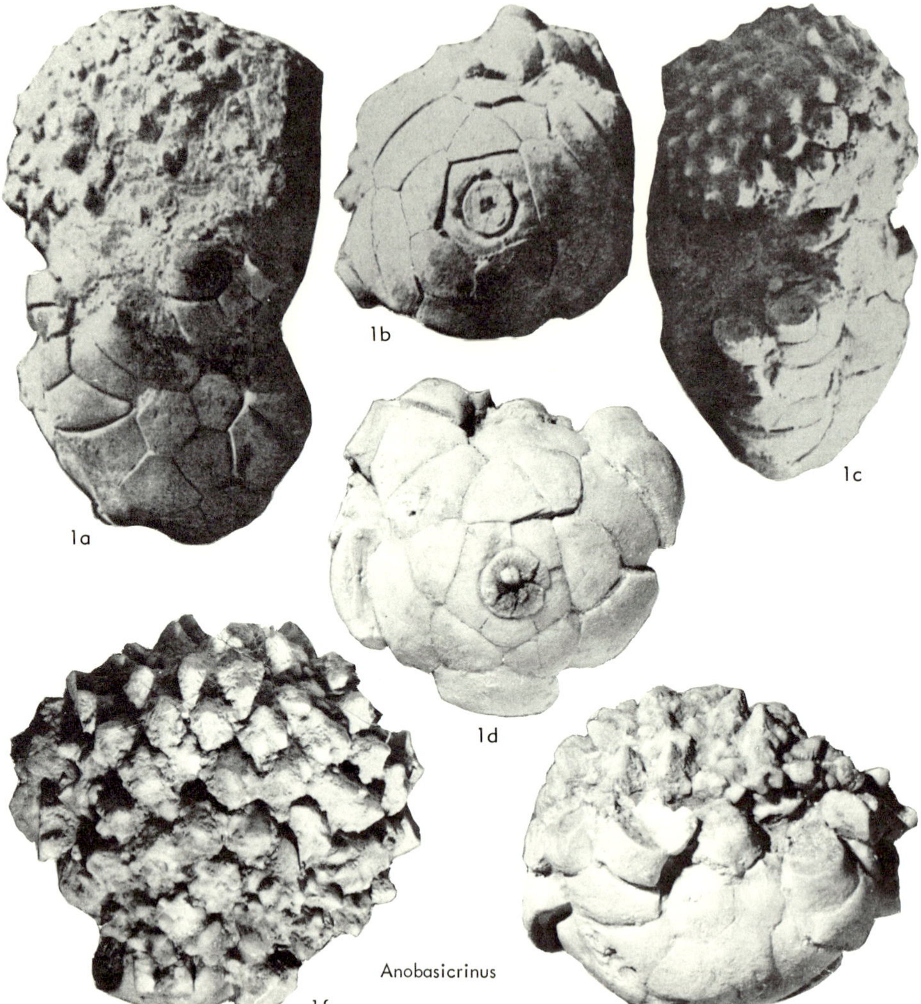

FIG. 448. Anobasicrinidae (p. *T679-T680*).

Desmoines.(Holdenville Sh.), near Beggs, Okmulgee Co., Okla.; *1a-c*, ant., dorsal (*CD* interray upward), and post. sides of crown, ×1 (Strimple, 1961d).——FIG. 448,*1d-f*. *A. brevis* STRIMPLE & MOORE, U.Penn.(Missour., LaSalle Ls.), Livingston Co., Ill.; *1d-f*, dorsal (*CD* interray upward), post., and ventral views of holotype, ×2 (Strimple & Moore, 1971a).

Synyphocrinus TRAUTSCHOLD, 1867, p. 1 [**S. cornutus*; OD]. Similar to *Haeretocrinus* except that anal sac composed of many vertical rows of rugose plates is broad balloon-like. *U.Carb.(Moscov.)-U.Perm.*, USSR(Moscow Basin)-Timor.——FIG. 438,*2a-e*. **S. cornutus*, U.Carb., USSR; *2a-c*, post., ant., and dorsal sides of cup, ×1; *2d,e*, arms and ext. surface of anal sac, ×0.5 (Yakovlev & Ivanov, 1956).——FIG. 438,*2f*. *S. trautscholdi* (WANNER), Perm., Indon.(Timor); diagram of holotype (radials black, anal plates stippled) (after Wanner, 1924).——FIG. 438,*2g,h*. *S. magnus* YAKOVLEV & IVANOV, U.Carb.(Moscov.), Myachkovo, USSR; edge view of broken balloon-like anal sac, int. view of broken anal sac, showing pores penetrating sac wall from exterior, ×1 (Yakovlev & Ivanov, 1956).

Terpnocrinus STRIMPLE & MOORE, 1971, p. 23 [**T. ocoyaensis*; OD]. Cup more globose than bowl shaped, midportion of infrabasal circlet sharply impressed, but distal tips curved upward, visible from side. Anal sac large, recurved as in *Haereto-*

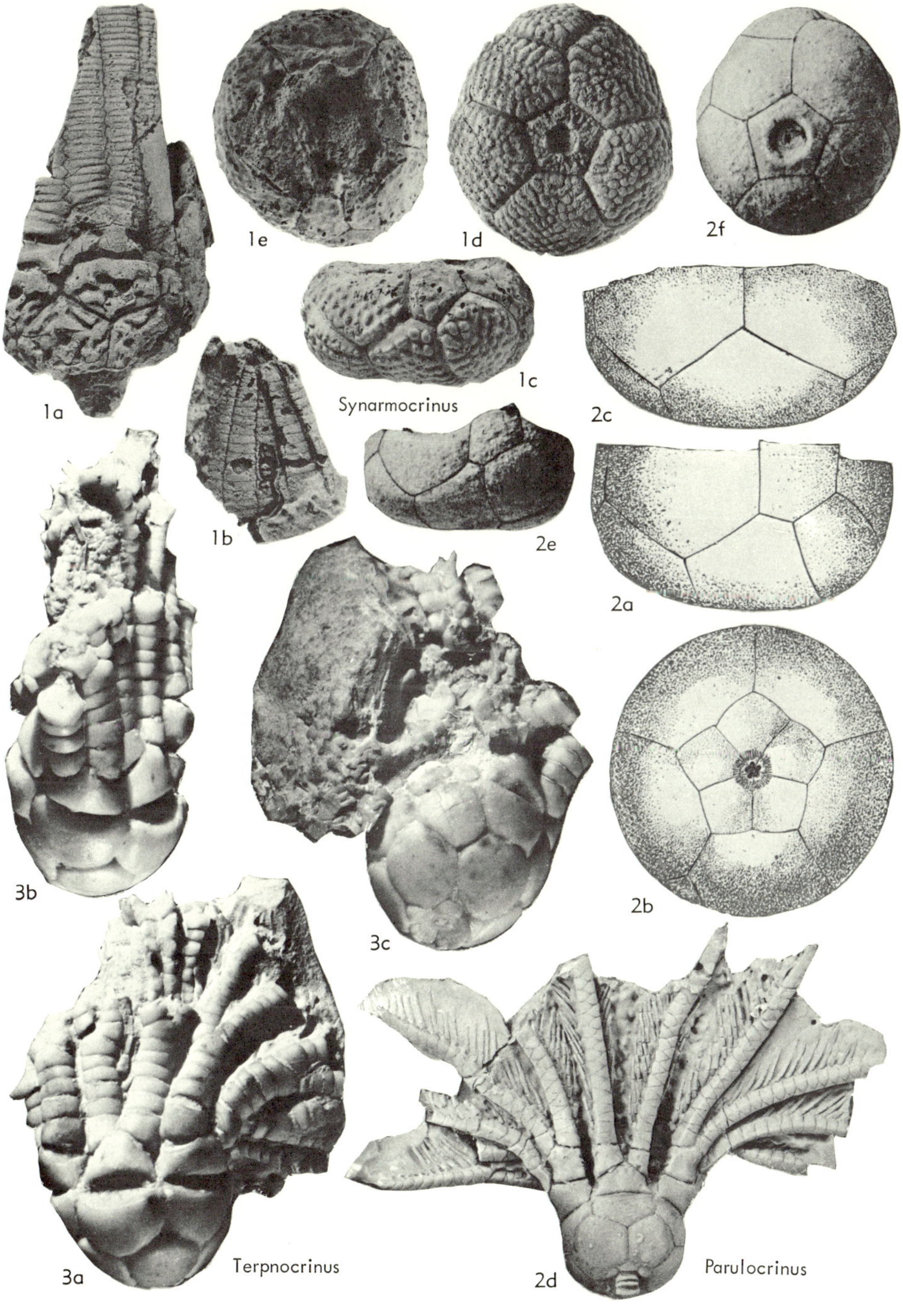

FIG. 449. Anobasicrinidae *(3)*; Cromyocrinidae *(1,2)* (p. *T*680-*T*681, *T*700-*T*701).

crinus. U.Penn.(Missour.), USA(Ill.-Neb.-Okla.).——FIG. 449,*3*. **T. ocoyaensis,* Missour.(LaSalle Ls.), Ill.; *3a-c,* holotype viewed from *AE* interray, *B* ray showing recurved anal tube, and oblique dorsal sides, ×2 (Strimple & Moore, 1971a).

Family AGASSIZOCRINIDAE S. A. Miller, 1889

[Agassizocrinidae S. A. MILLER, 1889, p. 214] [incl. Paragassizocrininae STRIMPLE & WATKINS, 1969, p. 186 (assigned to Ampelocrinidae)]

Crown ovoid to pyriform, medium sized, Cup subconical to semiovoid; infrabasals normally five, relatively large and clearly visible from side, or solidly fused into single round element, exceptionally (e.g., *Epipetschoracrinus*) much reduced in size, located within depressed stem attachment area, and not visible from side; basals five, large; radials five, wide and short, with plenary subhorizontal articular facets bearing rather poorly defined transverse ridge and ligament fossae; anal plates four to one in cup. Anal sac unknown. Arms generally ten, branching isotomously on first primibrach, uniserial, pinnulate; in some forms each arm bifurcating approximately on secundibrach *6*. Stem lacking except in juvenile stages and rarely in adult forms. *U.Miss.-L.Perm.*

Crinoids of this group are characterized by unusual thickness of their cup plates and large, mostly very massive infrabasal circlet, which commonly is fused into a single solid piece, with or without trace of stem impression. Radial articular facets are not extended inward by flangelike projection, as in many cladids.

Key to Genera of Agassizocrinidae

A. Lower part of cup convex, semiovoid ......
- I. Infrabasals 5, anal plates 3, hypertrophied arms in *B* and *E* rays .. *Anartiocrinus*
- II. Infrabasals 4, anal plates in cup 1 .... *Petschoracrinus*
- III. Infrabasals 1, comprising solidly fused circlet ....
  - a. Anal plates in cup 4, arms branching isotomously on first primibrachs .... *Agassizocrinus*
  - b. Anal plate in cup 1, arms branching isotomously on primibrachs *2* .. .. *Paragassizocrinus*

B. Cup with extremely narrow and small basal concavity, basals very large, single anal plate in cup .... *Epipetschoracrinus*

Agassizocrinus OWEN & SHUMARD, 1852, p. 93 [**A. conicus*; M] [=*Agassizocrinites* TROOST, 1850a, p. 420 *(nom. nud.)*; *Astylocrinus* ROEMER, 1854, p. 229 (type, *A. laevis*; OD)]. Crown moderately tall, pyriform. Cup truncate ovoid, with well-rounded base; infrabasals mostly fused into thick, solid piece, but may show complete or incomplete sutures dividing 5 plates; basals large and thick, forming approximately vertical sides of cup; radials wide and low, *C* and *D* radials commonly narrower than others, with subhorizontal articular facet equal in length to thickness of plate; anal plates normally 4, including obliquely oriented quadrangular radianal. Arms 10, unequal, uniserial, composed of externally rectangular well-rounded brachials bearing closely spaced pinnules, arms branching on axillary first primibrach, which may have distinctly unequal distal facets. Stemless in adults. *U.Miss.(Chester.)*, USA(Okla.-Ark.-Ill.-Ky.-Ala.).——FIG. 450,*2a*. **A. conicus*, Chester. (Okaw F.), Ill.; *?AB*-interray view of holotype, ×1 (Springer, 1926b).——FIG. 450,*2b-f*. *A. laevis* (ROEMER), Chester., USA(Ky.-Ill.); *2b, AB*-interray view of crown (ROEMER's type, much restored), from Okaw F., Ill., ×0.7 (Kirk, 1911); *2c,d, CD*-interray and *E*-ray views of ROEMER's holotype (unretouched photos) showing asymmetrical axillary primibrachs and stout unbranched arms, ×1 (Springer, 1926b); *2e,f, AB-* and *CD*-interray views of cups from Glen Dean Ls., Hardin Co., Ky., ×1 (Springer, 1926b).

Anartiocrinus KIRK, 1940, p. 47 [**A. lyoni*; OD]. Crown moderately tall, expanded. Cup tall, cone shaped, with upper portion subvertical; infrabasals 5, rising evenly from columnar attachment area; basals long; radials almost twice as wide as high, articular facets as wide as radials. Arms 10, rectiuniserial with well-rounded exteriors; primibrachs *1* axillary, short; some arms hypertrophied in right and left anterior rays. Column transversely round. *U.Miss.(Chester.)*, USA(Ky.-Ohio).——FIG. 450,*5*. **A. lyoni*, Glen Dean Ls., Ky.; *5a-c*, holotype, post., *E*-ray, ant. views; *5d*, post. view juvenile crown, all ×1.5 (Kirk, 1940b).

Epipetschoracrinus YAKOVLEV in YAKOVLEV & IVANOV, 1956, p. 81 [**E. borealis*; OD]. Basals much enlarged, forming entire lower half of cup, seemingly not incorporating infrabasals fused with them (as inferred by author), but infrabasal circlet concealed in very diminutive basal concavity or resorbed; single anal plate *(X)* in cup. *L.Perm.*, USSR(Pechora Basin).——FIG. 450,*3*. **E. borealis*; *3a,b, CD*-interray and dorsal views of cup, latter showing only large basals (*?CD* basal directed upward), ×1.5 (Yakovlev in Yakovlev & Ivanov, 1956).

Paragassizocrinus MOORE & PLUMMER, 1940, p. 340 [**Agassizocrinus tarri* STRIMPLE, 1938, p. 10; OD]. Cup high, cone or bullet shaped, long infrabasals usually fused about proximal remnant of column; radial articular facets wide and short; single anal plate *(X)* resting evenly on *CD* basal and followed by 2 tube plates. Arms 10, uniserial, pinnulate, with rectangular well-rounded exteriors, branching on primibrachs *2*. *L.Penn.(Morrow.)-U.Penn.(Missour.)*, N.Am.(USA).——FIG. 450, *4a,b*. **P. tarri* (STRIMPLE), U.Penn.(Missour.),

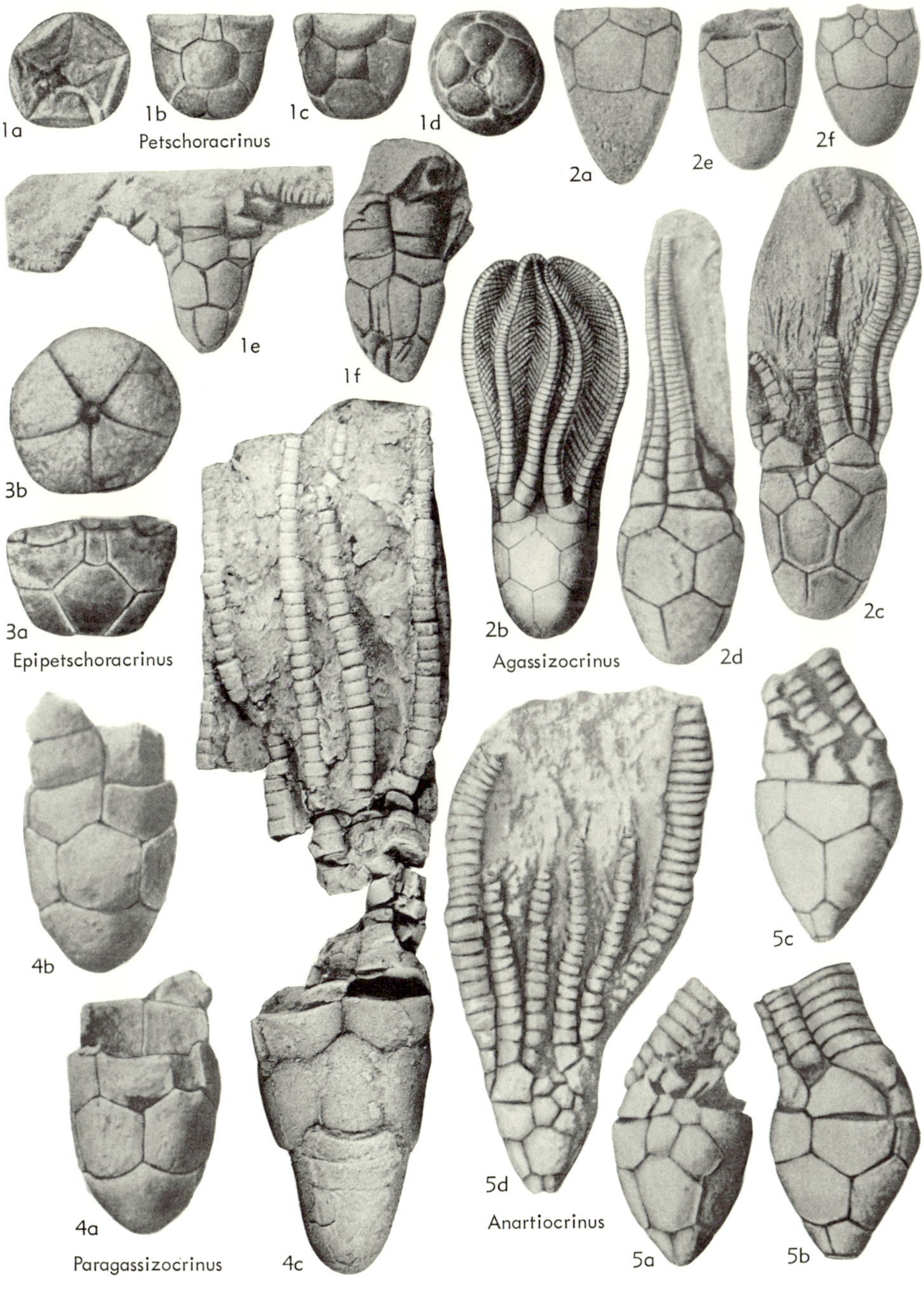

FIG. 450. Agassizocrinidae (p. *T*682-*T*684).

Kans.; cup from *D*-ray and *BC*-interray sides, ×1.5 (Moore & Plummer, 1940).——FIG. 450,*4c*. *P. kendrickensis* STRIMPLE & KNAPP, Morrow. (Kendrick Sh.), E.Ky.; hypotype crown from ant. *(AB)* side, ×1.4 (Strimple, n; Univ. Iowa SUI 42421).

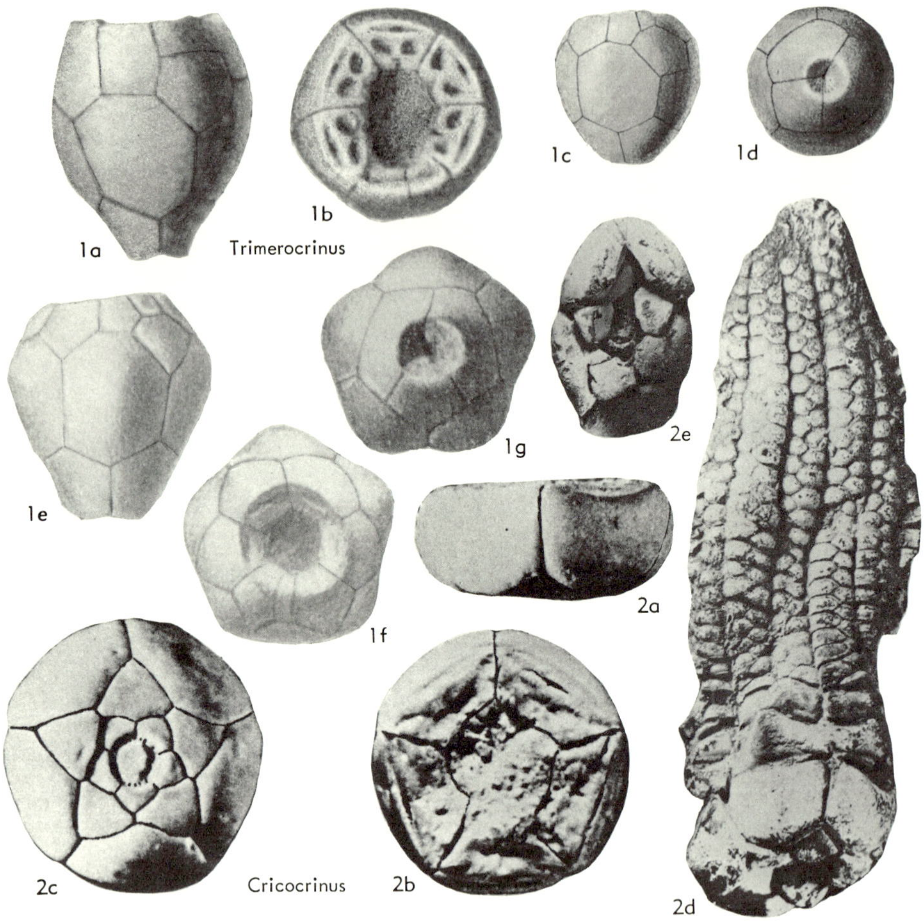

FIG. 451. Cricocrinidae *(2)*; Trimerocrinidae *(1)* (p. *T*684-*T*685).

Petschoracrinus YAKOVLEV, 1928, p. 5 [**P. variabilis*; OD]. Cup deep bowl shaped with 4 long infrabasals which tend to fuse but not solidly. Arms ?5. *L.Perm.*, USSR(Pechora Basin).——FIG. 450,*1*. **P. variabilis,* Pechora, Kojim R.; *1a-d,* cup from ventral, post., ant., and dorsal sides, ×1 (Yakovlev, 1928b); *1e,f,* side views of partial crown, ×1 (Yakovlev, 1934).

Family CRICOCRINIDAE Moore & Strimple, 1973

[Cricocrinidae MOORE & STRIMPLE, 1973, p. 21]

Crown tall, slender, with ten erect closely parallel biserial pinnulate arms. Cup low bowl shaped, round in outline as seen from below or above, base flat, lacking even shallow concavity, plates smooth, with sutures between them not impressed; radials with plenary articular facets; anal plate *(X)* in notch at top on inner side of *CD* interradial suture. *M.Penn.(Desmoines.).*

Cricocrinus STRIMPLE & WATKINS, 1969, p. 181 [**Paradelocrinus regulatus* STRIMPLE, 1949d, p. 325; OD]. Characters of family. *M.Penn.(Desmoines.),* USA(Okla.).——FIG. 451,*2*. **C. regulatus* (STRIMPLE), *2a-c,* Pumpkin Creek F., Love Co., Okla.; post., ventral, dorsal views of holotype, ×1.5 (Strimple, 1949d); *2d,e,* Pumpkin Creek F., Carter Co., Okla.; side and dorsal views of somewhat crushed crown, with displaced lower plates of cup, giving false appearance of basal concavity, ×1.4 (Frederickson & Waddell, 1960).

Family TRIMEROCRINIDAE Moore & Laudon, 1943

[Trimerocrinidae MOORE & LAUDON, 1943a, p. 62]

Cup globose, with narrow base and restricted in width at level of radials; infrabasals three, large; radials small, with plenary articular facets inclined obliquely outward-upward, restricting opening to body cavity, no interradial notches; three anal plates in cup. *L.Perm.-U.Perm.*

The Trimerocrinidae are a Permian family of very uncertain origin. The form of the cup is archaic, except for constriction toward the summit that produces an urn-shaped outline. The three anal plates of the cup are also a primitive character, whereas the steeply outward-upward sloping radial facets, which bear well-developed transverse ridges, are specialized features.

Trimerocrinus WANNER, 1916, p. 143 [**T. pumilus*; OD]. Cup bowl shaped to globose, expanding upward from narrow base and strongly restricted in width near top of cup; infrabasals 3, small one in *A* ray, upflaring and forming about one-third height of cup; basals 5, as high as wide; radials smaller than basals; 3 anal plates in cup, radianal 5-sided, quadrangular anal *X* on truncate upper edge of *CD* basal. Arms unknown. *L.Perm.-U.Perm.*, Indon.(Timor)-Eu.(USSR, Krasnoufimsk).——FIG. 451,*1a,b*. **T. pumilus*, Basleo, Timor; post. and ventral views of cup, ×4 (Wanner, 1916a).——FIG. 451,*1c,d*. *T. ventricosus* WANNER, Basleo, Timor; post. and dorsal views of cup, ×2 (Wanner, 1916a).——FIG. 451,*1e-g*. *T. pentagonus* WANNER, Basleo, Timor; post., ventral, and dorsal views of cup, ×2 (Wanner, 1916a).

Superfamily DECADOCRINACEA Bather, 1890

[*nom. transl.* STRIMPLE, 1975, p. 17 (*ex* Decadocrinidae BATHER, 1890b, p. 384)] [Materials for this superfamily prepared by R. C. MOORE and H. L. STRIMPLE]

Cup low conical to bowl shaped, sides curving inward near rim, radial arm facets commonly do not quite fill the width of radial, anal plates normally three in cup but may be reduced in number; arms ten, uniserial, pinnulate or bearing ramules, brachials usually zigzag in appearance; anal sac small, circular in outline. *M.Dev.-U.Penn.*

The Decadocrinidae was placed in the Erisocrinacea by MOORE, LANE, & STRIMPLE in MOORE & STRIMPLE (1973, p. 21), although the family is characterized by presence of three anal plates and the Erisocrinacea is characterized by one or no anal plate. Clathrocrinidae, whose members normally have three anal plates or rarely two, was placed in the Apographiocrinacea. Other apographiocrinaceans have one anal plate and relatively broad arms with delicate pinnules, quite unlike the arms of clathrocrinids or decadocrinids. The present grouping under a separate superfamily is deemed advisable.

Family DECADOCRINIDAE Bather, 1890

[Decadocrinidae BATHER, 1890b, p. 384] [incl. Decadocrininae JAEKEL, 1918, p. 63; Ramulocrinidae STRIMPLE & WATKINS, 1969, p. 199]

Crown slender. Cup widely expanded, truncate cone or bowl shaped with small basal concavity; five infrabasals with only distal tips at most visible in side view; five medium-sized basals; five radials with articular facets as wide as plates; one to three anals in cup; anal sac tall, slender. Arms ten, formed of cuneate uniserial brachials, branching isotomously on primibrachs *2* in geologically older forms and on primibrachs *1* in later ones, no further branching, arms sinuous or zigzag in appearance, pinnules stout, tending to resemble ramules. Stem preponderantly round transversely and noncirriferous (except *Aulocrinus*). [Similar forms from the Devonian lacking sinuous arms are included in the Scytalocrinidae.] *M.Dev.-U.Penn.*

Middle to Late Paleozoic genera assigned to the Decadocrinidae are characterized by having ten uniserial pinnulate arms which branch isotomously on the first or second primibrach in each ray. The arms have a noteworthy zigzag pattern resulting from the strongly cuneate form of the brachials. As a group these crinoids represent an intermediate evolutionary stage between uniserial arms composed of brachials having joint surfaces disposed at right angles to sides of the arms and a biserial arrangement of the brachials which is inferred to be an advanced structure.

Key to Genera of Decadocrinidae

A. Upflared or subhorizontal infrabasals, distal tips visible from side ....
  - I. Primibrachs *2* axillary ....
    - a. Small infrabasals .... *Decadocrinus*
    - b. Large infrabasals .... *Zostocrinus*
  - II. Primibrachs *1* axillary .... *Ramulocrinus*

B. Infrabasals in basal concavity, not visible from side; primibrachs *1* axillary ....
  I. Brachials decidedly staggered or zigzag .... *Glaukosocrinus*
  II. Brachials mildly staggered ....
    a. Elongate brachials ....
      1. Medium thick arms .... *Trautscholdicrinus*
      2. Thin arms .... *Eireocrinus*
    b. Short brachials ....
      1. Anal sac with specialized spout .. .. *Aulocrinus*
      2. Anal sac tall, cylindrical, lacking spout .... *Acylocrinus*

Decadocrinus Wachsmuth & Springer, 1880, p. 119 [**Poteriocrinus (Scaphiocrinus) scalaris* Meek & Worthen, 1861, p. 145; OD]. Calyx depressed, shallow truncate cone or bowl shaped; infrabasals small, usually visible in side view but may be hidden in concave base; radials and brachials with straight hinge line occupying entire width of plates. Arms invariably 10, branching isotomously on primibrach *2,* composed of angular wedge-shaped joints, zigzag, with their larger, longer sides alternating and projecting so as to support short, rounded pinnules, which may resemble ramules. Stem more or less pentangular in section. *M.Dev.-Miss.,* cosmop.——Fig. 452,*1a,b.* **D. scalaris* (Meek & Worthen), L.Miss., Osag.(Burlington Ls.), Iowa; *1a, C*-ray view of holotype crown, ×1 (Meek & Worthen, 1873); *1b,* plate diagram of cup (radials black, radianal cross ruled, anal plates stippled) and pinnulate arms (Moore, 1962b).——Fig. 452,*1c. D. tumidulus* Miller & Gurley, L.Miss.(Keokuk), Ind.; post. view of incomplete crown showing summit of anal sac, ×1 (Springer, 1926b).

Acylocrinus Kirk, 1947, p. 293 [**A. tumidus*; OD] [=*Tropiocrinus* Kirk, 1947, p. 299 (type, *Poteriocrinites (Scaphiocrinus) rudis* Meek & Worthen, 1869, p. 139)]. Crown erect, of medium height, erect arms separated laterally. Cup bowl shaped, with sharply depressed pit for attachment of stem, surface of plates granulose, infrabasals small, not visible from side; basals relatively small, posterior one truncate for support of anal plate; radials large; single large anal plate in posterior interray rising well above level of radials, distally lunate in section; anal sac tall, slender. Arms 10, relatively long and slender, branching isotomously on medially constricted primibrachs *1*; brachials strongly cuneate, dorsally nodose or carinate, with elevations appearing on pinnuliferous side of each, pinnules long, stout. Stem transversely circular, columnals forming well-defined nodal and internodal series; lumen obscurely pentagonal. *L.Miss. (Osag., Burlington Ls.),* USA(Iowa-Mo.).——Fig. 452,*2.* **A. tumidus,* Iowa; *2a-c, D-, A*-ray, and dorsal views of crown, ×1.5 (Kirk, 1947).

Aulocrinus Wachsmuth & Springer, 1897, p. 138 [**A. agassizi*; OD]. Monotypic genus founded on *A. agassizi,* figured without definition, or description, but based solely upon extraordinary form of the anal sac, which with its lateral spout exhibits a specialization almost without parallel among crinoids (exception, *Carcinocrinus* Laudon, 1941). Form otherwise like *Decadocrinus* as to cup and arms, but differing in stem and cirri. Type species strongly characterized by sharp prominent wrinkles on calyx plates, angular and keeled brachials, well-defined pores at sides of hexagonal anal tube plates, and sharp, longitudinal rows in which these plates are arranged. Stem distinctly pentagonal in section, with strong cirri. *L.Miss.(Osag.),* USA(Ind.).——Fig. 452,*3.* **A. agassizi,* Borden Gr., Ind.; *3a-c, A*-ray, *AB*-interray, and post. views of specimen showing anal sac; *3d, A*-ray view of another specimen; all ×1 (Springer, 1926b).

Eireocrinus Wright, 1951, p. 102 [**E. ornatus*; OD]. Cup shallow truncate cone shaped, discoid, with basal concavity. Uniserial arms keeled, branching on elongated primibrachs *1* (in 3 exposed rays of type species holotype); secundibrachs slightly elongated. Column large, transversely subpentagonal. *L.Carb.(Tournais.),* Eire.——Fig. 453,*3.* **E. ornatus*; side view of holotype crown, ×1.5 (Wright, 1951).

Glaukosocrinus Strimple, 1951, p. 191 [**Malaiocrinus parviusculus* Moore & Plummer, 1940, p. 100; OD] [=*Gloukosocrinus* Strimple, 1961d, p. 114 *(nom. null.)*]. Cup low, truncate bowl shaped with evenly rounded sides and small basal concavity; 3 anal plates in normal (primitive) position in broad posterior interray; radial articular facets wide (plenary) but short. Ten arms, branching evenly on primibrachs *1,* secundibrachs zigzag in appearance, bearing free armlets (pinnules) on alternate sides. Column small, circular in outline. *M.Penn.(Desmoines.)-U.Penn.(Virgil.),* USA (Okla.-Texas).——Fig. 452,*4a-c.* **G. parviusculus* (Moore & Plummer), M.Penn.(Millsap Lake F.), Parker Co., Texas; holotype cup from dorsal, post., and ventral sides, ×3 (Moore & Plummer, 1940).——Fig. 452,*4d. G. planus* Strimple & Moore, U.Penn.(Francis Sh.), near Ada, Okla.; post. view of holotype crown from obliquely below, ×1 (Strimple & Moore, 1971b).

Ramulocrinus Laudon, Parks, & Spreng, 1952, p. 557 [**R. nigelensis*; OD]. Cup widely flaring, 3 anals in cup. Arms not branching beyond axillary primibrachs *1,* unbranched in *A* ray; arms with characteristic zigzag appearance of *Decadocrinus.* [*Ramulocrinus* is thought to have developed from Devonian *Decadocrinus* with sinuous arms in the same manner as *Scytalocrinus* is presumed to have descended from smooth-armed decadocrinids without elimination of branching in any ray.] *L.Miss.(Osag.)-M.Penn.(Desmoines.),* Can.(Alta.), USA(Iowa-Texas).——Fig. 453,*4a.*

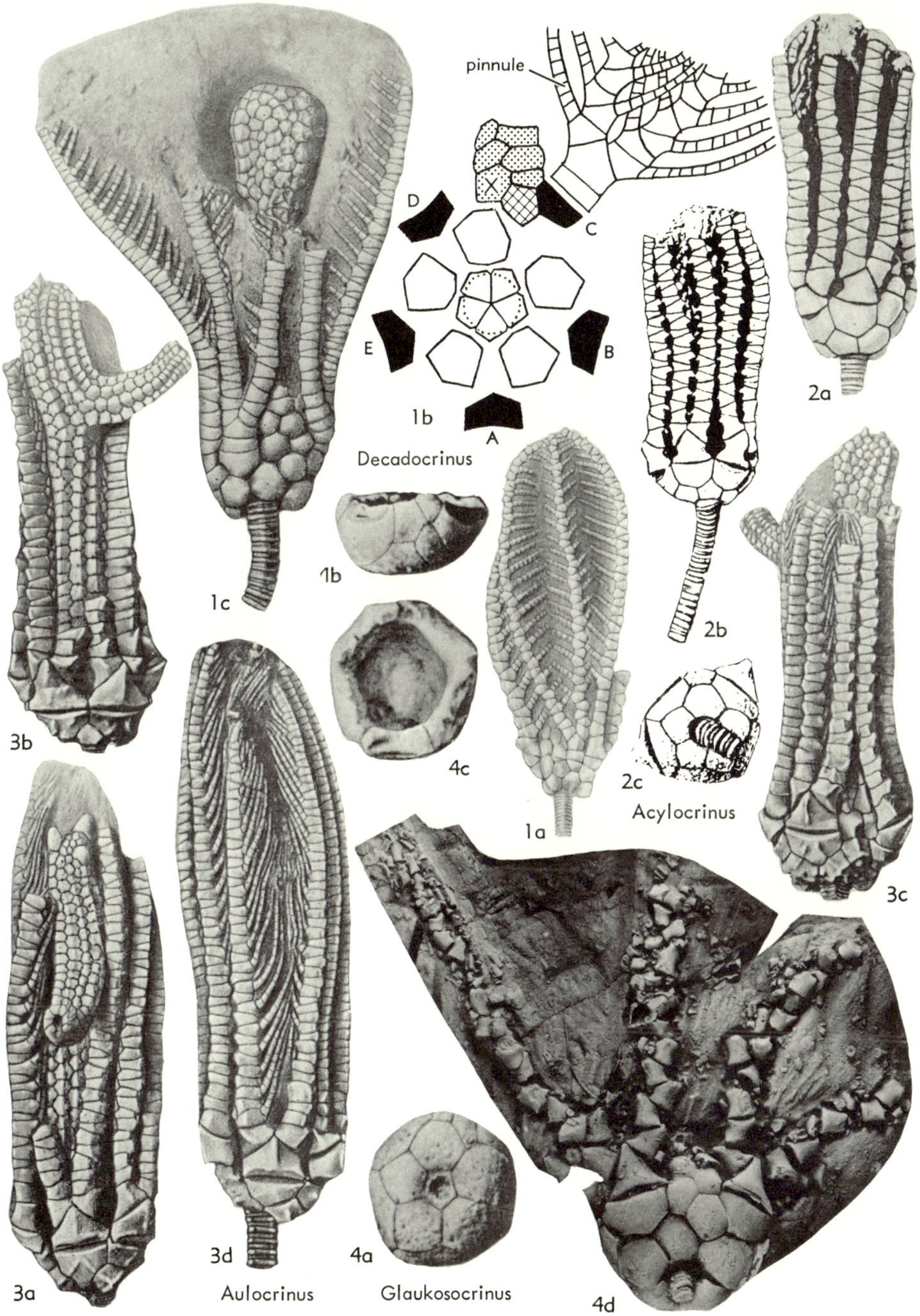

FIG. 452. Decadocrinidae (p. *T*686).

**R. nigelensis,* L.Miss.(Banff F.), Alta.; *A*-ray view, ×1.5 (Laudon, Parks, & Spreng, 1952).——

FIG. 453,*4b*. *R. halli* (HALL), L.Miss.(up. Burlington Ls.), Iowa; post. view of crown, ×1

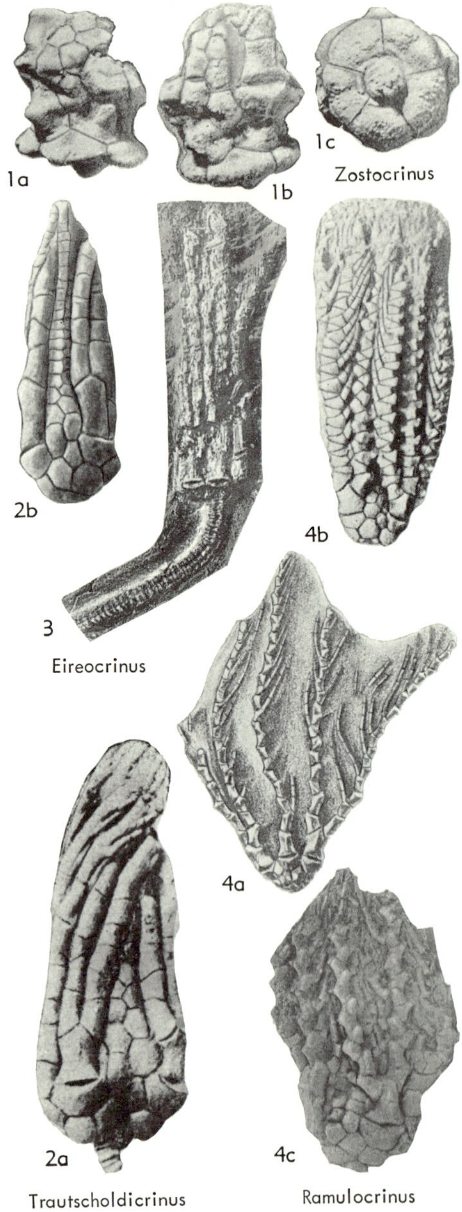

FIG. 453. Decadocrinidae (p. T686-T688).

(Springer, 1926b).——FIG. 453,*4c*. *R. consectatus* STRIMPLE & WATKINS, Desmoines.(Millsap Lake F.), Texas; post. view of holotype crown, ×1.5 (Strimple & Watkins, 1969).

Trautscholdicrinus YAKOVLEV & IVANOV in YAKOVLEV, 1939, p. 66 [**T. miloradowitschi*; OD] [=*Trautscholdicrinus* IVANOV, 1926 (no description or type species) (*non Trautscholdicrinus* MOORE & PLUMMER, in MOORE, 1939c, p. 195) (type, *T. jaekeli*, =*Synerocrinus incurvus* (TRAUTSCHOLD, 1867); OD)]. Crown cylindrical, arms not abutting. Cup expanded, with small basal concavity; infrabasals small, confined to concavity; basals large, extending to midheight of cup or above; radials large with well-developed subhorizontal articular facets having prominent adsutural slopes which extend to lateral cup walls; 3 large anal plates in normal arrangement followed above by large sac plates. Anal sac extending to top of crown (as preserved), distal portion composed in part of irregular plates but with at least one regular series bearing pore slits on lateral edges. Arms 10, elongate cuneate brachials with well-rounded exteriors, arranged uniserially, long side of each bearing stout pinnule; primibrachs *1* axillary, elongated, constricted in midsection. Column round in section, composed of alternatingly narrow and expanded columnals. *U.Carb.*, USSR(Moscow Basin).——FIG. 453,*2*. **T. miloradowitschi*, C_2(=Missour.); *2a,b*, post. views of 2 crowns, ×2 (Yakovlev & Ivanov, 1956).

Zostocrinus KIRK, 1948, p. 708 [**Z. ornatus*; OD]. Cup low, broad, with pronounced outward flare of radials and infrabasals, ornamented with high sharply defined ridges. Infrabasals hypertrophied, forming broad basal plane, flexing inward distally, basals short and small, radials large with wide articular facet; anal plates 3 in normal arrangement, anal *X* the largest element. Arms known from a few proximal brachials, branching with primibrach *2*, brachials long, deep and strongly keeled. Column pentagonal, lumen pentalobate. *M.Dev.(Hamilton)*, USA(N.Y.).——FIG. 453,*1*. **Z. ornatus; 1a-c*, hypotype, post., post. of another specimen showing proximal plates of anal tube, and base, ×3 (Goldring, 1954).

Family CLATHROCRINIDAE Strimple & Moore, 1971

[Clathrocrinidae STRIMPLE & MOORE, 1971a, p. 35]

Crown cylindrical to ovoid, with ten slender uniserial arms well separated from each other. Cup evenly rounded bowl shaped with small basal concavity, plates smooth or finely granulose, bounded by well-marked sutures which are impressed in some species but not impressed in others; infrabasals not visible in side views of cup, basals reach distally to midheight of cup, and proximal extremities of radials are above its basal plane; radial articular facets peneplenary, leaving distinct interradial notches; three anal plates occur in normal (primitive) arrangement. By far the most

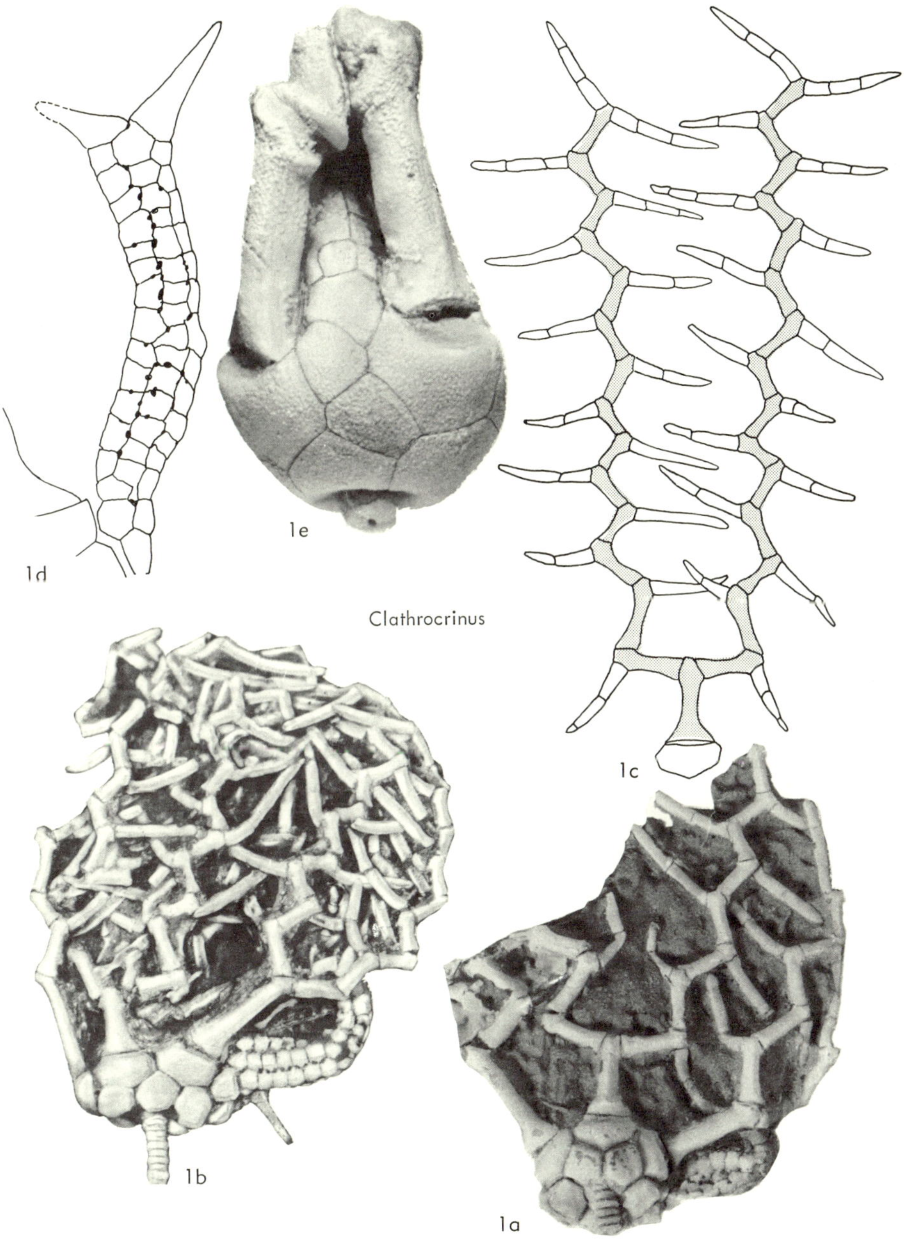

FIG. 454. Clathrocrinidae (p. *T690*).

striking feature of the family is arm structure, for all brachials are exceptionally long and slender, primibrachs and commonly secundibrachs being equal in length to height of the entire cup or greater than it; pinnules actually are ramules, little smaller in diameter than main arm components, which follow one another in a strongly zigzag manner, some with 90-degree deflections. Stem transversely circular. *U.Penn.-(Missour.).*

Clathrocrinus STRIMPLE & MOORE, 1971, p. 35 [*C. clathratus*; OD]. Characters of family. *U. Penn.(Missour.)*, USA(Ill.).——FIG. 454,*1a-d*. **C. clathratus*, Bond F.(LaSalle Ls.) near Pontiac, Livingston Co., Ill.; *1a, E*-ray view of paratype showing extremely long, slender brachials branching at right angle on primibrachs *1* and zigzag arrangement of secundibrachs, each bearing ramule-like pinnule on opposite sides at distal extremities, ×2.7; *1b, EA*-interray view of holotype with anal sac (lower right) and part of stem, ×2; *1c,d,* arms of single ray and anal sac with spinose top, ×2.7 (all Strimple & Moore, 1971a).——FIG. 454,*1e*. *C. clinatus* STRIMPLE & MOORE, from same source as type species, *CD* view of paratype showing primibrachs taller than cup and anal *X* not touching *CD* basal, ×5.3 (Strimple & Moore, 1971a).

Superfamily CROMYOCRINACEA Bather, 1890

[Cromyocrinacea BATHER, 1890b, p. 385] [*nom. transl.* MOORE & STRIMPLE, 1973, p. 21 (*ex* Cromyocrinidae BATHER, May, 1890b, *nom. transl.* JAEKEL, 1918, p. 65, *ex* series Cromyocrinites BATHER, 1890b)] [Materials for this superfamily prepared by R. C. MOORE and H. L. STRIMPLE]

Cup bowl shaped, with or without basal concavity, radials with plenary articular facets, two or three anals in cup. Anal sac not prominent. Arms uniserial or biserial, five to 20 or more, branching confined to primibrachs *1* and secundibrachs *1*. Stem transversely circular. *U.Miss.-Perm.*

Family EUPACHYCRINIDAE S. A. Miller, 1890

[Eupachycrinidae S. A. MILLER, 1890, p. 350]

Cup bowl shaped with narrow shallow to deep basal concavity; two or three anal plates in *CD* interray; infrabasals not visible in side view. Arms biserial, appressed when closed, branching isotomously on first primibrachs and also on first secundibrachs, two to four branches per ray. Column transversely round. *U.Miss.(Chester.)*.

The Eupachycrinidae differ from Cromyocrinidae in the more depressed form of the cup and presence of a well-developed basal concavity. These features indicate evolutionary specialization. The arms branch isotomously once above the first brachial and may branch again in any half ray on axillary secundibrach *1*. Brachials are biserial and pinnular. In that Eupachycrinidae are more specialized in cup form and in arm structure than most Cromyocrinidae, and are stratigraphically older, any relationship must stem from a common ancestor, probably Scytalocrinidae.

Key to Genera of Eupachycrinidae

A. Crown tall cylindrical, with erect biserial arms. Cup mostly with well-defined basal concavity, proximal tips of radials well above basal plane on outer sides of cup, plates smooth ....
  I. Concavity deep, sutures between plates impressed ....
    a. Infrabasals downflared, proximal parts of basals transversely concave, 3 anals in cup, arms 14 .... *Eupachycrinus*
    b. Basals not transversely concave, 2-3 anals in cup, arms 15-16 .... *Intermediacrinus*

Eupachycrinus MEEK & WORTHEN, 1865, p. 159 [**Graphiocrinus 14-brachialis* LYON, 1857, p. 477; OD]. Cup bowl shaped with deep basal concavity; infrabasals downflared; normally 3 anal plates in *CD* interray. Arms biserial, primibrachs *1* axillary in all rays, secundibrachs *1* axillary on right side of *B* and *C* rays and left side of *D* and *E* rays. Anal sac shorter than length of arms, with elongate spine at distal end. *U.Miss.(Meramec.-Chester.)*, USA(Ky.-Ind.).——FIG. 455,*3*. **E. quatuordecimbrachialis* (LYON); *3a-c*, specimen from Ste. Genevieve Ls.(Meramec.), dorsal, *DE*-interray, and *CD*-interray (post.) views of crown, ×0.93 (Kirk, 1937c); *3d*, specimen from Glen Dean Ls.(up. mid.Chester.), Ky.; radial articular facet, ×1.86 (Springer, 1911c).

Intermediacrinus SUTTON & WINKLER, 1940, p. 559 [**Eupachycrinus asperatus* WORTHEN, 1882, p. 34; OD]. Elongate, cylindrical crown. Cup medium sized, bowl shaped, with deep basal concavity; 2 or 3 anal plates in *CD* interray, *A* ray invariably bearing 2 biserial arms, in other rays secundibrachs *2* may be axillary, 4 arms in *C* ray. *U.Miss. (Chester.)*, USA(Ala.-Miss.-Ill.).——FIG. 455,*2*. *I. davidsoni* BURDICK & STRIMPLE, Jasper F., Ala.; *2a-c, B*-ray view (holotype), dorsal and *A*-ray views, ×1; *2d*, diagram of arm structure, axillary secundibrachs finely stippled, *X* and right tube plate coarsely stippled, radianal cross ruled, ×0.93 (all Burdick & Strimple, 1969).

Family PHANOCRINIDAE Knapp, 1969

[Phanocrinidae KNAPP, 1969 (March), p. 351] [=Phanocrinidae STRIMPLE & WATKINS, 1969 (July), p. 216]

Tall, comparatively narrow cylindrical crown with apposed arms. Cup bowl shaped, medium high or low, base concave or subhorizontal; radial articular facets plenary; three to one anal plates in cup.

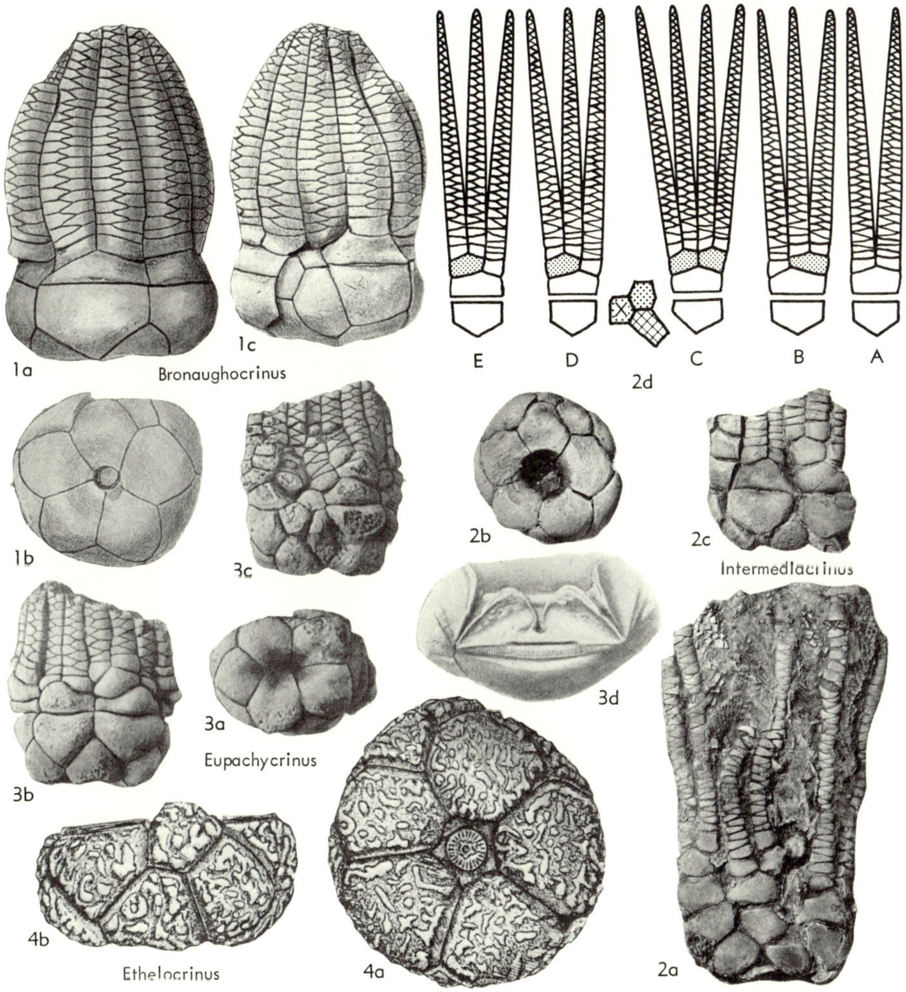

FIG. 455. Eupachycrinidae *(2,3)*; Phanocrinidae *(1)*; Cromyocrinidae *(4)* (p. *T690*, *T692*, *T694*).

Anal sac slender, shorter than arms. Arms five to ten with branching on primibrachs *1*, usually uniserial, but may become biserial (e.g., in later Chesterian) with gently convex outer surfaces and flattened sides, pinnulate. Stem transversely round. *U.Miss.(Chester.), L.Carb.(Visean)-U.Carb.(Namur.).*

Key to Genera of Phanocrinidae

A. Crown tall, cylindrical, with well-rounded erect arms. Cup bowl shaped, with or without basal concavity, plates smooth; anal plates in cup 1-3; arms 5-10 ....

I. Cup with very shallow to deep basal concavity; radials strongly rounded longitudinally to somewhat constricted summit of cup, 3 anals in cup; arms 10 ....

a. Sutures between plates moderately impressed; uniserial or incipiently biserial arms .... *Phanocrinus*

b. Sutures between plates not impressed; equibiserial arms .... *Bronaughocrinus*

II. Cup with shallow basal concavity, not constricted at summit, sutures between plates not impressed; 3 anals in cup ..

a. Arms 8, *C* and *D* first primibrachs

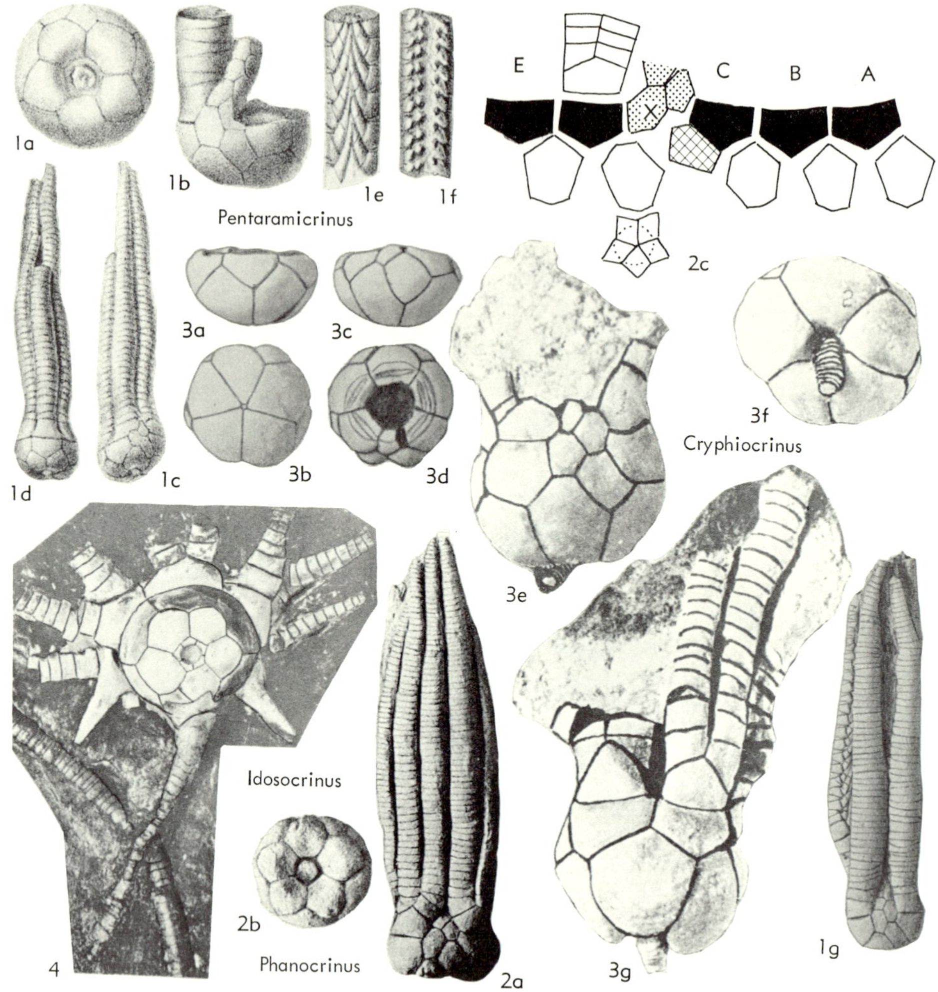

FIG. 456. Phanocrinidae (p. *T692-T693*).

spine bearing, arms of these rays unbranched .... *Idosocrinus*

b. Arms 5-10, no spinose first primibrachs .... *Pentaramicrinus*

III. Cup truncate ovoid, infrabasals small, basals very large; arms 10 .. *Cryphiocrinus*

B. Base of cup with small shallow concavity, infrabasals 5 or fused to 1, radial articular facets wide in *A, C,* and *D* rays, narrow in others .... *Hosieocrinus*

Phanocrinus KIRK, 1937, p. 602 [**Zeacrinus formosus* WORTHEN in MEEK & WORTHEN, 1873, p. 549; OD]. Cup low bowl shaped, like *Pentaramicrinus* except that cup is proportionally lower and constricted near summit, and proximal tips of radials tend to enter basal plane. Considerable variability in arrangements of anal plates may be found in large populations; basal invagination may be narrow and deep, wide and deep, or shallow to essentially nonexistent, all of which affects attitude of infrabasals. Arms well rounded, consistently 10, uniserial, brachials short, externally cuneate, with tendency toward biserial arrangement in *Phanocrinus cooksoni* LAUDON (1941), from the upper Chesterian (Pitkin Ls.) of Oklahoma. Anal sac slender, with long spine at distal end. *L.Carb.(Visean),* Eu.(Brit.-Eire)-N.Afr.(Morocco); *U.Miss.(Chester.),* USA(Ala.-Ky.-Ark.-Okla.-Ill.-Ind.-?Nev.).——FIG. 456,*2*. **P. formosus* (WORTHEN), Chester., Ky.; *2a,b,* post. and dorsal views of crown, ×1 (Kirk, 1937c); *2c,* plate diagram (radials black, radianal cross ruled, *X* and other anals stippled) (Moore, 1962b).

Bronaughocrinus STRIMPLE, 1951, p. 671 [**B. figuratus*; OD]. Crown pear shaped with height less than twice width, narrowing markedly above midheight. Cup bowl shaped, rounding to vertical near summit, with deep basal concavity, proximal tips of radials slightly above basal plane of cup or barely reaching it, surface of plates smooth, sutures between them not impressed; infrabasal circlet very small, covered by stem at center of basal concavity; basals and radials large, latter with plenary articular facets; anals in cup 3, radianal largest and small right tube plate barely touching *C* radial; anal sac unknown. Arms 10, all branching isotomously on very low and wide primibrachs *1*, closely appressed, with slightly rounded exteriors very wide, equibiserial. Stem round, transversely and relatively slender. *U.Miss.(Chester.)*, USA(Okla.-Ga.).——FIG. 455,*1*. **B. figuratus*, Pitkin Ls., Muskogee Co., Okla.; *1a-c*, ant., dorsal, and post. views of holotype crown, ×0.93 (Strimple, 1951d).

Cryphiocrinus KIRK, 1929, p. 153 [**C. girtyi*; OD]. Crown moderately tall, cylindrical. Cup truncate ovoid with very narrow basal invagination for stem attachment; infrabasals small, not visible from side (solid plug may cover infrabasals if stem is reduced); basals very large and thick; radials wide and low with subhorizontal articular facets no longer than thickness of plates; 3 anal plates in normal arrangement. Arms 10, uniserial, pinnulate, branching on primibrachs *1*. Stem very small, when present, circular in outline. *U.Miss.(Chester.)*, USA(Ky.-W.Va.-Okla.). ——FIG. 456,*3a-d*. **C. girtyi*, Chester.(Greenbrier Ls.) near Crellin, W.Va.; holotype, *A*-ray, dorsal, *CD*-interray, and ventral views, ×2 (Kirk, 1929c). ——FIG. 456,*3e-g*. *C. bowsheri* (STRIMPLE), Chester.(?Hindsville F.), near Locust Grove, Okla.; post., dorsal, and ant. *(AB)* views, ×2 (Strimple, 1949a).

Hosieocrinus WRIGHT, 1952, p. 137 [**Tribrachiocrinus caledonicus* WRIGHT, 1936, p. 402; OD]. Cup low bowl shaped, with narrow basal concavity, plates with rugose surface; infrabasals 5 or fused together in small basal concavity, not visible in side view; basals 5, large, convex; 3 large pentagonal radials *(A, C, D)* with wide articular facets, 2 small narrow *(B, E)* radials with narrow articular facets and small triangular primibrachs curved forward and fused to upper surface of cup; 3 anal plates in cup, radianal large. *L.Carb.(Visean)*, Eu.(Eng.).——FIG. 447,*2*. **H. caledonicus* (WRIGHT); *2a-c*, ventral, dorsal, and post. views of cup, ×2 (Wright, 1952).

Idosocrinus WRIGHT, 1954, p. 167 [**I. bispinosus*; OD]. Characters of *Phanocrinus* except that cup is not constricted at summit and primibrachs *1* in *C* and *D* rays are extended as long spines and bear single arms, making total of 8 arms. [The development of spinose primibrachs represents the first appearance which later is a common character of all primibrachs of most representatives of *Delocrinus*.] *L.Carb.(Visean)*, Eu.(Scot.).——FIG. 456,*4*. **I. bispinosus*; dorsal view, ×1.5 (Wright, 1954).

Pentaramicrinus SUTTON & WINKLER, 1940, p. 248 (*emend.* BURDICK & STRIMPLE, 1969, p. 9) [**Cromyocrinus gracilis* WETHERBY, 1880a; OD]. Crown tall, cylindrical, with slender rounded arms. Cup medium high, with small shallow basal concavity, not constricted toward summit; 3 anal plates in cup. Anal sac tubular, formed of small polygonal plates. Arms 5-10. [As originally defined the genus was confined to *Phanocrinus*-like forms bearing 5 arms. BURDICK & STRIMPLE modified the concept to include species having erect lateral cup walls with no appreciable constriction in distal portion, regardless of the number of arms. In specimens where the distal end of the anal sac has been observed 2 rows of small spines occur. Variability in the arrangement of the 3 anal plates is common and forms with 5, 9, and 10 uniserial arms have been observed. Species with 5 arms appear to be confined to upper middle Chesterian strata and are therefore thought to represent advanced, divergent species rather than basic or primitive stock.] *U.Miss. (Chester.)*, USA; *U.Carb.(Namur.)*, Eu.(Brit.). ——FIG. 456,*1*. **P. gracilis* (WETHERBY), Chester. (Glen Dean Ls.), USA(Ky.); *1a,b*, dorsal and post. views of cup, latter showing lower part of anal sac, ×2; *1c,d*, *D*- and *A*-ray sides of crown, ×1; *1e,f*, inside view of arms with and without attached pinnules, ×2 (Wetherby, 1880a); *1g*, post. view of crown, ×1 (Kirk, 1937c).

Family CROMYOCRINIDAE Bather, 1890

[*nom. transl.* JAEKEL, 1918, p. 65, *ex* series Cromyocrinites BATHER, 1890b, p. 385] [=Ethelocrinidae STRIMPLE, 1961d, p. 77]

Cup more or less globe shaped, including forms with upflared or subhorizontal infrabasals; mostly three anal plates in cup but exceptionally only two. Anal sac very short and small. Arms uniserial, unbranched (e.g., *Cromyocrinus*), or branched isotomously on primibrachs *1* in all rays or biserial and often branching a second time on secundibrachs *1* in some half rays. Column transversely round. *L.Carb.(U.-Miss.)-Perm.*

The Cromyocrinidae are closely related to the Ulocrinidae. The cup is globose, radial articular facets have strongly marked evidence of muscular articulation and are characterized by their relatively large size and horizontal attitude. Juveniles have a mildly convex to subhorizontal base and rather large infrabasals that changed with growth to shallowly concave to deeply concave basal invaginations and infrabasals of reduced size (Fig. 457). Anal plates were remarkably stable with a prominent radianal retained throughout the commonly included one or two additional plates. Arms were basically ten in number and progressed from uniserial to biserial. In many genera more arms were added with the second bifurcation on the first secundibrach. *Cromyocrinus* is the only genus in the family that bore five arms and the condition is considered to be divergent rather than primitive.

Key to Genera of Cromyocrinidae

A. Infrabasals slightly upflared ....
- I. Anal plates in cup 3 ....
  - a. Arms 10, uniserial ...... *Mantikosocrinus*
  - b. Arms 5, uniserial .... *Cromyocrinus*
- II. Anal plates in cup 2 or 3, arms 10, biserial .... *Metacromyocrinus*

B. Base of cup planate ....
- I. Anal plates in cup 3 ....
  - a. Arms 10, cuneate brachials .... *Dicromyocrinus*
  - b. Arms 10, uniserial low, rectangular brachials .... *Mooreocrinus*
- II. Anal plates in cup 2 but tip of right tube plate may enter cup, more than 10 biserial arms .... *Parulocrinus*
- III. One anal plate in cup, 10 uniserial arms .... *Moapocrinus*

C. Distal ends of infrabasals in basal plane but rest of infrabasal circlet forming an impressed base, anals 2 or 3 .... *Goleocrinus*

D. Shallow basal concavity with subhorizontal infrabasal circlet composed of large plates ..
- I. Anal plates 2 or 3, surface of cup plates markedly rugose ....
  - a. Arms 10, uniserial, 2 anal plates .... *Synarmocrinus*
  - b. Arms 10, biserial, 2 anal plates .... *Paracromyocrinus*
- II. Same as I, except surface of cup plates finely rugose or smooth ....
  - a. Surface of cup plates finely rugose, sutures strongly impressed, 2 or 3 anal plates, 16 biserial arms .... *Aglaocrinus*
  - b. Surface of cup plates smooth, sutures not impressed, 2 anal plates, 12 to 16 biserial arms .... *Parethelocrinus*

E. Deep basal concavity, downflared infrabasals, 2 anal plates, cup plates strongly ornamented, 16 to 18 biserial arms .... *Ethelocrinus*

Cromyocrinus TRAUTSCHOLD, 1867, p. 12 [**C. simplex*; OD]. Cup subglobular with strong suture lines between plates and distinctly convex base although gerontic specimens may flatten; surface of cup plates and arms smooth or nearly so; 5 large infrabasals clearly visible from side, basals and radials fairly large but thin; 3 anal plates below summit of radials. Anal sac very short, tapering rapidly to small anus near summit. Arms 5, long, uniserial, composed of short, wide brachials. *Penn.-Perm.*, N.Am.(Okla.-Ark.); *L.Carb.-U.Carb.*, USSR(Moscow Basin); *L.Carb.*, Eu.(Spain).——FIG. 458,*2*. **C. simplex*, U.Carb., Myachkova, USSR; *2a*, ant. view of crown, ×1 (Yakovlev & Ivanov, 1956); *2b-d*, topotype showing very short anal sac with attachment scars of gastropod (*Platyceras*), and specimens with *Platyceras* attached, ×0.6 (Yakovlev, 1956); *2e*, plate diagram (radials black, radianal cross ruled, anal *X* and higher anal sac plates stippled), ×1 (Moore, 1962b).

Aglaocrinus STRIMPLE, 1961, p. 86 [**Ethelocrinus magnus* STRIMPLE, 1949d, p. 12; OD] [=*Tarachiocrinus* STRIMPLE, 1962e, p. 135 (*nom. subst. pro Ataxiacrinus* STRIMPLE, 1961d, p. 89, *non Ataxiacrinus* LYON, 1869, p. 463) (type, *Ataxiacrinus multiramus* STRIMPLE, 1961d, p. 90)]. Cup broad, low bowl shaped, surface finely rugose, sutures typically impressed deeply in V-shaped notches, shallow basal concavity; infrabasals subhorizontal or slightly downflared, forming moderately large disc; 5 basals, and 5 radials large; 2 or 3 large anal plates. Arms 16, equibiserial; primibrachs *1* low, axillary, secundibrachs 1 may be axillary in right and left posterior arms and anterior arm; pinnulate. Stem transversely round with crenulae visible from exterior columnals alternatingly expanded. *M.Penn.(Desmoines.)-L.Perm.*, USA(Iowa-Kans.-Okla.-Texas).——FIG. 458,*3*. **A. magnus* (STRIMPLE), *3a-d*, Pumpkin Cr. Ls., Dornick Hills F., Love Co., Okla.; post., ant., dorsal, ventral views of holotype, ×1 (Strimple, 1949d); *3e*, M.Penn.(up.Desmoines., Holdenville F.), Okmulgee Co., Okla.; holotype

(*See facing page.*)

FIG. 457. Diagram showing comparison of thirteen genera of poteriocrinines having subglobular cups with two or three anal plates in the cup; *1)* those with ornamented outer surfaces: *Ethelocrinus* KIRK, *Aglaocrinus* STRIMPLE, *Paracromyocrinus* STRIMPLE, *Dicromyocrinus* JAEKEL, *Metacromyocrinus* STRIMPLE, *Mantikosocrinus* STRIMPLE; and *2)* those with smooth surfaces: *Parethelocrinus* STRIMPLE, *Parulocrinus* MOORE & PLUMMER, *Probletocrinus* STRIMPLE & MOORE, *Ulocrinus* MILLER & GURLEY, *Goleocrinus* STRIMPLE & WATKINS, *Cromyocrinus* TRAUTSCHOLD, and *Ureocrinus* WRIGHT & STRIMPLE (Strimple, n).

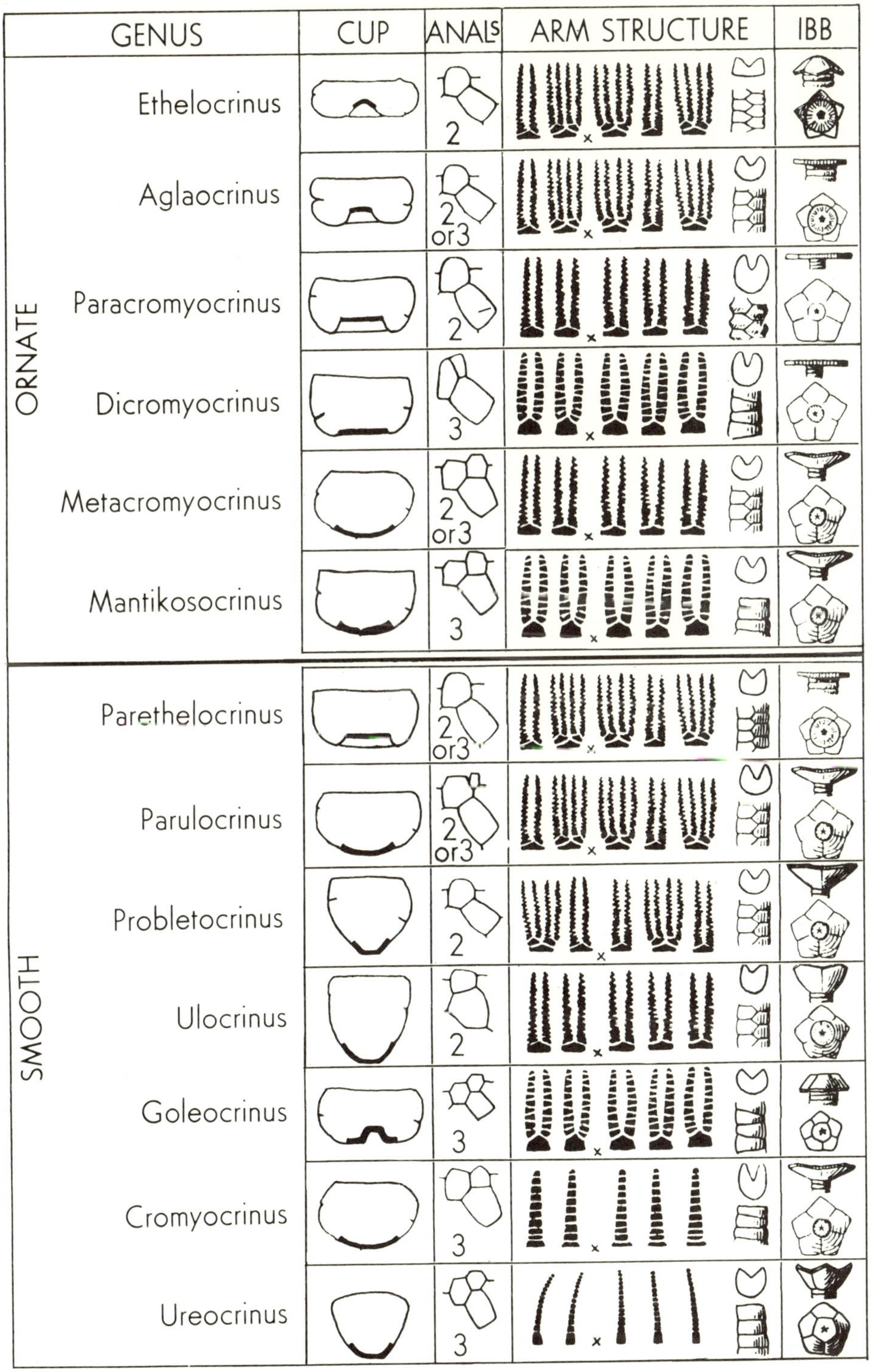
GENUS
CUP
ANALS
ARM STRUCTURE
IBB
ORNATE
Ethelocrinus
2
Aglaocrinus
2 or3
Paracromyocrinus
2
Dicromyocrinus
3
Metacromyocrinus
2 or3
Mantikosocrinus
3
SMOOTH
Parethelocrinus
2 or3
Parulocrinus
2 or3
Probletocrinus
2
Ulocrinus
2
Goleocrinus
3
Cromyocrinus
3
Ureocrinus
3

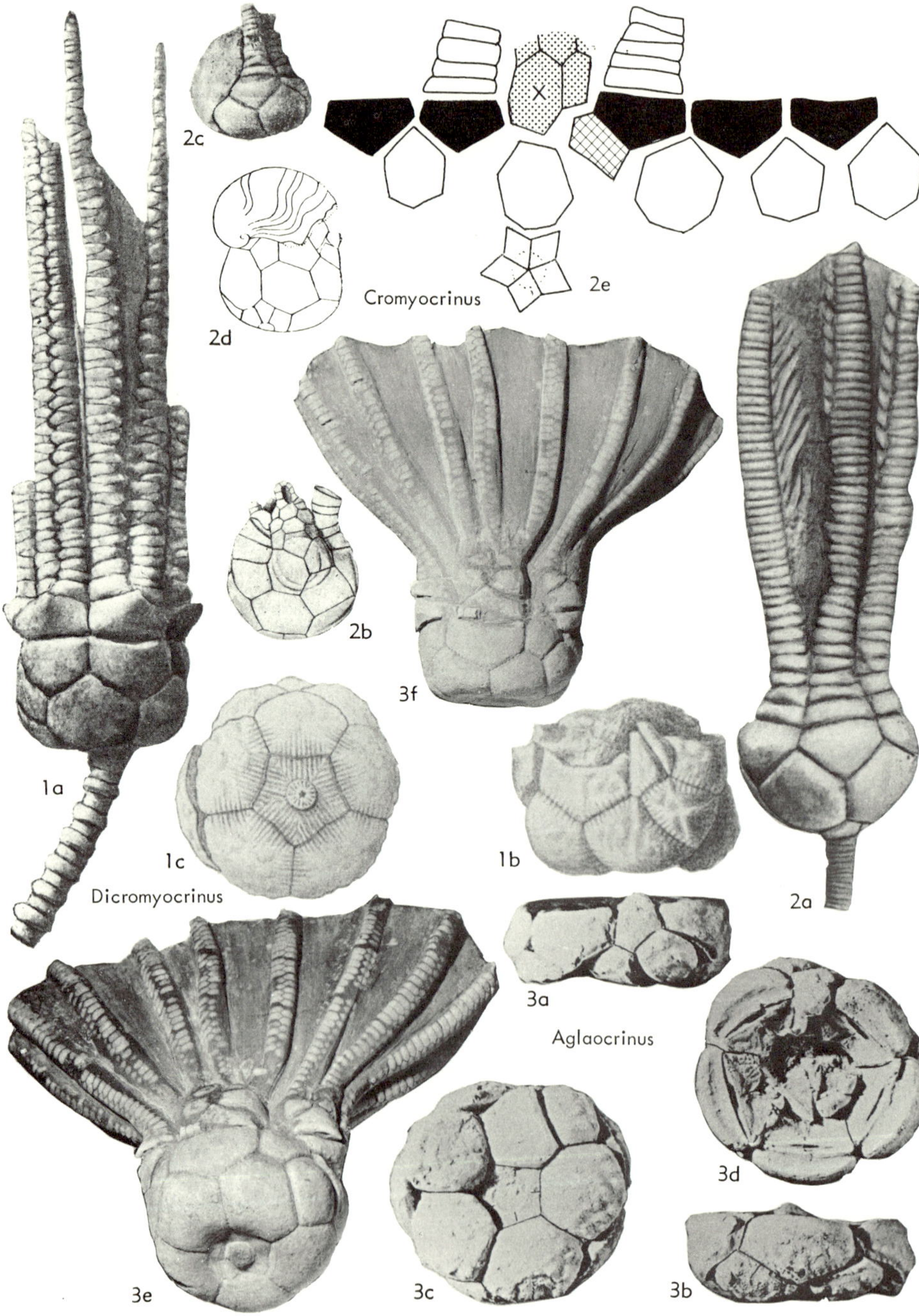

FIG. 458. Cromyocrinidae (p. *T694, T696-T698*).

crown, oblique dorsal view, ×0.8, and *3f,* left post. (*D*-ray) view, ×0.7 (Strimple, 1961d).——FIG. 459,*2*. *A. compactus* (MOORE & PLUMMER), Missour.(Brad F.), Palo Pinto Co., Texas; *2a-c,* dorsal, post., and ventral views, ×1 (Moore & Plummer, 1940).

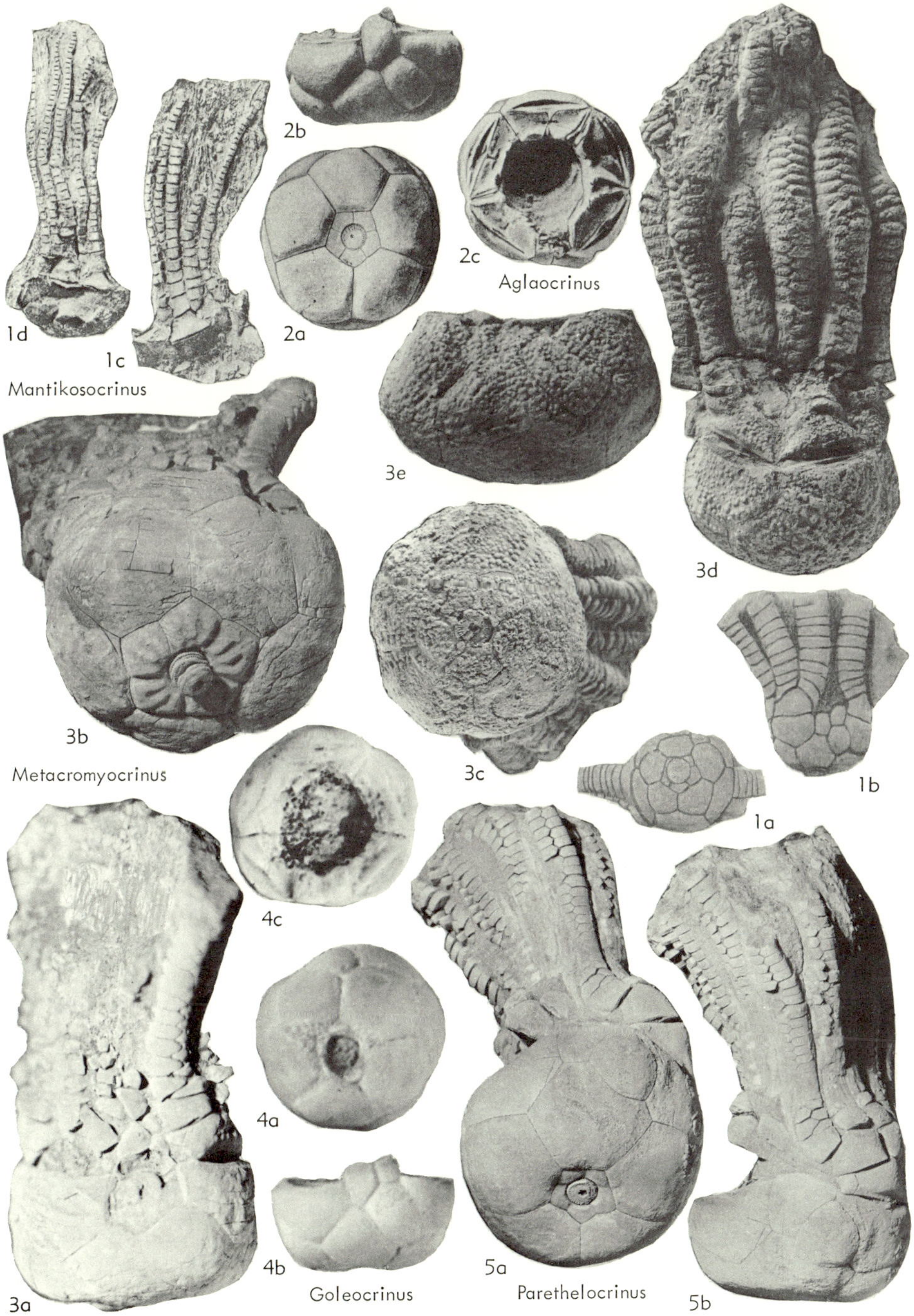

FIG. 459. Cromyocrinidae (p. *T694*, *T696*, *T698*, *T700*).

Dicromyocrinus JAEKEL, 1918, p. 66 [**Cromyocrinus ornatus* TRAUTSCHOLD, 1879, p. 121; SD MOORE & PLUMMER, 1940, p. 367]. Crown slender, tall and compact, arms abutting closely, entire outer surface granular and nodose. Cup bowl shaped, with broad base and nearly vertical sides; infrabasals forming large pentagonal horizontal disc; basals large, extending above midheight of cup; radials wider than long; 3 anal plates, radianal large, oblique, anal *X* in contact with posterior basal, extending well above cup summit, right tube plate narrowly in contact with radianal, extending above summit of anal *X*. Arms 10, long, slender, uniserial, typically with cuneate brachials, pinnulate. Stem transversely round. *L.Penn.-U.Penn.*, USA(Mont.-Texas-Okla.); *U.Carb.*, S.Am.(Brazil)-USSR-Afr.(Morocco).——FIG. 458,*1*. **D. ornatus* (TRAUTSCHOLD), U.Carb. (Moscov.), Myachkovo, USSR; *1a*, ant. view of large crown, ×1 (Yakovlev & Ivanov, 1956); *1b,c*, post. and dorsal views of cup, ×1 (Trautschold, 1867).

Ethelocrinus KIRK, 1937, p. 605 [**Eupachycrinus magister* MILLER & GURLEY, 1890, p. 4; OD]. Cup low basin shaped, with wide flattened base as viewed in profile and with tendency to constriction at summit of radial circlet; base concave, plates convex with surface covered by coarse granules and irregularly vermicular short ridges, sutures between plates impressed; basals large; radials broad and relatively low, articular facets occupying full width of radial with sharply defined transverse ridge and broad muscle fields, 2 anal plates in cup include large quadrangular radianal and hexagonal anal *X*. Arms biserial, commonly 16 (2 each in *B* and *E* rays, 4 each in others; an 18-armed specimen has 3 each in *B* and *E* rays and 4 in others). Stem transversely circular, proximal portion composed of alternate nodals and internodals; whorls of cirri in *Ethelocrinus magister* borne by column to within short distance of crown. *U.Penn.(Missour.)*, N.Am.——FIG. 455,*4*. **E. magister* (MILLER & GURLEY), Missour. (Lane Sh.), Kansas City, Mo.; *4a,b*, holotype cup from base and post. side, ×0.93 (Miller & Gurley, 1890).

?**Goleocrinus** STRIMPLE & WATKINS, 1969, p. 167 [**G. masonensis*; OD]. Cup medium bowl shaped, with broad base and central concavity, surfaces of plates smooth, sutures impressed; infrabasals downflared from impressed stem attachment scar to distal tips flexed sharply into basal plane; anals 2-3, radianal dominant. Arms 10, recti-uniserial. [The form described as *Phanocrinus? vadosus* WASHBURN (1968) from the Bridal Veil Falls Member, Oquirrh F.(Morrow.), of Utah is referred to *Goleocrinus vadosus* (WASHBURN), Moore & Strimple, 1973, p. 56.] *U.Miss.(Chester.)-M.Penn.(Desmoines.)*, USA(Utah-Texas).——FIG. 459,*4*. **G. masonensis*, M.Penn., Big Saline F. (Atokan), Mason Co., Texas; *4a-c*, holotype from dorsal, post., and ventral sides, ×1.5 (Strimple & Watkins, 1969).

Mantikosocrinus STRIMPLE, 1951, p. 673 [**M. castus*; OD]. Cup bowl shaped with gently convex base; infrabasals extending slightly beyond column, visible from side; basals medium sized, evenly curved to midheight of cup; proximal tips of radials well above basal plane; 3 anal plates in normal arrangement, radianal slightly larger than anal *X*, latter in broad contact with *CD* basal and right tube plate which rests on distal face of radianal. Arms 10, recti-uniserial with gently convex outer face, pinnulate; primibrachs *1* axillary, wider than high. [*Mantikosocrinus* is closely similar to *Mooreocrinus* WRIGHT & STRIMPLE (1945), from which it differs in having more evenly curved cup walls and lacking pronounced forefacets in distal portion of radials, which typify *Mooreocrinus*. The form from Scotland ascribed to *Dicromyocrinus geminatus* TRAUTSCHOLD has been named *D. wrighti* by YAKOVLEV & IVANOV (1956, p. 20) and referred to *Mantikosocrinus* by STRIMPLE (1966a, p. 5).] *U.Miss.(Chester.)*, USA (Okla.); *L.Carb.(Calciferous Ss. Series)*, Eu. (Scot.).——FIG. 459,*1a,b*. **M. castus*, U.Miss. (up. Fayetteville F.), Okla.(Braggs Mt., Muskogee Co.); dorsal and post. views of holotype crown, ×1 (Strimple, 1951d).——FIG. 459,*1c,d*. *M. wrighti* (YAKOVLEV & IVANOV), Calciferous Ss. Ser., near St. Andrews, Scot.; side views of 2 crowns, ×1 (Wright, 1951).

Metacromyocrinus STRIMPLE, 1961, p. 68 [**M. holdenvillensis*; OD]. Cup globe shaped with upflared infrabasals, exterior ornate, with pustules and granules; 5 large infrabasals forming low pentagonal cone visible in side view, base impressed for reception of proximal columnal; 5 large basals, almost as long as wide, strongly curved longitudinally and laterally; radials with slight forefacet, plenary articular facets large, marked by outer ligament pit, transverse ridge and moderately large muscle areas; 2 or 3 anal plates in cup, radianal large, right tube plate excluded from cup in later forms. Arms 10, equibiserial, stout, with well-rounded exteriors, pinnulate. *L.Penn.(Morrow.)-M.Penn.(Desmoines.)*, USA(Okla. Ark.).——FIG. 459,*3a,b*. **M. holdenvillensis*, up.Desmoines.(Holdenville F.), Beggs, Okla.; *3a*, holotype crown from post. side, ×1; *3b*, oblique dorsal view, ×1 (Strimple, n; Univ. Oklahoma 4125).——FIG. 459,*3c-e*. *M. gillumi* STRIMPLE, Morrow.(Bloyd F.), Sweetwater Creek, Ark.; *3c,d*, holotype crown from dorsal and ant. sides; *3e*, paratype cup from post. side, all ×2 (Strimple, 1966b; Univ. Iowa SUI 12279 and SUI 12280, paratype).

Moapocrinus LANE & WEBSTER, 1966, p. 30 [**M. rotundatus*; OD]. Crown tall, cylindrical; cup bowl shaped, base flat; single anal plate in cup; 10 wide, closely appressed uniserial arms, branching on primibrachs *1*. *L.Perm.(Wolfcamp.)*, USA (Nev.).——FIG. 460,*2*. **M. rotundatus*, Battleship Wash, Nev.; *2a-d*, paratype cup from ant., dorsal, post., and ventral sides, ×1 (Lane & Webster, 1966); *2e*, topotype, crown from right

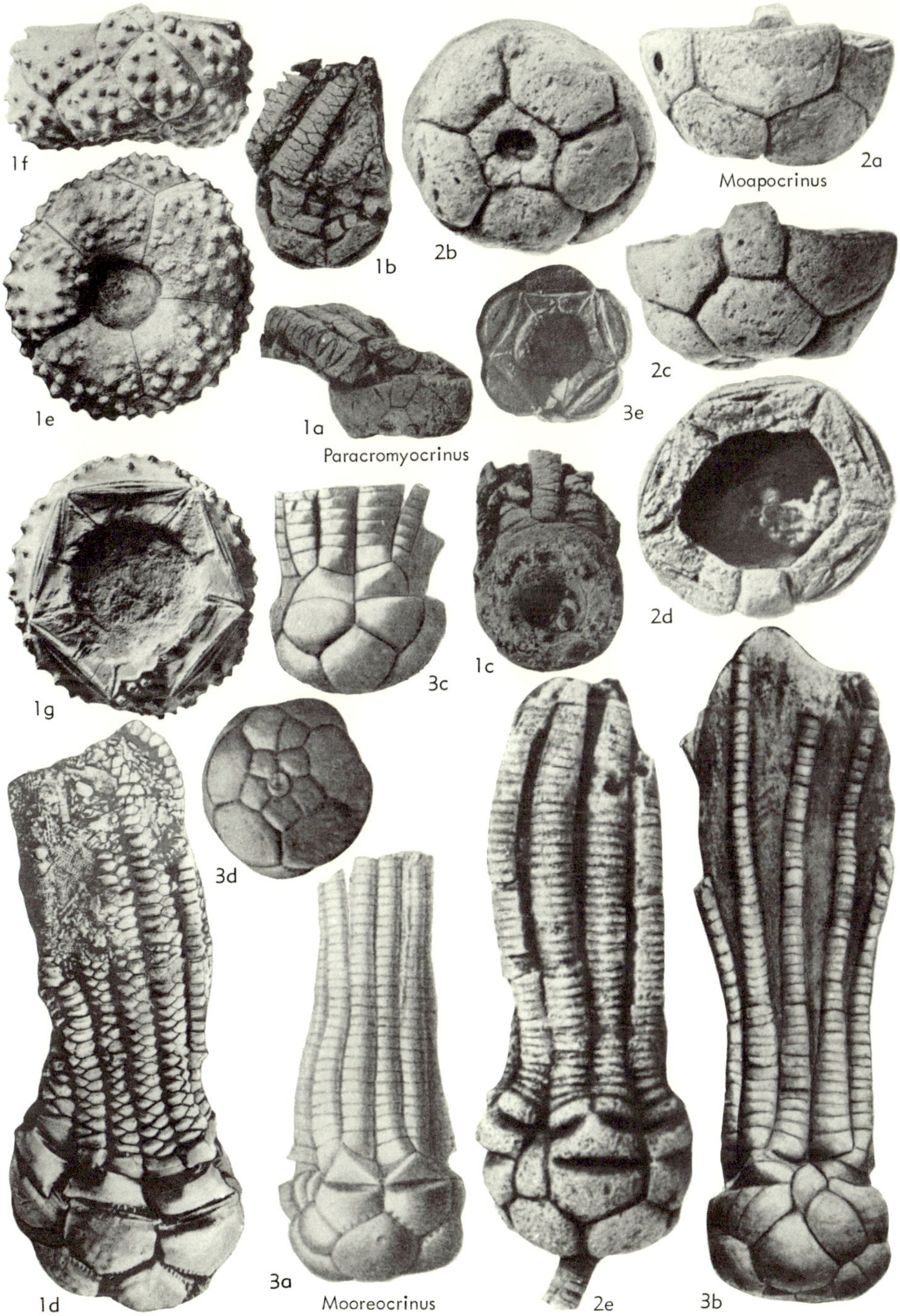

Fig. 460. Cromyocrinidae (p. *T*698-*T*700).

post. side, ×1 (Webster & Lane, 1967).
Mooreocrinus Wright & Strimple, 1945, p. 221 [**Cromyocrinus geminatus* Trautschold, 1867, p. 25; OD]. Crown tall, cylindrical, outer surface

devoid of pronounced ornamentation, arms abutting closely. Cup bowl shaped with broad, flat base and erect lateral walls, sutures impressed; infrabasals extending well beyond column, gently convex, distal tips visible in side view of cup; basals large with proximal portion in basal plane, flexing sharply to extend well up into lateral walls; radials wider than long, distal edges curved sharply inward to form a large forefacet (Fig. 460,*3e*); 3 anal plates normal or in advanced arrangement with radianal tending to migrate to posterior position and right tube plate barely notching cup summit. Arms 10, recti-uniserial, broad in proximal region but tapering to narrow width before attaining midheight, pinnulate; primibrachs *1* axillary, wide, with sharply inwardly sloped proximal surface. Stem relatively small. *U.Carb.*, USSR(Moscow Basin)-S.Am.(Brazil).——Fig. 460,*3*. **M. geminatus* (Trautschold), Moscov., Myachkovo, USSR; *3a*, *BC*-interray view of crown, ×1 (Trautschold, 1867); *3b-e*, topotype crowns from post. *(CD)* and ant. *(AB)* sides, cups from dorsal and ventral sides, all ×1 (Yakovlev & Ivanov, 1956).

Paracromyocrinus Strimple, 1966, p. 4 [**Parulocrinus vetulus* Lane, 1964b, p. 681; OD]. Crown long, cylindrical, arms abutting closely, surfaces ornate in most species ascribed to genus but typically smooth. Cup low bowl shaped, evenly curved sides, with slight or decided basal concavity; infrabasal disc broad, horizontal or mildly downflared; basals large, extending to midheight of cup; radials large, wider than high; 2 or 3 anal plates with tendency for very large radianal to dominate posterior interray. Arms 10, equibiserial, branching on low primibrachs *1*. [*Paracromyocrinus* has a more pronounced basal invagination than *Synarmocrinus* Lane, 1964, but main difference is in the arms, which are uniserial in the latter.] *L.Penn.(Morrow.)-L.Perm.(Wolfcamp.)*, USA(Okla.-Nev.-Texas-Kans.-Ark.-Colo.).——Fig. 460,*1a-c*. **P. vetulus* (Lane), L. Penn.(Bird Spring F.), Arrow Canyon, Clark Co., Nev.; holotype crown from post., ventral, and dorsal sides, ×1 (Lane, 1964b).——Fig. 460,*1d*. *P. marquisi* (Moore & Plummer); Missour.?, near Mason, Texas; holotype crown from ant. side, ×0.7 (Moore & Plummer, 1940).——Fig. 460, *1e-g*. *P. texasensis* (Moore & Plummer), L.Penn. (Morrow., Marble Falls F.), SW. of San Saba, Texas; holotype cup from dorsal, post., and ventral sides, ×1 (Moore & Plummer, 1940).

Parethelocrinus Strimple, 1961, p. 81 [**P. ellipticus*; OD]. Cup smooth, with nonimpressed sutures, broad, low bowl shaped with shallow basal concavity and subhorizontal infrabasal disc; 5 large basals, extending out of basal plane well into lateral cup walls; 5 large radials, wider than long, not extending to basal plane; 2 anal plates. Arms 12 to 16, equibiserial; primibrachs *1* low, axillary; secundibrachs *2* may be axillary in *C*, *D*, and *A* rays. Column transversely round. [*Parethelocrinus* is distinguished from closely similar *Parulocrinus* in having a basal concavity and much smaller infrabasal circlet. The associated *Aglaocrinus* has a more pronounced basal concavity, larger infrabasal disc, surface rugosity and deeply impressed sutures between cup plates.] *M.Penn.(Desmoines.)*, USA(Okla.-Texas).——Fig. 459,*5*. **P. ellipticus*, up.Desmoines.(Holdenville F.), near Beggs, Okmulgee Co., Okla.; *5a,b*, holotype crown in oblique post. and post. views, ×1 (Strimple, n; Univ. Oklahoma OU4135).

Parulocrinus Moore & Plummer, 1940, p. 360 [**Ulocrinus blairi* Miller & Gurley, 1893, p. 57; OD]. Crown moderately tall, having more than 10 long, biserial arms branching on primibrachs *1* in all rays and secundibrachs *1* in some rays. Cup medium sized, deep bowl shaped, without sharp constriction at cup summit, flat based or with very shallow basal concavity or weak convexity; infrabasals subhorizontal, not visible from side; typically with 2 anal plates in cup but proximal tip of right tube plate may enter cup. [*Parulocrinus* is closely similar to *Parethelocrinus* Strimple (1961d), the latter being distinguished by a proportionally smaller infrabasal circlet and a decided, though shallow, basal concavity, both of which are interpreted as more advanced characters than the wide, flat or faintly convex infrabasal circlet of typical *Parulocrinus*.] *U.Penn.(Missour.-Virgil.)-L.Perm.(Wolfcamp.)*, USA (Okla.-Kans.-Ill.-Mo.-Texas).——Fig. 449,*2a-c*. **P. blairi* (Miller & Gurley), Missour.(Argentine Ls.), SW. of Sedalia, Mo.; *2a-c*, post., dorsal, and right ant. *(AB)* views of holotype cup, ×1 (Miller & Gurley, 1893).——Fig. 449,*2d*. *P. pontiacensis* Strimple & Moore, Missour.(LaSalle Ls.), near Pontiac, Ill.; slightly oblique ant. view of hypotype crown, ×1.5 (Strimple & Moore, 1971a).——Fig. 449,*2e,f*. *P. beedei* (Moore & Plummer), Missour.(Palo Pinto Ls.), west of Strawn, Texas; post. and dorsal views of holotype cup, ×2 (Moore & Plummer, 1940).

Synarmocrinus Lane, 1964, p. 678 [**S. brachiatus*; OD]. Five radials, basals and subplanate infrabasals in cup; 2 anal plates; primibrachs *1* axillary, supporting 10 massive, wide, uniserial arms; proximal brachials laterally interlocking when closed; stem round. [Lane (1964, p. 678) stated: "proximal brachials fixed into dorsal cup by lateral interlocking with adjacent brachials," a premise which cannot be supported. Many Upper Paleozoic inadunates have similar interlocking devices in proximal brachials (e.g., *Dicromyocrinus ornatus* Trautschold), which operate simply for thorough closure from the exterior to protect the visceral mass from small predators.] *L.Penn.(Morrow.)-L.Perm.*, USA(Nev.-Utah-Okla.).——Fig. 449, *1a,b*. **S. brachiatus*, L.Perm.(Bird Springs F.), Indian Springs, Nev.; *1a*, left post. (*D*-ray) view of holotype, ×0.7; *1b*, arm tips of paratype, ×1

(Lane, 1964b).——Fig. 449,*1c-e*. *S. fundunus* Strimple, Atoka F., Clarita, Okla.; holotype cup from post. side, dorsal, and ventral, ×1.3 (Strimple, 1966a).

Family ULOCRINIDAE Moore & Strimple, 1973

[Ulocrinidae Moore & Strimple, 1973, p. 14]

Crown tall, arms not appressed. Cup globose, with convex base; infrabasals upflared, part outside of stem impression entirely visible from side; basals large; radials wider than high, with subhorizontal articular facets occupying full width of plates, no interradial notches; anals mostly two or three in cup but may be only one. Arms transversely well rounded, unbranched uniserial (e.g., *Ureocrinus*), or branching isotomously on first primibrachs mostly without higher bifurcations, ten arms in all (exceptionally 14-18, e.g., *Probletocrinus*), biserial, pinnulate (Fig. 457). *L.Carb.(Visean); M.Penn.-Perm.*

Key to Genera of Ulocrinidae

A. Infrabasals distinctly upflared, visible from side; 2 or 3 anal plates in cup ....
  I. Five uniserial arms .... *Ureocrinus*
  II. Ten biserial arms, commonly 2 anal plates .... *Ulocrinus*
  III. More than 10 biserial arms, 2 anal plates .... *Probletocrinus*

B. Cup moderately low bowl shaped; infrabasals subhorizontal, not visible from side; one anal plate .... *Tyrieocrinus*

Ulocrinus Miller & Gurley, 1890, p. 6 [**U. buttsi*; OD]. Crown tall, subcylindrical, with parallel erect arms not appressed. Globular to pyramidal calyx; infrabasals upflared weakly to strongly; basals as large or larger than infrabasal circlet; radials pentagonal, much wider than high, with subhorizontal broad articular facets filling full width of plates; single, large, quadrangular radianal resting obliquely between 2 basals, may reach *D* radial; anal *X* resting on truncate distal edge of *CD* basal or above radianal. Arms 10, long, well-rounded exteriors, secundibrachs cuneate in short proximal portions, equibiserial above. *M.Penn.(Atokan)-U.Penn.(Virgil.),* USA(Ill.-Mo.-Okla.-Kans.-Texas); *L.Perm.,* USA(Texas)-USSR (Krasnoufimsk); *U.Perm.,* Indon.(Timor, Basleo)-Pak.(Salt Ra.).——Fig. 461,*4a*. **U. buttsi,* Missour., Kansas City, Mo.; post. view of holotype cup, ×0.7 (Miller & Gurley, 1890).——Fig. 461,*4b*. *U. convexus* (Strimple), Missour.(LaSalle Ls.), Pontiac, Ill.; hypotype crown from *AB*-interray side, ×0.7 (Strimple & Moore, 1971a).

Probletocrinus Strimple & Moore, 1971, p. 26 [**P. curtus*; OD]. Crown tall, subcylindrical, with at least 14 (possibly 18) biserial arms. Cup somewhat pear shaped with posterior distal area protruded; infrabasals moderately large, gently upflared and readily visible in side view of cup; large basals decidedly curved longitudinally; radials large, 5-sided; 2 large anals, with radianal in oblique position and anal *X* plate in broad contact with posterior *(CD)* basal. Arms biserial, 2 arms preserved in *A* ray, 4 arms (with second branching on secundibrachs *1*) in *B* and *E* rays, other arms missing or unattached, but *C* and *D* rays almost certainly identical to *B* and *E* rays, total 18, arm exteriors mildly convex, pinnules imperfectly preserved. Anal sac not observed; columnar cicatrix round, crenulated, lumen subpentagonal. *U.Penn. (Missour.),* USA(Ill.).——Fig. 461,*3*. **P. curtus,* Missour.(LaSalle Ls.), near Pontiac, Ill.; *3a,b,* *B*-ray and post. views of holotype crown, ×0.7 (Strimple & Moore, 1971a).

?**Tyrieocrinus** Wright, 1945, p. 114 [**T. laxus*; OD]. Cup globose, much wider than high, plates rather thin, smooth, sutures flush; infrabasal circlet slightly convex to almost flat, just visible in side view and rather large in proportion to basals; radial circlet greatly constricted at summit, sloping upward-inward, plenary articular facets sloping slightly outward-upward, with strong articulating cross ridge; anal area occupied by single large plate resting on *CD* basal and extending slightly beyond line of radials. Arm structure unknown. *Up.L.Carb.,* Eu.(Scot.).——Fig. 461,*1*. **T. laxus,* Low.Ls.Gr.; *1a-c,* dorsal, post., ant. views of cup, ×1.5 (Wright, 1945).

Ureocrinus Wright & Strimple, 1945, p. 225 [**Poteriocrinus bockschii* Geinitz, 1845; OD]. Cup globular, subglobular, or somewhat elongated, infrabasal circlet plainly visible from side; radial circlet commonly constricted; anal area of advanced type with large radianal surmounted by 1 or 2 small plates, with some variation. Arms 5, uniserial, brachials cuneiform. *L.Carb.,* Eu.(Brit.-Ger.).——Fig. 461,*2*. **U. bockschii* (Geinitz); Eng.*(2a),* Scot.; *2a,* lat. view of crown, *2b-d,* series of post. views, ×1; *2e,* post. view with primitive type of anal area and with part of left and right post. unbranched arms attached, all ×1 (Wright, 1950-54).

Family CADOCRINIDAE Moore & Laudon, 1943

[Cadocrinidae Moore & Laudon, 1943a, p. 62]

Cup strongly bowl shaped, with narrow, deep basal concavity; infrabasals not visible from side; radial articular facets occupying full summit of these plates; anal plates of posterior interray exceptionally variable in size and arrangement, occupying large part

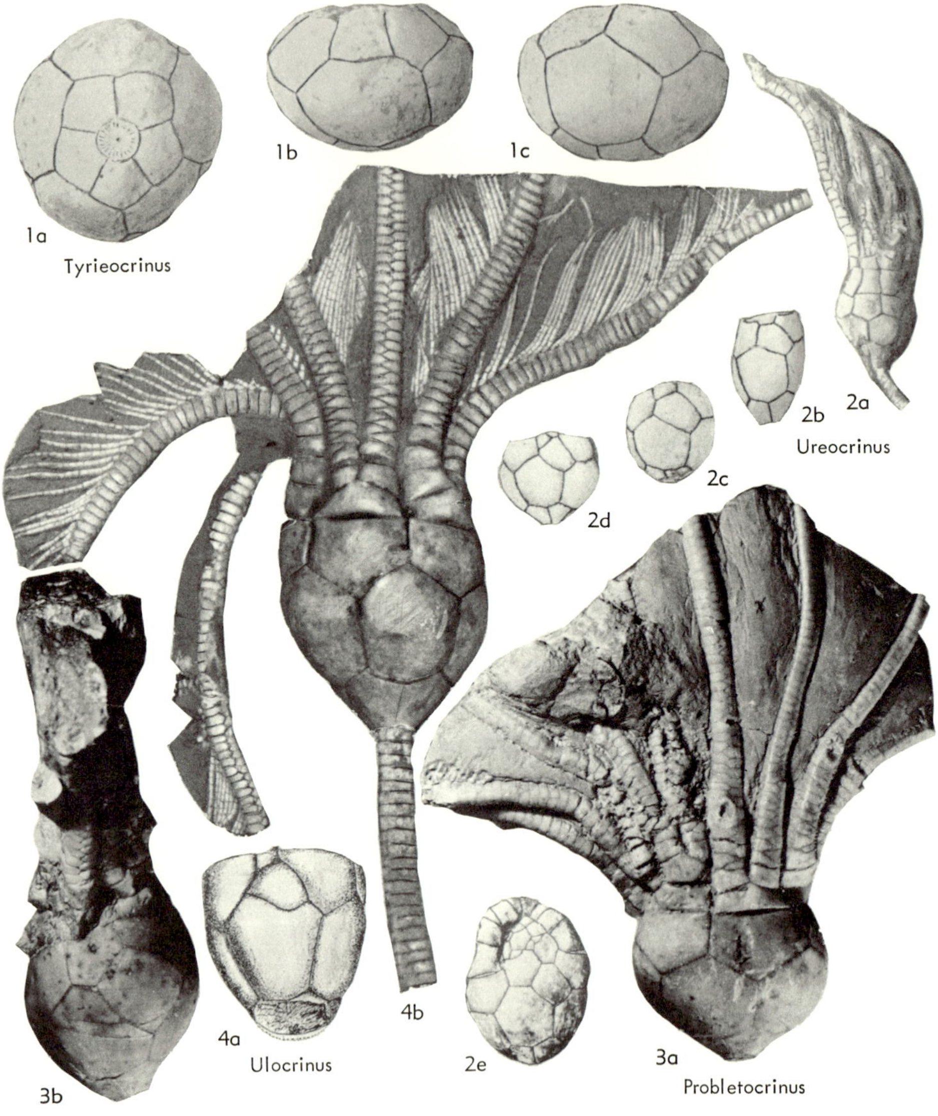

FIG. 461. Ulocrinidae (p. *T*701).

of cup or almost completely eliminated; and sac ?mushroom shaped. Arms unknown. Stem round, slender. *U.Perm.-(Basleo beds).*

The Upper Permian crinoids from Timor, described under the name *Cadocrinus,* are not satisfactorily referable to any other defined family and are placed in a separate group, the Cadocrinidae. Their globose cup suggests possible relationship with the cromyocrinids, but, if the anal sac that is supposed (WANNER, 1916a) to belong to this genus—not found attached to any cups—is correctly identified, the Cadocrinidae are far removed from the Cromyocrinidae. Variation of the anal plates in specimens belonging to the type species of *Cadocrinus* shows a recapitulation of the evolution of these plates, proceeding in some to the virtual exclusion of anals from the cup.

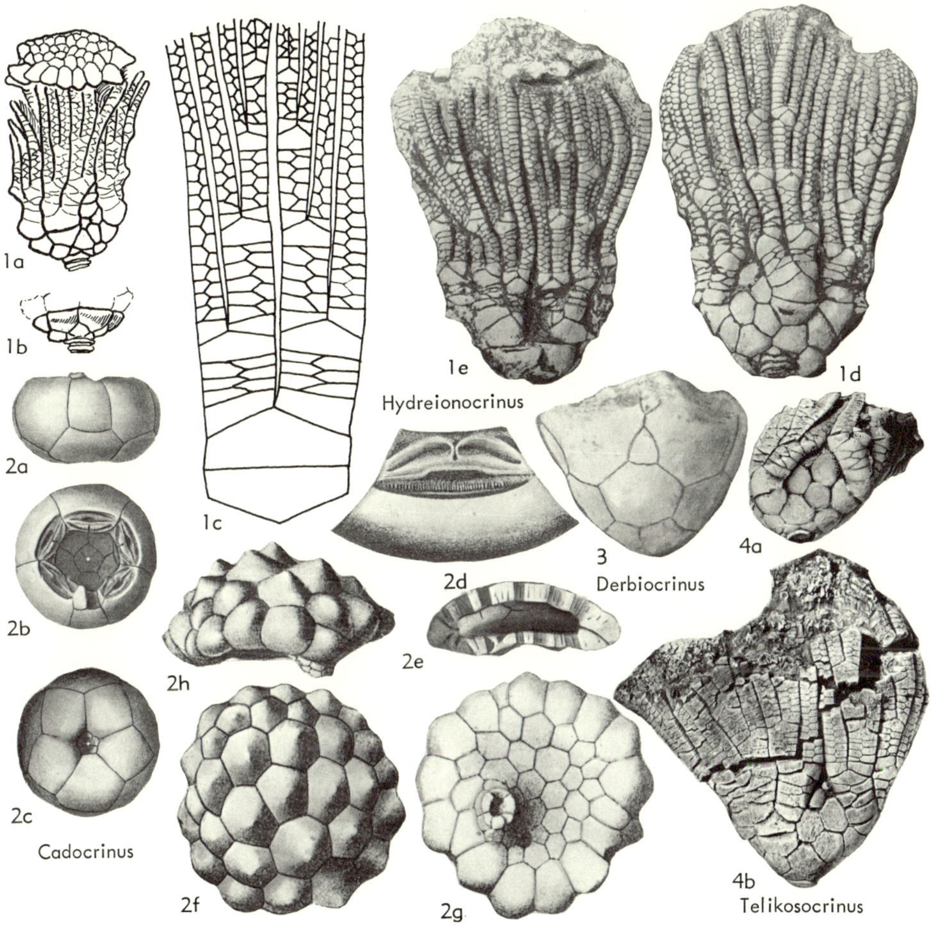

FIG. 462. Cadocrinidae *(2)*; Hydreionocrinidae *(1,3,4)* (p. T703-T704).

Cadocrinus WANNER, 1924, p. 250 [**Hydreionocrinus variabilis* WANNER, 1916a, p. 250; OD]. Cup bowl shaped, with very deep, narrow basal concavity; infrabasals small, confined to concavity; basals large and strongly arched; radials large, restricted in width and curved inward distally; anal plates 1 to 3 in cup, highly variable in size and position, radianal large and trapezoidal or moderate in size and quadrangular to small diamond shaped, anal *X* moderately large to quite small, either in or above cup; anal sac apparently mushroom shaped. Arms not known. Stem transversely round, slender. *U.Perm.(Basleo beds)*, Indon.(Timor).——FIG. 462,*2*. **C. variabilis* (WANNER); *2a-c*, post., ventral, and dorsal views of cup, ×1; *2d*, radial articular facet, ×10; *2e*, cross section of anal sac, ×1; *2f-h*, upper, lower, and side views of anal sac inferred to belong to *Cadocrinus*, ×1 (Wanner, 1916a).

Superfamily HYDREIONOCRINACEA Jaekel, 1918

[*nom. transl.* MOORE & STRIMPLE, 1973, p. 21 (*ex* Hydreionocrinidae JAEKEL, 1918, p. 63)] [Materials for this superfamily prepared by R. C. MOORE and H. L. STRIMPLE]

Cup low bowl shaped, sides flared outward at rim, plenary arm facets, arms biserial, exotomous; three anals in cup. *Up.L.Carb., U.Miss.*

Family HYDREIONOCRINIDAE Jaekel, 1918

[Hydreionocrinidae JAEKEL, 1918, p. 63]

Crown cylindrical or pear shaped with summit largely formed by bulged umbrellalike platform which caps anal sac. Cup shallow to deep bowl shaped; infrabasals five, flaring upward, distal parts visible from side; radial articular facets plenary, three anal plates in cup. Anal sac slightly taller than arms, mushroom shaped, with summit platform of small polygonal plates surrounded peripherally by horizontally directed spines. Arms branching isotomously on primibrachs *1*, with two to four exotomous divisions in each half-ray, brachials biserial, nonpinnulate. Stem transversely round. *Up.L.Carb., U.Miss.*

The Hydreionocrinidae are specialized in having suppressed isotomy of some branches, but this differs from that of the Cercidocrinidae in giving rise to an exotomous plan of branching, instead of an endotomous one. The brachials are arranged biserially, which is an advanced type in evolution. The low cup, which has a convex base, has three anal plates. The anal sac is modified into a mushroom shape, the expanded summit bearing a circle of laterally directed spines.

Hydreionocrinus deKoninck, 1858, p. 93 [**H. woodianus,* p. 97; SD Wachsmuth & Springer, 1879, p. 131]. Cup low bowl shaped with rounded convex base; *BC* and *CD* basals beneath cup anal plates as well as radials; 4 or 5 tube plates above radianal and anal *X* forming posterior rigid upward projection of cup, above radials as in *Zeacrinites*. Arms 40 or more, biserial, branching isotomously on first primibrachs and exotomously at higher levels. Pore-slits occur around edges of platform plates of the umbrella-like termination of anal sac. *Up.L.Carb.,* Eu.(Scot.-Isle of Man-Eng.-Belg.).——Fig. 462,*1a-c*. **H. woodianus,* Carb., Eng.; *1a,b*, *D*-ray view of crown, lat. view of cup, diagram., ×0.7 (Moore, 1938); *1c*, plan of arms, ×2 (Moore, n).——Fig. 462, *1d,e*. *H. amplus* Wright, Visean, Scot.; *C*-ray and *AB*-interray views of 2 crowns, ×0.7 (Wright, 1951).

Derbiocrinus Wright, 1951 [**D. diversus*; OD]. Cup deep bowl shaped or conical; infrabasal circlet distinctly visible from side, expanding rapidly from column facet; radials somewhat spreading and indented at interradial sutures, articular facets horizontal with strong transverse ridge; anal plates of cup extremely variable; anal sac and arms unknown. *Up.L.Carb.(Visean),* Eng.——Fig. 462,*3*. **D. diversus,* Derbyshire; post. view of type, ×1 (Wright, 1951).

Telikosocrinus Strimple, 1951, p. 261 [**T. caespes*; OD]. Crown broadly expanded. Cup low cone shaped, with infrabasals readily visible in side view, radial articular facets plenary; anal sac cylindrical, extending to summit, or near summit of crown, composed of small polygonal plates with pore-slits along margins, terminating with several small subvertical spines. Arms equibiserial, branching twice isotomously and thereafter heterotomously at least twice, primibrachs *1* axillary in all rays. Column round. *U.Miss.(Chester.),* USA (Okla.-Ark.).——Fig. 462,*4a*. **T. caespes,* Pitkin Ls., Muskogee Co., Okla.; holotype from post. side, ×1.2 (Strimple, n; National Museum Nat. History S-4970).——Fig. 462,*4b*. *T. residuus* Strimple, Pitkin Ls., Muskogee Co., Okla.; holotype from post. side, ×1.2 (Strimple, n; Natl. Museum Nat. History S-4973).

Superfamily ERISOCRINACEA Wachsmuth & Springer, 1886

[*nom. transl.* Moore & Strimple, 1973, p. 21 (*ex* Erisocrinites Wachsmuth & Springer, 1886, p. 115(191))] [incl. Arkacrinidae Knapp, 1969, p. 367 (=Arkacrininae); Diphuicrinidae Strimple & Knapp, 1966, p. 312; Encrinidae Austin & Austin, 1842 (*nom. correct.* Austin & Austin, 1843, p. 198, *pro* fam. Encrinoidea Austin & Austin, 1842, p. 108); Erisocrinidae S. A. Miller, 1890, p. 351; Graphiocrinidae Bather, 1899b, p. 922; Paradelocrinidae Knapp, 1969, p. 352; Protencrinidae Knapp, 1969, p. 353; Catacrinidae Knapp, 1969, p. 365; Stachyocrinidae Moore & Strimple, 1973, p. 22] [Materials for this superfamily prepared by R. C. Moore, H. L. Strimple, and N. Gary Lane]

Medium-sized crown commonly cylindrical and moderately tall, with pinnulate erect arms juxtaposed, mostly ten, branching on first primibrach. Cup low bowl shaped, with sides mostly formed by steep-sloping radials, base slightly convex, planate or with narrow to broad and shallow to deep concavity, radial articular facets large, plenary, without appreciable outward or inward slope, with transverse ridge, ligament fossae, and muscle fields well developed; single anal plate next above *CD* basal, or located in notch between posterior radial articular facets and not visible externally. Anal sac cylindrical, composed of small plates in vertical series, commonly with terminal spine or spines directed upward. Stem transversely round. *L.Carb.(Miss.)-U.Perm.; M.Trias.*

The Erisocrinacea are cladids of comparatively simple but advanced evolutionary type. The low truncate bowl- or saucer-shaped cup, in which the infrabasals are not visible from the side, is more specialized than in the Scytalocrinacea. As here interpreted, the assemblage includes genera having advanced structure of the posterior side, characterized by occurrence of a single anal or absence of an externally visible anal in the cup. The arms are like those of many Scytalocrinacea, except that in numerous genera evolution of brachials has advanced to a biserial arrangement; also, in some, suppression of isotomous arm division near the base of the rays produces unbranched (atomous) arms. The radial arm facets have marks of highly developed muscular articulation. The anal sac is thin plated and very weak, so that generally it is not preserved, or a sac may be entirely lacking.

Family ERISOCRINIDAE Wachsmuth & Springer, 1886

[*nom. transl.* S. A. MILLER, 1890, p. 351, *ex* section Erisocrinites WACHSMUTH & SPRINGER, 1886, p. 115(191)]

Crown subcylindrical. Cup truncate cone or bowl shaped, base planate, weakly convex, or mildly concave; infrabasals five, subhorizontal; basals five, medium sized; radials five, very prominent, with proximal tips well above basal plane of cup, long and wide (plenary) subhorizontal articular facets; no anal plate other than rudimentary remnant in notch between *C* and *D* radial articular facets. Anal sac small, short, recurved, composed of small polygonal plates. Arms ten, equibiserial, primibrach *1* axillary, arms with flattened outer surfaces, closely appressed when closed, each brachial bearing a thin pinnule. Stem transversely round, moderately large. *U.Miss.-L.Perm.*

Key to Genera of Erisocrinidae

A. Cup with faintly convex, flat, or broad shallow base; plenary planate radial articular facets ....
- I. Cup with flat or shallowly concave base
  - a. Small infrabasals, cup sides expanding .... *Erisocrinus*
  - b. Large infrabasals, cup sides erect .... *Sinocrinus*
- II. Cup with nearly flat, slightly convex base .... *Exaetocrinus*

Erisocrinus MEEK & WORTHEN, 1865, p. 174 [**E. typus*; OD] [=*Stemmatocrinus* TRAUTSCHOLD, 1867, p. 28 (type, *S. cernuus*; OD); *Libratocrinus* KNAPP, 1969, p. 354 (type, *Erisocrinus mediator* STRIMPLE, 1961d, p. 14; OD), M.Penn.(Desmoines.), Oologah Ls., NE.Okla.; *Pontotocrinus* KNAPP, 1969, p. 352 (type, *Paradelocrinus wapanucka* STRIMPLE, 1961e, p. 225; OD), L.Penn. (Morrow.), Wapanucka Ls., Pontotoc Co., Okla.; *Parerisocrinus* KNAPP, 1969, p. 358 (type, *Paradelocrinus obovatus* MOORE & PLUMMER, 1940, p. 326; OD), U.Penn.(Missour.), Texas-Okla.-Kans.]. Cup broadly truncate cone shaped, flattened base with little or no basal concavity, circular stem impression; cup outline typically pentagonal when viewed from above or below; 5 small infrabasals, or fused, confined to basal area together with proximal portions of 5 large basal plates; 5 radial plates forming most of lateral walls of cup; single anal plate confined to notch between radial articular facets. Arms 10, equibiserial, flattened exteriors, pinnulate, branching on short primibrach *1* on all rays. *L.Penn.(Morrow.)-L. Perm.(Wolfcamp.),* USA(Ill.-Ind.-Iowa-Mo.-Kans.-Neb.-Nev.-Okla.-Texas); *U.Carb.(Moscov.),* USSR.——FIG. 463,*1a-d*. **E. typus; 1a,b,* M.Penn.(Desmoines.), Sangamon Co., Ill.; syntype cup from post. and dorsal sides, ×2 (Moore & Plummer, 1940); *1c,* U.Penn.(Missour.), Bond F., Livingston Co., Ill.; lat. view of hypotype crown, ×0.7 (Strimple & Moore, 1971a); *1d,* long. sec. of cup through midline of *A* radial and *CD* interray, ×2 (Moore, n).——FIG. 463,*1e,f*. *E. longwelli* LANE & WEBSTER, L.Perm.(Bird Springs F.), Clark Co., Nev.; 2 views of holotype crown, ×0.7 (Lane & Webster, 1966).——FIG. 463,*1g-i*. *E. wapanucka* (STRIMPLE), L.Penn.(Morrow., Wapanucka F.), Pontotoc Co., Okla., type of *Pontotocrinus* KNAPP (1969, p. 352); ventral, dorsal, and post. views of holotype, ×2 (Strimple, 1961e).

Exaetocrinus STRIMPLE & WATKINS, 1969, p. 181 [**Stuartwellercrinus argentinei* STRIMPLE, 1949c, p. 125; OD]. Cup truncate cone shaped, with subhorizontal to slightly upflared infrabasals, visible from side, contour of cup pentagonal when viewed from summit or base, large, subhorizontal radial articular facets; anal *X* plate in notch between posterior radials at inner margin of their articular facets. [Differs from *Erisocrinus* in visibility of infrabasals from side.] *L.Carb.(Visean),* G.Brit.(Scot.)-*U-Carb.,* Afr.(Morocco); *U.Penn. (Missour.),* USA(Kans.-Mo.-Okla.); *L.Perm.,* USSR(Kashira).——FIG. 463,*2*. **E. argentinei* (STRIMPLE), Missour.(Argentine Ls.), Mo.; *2a-d,* dorsal, ventral, ant., and post. sides, ×2 (Strimple, 1949c).

Sinocrinus TIEN, 1926, p. 17 [**S. microgranulosus* TIEN, 1926, p. 17; OD]. Cup bowl shaped, with nearly flat base, impressed plate sutures, and pentameral symmetry; infrabasals 5, small, slightly upflared, forming pentagonal low cone basally hollowed for stem attachment; basals large, inflated; radials large, wider than high, with plenary horizontal articular facets; no anal plates in or above cup. Arms branching isotomously on large first primibrachs, which project outward as short or long spines; brachials biserial above primibrachs. *U.Carb.(Taiyuan),* China.——FIG. 463,*3*. **S. microgranulosus; 3a,b,* lat. and dorsal views of cup with proximal part of arms, ×1; *3c-f,* outer, upper, inner, and lat. views of radial plate, ×1; *3g,h,* lat. and articular views of stem, ×1 (Tien, 1926); *3i,* plate diagram of cup (planes of perfect bilateral symmetry marked by heavy black lines, radials black) (Moore, 1962b).

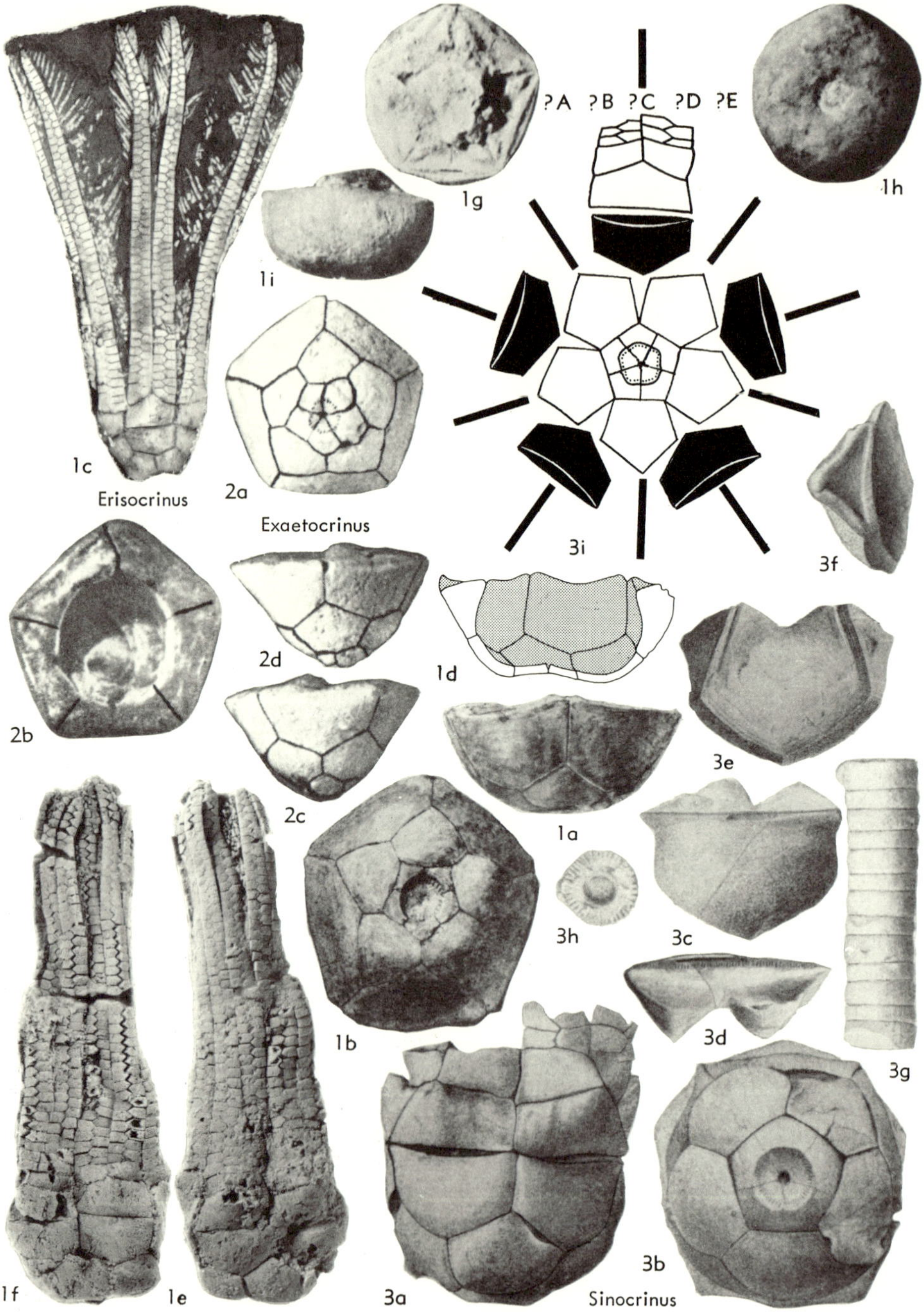

FIG. 463. Erisocrinidae (p. T705).

Family GRAPHIOCRINIDAE Wachsmuth & Springer, 1886

[*nom. transl.* BATHER, 1899b, p. 922 (*ex* series Graphiocrinites WACHSMUTH & SPRINGER, 1886, p. 115[191])] [=Scaphiocrinidae BATHER, 1899b, p. 922] [incl. Scaphiocrininae JAEKEL, 1918, p. 63]

Crown cylindrical, moderately slender or ovoid. Cup low, bowl or saucer shaped;

five small infrabasals, not visible in side view, may be in basal plane or in mild concavity; basals medium sized, with distal portions visible in side view; radials moderately large, wider than long, articular facets filling distal face of plates; single anal plate (*X*) in cup or absent in late genera. Arms ten, wide and usually recti-uniserial, branching on large primibrach *1* in all rays. Column transversely round. *Low.L.Carb.-(Miss.); M.Penn.-U.Penn.; U.Perm.*

The graphiocrinids were probably derived from Devonian decadocrinids. They retained the sinuous more or less zigzag arms with large pinnules and a pentagonal stem in geologically older forms, but eliminated all but one anal plate from the cup. Rather than progressing from cuneate uniserial brachials to biserial they reduced size of the pinnules and shifted attachment facets for pinnules toward the interior so that the arms were able to abut closely when closed. In several forms the brachials widened and became recti-uniserial or slightly cuneate.

Key to Genera of Graphiocrinidae

A. Crown slender cylindrical, small, with appressed erect arms. Cup bowl shaped; radial with wide (plenary) articular facets; 1 or 0 anal plate in cup. Arms 10, recti-uniserial, branching isotomously on primibrachs *1* in all rays ....

- I. Primaxils and anal *X* elongate, or absent, basals small ....
  - a. Anal small, on truncate *CD* basal, basals lozenge shaped, arms medium in height .... *Graphiocrinus*
  - b. Anal pentagonal, not touching *CD* basal .... *Euerisocrinus*
  - c. Anal plate absent .... *Permiocrinus*
- II. Primaxils short to medium in height, not projecting appreciably sideward, basals large ....
  - a. Anal *X* short, narrow .... *Contocrinus*
  - b. Anal *X* large, wide .... *Holcocrinus*

Graphiocrinus DE KONINCK & LE HON, 1854, p. 115 [**G. encrinoides*; M] [=*Scaphiocrinus* HALL, 1858b, p. 550 (type, *S. simplex*; OD)]. Greatest width of crown just above primibrachs. Anal plate long and slender, extending well above summit of radials, distal end terminating in an apex between sides of posterior primibrachs *1* which are closely abutting and closed over distal tip of anal. Ten arms composed of rather wide, gently convex segments with width only slightly greater than length, arms in close contact laterally. *L.Carb.(Tournais.),* Eu.(Belg.-G.Brit.-?Eire).——FIG. 464,*3*. **G. encrinoides,* Tournais., Belg.; *3a,b,* post. views of crown and cup with proximal portion of arms attached, ×2 (Springer, 1911c).

Contocrinus KNAPP, 1969, p. 355 [**Graphiocrinus stantonensis* STRIMPLE, 1939a, p. 12; OD]. Like *Graphiocrinus* except that basals are relatively large, anal *X* short and commonly truncated for contact with single tube plate, base of cup is typically planate and radial articular facets fill full width of plate. Arms uniserial, in close contact laterally. *M.Penn.(Desmoines.)-U.Penn.(Virgil.),* USA(Okla.-Kans.-Texas).——FIG. 464,*1a-c*. *C. lineatus* (STRIMPLE), Missour.(Wann F.), Okla.; *D*- and *B*-ray sides and base of crown, ×2 (Strimple, 1963c).——FIG. 464,*1d-f*. **C. stantonensis* (STRIMPLE), Missour.(Wann F.), near Ochelata, Okla.; hypotype crown from *C*-ray and opposite sides, cup from summit, *1d,e,* ×1.7; *1f,* ×2.8 (Strimple, 1962f).——FIG. 464,*1g,h*. *C. kingi* (MOORE & PLUMMER), M.Penn.(Mineral Wells F.), Palo Pinto Co., Texas; holotype crown from post. and ant. sides, ×1.5 (Moore & Plummer, 1940).——FIG. 464,*1i-k*. *C. delicatulus* (MOORE), U.Penn.(Virgil., Brownville Ls.), Osage Co., Okla.; dorsal, ventral, post. views, ×2.7 (Moore, 1939c).

Euerisocrinus STRIMPLE, 1939, p. 13 [**E. waysidensis*; OD]. Cup bowl shaped, with flat base or small basal concavity; infrabasals covered by stem; basals large, mostly visible from side; proximal ends of radials well above basal plane, articular facets filling distal face of radials, planate; anal *X* pentagonal, not touching posterior basal, extending well above cup summit, faceted for one tube plate. Anal sac and arms unknown. *U.Penn.(Missour.),* USA(Kans.).——FIG. 464,*4*. **E. waysidensis,* Stanton F., Montgomery Co., Kans.; *4a-c,* dorsal, post., and ant. views of holotype cup, ×2 (Strimple, n; Natl. Museum Nat. History S-4309).

Holcocrinus KIRK, 1945, p. 517 [**Graphiocrinus longicirrifer* WACHSMUTH & SPRINGER, 1890, p. 193; OD]. Crown very high, compact, arms 10 to 17 times height of cup. Cup broadly bowl shaped; 5 small infrabasals, barely visible in side view; 5 medium-sized basals; 5 large radials, with articular facets full width of plates; single large anal *X* plate resting on truncated distal end of *CD* basal, extending well above summit of radials and bearing single large tube plate on its distal face. Anal sac tall, extending nearly to tip of arms or beyond them. Arms 10, relatively slender with cuneate brachials which may bear spinose processes in later species; primibrachs *1* axillary, low. Column pentagonal in section in early species and pentagonal with concave faces in later forms. *L.Miss.(Kinderhook.-Osag.),* USA(Iowa-Mo.); *L.Carb.,* Eire.——FIG. 464,*2a,b*. **H. longicirrifer* (WACHSMUTH & SPRINGER), Kinderhook

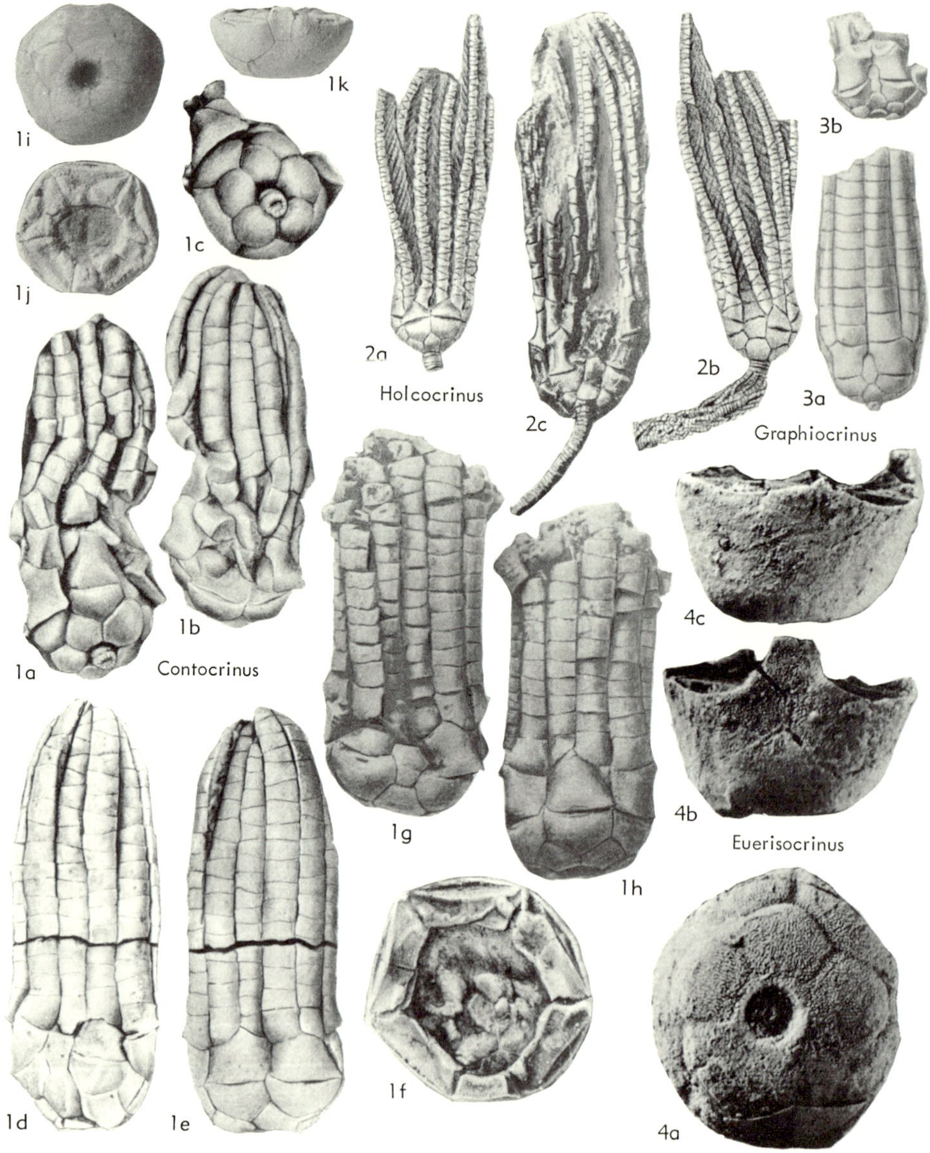

FIG. 464. Graphiocrinidae (p. *T707-T708*).

(Hampton F.), LeGrand, Iowa; holotype crown from ant. and post. sides, ×1 (Wachsmuth & Springer, 1890).——FIG. 464,*2c*. *H. smythi* (WRIGHT), Tournais.(Supra-dolomite beds), Eire (Hook Head, Wexford Co.); crown from post. side, ×1.5 (Wright, 1951-54).

Permiocrinus WANNER, 1949, p. 53 [**P. immaturus*; OD]. Crown small, cylindrical. Cup low, saucer shaped, with broad basal concavity; infrabasals small, not visible in side view, mostly covered by stem; basals small, convex, proximal parts forming edge of basal concavity; radials low, broad, convex; anal plates absent; anal sac not known. Arms uniserial, primibrach *1* axillary, very high in *A*-ray, shorter in other rays; secundibrachs high and narrow, distal parts of arms not known. Stem round. *U.Perm.*, Indon.(Timor).——FIG. 465,*1*. **P. immaturus*, Basleo; *1a,b*, *A*-ray and *C*-ray views of partial crown, ×1.5 (Wanner, 1949b).

Family PARADELOCRINIDAE Knapp, 1969

[Paradelocrinidae KNAPP, 1969, p. 352] [incl. Atokacrininae KNAPP, 1969, p. 357]

Base of cup broadly to narrowly concave, moderate in depth or shallow; infrabasals five, downflared steeply or gently, mostly covered by stem; proximal part or all of basals included in concavity; radials wide, with planate facets occupying full width of radials; anal *X* in notch between posterior radials, generally not visible from side. Arms ten, biserial except in lower portions, branching on axillary first primibrachs. Stem transversely round. *U.Carb., Penn.-(Morrow.); U.Perm.(Basleo beds).*

Paradelocrinids are somewhat smaller than associated catacrinids. They differ from Erisocrinidae, which also lack an externally visible anal plate, in smaller height of the cup, longitudinally rounded form of the radials, and narrower, deeper basal concavity.

Key to Genera of Paradelocrinidae

A. Cup low bowl shaped, with narrow moderately deep basal concavity, anal *X* mostly not visible from exterior of cup ....
- I. Cup plates smooth, sutures between plates impressed .... *Neocatacrinus*
- II. Cup plates smooth, sutures between plates not impressed ....
  - a. Basals transversely concave .... *Sublobalocrinus*
  - b. Basals not transversely concave .... *Paradelocrinus*
- III. Cup plates with rugose surface ....
  - a. Coarse nodes covering plates, 5 infrabasals .... *Atokacrinus*
  - b. Fine tubercles, aligned in longitudinal ridges, 3 infrabasals .. *Lopadiocrinus*

Paradelocrinus MOORE & PLUMMER, 1938, p. 294 [**P. aequabilis*; OD]. Characters of family, with smooth plates, arms unknown, and anal *X* in notch between posterior radial articular facets; cup very shallow, radials reaching basal plane but not in concavity. *L.Penn.(Morrow.)-U.Penn.(Missour.),* USA(Texas-Okla.-Ark.-Kans.).——FIG. 466,*4*. **P. aequabilis,* Morrow.(Bloyd F.), Ark.; *4a-d,* holotype cup from ant., post., ventral (*A* ray upward), and dorsal (*A* ray downward) sides, ×1.5 (Moore & Plummer, 1938); *4e,* long. sec. of cup through midline of *A* radial and *CD* interray (portion of cup beyond section shaded), ×1.5 (Moore, n).

[BREIMER (1962, p. 172) reported two specimens from Spain as *Paradelocrinus* spec. 1 from the Upper Visean strata and as *Paradelocrinus* spec. 2 from Moscovian strata. *Paradelocrinus* spec. 1 is apparently an advanced *Cyathocrinites* that has eliminated the anal plate from the cup. *Paradelocrinus* spec. 2 is considered here as *Atokacrinus* sp.]

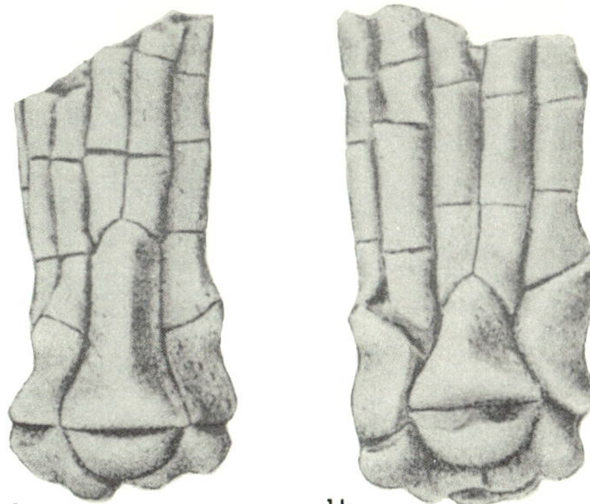

1a 1b

Permiocrinus

FIG. 465. Graphiocrinidae (p. *T*708).

Atokacrinus KNAPP, 1969, p. 358 [**A. obscurus*; OD]. Differs from *Lopadiocrinus* in having 5 infrabasals and proximal tips of radials above basal plane. Arms unknown. [This genus appears to be a direct derivative of *Palmerocrinus,* in which anal plate has migrated to position notching cup between radial articular facets.] *M.Penn.(Atokan-Desmoines.),* USA(Okla.-Mo.); *U.Carb.,* Eu.(Spain).——FIG. 466,*3a-c*. **A. obscurus,* Atokan(Burgner F.), Mo.; holotype cup from dorsal, post., and ventral sides, ×1.5 (Knapp, 1969).——FIG. 466,*3d-f*. *A. decorus* STRIMPLE & MOORE, Desmoines.(Lester F.), Love Co., Okla.; dorsal, ventral, and post. views of holotype cup, ×1.7 (Strimple & Moore, 1971c).

Lopadiocrinus WANNER, 1916, p. 227 [**L. granulatus*; OD]. Cup shallow bowl shaped, with broad base and large basal concavity; infrabasals 3, small, almost covered by stem impression; basals narrow and moderately long but only distal tips visible in side view; radials large, with proximal portions entering basal plane, forefacet present near summit and pronounced outer ligament pit extending full width of plate, transverse ridge prominent and expanded in midportion of facet, muscle areas large; facet for rudimentary anal plate in built-up area between *C* and *D* radials toward inner edge of facets and sloping outward-upward (sursumate). Entire outer surfaces of cup covered by large granulations which may coalesce on radials and form longitudinal ridges. *U.Perm.(Basleo beds),* Indon.(Timor).——FIG. 466,*2*. **L. granulatus,* Basleo; *2a-c,* holotype from ant., dorsal, and ventral sides, ×1.5 (Wanner, 1916a).

Neocatacrinus KNAPP, 1969, p. 366 [**Paradelocrinus protensus* MOORE & PLUMMER, 1940, p. 324;

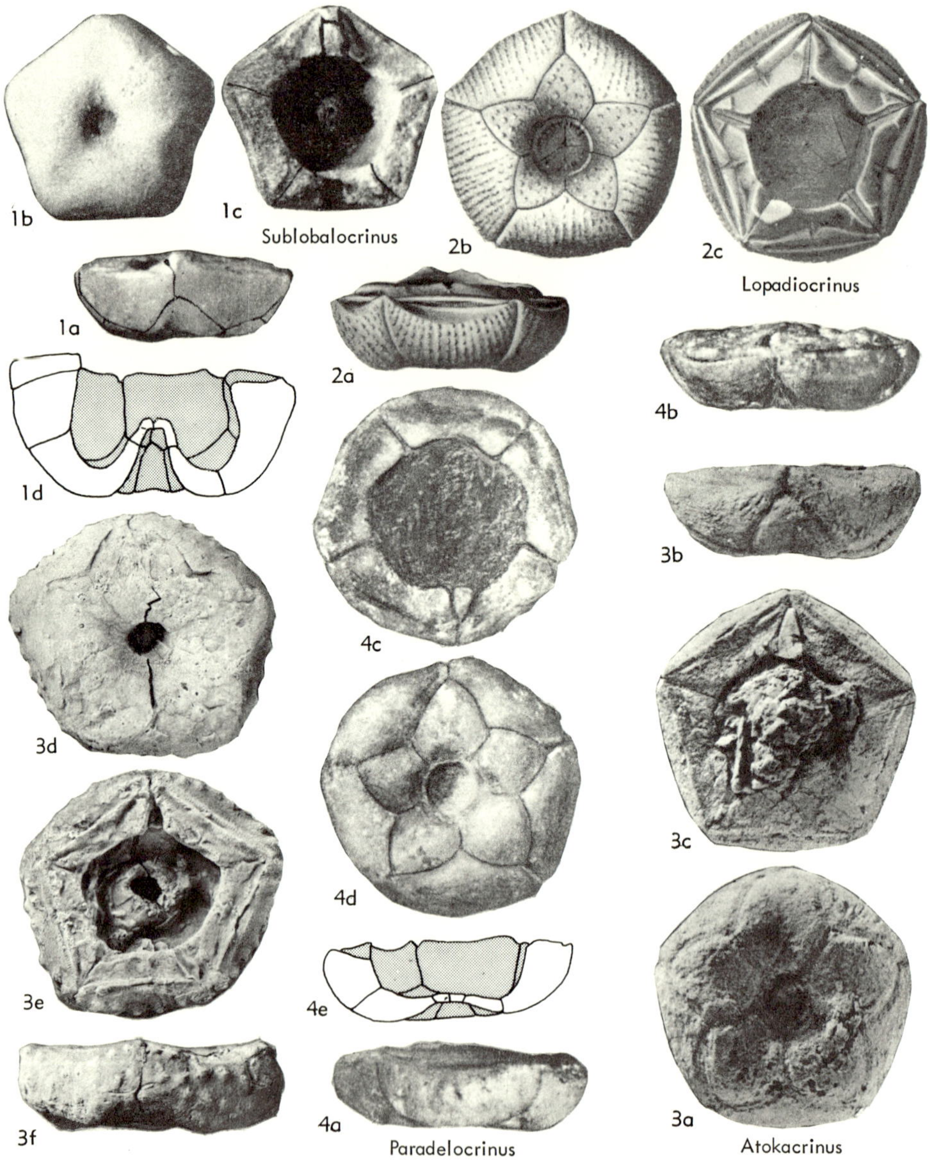

FIG. 466. Paradelocrinidae (p. T709-T710).

OD]. Cup low bowl shaped, with moderately deep basal concavity, plates smooth, sutures impressed; anal *X* confined to notch at inner side of *CD* interradial suture. Arms unknown. [Distinguished from *Paradelocrinus* by appreciably greater depth of the basal concavity and impressed nature of sutures between cup plates.] *U.Penn.(Missour.)*, USA(Texas).——FIG. 467,*1*. **N. protensus* (MOORE & PLUMMER), Graford F., Palo Pinto Co., Texas; *1a-c*, dorsal, post., ventral views of holotype, ×2 (Moore & Plummer, 1940); *1d*, long. sec. of cup through midline of *A* radial and *CD* interray (portions of cup beyond section shaded), ×3.5 (Moore, n).

Sublobalocrinus KNAPP, 1969, p. 365 [**Paradelocrinus iolaensis* STRIMPLE, 1949c, p. 124; OD]. Cup small, nearly straight-sided pentagonal with narrowly rounded corners as seen from above and below, shallow bowl shaped in side profile, with very narrow deep basal concavity, plates smooth, without impressed sutures; basals transversely concave as in *Lobalocrinus* (Catacrinidae); radials extending to basal plane of cup but not into concavity; anal *X* in notch between *C* and *D* radials, its proximal part visible externally, well separated from *CD* basal, its medial and distal parts recumbent inward and nearly horizontal. Arms unknown. *U.Penn.(Missour.)*, USA(Kans.).——FIG. 466,*1*. **S. iolaensis* (STRIMPLE), Iola Ls., Allen Co., Kans.; *1a-c*, post., dorsal, ventral views of holotype, ×2 (Strimple, 1949c); *1d*, long. sec. of cup through midline of *A* radial and *CD* interray (portions of cup beyond section shaded), ×2.8 (height exaggerated) (Moore, n).

Family ARKACRINIDAE Knapp, 1969

[*nom. transl.* MOORE & STRIMPLE, 1973, p. 22 (*ex* Arkacrininae KNAPP, 1969, p. 356)]

Cup bowl shaped, with deep basal concavity and constriction at upper margin, plates smooth and exceptionally thick; infrabasals five, distally downflared steeply; basals five, confined to concavity and also downflared steeply, concave transversely; radials five, subequal, broadly flattened medially and distally, with proximal tips reaching basal plane of cup, anal plate in notch between *C* and *D* radials, mostly at inner margins of their articular facets. Arms unknown. *L.Penn.(Morrow.)*.

Arkacrinus KNAPP, 1969, p. 357 [**Delocrinus dubius* MATHER, 1915, p. 105; OD]. Characters of family. Arms unknown. *L.Penn.(Morrow.)*, USA(Okla.-Ark.).——FIG. 468,*1*. **A. dubius* (MATHER), Brentwood Ls., Ark.; hypotype, ant., ventral, and dorsal views, ×2 (Moore & Plummer, 1938); *1d*, long. sec. of cup through midline of *A* radial (portions of cup beyond section shaded), ×2 (Moore, n).

Family DIPHUICRINIDAE Strimple & Knapp, 1966

[Diphuicrinidae STRIMPLE & KNAPP, 1966, p. 312] [incl. Graffhamicrininae KNAPP, 1969, p. 362]

Bowl-shaped cup with deep basal invagination, plates rugose and very thick; five infrabasals, steeply downflared; large basals and radials dominant in side view; one anal plate in contact with posterior basal or essentially eliminated from cup exterior, followed by either one or two tube plates. Arms ten, cuneate uniserial, branching on axillary primibrachs *1*. Stem small, transversely round. *L.Penn.(Morrow.)-L.Perm.*

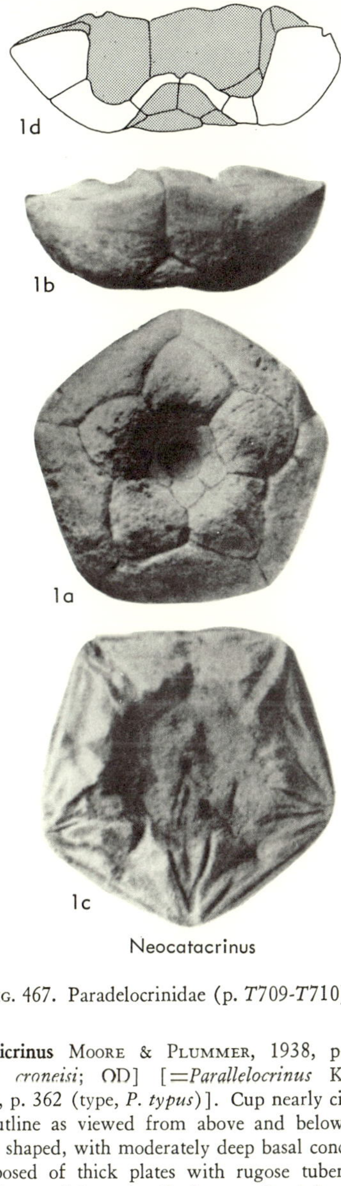

FIG. 467. Paradelocrinidae (p. *T709-T710*).

Diphuicrinus MOORE & PLUMMER, 1938, p. 307 [**D. croneisi*; OD] [=*Parallelocrinus* KNAPP, 1969, p. 362 (type, *P. typus*)]. Cup nearly circular in outline as viewed from above and below, low bowl shaped, with moderately deep basal concavity, composed of thick plates with rugose tuberculate exteriors and irregular finely crenulated interplate facets; subhorizontal radial articular facets with prominent crenulated transverse ridge and narrow deep ligament fossae; narrow but distinct forefacet at distal edge of radials; distal portion of single anal plate *(X)* elongated, recumbent inward, faceted distally for 2 sac plates. Arms uniserial, composed of cuneate brachials, pinnulate. Stem transversely round. *L.Penn.(Morrow.)-U.Penn. (Virgil.)*, USA(Texas-Okla.-Mo.-Ark.-Ohio-Ky.-Kans.).——FIG. 469,*1a-j*. **D. croneisi*, Morrow. (Bloyd F.), Ark.; *1a-c*, holotype cup from ant.,

FIG. 468. Arkacrinidae (p. *T711*).

dorsal, and post. sides; *1d-g,* radial plate from summit, ext., int., and dorsal sides; *1h,* ventral view showing radial forefacets; *1a-g,* ×1.5, *1h,* ×2.4 (Moore & Plummer, 1938); *1i,j,* post., ant. views of hypotype crown, ×1.2 (Strimple & Knapp, 1966).——FIG. 469,*1k-m. D. pegmus* STRIMPLE & MOORE, Desmoines.(Frensley F.), Carter Co., Okla.; post., ventral, and dorsal views of holotype cup, ×3 (Strimple & Moore, 1971c).——FIG. 469,*1n-q. D. dovelyensis* STRIMPLE & MOORE, Desmoines. (Savanna F.), Roger Co., Okla.; dorsal, ant., post., and ventral sides of holotype cup, ×1.6 (Strimple & Moore, 1961c).——FIG. 470,*1. D. typus,* M.Penn.(Burgner F.), Jasper Co., Mo.; long. sec. of cup through midline of *A* radial and *CD* interray (portions of cup beyond section shaded), ×2 (Moore, n).

Graffhamicrinus STRIMPLE, 1961, p. 123 [**G. acutus*; OD] [=*Tholiacrinus* STRIMPLE, 1962e, p. 136 (type, *Corythocrinus undulatus* STRIMPLE, 1961d, p. 129; OD), M.Penn.(Desmoines.)-L. Perm.(Wolfcamp.), USA(Neb.-Kans.-Okla.-Texas)]. Cup similar to that of *Delocrinus* but surface of plates rugose. Anal plate typically erect, faceted for one tube plate, resting on truncated distal end of *CD* basal. Arms 10, equibiserial, ornate for at least half of their length. *M.Penn.(Atokan-Desmoines.)-L.Perm.(Wolfcamp.),* USA(Mo.-Okla.-Kans.-Neb.-Texas-Iowa).——FIG. 471,*1a-c.* **G. acutus,* Desmoines.(Holdenville F.), Okla.; *1a,b,* paratype from post. and dorsal sides; *1c,* holotype crown from ant. side, ×1.2 (Strimple, 1961d).——FIG. 471,*1d-g. G. magnificus* (STRIMPLE), U.Penn.(Haskell Ls., Virgil.), Kans.; *1d,* post. view of holotype crown slightly oblique from below, ×1; *1e-g,* post., dorsal, and ventral views of paratype cup, ×1 (Strimple, 1947).——FIG. 471, *1h-o. G. granulosus* (MOORE & PLUMMER); *1h-k,* Desmoines.(Mineral Wells F.), Palo Pinto Co., Texas, dorsal, post., ventral, and ant. views of holotype cup, ×1.5 (Moore & Plummer, 1940); *1l-o,* Atokan, Okla., *1l,* int. mold showing form of anal tube, ×1; *1m,n,* restoration and cross section of anal sac, ×2; *1o,* rubber cast from ext. mold, ×1 (Moore & Strimple, 1941).

Family PROTENCRINIDAE Knapp, 1969

[Protencrinidae KNAPP, 1969, p. 353]

Crown tall, cylindrical. Cup saucerlike, smooth plated, with wide, shallow basal concavity; basals small, mostly confined to concavity; radials large, with proximal tips in basal plane of cup or extending into concavity; anal plate in notch at upper-inner edge of *CD* interradial suture, not visible externally. Arms ten, biserial, branching isotomously on primibrachs *1*. Stem round transversely. *U.Carb.(Moscov.), M.Penn.-(Atokan)-L.Perm.(Artinsk.).*

Protencrinus JAEKEL, 1918, p. 66 [**P. moscoviensis*; OD]. Cup very low; radials forming part of basal concavity; basals very small and tending to be separated laterally by infrabasals; proximal

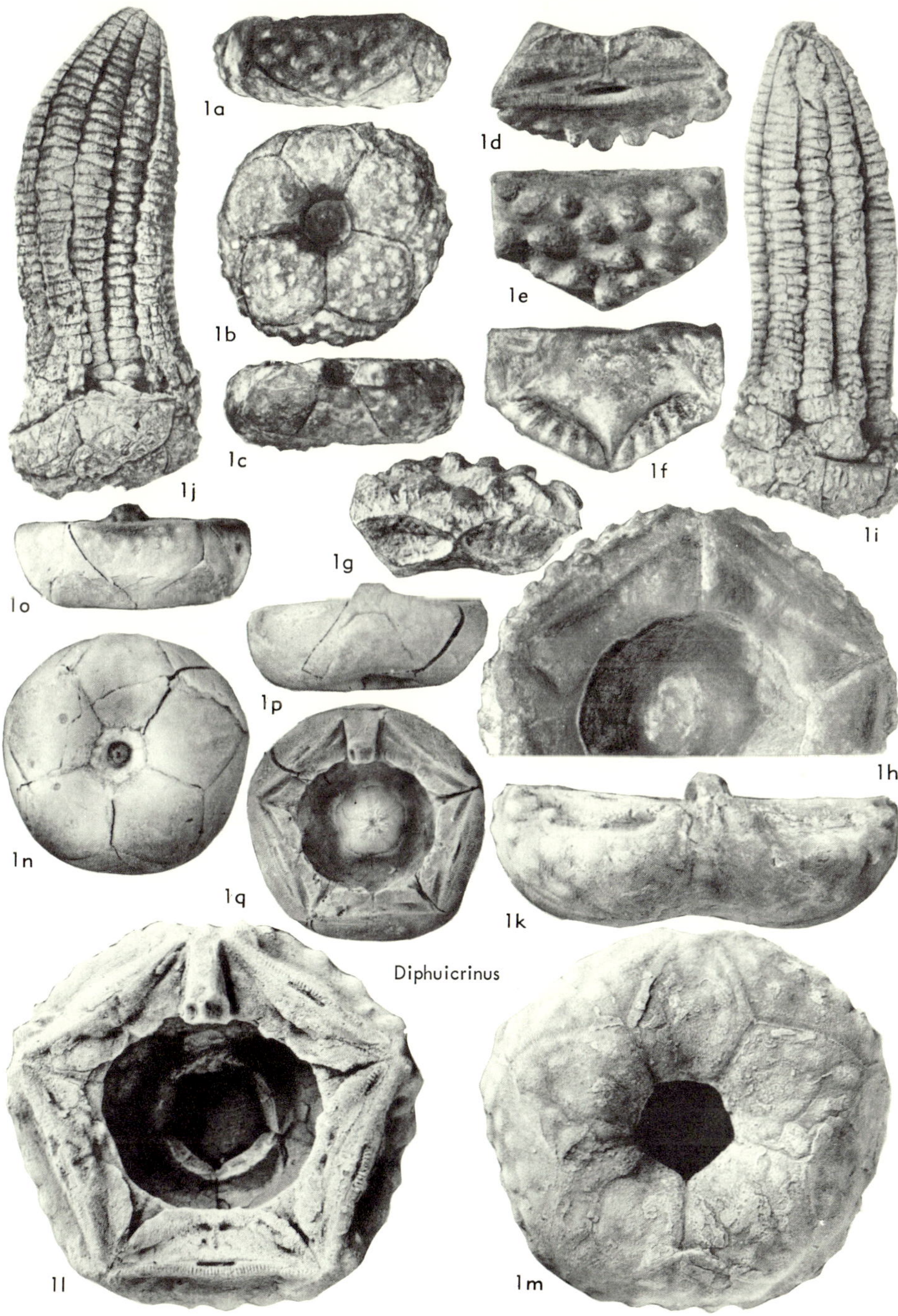

FIG. 469. Diphuicrinidae (p. *T711-T712*).

tips of radials touching infrabasals; no externally visible anal plate in cup. Arms 10, biserial except for proximal segments. *M.Penn.(Atokan-Desmoines.)*, USA(Okla.-Texas); *U.Carb.(Moscov.)*-

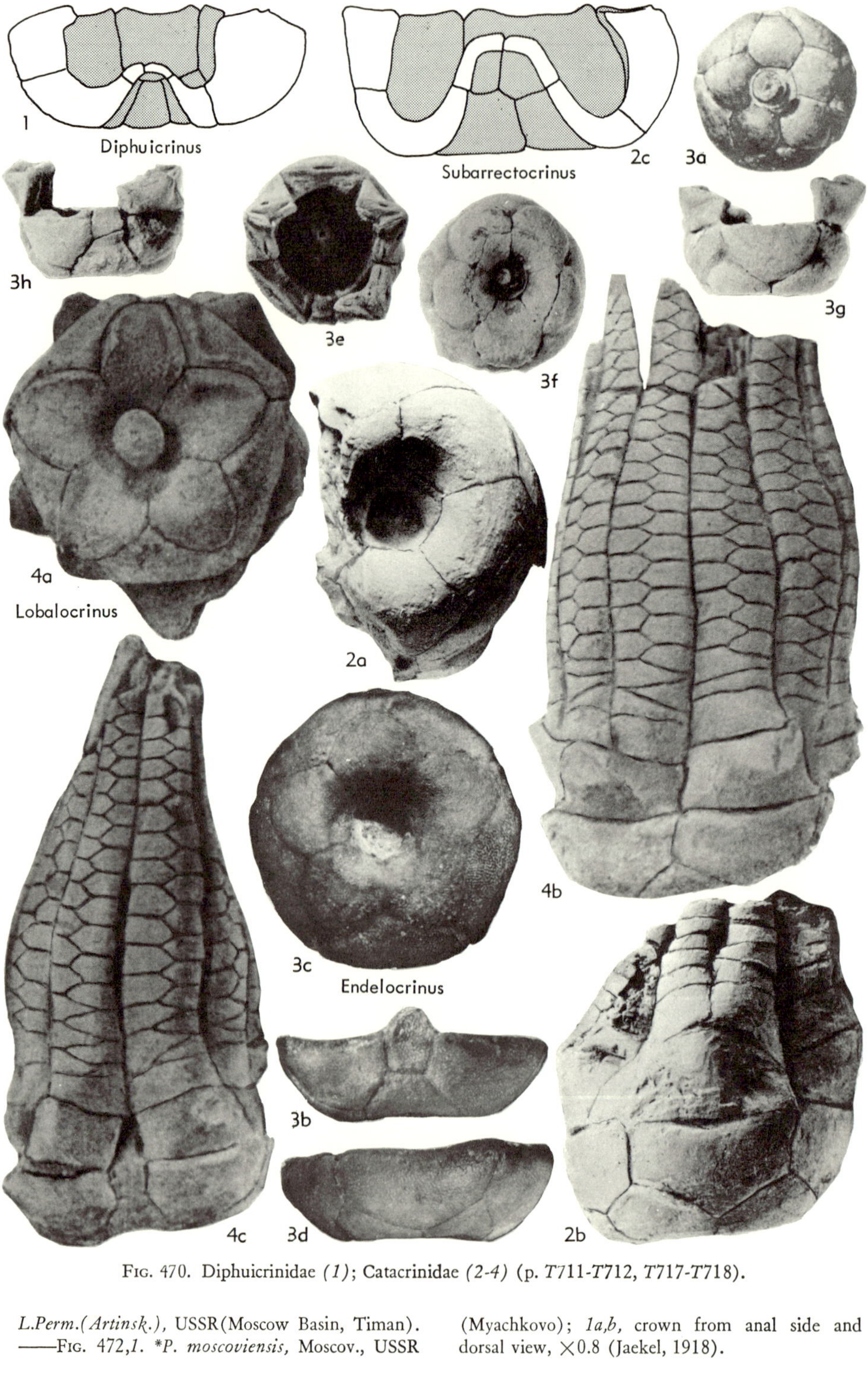

FIG. 470. Diphuicrinidae *(1)*; Catacrinidae *(2-4)* (p. T711-T712, T717-T718).

L.Perm.(Artinsk.), USSR(Moscow Basin, Timan).——FIG. 472,*1*. **P. moscoviensis*, Moscov., USSR (Myachkovo); *1a,b,* crown from anal side and dorsal view, ×0.8 (Jaekel, 1918).

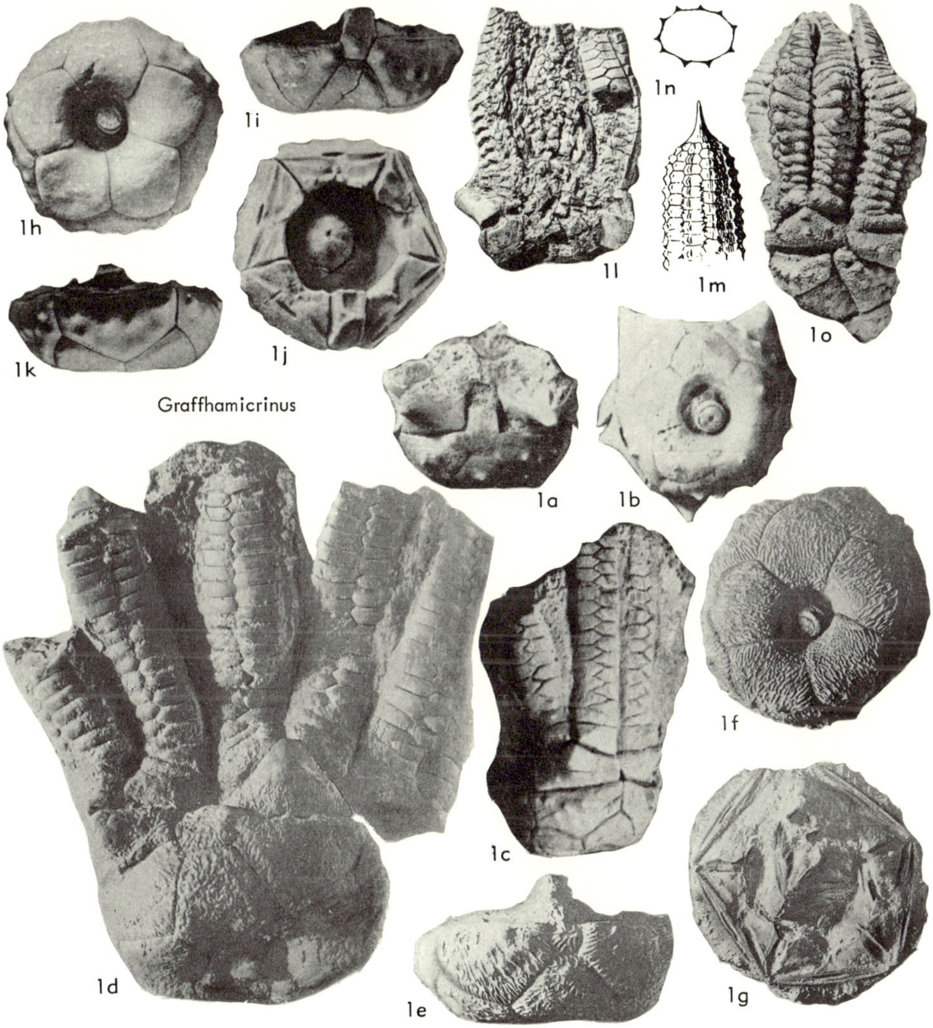

FIG. 471. Diphuicrinidae (p. T712).

Neoprotencrinus KNAPP, 1969, p. 354 [**Paradelocrinus subplanus* MOORE & PLUMMER, 1940, p. 321; OD]. Crown tall, cylindrical. Base of cup planate or mildly concave; proximal tips of radials in basal plane of cup but not entering basal concavity, well separated from infrabasals; anal plate in notch on inner side of *C* and *D* radial articular facets. Arms 10, equibiserial. *M.Penn. (Desmoines.),* USA(Okla.-Texas).——FIG. 472, *2a-d.* **N. subplanus* (MOORE & PLUMMER); *2a,b,* Strawn Gr.(Millsap Lake F.), Hood Co., Texas, holotype cup from dorsal and ventral, ×2.4 (Moore & Plummer, 1940); *2c,d,* Oologah F., Okla.; *2c,* hypotype crown, ×1.8 (Strimple, 1962a); *2d,* long. sec. of holotype through mid-line of *A* radial and *CD* interray (portions of cup beyond section shaded), ×4 (Moore, n).——FIG. 472,*2e.* *N. brachiatus* (MOORE & PLUMMER), Millsap Lake F., Hood Co., Texas; side view of holotype crown, ×1.2 (Moore & Plummer, 1940).

Family CATACRINIDAE Knapp, 1969

[*nom. transl.* MOORE & STRIMPLE, 1973, p. 22 (*ex* Catacrininae KNAPP, 1969, p. 365)] [incl. Arrectocrininae KNAPP, 1969, p. 363; Palmerocrininae KNAPP, 1969, p. 360]

Crown tall, cylindrical. Cup low bowl shaped, smooth plated or rugose, with moderately deep and narrow basal concavity, proximal tips of radials above basal plane

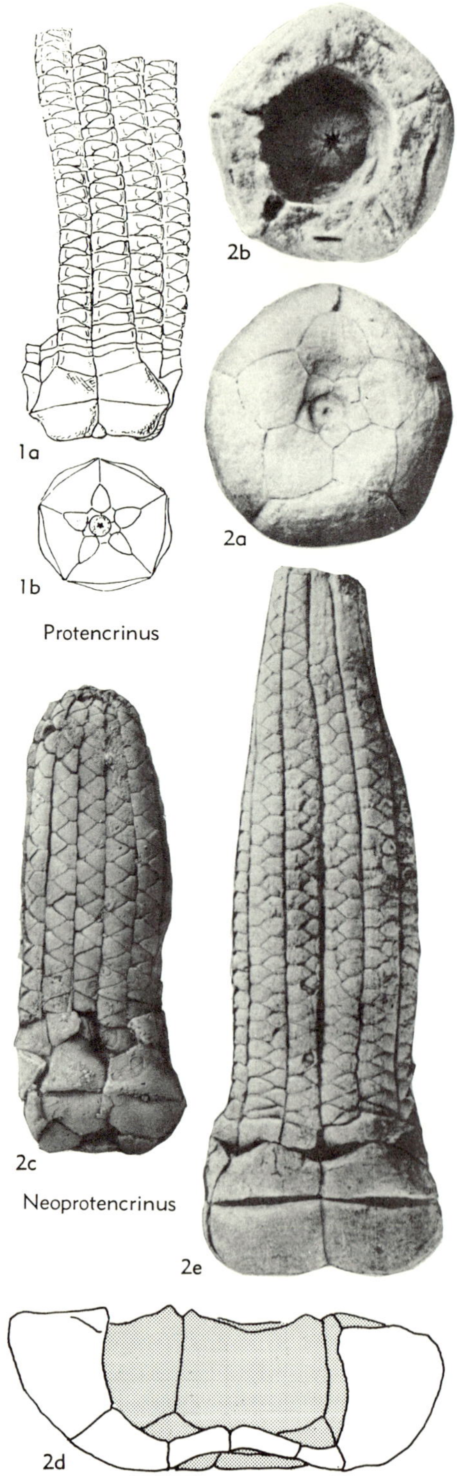

FIG. 472. Protencrinidae (p. *T712-T715*).

of cup or barely reaching it; articular facets of radials subhorizontal, occupying full width of plates; single upright anal plate (*X*) in cup extending above radial summits and distally recumbent inward in some genera. Anal sac cylindrical, not reaching above arms, composed of small polygonal plates in vertical rows, pointed at summit. Arms ten, biserial, branching isotomously on primibrachs *1*, which uncommonly is produced laterally as stout long spine, pinnulate. Stem transversely round, homeomorphic. *L.Penn.(Morrow.)-U.Perm.(Basleo beds).*

Key to Genera of Catacrinidae

A. Cup bowl shaped, with medium to deep basal concavity, radial articular facets planate and plenary; anal *X* plate on truncate distal edge of *CD* basal; arms 10, biserial; stem transversely round ....
- I. Cup plates smooth ....
  - a. Sutures between plates impressed ..
    1. Cup low, infrabasals 5, basals transversely concave ...... *Lobalocrinus*
    2. Cup low, infrabasals 3, basals not concave .... *Paraplasocrinus*
    3. Cup medium in height, basals not concave, pits at plate angles .... *Endelocrinus*
  - b. Sutures between plates not impressed
    1. Cup low, exceptionally deep basal concavity .... *Subarrectocrinus*
    2. Cup low, deep basal concavity .... *Delocrinus*
    3. Cup low, shallow basal concavity .... *Pyndaxocrinus*
    4. Cup high, nearly straight sided longitudinally .... *Arrectocrinus*
- II. Cup plates granulose to coarsely rugose. Sutures between plates distinctly impressed, surface granulose .... *Palmerocrinus*

Delocrinus MILLER & GURLEY, 1890, p. 9 [**Poteriocrinus hemisphericus* SHUMARD, in SHUMARD & SWALLOW, 1858, p. 26; OD) (neotype, ICZN Op. 1006, 1974, *D. hemisphericus* MILLER & GURLEY, 1890, p. 9, Univ. Chicago spec. 6234 in Field Museum Nat. History, Chicago, Lane Sh., Missour., Kansas City, Mo.)] [=*Catacrinus* KNAPP, 1969, p. 365 (type, *Delocrinus subhemisphericus* MOORE & PLUMMER, 1940, p. 258, obj. syn. of *D. hemisphericus* MILLER & GURLEY, 1890, with same type specimens; OD); *Asaccocrinus* WANNER, 1949b, p. 55 (type, *Delocrinus? malaianus* WANNER, 1916, p. 202; OD), U.Perm., Timor; *Wewokacrinus* KNAPP, 1969, p. 361 (type, *Delocrinus wewokaensis* STRIMPLE, 1940b, p. 7; OD), M. Penn.(Desmoines.), USA(Okla.); *Palmatocrinus*

KNAPP, 1969, p. 363 (type, *Delocrinus ponderosus* STRIMPLE, 1949, p. 341; OD), U.Penn., Virgil., USA(Okla.-Texas); *Cathetocrinus* KNAPP, 1969, p. 361 (type, *Delocrinus stullensis* STRIMPLE, 1947, p. 5; OD), U.Penn.(Virgil.), USA(Kans.)]. Cup low bowl shaped, with basal concavity moderately narrow and deep; infrabasals downflared, basals large, proximal ends of radials well above basal plane of cup or barely reaching it; erect or distally somewhat recumbent anal *X* resting on truncated tip of *CD* basal, typically faceted for 2 sac plates but commonly for single plate; anal sac slender, cylindrical, terminating in single spine. Arms 10, biserial, branching isotomously on first primibrach which may be projected as spine. Stem transversely round. *M.Penn.(Desmoines.)-U.Perm. (Basleo beds),* cosmop.——FIG. 473,*2a-g*. **D. hemisphericus* (MILLER & GURLEY), U.Penn.(Missour., Lane Sh.), Mo.; *2a-d,* dorsal, ventral, post., and ant. views of neoparatype, ×2; *2e-g,* dorsal, ventral, and inner end articular views of neoparatype axillary primibrach *1,* all ×2 (Moore & Strimple, 1970).——FIG. 473,*2h*. *D.* sp., U.Penn. (Missour., Wann F.), Okla.; slender anal sac with summit spine, ×2 (Moore & Strimple, 1941).——FIG. 473,*2i,j*. *D.* sp., U.Perm.(Word F.), Brewster Co., Texas; interiors of radial and *CD* basal showing growth lines and ridges running to plate corners, ×2 (Moore, n).

Arrectocrinus KNAPP, 1969, p. 364 [**Delocrinus abruptus* MOORE & PLUMMER, 1940, p. 289; OD] [=*Metarrectocrinus* KNAPP, 1969, p. 364 (type, *Delocrinus major* WELLER, 1909b, p. 627; OD), *U.Penn.(Virgil.)-L.Perm.(Wolfcamp.)*]. Cup distinguished from that of *Delocrinus* by its exceptional height and longitudinally nearly straight sides; basal concavity deep. *U.Penn.(Virgil.)-L. Perm.(Wolfcamp.),* USA(Kans.-Neb.-Texas).——FIG. 473,*1*. **A. abruptus* (MOORE & PLUMMER); *1a,b,* holotype from Florena Sh., Cowley Co., Kans., dorsal and post. views, ×1.5; *1c,d,* paratype from Moran F., Callahan Co., Texas, dorsal and ventral views, ×1.5; *1e,* long. sec. through midline of *A* radial and *CD* interray (portions of cup beyond section shaded), ×2.25 (*1a-d,* Moore & Plummer, 1940; *1e,* Moore, n).

Endelocrinus MOORE & PLUMMER, 1940, p. 296 [**Eupachycrinus fayettensis* WORTHEN, in MEEK & WORTHEN, 1873, p. 565; OD]. Characters of family with anal plate resting on truncated tip of *CD* basal, plates moderately tumid, with dimpled depressions at plate angles, arms 10, biserial, branching isotomously on first primibrachs. *L. Penn.(Morrow.)-L.Perm.(Wolfcamp.),* USA(Okla.-Kans.-Texas-Neb.-Ill.-Mo.-Nev.-Pa.-Ariz.-Ohio).——FIG. 470,*3a*. **E. fayettensis* (WORTHEN), U. Penn.(Missour.), Ill.; holotype cup from base, ×2.5 (Moore & Plummer, 1940).——FIG. 470, *3b-d*. *E. matheri* (MOORE & PLUMMER), L.Penn. (Morrow., Brentwood Ls.), Okla.; holotype, post., dorsal, ant. views, ×2.5 (Moore & Plummer, 1938).——FIG. 470,*3e-h*. *E. solus* (WEBSTER & LANE), M.Penn.(Desmoines., Naco F.), Ariz.; ventral, dorsal, ant., and post. views of holotype, ×1.5 (Webster & Lane, 1970).

Lobalocrinus KNAPP, 1969, p. 366 [**Delocrinus wolforum* MOORE & PLUMMER, 1940, p. 278; OD]. Cup distinguished from that of *Delocrinus* by petaloid transversely concave basals; anal *X* erect and tall, resting on truncate distal edge of *CD* basal. Arms 10, biserial. *U.Penn.(Virgil.),* USA (Texas).——FIG. 470,*4*. **L. wolforum* (MOORE & PLUMMER), Graham F., Cisco Gr., Brown Co., Texas; *4a-c,* dorsal, *AB*- and *CD*-interray views of holotype, ×1.3 (Moore & Plummer, 1940).

Palmerocrinus KNAPP, 1969, p. 360 [**P. comptus*; OD]. Basal concavity of cup narrow and deep, surface of plates covered by granules, sutures between them impressed; proximal tips of radial plates above basal plane, their distal parts curved sharply inward; anal *X* recumbent inward at low angle to horizontal. Arms 10, obliqui-uniserial in proximal portion, biserial distalward. *L.Penn. (Morrow.)-M.Penn.(Atokan),* USA(Okla.-Mo.-Utah).——FIG. 473,*3*. **P. comptus,* Atokan (Burgner F.), Mo.; *3a-c,* holotype cup from post., dorsal, and ventral views, ×1.5 (Knapp, 1969).

Paraplasocrinus MOORE & PLUMMER, 1938, p. 306 [**Cibolocrinus transitorius* WANNER, 1916a, p. 208; OD]. Basal concavity deep, including 3 planate or slightly downflared infrabasals and proximal portions of basals, small azygous infrabasal in *C* ray; all plates smooth with sutures not impressed, or moderately tumid with impressed sutures; anal *X* on truncate tip of *CD* basal or only notching inner edges of posterior radials; radial articular facets plenary, planate, with muscular articulations for first primibrachs. *U.Perm. (Basleo beds),* Indon.(Timor).——FIG. 474,*2*. **P. transitorius* (WANNER), Basleo; *2a-c,* post., ventral, and dorsal views of holotype cup, ×1 (Wanner, 1916a).

Pyndaxocrinus KNAPP, 1969, p. 365 [**Delocrinus separatus* STRIMPLE, 1949d, p. 26; OD]. Similar to *Delocrinus* in having a shallow cup with proximal tips of radials reaching basal plane, but differing in having very shallow basal concavity. Arms unknown. Stem small, transversely round. *U. Penn.(Virgil.)-L.Perm.(Wolfcamp.),* USA(Kans.-Neb.-Nev.).——FIG. 474,*1*. **P. separatus* (STRIMPLE), Virgil.(Stull Sh.), Kans.; *1a-c,* holotype cup from post., ventral, and dorsal sides, ×1.3 (Strimple, 1949d).

Subarrectocrinus KNAPP, 1969, p. 365 [**Delocrinus? perexcavatus* MOORE & PLUMMER, 1940, p. 284; OD]. Cup high bowl shaped with very deep basal concavity, its bottom only little below summit of cup; plates smooth, sutures not impressed. Arms unknown above uniserial cuneate proximal secundibrachs, doubtless biserial distally. *U.Penn. (Virgil.),* USA(W.Texas-Kans.).——FIG. 470,*2*. **S. perexcavatus* (MOORE & PLUMMER), Gaptank

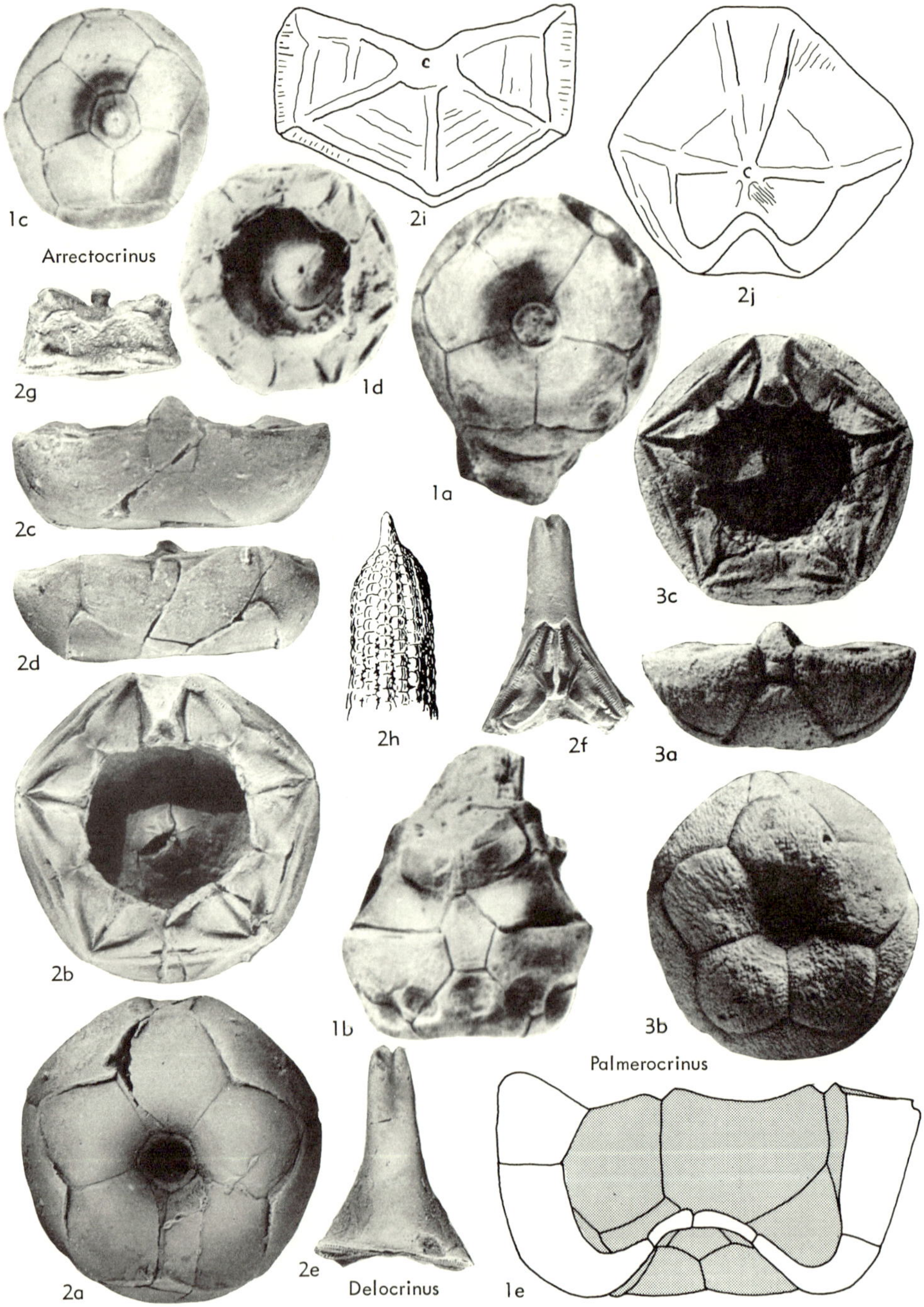

FIG. 473. Catacrinidae (p. *T716-T717*).

F., Brewster Co., Texas; *2a,b,* dorsal and ant. views of holotype, ×2 (Moore & Plummer, 1940); *2c,* long. sec. of cup through midline of *A* radial and *CD* interray (portion of cup beyond section shaded), ×2.25 (Moore, n).

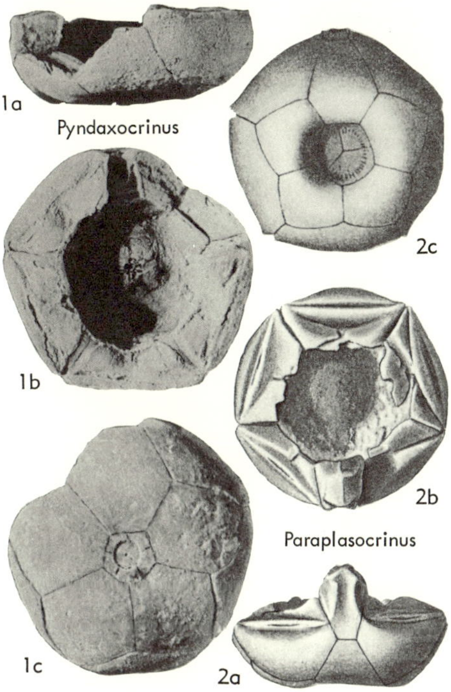

FIG. 474. Catacrinidae (p. T717).

Family STACHYOCRINIDAE Moore & Strimple, 1973

[Stachyocrinidae MOORE & STRIMPLE, 1973, p. 22]

Crown cylindrical to ovoid. Cup low bowl or saucer shaped, with shallow basal concavity, proximal tips of radials reaching basal plane of cup or well above it; anal in notch between *C* and *D* radial articular facets, not visible from side. Arms ten, uniserial, brachials rectangular in side view. Stem impression circular. *U.Perm.(Basleo beds).*

Stachyocrinus WANNER, 1916, p. 233 [**S. zea*; OD]. Cup shallow bowl shaped with pronounced but narrow basal concavity; infrabasals covered by stem; basals small, narrow, confined to basal area; radials large, with proximal portions in basal plane of cup, small forefacet near summit; anal plate not visible when arms are in place. Arms 10, rectilinear, convex externally, abutting closely, primibrachs *1* axillary, very large and tumid, some assisting in support of secundibrachs belonging to arms of adjacent rays. *U.Perm.(Basleo beds),* Indon.(Timor).——FIG. 475,*2*. **S. zea,* Noil Fatoe; *2a-c,* dorsal and two lat. views of crown, ×2 (Wanner, 1916a).

Parastachyocrinus WANNER, 1949, p. 44 [**Erisocrinus malaianus* WANNER, 1924, p. 42; OD]. Crown medium in height, cylindrical except for strongly protruded axillary primibrachs *1*. Cup

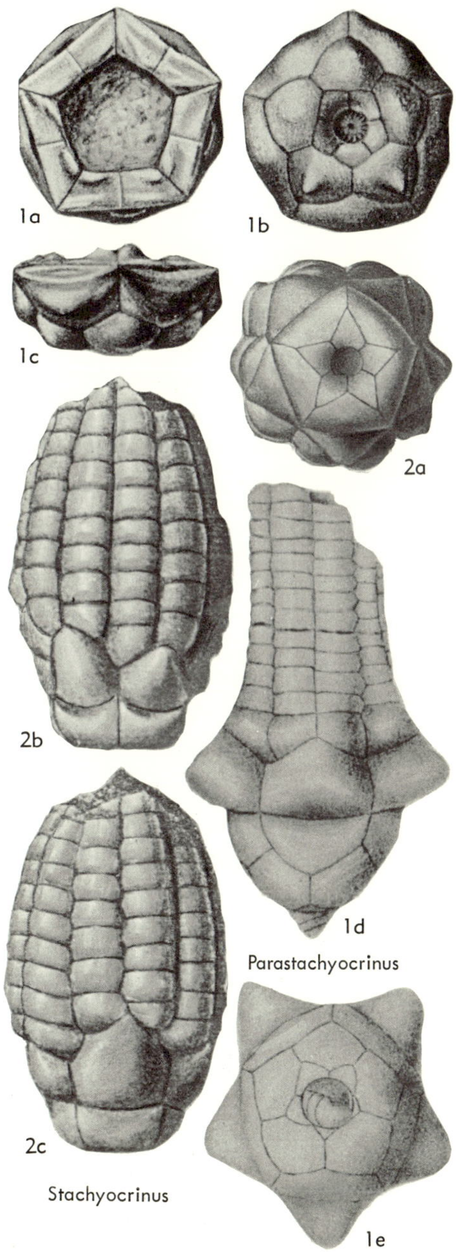

FIG. 475. Stachyocrinidae (p. T719-T720).

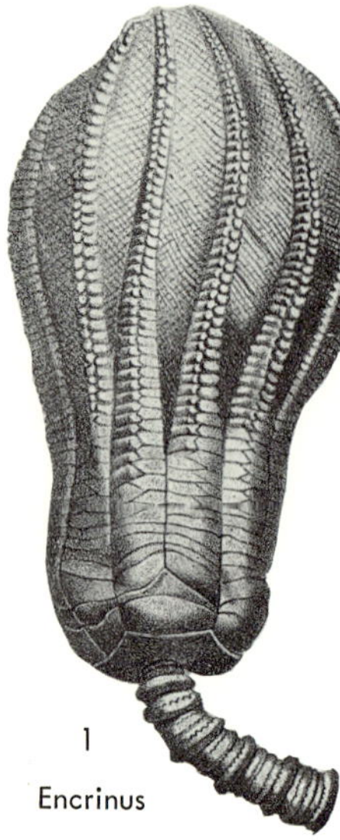

FIG. 476. Encrinidae (p. T720).

truncate bowl shaped with slight basal concavity; anal *X* excluded from exterior of cup but marked by notch on inner side between posterior radial articular facets, not visible externally from side. Ten arms, closely abutting, primibrachs *1* axillary, secundibrachs recti-uniserial. Column transversely round. *U.Perm.(Basleo beds),* Indon.(Timor).——FIG. 475,*1*. **P. malaianus* (WANNER), Basleo; *1a-c,* ventral, dorsal, and post. sides of holotype cup, ×1.5 (Wanner, 1924); *1d,e,* lat. and dorsal views of crown, ×1.5 (Wanner, 1949b).

Family ENCRINIDAE Dujardin & Hupé, 1862

[*nom. correct.* ICZN, 1962 (Bull. Zool. Nomencl., v. 19, pt. 5, p. 263) (*pro* Encriniens DUJARDIN & HUPÉ, 1862, p. 161)]

Characters of *Encrinus. M.Trias.*

Encrinus LAMARCK, 1801, p. 379 [**E. liliiformis* LAMARCK, 1801, p. 379; SD ICZN, 1962 (plenary powers, Op. 636)]. Cup low basin shaped with moderate basal concavity, distinguished by perfect pentameral symmetry in arrangement of three plate circlets, absence of anal or anals, and 10 biserial, highly pinnulate arms. *M.Trias.,* Eu. (Ger.).——FIG. 476,*1*. **E. liliiformis*; lat. view of crown and proximal part of stem, ×0.7 (Bronn & Roemer, 1851-56). [=*Encrina* BRONN, 1848 *(nom. null.)*.] [=*Chelocrinus* VON MEYER, 1837, p. 316; *Flabellocrinites* KLIPSTEIN, 1845 (type, *F. cassianus*; M); *Chelocrinites* GEINITZ, 1846, p. 540 *(nom. van pro Chelocrinus); Cassianocrinus* LAUBE, 1865, p. 274 (type, *?Encrinus cassianus* LAUBE); *Porocrinus* DITTMAR, 1866, p. 392 (*non* BILLINGS, 1857, p. 279); *Traumatocrinus* WÖHRMANN, 1889, p. 290; *Beyrichocrinus* JAEKEL, 1918, p. 67 (*non* WAAGEN & JAHN, 1899) (type, *Encrinus carnalli* BEYRICH, M; *Jenaicrinus* JAEKEL, 1918, p. 67.]

Superfamily APOGRAPHIOCRINACEA Moore & Laudon, 1943

[*nom. transl.* MOORE & STRIMPLE, 1973, p. 22 (*ex* Apographiocrinidae MOORE & LAUDON, 1943a, p. 55)] [Materials for this superfamily prepared by R. C. MOORE and H. L. STRIMPLE]

Cup bowl shaped with narrow basal concavity and steep to incurved sides near rim, arm facets peneplenary, anal on squarely truncate *CD* basal only cup anal, arms five or 10, uniserial. *L.Penn.-U.Perm.*

Family APOGRAPHIOCRINIDAE Moore & Laudon, 1943

[Apographiocrinidae MOORE & LAUDON, 1943a, p. 55]

Crown medium sized, slender and subcylindrical. Cup low bowl shaped, composed of gently tumid smooth plates with slightly indented sutures; base gently concave, with horizontal infrabasals not visible from side; radials with facets less than greatest width of plates, leaving small interradial prongs, usually with finely tuberculose extrafacetal areas surrounded by fine thin line; single anal *X* plate resting on squarely truncate tip of posterior *(CD)* basal and followed by two equidimensional plates well above cup summit. Arms ten (five in one late genus), branching on slightly elongated primibrachs *1,* brachials rectiuniserial or slightly cuneate, pinnulate. Anal sac small, cylindrical, extending slightly above midheight of arms. Stem transversely circular, noncirriferous, heteromorphic. *L.Penn.-(Morrow.)-U.Perm.*

The Apographiocrinidae are similar to poteriocrinitid stock in having peneplenary radial articular facets, but in other respects are more closely aligned with scytalocrinid-decadocrinid stock. The radial articular facets are advanced in evolutionary status, as indicated by their slope, so that the pronglike divisions bordering the facets are thought to be a specialized advanced feature rather than retention of the primitive structure found in Poteriocrinitidae. This is based on their being functional with the proximal sides of the primibrachs resting on them and the anal sac attached to their inner facets. A single anal plate occurs in the cup. Presence of numerous pinnules is considered advanced, but the uniserial arrangement of the brachials is primitive. Reduction in number of arms from ten to five in one late genus is considered to be specialization but could be construed as atavistic.

Apographiocrinus Moore & Plummer, 1940, p. 115 [**A. typicalis*; OD]. Crown relatively tall and slender, subcylindrical, with arms generally held closely together. Cup small, slightly truncate bowl shaped, with small well-defined basal concavity; plates of cup slightly bulbous in typical examples; 5 small infrabasals, forming pentagon at bottom of basal concavity, nearly covered by stem; basals 5, subequal, except posterior basal, which is distinctly larger than others and truncated distally for contact with anal *X*, only very small proximal part of basals involved in basal concavity of cup; radials 5, pentagonal, length about two-thirds of width; transverse convexity giving scalloped appearance to summit of radials in ventral or dorsal view of cup; single anal plate in cup between posterior radials, resting on truncated tip of posterior basal, about 0.5 of its height generally raised above summit of radials followed by 2 equidimensional sac plates. Ten arms, composed of quadrangular or very slightly cuneate uniserial segments approximately uniform in size throughout length, branching isotomously on axillary primibrachs, those of left and right anterior rays slightly shorter than others, outer surface gently rounded but flat where sides of arms fit together, although some segments are marked by fine grooves and ridges for firmer union between arms, sharp angulation separating sides from outer surface of arms; pinnules not seen on lowermost parts of arms, but observed in some species throughout most of arm length, radial articular facets distinctly less than greatest width of radials in all North American forms, and extension of outer face of radials present along interradial sutures to inner border of facets where they are joined by small covering plates of tegmen, outer ligament area wide but short and rather strongly depressed, transverse ridge straight and well defined, inner ligament area sloping strongly outward-upward. Anal sac small, cylindrical, composed of irregular polygonal plates, extending slightly above midheight of arms. *L.Penn.-L.Perm.*, central USA; *U.Perm.*, USSR-Indon.(Timor).——Fig. 477,*1a-c*. **A. typicalis*, U.Penn.(Missour.); *1a*, from Plattsburg Ls., Wilson Co., Kans.; holotype crown from ant. (*AB* interray) side, ×1.8 (Moore & Plummer, 1940); *1b*, plate diagram (radials black, anal *X* stippled), enl.(Moore, n); *1c*, hypotype crown from LaSalle Ls., near Pontiac, Livingston Co., Ill., showing pinnulate arms and attached stem, ×1.2 (Strimple & Moore, 1971a). ——Fig. 477,*1d-f*. *A. exculptus* Moore & Plummer, U.Penn.(Missour., Graford F.), NE.Texas; ventral, post., and dorsal views of holotype cup, ×1.8 (Moore & Plummer, 1940).

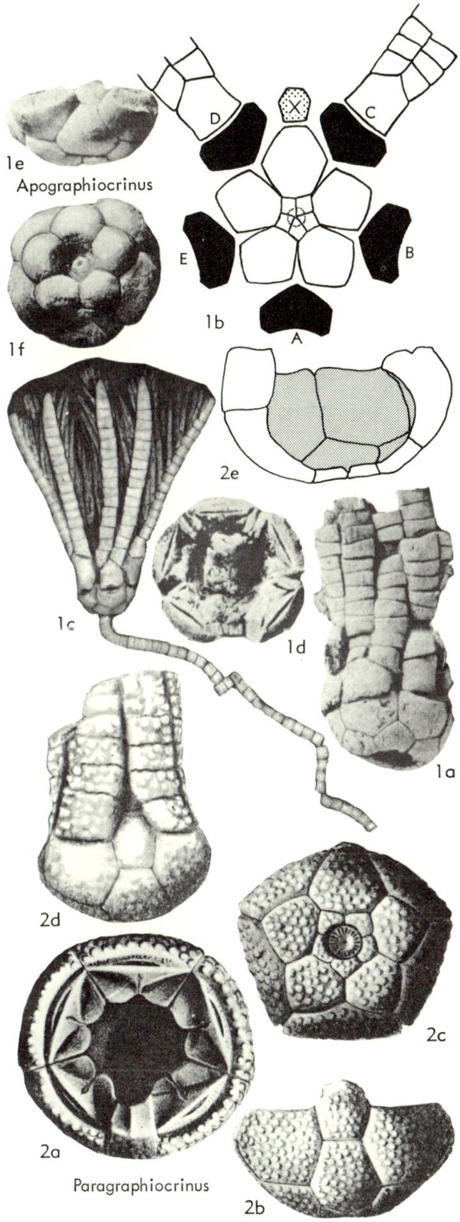

Fig. 477. Apiographiocrinidae (p. *T721-T722*).

Paragraphiocrinus WANNER, 1937, p. 173 [**Graphiocrinus exornatus* WANNER, 1916a, p. 172; OD]. Cup either smooth of highly ornate, according to WANNER, shape and plate tumidity similar to *Apographiocrinus* except in lacking projections from the exterior between radial articulating facets. Five robust arms. *U.Perm.*, Indon.(Timor).——FIG. 477,2. **P. exornatus* (WANNER), Basleo beds, Timor; *2a-c,* ventral, post., dorsal views of holotype, ×2 (Wanner, 1916a); *2d,* post. view of crown, ×1.3 (Wanner, 1937); *2e,* vert. sec. of cup through midline of *A* radial (portion of cup beyond section shaded), ×1.8 (Moore, n).

Superfamily PIRASOCRINACEA Moore & Laudon, 1943

[*nom. transl.* MOORE & STRIMPLE, 1973, p. 22 (*ex* Pirasocrinidae MOORE & LAUDON, 1943a, p. 58)] [Materials for this superfamily prepared by R. C. MOORE and H. L. STRIMPLE]

Crown pyriform; cup very low bowl shaped, with mostly deep basal concavity and rounded sides incurved at rim, arm facets peneplenary, interradial notches at rim, three anals in cup, anal sac tall, typically mushroomlike with girdle of horizontal spines around summit platform. *L.Miss.-U.Perm.*

Family PIRASOCRINIDAE Moore & Laudon, 1943

[Pirasocrinidae MOORE & LAUDON, 1943a, p. 58] [incl. Aatocrininae KNAPP, 1969, p. 374; Affinocrininae KNAPP, 1969, p. 372; Exterocrininae KNAPP, 1969, p. 380; Lasanocrininae KNAPP, 1969, p. 369; Pirasocrininae MOORE & LAUDON, 1943a (KNAPP, 1969, p. 370); Psilocrininae KNAPP, 1969, p. 375; Sciadiocrininae KNAPP, 1969, p. 377; Triceracrininae KNAPP, 1969, p. 382]

Crown compact. Cup low; base slightly to deeply concave; infrabasals five, not visible from side, radial facets wide (peneplenary) and moderately long, sloping gently outward-downward (declivate), having distinct transverse ridge, muscle, and ligament areas; three anals in cup. Anal sac prominent, mushroom shaped. Arms uniserial, branching twice or more isotomously, beginning on first primibrachs. *U.Miss.-U.Perm.*

The pirasocrinids are distinguished by their very low, almost dicoid cups with outward-downward sloping radial facets and multibranched uniserial arms, combined wherever known with a prominent flat-topped anal sac reaching to or above tips of the arms. The basal concavity of the cup may be broad and shallow or somewhat narrow and moderately to very deep. The nature of the low cup, arms, and mushroom-shaped anal sac are chiefly important characters. The horizontally outspread spines surrounding the summit of the anal sac range from short to exceptionally elongate (e.g., *Eirmocrinus* STRIMPLE & WATKINS, 1969, from the Middle Pennsylvanian of Texas).

KNAPP (1969, p. 369) stated: "Presumably, the ancestry of the pirasocrinids is in a phanocrinid-like dorsal cup, and by multiplication of the arms, lowering of the dorsal cup, and by decreasing the downflaring of the infrabasal plates, the Pirasocrinidae were established." The phanocrinid lineage is remarkably stable and, as demonstrated by KNAPP, continued with ten-armed derivatives into the Permian. It is not involved in the pirasocrinid lineage. We find no valid reason for questioning *Zeusocrinus* STRIMPLE as the probable progenitor of the pirasocrinids, as proposed by STRIMPLE (1961d, p. 21). *Zeusocrinus* has a low cup with multiple arms and a summit umbrellalike platform on the anal sac.

The family Pirasocrinidae appears to be polyphyletic, but the exact affinities of the included lineages are obscured. *Dasciocrinus,* of late Mississippian age, apparently gives rise to *Stenopecrinus rugosus* STRIMPLE of early Pennsylvanian age. *S. rugosus* has been assigned to the genus *Anchicrinus* by MOORE & STRIMPLE (1973), but has some atypical morphologic features (e.g., the cup of typical *Anchicrinus* has an almost imperceptible basal concavity and the tegmen spines have a broad, flattened base). The tegmen spines of *S. rugosus* have a narrow, rounded base and the basal concavity is pronounced. Narrow rounded tegmen spines are deemed to be primitive (i.e., like *Dasciocrinus*), and a pronounced basal concavity usually represents a more advanced state than a shallow basal concavity.

Zeusocrinus of late Mississippian age has a pronounced basal concavity (advanced?), flattened tegmen spines surrounding a small platform of polygonal plates, and agrees in all respects with Pennsylvanian genera assigned to the family.

The family Laudonocrinidae is closely related to the Pirasocrinidae. Basic stock of Laudonocrinidae is thought to have upflared infrabasals as represented by *Paianocrinus* STRIMPLE of late Mississippian age.

The cup is essentially flared cone shaped and is transformed to a planate base and then to a shallow basal concavity. Some species of *Laudonocrinus* of Missourian age can hardly be separated from young specimens of species of *Plaxocrinus,* which in turn casts some doubt on the viability of the family Laudonocrinidae.

Structural placement of plates belonging to successive circlets—infrabasals, basals, radials—of the cups of the Pirasocrinidae and similar crinoids having shallow to deep concave bases, as well as some broadly flat-bottomed forms (e.g., Anobasicrinidae), is important for purposes of taxonomic classification. This can be expressed most simply and explicitly in terms of the position of their proximal tips and somewhat less so of their distal extremities, both considered in relation to the basal plane of the cup and both determined by points on the exterior surface of the plates. Terms used are 1) **infral,** above basal plane within basal concavity of the cup, 2) **basiplanal,** in or tangent to basal plane of the cup, and 3) **supral,** above basal plane on outer side of the cup. In the *Treatise* these terms are given uniformly in the sequence: infrabasal plate, basal plate, and radial plate, which is outward from center of the stem impression. Symbols may be employed for most compact statement, minus sign (−) for infral, zero (0) for basiplanal, and plus sign (+) for supral.

Unless stated otherwise, terms and symbols for crinoid plate positions refer to proximal extremities because these seem to be more significant than distal ones. Finally, points for basal plates are located in an interradial plane at an approximate 36° angle from that of a neighboring radial plane containing extremities of infrabasal and radial plates. Points for plates of successive circlets are equally well given in longitudinal sections of cups through the midline of the *D* radial and *CD* interray or *C* radial and *AB* interray (Fig. 436). Irregularities introduced by the insertion of anal cup plates on the posterior side of the cup are ignored.

Examples of plate-position recording (all proximal tips) as explained here are for 1) *Pirasocrinus,* with very deep basal invagination (infral-infral-infral, − − −), 2) *Perimestocrinus,* with small, moderate basal concavity (infral-infrabasiplanal, − − 0), 3) *Athlocrinus,* with flat base, except for shallowly concave stem impression (infral-basiplanal-basiplanal, − 0 0), and 4) *Laudonocrinus,* with nearly flat base (infral-basiplanal-supral, − 0 +) (Fig. 436).

Key to Genera of Pirasocrinidae

A. Cup with proximal extremity of plates of all circlets confined to basal concavity (− − −) ....
- I. Concavity very deep, only radials and 3 anal plates in cup visible from side ..
  - a. Axillary primibrach *1* bulbous .... *Pirasocrinus*
  - b. Axillary primibrach *1* spinose .... *Eirmocrinus*
- II. Concavity moderately deep, only radials, *CD* basal and 3 cup anals visible from side .... *Sciadiocrinus*
- III. Concavity moderately deep, all basals partly visible from side; plates of cup and arms smooth, interradial notches prominent .... *Aatocrinus*

B. Cup with proximal extremities of infrabasals and basals confined to basal concavity, those of radials to basal plane (− − 0)
- I. Concavity moderately deep and wide, proximal part of basals transversely concave ....
  - a. Plates tumid, sutures impressed .... *Affinocrinus*
  - b. Plates not tumid, sutures not impressed .... *Platyfundocrinus*
- II. Plates tumid, sutures impressed, interradial notches narrow ....
  - a. Basals thick throughout their length, with proximal edges markedly overhanging infrabasals .... *Stenopecrinus*
  - b. Proximal margins of basals not overhanging infrabasals .... *Vertigocrinus*
- III. Concavity shallow and wide, plates not tumid, sutures not impressed ....
  - a. Interradial notches prominent, *CD* interray wide .... *Metaffinocrinus*
  - b. Interradial notches and *CD* interray narrow, rounded protuberances at middle of radials .... *Lasanocrinus*
  - c. No interradial notches, *CD* interray narrow, rounded median protuberances on both basals (as in *Utharocrinus*) and radials (as in *Lasanocrinus*) .... *Metutharocrinus*
- IV. Concavity shallow and narrow, *CD* interray narrow ....
  - a. Plates tumid, sutures impressed .... *Schedexocrinus*
  - b. Plates not tumid, sutures not im-

pressed ....
1. Interradial notches wide and prominent .... *Retusocrinus*
2. Interradial notches narrow but well defined, *CD* interray narrow .... *Simocrinus*

C. Cup with proximal extremities of infrabasals and basals confined to basal concavity, those of radials above basal plane on outer side of cup (− − +) ....
I. Cup moderately high bowl shaped ....
a. Concavity deep and wide, plates tumid, *CD* interray wide .... *Exterocrinus*
b. Concavity shallow and narrow ......
1. Plates tumid, *CD* interray narrow
a.) Interradial notches well marked, plates lacking protuberances .... *Separocrinus*
b.) Interradial notches small, tumid plates .... *Metaperimestocrinus*
c.) Interradial notches small and weak, prominent rounded protuberances on radials .... *Triceracrinus*
2. Plates not tumid, *CD* interray wide, interradial notches prominent .... *Perimestocrinus*
II. Cup low, saucer shaped, plates not tumid, concavity wide ....
a. Concavity wide and very shallow, sutures impressed, base of cup nearly flat, interradial notches wide .... *Plaxocrinus*
b. Concavity wide and moderately deep, *CD* interray narrow ...... *Polygonocrinus*
c. Concavity moderate or narrow and shallow, small interradial notches ..
1. Distal spine plates of anal sac meet at base .... *Dasciocrinus*
2. Distal spine plates of anal sac surround small platform of polygonal plates .... *Zeusocrinus*
d. Concavity wide to somewhat narrow and shallow, small interradial notches, *CD* interray narrow, pointed protuberances directed obliquely downward developed on basals .... *Utharocrinus*
III. Cup low, plates not tumid, concavity narrow and shallow, well-defined interradial notches, wide *CD* interray .... *Psilocrinus*

Pirasocrinus Moore & Plummer, 1940, p. 235 [**P. scotti*; OD]. Crown tall, relatively slender, and widest at height of secundaxils, total height about 3 times greatest width; isotomous branching at uniform heights in each ray, combined with strongly bulbous nature of axillary brachials, provides a pagodalike, turreted appearance; horizontally placed sac spines spread outward above arm tips at crown summit. Cup very low bowl shaped with strongly concave base (formula for proximal tips of cup plates, − − −); except in posterior interray only radials visible in side view of cup; height of cup only about 0.1 that of crown and, because proximal brachials are considerably more bulky than any cup plates and project well beyond outer edges of radials, cup appears relatively small. Infrabasals almost or entirely covered by stem at base of deep central concavity, 5 subequal basals strongly downflared throughout their length and all pentagonal except posterior basal, which is hexagonal and truncated distally for contact with anal *X*, distal tip of *CD* basal visible in side view. Radials moderately convex transversely and extremely so longitudinally, with proximal parts sloping steeply into basal concavity and distal parts forming subvertical sides of cup; articular facets occupying slightly less than full width of radials, their plane sloping distinctly outward-downward; 3 anal cup plates in normal position, elongate radianal lying parallel to *CD* basal and narrowly touching *BC* basal, large anal *X* and right tube plate, former touching distal tip of *CD* basal. Anal sac formed of relatively strong plates rising to summit of crown and roofed by few broad plates that project laterally in form of spines. Arms long, rounded, branching isotomously at regular intervals, composed of uniserial externally rectangular segments that are much wider than long; axillary brachials bulbous and much thicker than succeeding segments; arms in each ray bifurcating typically 4 times, so that at top of crown maximum of 80 slender branches may occur; first 3 divisions of each ray very uniform, but fourth less regular; some branches apparently undivided above third dichotom, slender pinnules borne by at least upper arm branches; all specimens show arms parallel and closely adjoining. Stem transversely round. *M.Penn.(Desmoines.)*, USA(Texas).——Fig. 436, *1*; 478,*2*. **P. scotti*, Millsap Lake F., Hood County, Texas; 436,*1a,b*, diagram. secs. of cup through midlines of ant. and left post. *(D)* radials (portions of cup beyond section shaded), ×1 (Moore, n); 478,*2a*, plate diagram of cup (radials black, radianal cross ruled, anal *X* stippled) (Moore, n); 478,*2b,c*, *C*-ray and dorsal side of holotype crown and paratype cup, with *CD* interray directed upward, ×1 (Moore & Plummer, 1940).

Aatocrinus Moore & Plummer, 1940, p. 213 [**Zeacrinus? robustus* Beede, 1900b, p. 22; OD]. Cup very low, with inner margin of radial articular facets about as high above outer summit of radials as latter are above basal plane of cup; central part of base strongly depressed but distal portions of basals and radials flare upward and are not involved in basal concavity, plates somewhat tumid (formula for proximal tips of cup plates, − − −); infrabasals subhorizontal or slightly

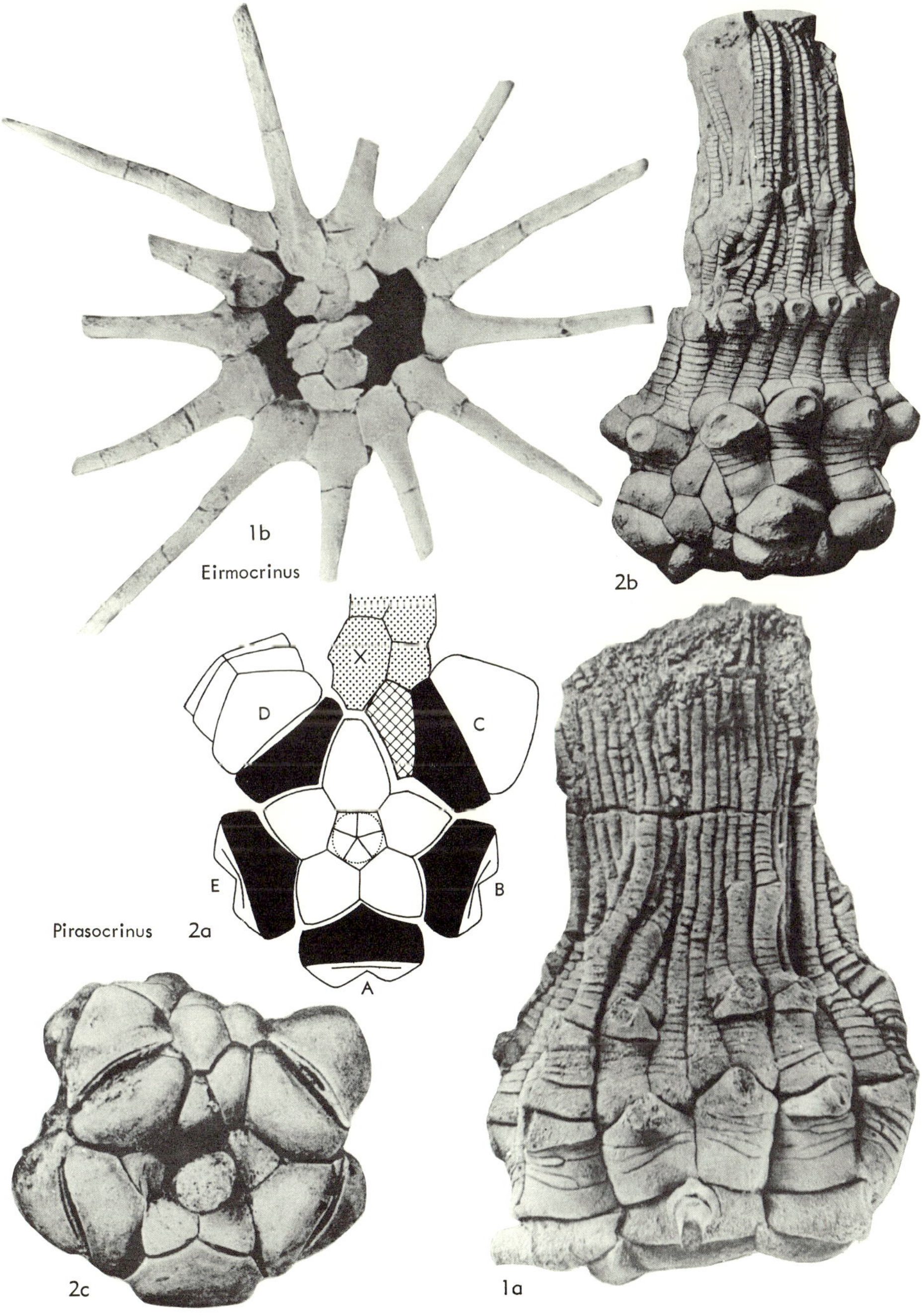

FIG. 478. Pirasocrinidae (p. *T724*, *T726*).

downflared forming small pentagon covered largely by stem; 5 basals subvertical near proximal margin, curving outward-downward to subhorizontal beyond midlength and then curving upward, strongly convex in longitudinal profile and moderately convex transversely; distal parts of these plates visible in side view; *CD* basal equal in width to others but longer, *BC* basal largest and distinctly widest, radials relatively large, width about twice length, proximal portion slightly downflaring in longitudinal profile, curved upward to margin of facetal ligament area in distal portion; articular radial facets slightly less than maximum width of plates, but because of lateral projection of radials beyond points of union with neighbors well-defined notches occur between plates, and outer surface of cup is confluent with space along inner articular suture not covered by facets, which are declivate at angle of 30° to 40°; *CD* interray narrow and rather deeply indented, containing 3 anal plates, with radianal in contact with *BC* basal, anal *X* resting on truncated tip of *CD* basal, about 0.5 its length extending above facet of *D* radial, with right tube plate on *X*, radianal reaching only slightly below *C* radial. Anal sac unknown. Stem transversely round. [This genus differs from *Pirasocrinus* in its shallower basal concavity and from *Perimestocrinus* in its much deeper concavity.] *M.Penn.(Desmoines.)-U.Penn.(Missour.),* N.Am.(Midcontinent region).——FIG. 436,*4*; 479,*3*. **A. robustus* (BEEDE), U.Penn.(Missour.), Kansas City, Mo.; 436,*4*, vert. sec. through midline of *A* radial (portions of cup beyond section shaded) (Moore & Plummer, 1940, mod.); 479,*3a-c*, dorsal, post., and ventral views of cup; ×1 (Moore & Plummer, 1940).

Affinocrinus KNAPP, 1969, p. 372 [**A. concavus*; OD]. Cup low bowl shaped, with smooth tumid plates and impressed sutures; proximal extremities of infrabasals and basals confined to moderately deep and wide basal concavity, those of radials in basal plane of the cup (formula, – – 0); proximal parts of basals transversely concave, their distal tips visible in side view; summit outer parts of radials bearing arcuate impressed fore-facet, planes of articular facets moderately declivate; interradial notches shallow; 3 anal plates in narrow *CD* interray. Arms unknown. Stem impression circular. *L.Penn.-M.Penn.(Atokan-Desmoines.),* USA(Mo.).——FIG. 480,*2*. **A. concavus,* Atokan(Burgner F.), Jasper Co., Mo.; long. sec. of cup through midline of *D* ray (portion of cup beyond section shaded), ×2 (Knapp, 1969).

Dasciocrinus KIRK, 1939, p. 472 [**Cyathocrinus florialis* YANDELL & SHUMARD, 1847, p. 24; OD]. Crown subcylindrical. Cup low, bowl or saucer shaped with invaginated base; infrabasals small, concealed by column within small basal pit; basals relatively small, proximal portions taking part in basal pit; radials large; articular facet extending full width of radial, linear, gaping; 3 anal plates in cup, radianal large, elongate, penetrating cup deeply and resting upon *BC* basal, anal *X* relatively small, meeting *CD* basal narrowly. Anal sac taller than arms with spinose processes (mostly 3 or 4) common on distal plates; anal opening lateral near distal end of sac. Arms uniserial, long, slender, endotomous, with few divisions; primibrachs *1* axillary in all rays, tending to be nodose or spinose; next branching high above primaxil, so that secundibrachs are numerous; all axillaries nodose and other brachials cuneate. *U.Miss.(Chester.),* USA(Ill.-Ky.-Okla.).——FIG. 446,*1*. **D. florialis* (YANDELL & SHUMARD), Ky.; *1a,b,* complete crown showing distal end of sac above arm tips and distal position of sac with spines and anal opening, ×1 (Springer, 1926b).

Eirmocrinus STRIMPLE & WATKINS, 1969, p. 204 [**E. grossus*; OD]. Crown robust, pear shaped, widest at about 0.25 of its height. Cup extremely low, saucerlike; deeply concave at base, only radials and 3 long, narrow anal plates visible from side. Arms uniserial (except for unevenly biserial arrangement of secundibrachs), branching isotomously on primibrachs *1*, which project laterally in stout spines, similar branching on secundibrachs *4-6* and endotomously on outer sides of each half-ray at 4 or 5 higher levels, all arms in upper part of crown very slender. Anal sac mushroom shaped, with expanded nearly flat summit above arm tips, approximately 13 long slender spines directed outward from central area composed of moderately large polygonal plates. *M. Penn.(Desmoines.),* USA(Texas).——FIG. 478,*1*. **E. grossus,* Millsap Lake F.(Strawn Gr.), Hood Co., Texas; *1a,b, B*-ray view of crown and summit of anal sac from above, ×0.8 (Strimple & Watkins, 1969).

Exterocrinus KNAPP, 1969, p. 380 [**Perimestocrinus pumilis* MOORE & PLUMMER, 1938, p. 281; OD]. Cup moderately deep bowl shaped with wide and deep basal concavity, plates tumid, proximal tips of infrabasals and basals confined to concavity and of radials distinctly above basal plane on outer side of cup (formula, – – +); articular radial facets strongly declivate; 3 cup anal plates in wide *CD* interray. Anal sac and arms unknown. Stem impression circular. *L. Penn.(Morrow.),* USA(Okla.).——FIG. 480,*3*. **E. pumilis* (MOORE & PLUMMER), Bloyd F.; *3a,b,* dorsal and post. views, ×2 (Knapp, 1969); *3c,* long. sec. of cup through midline of *D* radial (portion of cup beyond section shaded), ×2 (Moore, n).

Lasanocrinus MOORE & PLUMMER, 1940, p. 181 [**Hydreionocrinus daileyi* STRIMPLE, 1940a, p. 5; OD]. Cup very low, almost discoidal, outline distinctly pentagonal in dorsal or ventral view but unlike most with angles of pentagon in midportion of the radials, greatest width of cup below midheight or at basal plane, outline of cup in side view subrectangular to trapezoidal, depending on orientation; sutures not impressed but distinct, surfaces of adjoining plates smoothly confluent; base of cup broadly and shallowly concave or marked by central moderate deep concavity that involves infrabasals and proximal parts of basals, proximal extremities of radials in basal plane of

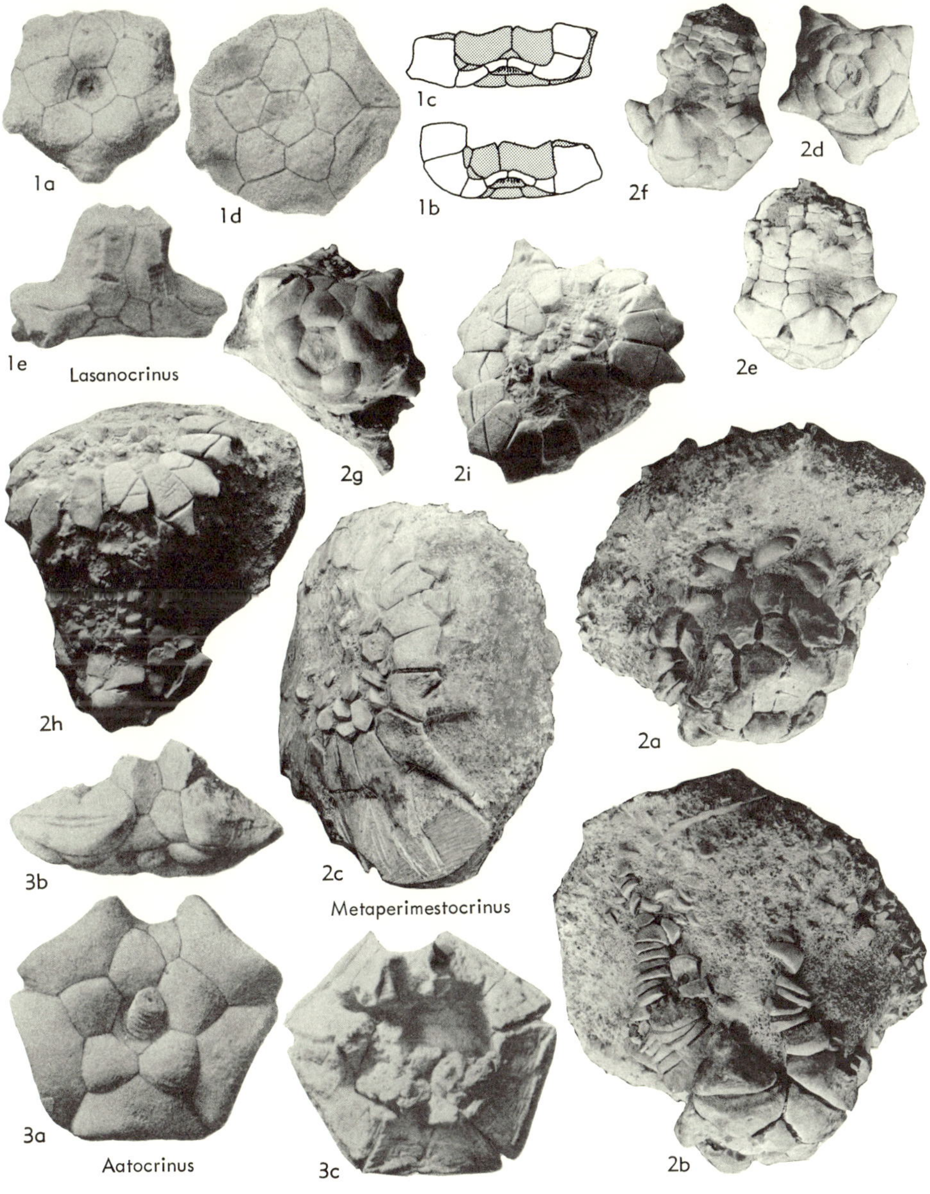

FIG. 479. Pirasocrinidae (p. *T724-T729*).

cup (formula, – – 0); infrabasals forming regular pentagon twice stem impression in diameter; subequal basals with subhorizontal midportions and distal parts gently upflared, *CD* basal slightly longer than others and truncated distally, for contact with anal *X*; radials strongly bulbous or spinose, articular facets sloping outward-downward, width of facets slightly less than maximum width of radials, transverse ridge and ligament areas well defined on facets; anal plates 3, elongate radianal with narrower end typically in contact with *BC* basal. Anal *X* on truncated distal extremity of *CD* basal obliquely above radianal; right tube plate relatively large and in normal contact with anal *X* and radianal. Arm structure unknown. Stem unknown but impression of it round. [Distinguished from other genera of low discoidal form by distinctive shape of cup and

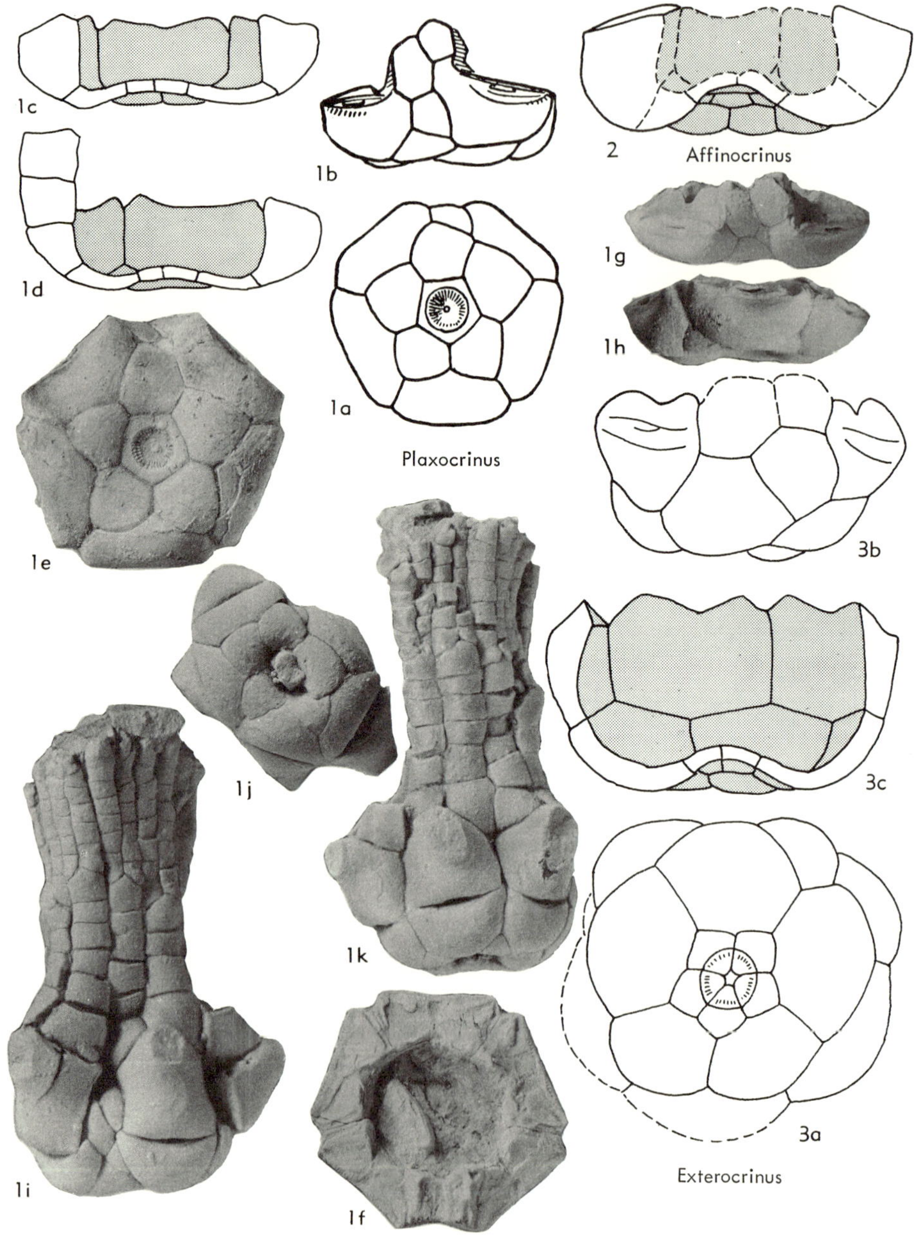

FIG. 480. Pirasocrinidae (p. *T726, T731*).

nearly horizontal attitude of radial facets.] *L. Penn.(Morrow.),* USA(Okla.-Texas).——FIG. 479, *1a-c.* **L. daileyi* (STRIMPLE), Okla.; *1a,* dorsal view of cup, ×2; *1b,c,* long. secs. of cup through midlines of *A* and *D* radials; ×2 (Moore & Plummer, 1940).——FIG. 479,*1d,e.* *L. cornutus* MOORE & PLUMMER, Marble Falls Ls., San Saba Co., Texas; holotype, dorsal and post. views, ×2 (Moore & Plummer, 1940).

Metaffinocrinus KNAPP, 1969, p. 373 [**Plaxocrinus perundatus* MOORE & PLUMMER, 1940, p. 193; OD]. Cup low basin shaped, with shallow wide basal concavity, proximal extremities of infrabasals and basals within concavity, those of radials in basal plane of cup (formula, – – 0), plates not tumid and sutures not impressed; interradial notches prominent; 3 anal plates in wide *CD* interray. Arms uniserial, branching isotomously on large, convexly rounded first primibrachs, secundibrachs *2*, and possibly at higher levels but unknown. Stem transversely round. *M.Penn. (Desmoines.),* USA(Okla.-Texas).——FIG. 481,*1*. **M. perundatus* (MOORE & PLUMMER), Millsap Lake F.(Strawn Gr.), Hood Co., Texas; *1a,b,* long. secs. of cup through midlines of *A* and *D* radials (portion of cup beyond section shaded), ×1.8; *1c-f,* dorsal, post., ant., ventral views of holotype; *1g,* post. side of paratype obliquely from below, showing proximal part of arms, all ×1.2 (Moore & Plummer, 1940).

Metaperimestocrinus STRIMPLE, 1961, p. 37 [**M. spiniferus*; OD]. Cup low, truncate bowl shaped with mildly depressed basal area; infrabasals forming pentagonal, subhorizontal disc; proximal edges of basals within depression, only distal tips visible in side view; radials wider than long, entirely outside basal area (formula, – – +); anals usually 3 within cup, in normal arrangement (primitive type); anal sac umbrella shaped with wide summit composed of large flattened spines surrounding elliptical central area of many small plates. Arms uniserial, with axillary first primibrachs protruded as short, blunt spines, subsequent brachials externally rectangular, rounded in lower arms but rather flattened in higher ones; second bifurcation on secundibrachs *4* to *6*, which may be nodelike or bear short spine; other bifurcations on tertibrachs *7* to *10* in outer arms of each ray indicating minimum of 30 endotomous arms. Stem attachment circular in outline and crenulated. [Only closely comparable genus is *Stenopecrinus* which has a deeper and narrower basal concavity, proximal brachials extended as long slender spines, and terminating spines of anal sac less numerous.] *M.Penn.(Desmoines.),* N.Am.(Midcontinent region).——FIG. 479,*2*. **M. spiniferus,* up.Desmoines.(Holdenville F.), near Beggs, Okla.; *2a-c,* oblique post., side, and ventral views of large paratype, ×1; *2d-f,* dorsal, ant., and post. views of small paratype, ×1; *2g-i,* dorsal, lat., and ventral views of holotype, ×1 (Strimple, n; Univ. Oklahoma OU 4015, holotype; OU 4013, paratype).

Metutharocrinus MOORE & STRIMPLE, 1973, p. 71 [**M. cockei*; OD]. Cup and median protuberances on basals as in *Utharocrinus* combined with similar protuberances on radials as in *Lasanocrinus*; 3 anal plates in narrow gently convex *CD* interray; no interradial notches. *L.Penn.(Morrow.),* USA(Ark.-Okla.).——FIG. 482,*5*. **M. cockei,* Bloyd F., Crawford Co., Ark.; *5a-d,* holotype, dorsal, ant., post., ventral views, ×4.5 (Moore & Strimple, 1973).

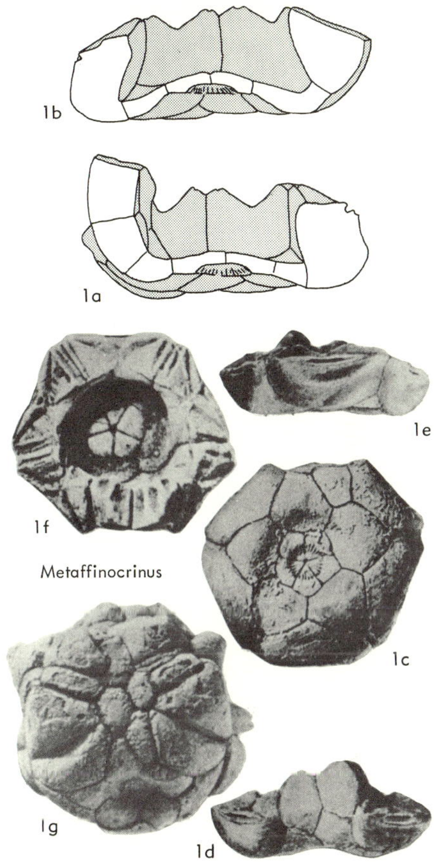

FIG. 481. Pirasocrinidae (p. *T728-T729*).

Perimestocrinus MOORE & PLUMMER, 1938, p. 280 [**Hydreionocrinus nodulifer* MILLER & GURLEY, 1894b, p. 41; OD]. Crown moderately tall, cylindrical. Cup low, bowl shaped with flaring sides, base with sharply impressed concavity but shallow; infrabasals mostly small, curved by stem impression at bottom of basal concavity; basals subequal, proximally included in basal concavity but main part forming outer lower slope of cup; radials large, articular facets peneplenary, declivate, small interradial notches (formula for proximal plate extremities of cup circlets, – – +); anals 3. Anal *X* and right tube plate projecting well above summit of radials in wide *CD* interray. Arms poorly known; first primibrachs axillary, secundibrachs externally rectangular, uniserial, no branching observed above primibrachs. Stem transversely round. *M.Penn.(Desmoines.)-L.Perm.,* USA(Kans.-Mo.-Ill.-Okla.-Texas).——FIG. 436,*2*. *P. excavatus* (WELLER), L.Perm.(Wolfcamp., Cibolo Ls.), Presidio Co., Texas; *2a,b,* vert. secs. of cup through midlines of *A* and *D* radials (portions of cup

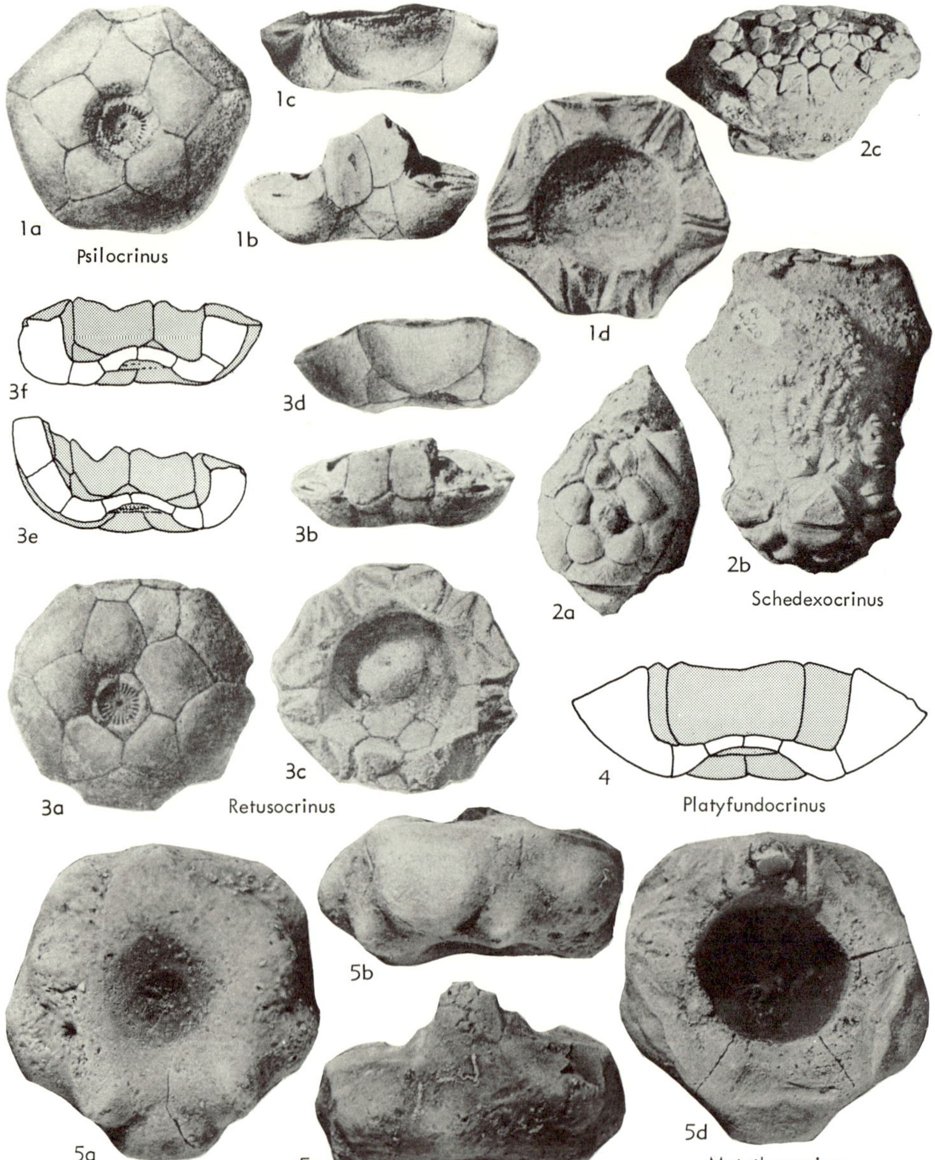

FIG. 482. Pirasocrinidae (p. T729-T732).

beyond section shaded) (Moore & Plummer, 1940, mod.).——FIG. 483,*1a-c*. **P. nodulifer* (MILLER & GURLEY), Missour.(Argentine Ls.), Kansas City, Mo.; post., ant., dorsal views, ×2 (Moore & Plummer, 1938).——FIG. 483,*1d-f*. *P. formosus* MOORE & PLUMMER, Missour.(Martins Lake Ls., Palo Pinto F.), Palo Pinto Co., Texas; dorsal, post., ventral views of cup, ×2 (Moore & Plummer, 1940).——FIG. 483,*1g-j*. *P. impressus* MOORE & PLUMMER, Desmoines.(Mineral Wells F.), Palo Pinto Co., Texas; dorsal, post., ventral, ant. views of cup, ×2 (Moore & Plummer, 1940). [=*Peremistocrinus* STRIMPLE, 1948a (*nom. null.*).]

Platyfundocrinus KNAPP, 1969, p. 370 [**P. typus*; OD]. Similar to *Affinocrinus* except that cup plates are not tumid or sutures impressed, basal concavity wide and shallow (proximal extremities of circlet plates – – 0). *M.Penn.(Atokan),* USA (Mo.).——Fig. 482,*4*. **P. typus,* Burgner F., Jasper Co., Mo.; long. sec. of cup through midline of *D* radial (portion of cup beyond section shaded), ×2 (Moore, n).

Plaxocrinus Moore & Plummer, 1938, p. 277 [**Hydreionocrinus crassidiscus* Miller & Gurley, 1894, p. 43; OD]. Crown moderately tall, cylindrical. Cup very low, saucerlike; with wide, shallow basal concavity; proximal extremities of infrabasals and basals within concavity, those of radials on outer side of cup above basal plane (formula, – – +); infrabasals largely covered by stem impression but not strongly depressed; basals subequal, interbasal sutures subhorizontal, distal parts of basals nearly horizontal; radials moderately convex, articular facets less than width of plates, declivate; interradial notches distinct; anal plates 3 in cup, thick, forming rather prominent protuberance at posterior side of cup. Arms branching isotomously on primibrachs *1* and secundibrachs *4-6,* axillary primibrachs spine-bearing. Stem transversely round. *M.Penn.(Atokan)-U.Penn.(Virgil.),* N.Am.——Fig. 480,*1a-d*. **P. crassidiscus* (Miller & Gurley), U.Penn.(Missour., Argentine Ls.), Kansas City, Mo.; *1a,b,* dorsal, post. views of cup, ×1.3 (Moore, 1939c); *1c,d,* long. secs. of cup through midlines of *D* and *A* radials (portion of cup beyond section shaded), ×2 (Moore, n).——Fig. 480,*1e-h*. *P. modestus* Moore, Virgil.(Brownville Ls.), Osage Co., Okla.; dorsal, ventral, post., ant. views of cup, ×1.35 (Moore, 1939c).——Fig. 480,*1i-k*. *P. politus* Moore, U.Penn.(Barnsdall F.), Mo.; post., dorsal, ant. views of crown, ×2.7 (Moore, 1939c).

Polygonocrinus Strimple, 1961, p. 59 [**P. multiextensus*; OD]. Crown widely expanded and relatively short. Cup low, broad, truncate bowl shaped, with deep basal concavity; infrabasals forming subhorizontal disc at base of concavity; basals confined to concavity although their distal tips may be visible in side view, owing to deep, V-shaped depressions between radials; anal plates 3 (anal *X,* radianal, right tube plate) in normal primitive arrangement. Anal sac with wide summit platform higher than arms (in type species composed of 32 small, polygonal plates surrounded by 14 large, flattened, laterally directed spines). Arms as many as 70, uniserial or partly biserial, with isotomous branching, large axillary first primibrachs; secundibrachs *2* and *3* normally low and interlocking (biserial); second bifurcation usually on fourth secundibrachs, and another regular branching on about tertibrachs *10* in most rays and further branching in some rays in irregular pattern; axillary brachials above first division mildly tumid. Stem transversely round. *M. Penn.(Desmoines.),* USA(Okla.).——Fig. 483,*2*. **P. multiextensus,* Holdenville F., Okmulgee Co., Okla.; *2a,b,* dorsal view of cup, summit of anal sac, ×0.4 (Strimple, 1961d).

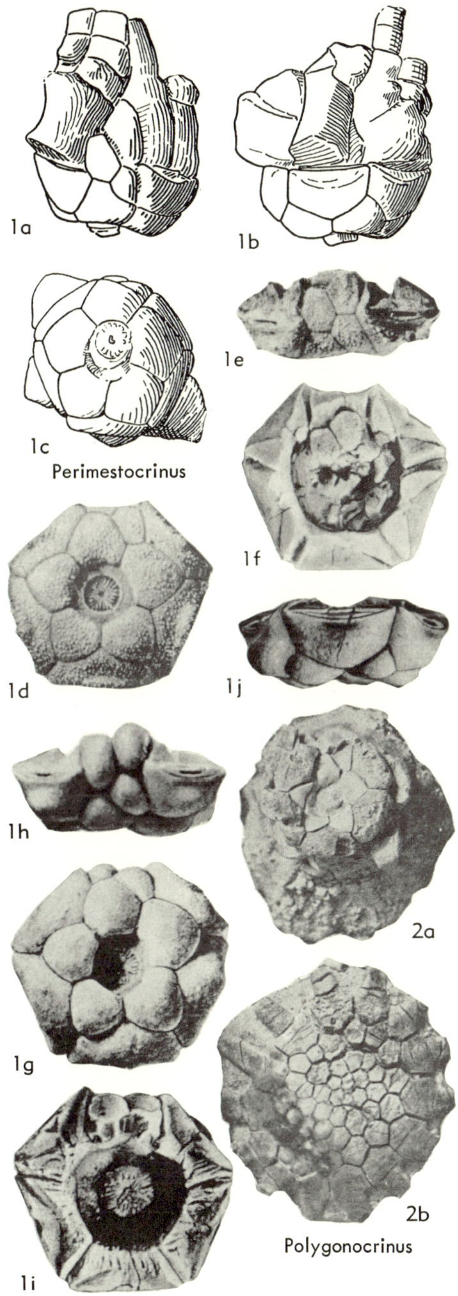

Fig. 483. Pirasocrinidae (p. *T*729-*T*731).

Psilocrinus KNAPP, 1969, p. 375 [**Plaxocrinus omphaloides* MOORE & PLUMMER, 1940, p. 194; OD]. Cup low, saucerlike, hexagonal outline in basal and summit views, plates smooth and sutures not impressed; narrow, shallow basal concavity containing small infrabasal circlet and proximal tips of basals, radials entirely above basal plane of cup; peneplenary articular facets on radials, narrow but distinct interradial notches; 3 large anal plates in cup, *CD* interray wide. Arms and anal sac unknown. Stem transversely circular. *M. Penn.(Desmoines.),* USA(Texas).——FIG. 482,*1*. **P. omphaloides* (MOORE & PLUMMER), Mineral Wells F., McCulloch Co., Texas; *1a-d,* dorsal, post., ant., ventral views of holotype, ×2 (Moore & Plummer, 1940).

Retusocrinus KNAPP, 1969, p. 375 [**Plaxocrinus lobatus* MOORE & PLUMMER, 1940, p. 196; OD]. Cup low bowl shaped, with shallow wide basal concavity, hexagonal outline in dorsal or ventral view owing to wide *CD* interray, plates not tumid, interradial notches wide and prominent; proximal tips of radials in basal plane of cup, those of other plate circlets without basal concavity (formula, – – 0); 3 anals in cup. Stem impression circular, arms unknown. *M.Penn.(Desmoines.),* USA(Texas).——FIG. 482,*3*. **R. lobatus* (MOORE & PLUMMER), Mineral Wells F.(Strawn Gr.), McCulloch Co., Texas; *3a-d,* dorsal, post., ventral, ant. views of holotype; *3e,f,* long. secs. of cup through midlines of *A* and *D* radials (portion of cup beyond section shaded), all ×2 (Moore & Plummer, 1940).

Schedexocrinus STRIMPLE, 1961, p. 27 [**S. gibberellus*; OD]. Cup low, truncate, bowl shaped, with narrow, shallow basal concavity; infrabasals in bottom and basals forming sides of concavity, with distal parts of latter commonly curved upward enough to be visible in side view; wide radials with erect distal portions (formula for proximal tips of circlet plates, – – 0); anal plates 3. Anal sac mushroomlike, extending above arms, summit typically composed of 15 long, flat spines directed outward and surrounding platform of about 32 small polygonal plates. Arms normally 40 or more, endotomous, uniserial or partially biserial. Stem circular. *M.Penn.(Desmoines.),* N. Am.(Midcontinent region).——FIG. 482,*2*. **S. gibberellus,* Holdenville F., near Beggs, Okla.; *2a-c,* dorsal, ant. views of crown, and summit of anal sac (holotype), ×0.45 (Strimple, 1961d).

Sciadiocrinus MOORE & PLUMMER, 1938, p. 275 [**Zeacrinus (Hydreionocrinus) acanthophorus* MEEK & WORTHEN, 1870, p. 28; OD]. Crown short, somewhat expanded. Cup very low, with moderately deep basal concavity; similar to *Retusocrinus* except in confinement of radial proximal extremities to concavity and in having only distal tip of *CD* basal visible from side of cup, (formula, – – –); interradial notches distinct; anal sac mushroom shaped, cylindrical below, with large anal vent, summit wide and flat topped, composed of numerous small polygonal plates peripherally bordered by laterally directed marginal spines. Arms uniserial, rounded, branching isotomously, twice or more in each ray. Stem transversely round. *L.Penn.-U.Penn.(Missour.),* N.Am.——FIG. 484,*1a-f*. **S. acanthophorus* (MEEK & WORTHEN), U.Penn.(Nodaway Ls.), Ill.; *1a,b,* dorsal, post. sides of crown (radials black), ×1.5; *1c,d,* ant. view of same showing mushroom-shaped anal sac with large opening (radials black) and widely expanded summit, ×1.5; *1e,f,* long. secs. through midlines of *A* and *D* radials (radials black) (Moore & Plummer, 1940).——FIG. 484,*1g-i*. *S. disculus* MOORE & PLUMMER, Missour.(Merriman Ls., Graford F.), Palo Pinto Co., Texas; dorsal, post., ventral views of cup, ×1.5 (Moore & Plummer, 1940).——FIG. 484,*1j,k*. *S. harrisae* MOORE & PLUMMER, Desmoines.(Millsap Lake F.), Hood Co., Texas; post., dorsal views of cup, ×1.5 (Moore & Plummer, 1940).——FIG. 484, *1l,m*. *S. confertus,* Desmoines.(Millsap Lake F.), Parker Co., Texas; summit of anal sac, arms in typical ray, ×0.75 (Moore & Plummer, 1940).

Separocrinus KNAPP, 1969, p. 382 [**Plaxocrinus praevalens* MOORE, 1939c, p. 229; OD]. Cup low bowl shaped, with narrow and shallow basal concavity, tumid plates and well-marked interradial notches, *CD* interray narrow; proximal tips of radials above basal plane, on outer side of cup, those of other plate circlets within basal concavity (formula, – – +); anal plates in cup 3; anal sac and arms unknown. Proximal columnals of stem circular. *U.Penn.(Virgil.),* USA(Okla.).——FIG. 484,*2a-e*. **S. praevalens* (MOORE), Brownville Ls., Osage Co., Okla.; *2a-c,* dorsal view of holotype, dorsal and ventral views of primibrach spine; all ×1.5 (Moore, 1939c); *2d,e,* long. secs. through midlines of *A* and *C* radial and primibrach (portions of cup beyond section shaded), ×1.5 (Moore, 1939c).

Simocrinus KNAPP, 1969, p. 375 [**Plaxocrinus modestus* MOORE, 1939c, p. 227; OD]. Cup saucer shaped, plates not tumid; radials large, with peneplenary declivate articular facets, proximal tips on outer side of cup above basal plane, those of other plate circlets within concavity (formula, – – +); interradial notches distinct; *CD* interray narrow and depressed, containing 3 anal plates, anal *X* largest. Anal sac and arms unknown. Columnals of stem circular. *U.Penn.(Virgil.),* USA(Okla.-Kans.).——FIG. 485,*3*. **S. modestus* (MOORE), Brownville Ls., Osage Co., Okla.; *3a-d,* dorsal, post., ventral, ant. views of holotype; *3e,f,* long. secs. of cup through midlines of *A* and *D* radials (portions of cup beyond section shaded), all ×1.5 (Moore, 1939c).

Stenopecrinus STRIMPLE, 1961, p. 40 [**Perimestocrinus planus* STRIMPLE, 1952, p. 787; OD].

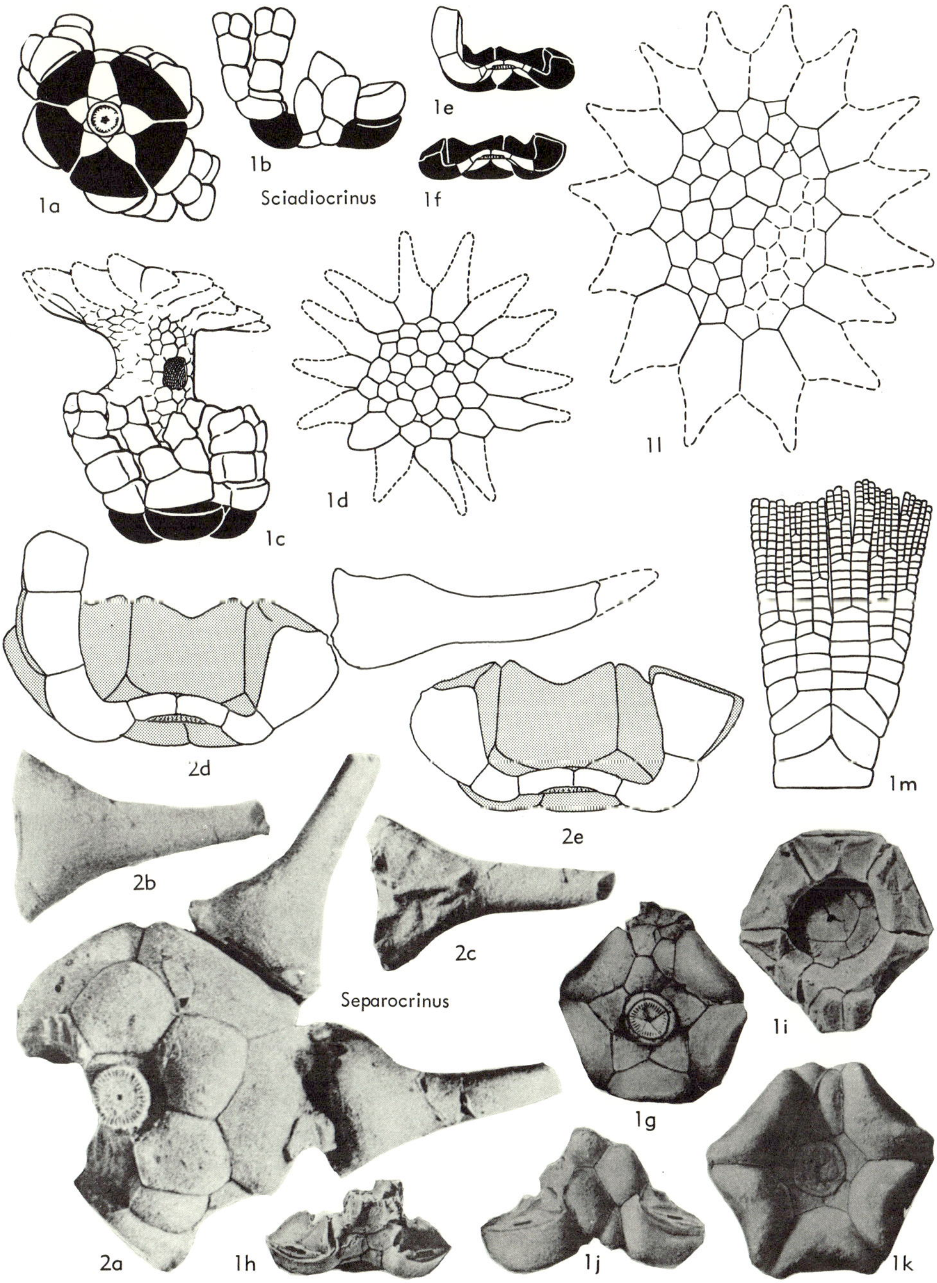

FIG. 484. Pirasocrinidae (p. *T732*).

Crown cylindrical, height medium. Cup low, bowl shaped, with rather deep, narrow basal concavity, plates tumid; infrabasals and proximal portions of basals restricted to concavity; large pentagonal radials with distal portions upright, and proximal extremities barely reaching basal plane of cup, notches at summit of interradial sutures not pronounced; cup anal plates 3. Anal sac with

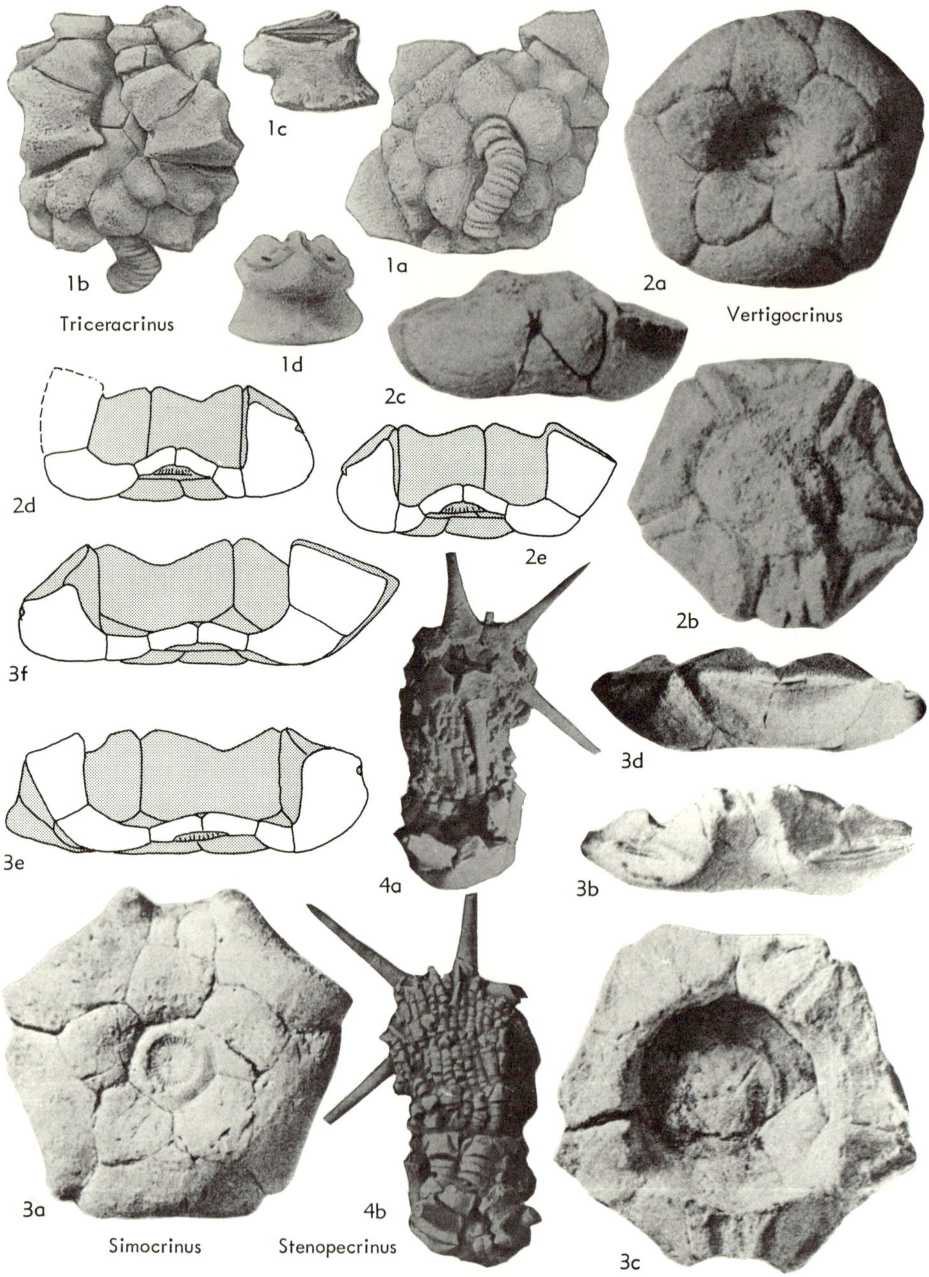

FIG. 485. Pirasocrinidae (p. *T732-T735*).

small platform at summit, bordered by very long flattened spines directed outward. Arms uniserial, endotomous, with curved exterior surfaces in distal portions, first primibrachs axillary, higher axillaries spinose. Stem transversely rounded and crenulate. *L.Penn.(Morrow.)-U.Penn.(Missour.),* N.Am.(Midcontinent region).——FIG. 485,*4*. **S. planus* (STRIMPLE), Missour.(Barnsdall F.), Washington Co., Okla.; *4a,b,* post., ant. sides of crown showing long terminal anal sac spines, ×1 (Strimple, 1961d).

Triceracrinus BRAMLETTE, 1943, p. 550 [**T. moorei*; OD]. Crown and cup resembling *Perimestocrinus*, but with cup plates much more protuberant; basal concavity narrow, proximal extremities of circlet plates disposed as in *Plaxocrinus* and *Perimestocrinus* (formula, – – +); with peneplenary articular facets leaving small interradial notches; anal plates in cup 3. Anal sac unknown. Arms uniserial, composed of laterally projecting longitudinally keeled thick brachials and especially characterized by robust axillary first primibrachs, which have broadly flattened lower surface next to gaping suture above radials. *L.Perm.(Wolfcamp.)-U.Perm.*, USA(Texas-Okla.-Kans.-Iowa).——FIG. 485,*1*. **T. moorei*, L.Perm., McCulloch Co., Texas; *1a,b*, dorsal and post. views of incomplete crown; *1c,d*, lat. and front views of first primibrach; all ×2 (Bramlette, 1943).

Utharocrinus MOORE & PLUMMER, 1938, p. 285 [**Delocrinus pentanodus* MATHER, 1915, p. 106; OD]. Cup small, low, basin shaped, with broad to somewhat narrow and shallow concave base and no constriction at upper margin of radials; infrabasals 5, suture lines scarcely discernible, projecting well beyond shallow circular depression of stem attachment; basals 5, pentagonal, except hexagonal *CD* basal, gently to strongly convex transversely, each plate bearing single massive subcentral horn-shaped node protruding outward-downward; radials pentagonal, broader than high, comparatively thick, strongly convex longitudinally; interradial notches distinct; narrow *CD* interray containing elongate radianal and 1 or 2 small additional plates. Arm structure and anal sac unknown. *L.Penn.(Morrow.)-U.Penn.(Virgil.)*, USA(Okla.-Kans.).——FIG. 486,*1a-d*. **U. pentanodus* (MATHER), Morrow., NW.Ark.; dorsal, *D*-ray, ant., post. views of cup, ×2 (Moore & Plummer, 1938).——FIG. 486,*1e-g*. *U. quinquacutus* MOORE, Virgil.(Brownville Ls.), Osage Co., Okla.; dorsal, ventral, post. views of cup, ×2.7 (Moore, 1939c).

Vertigocrinus KNAPP, 1969, p. 377 [**Perimestocrinus subtilis* MOORE, 1939c, p. 235; OD]. Cup low, basally inverted bowl shaped with tumid plates and impressed sutures; proximal extremities of infrabasals and basals confined to basal concavity, those of radials in basal plane of cup (formula, – – 0), distal edges of basals not overhanging infrabasals; interradial notches small but clearly evident; large radianal, other 2 anals in cup very small. Arms and anal sac unknown. Stem transversely round. *U.Penn.(Virgil.)*, USA (Okla.).——FIG. 485,*2*. **V. subtilis* (MOORE), Brownville Ls., Osage Co., Okla.; *2a-c*, dorsal, ventral, post. views of holotype; *2d,e*, long. secs. of cup through midlines of *A* and *D* radials (portions of cup beyond sections shaded), all ×1.5 (Moore, 1939c).

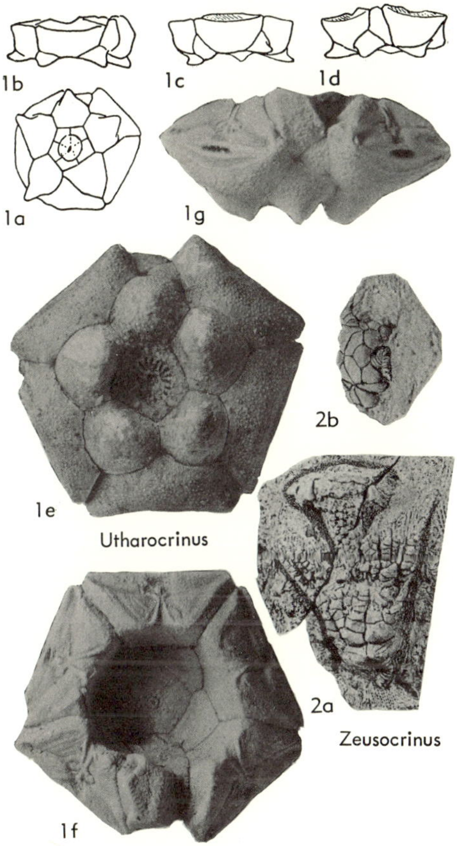

FIG. 486. Pirasocrinidae (p. *T735*).

Zeusocrinus STRIMPLE, 1961, p. 21 [**Tholocrinus foveatus* STRIMPLE, 1951d, p. 674; OD]. Diminutive with upflared compact crown. Cup low, bowl shaped with narrow basal concavity; infrabasals entirely obscured, within concavity; basals curving sharply out of concavity, readily visible in side view; radials wide, pentagonal, with proximal edges above basal plane of cup (formula, – – +); anal plates 3, in normal (primitive) arrangement; anal sac tall, mushroomlike, summit platform above arm tips, with circlet of horizontally directed, flat, spinose plates surrounding small polygonal plates. Arms uniserial, axillary brachials spinose, first bifurcation on primibrachs *1* and one other isotomous division known in some rays. Stem transversely round with alternating expanded columnals. *U.Miss.(Chester.)*, USA (Okla.).——FIG. 486,*2*. **Z. foveatus* (STRIMPLE), near Ft. Gibson, Okla.; *2a,b*, *AB*-interray view of crown and cup, ×1 (Strimple, 1951d).

FIG. 487. Adinocrinidae *(2)*; Texacrinidae *(3,4)*; Galateacrinidae *(5)*; Sellardsicrinidae *(1)* (p. *T737-T738*).

Family ADINOCRINIDAE Strimple, 1961

[Adinocrinidae STRIMPLE, 1961d, p. 63]

Crown tall, expanded upward. Cup low, bowl shaped, pentagonal outline; infrabasals in wide shallow basal concavity; basals small, confined to basal concavity and separated by proximal ends of large radials; *CD* interray narrow; 3 anal plates; anal sac unknown. Primibrachs *1* axillary in all rays except *A* ray which has 5 primibrachs, arms branch twice in most rays. Column round. *L.Miss.*

The Adinocrinidae are as advanced in cup shape and structure as the Pennsylvanian (Missourian) species *Hydreionocrinus pentagonus* MILLER & GURLEY (1890), ascribed to the genus by MOORE & LAUDON, 1941. The Pennsylvanian species is considered here to belong to the pirasocrinids *(Sciadiocrinus)*, none of which are as advanced in cup structure as the adinocrinids.

Adinocrinus KIRK, 1938, p. 161 [**Zeacrinus nodosus* WACHSMUTH & SPRINGER, 1885, p. 243; OD]. Crown expanding upward, compact. Cup depressed basin shaped, small in comparison to size of crown, sharply pentagonal in outline; proximal extremities of plates in all circlets within basal concavity (formula, ---); infrabasals in bottom of concavity covered by column or slightly projecting beyond it; basals relatively very small, typically appearing as triangles separated by radials, which may touch infrabasals, *CD* basal larger than others; radials proportionally very large, with articular facets extending nearly full width of radials, suture gaping; plane of facets sloping strongly outward-downward; radianal low and narrow, meeting *CD* basal on long lateral face; anal *X* usually high in cup, separated from *CD* basal and lying on radianal, right tube plate on sloping upper right shoulder of *X*. Anal sac unknown. Arms uniserial, primibrachs 5 in anterior ray in 3 observed specimens, primibrachs *1* axillary in all other rays; branching isotomously, with 2 divisions above main dichotom; rami very stout, with rounded backs and composed mostly of externally rectangular brachials. All axillaries nodose, and proximal brachial of each series larger than succeeding ones and strongly convex to nodose; below tertaxils processes on lateral margins of brachials interlock with similar processes of juxtaposed arms. *L.Miss.(Osag.),* USA(Ky.-Tenn.).——FIG. 487,*5*. **A. nodosus* (WACHSMUTH & SPRINGER), Keokuk, Tenn.; *A*-ray view of holotype crown, ×1 (Wachsmuth & Springer, 1886).

Superfamily TEXACRINACEA Strimple, 1961

[*nom. transl.* MOORE & STRIMPLE, 1973, p. 22 (*ex* Texacrinidae STRIMPLE, 1961d, p. 92)] [Materials for this superfamily prepared by R. C. MOORE and H. L. STRIMPLE]

Crown tall and slender, cup bowl shaped with basal concavity and steep sides near rim, arm facets plenary, one to three anals in cup, anal sac tall, composed of longitudinal rows of plates, arms long, commonly many (to 40) but five or 10 in some. *L.Miss.-U.Perm.*

Family TEXACRINIDAE Strimple, 1961

[Texacrinidae STRIMPLE, 1961d, p. 92] [incl. Texacrininae KNAPP, 1969, p. 381]

Crown tall, somewhat expanded upward, arms not appressed. Cup truncate cone or bowl shaped, usually with slightly constricted summit; infrabasals obscured in basal concavity and mostly covered by stem; five subequal, pentagonal basals except for *BC* basal which is hexagonal; five radials with length about 0.7 of width, articular facets equal to greatest width of radials; anal plates three, large radianal resting on sloping edges of *BC* and *CD* basals, also touching *CD* radial, other two plates barely reaching below summit of radials; double series of anal sac plates visible above these. Arms uniserial, branching isotomously in each ray on first primibrach, and with two or three exotomous higher divisions in each half-ray; long, slender pinnules borne by alternating brachials. [General appearance resembles *Pachylocrinus* but differs in arm structure with exotomous pattern of branching accentuated by each higher bifurcation of arms in half-rays and absence of pronounced indentations between radials.] *L.Penn.-L. Perm.*

Texacrinus MOORE & PLUMMER, 1940, p. 143 [**T. gracilis*; OD]. Characters of family. Holotype has 30 slender delicate arms with moderately long brachials. Some species have broad arms with very short brachials. *M.Penn.(Desmoines.-Missour.)-L.Perm.(Wolfcamp.),* USA(Texas-Okla.-Nev.).——FIG. 487,*1e,f*. **T. gracilis,* M.Penn. (Millsap Lake F.), Parker Co., Texas; *3e,f,* ant., post. sides of holotype crown, ×1.5 (Moore & Plummer, 1940).——FIG. 487,*1a-d*. *T. associatus* STRIMPLE, Desmoines.(Oologah Ls.), Okla.; *1a-d,* post., ant., dorsal, ventral sides of holotype cup, ×1.5 (Strimple, 1952d).

Marathonocrinus MOORE & PLUMMER, 1940, p. 240 [**M. bakeri*; OD]. Crown pyriform. Cup apparently very low, with radials about twice as wide as long and strongly convex longitudinally; cup anal plates and anal sac unknown. Arms uniserial, pinnulate, isotomously branched on first primibrach and exotomously at higher levels. *L.Penn. (?Morrow.-Bend.),* USA(Texas).——FIG. 487,*4*. **M. bakeri*; side view of crown, ×1 (Moore & Plummer, 1940).

Family GALATEACRINIDAE Knapp, 1969

[*nom. transl.* MOORE & STRIMPLE, 1973, p. 22 (*ex* Galateacrininae KNAPP, 1969, p. 382)]

Crown low, outspread laterally, with numerous, well-separated, cuneate uniserial arms, which are pinnulate. Cup discoid, mainly characterized by broadly flangelike nature of radials, infrabasals in concavity surrounded by bulbous basals, not visible from side. Stem transversely circular. *M. Penn.(Desmoines.)-U.Penn.(Virgil.).*

Galateacrinus Moore, 1940, p. 45 [**G. stevensi*; OD]. Radials projecting laterally very prominently with interradial notches, rather sharply separated from basals, which are tangent to basal plane of cup at about their midlength. Radials extremely convex in longitudinal profile, their proximal and distal areas nearly horizontal, articular facets subhorizontal, short, distinctly narrower than greatest width of radials, but with ends of neighboring transverse ridges almost touching. Three anal plates in cup, radianal obliquely above *CD* basal and supporting large *X* plate, which also rests on *CD* basal; anal sac unknown. Arms with axillary first primibrachs in all rays, 30 or 40 endotomous arms present, brachials cuneate uniserial. Stem round transversely. *M.Penn.(Desmoines.)-U.Penn.(Virgil.),* USA(Kans.-Okla.-Texas-Iowa-Neb.).——Fig. 487,*2a-d*. **G. stevensi,* Desmoines.(Oologah Ls.), NE.Okla.; *2a-d,* dorsal (*CD* interray up), post., ventral (*CD* interray down), and ant. views of holotype, ×2 (Moore, 1940a).——Fig. 487,*2e*. *G. ornatus* Strimple & Moore, Missour.(Francis Sh.), near Ada, Okla.; hypotype crown (*A* ray directed upward), ×2 (Strimple & Moore, 1971b).

Family SELLARDSICRINIDAE Strimple & Watkins, 1969

[Sellardsicrinidae Strimple & Watkins, 1969, p. 169]

Crown elongate. Cup low, truncate bowl shaped with slight basal concavity; plates tumid, sutures impressed; five infrabasals confined to concavity; large basals; prominent radials, with subhorizontal articular facets less than full summit width of radials, moderately short and lacking distinct flangelike inward extension; three large anal plates in normal (primitive) position followed by two series of prominent anal sac plates. Arms 40, biserial, pinnulate, with rounded exteriors, not appressed, unusually slender except for proximal brachials; three isotomous branchings, first on large, low primibrachs *1,* second on secundibrachs *3-10,* and third high in crown. *M.Penn.(Desmoines.).*

Sellardsicrinus Moore & Plummer, 1940, p. 357 [**S. marrsae*; OD]. Characters of family. *M. Penn.(Desmoines.),* USA(Texas).——Fig. 487,*3*. **S. marrsae,* Millsap Lake F., Parker Co., Texas; side view of holotype crown, ×1 (Moore & Plummer, 1940).

Family CYMBIOCRINIDAE Strimple & Watkins, 1969

[Cymbiocrinidae Strimple & Watkins, 1969, p. 188]

Cup ranging from basally impressed bowl of medium height to very low with deep basal invagination; five infrabasals usually confined to basal concavity; five large to small basals; five medium to large radials, with articular facets filling width of these plates, sloping inward and short (except *Aesiocrinus* which has relatively long facets); one anal plate in line with radials normally extended above cup summit (except *Aenigmocrinus* which has two subequal anal plates on *CD* basal), followed by two tube plates or by single tube plate in some later forms. Anal tube slender and recurved in older forms but usually long and straight in later ones. Arms mostly ten, but may be five *(Allosocrinus),* uniserial, primibrachs *2* normally axillary (primibrachs *1* and *2* may fuse in later forms), brachials may be cuneate, and syzygial pairs are not uncommon. Column transversely subpentagonal to pentagonal, or may become round. *U.Miss.-U.Penn.*

Key to Genera of Cymbiocrinidae

A. Cup large, bowl shaped, broad basal concavity ....
- I. Ten large arms, primibrachs *2* axillary, radianal followed by one tube plate .... *Proallosocrinus*
- II. Five large arms, brachials include many syzygial pairs .... *Allosocrinus*

B. Cup medium, bowl shaped, anal plate followed by two tube plates ....
- I. Ten narrow, elongated, pinnulated arms, no syzygial paired brachials except for the two primibrachs .... *Aesiocrinus*
- II. Same as BI except hyperpinnulated (one pinnule on each side of a brachial).... .... *Paracymbiocrinus*

C. Cup shallow, saucer shaped ....
- I. *CD* basal followed by two equidimensional anal plates .... *Aenigmocrinus*
- II. Large radianal in *CD* position followed by two tube plates which extend slightly below cup summit .... *Cymbiocrinus*

D. Cup shallow, essentially discoid; primibrachs *2* axillary ....

I. Anal plate (radianal) followed by three tube plates, some arms hypertrophied .. .. *Lecobasicrinus*

II. Radianal narrow, followed by one tube plate .... *Oklahomacrinus*

Cymbiocrinus KIRK, 1944, p. 233 [**C. romingeri*; OD]. Crown compact, subcylindrical. Cup shallow, patelliform or basin shaped, with flattened base; plates typically thick and convex; infrabasals small, lying within basal pit but usually extending well beyond column; basals relatively small, mostly lying within flattened base, but distal portions curving upward and clearly visible in lateral view, posterior basal truncate, usually supporting single anal (radianal) on its distal face, but in type species followed by 2 or 3 sac plates; radials very large, forming major part of cup, with strong sharply defined fulcral ridge extending full width of plates and dorsal ligament fossae narrow but distinct, large ligament pits cutting into and under fulcral ridge, pair of small, interarticular ligament fossae and pair of large muscle fields separated by intermuscular furrow. Anal sac formed by many small thin plates. Arms relatively short, composed of 2 stout primibrachs constricted medially, union of brachials strong, approaching anchylosis, except in earlier species (St. Louis and Ste. Genevieve), which have much looser union, junction of radials and primibrachs linear and gaping, first 2 secundibrachs closely united, appearing in some specimens as high, completely fused unit although sutures are clearly visible, especially in weathered juvenile specimens; distalward brachials have sloping articular faces, long pinnules of successive brachials or syzygial brachial pairs alternately on opposite sides. Stem transversely circular or subpentagonal with pentagonal lumen, unusually slender and weak, bearing whorls of short, slender cirri. *U.Miss.(Chester.),* USA(Mo.-Ill.-Ala.-Ky.-Okla.); *L.Penn.(Morrow.),* USA(Texas).——FIG. 488,*1a-c*. **C. romingeri,* U.Miss.(Ste. Genevieve Ls.), near Huntsville, Ala.; *BC* interray view of crown, dorsal and post. views of cup, ×1.5 (Kirk, 1944e).——FIG. 488,*1d,e*. *C. contentus* STRIMPLE & WATKINS, L.Penn.(Marble Falls F.), Espey Creek, San Saba Co., Texas; *C*-ray, *DE*-interray views of holotype crown, ×1.3 (Strimple & Watkins, 1969).

Aenigmocrinus STRIMPLE, 1973, p. 16 [**Poteriocrinus anomalus* WETHERBY, 1880b, p. 158 (*nom. correct.* STRIMPLE, 1973c, *pro P. anomalos* WETHERBY, 1880b); OD]. Crown rather compact but arms not abutting each other. Cup very shallow, basin shaped, with broad base containing small invagination; infrabasals small and mostly covered by stem; basals small and confined to basal concavity except *CD* basal which is exceptionally large, its distal apex almost reaching cup summit; radials large, with proximal ends entering basal plane of cup; 2 subequal plates on *CD* basal extending well above cup (plate at left may be termed anal *X* and that at right radianal, although the arrangement is rather unique); cup anal plates followed above by 2 series of tube plates; anal sac 0.8 or more height of arms, recurved, with large flattened, stout spines at summit. Arms 9 or 10, depending on whether anterior ray branches, stout, with well-rounded exteriors; primibrach *2* axillary in all except *A* ray which may branch on primibrach *6-7* or remain unbranched, brachials other than axillaries outwardly cuneate with constricted midsections, long sides bearing stout pinnules. Column round or subpentagonal. *U. Miss.(mid.Chester.),* USA(Ill. basin).——FIG. 489, *1*. **A. anomalus* (WETHERBY), Fraileys F., Ill.; *1a,b,* ant. view showing large, recurved anal sac, and ant. view with large plates of distal termination of anal sac, ×4 (Strimple, 1973c).

Aesiocrinus MILLER & GURLEY, 1890, p. 14 [**A. magnificus*; SD BASSLER, 1938, p. 34] [=*Pentadelocrinus* STRIMPLE, 1939a, p. 11 (type, *P. twinensis*; OD); *Phyalocrinus* TRAUTSCHOLD, 1870, p. 7 *(nom. nud.)*; *Phialocrinus* TRAUTSCHOLD, 1879, p. 14 (type, *P. patens*; M) (*non* EICHWALD, 1856)]. Low bowl shaped cup with shallowly concave or planate base; radial facets wide, relatively long, with articular facets sloping downward-inward; radianal typically large, extending above cup summit and followed by 2 sac plates. Anal sac tall, tapering upward near summit, composed of vertical rows of rugose hexagonal plates with prominent pore slits, anus at distal end. Arms 10, long slender, with well-rounded exteriors, short nearly rectilinear pinnules. Stem transversely pentagonal. *L.Penn.(Morrow.)-U.Penn.(Missour.-Virgil.),* USA(Mo.-Kans.-Okla.-Utah); *M.Carb.,* USSR.——FIG. 488,*2a,b*. **A. magnificus,* U.Penn.(Missour., Lane Sh.), Kansas City, Mo.; *2a,* side view of cup and anal sac, ×1 (Strimple, n); *2b,* crown showing oblique side view of crown from left (*D* ray), pinnulate outspread arms, rugose anal sac, and part of attached cirriferous stem (radials black, anal plates stippled), ×1 (Moore, 1952).——FIG. 488,*2c-h*. *A. dilatatus* MOORE, U.Penn.(Virgil., Brownville Ls.), Osage Co., Okla.; *2c,* drawing showing cross section of cup (portions beyond section shaded), *2d-h,* drawing of *A* radial plate (holotype), ext., int., summit, side and median section, all ×2.5 (Moore, 1939c).——FIG. 488,*2i*. *A. secundus* WASHBURN, L.Penn.(Morrow., Oquirrh F.), Utah; post. view of crown, ×1.4 (Washburn, 1968).

[WASHBURN (1968) described *Aesiocrinus secundus* based on a large specimen with an abraded base. In the same paper he proposed *Cymbiocrinus anatonus* and *C. cuneatus* for somewhat smaller specimens with better-preserved cups having a definite basal concavity. *C. anatonus* and *C. cuneatus* are considered here to be conspecific with *A. secundus*, and are thought to be progenitors of *Allosocrinus*.]

Allosocrinus STRIMPLE, 1949, p. 18 [**A. bronaughi*; OD]. Cup medium sized bowl shaped, with small deep basal concavity, basals large; anal plate

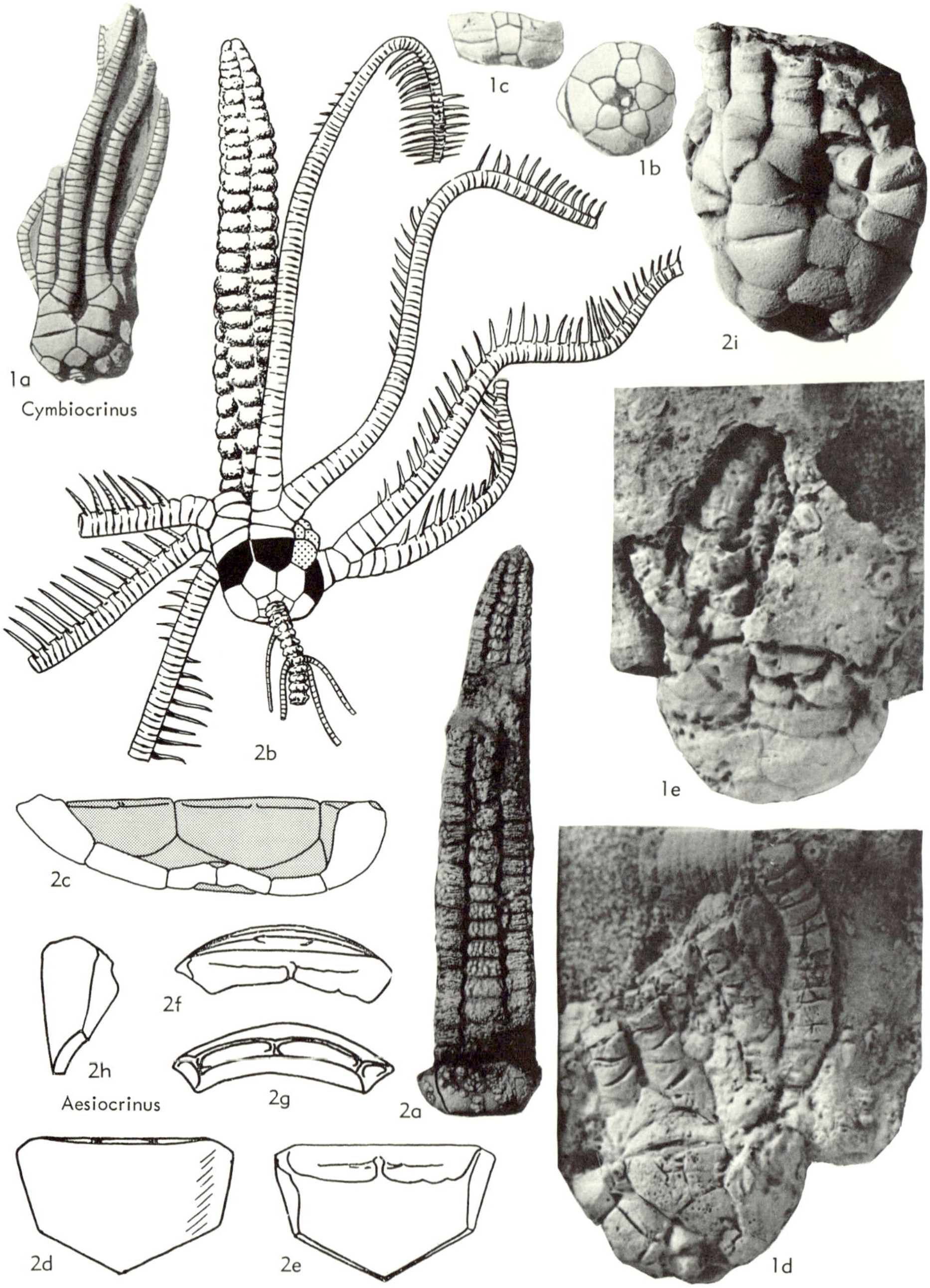

FIG. 488. Cymbiocrinidae (p. *T739*).

medium, followed by 2 sac plates. Arms 5, brachials cuneate, syzygially paired brachials common. Column may be transversely subpentagonal at cup but otherwise essentially round. *M.Penn.(Desmoines.)-U.Penn.(Missour.),* USA(Okla.-Ill.).——FIG. 490,*2*. **A. bronaughi; 2a-d,* U.Penn., Missour., Wann F., near Bartlesville, Okla.; holotype from dorsal, ventral, *B*-ray, and post. sides, ×1.5 (Strimple, 1949e); *2e,* LaSalle Ls., near Pontiac, Livingston Co., Ill.; hypotype crown, ×1.4 (Strimple & Moore, 1971a).

Lecobasicrinus Strimple & Watkins, 1969, p. 191 [**L. kickapooensis*; OD]. Crown very tall, cylindrical. Cup shallow bowl shaped with flat to faintly concave base, small infrabasals mostly covered by stem; small basals subtriangular (except distally truncate *CD* basal) barely meeting neighbors so as to separate radials from infrabasals; large radianal only plate in interray. Anal sac unknown. Ten recti-uniserial pinnulate arms; isotomously branched in each ray on primibrachs *2*. *M.Penn.(Atokan)-U.Penn.(Virgil.)*, USA(Texas-Kans.-Nebr.).——Fig. 490,*1*. **L. kickapooensis*, Desmoines.(Millsap Lake F.), Hood Co., Texas; *1a,b*, ant. and post. views of holotype crown, ×1.2 (Strimple & Watkins, 1969).

Oklahomacrinus Moore, 1939, p. 255 [**O. supinus*; OD]. Cup extremely depressed, consisting of nearly flat, straight-sided pentagonal disc with hollowed base; infrabasals forming slightly stellate pentagon in deepest part of basal concavity, distal portions flaring outward-downward, circlet appearing as cone on interior of cup that in some specimens is distinctly higher than radial facets; basals subequal and hexagonal in outline, except *CD* basal which is heptagonal, proximal portion of these plates flaring strongly outward-downward, distal tips visible in side view. Radials essentially horizontal, subequal pentagonal plates with mid-portions tangent to basal plane of cup; single quadrangular anal plate much longer than wide between posterior radials resting on truncated extremity of *CD* basal. Anal sac tubular, composed of plate rows with respiratory pores along sutures. Arms 10, curving outward and downward from nearly flat cup, composed of uniserial brachials with wide, short, quadrangular primibrachs *1* in each ray followed by triangular axillary primibrachs *2*, these plates joined by nearly immovable sutures and evidently forming a single functional unit commonly found attached to one another and cup; secundibrachs somewhat trapezoidal or rectangular in external view and arranged in structural pairs with smooth articular surfaces between members of pairs, whereas joints between successive pairs are marked by transverse ridges, ligament pits, and muscle areas. Radial articular facets plenary, with transverse strong ridge on outer margin of facets, outer ligament area deeply excavated and in subvertical position, inner ligament area short and gently declivate. Stem composed of thin subpentagonal columnals, heteromorphic, cirriferous. *M.Penn.(Desmoines.)-U.Penn. (Virgil.)*, USA(Okla.-Kans.-Texas).——Fig. 491, *1a-c*. **O. supinus*, Virgil.(Brownville Ls.), west of Strohm, Osage Co., Okla.; *1a,b*, dorsal and ventral views of holotype (*A* ray downward in *1a*, upward in *1b*), ×2; *1c*, oblique view of exterior from *B*-ray side, ×1.5 (Moore, 1939c).——Fig. 491,*1d-f*. *O. loeblichi* Moore, Missour. (Francis Sh.), Ada, Pontotoc Co., Okla.; *1d,e,* holotype from dorsal side (*A* ray downward), ×2, and oblique side view from below (*E* ray at right), ×1.5 (Moore, 1939c); *1f*, long. sec. of holotype cup (*A* ray at right), ×2 (Moore, 1939c).——Fig. 491,*1g-k*. *O. loeblichi* Moore(?); dissociated radial plate from outer, inner, upper, lower, and left sides, ×1.9 (Moore, 1939c).

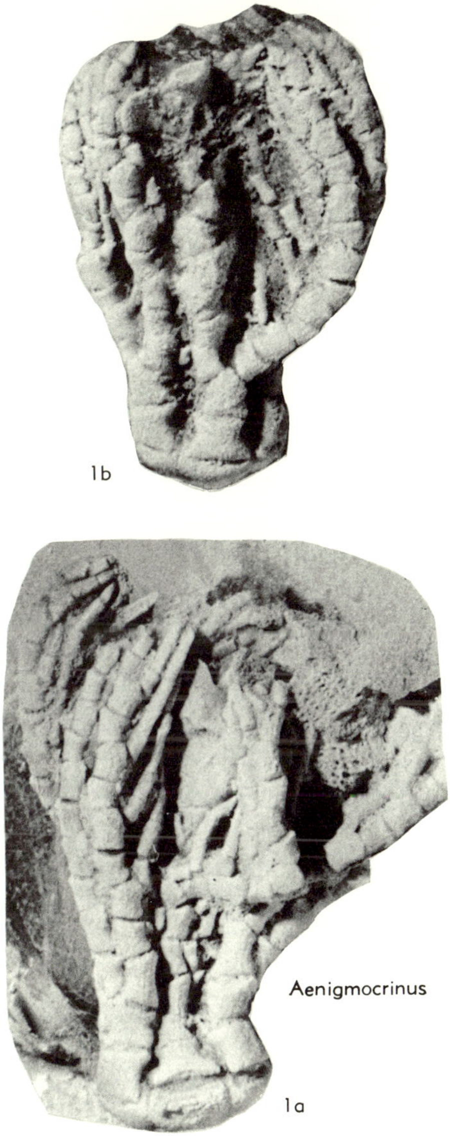

Fig. 489. Cymbiocrinidae (p. *T*739).

Paracymbiocrinus Burdick & Strimple, 1973, p. 239 [**P. ormondi*, p. 240; OD]. Cup low bowl shaped with shallow basal invagination, plate sutures slightly impressed. Infrabasals small, down-

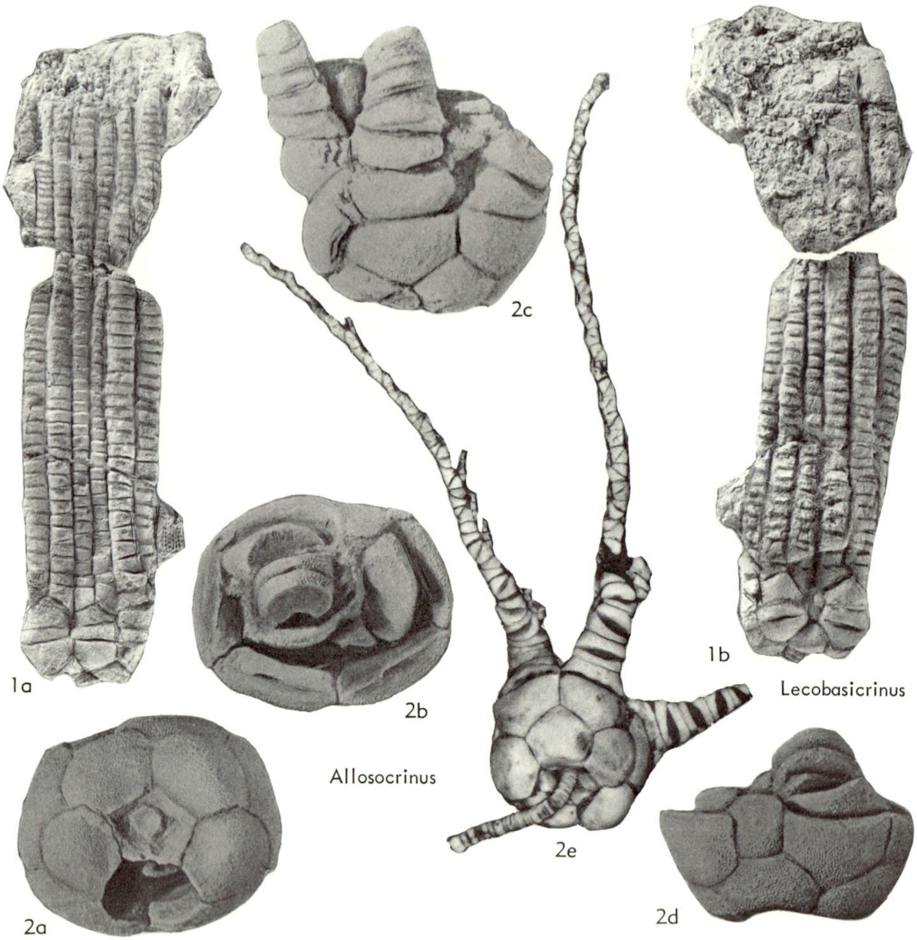

FIG. 490. Cymbiocrinidae (p. T739-T741).

flared, confined to basal concavity; basals large, *CD* basal truncated for reception of single anal plate; radials large with proximal ends almost reaching basal plane, articular facets wide; anal plate extends slightly above cup summit and is faceted for 2 tube plates, the larger one being to the right. Arms moderately long (10 times height of cup), dividing on primibrach 2; secundibrachs quadrangular when viewed from exterior and bearing a pinnule on each side of ambulacral groove. Column pentagonal. *U.Miss.(up.Chester.),* USA(Okla.-Ark.).——FIG. 492,*1*. **P. ormondi,* Ark.; *1a-c,* basal, post., ant. views of crown, ×2.7 (Burdick & Strimple, 1973b).

Proallosocrinus MOORE & STRIMPLE, 1973, p. 77 [**P. glenisteri*; OD]. Like *Allosocrinus* but transversely nearly circular arms more robust and shorter, composed of thin wide brachials. Arms 10, branching isotomously on primibrachs 2, pinnulate. *L.Penn.(Morrow.),* USA(Okla.).——FIG. 491,*2*. **P. glenisteri,* Gene Autry F., Carter Co., Okla.; *2a,b,* post. and dorsal views of holotype crown, ×1.5; *2c-e,* articular surface of secundibrach, lat. and ventral views of radial, ×3 (Moore & Strimple, 1973a).

Family STAPHYLOCRINIDAE Moore & Strimple, 1973

[Staphylocrinidae MOORE & STRIMPLE, 1973, p. 31]

Crown tall, slender; cup bowl shaped, with basal concavity or exceptionally with fused infrabasal circlet gently convex and visible from side, arm facets plenary, uni-

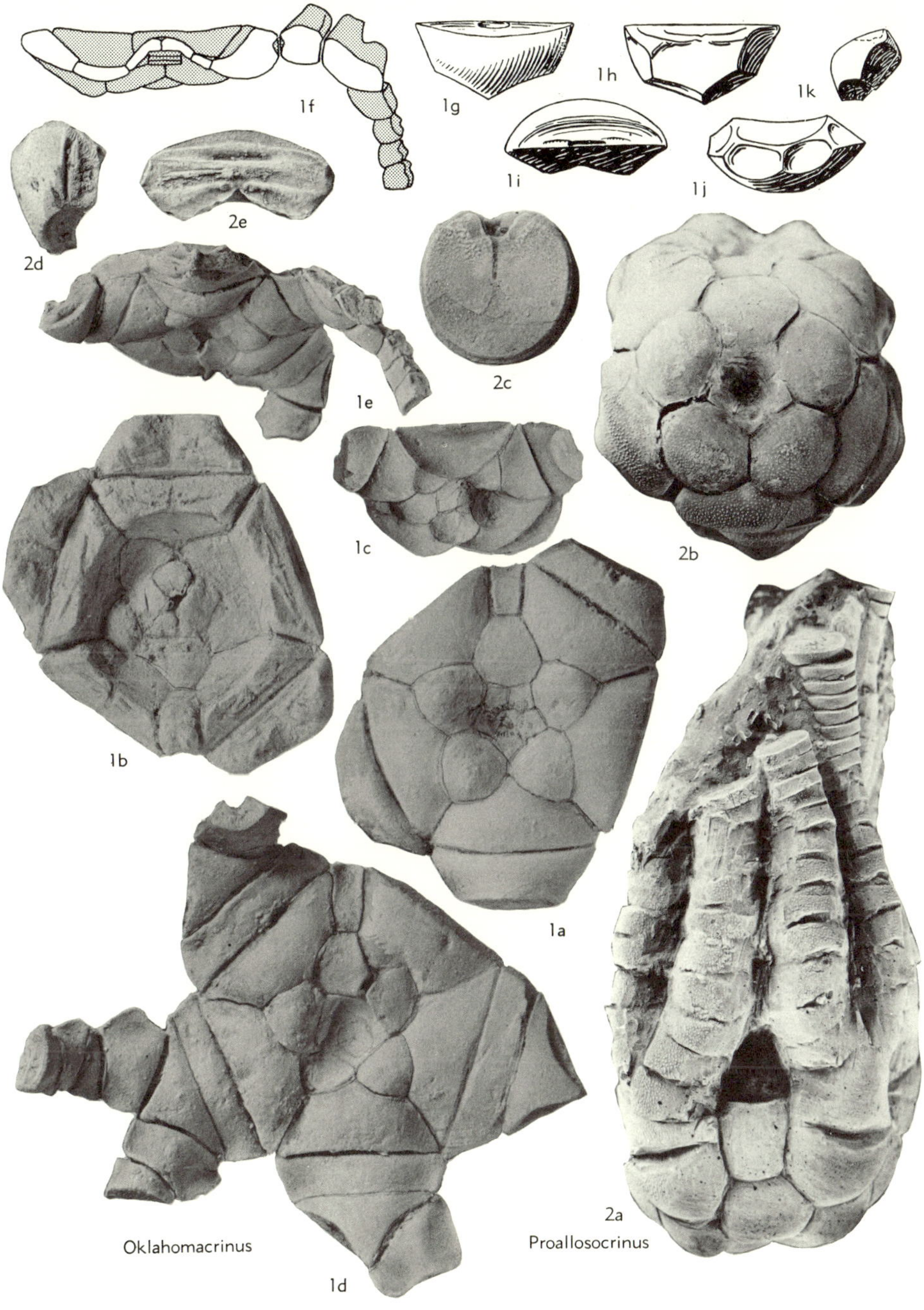

FIG. 491. Cymbiocrinidae (p. *T741-T742*).

serial arms 20 to 80 or more, two to three anals in cup. *L.Miss.-U.Miss.; M.Penn.-U. Perm.*

Staphylocrinus BURDICK & STRIMPLE, 1969, p. 9 [**S. bulgeri*; OD]. Cup bowl shaped, medium in height, composed of very thick tumid plates with tumidity extending to their margins, sutures

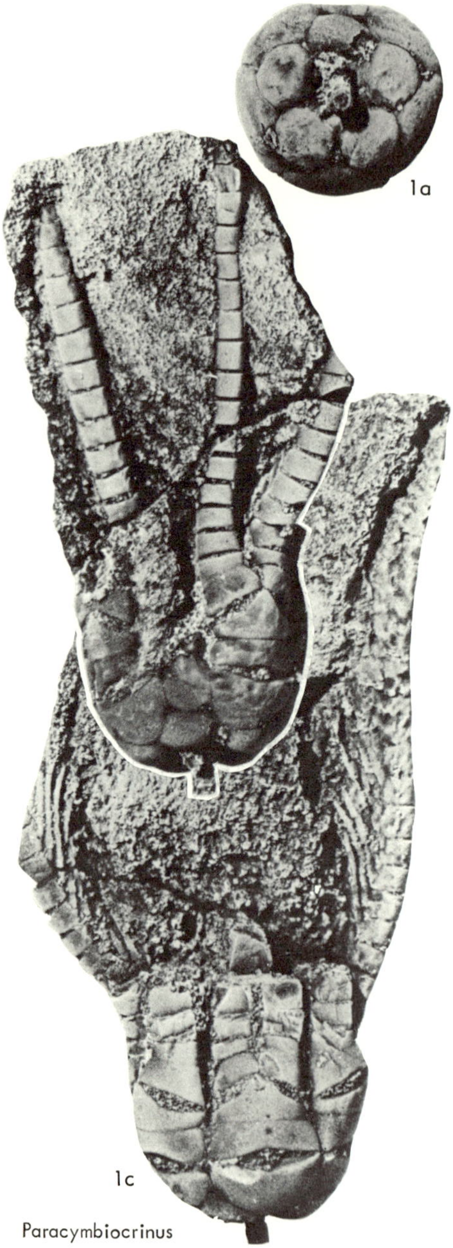

FIG. 492. Cymbiocrinidae (p. T741-T742).

impressed; infrabasal circlet fused in adults, visible from side as low element smoothly rounded below except for small central concavity marking attachment of vanished stem; 5 infrabasals separate in juvenile individuals; 3 thick, tumid anals in cup. Arms uniserial, composed of stout low brachials, branching isotomously on primibrachs *1* in all rays and again on axillary secundibrachs, generally secundibrach *6*. *U.Miss.(Chester.)*, USA(Ala.).——FIG. 493,*4*. **S. bulgeri*, Monteagle Ls.(Gasper); *4a,b*, *CD*-interray view of paratype and *AB*-interray view of holotype obliquely from below, ×0.8; *4c*, diagram vertical sec. of cup (portions of cup beyond section shaded), *A* radial at upper left, ×0.8 (all Burdick & Strimple, 1969).

Abrotocrinus MILLER & GURLEY, 1890, p. 30 [**A. cymosus* (=*Scaphiocrinus unicus* HALL, 1861a, p. 313); OD]. Crown tall, subcylindrical. Cup low, bowl shaped; infrabasals 5, within shallow concavity; basals prominent in side view; radials 5, larger than basals, proximal ends well above basal plane, distal portion flared outward; posterior interray wide, anals 3 in normal (primitive) arrangement. Anal sac cylindrical, reaching above arm tips, composed of small plates and culminating at top in one or more spines. Arms uniserial, bifurcating twice or more, pinnulate, primibrachs *1* axillary in all except anterior ray which has axillary primibrach *2*, brachials cuneate, rounded exteriors. Column obscurely pentagonal in section next to cup, becoming round below. *L.Miss.(Keokuk)*, USA(Ind.-Ill.-Iowa-Ala.).——FIG. 493,*1*. **A. unicus* (HALL), Keokuk, Ind.; *1a*, *E*-ray view of crown and attached part of pentagonal stem, ×1 (Springer, 1926b); *1b*, *?E*-ray view of crown and stem, ×0.7 (Miller & Gurley, 1890).

Agnostocrinus WEBSTER & LANE, 1967, p. 15 [**A. typus*; OD]. Crown moderately tall cylindrical. Cup bowl shaped with shallow basal concavity, plates bulging; radial articular facets peneplenary, sutures beneath primibrachs gaping; one large tumid anal plate. Anal tube long, extending above summit of arms, cylindrical and narrow. Arms 16, uniserial, branching isotomously on first primibrachs in all rays and on first secundibrachs in *A*, *C* and *D* rays only. *L.Perm.-U.Perm.*, USA (Nev.-Texas).——FIG. 494,*2*. **A. typus*, Bird Springs F., Nev.; *2a,b*, holotype crown from ant. *(EA)* and post. sides (*D* ray at left), ×1 (Webster & Lane, 1967).

Dinotocrinus KIRK, 1941, p. 513 [**D. compactus*; OD]. Crown subcylindrical, expanding slightly to about 0.5 of its height, then contracting. Cup broadly truncate low conical with basal depression; infrabasals small, fused into entirely concealed flat pentagonal plate; basals large, forming lower part of cup as seen in side view; radials wider than high, large, articular facets linear, occupying full width of radials; anal plates in cup 3, radianal large, entering deeply between *CD* and *BC* basals, anal *X* and right tube plate small, but extending above radials. Anal sac slender, reflexed. Arms uniserial, relatively short and stout, branching isotomously on primibrach *1* in each ray, typically one isotomous division well above main dichotom;

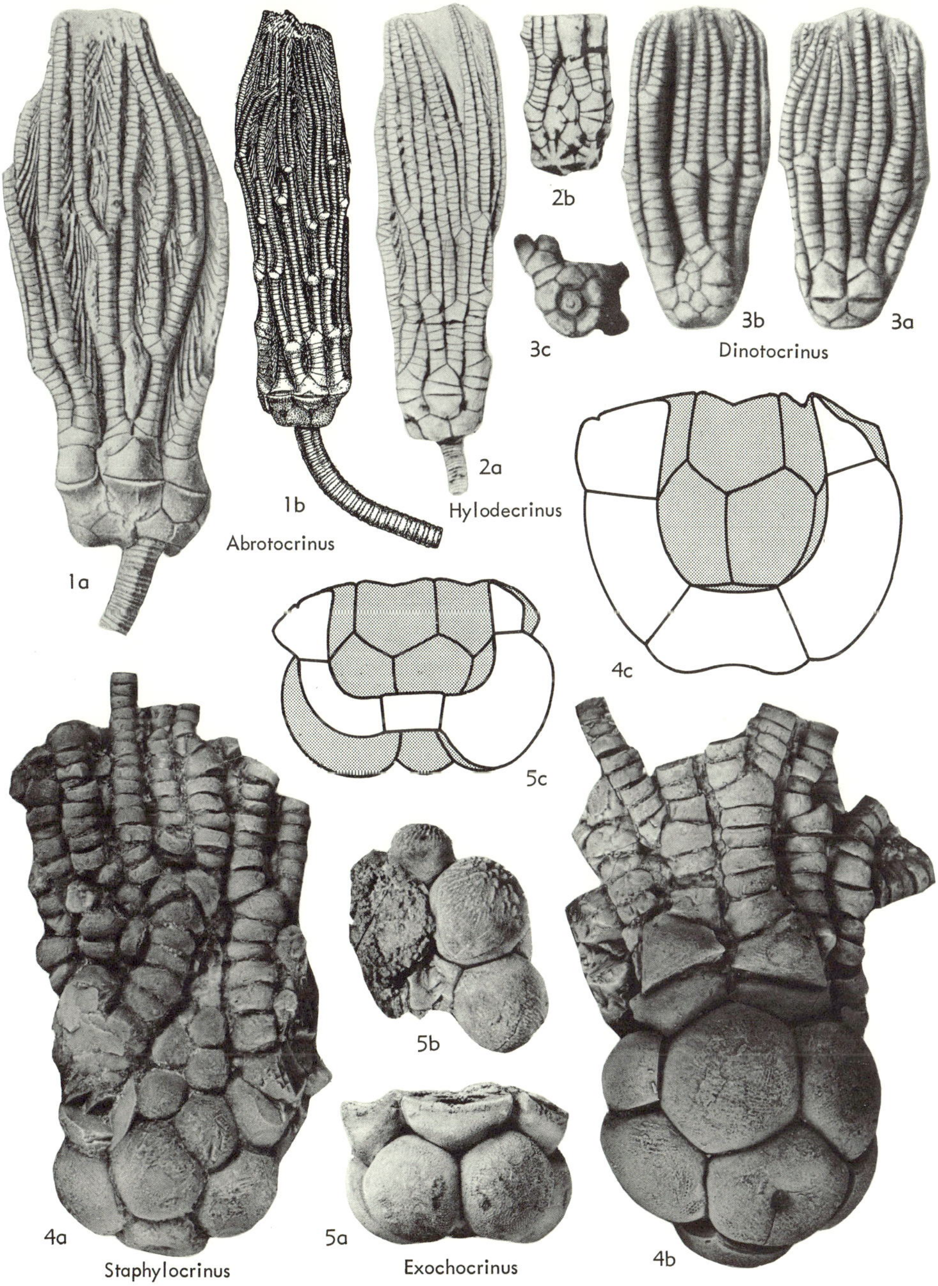

FIG. 493. Staphylocrinidae (p. *T743-T747*).

pinnules relatively long and stout. Stem transversely circular, with prominent nodals and internodal series. *L.Carb.*, G.Brit.(Eng.); *L.Miss.-U. Miss.*, N.Am.(Can.-USA).——FIG. 493,*3*. **D. compactus*, U.Miss.(Ste. Genevieve), Ala.; *3a-c*, holotype, ant., post., and dorsal views, ×1.5 (Kirk, 1941).

Exochocrinus BURDICK & STRIMPLE, 1969, p. 12 [**Eupachycrinus tumulosus* S. A. MILLER, 1892b, p. 680; OD]. Cup moderately deep, truncate bowl

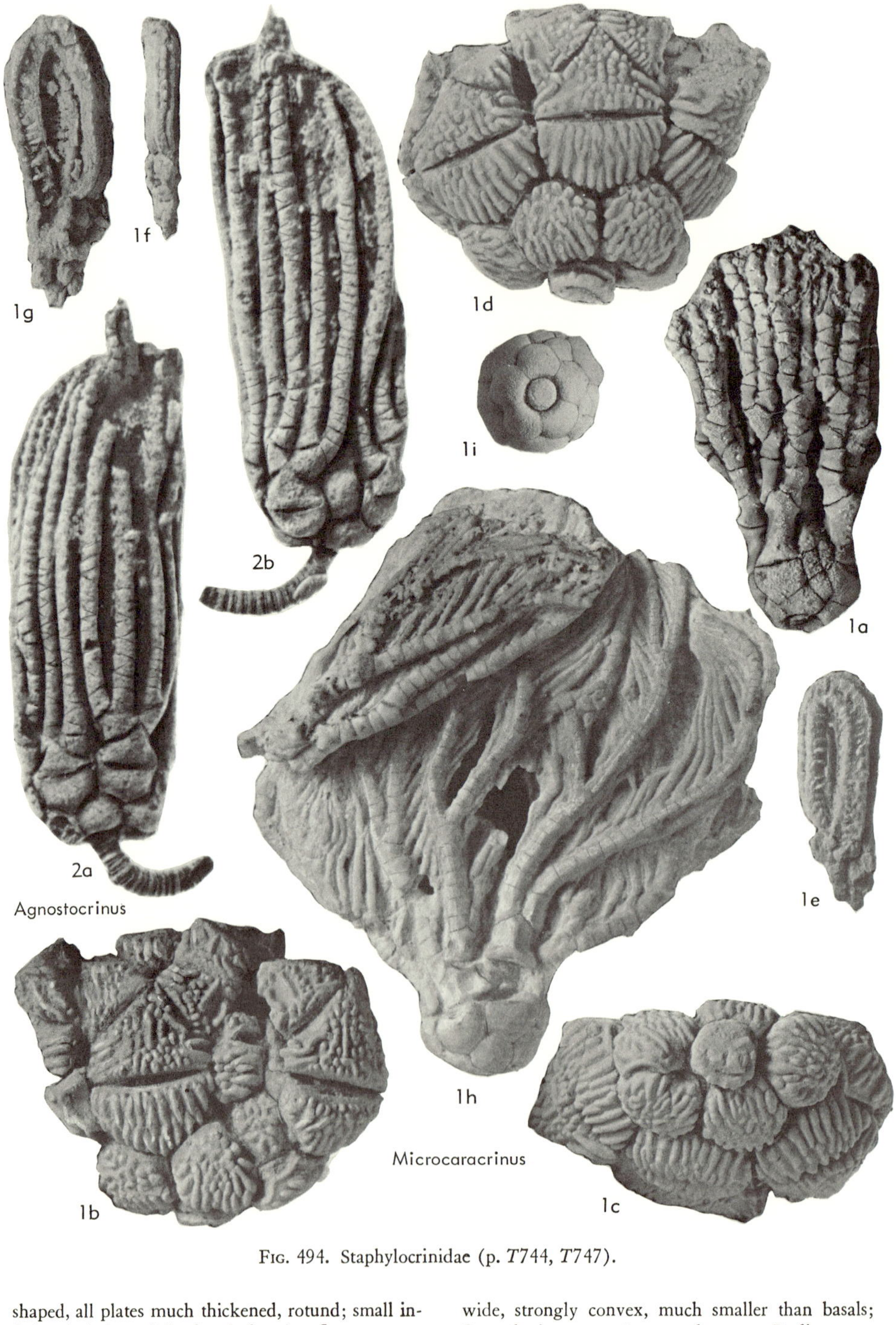

FIG. 494. Staphylocrinidae (p. *T744*, *T747*).

shaped, all plates much thickened, rotund; small infrabasal circlet solidly fused, forming flat pentagon disposed horizontally, located in concavity at base of cup, sides of which consist of upflared proximal parts of extremely tumid basals; radials low, wide, strongly convex, much smaller than basals; 3 anals in cup. Arms unknown. Rudimentary stem indicated by diminutive impression on infrabasal circlet. *U.Miss.(Chester.)*, USA(Ill.-Ky.-Ala.).——FIG. 493,5. **E. tumulosus* (MILLER);

5a, A-ray view of holotype, Ky.(Breckinridge Co.), ×1.3; *5b,* dorsal view of cup from Ill.(Randolph Co.), ×0.8; *5c,* diagram. vert. sec. of cup (portions beyond section shaded) (based on holotype) (Burdick & Strimple, 1969).

Hylodecrinus Kirk, 1941, p. 85 [**H. sculptus*; OD]. Crown compact, subcylindrical, arms approximately 9 or 10 times height of cup. Cup cyathiform, with flattened base, formed of heavy tumid plates; infrabasals small, not visible in lateral view, mostly covered by column; basals large, forming lower part of cup as seen in side view; radials large, articular facets full width of radial, sutures gaping; anal plates 3, radianal pentagonal, resting almost equally on *BC* and ***CD*** basals, anal *X* large, rising above radials, right tube plate reaching well up on *C*-ray primaxil. Anal sac relatively slender, reaching nearly to tips of arms, recurved, with cluster of small spines at flattened apex. Stem stellate in section in proximal part, becoming pentagonal below. Arms uniserial, stout, endotomous, with 2 primibrachs in each ray and 2 bifurcations above primaxils except in anterior ray which usually has single isotomous division above first dichotom, axillaries nodose, and usually small node on each brachial, giving crown a rough, uneven appearance; brachials with slightly sloping articular faces, and long, stout pinnules. [Resembles *Pachylocrinus* in general habit but has stout arms with few divisions, a long ventral sac, and a pentastellate to pentagonal stem.] *L.Miss.(up. Borden Gr.),* USA(Ind.-Ky.).——Fig. 493,*2*. **H. sculptus,* Keokuk, Ind.; *2a,b,* ant. (*B*-ray) and post. views of crown, ×0.7 (Kirk, 1941b).

Microcaracrinus Strimple & Watkins, 1969, p. 201 [**M. delicatus*; OD]. Crown tall, slender. Cup shallow bowl shaped, with narrow basal concavity; infrabasals small, confined to basal concavity; large basals tangent to basal plane of cup; large radials wider than high, articular facets filling distal face of radials; 3 anals, radianal in posterior position, anal *X* and right tube plate above with confluent distal surfaces. Anal sac short, sharply looped. Arms slender, uniserial, with well-rounded exteriors, not abutting, pinnules stout; first branching on primibrachs *1* in all rays and a second branching higher, a third bifurcation may take place in some inner half-rays only (exotomous). Column transversely round. *M.Penn. (Atokan)-L.Perm.(Wolfcamp.),* USA(Midcontinent reg.); *M.Carb.,* USSR.——Fig. 494,*1a*. *M. conjugulus* Strimple & Moore, U.Penn.(Missour., LaSalle Ls.), near Pontiac, Livingston Co., Ill.; holotype crown from post. side, ×4 (Strimple & Moore, 1971a).——Fig. 494,*1b-d*. *M. bellirugosus* (Moore), U.Penn.(Virgil., Wabaunsee Gr.), near Emporia, Kans.; holotype from post., dorsal, and ant. sides, ×2.7 (Moore, 1939c).——Fig. 494,*1e-i*. *M. colubrosus* (Moore), L.Perm.(?Ft. Riley Ls.), near Augusta, Kans.; *1e,f,* anal sac of paratype from left and ant. sides, ×2.7; *1g,* same from right side, ×3.2; *1h,i, AB*-interray and dorsal (anal series directed upward) views of holotype crown, ×2.7 (all Moore, 1939c).

Superfamily ZEACRINITACEA Bassler & Moodey, 1943

[*nom. transl.* Moore & Strimple, 1973, p. 31 (*ex* Zeacrinitidae Bassler & Moodey, 1943, *nom. correct. pro* Zeacrinidae Kirk, 1942a)] [Materials for this superfamily prepared by R. C. Moore, N. Gary Lane, & H. L. Strimple]

Concave to flat-based low saucer-shaped cup with one to three anal plates, mostly mushroom-shaped tall anal sac and multiple uniserial arms. *L.Miss-U.Perm.*

Family ZEACRINITIDAE Bassler & Moodey, 1943

[Zeacrinitidae Bassler & Moodey, 1943, p. 17 (*nom. correct. pro* Zeacrinidae Kirk, 1942a, p. 382)]

Crown compact, cylindrical or flaring upward. Cup low, with prominent but shallow basal concavity; radial articular facets wide, sloping outward-downward (declivate); three anals in cup, variable in relative size and position, solidly joined to form prominent upward projection of cup summit. Anal sac conspicuous, variable in form, pyramidal, cylindrical or mushroom shaped. Arms uniserial, branching endotomously, bi-endotomously, or isotomously in early forms; brachials pinnulate or hyperpinnulate, with two or four pinnules to each brachial except axillaries. Stem round transversely. *L.Miss.-U.Perm.*

Possibly related to the pirasocrinids and comparable to them in evolutionary specialization, the Zeacrinitidae are differentiated primarily by suppression of some isotomous arm bifurcations so as to produce an endotomous structure, like that of the Cercidocrinidae. The zeacrinitids are thought to be derivatives of the cercidocrinids. A peculiarity of many crinoids of this family is the firm attachment to the cup of some anal plates above radial summits.

Key to Genera of Zeacrinitidae

A. Crown subcylindrical. Cup saucer shaped, with basal concavity, arms uniserial ....
   I. Concavity wide, including proximal extremities of plates in all circlets (formula in terms given for proximal tips of plates in Pirasocrinidae, – – –), arms branching endotomously ....
      a. Anal sac short, pyramidal; arms

narrow .... *Zeacrinites*
b. Very numerous wide arms concealing anal sac .... *Parazeacrinites*
c. Anal sac tall, club shaped .... *Alcimocrinus*

II. Concavity wide, including proximal extremities of infrabasals and basals, those of radials in basal plane of cup (formula, – – 0) ....
a. Anal sac tall, club shaped .. *Eratocrinus*
b. Anal sac slender, planospirally coiled .... *Linocrinus*
c. Anal sac short and stout ...... *Sarocrinus*
d. Anal sac mushroom shaped, with laterally directed spines bordering widened summit .... *Tholocrinus*

III. Cavity chiefly consisting of sharply depressed stem attachment area on infrabasals but including proximal tips of basals, those of radials above basal plane on outer side of cup (formula, – – +); arms branching endotomously .... *Neozeacrinus*

Zeacrinites TROOST in HALL, 1858, p. 544 [**Z. magnoliaeformis* TROOST in HALL, 1858b, p. 684; SD WACHSMUTH & SPRINGER, 1880, p. 128] [=*Zeacrinus* HALL, 1858b, p. 546 *(nom. van.)*]. Crown compact. Cup low, broad, with wide basal concavity; infrabasals small, mostly covered by stem; basals small, spear shaped, rather widely separated by proximal parts of large radials which form solidly joined distally projecting part of cup; radianal may separate *CD* basal and anal *X*. Anal sac short, pyramidal, tapering distally. Arms broad and closely appressed; uniserial, primibrach *1* axillary in all but *A* ray, which may have 2 or, less commonly, 3 primibrachs; 2 to 4 higher divisions of arms endotomous; each brachial with 2 pinnules, one on each side. Stem transversely round. *U.Miss.(Chester.), ?M.Penn.,* USA(Ill.-Ky.-Ala.-Ind.-Va.-Mo.-Texas).——FIG. 495,*1a-f*. **Z. magnoliaeformis,* U.Miss., Ala.; post. and ant. views of crown, 2 dorsal, oblique, and ventral views of cup, ×1 (Springer, 1926b).——FIG. 495,*1g*. *Z. wortheni* (HALL), U.Miss., Ky.; post. view of crown, ×0.7 (Springer, 1926b).

Alcimocrinus KIRK, 1938, p. 162 [**Zeacrinus girtyi* SPRINGER, 1926b, p. 83; OD]. Crown subcylindrical, slightly expanding upward. Cup low, with broad, shallow basal concavity; infrabasals small, covered by stem; basals and proximal edges of radials within basal concavity; anal sac elongate, somewhat club shaped. Arms stout, uniserial brachials low and broad; primibrachs *2* axillary; 6 to 8 endotomous branches in each half-ray above primaxil. *U.Miss.(Chester.)-M.Penn.(Atokan),* USA (Okla.).——FIG. 495,*3*. **A. girtyi* (SPRINGER), L.Penn., Okla.; *3a,b,* post. and ant. views of crown, ×1 (Springer, 1926b).

Eratocrinus KIRK, 1938, p. 165 [**Zeacrinus elegans* HALL, 1858b, p. 547; OD]. Crown tall, slightly pear shaped. Cup shallow saucer shaped; basal concavity narrow and shallow; small infrabasals confined to center of concavity; distal parts of basals, and radials, visible in side view; 3 anal plates in cup. Anal sac elongate, and club shaped, some species with large polygonal plates in vertical row on posterior side and at summit. Arms uniserial, appressed; with large first primibrachs axillary in all but *A* ray, which may have 2 to 5 primibrachs; 3 or 4 endotomous divisions in each half-ray. Stem subpentagonal in outline proximally, becoming round distally. *L.Miss.(Osag.)-U.Miss.(Warsaw),* USA(Iowa-Mo.-Ind.-Ill.).——FIG. 495,*2a,b*. **E. elegans* (HALL), L.Miss., Iowa; *C*-ray and *EA*-interray views of crown, ×1 (Springer, 1926b).——FIG. 495,*2c,d*. *E. commaticus* (S. A. MILLER), U.Miss.(Chester.), Mo.; post. and *E*-ray views of crown showing pinnulate arms and anal sac, ×1 (Springer, 1926b).

Linocrinus KIRK, 1938, p. 168 [**L. wachsmuthi*; OD]. Crown subcylindrical. Cup low, broad, with shallow basal concavity; infrabasals concealed by stem; basals angularly tumid, radials with strong radiating ridges, sutures with first primibrachs gaping; anal sac large, elongate, planispirally coiled distally, with anal opening on anterior side below spiral. Arms uniserial, composed of wedge-shaped, nodose brachials; first primibrachs large, with strong median keel, axillary in all but anterior ray, which has 3 to 5 primibrachs; arms branching endotomously 2 or 3 times above primaxils. Stem round transversely. *L.Miss.(Kinderhook.)-U.Miss.(Chester.),* USA(Ind.-Iowa-Ill.-Ala.-Okla.-Ga.-Mo.-Mont.)-Can.(Alta.).——FIG. 496, *2a-c*. **L. wachsmuthi,* U.Miss., Ala.; *B-, C-,* and *B*-ray views of crown, ×1.3 (Springer, 1926b).——FIG. 496,*2d*. *L. arboreus* (WORTHEN), U. Miss., Ala.; *D*-ray view of partial crown showing spirally coiled distal part of anal sac, ×1.8 (Springer, 1926b).

Neozeacrinus WANNER, 1937, p. 144 [**N. peramplus*; OD]. Crown moderately tall, ovoid. Cup low, saucer shaped, with narrow basal concavity consisting of stem impression, confined to infrabasals; basals and radials slightly tumid; anal sac not known. Arms broad and compact, uniserial brachials wide and low, primibrachs *1* axillary in all rays; arms branching isotomously on secundibrachs and endotomously 3 or 4 times above secundaxils, producing bi-endotomous arms; brachials with 4 pinnules each, 2 on each side. *M. Penn.-U.Perm.,* USA(Texas)-Indon.(Timor).——FIG. 497,*1a,b*. *N. uddeni* (WELLER), L.Perm., Texas; post. and dorsal (*CD* interray directed upward) views of cup, ×1 (Moore and Plummer, 1940).——FIG. 497,*1c*. *N. praecursor* MOORE & PLUMMER, M.Penn.(Desmoines.), Texas; *D*-ray view of crown, ×0.7 (Moore & Plummer, 1940).——FIG. 497,*1d-g*. **N. peramplus,* U.Perm., Basleo, Timor; *1d-f,* post., dorsal, and ventral

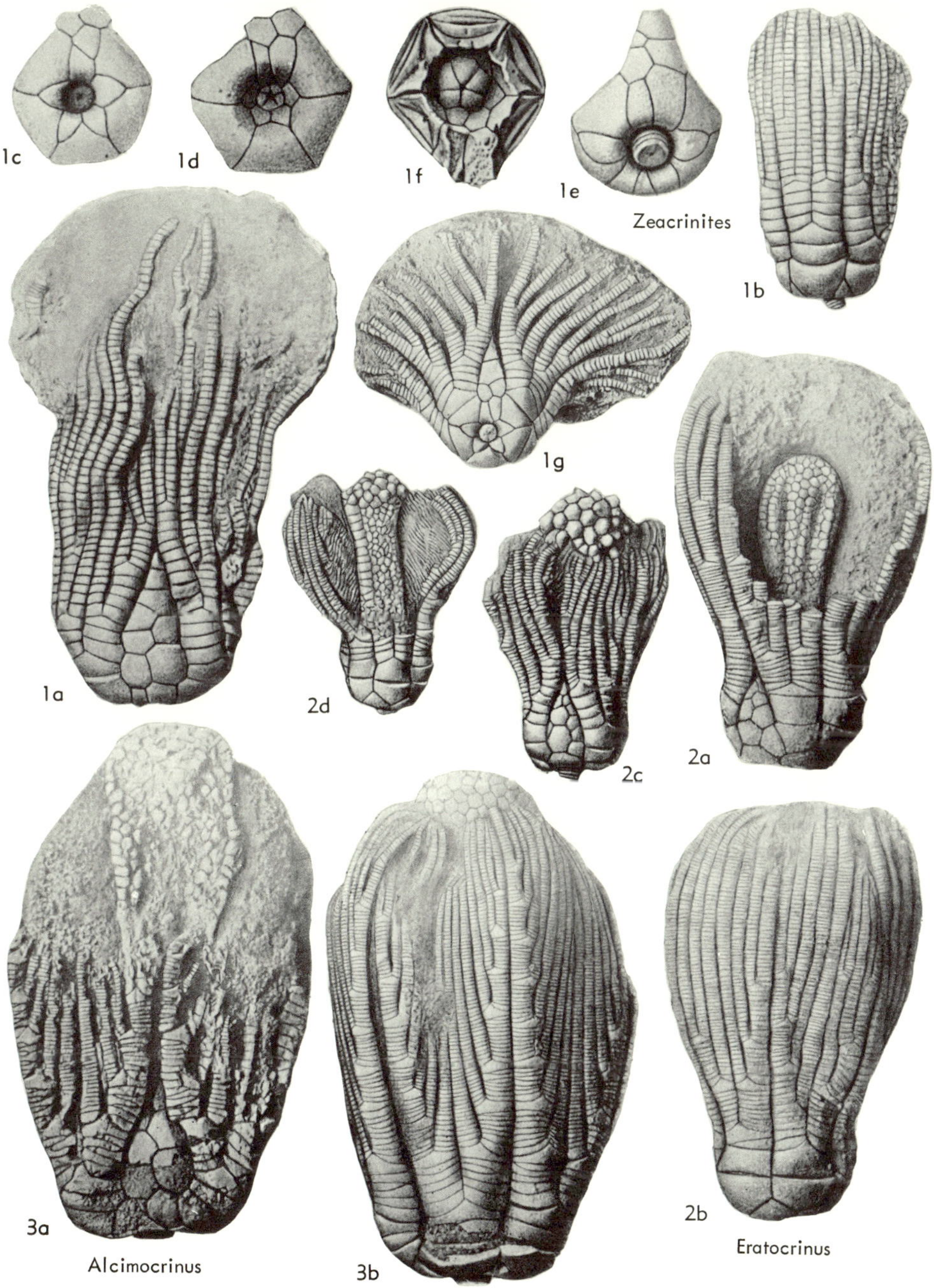

FIG. 495. Zeacrinitidae (p. *T*748).

views of holotype crown, ×1; *1g*, lat. view of *B* arm from primibrach *2* to tertibrach *4* with pinnules, ×2 (Wanner, 1937).

Parazeacrinites BURDICK & STRIMPLE, 1971, p. 25 [**Zeacrinus konincki* BATHER, 1912, p. 73; OD]. Crown pear shaped, narrowing upward. Like *Zeacrinites* except that brachials are thinner and each bears single pinnule on alternate sides of successive brachials. *L.Carb.(Visean)*, Eu.(Eng.). ——FIG. 496,*4*. **P. konincki* (BATHER); *4a-c, E-*

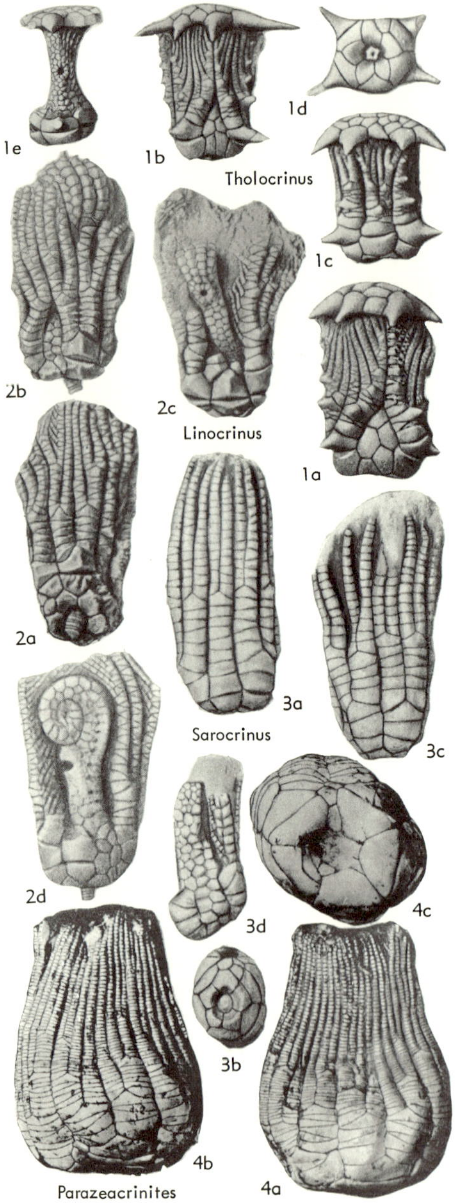

FIG. 496. Zeacrinitidae (p. T748-T750).

ray, *C*-ray, and dorsal views of crown, ×0.9 (Wright, 1952a).

Sarocrinus KIRK, 1942, p. 382 [**S. nitidus*; OD]. Crown small, cylindrical. Cup very low saucer shaped with deep, narrow basal concavity confined to infrabasals and proximal edges of basals. Anal sac short, stout, composed of numerous small polygonal plates. Arms uniserial, branching on primibrachs 2 in all but *A* ray, which has 3 to 8 primibrachs; all branches isotomous, with tertibrachs present in all but *A* ray, which has a single division. *L.Miss.(Osag.),* USA(Iowa-Ind.-Mo.).——FIG. 496,*3*. **S. nitidus,* Ind.; *3a-d, A*-ray, dorsal, *BC*-interray, and post. views of crown, ×1.3 (Kirk, 1942a).

Tholocrinus KIRK, 1939, p. 479 [**Hydreionocrinus spinosus* WOOD, 1909; p. 93; OD] [=*Xystocrinus* MOORE & PLUMMER, 1938, p. 27 (type, *Zeacrinus depressus* HALL, 1858b, p. 546; OD)]. Crown short and compact. Cup rounded saucer shaped, with narrow, deep basal concavity; infrabasals concealed by stem; anal sac elongate, higher than arms, narrowed medially, composed of numerous small plates below large, expanded summit composed of large plates, those of central area polygonal and of periphery comprising long, laterally directed spines; anal opening at midheight of sac. Arms composed of uniserial strongly wedge-shaped brachials proximally, becoming biserial distally; distal tips of arms preserved under expanded distal part of anal sac; primibrachs *1* axillary in all but *A* ray, which has several primibrachs; all axillary plates strongly nodose to spinose; arms branching isotomously on primaxils and endotomously 2 to 6 times above primaxils. *U.Miss.(Chester.),* USA(Ill.-Ky.).——FIG. 496,*1*. **T. spinosus* (WOOD), U.Miss., Ky.; *1a-e,* post. views of 2 crowns, *A*-ray and dorsal (*CD* interray upward) views of crowns, and ant. view of theca, ×0.9 (Springer, 1926b).

Family EXOCRINIDAE Strimple & Watkins, 1969

[Exocrinidae STRIMPLE & WATKINS, 1969, p. 183]

Small crinoids with multiple slender uniserial arms and low bowl-shaped cup which is flat to faintly convex or concave at base, main cup plates being radials with plenary articular facets; 3 anals in cup. Arms 30 to 40, branching isotomously on primibrachs *1* and at higher levels, pinnulate; brachials commonly keeled longitudinally. *L.Penn.(Morrow.)-L.Perm.(Wolfcamp.).*

Exocrinus STRIMPLE, 1949, p. 9 [**E. multirami*; OD]. Characters of family with pronounced tendency to fusion of successive brachials with pinnules borne on opposite sides of arms as though the individual brachials were still divided; radianal directly above *CD* basal followed by anal *X* and right tube plate with confluent distal edges. Anal sac short, cylindrical, composed of longitudinal rows of polygonal plates. Arms branching on elongate primibrachs *1,* secundibrachs *2* or *3,* and tertibrachs *3-5.* Stem transversely round, noncirriferous. *L.Penn.(Morrow.)-L.Perm.(Wolf-*

camp.), USA(Kans.-Okla.-Texas-Ill.-Neb.-Nev.). ——FIG. 498,*1*. *E. wanni* STRIMPLE, U.Penn. (Missour., LaSalle Ls.), Livingston Co., Ill.; *1a*, *C*-ray view of crown, ×2.4; *1b*, *B*-ray view of crown, ×2; *1c*, *C*-ray camera lucida drawing of crown, ×4 (Strimple & Moore, 1971a).

Oxynocrinus STRIMPLE & WATKINS, 1969, p. 184 [**O. spicata*; OD]. Crown short and broad. Cup low bowl shaped with gently convex base; radial articular facets plenary; 3 anal plates in normal (primitive) arrangement in cup. Arms approximately 40, branching isotomously on primibrachs *1*, secundibrachs *2* or *3*, and tertibrachs *4* or *5*, pinnulate. Stem transversely round, noncirriferous. *L.Penn.(Morrow.),* USA(Texas).——FIG. 498,*2*. **O. spicatus,* Marble Falls Ls., Lampasas Co., Texas; post. view of holotype crown, ×1.85 (Strimple & Watkins, 1969).

Petalambicrinus STRIMPLE, 1976, p. 867 [**P. craddocki*; OD]. Shallow cup with shallow basal invagination; anal plates advanced, as in *Exocrinus*; radial articular facets large, plenary; proximal columnal large, transversely round. *U.Penn.(Missour.),* USA(Texas).

Family TIMORECHINIDAE Jaekel, 1918

[Timorechinidae JAEKEL, 1918, p. 63] [=Timorocrinidae WANNER, 1916, p. 283]

Crown short and compact. Cup low, basally concave, infrabasals and basals small, not visible in side view; radials large, with articular facets occupying full width of plates; one small anal plate in or above cup. Anal sac stout and specialized, where known. Arms composed of uniserial medium high or elongate brachials, short, stout, primibrachs *1* axillary, heterotomous. *U.Perm.(Basleo beds).*

Rather strangely specialized crinoids of zeacrinitid type are included in the Timorechinidae, of Upper Permian age. The cup is very low and basally concave. Advanced evolution of the anal series is indicated by absence of a radianal, unless the so-called *X* plate that occurs in line with the radials of some genera is actually the radianal. In some genera all anals have been expelled from the cup. The anal sac is variously modified in form; in *Timorechinus* lateral projections of the sac make partitions between the arms in a manner strongly suggesting the separate arm compartments of *Eucalyptocrinites,* among the camerates. The strange calyx of *Timorechinus* was not at first suspected of being that of a crinoid.

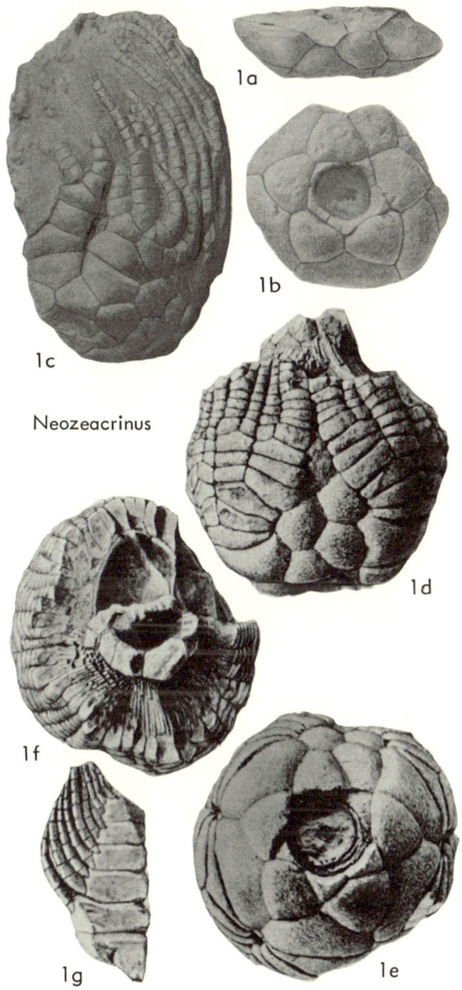

FIG. 497. Zeacrinitidae (p. *T*748-*T*749).

Key to Genera of Timorechinidae

A. Base of cup flat, lacking concavity, anal sac large, longitudinally fluted, with arms fitting in its grooves .... *Timorechinus*

B. Basal concavity of cup wide and shallow, numerous arms branching endotomously .. .. *Parabursacrinus*

C. Basal concavity of cup wide and deep ....
   - I. Radials longitudinally ridged, *CD* basal very elongate .... *Benthocrinus*
   - II. Radials smooth, *CD* basal short .... *Notiocrinus*

D. Basal concavity of cup narrow and deep, anal sac strongly bulging posteriorly .... *Prolobocrinus*

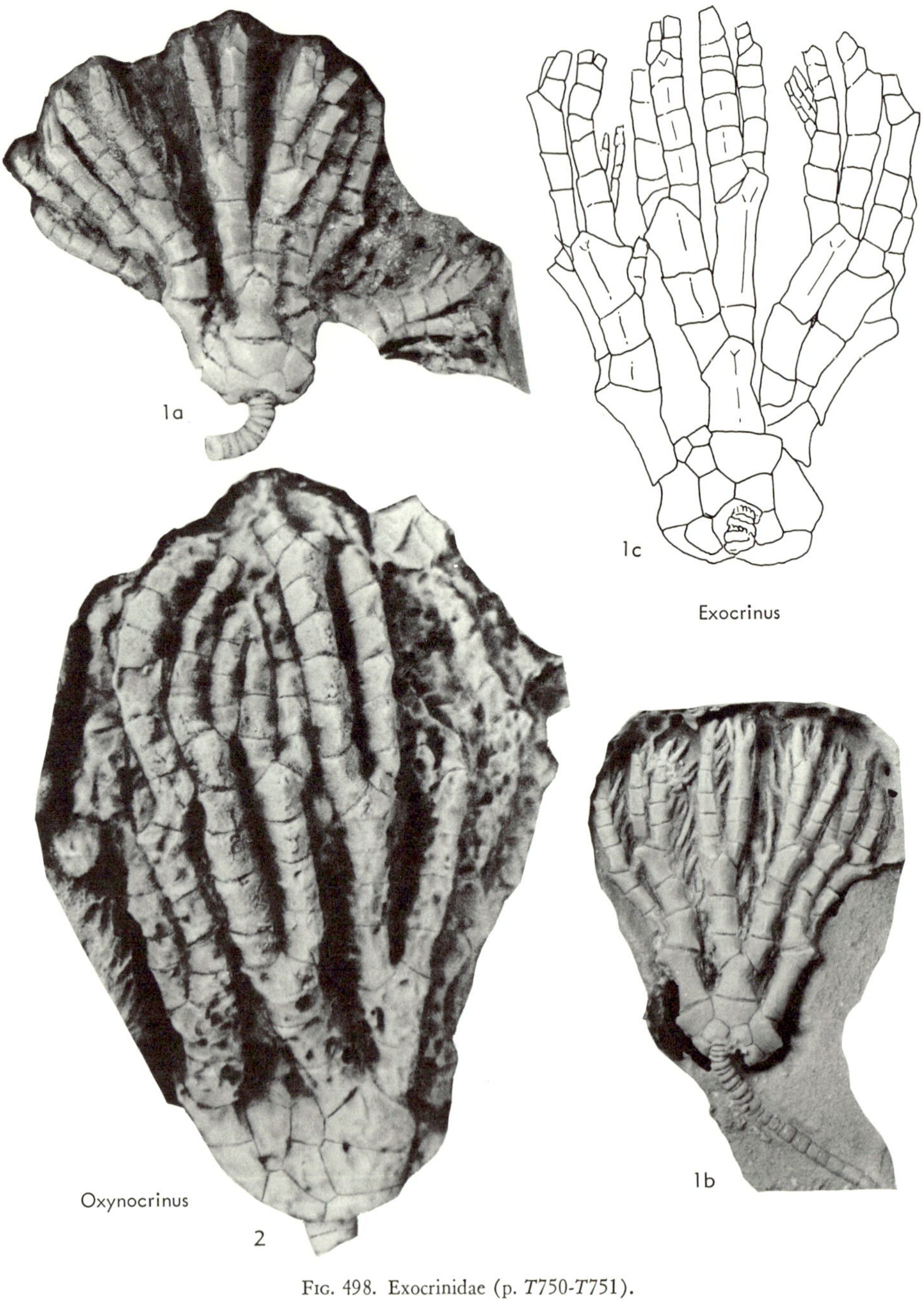

FIG. 498. Exocrinidae (p. *T*750-*T*751).

Timorechinus WANNER, 1911, p. 125 [**T. mirabilis*; OD] [=*Timorocystis* LAMBERT, 1911, p. 185 (obj.); *Timorocrinus* WANNER, 1912, p. 599 (obj.)]. Crown ovoid, with greatest width at midheight of arms. Cup low, bowl shaped, with flat base; infrabasals small, confined to stem impression, fused into single plate; basals spear shaped except for *CD* basal, which is very large and projects upward to top of cup; radials small, wider than high; articular facets occupying full width of radials; anal sac massive, with large plate directly above *CD* basal supporting several small

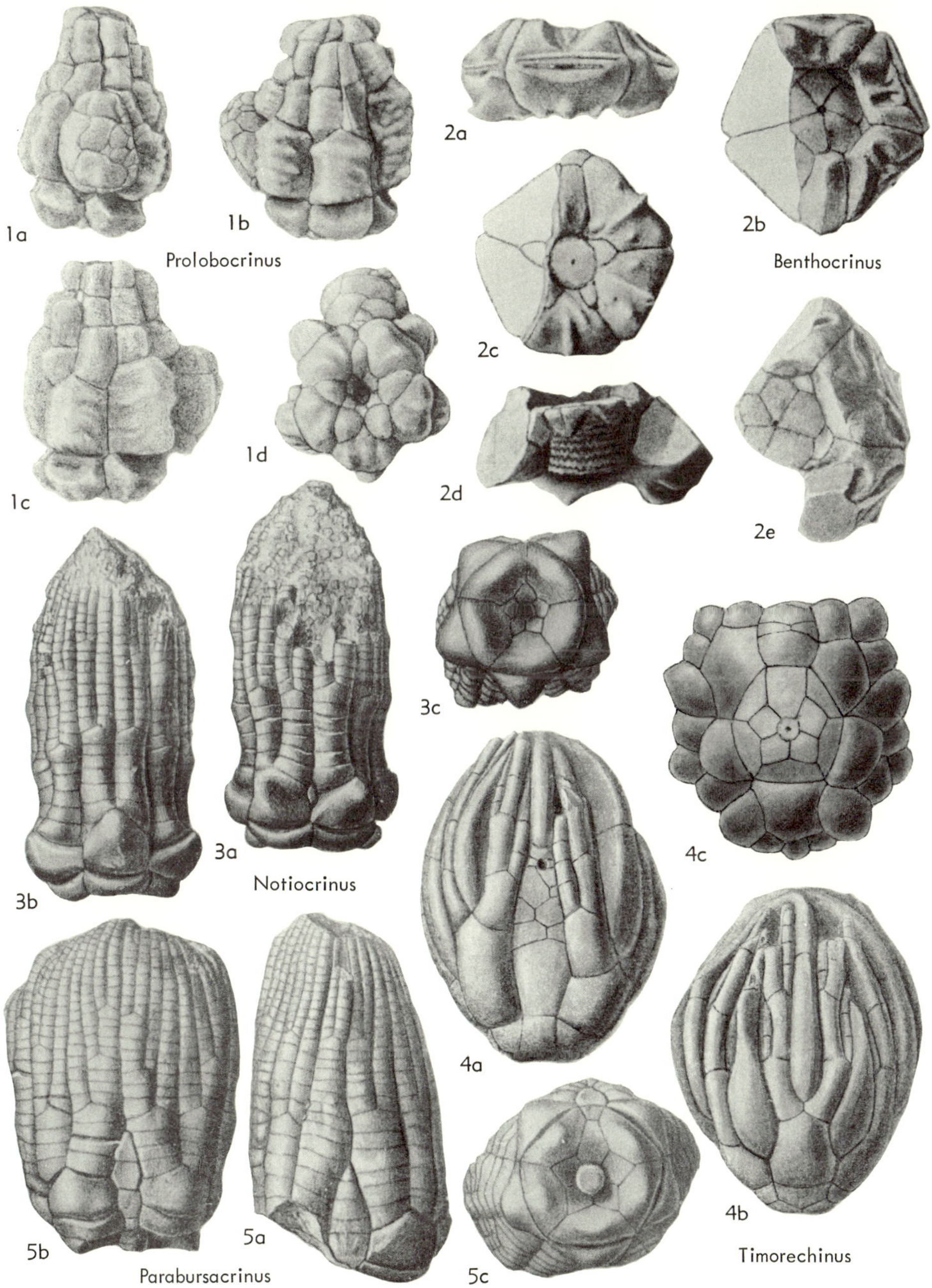

Fig. 499. Timorechinidae (p. *T752-T754*).

plates that surround a small, laterally directed anal opening; summit of sac and interradial and radial areas occupied by large plates that project between the arms and form niches within which arms are placed. Arms short, stout, with large, convex axillary first primibrachs; higher brachials elongate, heterotomous, with ramule on alternate sides of each brachial; tegminal plates separating arms of each ray and brachials of each half-ray. *U.Perm.(Basleo beds),* Indon.(Timor).——FIG. 499,*4*. **T. mirabilis; 4a-c,* post., ant., and dorsal (*CD* interray up) views of crown, ×1.5 (Wanner, 1911).

Benthocrinus WANNER, 1937, p. 187 [**B. cryptobasalis*; OD]. Cup bowl shaped, with wide, deep basal concavity, stem impression almost at height of radial articular facets; infrabasals small, thin, covered by stem impression; basals small, spear shaped, separated by tips of radials, except for *CD* basal, which is quadrangular, elongate, and separates posterior radials, extending almost to their summits; single elongate anal plate in cup, directly above *CD* basal and separating radials; radials large, proximally forming edge of basal concavity, ornamented with nodes and longitudinal ridges; articular facets as wide as radials, steeply inclined downward-outward with prominent transverse ridge and short ligament pit. *U.Perm. (Basleo beds),* Indon.(Timor).——FIG. 499,*2*. **B. cryptobasalis; 2a-e, D*-ray, ventral (*CD* interray downward), and dorsal (*CD* interray upward) views of incomplete cup, lat. view of broken cup showing deep basal concavity, ventral view of cup showing infrabasals and small basals, ×1.5 (Wanner, 1937).

Notiocrinus WANNER, 1924, p. 181 [**N. timoricus*; OD]. Crown cylindrical. Cup very low, with wide, deep basal concavity; infrabasals and basals not visible in side view; radials strongly convex, pentagonal, with articular facets that occupy full width of radials; single small anal plate in notch between posterior primibrachs. Arms stout, brachials rounded; first primibrachs axillary, convex, and protruding outward; secundibrachs *5* or *6* axillary, isotomous; higher divisions of arms with tertaxils occurring lower on abmedial side of each half-ray than on admedial sides. *U.Perm.(Basleo beds),* Indon.(Timor).——FIG. 499,*3*. **N. timoricus; 3a-c,* post., ant., and dorsal (*CD* interray upward) views of crown, ×1 (Wanner, 1924).

Parabursacrinus WANNER, 1924, p. 180 [**Bursacrinus procerus* WANNER, 1916a, p. 181; OD]. Crown wide, cylindrical. Cup very low and broad, with wide shallow basal concavity; infrabasals small, largely covered by stem; basals small, not visible in side view, spear shaped except for *CD* basal, which is elongate and projects to mid-height of radials, which are large, convex plates, each with transversely elongate node, articular facets as wide as radials; single small anal plate between posterior radials, supporting longitudinal row of small sac plates. Arms robust, first primibrachs axillary, large, convex and nodose; several endotomous divisions in each half-ray, with admedial branches narrow, elongate, and without higher divisions. *U.Perm.(Basleo beds),* Indon. (Timor).——FIG. 499,*5a*. **P. procerus* (WANNER); post. views of arms and lower anal sac plates, ×1 (Wanner, 1924).——FIG. 499,*5b,c*. *P. nefotassiensis* WANNER; post. and dorsal (*CD* interray upward) views of crown, ×1 (Wanner, 1924).

Prolobocrinus WANNER, 1937, p. 184 [**P. permicus*; OD]. Crown short and massive. Cup low, bowl shaped, with narrow, deep basal concavity; infrabasals small; basals small, not visible in side view, almost separated by radials and infrabasals; radials large, convex, fluted longitudinally; anal plate small, occupying notch between edges of posterior radials; anal sac swollen globose, protruding backward between posterior arms; short, and composed of numerous irregular plates; anal opening directed laterally toward *C* ray. Arms short, stout, with large gap between massive axillary first primibrachs and radials; several quadrangular, stout, elongate secundibrachs known above primaxils. *U.Perm.(Basleo beds),* Indon. (Timor).——FIG. 499,*1*. **P. permicus; 1a-d,* post., *B*-ray, *DE*-interray, and dorsal (*CD* interray upward) views of crown, all showing bulged anal sac, ×2 (Wanner, 1937).

Family SCOTIACRINIDAE Moore & Strimple, 1973

[Scotiacrinidae MOORE & STRIMPLE, 1973, p. 31]

Like *Zeacrinites* but cup moderately deep bowl shaped and plates tumid, uniserial arms branching on first primibrachs; arm facets plenary. *Up.L.Carb.*

Scotiacrinus WRIGHT, 1945, p. 119 [**Pachylocrinus tyriensis* WRIGHT, 1937, p. 406; OD]. Cup bowl shaped, with narrow basal concavity, plates of cup ridged or corrugated where they join one another, sutures strongly impressed, surface finely granular to coarsely vermiculate; infrabasals small in basal cavity, not visible from side; basals very tumid; radials rounded, not flaring, articular facets occupying full width of radials; anal area normally of 3 plates, radianal, anal *X,* and right tube plate. Arms rather long, uniserial, branching isotomously on primibrachs *1,* and subsequently 2 or 3 times; brachials cuneate, short to rather long, bearing stout pinnules. *Up.L.Carb.,* Eu.(Scot.).——FIG. 500,*1*. **S. tyriensis* (WRIGHT); *1a,* holotype, *C*-ray view of crown, ×0.7; *1b,* paratype, *CD*-interray view oblique from below, ×1.5 (Wright, 1951-54).

Superfamily CALCEOLISPONGIACEA Teichert, 1954

[*nom. transl.* TEICHERT, in MOORE & STRIMPLE, 1973, p. 31 (*ex* Calceolispongiidae TEICHERT, 1954, p. 70)] [Materials for this superfamily prepared by N. GARY LANE]

Cup large to small, base concave to flat, distinguished mainly by prominent spine- or spadelike projections of basals in most species, arms five obliqui-uniserial. *L.Perm.-U.Perm.*

Family CALCEOLISPONGIIDAE Teichert, 1954

[Calceolispongiidae TEICHERT, 1954, p. 70]

Cup large, bowl shaped, with shallow basal invagination; infrabasals small, concealed by stem, not visible from side; basals large, spinose or tuberculate; radial arm facets wide, inclined outward-upward; one anal plate in cup, separating radials. Arms uniserial, unbranched, pinnulate. Stem small relative to cup size. *L.Perm.(Sakmar.-Artinsk.)-U.Perm.(Word.equiv.).*

Among all cladid inadunates most weird examples of aberrant specialization are found among the Calceolispongiidae, the basals of which were originally misidentified as sponges. Some species are quite large, with the cup formed by extremely massive basals associated with comparatively normal circlets of infrabasals and radials, the latter interrupted by an anal *X* plate on the posterior side. The basals are prolonged laterally by huge horn- or spadelike processes. At present these crinoids are known only from Permian rocks of Australia, Timor, and India.

Calceolispongia ETHERIDGE, 1915, p. 9 [**C. hindei*; M] [=*Dinocrinus* WANNER, 1916a, p. 313 (type, *D. cornutus*)]. Cup large to small, bowl shaped with flat base or shallow basal concavity, and indented basal-radial sutures; infrabasals small, diamond shaped, not visible or barely visible from side; basals very large, hexagonal except for *CD* basal which is truncated by anal *X*, ornamented by large horn-like, mammillate, or spade-shaped spines; radials pentagonal and shorter than basals; separated by anal *X* in *CD* interray; radial arm facets occupying entire width of plates, inclined outward-upward, muscle area long. Arms uniserial, pinnulate; proximal 2 primibrachials large, trapezoidal, distal brachials cuneiform. Stem very small relative to cup size. *L.Perm.(Sakmar.-Artinsk.)-U.Perm.(Word. equiv.),* Australia(W. Australia-Queensl.-New S. Wales-Tasm.)-India (C. peninsular and Assam Himalayas)-Indon.(Timor).——FIG. 501,*1a,b*. *C. elegantula* TEICHERT, L. Perm., W.Australia; ventral and dorsal views of cup, ×1 (Teichert, 1949).——FIG. 501,*1c-e*. *C. spectabilis* TEICHERT, L.Perm., W.Australia; inner, ventral, and lat. views of basal spine plate, ×1 (Teichert, 1949).——FIG. 502,*1a,b*. *C. robusta* TEICHERT, L.Perm., W.Australia; dorsal and *CD*-interray views of crown, ×0.5 (Teichert, 1949).

Jimbacrinus TEICHERT, 1954, p. 71 [**J. bostocki*; OD]. Cup large, bowl shaped with shallow invagination below; basals very large, hexagonal, mammillate; ornamented with 1 to 3 tubercles; radials pentagonal, equal, smaller than basals, bearing one or more tubercles, separated in *CD* interray by anal *X*, radial arm facets form entire upper surface of plates, inclined outward-upward; anal *X* quadrangular, height equal to lateral radial facets, with one to several tubercles. Arms uniserial, pinnulate, primibrachs *1* and *2* trapezoidal, weakly tuberculate, distal brachials cuneiform. Stem small relative to cup size, with alternating long and short columnals. *L.Perm.(Artinsk.),* W.Australia.——FIG. 501,*2*. **J. bostocki*; *2a-c*, *C*-ray, dorsal and *CD*-interray views of crown, ×1 (Teichert, 1954).

FIG. 500. Scotiacrinidae (p. *T*755).

Superfamily and Family UNCERTAIN

[Materials for this superfamily prepared by H. L. STRIMPLE]

Aulodesocrinus WRIGHT, 1942, p. 278 [**A. parvus*, p. 279; OD]. Cup judged to be dicyclic based on proximal sutures on long slender basal plates, with 5 elongate radials, flared distalward, and single elongated anal plate as long as *C* and *D* radials. *L.Carb.(Visean),* Eu.(Scot.).——FIG. 503,*1*. **A. parvus*; *1a,b*, cup from *AB* interray

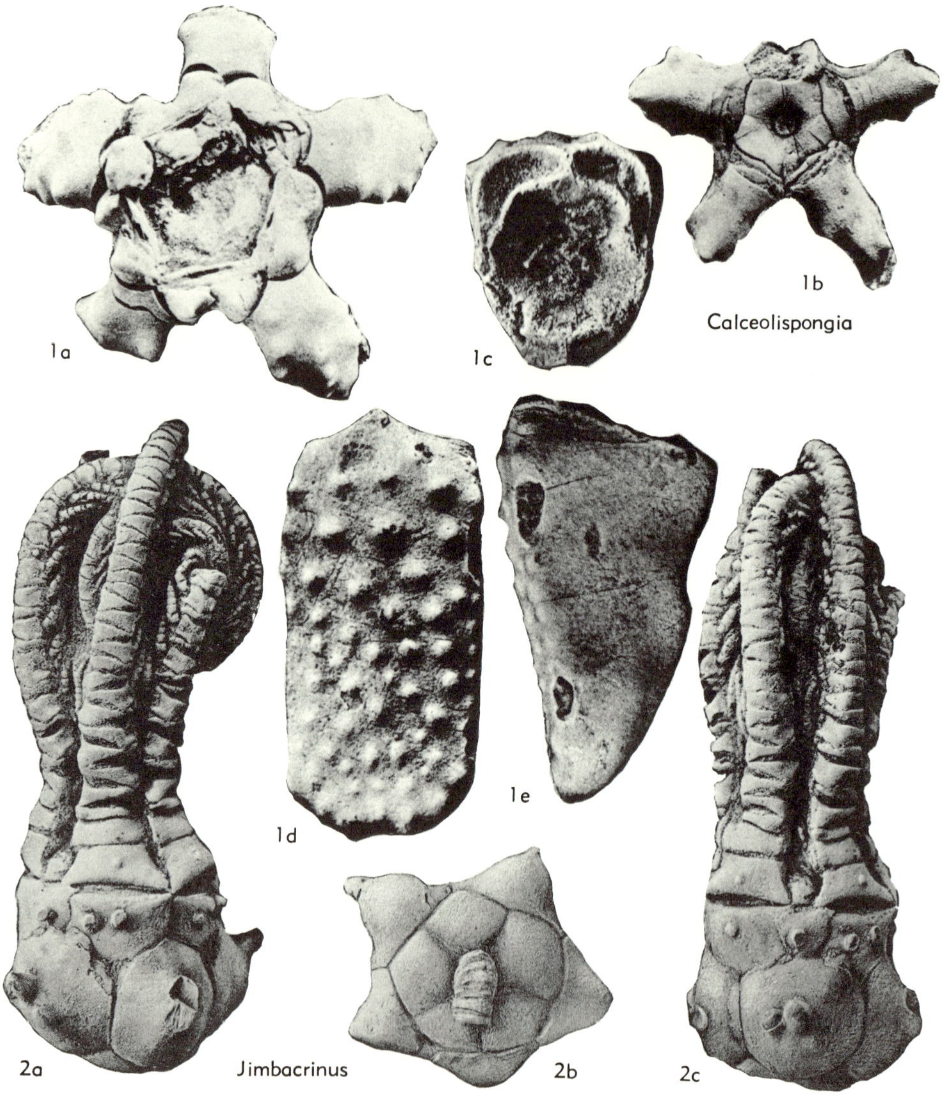

FIG. 501. Calceolispongiidae (p. *T755*).

and from *D* ray with anal plate to right, ×16.5 (Wright, 1952a).

Carlopsocrinus WRIGHT, 1933, p. 198 [**C. bullatus*; OD]. Crown subglobose; cup with swollen fused infrabasal circlet, minute basals, and large radials in contact with infrabasals, lacking anal plates. Each radial bears large, nonaxillary primibrach *1*, which reduces width and curves inward distally. Columnar facet circular, pierced by minute round lumen. *L.Carb.(Visean)*, Eu.(Scot.).——FIG. 503,*2*. **C. bullatus; 2a,b,* holotype crown from base and side, ×15 (Wright, 1952a).

Order and Superfamily UNCERTAIN

Family PARACTOCRINIDAE Jaekel, 1918

[Paractocrinidae JAEKEL, 1918, p. 25] [Materials for this family prepared by H. L. STRIMPLE]

Small crinoid cups composed of three or more circlets of plates which are only slightly offset rather than alternating. There

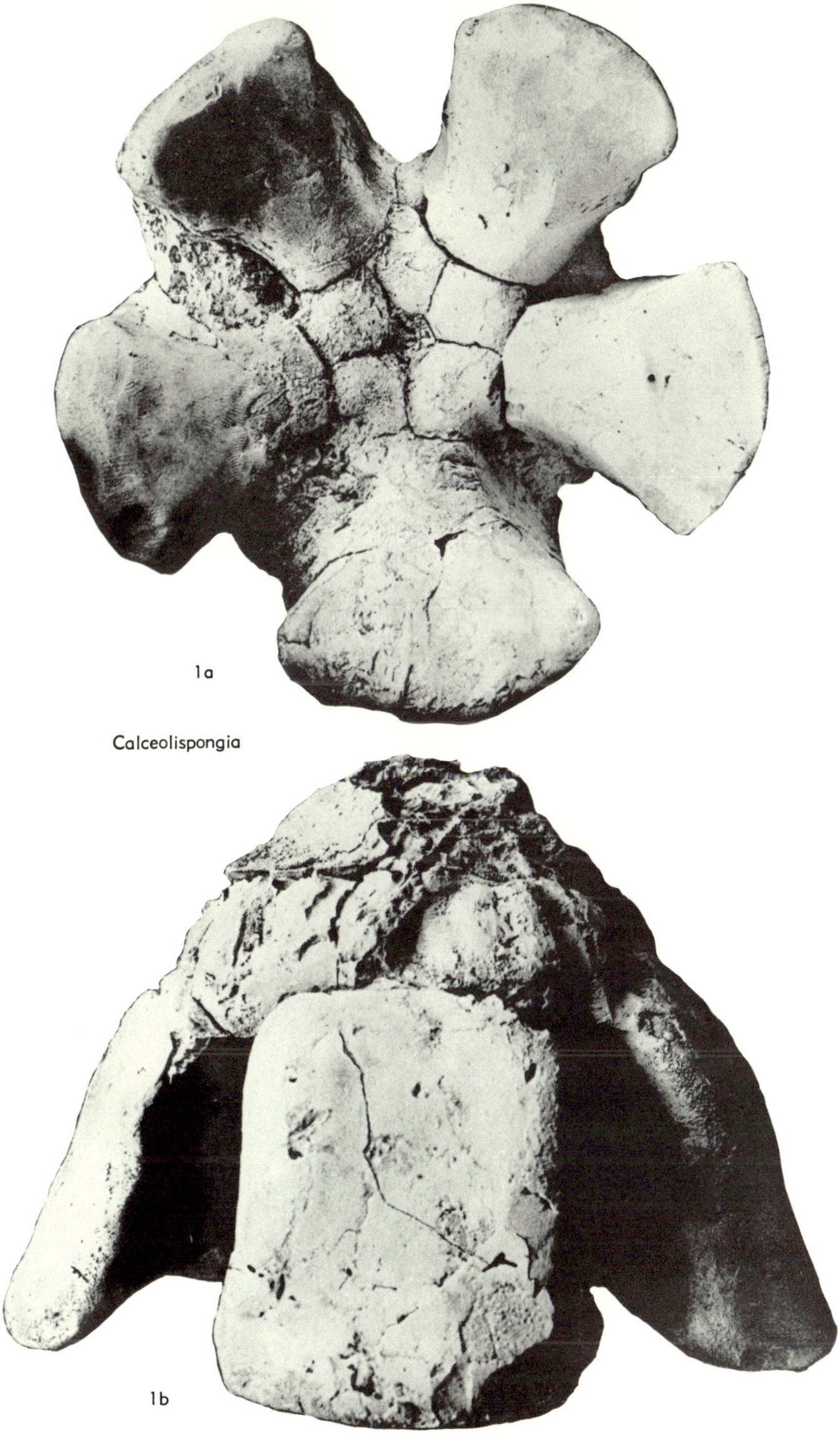

Fig. 502. Calceolispongiidae (p. *T*755).

are no anal plates and in at least some species the proximal columnal has a cone shape with no facet for additional columnals. *Ord.*, USSR.

Paractocrinus JAEKEL, 1918, p. 25 [**P. tuberculatus*, p. 22; OD]. Cup erect, almost cylindrical but widened slightly distalward. Lowermost circlet of plates considered to be infrabasals, with broad circular columnar cicatrix; next circlet comprised of slightly larger plates, thought to be basals; third circlet in summit view shows no arm facets to indicate radials; in fact, there is no body cavity, but instead a lumen in midportion as evidence of at least one more circlet of cup plates. The poorly preserved specimen described as *Paractocrinus laevis* JAEKEL, 1918, has a cuplike base indicating the presence of a cone-shaped proximal columnal as in *Parorthocrinus*. *Ord.* (*?Vaginatenkalk*), NW.USSR(Leningrad area).——FIG. 504,*1a-c*. **P. tuberculatus; 1a,c,* cup from side and summit, ×2; *1b,* from base, ×1 (Jaekel, 1918).——FIG. 504,*1d*. **P laevis* JAEKEL; diag., ×0.7 (Jaekel, 1918).

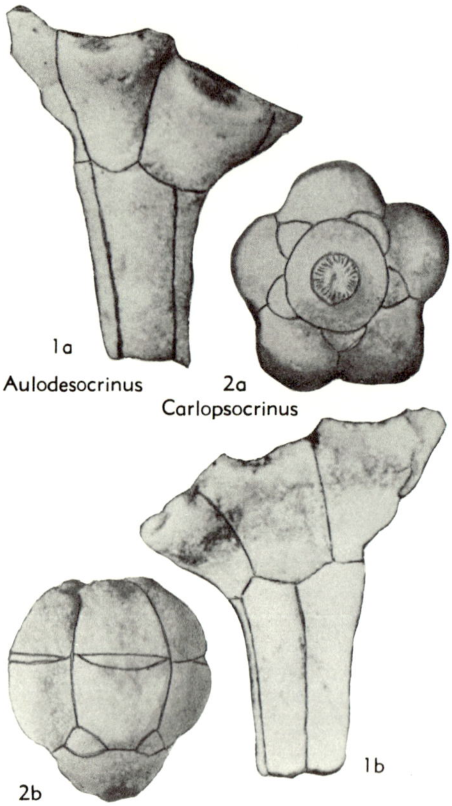

FIG. 503. Superfamily and Family Uncertain (p. T755-T756).

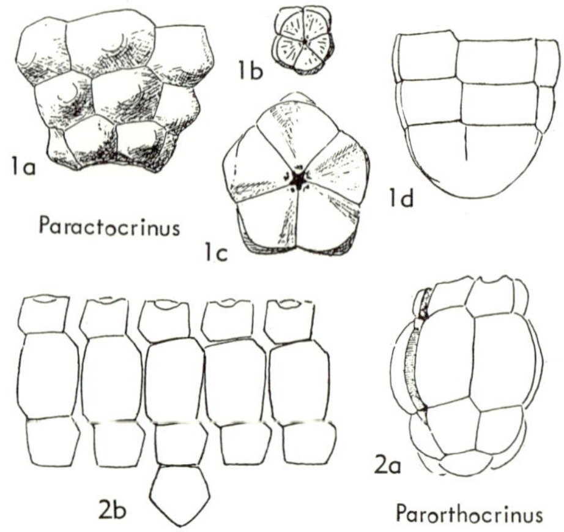

FIG. 504. Inadunata Order Uncertain, Paractocrinidae (p. T758-T759).

Parorthocrinus JAEKEL, 1918, p. 25 [**P. liber*; M]. Cup erect, cylindrical with slightly expanded midsection, and small cuplike basal element (termed a "centrobasal" by JAEKEL), which is probably homologous with the element previously considered to be a cone-shaped proximal columnal in *Paractocrinus*. Basals proportionately longer than in *Paractocrinus*; radial circlet appears to have facets for reception of arms. *Ord.*, NW.USSR(Leningrad area).——FIG. 504,*2*. **P. liber; 2a,b,* side view

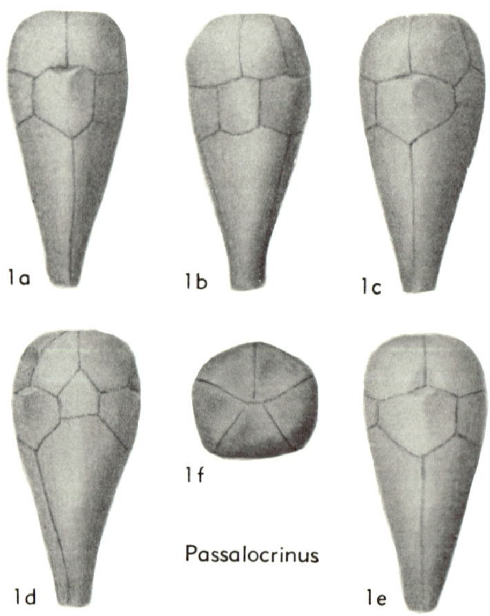

FIG. 505. Inadunata Order and Family Uncertain (p. T759).

of cup, diagram of cup (Jaekel, 1918).

[While this volume was in press, YU. A. ARENDT (1976) proposed a new class of Crinoidea that included the above-mentioned taxa in the following classification:

Class Hemistreptocrinoidea Arendt, 1976
  Order Hemistreptocrinida Arendt, 1976
    Family Hemistreptocrinidae Arendt, 1976
      Hemistreptocrinus Arendt, 1976
        L.Ord., USSR (Leningrad area)
      Nonparactocrinus Arendt, 1976
        L.Ord. or M.Ord., USSR (Leningrad area or Estonia)
      Parorthocrinus Jaekel, 1918
        L.Ord., USSR (Leningrad area)
      Tetractocrinus Jaekel, 1918
        L.Ord., USSR (Leningrad area)
    Family Paractocrinidae Jaekel, 1918
      Paractocrinus Jaekel, 1918
        L.Ord. or M.Ord., USSR (Leningrad area or Estonia).—Eds.]

Family UNCERTAIN

Passalocrinus PECK, 1936, p. 292 [**P. triangularis*; OD]. Small, monocyclic, theca shaped like old-fashioned ice cream cone. Basals 3, elongated, equidimensional; radials 5, short, variable sizes and shapes with 3 largest bearing articular facets (*A, B,* and *D* rays); orals 5, *CD* oral the largest. No anal plates or anal opening known. Arms and column unknown. [Specimens described by CRONEIS & GEIS, 1940, purported to be immature representatives of the blastoids *Mesoblastus glaber* and *Pentremites princetonensis*, are remarkably similar to *Passalocrinus*. The ontogeny presented did not show a transition from a *Passalocrinus*-like stage to a recognizable blastoid stage and the matter has not been resolved to date.] *L.Miss.-M.Miss.,* USA (Mo.-Ky.-Okla.-N.Mexico).——FIG. 505,*1*. **P. triangularis*, L.Miss., Mo.; *1a-f*, views of theca from *A, E, D, C,* and *B* rays, and ventral, ×45 (Peck, 1936). [STRIMPLE]

FLEXIBILIA[1]

By RAYMOND C. MOORE
[University of Kansas]

GENERAL MORPHOLOGY

Crinoids included in the class Flexibilia display from beginning to end of their long existence (M.Ord.-U.Perm.) a remarkable unity of important morphological characters, which serves to set them well apart from other crinoid groups. Among the features that appear in almost every taxonomic category the foremost are the following:

1) An aboral cup composed of three infrabasal plates (uncommonly reduced to two plates or fused to a single one) combined with five basals and five radials.

2) The infrabasal circlet typically is composed of a small plate (termed *azygous*, unyoked) and two larger ones each formed by fusion of a pair of antecedent small plates (hence termed *zygous*, yoked), uncommonly, all infrabasals are fused together.

3) In all but a few specimens of one genus (*Forbesiocrinus*) the azygous infrabasal is located in the *C* ray.

4) In simpler forms (e.g., lecanocrinids) the basals and radials form the main part of the aboral cup with summits of the radials mostly forming an even horizontal boundary at the arm bases.

5) Relatively large or diminutive anal plates consisting of radianal and anal *X* plate, or of the latter alone, compose posterior parts of most aboral cups, in one group (Taxocrinidae) the linearly arranged anal plates comprise a tube not firmly joined to the posterior basal or either arm of the posterior rays.

6) Observed tegmens of flexible crinoids are mainly built of extremely numerous small plates that in life presumably were buried in a leathery integument that covered and protected the viscera; ambulacral and interambulacral tracts are differentiated, the former leading between medium-sized to large oral plates surrounding the mouth; the posterior oral, exceeding the others in size, is a porous plate that functioned as a madreporite.

7) Position of the tegmen is at the level of radial plate summits in small rotund forms but well above it in larger, many-plated crowns (e.g., *Sagenocrinites*,

[1] In the chapter on Flexibilia the Wenlockian and Ludlovian of Sweden (Gotl.) are bracketed as undifferentiated Upper Silurian, whereas in other parts of this volume the Wenlockian is identified as Middle Silurian.

Forbesiocrinus, Onychocrinus, and many others); wherever precisely known, it is at the height of the so-called Y-line of the crown situated at the boundary between laterally united arm plates (associated in many forms with interlocked interradial and interbrachial plates) at the lowermost limit of free arms.

8) Arms universally are uniserial and lacking pinnules, their brachials generally much wider than high, with mostly flattened exterior surfaces. A patelloid process is present on the midline of the brachials of many genera.

9) Arms of a few genera are unbranched but generally they divide one or more times isotomously; heterotomous branching (mostly of biendotomous type) characterizes several forms.

10) Strong incurving of arms at summit of the crown is general.

11) Interradial and interbrachial plates characterize many genera; they exhibit much variation in size and number, as well as in being loosely or very firmly joined to brachials of rays.

12) Except in a very few aberrant forms, the stem is transversely circular and relatively robust, never pentagonal; where moderate to considerable lengths of it are known, its diameter is substantially uniform, but distal tapering characterizes several forms, especially just below the calyx. A differentiated proxistele of wide, thin columnals is present in many genera.

FORM AND STRUCTURE OF CROWN

A large majority of flexible crinoids have a rotund crown, either 1) subspherical, with width and height subequal (most lecanocrinids, nipterocrinids such as *Pycnosaccus* and *Hormocrinus,* dactylocrinids such as *Aexitrophocrinus, Paramphicrinus, Ainacrinus, Amphicrinus,* icthyocrinids such as *Clidochirus* and *Metichthyocrinus,* and taxocrinids such as *Parichthyocrinus, Synerocrinus,* and *Euonychocrinus*); 2) elongate, with width modestly to very distinctly less than height (*Lecanocrinus* and *Miracrinus* among lecanocrinids, *Nipterocrinus* and *Cholocrinus* among nipterocrinids, most homalocrinids, dactylocrinids, sagenocrinitids, icthyocrinids, and taxocrinids); and 3) low rotund, with width greater than height (*Caldenocrinus, Zenocrinus,* and *Dieurycrinus* among dactylocrinids). Illustrations and structural diagrams of all these forms are given with the systematic part of the text.

An examination of the structure of flexible crinoid crowns of all types reveals the anomalous nonflexible nature prevailing in their construction. Most plates of the calyx, including those of the more or less morphologically differentiated aboral cup and lower part of the arms, with associated interradial and interbrachial plates where present, are rigidly united together. Movement between them therefore is nil or at most insignificant. Only distal parts of arm appendages are capable of appreciable inward and outward movements, which can be defined as flexible. Thus, a comparative equivalence in the crown structure of all or nearly all crinoids clearly links the Flexibilia with Camerata, Inadunata, and Articulata. It is prevalence of interplate articulation by means of very short ligament fibers, accompanied by lack of structures providing for muscular articulations of any sort, which separate the flexible crinoids from others. In general, the Flexibilia comprise a readily recognizable assemblage of crinoids distinguished both by their crown structure and by all-but-constant morphological features of their aboral cup and arm construction.

ELEMENTS OF ABORAL CUP AND THEIR RELATIONSHIP TO OTHER PARTS OF WHOLE CRINOID

INFRABASALS

Morphologic features of the infrabasal plates at the base of aboral cups of flexible crinoids are described below according to the inclusion in six arbitrarily differentiated groups, in those numbered (3) and (4) entirely covered by the topmost columnal of the stem, but in others with a small or relatively large part of their distal extremities visible outside of the stem (Fig. 506). It is evident that the different groups lack taxonomic significance, for representatives

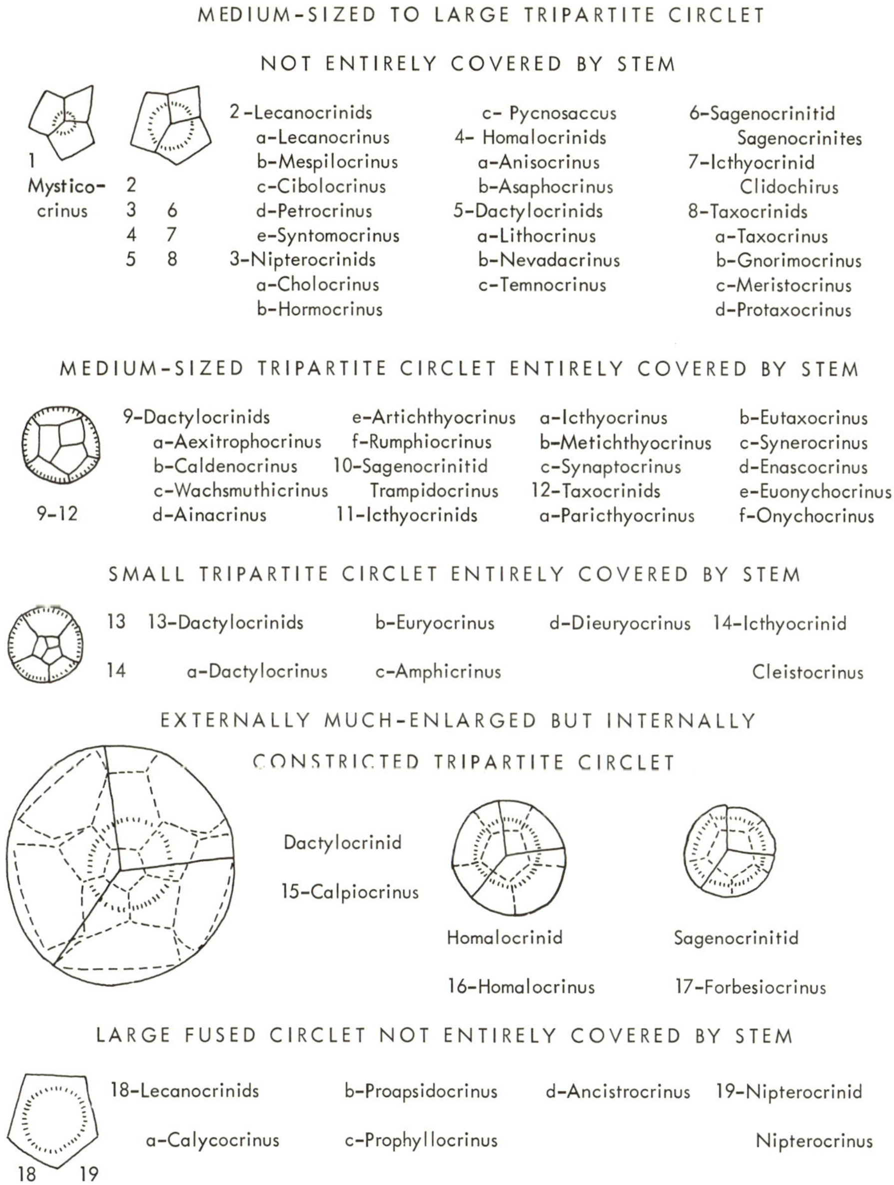

FIG. 506. Diagrams showing morphologic features of the infrabasal circlet of flexible crinoids (not to scale).

are distributed among different families.

1) Relatively large tripartite infrabasal circlet with most of plates visible beyond small stem impression. Such an arrangement characterizes *Mysticocrinus* (Lecanocrinidae) (Fig. 506,*1*), distinguishing it from all other flexible crinoids.

2) Medium-sized tripartite infrabasal circlet not entirely covered by relatively large stem. Crinoids of this group resemble the first except that the stem covers most but not all of the plates, their distal extrem-

ities being clearly visible beyond the stem impression. Here belong lecanocrinids (Fig. 506,*2a-e*), nipterocrinids (Fig. 506,*3a-c*), homalocrinids (Fig. 506,*4a,b*), dactylocrinids (Fig. 506,*5a-c*), sagenocrinitid (Fig. 506,*6*), icthyocrinid (Fig. 506,*7*), and taxocrinids (Fig. 506,*8a-d*). Orientation of the *azygous* plate is constantly in the *C* ray.

3) Medium-sized tripartite infrabasal circlet entirely concealed by stem. The circlet occupies most but not quite all of the stem impression. Here belong dactylocrinids (Fig. 506,*9a-f*), a sagenocrinitid (Fig. 506,*10*), icthyocrinids (Fig. 506, *11a-c*), and taxocrinids (Fig. 506,*12a-f*).

4) Small tripartite infrabasal circlet entirely concealed by stem which covers also proximal part of basals. These are mostly dactylocrinids (Fig. 506,*13a-d*). *Cleistocrinus* (Fig. 506,*14*) represents the icthyocrinids.

5) Externally much enlarged but internally constricted tripartite infrabasal circlet. This is an exceptional arrangement found in three flexible crinoids. The stem covers only proximal parts of exposed infrabasals in *Calpiocrinus* (Fig. 506,*15*) and *Homalocrinus* (Fig. 506,*16*) but most of these plates in *Forbesiocrinus* (Fig. 506,*17*). In *Calpiocrinus* the infrabasals are so enlarged that they cover not only basals but radial plates.

6) Large, fused infrabasal circlet mostly but not entirely covered by stem. In this group are lecanocrinids (Fig. 506,*18a-d*) and *Nipterocrinus* (Fig. 506,*19*).

BASALS

Morphologically viewed, plates of the basal circlet of flexible crinoids may be divided into two groups, based on relative importance as elements of the aboral cup.

1) Basals forming important, even major part of aboral cup, entirely outside and above area of stem impression. In numerous genera they form part of the base and the main part of the sides of the aboral cup. In others they are confined to the sides. Their size is relatively large, little if any smaller than the radials. Distally they invariably slope more or less steeply upward.

Basals of this group especially characterize lecanocrinids, as seen in *Mysticocrinus* (Fig. 507,*1a*), *Lecanocrinus* (Fig. 507,*1b*), *Miracrinus* (Fig. 507,*1c*), *Geroldicrinus* (Fig. 507,*1d*), *Mespilocrinus* (Fig. 507,*1e*), *Cibolocrinus* (Fig. 507,*1f*), and *Calycocrinus* (Fig. 507,*1g*). Among nipterocrinids typical representatives of the group are seen in *Cholocrinus* (Fig. 507,*2a*), *Pycnosaccus* (Fig. 507,*2b*), and *Hormocrinus* (Fig. 507,*2c*). Among homalocrinids these relatively large basals occur in *Anisocrinus* (Fig. 507,*3a*), and *Asaphocrinus* (Fig. 507, *3b*); in dactylocrinids they are well defined in *Lithocrinus* (Fig. 507,*4a*), *Nevadacrinus* (Fig. 507,*4b*), and *Temnocrinus* (Fig. 507,*4c*). The sagenocrinitids are represented by the single genus *Sagenocrinites* (Fig. 507,*5*) and icthyocrinids by the single genus *Clidochirus* (Fig. 507,*6*). Among taxocrinids are four genera with large basals outside of the stem impression area, *Taxocrinus* (Fig. 507,*7a*), *Meristocrinus* (Fig. 507,*7b*), *Protaxocrinus* (Fig. 507,*7c*), and *Gnorimocrinus* (Fig. 507,*7d*).

2) Basals forming minor part of aboral cup, largely or entirely covered by stem except for distal extremity of the posterior basal. In this group the aboral cup is not a prominent part of the calyx, being differentiated only by morphologic distinctions.

Most dactylocrinids have basals belonging to this second group, *Amphicrinus* (Fig. 508,*1a*), *Dactylocrinus* (Fig. 508,*1b*), *Aexitrophocrinus* (Fig. 508,*1c*), *Ainacrinus* (Fig. 508,*1d*), *Euryocrinus* (Fig. 508,*1e*), *Dieuryocrinus* (Fig. 508,*1f*), *Artichthyocrinus* (Fig. 508,*1g*), *Rumphiocrinus* (Fig. 508,*1h*), *Caldenocrinus* (Fig. 508,*1i*), and *Wachsmuthicrinus* (Fig. 508,*1j*). Sagenocrinitids are represented by *Forbesiocrinus* (Fig. 508, *3a*), and *Trampidocrinus* (Fig. 508,*3b*); icthyocrinids by *Icthyocrinus* (Fig. 508,*2a*), *Synaptocrinus* (Fig. 508,*2b*), and *Metichthyocrinus* (Fig. 508,*2c*). Five genera of taxocrinids have basals almost entirely covered by the stem; they are *Parichthyocrinus* (Fig. 508,*4a*), *Enascocrinus* (Fig. 508,*4b*), *Eutaxocrinus* (Fig. 508,*4c*), *Euonychocrinus* (Fig. 508,*4d*), *Synerocrinus* (Fig. 508,*4e*), and *Onychocrinus* (Fig. 508,*4f*).

ANALS

Anal plates of flexible crinoids range from a single one to many. They are more or less prominent elements of the posterior side of the aboral cup, also present only

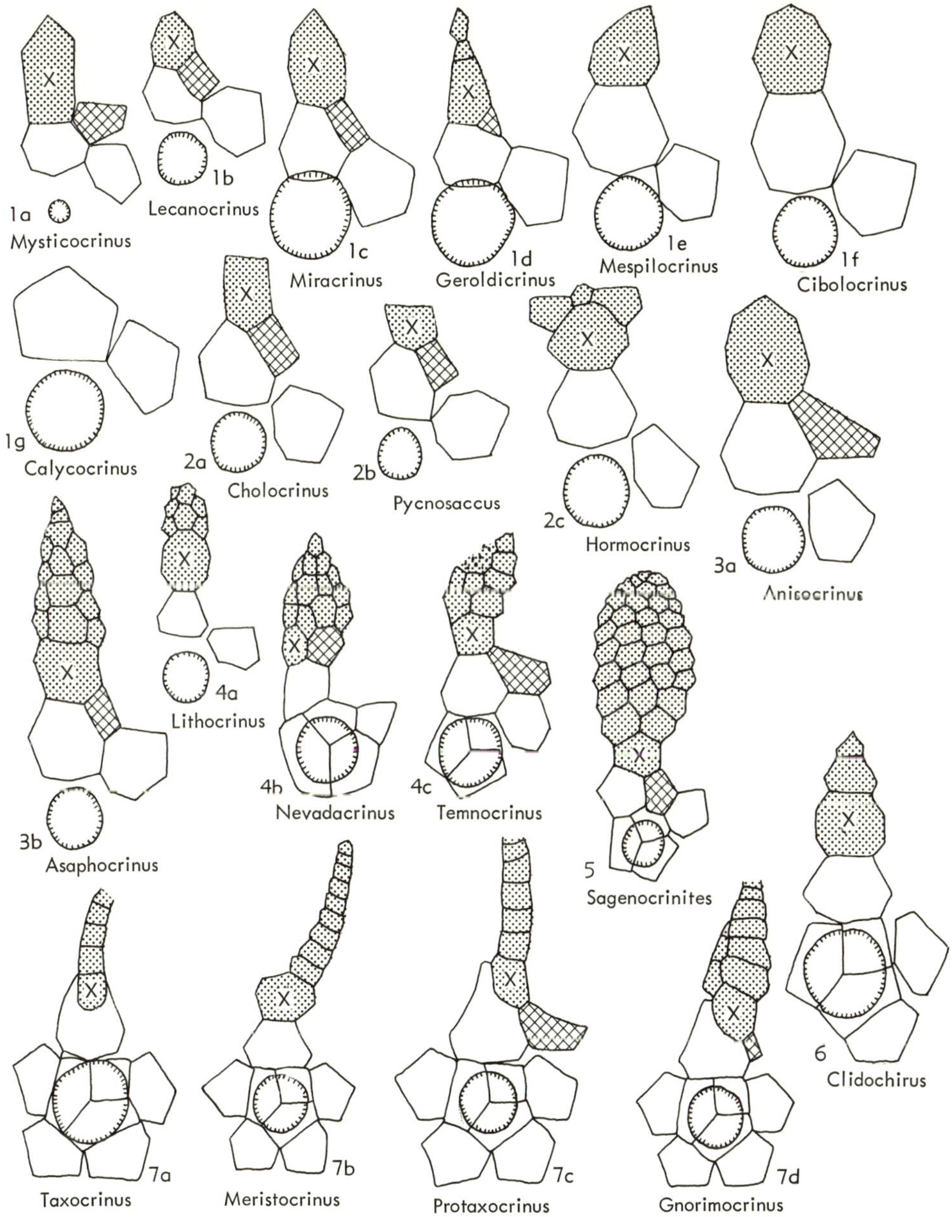

FIG. 507. Diagrams showing *BC* and *CD* basals or entire circlet in relation to stem impression, also anal-plate series (radianal plate cross ruled, anal series stippled) (not to scale).

above the summit of the cup. They form a generally significant part of the calyx. The anals are distinguishable as a radianal (absent in many forms); anal *X* plate and calyx or tube plates above the *X*.

Crinoids having a radianal are divisible into three groups differentiated as follows:

1) Radianal primitive, occurring directly below the *C* radial. Here belong the lecanocrinid *Mysticocrinus* (Fig. 507,*1a*), the homalocrinid *Anisocrinus* (Fig. 507,*3a*), the dactylocrinid *Temnocrinus* (Fig. 507,*4c*),

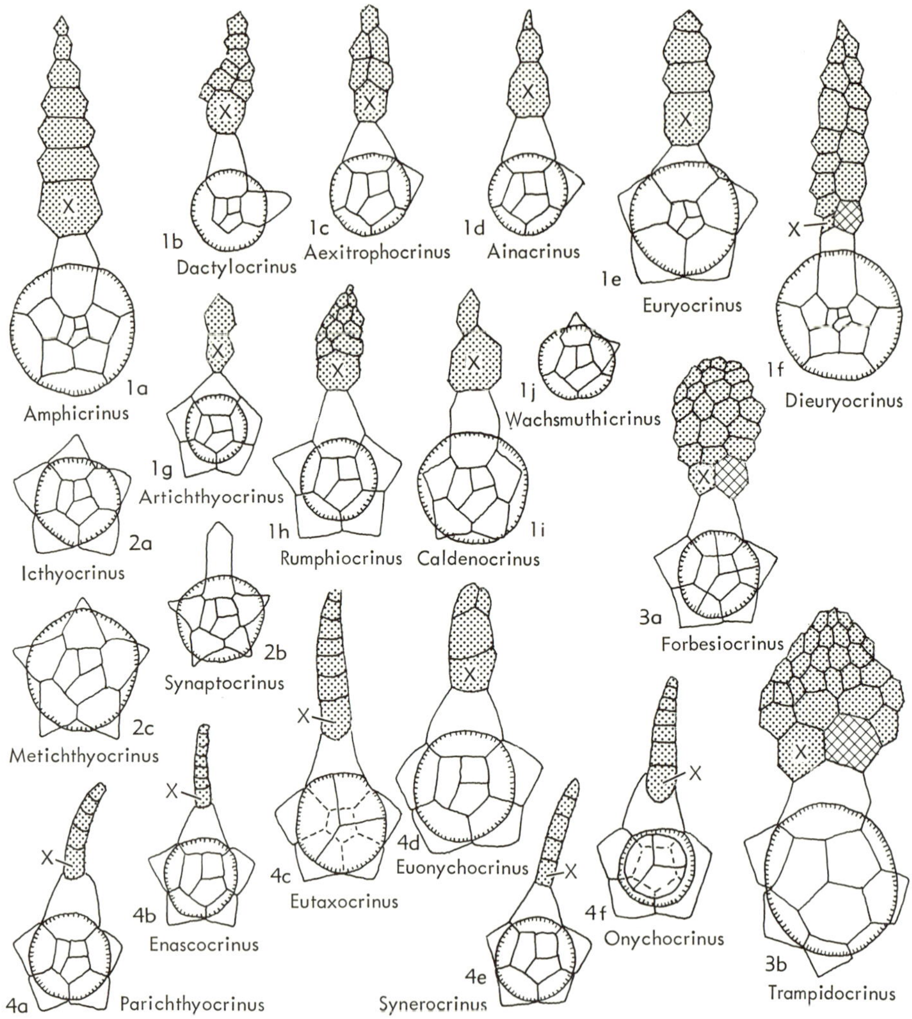

Fig. 508. Diagrams of same morphologic features given in Fig. 507 (patterns as in Fig. 507).

the sagenocrinitid *Sagenocrinites* (Fig. 507, *5*), and the taxocrinid *Protaxocrinus* (Fig. 507,*7c*).

2) Radianal rhombic (or modified to subtriangular, pentagonal, or hexagonal) obliquely at left below the *C* radial. In this group belong the lecanocrinids *Lecanocrinus* (Fig. 507,*1b*), *Miracrinus* (Fig. 507,*1c*), and *Geroldicrinus* (Fig. 507,*1d*); the nipterocrinids *Cholocrinus* (Fig. 507,*2a*), and *Pycnosaccus* (Fig. 507,*2b*); the homalocrinid *Asaphocrinus* (Fig. 507,*3b*); and the taxocrinid *Gnorimocrinus* (Fig. 507,*7d*).

3) Radianal displaced upward to adjoin the anal *X* plate on right. Here four genera are recognized: among dactylocrinids, *Nevadacrinus* (Fig. 507,*4b*) and *Dieuryocrinus* (Fig. 508,*1f*); sagenocrinitids include *Forbesiocrinus* (Fig. 508,*3a*) and *Trampidocrinus* (Fig. 508,*3b*). A radianal plate is lacking in all other flexibles.

Crinoids having no anal plates at all include the lecanocrinid *Calycocrinus* (Fig. 507,*1g*), the dactylocrinid *Wachsmuthicrinus* (Fig. 508,*1j*), and the icthyocrinids *Icthyocrinus* (Fig. 508,*2a*), *Synaptocrinus*

(Fig. 508,*2b*), and *Metichthyocrinus* (Fig. 508,*2c*). The orientation of cups and calyces belonging to these genera is determined by arrangement of plates in the infrabasal circlet.

RADIAL AND ARM PLATES

Radial plates of nearly all flexible crinoids have straight to very gently curved summits. In a few genera, however, the summit is moderately to strongly arcuate and asymmetrical. This latter group, all lecanocrinids, include *Mysticocrinus* (Fig. 509,*1a*), *Proapsidocrinus* (Fig. 509,*1b*), *Prophyllocrinus,* and *Ancistrocrinus.*

Construction of the arms, combined with other morphological features, has considerable taxonomic significance. Three groups of arm types can be recognized among flexible crinoids, with several subgroups.

1) Arms unbranched. The lecanocrinids *Palaeoholopus* (see Fig. 525,*1*) and *Permobrachypus* (see Fig. 525,*4*) lack arm bifurcations.

2) Flexible crinoids characterized by isotomous bifurcation of the arms comprise a second group, which is divisible into subgroups as follows:

2a) Single bifurcation in each ray. Only the lecanocrinids *Petrocrinus* (see Fig. 521,*1*) and *Mysticocrinus* (Fig. 509,*1a*), *Syntomocrinus* (see Fig. 521,*2*), *Gaulocrinus* (see Fig. 525,*5*), and *Ancistrocrinus* (see Fig. 528,*4*), appear to belong in this subgroup.

2b) Two even bifurcations in each ray. All structures of this type are seen in the lecanocrinids *Proapsidocrinus* (Fig. 509, *1b*), *Mespilocrinus* (Fig. 509,*2c*), and *Cibolocrinus* (Fig. 509,*2d*). Other flexibles of the subgroup include the nipterocrinids *Nipterocrinus* and *Cholocrinus,* the dactylocrinids *Artichthyocrinus* and *Nevadacrinus,* and the icthyocrinid *Metichthyocrinus.*

2c) Three uneven bifurcations in all branches of each ray. Structures of this type are seen in the nipterocrinids *Hormocrinus* and *Pycnosaccus* (Fig. 509, *3a,b*), in the dactylocrinid *Temnocrinus* (Fig. 510,*1a*), in the sagenocrinitids *Trampidocrinus, Sagenocrinites,* and *Forbesiocrinus,* in the icthyocrinids *Synaptocrinus* and *Clidochirus,* and in the taxocrinids *Taxocrinus, Meristocrinus, Parichthyocrinus,* and *Eutaxocrinus* (Fig. 510,*1b*).

2d) Two or more even bifurcations in some but not all branches of each ray. Structures of this type are found in the lecanocrinids *Lecanocrinus* (Fig. 509,*2a*), *Miracrinus* (Fig. 509,*2b*), and *Geroldicrinus,* in the dactylocrinids *Euryocrinus* (Fig. 510,*2a*), *Amphicrinus* (Fig. 510, *2b*), and *Caldenocrinus,* in the icthyocrinid *Icthyocrinus,* and in the taxocrinids *Gnorimocrinus* and *Protaxocrinus.*

3) Main arm branches bifurcating isotomously, then each branch dividing heterotomously. This readily differentiated assemblage is divisible into three subgroups.

3a) Uniendotomous (only two main branches). Here the main branches bear small ones facing each other on inner sides of the main ones, as in Homalocrinus (Fig. 509,*4a*).

3b) Biendotomous (four main branches). The two pairs of main branches bear small arms facing each other so as to form a double-combed pattern. This predominates in the dactylocrinids as illustrated by *Aexitrophocrinus* (Fig. 509,*4b*), *Wachsmuthicrinus* (Fig. 509, *4c*), *Dactylocrinus* (Fig. 509,*4d*), *Calpiocrinus* (Fig. 509,*4e*), *Paramphicrinus* (Fig. 510,*3a*), *Lithocrinus, Calpiocrinus,* and *Zenocrinus.* It is also seen in the taxocrinids *Synerocrinus* (Fig. 510,*3b*), *Enascocrinus, Euonychocrinus,* and *Onychocrinus.*

3c) Uniexoendotomous (two main branches with ramules on both sides of branches). This distinctive type of arm structure is found only in *Onychocrinus* of the Taxocrinidae.

Subclass FLEXIBILIA Zittel, 1895

[*nom. transl.* MOORE & LAUDON, 1943a, p. 64 (*ex* order Flexibilia ZITTEL, 1895, p. 165)] [=Articulata ZITTEL, 1879, p. 345, *non* MILLER, 1821, p. 13; suborder Articulosa JAEKEL, 1894, p. 118; Articulata Impinnata WACHSMUTH & SPRINGER, 1885, p. 304(82) *(partim)*; grade Impinnata BATHER, 1899b, p. 922; order Flexibilida PEARSE, 1947, p. 10] [Materials for this subclass prepared by R. C. MOORE, with additional contributions as noted in text]

Crown ovoid to globose, consisting mainly of closely abutting uniserial arms incurved distally, distinctly differentiated from cup or with proximal brachials and interarm plates merging uninterruptably with radials

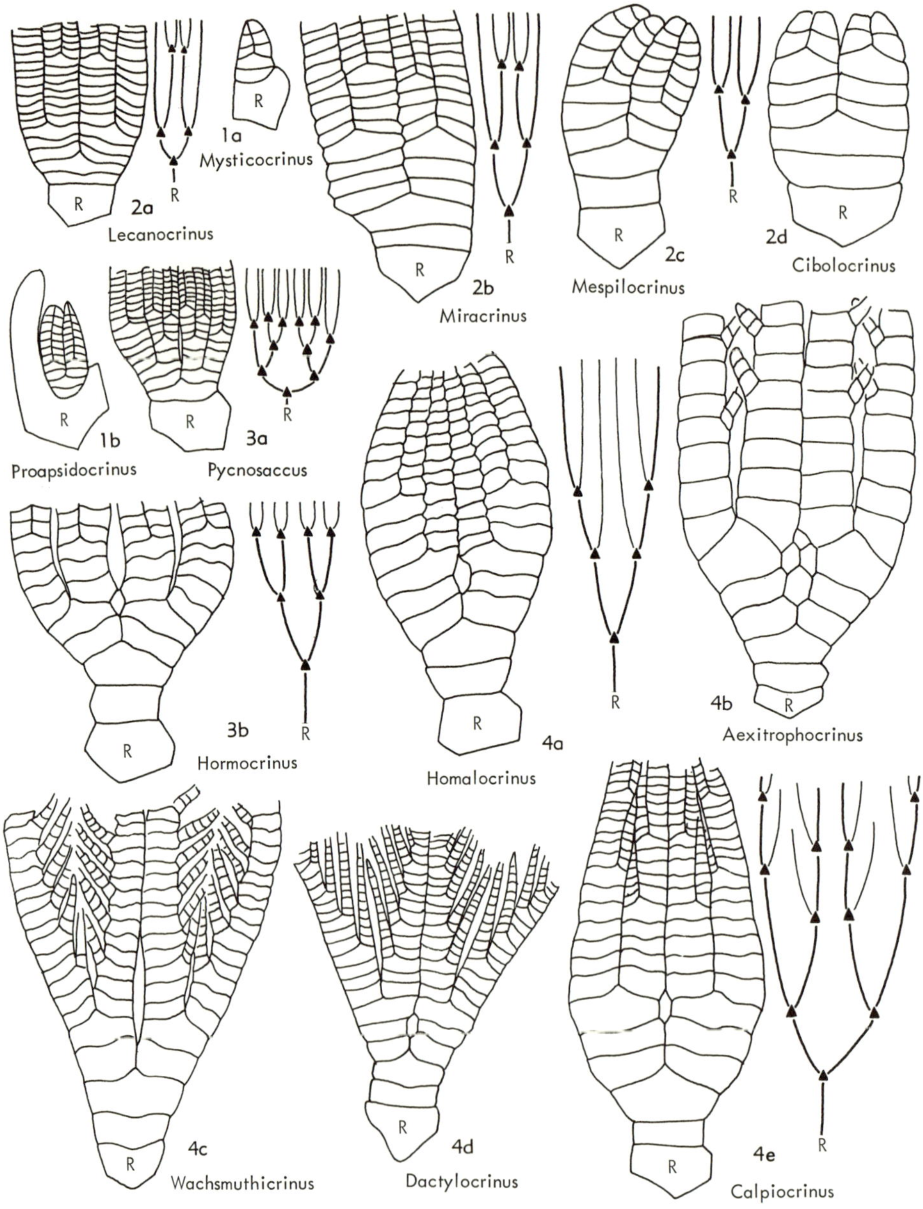

FIG. 509. Diagrams of radial *(R)* and arm structure of *C* ray observed in flexible crinoids, some accompanied by drawings designed to portray arm-branching pattern in simplified form (black triangles denote axillary brachials and heavier lines main arm branches; *R*, radial) (not to scale).

and two subjacent plate circlets, nearly all joined together by ligaments rather than by solid union. Infrabasals three, small azygous one in *C* ray and large ones in *AB* and *DE* rays, or rarely fused into single disc or entirely resorbed. Five basals and five radials subequal in each circlet. Radianal commonly present, either in primitive position directly below *C* radial or shifted obliquely left. Anal *X* plate inserted be-

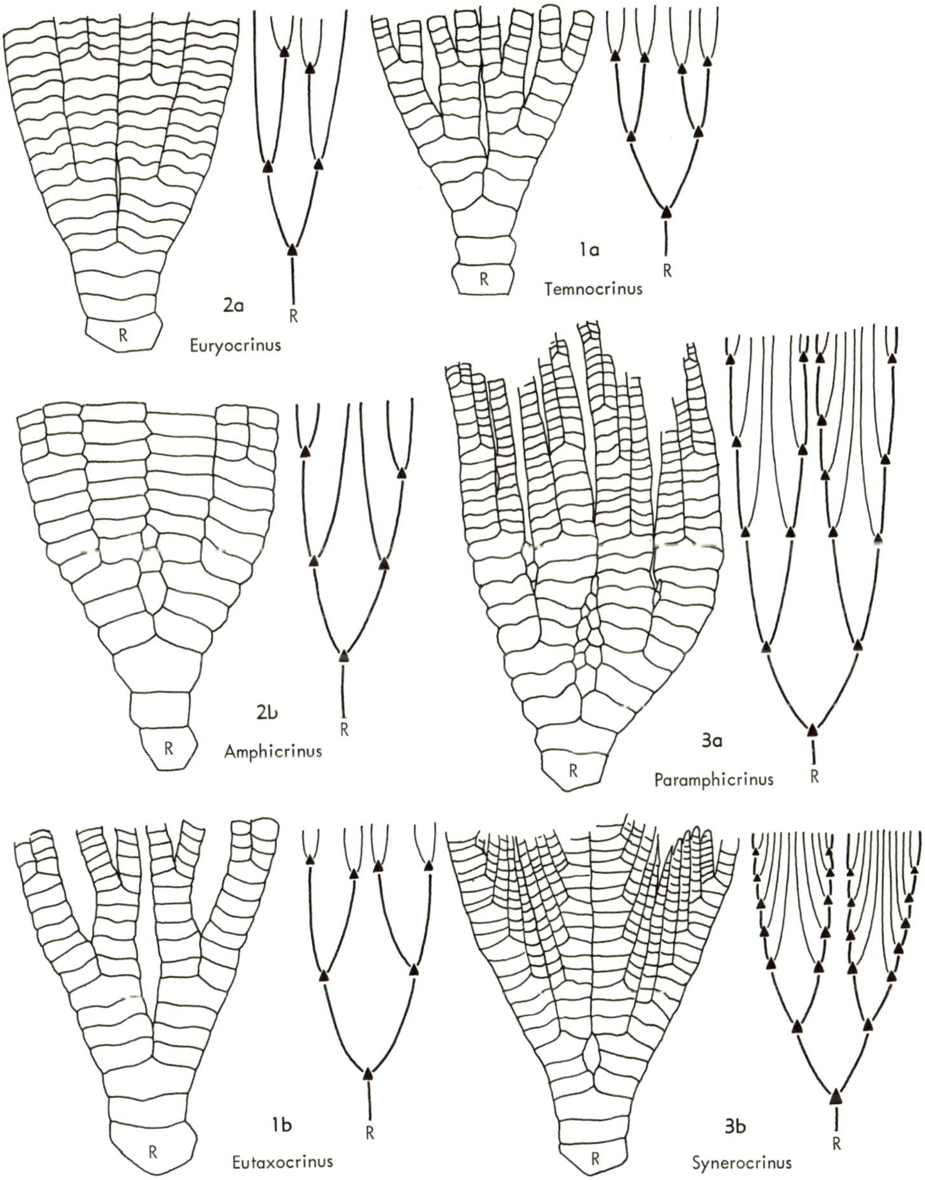

FIG. 510. Diagrams of same morphologic features given in Fig. 509.

tween posterior radials. Arms nonpinnulate generally branching isotomously or both isotomously and heterotomously, exceptionally atomous. Supplementary small, medium-sized or large polgonal plates may be developed interradially as well as between arm branches (interbrachials). Tegmen flexible, lacking anal sac but in many genera containing five distinct orals (posterior one largest), rows of ambulacrals leading from arm bases to exposed mouth, and large areas of small interambulacral plates. Stem transversely round, composed of thin columnals, or very rarely crescentic *(Ammonicrinus)*; noncirriferous. *M.Ord.-U.Perm.*

Order TAXOCRINIDA Springer, 1913

[*nom. correct.* MOORE in MOORE, LALICKER, & FISCHER, 1952, p. 614 (*pro* order Taxocrinoidea SPRINGER, 1913, p. 205)] [=*Taxocrinites* JAEKEL, 1918, p. 78]

Crown elongate ovoid, with well-rounded arms not abutting next adjacent neighbors and relatively weak calyx lacking distinct cup, anal *X* and succeeding longitudinal row of anal plates separated from posterior radials and primibrachs by small-plated flexible perisome instead of joined to them and *CD* basal by close suture. *M.Ord.-U. Penn.*

Representatives of the Taxocrinida are the only Flexibilia known from pre-Silurian deposits anywhere in the world. These most ancient flexible crinoids resemble contemporaneous inadunates in having a steeply conical cup with strongly upflaring infrabasals, large radianal directly beneath the *C* radial, well-developed anal *X* and row of following anal plates, and slender round arms which are long and isotomously branched at various heights in the crown. Interbrachial plates generally are present but not abundant.

Silurian and youngest crinoids of the Taxocrinida are larger than oldest ones, have more gently upflared cups, display a leftward shift of the radianal or disappearance of this plate (e.g., *Eutaxocrinus*), and diminution of interbrachials without close juxtaposition of the arms. No evolutionary tendency toward greater simplicity of structure or decrease in size is observable.

Superfamily TAXOCRINACEA Angelin, 1878

[*nom transl.* MOORE & STRIMPLE, 1973 (*ex* Taxocrinidae ANGELIN, 1878, p. 8)] [incl. Onychocrinidae JAEKEL, 1918, p. 79]

Crown usually elongate, with rays above radials partly or wholly separated all around; posterior interray invariably differentiated by presence of anal plates in tubelike series not incorporated in calyx; branching of arms isotomous, heterotomous, or both. *M.Ord. (Trenton.)-U.Penn.(Missour.).*

Family TAXOCRINIDAE Angelin, 1878

[Taxocrinidae ANGELIN, 1878, p. 8]

Branching of arms isotomous; radianal present or absent. *M.Ord.(Trenton.)-U. Miss.(Chester.).*

Taxocrinus PHILLIPS in MORRIS, 1843, p. 59 [**Cyathocrinus? macrodactylus* PHILLIPS, 1841, p. 29 (=*Isocrinus macrodactylus* PHILLIPS, 1841, p. 30); SD WORTHEN in MEEK & WORTHEN, 1866a, p. 271] [=*Isocrinites* PHILLIPS, 1841, p. 30 (group) (*non* VON MEYER, 1836); *Isocrinus* PHILLIPS, 1841, p. 30 (obj.) (*non* VON MEYER, 1837); *Cladocrinites* AUSTIN & AUSTIN, 1843, p. 197 (type, *Poteriocrinites egertoni* PHILLIPS, 1836); *Cladocrinus* MORRIS, 1843, p. 59 *(nom. van. pro Cladocrinites); Euryalecrinus* AUSTIN & AUSTIN, 1846, p. 66 *(nom. nud.); Taxocrinites* STEININGER, 1853, p. 37 *(nom. van. pro Taxocrinus); Euryalocrinus* DELAGE & HÉROUARD, 1904, p. 477 *(nom. van. pro Euryalecrinus)*]. Crown elongate with rays not abutting above interray areas, arms divergent; infrabasals subhorizontal, mostly or entirely concealed by column; posterior basal elongate, with distal extremity hollowed for reception of anal *X* followed by anal tube; radianal only in upper oblique position when present; interprimibrachs variable, usually well defined between primibrach series of all rays except *CD* interray, distal margin of these plates crescentic, rising toward rays. Primibrachs 3, followed by arm branches which bifurcate additionally above axillary primibrach. Column usually enlarged proximally. *M.Dev.,* USA(Iowa-Wis.-N.Y.); *U.Dev.,* Eu.(Eng.); *L.Miss.(Kinderhook.-Osag.),* USA (Iowa-Mo.-Ill.-Ind.-Ohio-Ky.); *U.Miss.(Chester.),* USA(Mo.-Ill.-Ky.-Ala.-Okla.); *L.Carb.(Tournais.),* Eu.(Belg.-Eng.-Eire-Ger.).——FIG. 511,*1*; 512,*5c*. **T. macrodactylus* (PHILLIPS), L.Carb.(Pilton beds), Eng.(Devon.); 511,*1*, plate diagram of cup (radials black, *X* and higher anals stippled), part of *C*- and *D*-ray arms, anal-plate series, and interprimibrachs of *BC* and *DE* interrays (mod. from Springer, 1920); 512,*5c,* lat. view of holotype, rays unspecified, ×0.5 (Springer, 1920).——FIG. 512,*4*. *T. colletti* WHITE, L.Miss.(Osag.), USA (Ind.); partial plate diagram with edge views of thick brachials (Springer, 1920).——FIG. 512,*5a,b*. *T. intermedius* WACHSMUTH & SPRINGER, L.Miss. (Kinderhook.), USA(Iowa); tegminal view of holotype showing ambulacral series reaching between large orals, and *D*-ray view of another specimen, ×1 (Springer, 1920).——FIG. 512,*5d*. *T. ornatus* SPRINGER, L.Miss.(Osag., Burlington Ls.), USA(Iowa); *EA*-interray view of holotype showing strongly incurved distal parts of arms, ×1 (Springer, 1920). [=*Chladiocrinus* DELAGE & HEROUARD, 1903 *(nom. null.)*.]

Eutaxocrinus SPRINGER, 1906, p. 493 [**Taxocrinus affinis* MÜLLER, 1856, p. 353; SD SPRINGER, 1920,

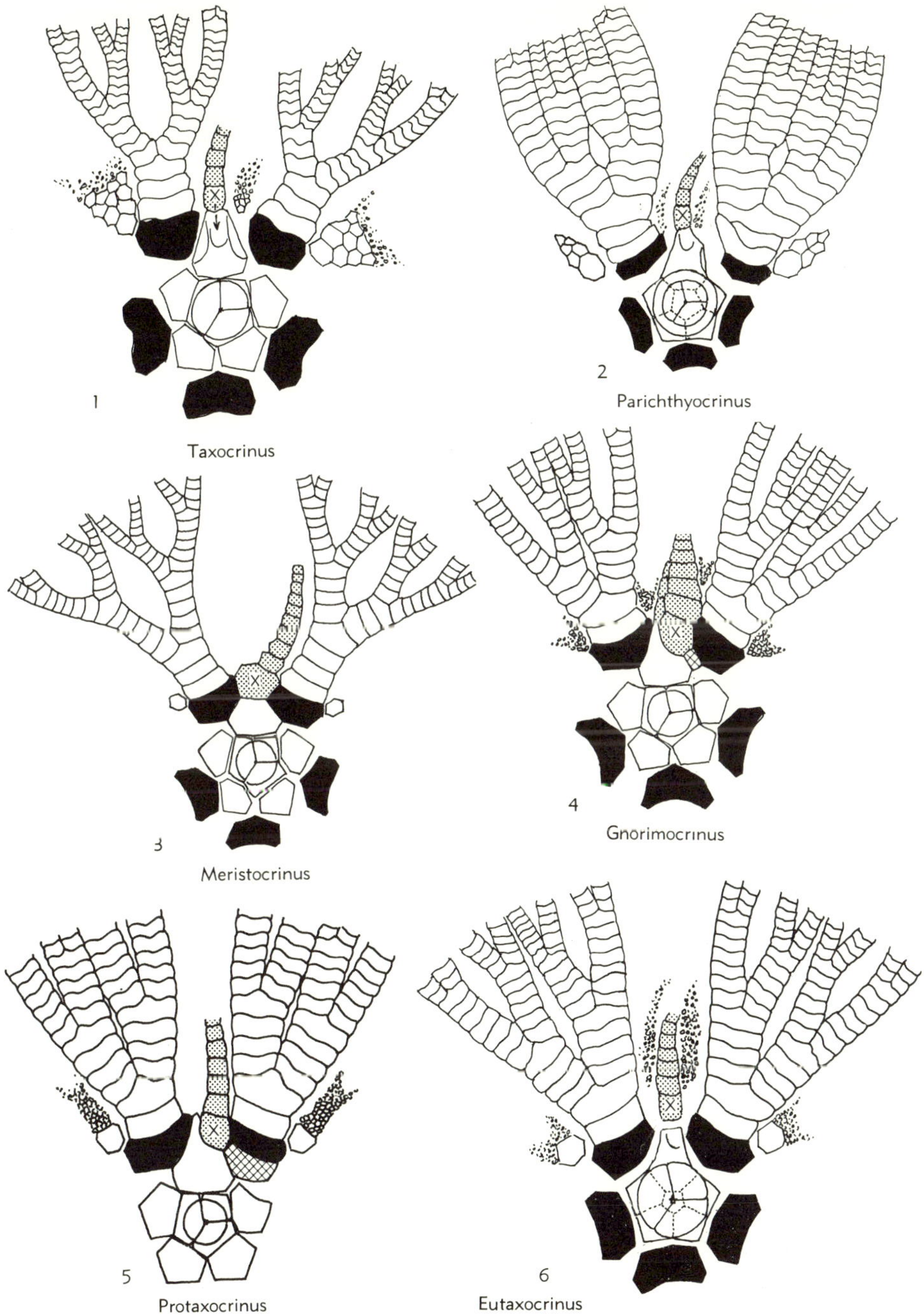

FIG. 511. Taxocrinidae (radials black, *X* and higher anals stippled; radianal cross ruled) (p. *T*768-*T*773).

p. 357] [*Taxocrinus oblongatus* ANGELIN, 1878, and *T. rigens* ANGELIN, 1878, only originally mentioned species, excluded from eligibility as type species because then considered incertae sedis (*Code,* Art. 67h)]. Very similar to *Taxocrinus* but with only 2 primibrachs in each ray and with interprimibrachs very few or lacking. *U.Sil.,* Eu.(Sweden); *L.Dev.,* Eu.(Ger.); *M.Dev.,* Eu.

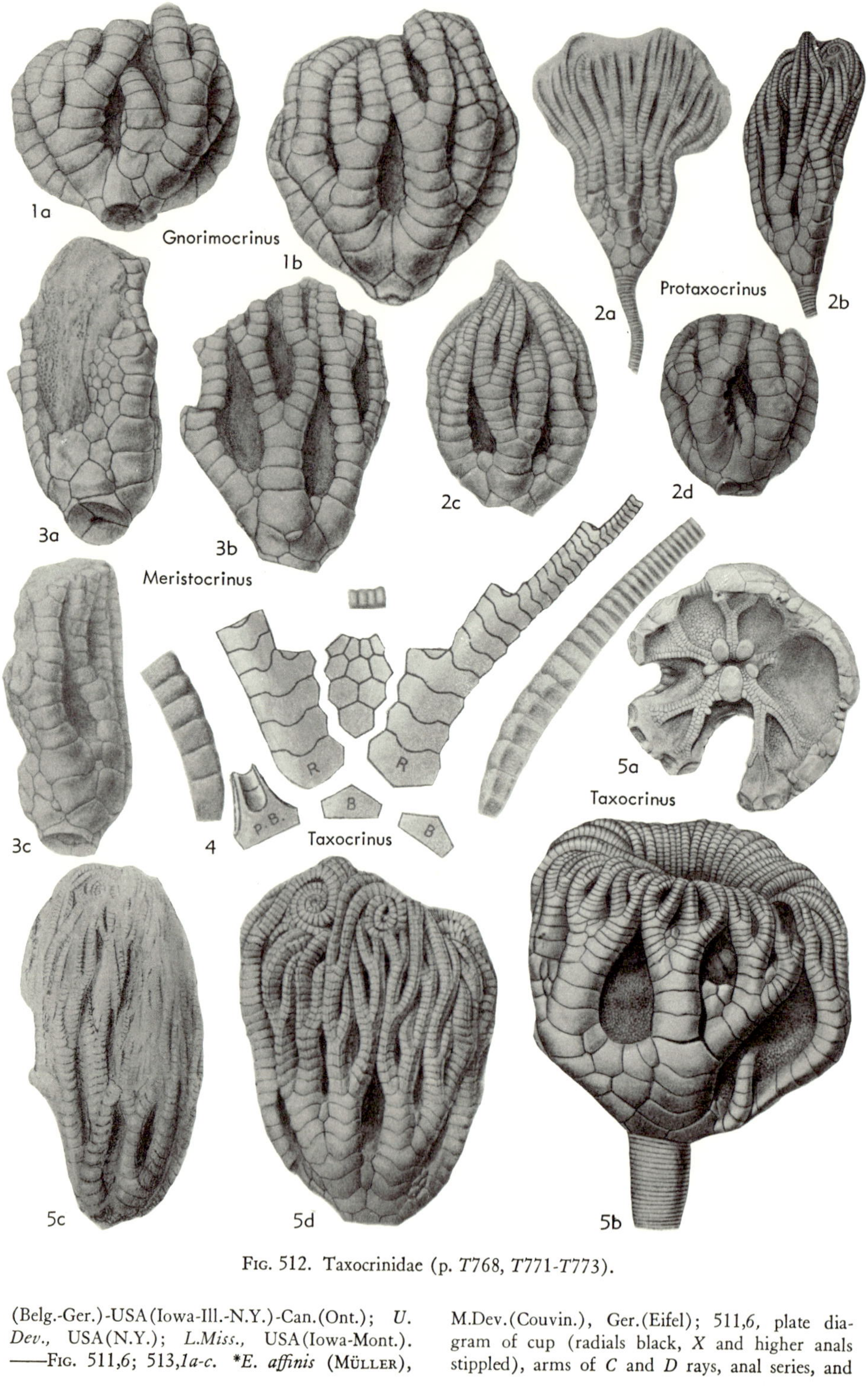

FIG. 512. Taxocrinidae (p. *T768*, *T771-T773*).

(Belg.-Ger.)-USA(Iowa-Ill.-N.Y.)-Can.(Ont.); *U. Dev.*, USA(N.Y.); *L.Miss.*, USA(Iowa-Mont.).——FIG. 511,*6*; 513,*1a-c*. *E. *affinis* (MÜLLER), M.Dev.(Couvin.), Ger.(Eifel); 511,*6*, plate diagram of cup (radials black, *X* and higher anals stippled), arms of *C* and *D* rays, anal series, and

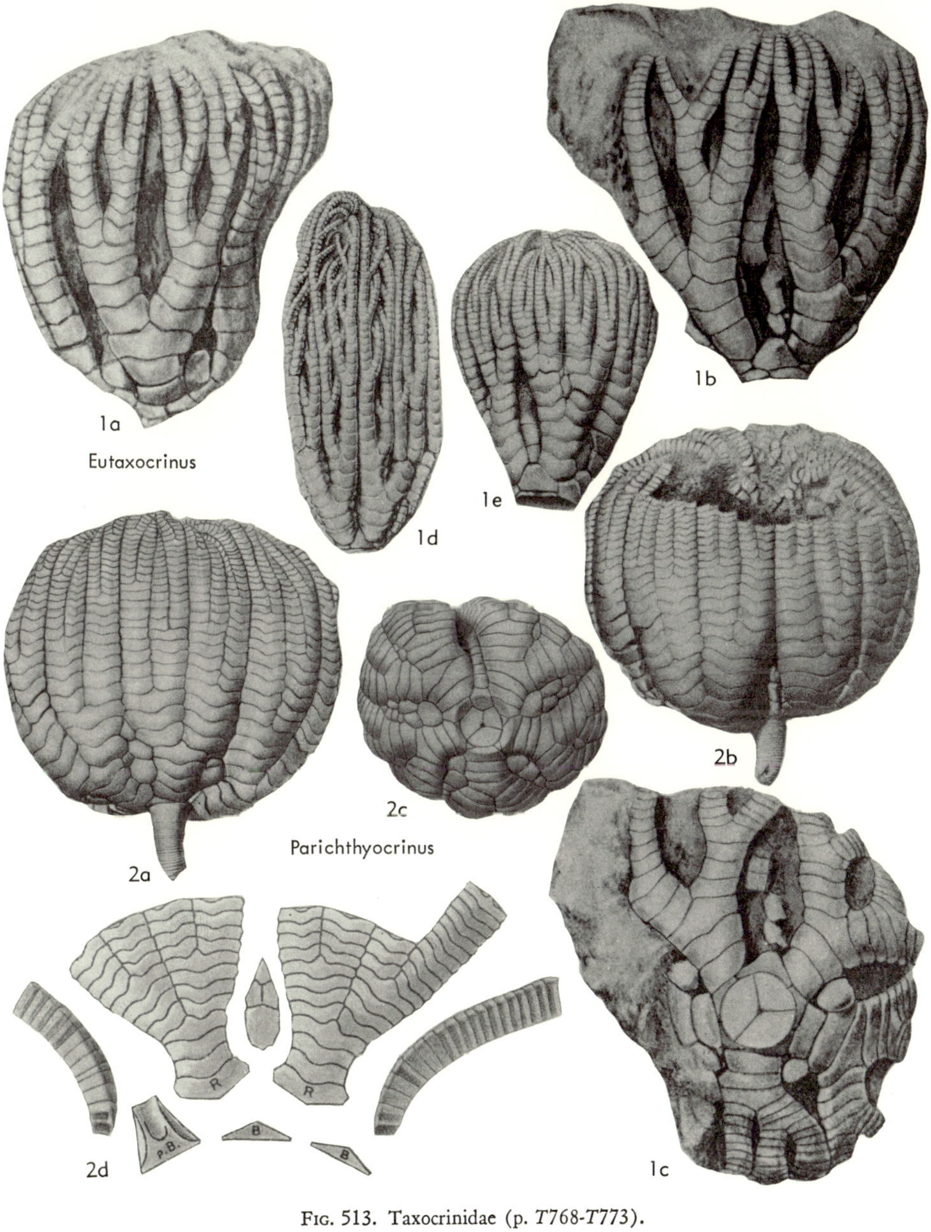

FIG. 513. Taxocrinidae (p. *T768-T773*).

BC and *DE* interrays (mod. from Springer, 1920); 513,*1a-c,* ant., post., and dorsal views, ×1.5 (Springer, 1920).——FIG. 513,*1d*. *E. oblongatus* (ANGELIN), U.Sil.(Wenlock.), Sweden(Gotl.); *DE*-interradius view of holotype showing biheterotomous branching of arms, ×1 (Springer, 1920).——FIG. 513,*1e*. *E. fletcheri* (WORTHEN), L.Miss. (Kinderhook.), USA(Iowa); *C*-ray view of crown, ×1 (Springer, 1920).

Gnorimocrinus WACHSMUTH & SPRINGER, 1880, p. 273 [**Taxocrinus expansus* ANGELIN, 1878, p. 9; SD BATHER, 1899b, p. 923]. Crown low, rotund; infrabasals usually extending slightly beyond stem impression; rhombic radianal obliquely at left beneath *C* radial. Rays as in *Eutaxocrinus* with 2 primibrachs followed by divergent arm branches which bifurcate isotomously; interprimibrachs few or lacking. Stem short, not enlarging near calyx,

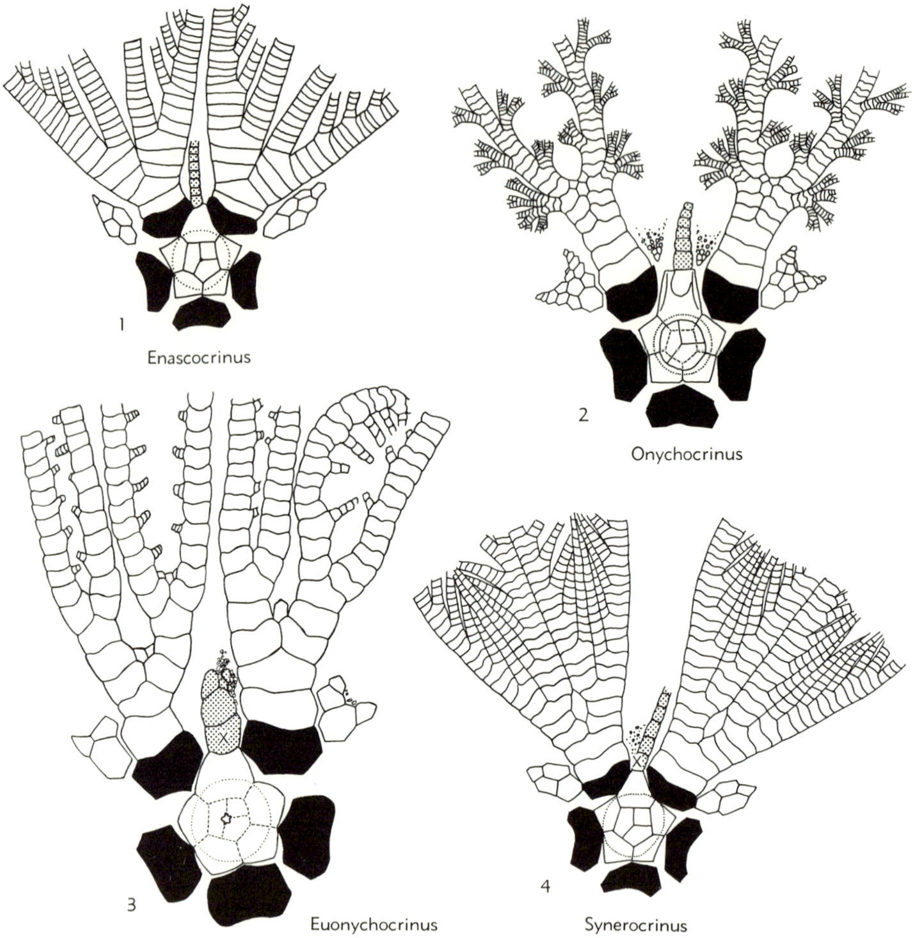

FIG. 514. Synerocrinidae (radials black, *X* and higher anals stippled) (p. *T773-T775*).

terminating distally in branched root. *U.Sil.*, Eu. (Sweden)-USA(Tenn.).——FIG. 511,*4*; 512,*1*. **G. expansus* (ANGELIN), Gotl.; 511,*4*, plate diagram of cup (radials black, *X* and higher anals stippled, radianal cross ruled), arms of *C* and *D* rays, anal series, and minute interprimibrachs (mod. from Springer, 1920); 512,*1a,b*, post. view of holotype and *EA*-interray view of another specimen, ×3 (Springer, 1920).

Meristocrinus SPRINGER, 1906, p. 515 [**Taxocrinus (Gnorimocrinus) loveni* WACHSMUTH & SPRINGER, 1880, p. 273 (=*Cyathocrinus interbrachiatus* ANGELIN, 1878, p. 23); OD]. Crown elongate, expanding above radials, with divergent isotomous arm branches; infrabasal slightly exposed beyond stem impression; posterior basal somewhat larger than others and truncated for contact with large anal *X*, which is followed by plates in anal tube bordered on left side by minute platelets of perisome; radianal usually in form of inferradial directly beneath *C* radial. Primibrachs 3 each ray; interprimibrachs very few and small, confined to lower part of interray areas, or lacking. Stem large, expanding toward calyx. *U.Sil.*, Eu.(Sweden-Eng.).——FIG. 511,*3*; 512,*3*. **M. interbrachiatus* (ANGELIN), Sweden(Gotl.); 511,*3*, plate diagram of cup (radials black, *X* and higher anals stippled), arms of *C* and *D* rays, and anal series (mod. from Springer, 1920); 512,*3a-c*, post., *B*-ray, and *C*-ray views of holotype, ×1.5 (Springer, 1920).

Parichthyocrinus SPRINGER, 1902, p. 94 [**Ichthyocrinus nobilis* WACHSMUTH & SPRINGER, 1879, p. 254; M]. Crown elongate ovoid, characterized by closely abutting nature of isotomous arms above radials; infrabasals prone, resembling columnal but not surrounded by ring of basals; posterior basal elongate; radianal only in upper oblique position if present. Primibrachs 3; interprimibrachs few and confined to lower part of interray areas. Column enlarging beneath calyx. *L.Miss. (Osag.)*, USA(Iowa-Ill.-Ind.).——FIG. 511,*2*; 513, *2*. **P. nobilis* (WACHSMUTH & SPRINGER), Burlington Ls., USA(Iowa); 511,*2*, plate diagram of cup (radials black, *X* and higher anals stippled), arms of *C* and *D* rays, anal series, and 2 interrays (mod. from Springer, 1920); 513,*2a,b*, ant. and post. views of nearly complete crown and small part of attached stem, ×1 (Springer, 1920).——FIG. 513,*2c*. *P. meeki* (HALL), Keokuk Ls., USA(Ill.); dorsal view of complete crown with stem removed, ×0.7 (Springer, 1920).——FIG. 513,*2d*. *P. subovatus* (MILLER & GURLEY), Borden Gr., USA(Ind.); partial plate diagram with edge view of thick brachials (Springer, 1920).

Protaxocrinus SPRINGER, 1906, p. 515 [**Taxocrinus ovalis* ANGELIN, 1878, p. 8; SD SPRINGER, 1920, p. 345]. Crown elongate, with isotomous arms usually not abutting laterally; infrabasals low, with tips exposed beyond stem impression; posterior basal elongate; radianal in primitive position resembling inferradial directed beneath *C* radial; anal tube plates tending toward sutural connections with *C* brachials. Primibrachs 2; interprimibrachs very few, small or lacking. Stem enlarging next to calyx. *M.Ord.*, Can.(Ont.); *U.Ord.*, USA(Mo.-Ill.); *L.Sil.*, USA(Mo.); *M.Sil.-U.Sil.*, USA(Tenn.)-Eu.(Sweden-Eng.)-Can.(Quebec); *L.Dev.*, USA (W.Va.).——FIG. 511,*5*; 512,*2c,d*. **P. ovalis* (ANGELIN), M.Sil.(Wenlock.), Gotl.; 511,*5*, plate diagram of part of cup (radials black, *X* and higher anals stippled, radianal cross ruled), arms of *C* and *D* rays, anal series, and *BC* and *DE* interbrachial areas (mod. from Springer, 1920); 512,*2c,d*, ant. and post. views of holotype, ×2 (Springer, 1920).——FIG. 512,*2a*. *P. elegans* (BILLINGS), M.Ord.(Trenton.), Can.(Ont.); ant. view of holotype, ×1 (Springer, 1920).——FIG. 512,*2b*. *P. laevis* (BILLINGS), M.Ord.(Trenton.), Can.(Ont.); post. view of complete crown, ×1.5 (Springer, 1920).

Family SYNEROCRINIDAE Jaekel, 1918

[Synerocrinidae JAEKEL, 1918, p. 78)] [incl. Onychocrinidae JAEKEL, 1918, p. 79]

Taxocrinids distinguished by distinctly heterotomous branching of arms above isotomous division on primibrachs *2*. *L. Miss.-U.Penn.; L.Carb.-U.Carb.*

Synerocrinus JAEKEL, 1898, p. 47 [**Forbesiocrinus incurvus* TRAUTSCHOLD, 1867, p. 31 (=*Talanterocrinus jaekeli* MOORE & PLUMMER, 1940, p. 196); M] [*non Synerocrinus* MOORE & PLUMMER in MOORE, 1939c, p. 195; *nec* MOORE & PLUMMER, 1940, p. 90]. Crown elongate ovoid, expanding upward from radials; infrabasals and proximal part of basals covered by stem; posterior basal larger than others and truncate distally for support of anal *X* which is followed by anal tube; no radianal. Arms in each ray branching isotomously on primibrachs *2* and secundibrachs *3*, thereafter with strongly heterotomous branching

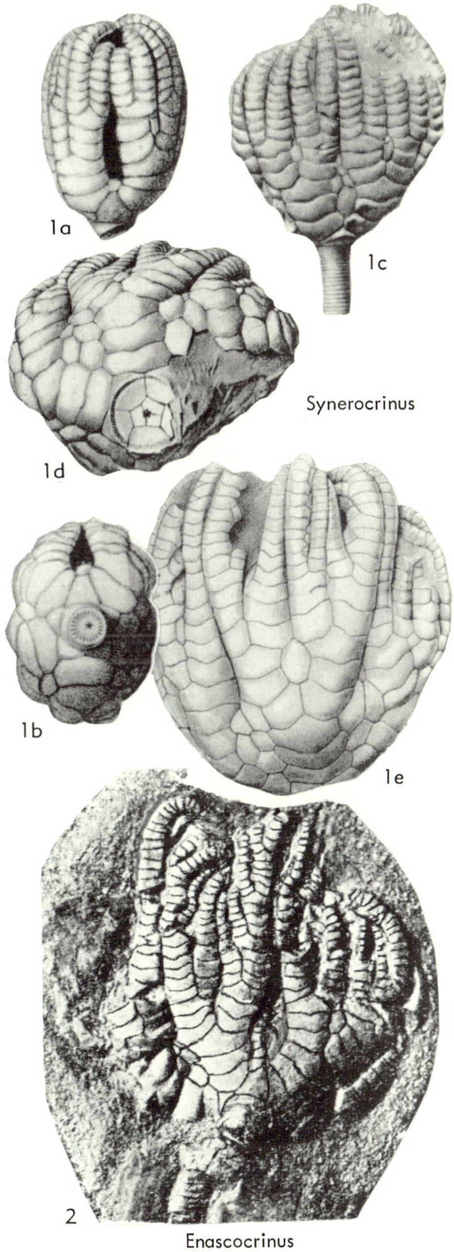

FIG. 515. Synerocrinidae (p. *T773-T775*).

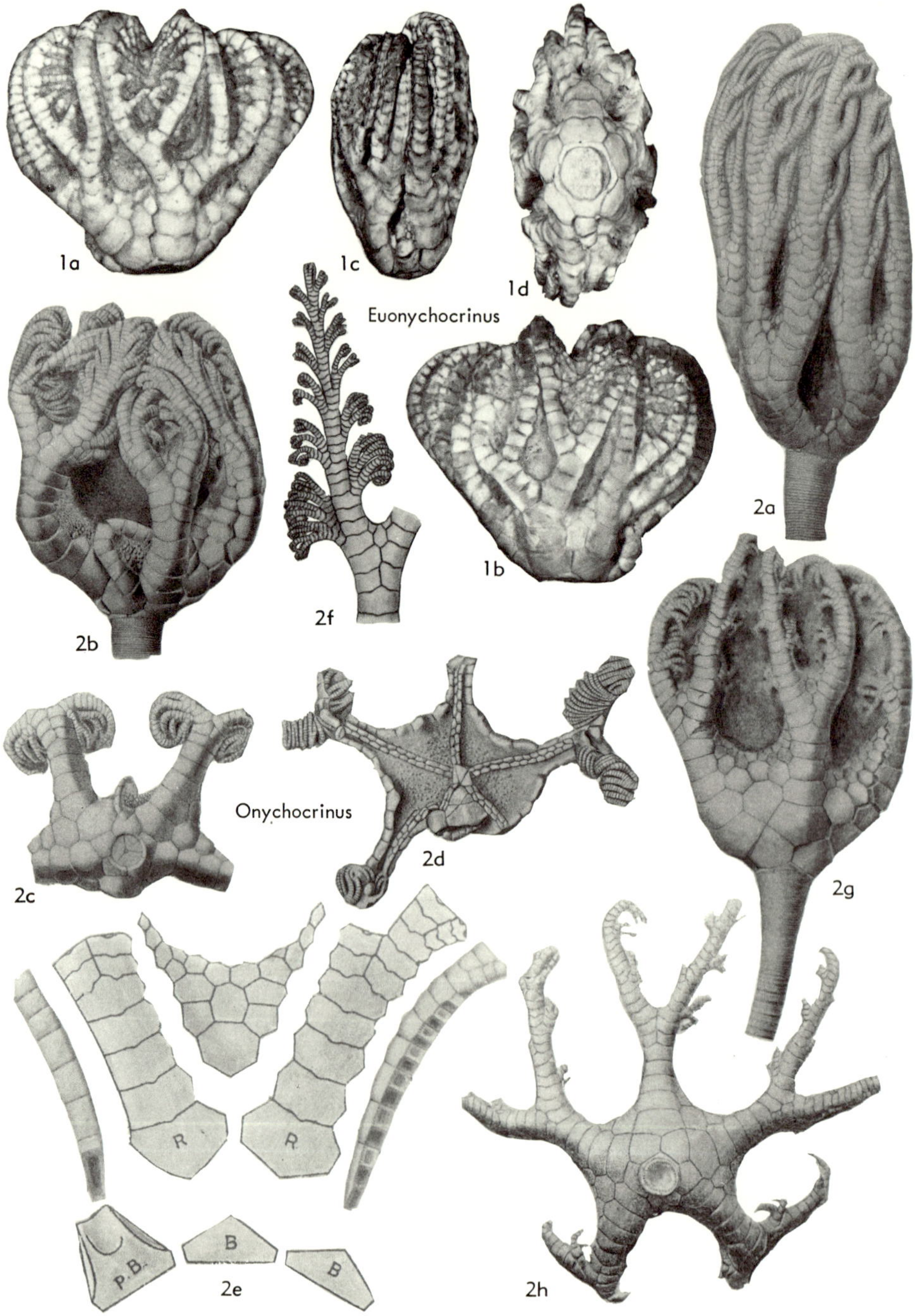

Fig. 516. Synerocrinidae (p. *T775*).

of bi-endotomous type, ramules unbranched, interprimibrachs few. Column large, expanding toward calyx. *U.Carb.*, Eu.(USSR); *Penn.*, USA(Okla.).——Fig. 514,*4*; 515,*1*. *S. *incurvus* (Traut-

SCHOLD), Moscov., Moscow basin; 514,*4*, plate diagram of cup (radials black, *X* and higher anals stippled), arms of *C* and *D* rays, anal series, and *BC* and *DE* interprimibrachs (mod. from Springer, 1920); 515,*1a,b*, post. and dorsal views of syntype, ×1 (Springer, 1920); 515,*1c*, post. view of another syntype, ×1 (Springer, 1920); 515,*1d,e*, dorsal view (*B* ray directed upward) and *B*-ray view of typical crown, ×1 (Springer, 1920). [=*Forbesiocrinus* TRAUTSCHOLD, 1867 (*non* DE KONINCK & LEHON, 1854); *Forbsiocrinus* MEEK & WORTHEN, 1860 (*nom. null.*).]

Enascocrinus STRIMPLE & WATKINS, 1969, p. 231 [**Talanterocrinus redesdalensis* WRIGHT, 1952c, p. 320; OD]. Very similar to *Synerocrinus* but crown smaller and heterotomous branching of arms in each ray of endotomous instead of bi-endotomous type. *L.Carb.(Visean)-U.Carb.(Namur.)*, Eu.(Scot.-Eng.).——FIG. 514,*1*; 515,*2*. **E. redesdalensis* (WRIGHT), Visean, Scot.; 514,*1*, plate diagram of cup (radials black, *X* and higher anals stippled), arms of *C* and *D* rays, anal series, and *BC* and *DE* interbrachials (Moore, n); 515,*2*, *D*-ray view of crown, ×1 (Wright, 1954).

Euonychocrinus STRIMPLE, 1940, p. 101 [**E. dubius*; OD]. Crown elongate ovoid with 20 well-separated, transversely rounded arms rising in forked manner from axillary secundibrachs, arm branching isotomous to this level, above which arm pairs give off short rounded unbranched ramules on sides facing each other, thus producing very simple sort of bi-endotomous heterotomy, each ramule-bearing tertibrach commonly separated by one without ramule; like *Synerocrinus* in having 2 primibrachs and 3 secundibrachs in each arm branch; each interray with small number (3-5) moderately large interprimibrachs; only tips of basals appearing beyond stem impression, posterior one longer than others and supporting anal *X* followed by transversely rounded anal tube not laterally united to brachials. *U.Penn.(Missour.)*, USA(Kans.-Ill.).——FIG. 514,*3*; 516,*1*. *E. simplex* STRIMPLE & MOORE, LaSalle Ls., USA(Ill.); 514,*3*, plate diagram of cup (radials black, *X* and higher anals stippled), arms of *C* and *D* rays, anal series, and *BC* and *DE* interprimibrachs (Strimple & Moore, 1971a); 516,*1a-d*, *BC*-interray, *DE*-interray, post., and dorsal (anal series directed downward) views of holotype, ×1 (Strimple & Moore, 1971a).

Onychocrinus LYON & CASSEDAY, 1860, p. 77 [**O. exsculptus*; M] [=*Oligocrinus* SPRINGER, 1906, p. 519 (type, *Forbesiocrinus asteriaeformis* HALL, 1861a, p. 320; M); *Origocrinus* JAEKEL, 1918, p. 79 (*nom. null.*)]. Crown low with pentamerous outward extension of rays resembling starfish in shape, or rotund to elongate ovoid, especially distinguished by wide separation of rays above areas of small interprimibrachs adjacent to calyx; infrabasals low, with outward extremities commonly visible beyond stem; posterior basal elongate, with distal portion hollowed for reception of anal *X* which is followed by anal tube; radianal only in upper oblique position, if present. Each ray with 3 or more primibrachs, arms branching isotomously above axillary primibrach, thereafter with heterotomous branching, with ramules on both sides of 10 main arm trunks, these ramules commonly with 2 to several bifurcations, producing a unique, highly distinctive type of arm structure. Column enlarging next to calyx. *L.Miss.(Osag.)*, USA(Iowa-Ill.-Ind.-Ky.)-Can.; *U.Miss.(Meramec.-Chester.)*, USA(Ill.-Ind.-Ky.-Ala.-Okla.); *L.Carb.*, Eu.(Eire-Ger.-Scot.).——FIG. 514,*2*; 516,*2f-h*. **O. exsculptus*, L.Miss.(Borden Gr.), USA(Ind.); 514,*2*, plate diagram of cup (radials black, *X* and higher anals stippled), arms and ramules of *C* and *D* rays, anal series, and *BC* and *DE* interprimibrachs (mod. from Springer, 1920); 516,*2f-h*, diagram of arm with ramules, *A*-ray view of crown, and dorsal view of crown with spread-out arms (post. side directed downward), ×0.7, ×0.7, ×1 (Springer, 1920).——FIG. 516,*2a*. *O. ramulosus* (LYON & CASSEDAY), L.Miss.(Borden Gr.), USA (Ind.); post. view of nearly perfect crown with attached part of stem, ×0.5 (Springer, 1920).——FIG. 516,*2b-e*. *O. ulrichi* MILLER & GURLEY, L. Miss.(Borden Gr.), USA(Ind.); 516,*2b-d*, post., dorsal (post. side upward), and ventral (post. side downward) views of crown, ×0.7 (Springer, 1920); 516,*2e*, partial plate diagram showing edges of thick brachials (Springer, 1920).

Order SAGENOCRINIDA Springer, 1913

[*nom. correct.* MOORE in MOORE, LALICKER, & FISCHER, 1952, p. 614 (*pro* order Sagenocrinoidea SPRINGER, 1913, p. 203)]

Crown globose to elongate ovoid with plates of calyx and commonly those of arms, interradial and interbrachial areas united rather firmly together, anal plates of posterior interray (radianal, if present, anal *X*, and higher anal plates) joined firmly to adjoining basals, radials, and brachials instead of marked off by flexible perisome as in Taxocrinida. Cup low bowl shaped, mostly lacking well-defined separation from lower arm and interarm plates; infrabasals visible beyond stem or entirely concealed by it; basals and radials relatively large, upflaring gently or in distal part of radials moderately to steeply; radianal in primitive position directly below *C* radial shifted leftward and upward, or absent; anal *X* mostly large and reaching above radial summits but may be small or lacking entirely. Arms generally wide, externally little rounded, composed of uniserially arranged short brachials, closely adjoining one another laterally or separated by interbrachials confluent with arm exterior surfaces; distal extremities of arms typically curved inward strongly. Stem transversely circular and tending to taper distally, very rarely columnals crescentic in outline. *L.Sil.-U. Perm.(Basleo beds).*

Comparatively numerous sagenocrinid genera which abruptly appear in Upper Silurian deposits of Europe and North America show such wide diversity of morphological characters that relationships to one another and to ancestors probably belonging among Taxocrinida are highly conjectural. For example, *Lecanocrinus* and *Mysticocrinus* (the latter misinterpreted at first by SPRINGER as an inadunate crinoid) possess few features of contemporary *Sagenocrinites, Icthyocrinus, Calpiocrinus,* and *Lithocrinus*—not even in content and arrangement of cup anal plates—yet all are undoubtedly non-taxocrinid flexible crinoids. The tripartite pentagonal infrabasal circlet is horizontal. Branching of the arms is exclusively isotomous; although in some later sagenocrinids, like several taxocrinids, it is heterotomous. Interbrachial plates are found only in *Sagenocrinites,* among the Silurian genera mentioned. *Icthyocrinus* has no cup anal plates, *Calpiocrinus* and *Lithocrinus* an *X* plate only, and the others both radianal and anal *X*. These similarities and differences sufficiently characterize Sagenocrinida in general and indicate presence in the group of more than a single lineage.

Superfamily LECANOCRINACEA Springer, 1913

[*nom. transl.* MOORE & STRIMPLE, 1973, p. 15 (*ex* Lecanocrinidae SPRINGER, 1913, p. 203)]

Crown typically short, subglobular (exceptionally biconvex, e.g., *Ammonicrinus*), with stoutly built cup well delimited from arms, which are composed of very low and wide brachials and free above radials, arms unbranched or with one to three isotomous bifurcations, their distal extremities curved inward to conceal tegmen and with or without clockwise torsion. Interbrachials generally lacking, but present in Nipterocrinidae. Infrabasals sloping upward and visible from side or forming flat base of cup with extension beyond stem impression, rarely in basal concavity, flat and down-sloping circlets not visible from side, normally with two large plates (*AB, DE*) and small one in *C* ray but solidly fused together in several genera. Basals relatively large, with surface area little smaller than large radials, which have straight or gently concave articular facets occupying entire width of plate or uncommonly only small part of it, facets short transversely, approximately equal to thickness of radials, and lacking well-defined ridge and ligament fossae. Radianal obliquely left beneath *C* radial and moderately large anal *X* above squarely truncate distal edge of *CD* basal, or lacking radianal. Arms composed of very wide short brachials, branching on primibrachs *2,* secundibrachs *2* to *4,* and higher in some genera. Stem mostly subcylindrical, composed of hetero- or homeomorphic low columnals with small axial canal, without cirri; stem may be distinctly xenomorphic and in forms with enrolled column moderately to strongly crescentic in transverse shape. *U.Sil.-U. Perm.*

Family LECANOCRINIDAE Springer, 1913

[Lecanocrinidae SPRINGER, 1913, p. 203 (incl. Lecanocrininae BOWSHER, 1953, p. 1)] [=Cyrtidocrinidae ANGELIN, 1878, p. 20]

Cup low bowl shaped with flaring sides and strongly defined even summit formed by straight facets of radials, which are subequal in size and adjoin one another except at *CD* interray where large anal *X* plate generally occurs; radianal typically rhombic in outline obliquely at left beneath *C* radial; anal *X* may extend high enough above rim of cup to separate proximal secundibrachs. Infrabasals sloping upward or flat to slightly down-sloping in very shallow basal concavity. Arms formed by very low and wide brachials, primibrachs fully equal in width to radials; arms with two or three isotomous divisions, except in one genus *(Miracrinus)* with close sutures joining rays and branches laterally in proximal region to make rigid calyx, distal parts of arms free and strongly incurved over tegmen. Column transversely circular. *U. Sil.-M.Dev.*

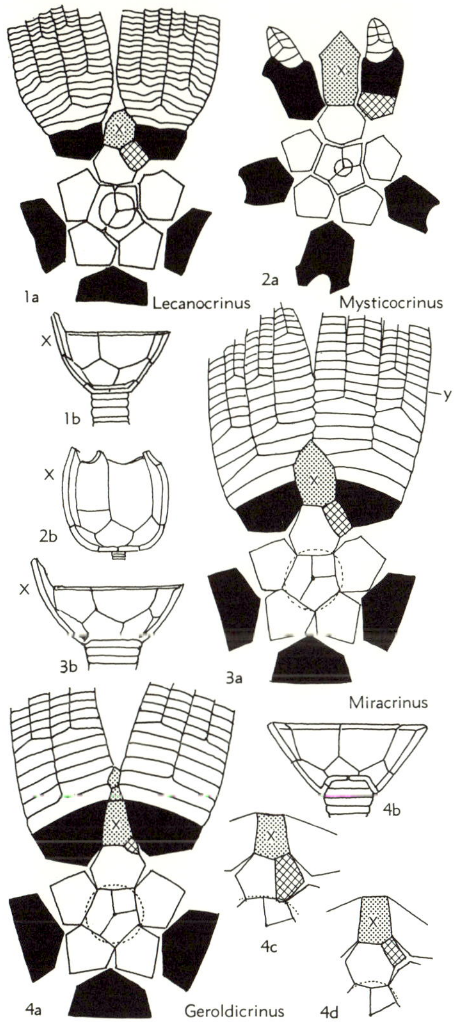

FIG. 517. Lecanocrinidae (radials black, *X* and higher anals stippled; radianal cross ruled) (p. T777-T779).

Lecanocrinus HALL, 1852, p. 199 [**L. macropetalus*; SD WACHSMUTH & SPRINGER, 1880, p. 40] [=*Cyrtidocrinus* ANGELIN, 1878, p. 20 (type, *C. facietatus*; M)]. Infrabasals sloping slightly upward, visible in side view of cup, which has even summit defined by straight, wide articular facets of radials; medium-sized to large rhombic radianal obliquely beneath left margin of *C* radial; anal *X* large, may or may not extend above cup summit. Arms wide, free above radials, composed of very low brachials not united by synostosial sutures, with 2 or 3 isotomous bifurcations in each ray, tips of arms curved inward so as to conceal tegmen. Column circular in section, formed of moderately low heteromorphic columnals with crenulate articula and small circular axial canal. *U.Sil.-L.Dev.*, N.Am.-Eu.-Australia. ——FIG. 517,*1*. *Lecanocrinus* sp.; *1a,* plate diagram of cup (radials black, radianal cross ruled, anal *X* stippled) and post. arms; *1b,* long. sec. of cup showing nearly horizontal infrabasals (mod. from Springer, 1920).——FIG. 518,*1a-c*. **L. macropetalus,* U.Sil.(Rochester Sh.), USA(Lockport, N.Y.); *1a,b,* side views of crown, ×1; *1c,* slightly oblique dorsal view of cup, ×1 (all Springer, 1920). FIG. 518,*1d,e*. *L. facietatus* (ANGELIN), U.Sil.(Wenlock.), Sweden(Gotl.); *1d,e,* side views of cup and incomplete crown, ×2 (Springer, 1920).

Geroldicrinus JAEKEL, 1918, p. 80 [**Lecanocrinus roemeri* SCHULTZE, 1867, p. 153; M]. Like *Lecanocrinus* but with evenly sloping sides of low truncate conical cup and with infrabasal circlet in shallow concavity entirely covered by stem; radianal variable in shape and size, subtriangular, rhombic, or pentagonal; anal *X* not reaching above summit of radials, followed by one or 2 small anal plates between primibrachs. Column circular, with finely crenulate articula on columnal facets. *M.Dev.*, Eu.(Ger.).——FIG. 517,*4*; 518,*2*. **G. roemeri* (SCHULTZE), Eifel; 517,*4a,* plate diagram of cup (radials black, radianal cross ruled, *X* and higher anals stippled) and arms of *C* and *D* rays; 517,*4b,* vertical section of cup through *A* radial and anal *X* showing attitude of infrabasals; 517,*4c,d,* partial plate diagrams showing variations in shape of radianal (all Moore, n); 518,*2a-c,* side, oblique dorsal, and ventral views of typical crown, ×1; 518,*2d,* dorsal view of cup showing rhombic radianal and infrabasals in steep-sided concavity, ×1 (all Springer, 1920).

Miracrinus BOWSHER, 1953, p. 3 [**Lecanocrinus (Miracrinus) perdewi*; OD]. Resembles *Geroldicrinus* in having infrabasal circlet in concavity entirely covered by proximal part of column, but differs from this genus and *Lecanocrinus* in firmly sutured brachials in all rays to height of nearly twice that of cup, thus producing stout calyx above which free arms curve inward over tegmen; radianal subrhombic obliquely beneath *C* radial and large anal *X* which projects well above summit of cup enough to separate proximal secundibrachs of *C* and *D* rays. Stem transversely circular, articular facets of columnals with even short crenulae surrounding wide smooth areola, axial canal small, circular. *L.Dev.(Helderberg.)*, USA (Md.).——FIG. 517,*3*; 518,*4*. **M. perdewi* (BOWSHER), New Scotland Ls., near Cumberland, Md.; 517,*3a,b,* plate diagram (radials black, radianal cross ruled, *X* and anals stippled) and vertical section through cup (Moore, n); 518,*4a-c,* ant., post., and summit views of crown (holotype) showing position of boundary between fixed and free brachials (*Y, Z,* primibrach *1*), ×2 (Bowsher, 1953).

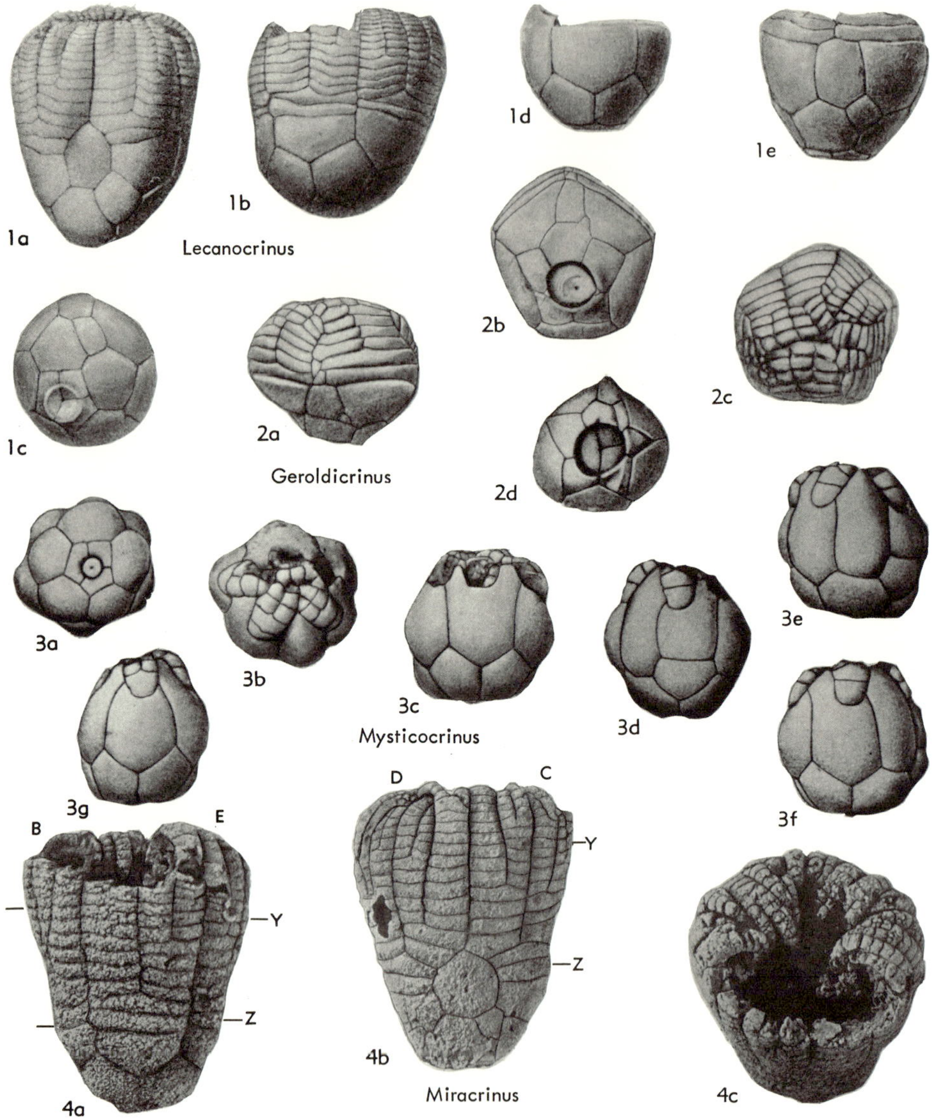

FIG. 518. Lecanocrinidae (p. *T777-T779*).

Mysticocrinus SPRINGER, 1918, p. 666 [**M. wilsoni*; OD]. Calyx globose, with rigidly united plates and cup with flat base and roundly curved sides; infrabasal circlet subpentagonal, composed of 3 plates, small one in *C* ray; radials with height greater than width, curved articular facets occupying less than full width of plates, 1 or 2 shoulders of some projected upward in rounded elevation; radianal directly beneath *C* radial; relatively large and very tall anal *X* distally pointed and rising above summit of radials. Arms very short, bifurcating isotomously once on primibrach 2, each branch typically with 3 secundibrachs; brachials wide and short, with outer face distinctly curved. [This genus was considered by SPRINGER to be an inadunate, and, in spite of its dicyclic cup and tripartite infrabasal circlet, most closely related to pisocrinids. Such interpretation is erroneous. In the nature of its radials and arms *Mysticocrinus* is most suggestive of undoubted flexible crinoids classified in the family Prophyllocrinidae of the Lecanocrinacea but its possession of a radianal removes it from possible placement there. The genus differs from other members of the Lecanocrinidae in shape and width of the radial articular facets, shorter and narrower arms, and primitive nature of the radianal, directly below the *C* radial.] *U.Sil.*, USA(Ind.).——FIG. 517,*2*; 518,*3*. **M. wilsoni*, Laurel Ls.; 517,*2a,b*, plate diagram (symbols as in Fig. 517,*1a*) and long. sec. of cup (anal *X* at left) (Moore, n); 518,*3a-g*, dorsal (*A* ray down), ventral (*A* ray up), *A*-ray, *C*-ray, *CD*-interray, *D*-ray, and *E*-ray views, ×4 (Springer, 1926a).

Family NIPTEROCRINIDAE Jaekel, 1918

[Nipterocrinidae JAEKEL, 1918, p. 80]

Crown ovoid, moderately tall, characterized especially by prominence of line at summit of cup (except in *Hormocrinus*) marking division between radials and free arms; cup typically low bowl shaped or conical with evenly flaring sides, but in both types height may equal diameter; infrabasal circlet sloping upward and visible from side or flat and not visible except in dorsal view, normally tripartite with small infrabasal in *C* ray, but fused to solid pentagonal plate in type genus; basals five, moderately large pentagonal plates forming important part of cup; radials also large, mostly with width exceeding height, gently curved articular facets mostly not extending to interradial sutures, thus providing shoulders which meet interradial plates; radianal and anal *X* plates present or absent. Arms composed of brachials which are wider than high, with isotomous bifurcations on primibrachs *1*, *2*, or *3*, and above this isotomous to moderately heterotomous additional branching. Interradial plates present between all rays mostly very diminutive and irregular; interbrachials also may be present but consist of numerous platelets only in type genus. Column medium sized, transversely circular, composed of homeomorphic to heteromorphic columnals and in some species exhibiting distinct xenomorphy. [The family name, determined by priority, is derived from genus modified by evolution and considered less representative of the assemblage than others.] *U.Sil.-L.Dev.; U.Dev.-L.Miss.*

Nipterocrinus WACHSMUTH in MEEK & WORTHEN, 1868, p. 341 [**N. wachsmuthi* MEEK & WORTHEN, 1868; M]. Crown relatively tall and marked by prominent differentiation of low bowl-shaped, large-plated cup from well-rounded moderately slender branching arms separated by interradial and interbrachial platelets; infrabasals fused into pentagonal plate partly visible from sides; basals slightly to distinctly smaller than large radials which have articular facets confined to middle portion of summit, forming scalloped indentation below prominent, subhorizontal shoulders; lack of any anal plates and undivided infrabasal circlet make cup perfectly symmetrical radially preventing identification of CARPENTER-lettered rays. Well-rounded relatively slender arms formed by low brachials with straight to somewhat flexuous sutures; isotomous bifurcation on primibrachs *3* followed by isotomous branching at higher levels in each ray; extremely numerous irregular platelets in interray areas denote lofty flexible tegmen. Robust stem of proximal region composed of very low circular columnals, those of intermediate and distal parts of stem somewhat narrower and distinctly taller. *L.Miss.(Osag.)*, USA(Iowa).——FIG. 519,*2*; 520,*3b,c*, **N. wachsmuthi* MEEK & WORTHEN, U.Burlington Ls.; 519,*2*, plate diagram (radials black), *CD* interray upward (Springer, 1920, mod.); 520,*3b*, lat. view of distorted holotype crown, ×1; 520,*3c*, same (reconstr.), ×1 (both Springer, 1920). FIG. 520,*3a*. *N. arboreus* WORTHEN, L.Burlington Ls.; lat. view of crown, ×2 (Springer, 1920).

Cholocrinus SPRINGER, 1906, p. 517 [**Forbesiocrinus obesus* ANGELIN, 1878, p. 9; M]. Crown elevated ovoid with prominent line between the bowl-shaped cup and transversely rounded arms separated by wide areas of interradial platelets and less prominent interbrachial ones; flat, fairly large infrabasal circlet with 3 plates, small one in *C* ray, not visible from side; large basals and radials similar to those of *Nipterocrinus*; rhombic radianal obliquely beneath *C* radial at left, followed above by large anal *X*, its summit evenly confluent with shoulders of adjacent radials. Arms branching repeatedly, mostly in isotomous manner but partly heterotomously, initial bifurcation on primibrach *2*. *U.Sil.*, Sweden(Gotl.).——FIG. 519,*3*; 520,*4*. **C. obesus* (ANGELIN); 519,*3*, plate diagram (radials black, radianal cross ruled, *X* and higher anals stippled), *CD* interray upward (Springer, 1920, mod.); 520,*4a,b*, *A*- and *B*-ray views of holotype crown, ×1.5 (Springer, 1920); 520,*4c*, part of same view as *4b* showing diminutive platelets of *AB* and *BC* interrays, ×2 (Springer, 1920).

Hormocrinus SPRINGER, 1920, p. 166 [**Centrocrinus tennesseensis* WORTHEN, 1890, p. 95; OD]. Crown ovoid, with width typically greater than height, with prominent division between large-plated calyx and part of crown formed by rounded branching arms and small-plated interradial areas, but actually this division is above Z line, being formed by summit of large interradial plates and by division between fixed and free brachials, at distal edges of primibrachs *2* (therefore, this

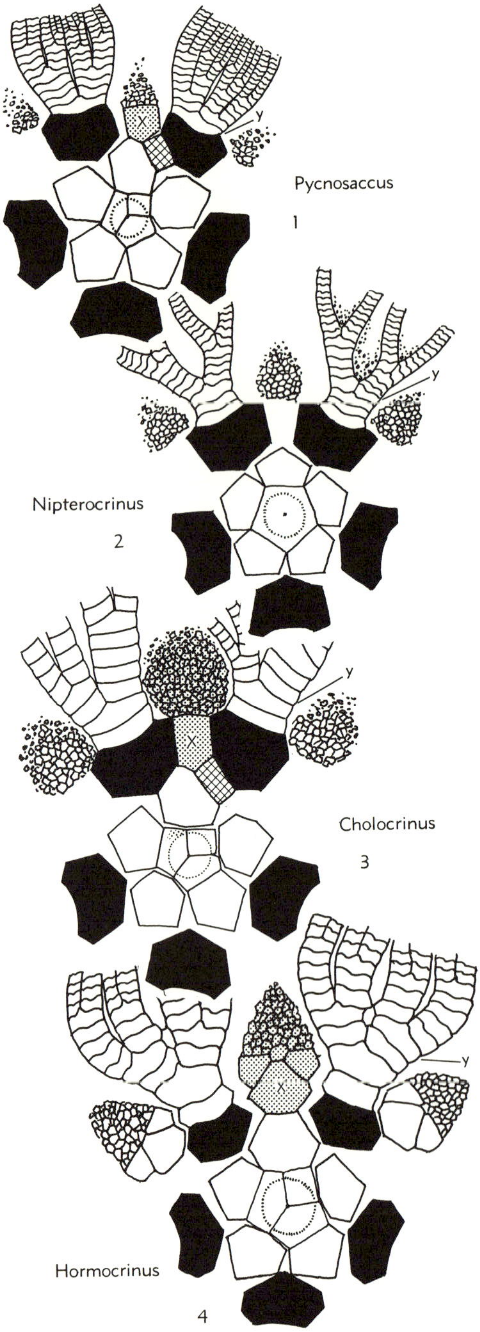

FIG. 519. Nipterocrinidae (radials black, *X* and higher anals stippled; radianal cross ruled) (p. *T779-T780*).

boundary denominated as Y line); posterior interray with large anal *X* resting on truncated extremity of posterior basal, not accompanied by radianal. Arms branching isotomously at 2 or 3 levels. Column distinctly heteromorphic, nearly uniform in diameter in proximal and intermediate areas but narrowing distally, with noticeable increase in height of columnals. *U.Sil.*, Eu.(Eng.-Sweden)-USA(Tenn.).——FIG. 519,*4*; 520,*2*. **H. tennesseensis* (WORTHEN), Brownsport Ls., USA (Tenn.); 519,*4*, plate diagram (symbols as in Fig. 519,*3*), *CD* interray upward (Springer, 1920, mod.); 520,*2a,b*, oblique dorsal (*E* ray up) and *E*-ray views of 2 crowns, ×2 (Springer, 1920); 520,*2c*, somewhat oblique *EA*-interray view of crown with attached long tapering heteromorphic stem, ×1 (Springer, 1920).

Pycnosaccus ANGELIN, 1878, p. 13 [**Cyathocrinites scrobiculatus* HISINGER, 1840, p. 6; SD SPRINGER, 1920, p. 180] [=*Oncocrinus* BATHER, 1890b, p. 387 (type, *O. bucephalus*; M)]. Similar to *Cholocrinus* but cup more conical with upsloping infrabasals clearly visible from side, large basals and radials smooth or marked by ridges crossing plates normal to sutures between them; rhombic radianal and anal *X* moderately large. Arms well-rounded laterally, typically slender, branching isotomously, with initial one on primibrachs *1*; interradial areas filled by minute plates. Stem resembles that of *Hormocrinus* in heterotomous arrangement of columnals, thicker ones with well-rounded peripheries. *U.Sil.*, Eu.(Eng.-Sweden)-USA(Ill.-Ind.-Tenn.-N.Y.); *L.Dev.*, USA (W.Va.); *U.Dev.*, NW.Can.——FIG. 519,*1*; 520,*1d-f*. **P. scrobiculatus* (HISINGER), U.Sil., Sweden; 519,*1*, plate diagram (symbols as in Fig. 519,*3*), *CD* interray upward (Springer, 1920, mod.); 520,*1d,e*, *E*-ray and *CD*-interray views of holotype cup, ×1.5 (Springer, 1920); 520,*1f*, facetal view of *B* radial (center) showing impressions of minute interradial plates on shoulders next to sutures, ×2 (Springer, 1920).——FIG. 520,*1a*. *P. nodulosus* ANGELIN, U.Sil., Sweden(Gotl.); *CD*-interray view of crown with single primibrach in each ray, ×1.5 (Springer, 1920).——FIG. 520,*1b,c*. *P. tenuibrachiatus* SPRINGER, L.Dev.(Keyser Ls.), USA(W.Va.); *C*- and *E*-ray views of crown, ×1.5 (Springer, 1920).——FIG. 520,*1g*. *P. patei* SPRINGER, U.Sil.(Brownsport Ls.), USA(Tenn.); *B*-ray view of crown with stem, ×1 (Springer, 1920).

Family MESPILOCRINIDAE Jaekel, 1918

[Mespilocrinidae JAEKEL, 1918, p. 80]

Generally similar to Lecanocrinidae, differing mainly in absence of radianal in cup, flat infrabasal circlet extending beyond stem impression, commonly tripartite but may be solidly fused. Arms typically very short and crown rotund, with or without tendency toward clockwise torsion of rays. Stem transversely circular, medium sized to slen-

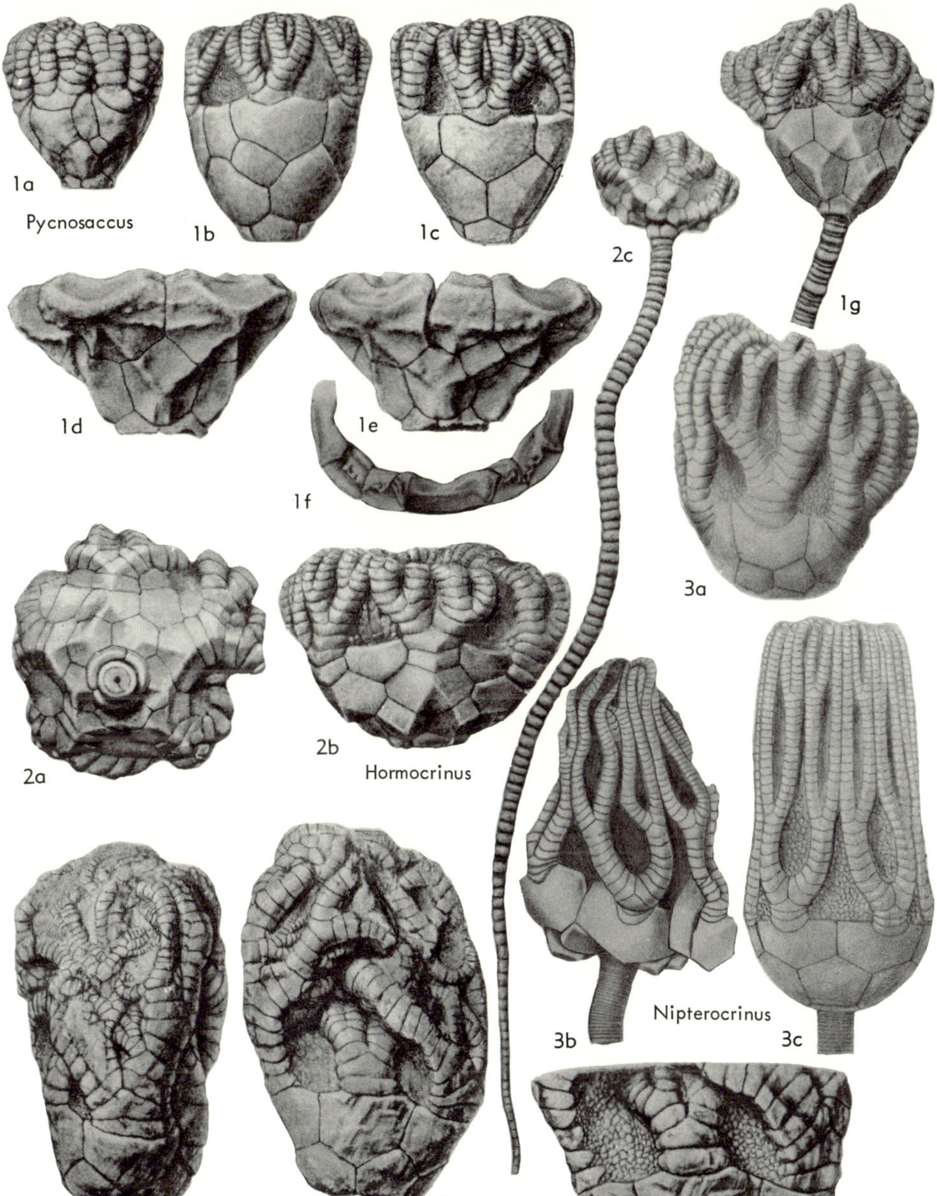

FIG. 520. Nipterocrinidae (p. *T779-T780*).

der, commonly heteromorphic and xenomorphic, cirri lacking. *L.Miss.(low.L.Carb.)-U.Perm.*

Mespilocrinus DE KONINCK & LE HON, 1854, p. 111 [**M. forbesianus*; SD S. A. MILLER, 1889, p. 262]. Small subglobular crown with short arms twisted spirally clockwise, isotomously bifurcating on primibrach *2* and secundibrachs *2* or *3*, tips curved strongly inward; *CD* basal distinctly larger than others and followed by large anal *X* with distally pointed extremity reaching level of proximal secundibrachs; infrabasal circlet with width 1.5 to 2 times that of circular stem impression. Stem nearly homeomorphic to distinctly heteromorphic and typically xenomorphic, with quite dissimilar proximal, intermediate, and distal portions. *L.Miss.(Osag.)*, N.Am.; *L.Carb.*

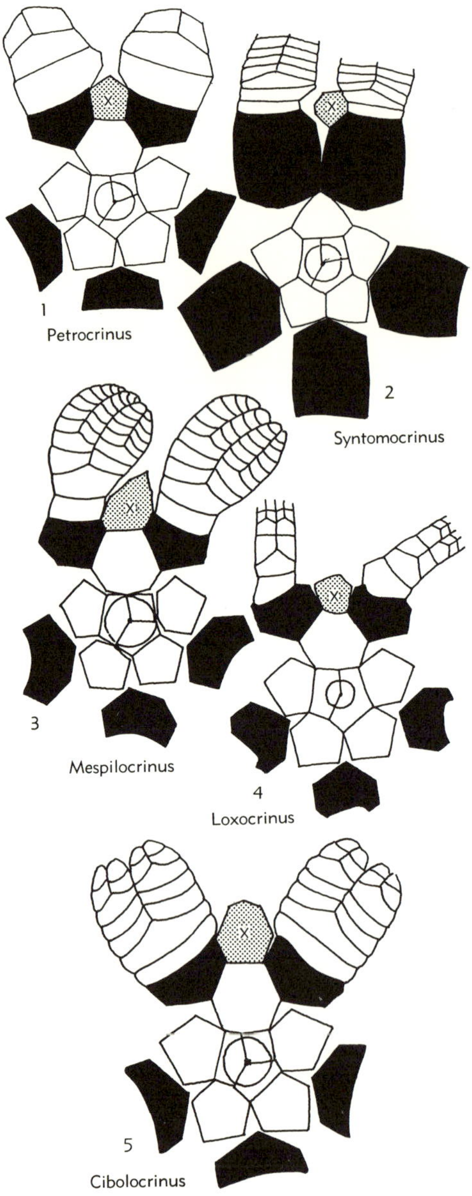

FIG. 521. Mespilocrinidae (radials black, anal *X* stippled) (p. *T*781-*T*783).

(Tournais.-Visean), Eu.——FIG. 521,*3*; 522,*2c-f*. **M. forbesianus*, Belg.(Tournai); 521,*3*, plate diagram of cup (radials black, *X* stippled) and post. arms (Moore, n); 522,*2c-e*, *D*-ray, *CD*-interray, and dorsal (*A* ray down) views, ×1.5 (Springer, 1920); 522,*2f*, lat. views of crown and stem, ×1.5 (Springer, 1920).——FIG. 522, *2a,b*. *M. konincki* HALL, Burlington Ls., USA (Iowa); *2a,b*, *B*- and *D*-ray views of crown with attached part of stem, ×2 (Springer, 1920).

Cibolocrinus WELLER, 1909, p. 630 [**C. typus*; OD] [*non Cibolocrinus* WELLER, 1909b, p. 632-633; *nec* WANNER, 1916a, p. 211; *nec* WANNER, 1924, p. 257; *nec* WANNER, 1934, p. 260; *nec* WANNER, 1937, p. 190; *nec* YAKOVLEV, 1930, p. 377; *nec* YAKOVLEV, 1934, p. 278]. Crown subglobular, with shallow to moderately high bowl-shaped cup forming half or more of crown height; infrabasals 3, small one normally in *C* ray but found in *A* ray in type species; medium-sized to large anal *X* extending well above radial summits separating primibrachs and even proximal secundibrachs of *C* and *D* rays. Arms short, branching isotomously on primibrach *2* and secundibrachs *3* to *5*, brachials very low and wide. Stem transversely circular, facets of columnals with narrow crenulate rim. *L.Penn.-L.Perm.*, N.Am.-?Sib.-S. Am.(Bol.).——FIG. 521,*5*; 522,*1d-g*. *C. punctatus* MOORE & PLUMMER, M.Penn.(Atokan, Marble Falls Ls.), USA(Texas); 521,*5*, plate diagram of cup (radials black, *X* stippled) and post. arms (Moore & Plummer, 1940); 522,*1d-g*, dorsal (*A* ray down), *CD*-interray, *A*-ray, and ventral (*E* ray up) views of crown, ×1.5 (Moore & Plummer, 1940).——FIG. 522,*1a-c*. **C. typus*, L.Perm. (Wolfcamp.), USA(W.Texas); *1a-c*, dorsal (*A* ray down), *CD*-interray, and ventral (*A* ray up) views of cup, ×1 (Moore & Plummer, 1940).——FIG. 522,*1h-l*. *C. banioni* MOORE, U.Penn.(Virgil., Brownville Ls.), USA(Okla.); *1h-l*, long. sec. of cup (*CD* interray at left) and ext., int., proximal, and distal views of radial plate, ×1.5 (Moore, 1939c).

Loxocrinus WANNER, 1916, p. 251 [**L. globulus*; OD]. Crown small, globular, with bowl-shaped cup forming more than half its height, plates of cup massive; infrabasals 2 in type species but may be 3 in others or fused entirely; posterior basal larger than others, with distal curved truncation for reception of medium-sized anal *X*; radials with slightly inclined articular facets occupying almost but not quite entire width of plate, bearing short arms twisted clockwise as in *Mespilocrinus*, with isotomous bifurcations on primibrachs *2* and secundibrachs *2*. Stem transversely circular. *U. Perm.*, Indon.(Timor).——FIG. 521,*4*; 523,*1*. **L. globulus*, Basleo beds; 521,*4*, plate diagram of cup (radials black, anal *X* stippled) and post. arms (Wanner, 1916a, mod.); 523,*1a-d*, *CD*-interray, *A*-ray, ventral (*A* ray up), and dorsal (*A* ray down) views of crown, ×2 (Wanner, 1916a); 523,*1e-g*, *CD*-interray, ventral (*A* ray up), and dorsal (*A* ray down) views of cup, ×2 (Wanner, 1916a).

Petrocrinus WANNER, 1924, p. 92 [**P. beyrichi*; OD]. Crown relatively large, ovoid; cup conical to bowl shaped, fused infrabasal circlet flaring upward peripherally and visible from side; radials

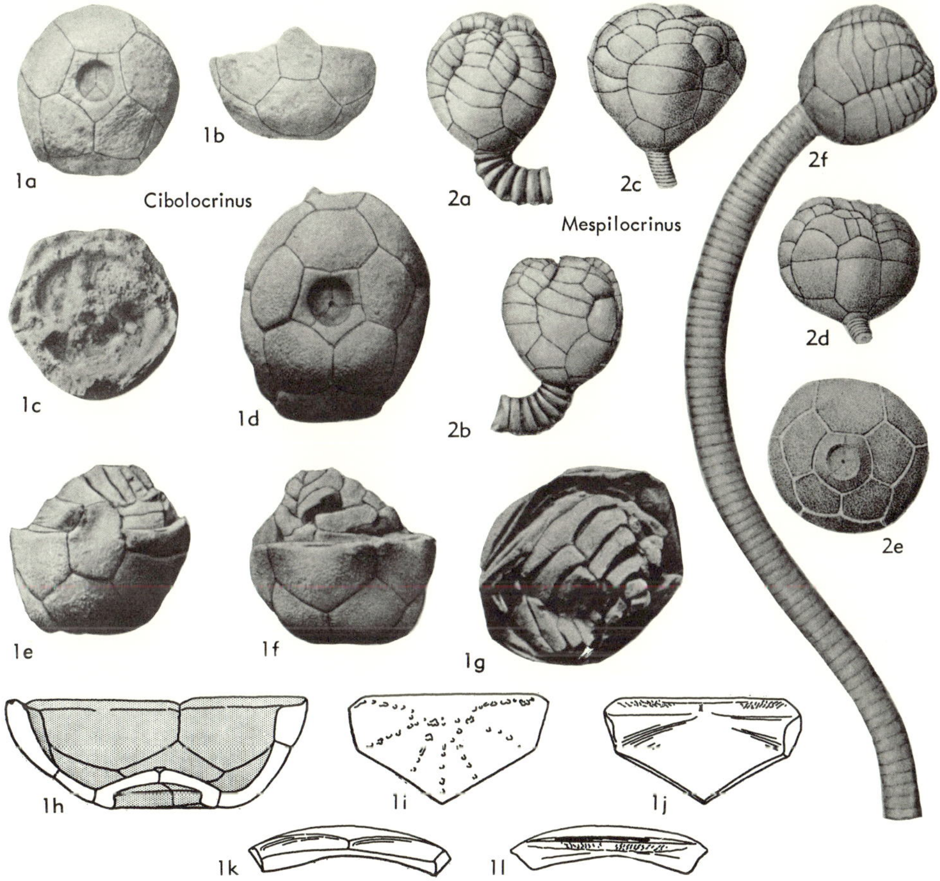

FIG. 522. Mespilocrinidae (p. *T*781-*T*782).

with wide straight articular facets forming very even summit of cup; anal *X* barely rising above radials. Arms short, composed of wide and unusually tall brachials, sloping evenly upward to pointed tip of crown, arms branching isotomously in each ray on primibrachs 2, which is followed by only single pairs of secundibrachs, so that total in crown is 10. Stem transversely circular. *U.Perm.*, Indon.(Timor).——FIG. 521,*1*; 523,*2*. **P. beyrichi*, Basleo beds; 521,*1*, plate diagram of cup (radials black, anal *X* stippled) and post. arms (Wanner, 1924, mod.); 523,*2*, *CD*-interray view of holotype, ×1 (Wanner, 1924).

Syntomocrinus WANNER, 1916, p. 256 [**S. sundaicus*; OD]. Crown pear shaped, narrowing upward from very even summit of cup, which is nearly straight sided and gently concave at base, only distal tips of basals visible from side, if at all, small pentagonal infrabasal circlet largely covered by circular stem. Arms composed of wide, short brachials, branching isotomously on primibrachs *2* or *3*, secundibrachs *3* or *4*, uppermost arm branches having 3 or 4 tertibrachs, but rarely preserved. *U.Perm.*, Indon.(Timor).——FIG. 521, *2*; 523,*3*. **S. sundaicus*, Basleo beds; 521,*2*, plate diagram of cup (radials black, anal *X* stippled) and post. arms (Wanner, 1916a, mod.); 523,*3a,b*, *A*-ray and dorsal (*A* ray down) views of crown, ×1.5 (Wanner, 1916a); 523,*3c,d*, dorsal (*A* ray down) and ventral (*A* ray up) views of cup, ×1.5 (Wanner, 1916a); 523,*3e*, part of specimen figured in *3c,d*, showing *A*-ray facet, ×2 (Wanner, 1916a).

Family CALYCOCRINIDAE Moore & Strimple, 1973

[Calycocrinidae MOORE & STRIMPLE, 1973, p. 34]

Lecanocrinids chiefly characterized by development of bilateral symmetry in plane bisecting *CD* interray and on anterior side

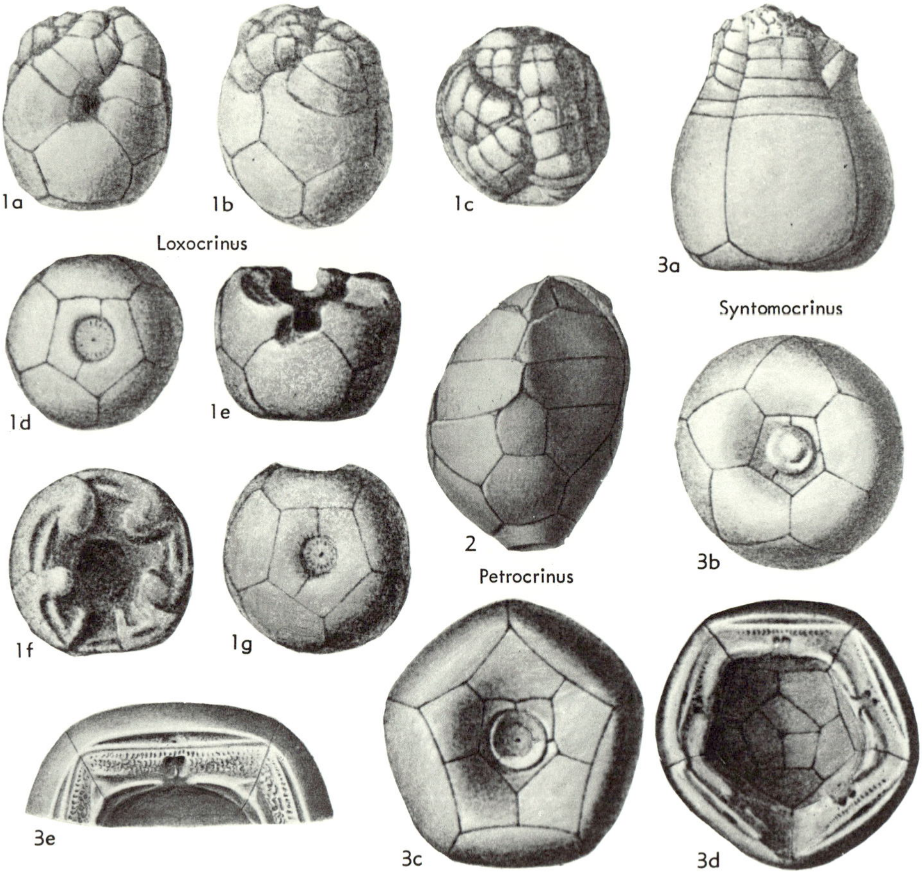

FIG. 523. Mespilocrinidae (p. T782-T783).

of crown either bisecting *A* ray or *AE* interray, this bisymmetry best developed in forms with crown distinctly bent on stem or stem coiled around crown; cup low to tall conical or bowl shaped, with shortest and tallest sides in plane of symmetry; infrabasal circlet reduced to two plates or fused into single plate largely or entirely covered by proximal columnal of stem; basals with distinct tendency for *CD* basal to be larger than others or reduced to three (*AE* and *AB* basals lacking) or with *CD* basal standing alone, radials subequal in shape and size or distinctly unequal, with margin or articular facets straight or strongly curved; radianal and anal *X* present or absent. Arms formed by wide short brachials, branching isotomously once or more in each ray. Stem transversely circular in section or unevenly elliptical. *M.Dev.; L.Penn.; U.Perm.*

Calycocrinus WANNER, 1916, p. 260 [**C. curvatus*; OD]. Crown low to moderately tall, subconical to bowl shaped with rounded summit formed by incurved arms; cup with similar sides all around or with posterior and anterior sides quite dissimilar in height and curvature, with posterior side longitudinally convex and anterior side concave; infrabasals solidly fused into pentagonal plate largely covered by stem impression; basals 5, pentagonal in outline and equal in size or with anterior one smaller than *CD* basal; radials 5 with nearly straight margin of articular facet; without radianal and anal *X* (Fig. 524). Arms short, formed by wide short brachials, branching isotomously, normally on primibrach *3*, and with additional bifurcation in some branches. Stem transversely circular, columnal articula with short peripheral crenulae. *L.Penn.(Morrow.),* USA(Okla.); *U. Perm.,* Indon.(Timor).——FIG. 525,*2*; 526,*3c-f*. **C. curvatus curvatus* (=*C. curvatus typus* WANNER, 1924, p. 96), U.Perm., near Basleo; 525,*2*, plate diagram of cup (radials black) and post. arms showing bilateral symmetry through *A* ray and *CD* interray (Wanner, 1924, mod.); 526,*3c-f*, *CD*-interray, *A*-ray, *DE*-interray, and ventral (*A* ray down) views of holotype crown, ×2 (Wanner, 1916a).——FIG. 526,*3a,b*. *C. curvatus subturbinatus* WANNER, U.Perm., Basleo; *3a,b*, *DE*-interray and ventral (*A* ray down) views of crown, ×1 (Wanner, 1924).

Ammonicrinus SPRINGER, 1926, p. 22 [**A. wanneri*; OD]. Crown short, rounded asymmetrically, incurved strongly in plane bisecting *AE* and *CD* interrays; infrabasals reduced to 2 subequal, symmetrically disposed plates which are larger than any of 3 basals adjoining them on posterior side (*AB* and *EA* basals lacking); *A* and *E* radials symmetrically disposed and distinctly larger than others, with margins of articular facets rather strongly curved; rhombic radianal obliquely at left below *C* radial (apparently doubled in some specimens); large anal *X* above *CD* basal and followed by several smaller anal plates. Arms formed by wide, short and straight or laterally somewhat curved brachials, branching isotomously on primibrachs *6* to *8* with up to 10 secundibrachs in some branches, followed by at least some tertibrachs. Stem coiled almost completely around crown, with distinct impressions of columnals on cup; columnals throughout midlength of stem strongly arched, with crescentic cross section; articular facets with transverse ridge and otherwise with nearly smooth surface, indicating synostosial articulation; distal part of stem narrowing to point without holdfast or prolonged in elongate cylindrical columnals; axial canal very diminutive, circular. [SPRINGER was uncertain whether *Ammonicrinus* belonged to the Inadunata or Camerata, but judged that it most probably was a hexacrinid camerate. A painstaking study by UBAGHS (1953) that included dissection of a well-preserved specimen of the type species has served to demonstrate beyond doubt that this crinoid is a lecanocrinacean flexible.] *M.Dev.,* Eu.(Ger.).——FIG. 525,*3*; 526,*5*. **A. wanneri,* Eifel; 525,*3*, plate diagram (radials black, radianals cross ruled, *X* and higher anals stippled), *CD* interray upward (Ubaghs, 1952); 526,*5a-c,* opposite sides and edge of enrolled column, distal extremity pointed upward, ×1.5 (Springer, 1926b); 526,*5d,e,* edge and side views of larger specimen, ×1.5 (Springer, 1926b).

Plagiocrinus WANNER, 1924, p. 110 [**P. torynocrinoides*; OD]. Crown asymmetrically pear shaped, with nearly straight anterior side, rounded summit, and longitudinally strongly convex posterior side; cup obliquely inclined from stem or bent almost at right angles to it; infrabasal circlet reduced to small subcrescentic fused plate on posterior side of stem impression, followed on this side by single *CD* basal (no others in cup); radials 5, quadrangular to pentagonal in outline, with straight margin of articular facet, *A* and *E* radials in contact with infrabasal circlet beneath proximal columnal. Arms varying in width, composed of low, wide brachials, branching isotomously on primibrach *2*, and with 1 or 2 additional bifurcations in different branches. Stem unknown but impression at base of cup unevenly elliptical. [*Plagiocrinus* resembles *Ammonicrinus* in disposition of its plane of near-bilateral symmetry. It differs from strongly oblique species of *Calycocrinus* mainly in having only a single basal plate.] *U.Perm.,* Indon.(Timor, Basleo).——FIG. 526,*6*; 526,*4d*. **P. torynocrinoides,* Basleo; 525,*6*, plate diagram (radials black), *CD* interray upward (Wanner, 1924, mod.); 526,*4d*, *E*-ray view of holotype crown, ×2 (Wanner, 1924).——FIG. 526,*4a-c*. *P. jaekeli* WANNER, Basleo; *4a-c*, *DE*-interray, *CD*-interray, and ventral (*A* ray down) views of holotype cup, ×1.5 (Wanner, 1924).

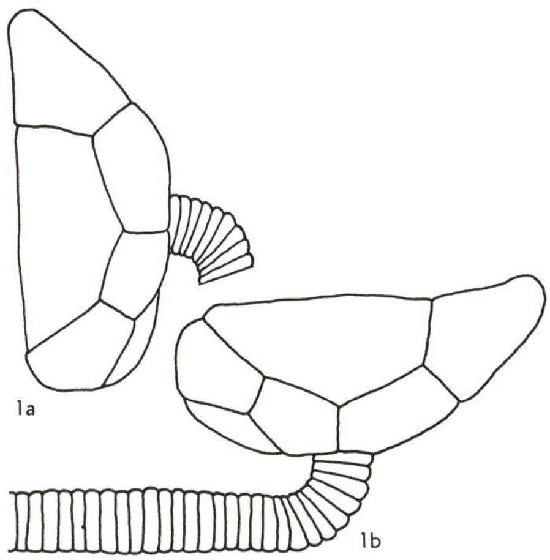

FIG. 524. Interpretations of orientation of cup and stem of *Calycocrinus* during life; *1a,* erect attitude like that of an air ventilator, postulated by YAKOVLEV; *1b,* prone attitude on sea bottom, suggested by MOORE & STRIMPLE (Moore & Strimple, 1973).

Family GAULOCRINIDAE Moore & Strimple, 1973

[Gaulocrinidae MOORE & STRIMPLE, 1973, p. 15]

Cup low bowl shaped , with flat to slightly concave base and longitudinally convex sides, becoming subvertical in upper part and curving inward slightly at margins of

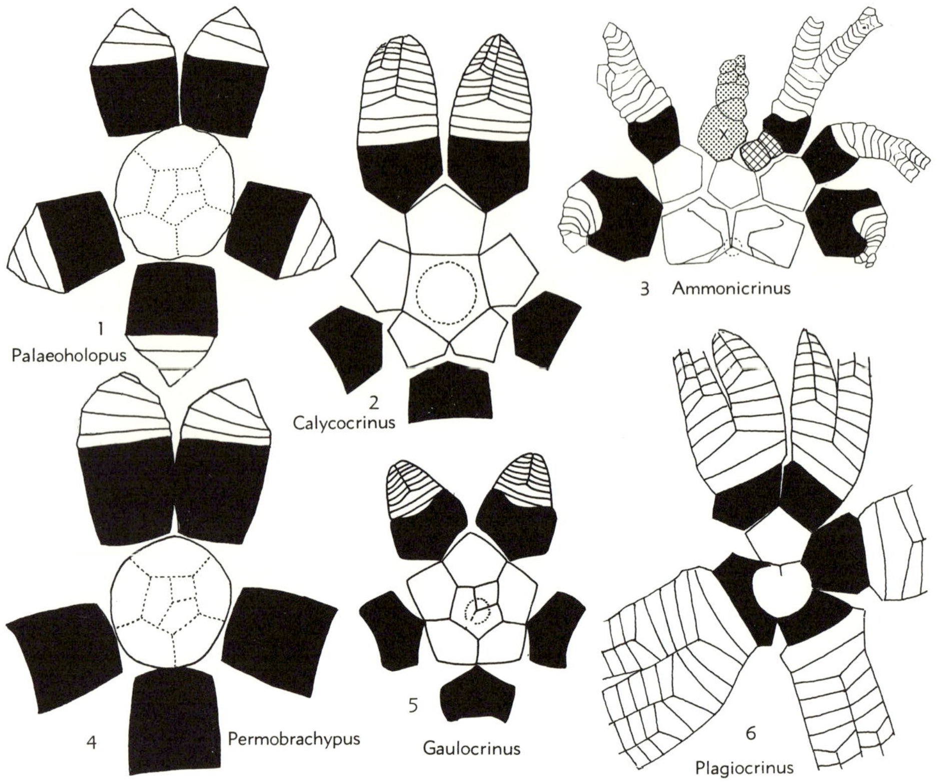

FIG. 525. Calycocrinidae *(2,3,6)*; Gaulocrinidae *(5)*, Palaeoholopodidae *(1,4)* (radials black, anal series stippled, radianals cross ruled) (p. *T784-T786, T789*).

radials; infrabasal circlet angularly pentagonal, with small infrabasal in *C* ray or possibly all plates fused, approximately half of its diameter occupied by shallowly concave stem impression; basals large and very thick, strongly convex both longitudinally and transversely, all pentagonal in outline; radials wider than high, also exceptionally thick, with broad, nearly featureless facets with faint transverse ridge not quite reaching interradial sutures. Arms composed of very wide short brachials, branching isotomously on primibrachs 2, succeeded in each branch by five or more secundibrachs, tertibrachs unknown; arms slope inward with gentle convexity to meet at center of crown. No radianal or anal *X* in cup. Column transversely circular but otherwise unknown. *L.Miss.*

Gaulocrinus KIRK, 1945, p. 180 [**Stemmatocrinus trautscholdi* WACHSMUTH & SPRINGER, 1885, p. 180; OD]. Characters of family. *L.Miss.(Osag.)*, USA(Ind.-Tenn.-?Ky.).——FIG. 525,*5*; 527,*1*. **G. trautscholdi* (WACHSMUTH & SPRINGER), New Providence F., White's Creek Springs, Tenn.; 525,*5*, plate diagram (radials black), *CD* interray upward (Moore, n); 527,*1a,b*, ventral (presumed *A* ray up) view showing very thick radials and nature of facets, and part of dorsal view (presumed *BC* interray down), ×1.5 (Moore, n; U.S. Natl. Museum); 527,*1c-g*, *A* to *E* radials from exterior showing near identity in appearance and broadly scalloped profile of articular facets, ×1.5 (Moore, n; U.S. Natl. Museum); 527,*1h*, side view of incomplete crown (rays unidentified) showing typical sort of flexible arm structure, ×1.5 (Moore, n; U.S. Natl. Museum).

Family PROPHYLLOCRINIDAE Moore & Strimple, 1973

[Prophyllocrinidae MOORE & STRIMPLE, 1973, p. 15]

Crown moderately small, globose; cup with flat or somewhat concave base formed by pentagonal completely fused infrabasal circlet, surrounded by five equal pentagonal basals, strongly curved longitudinally as well as transversely; radials with articu-

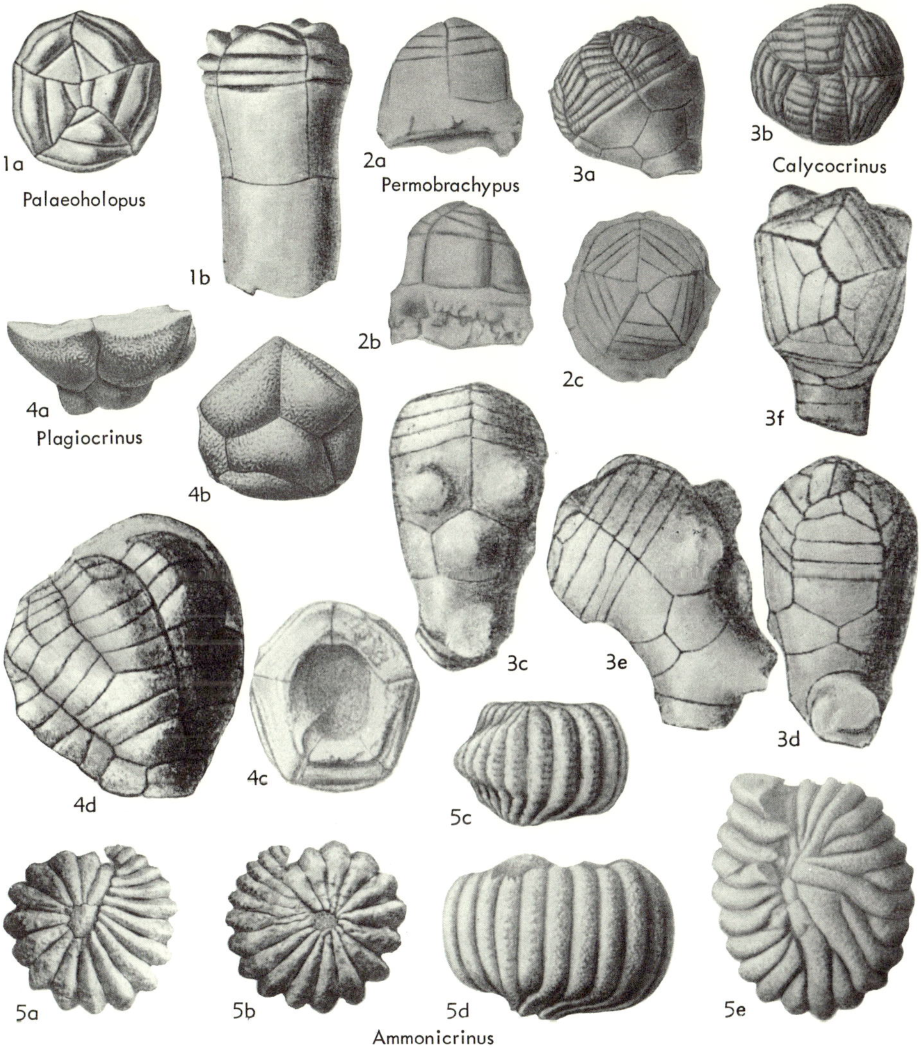

FIG. 526. Calycocrinidae *(3-5)*, Palaeoholopodidae *(1,2)* (p. *T*784-*T*785, *T*789).

lar arm facets much narrower than width of plates, bordered on one or both sides by pronglike projections, which may be symmetrical or decidedly asymmetrical, with projection on left side of facet most extended; quadrangular radianal directly below *C* radial, pentagonal anal *X* on distally truncate *CD* basal or both of these plates lacking. Arms short, narrow, bifurcating isotomously once or twice. *U.Perm.*

Prophyllocrinus WANNER, 1916, p. 273 [**P. dentatus*; OD]. Cup subglobular, with flattened or concave base, formed of massive plates; pentagonal infrabasal circlet entirely fused; radials with symmetrical or nearly symmetrical upward projections on either side of medially placed articular facet approximately 0.3 width of radial. Arms formed of wide, low brachials with transversely curved faces, branching isotomously on primibrachs *2* or *3* (rarely *4*) followed by 2 to 6 secundibrachs in each branch and 0 to 8 tertibrachs; arms bent strongly inward concealing tegmen. Stem transversely circular, diameter 0.4 to 0.8 width of infrabasal circlet, facets with peripheral girdle of short, fine crenulae, axial canal very small, circular.

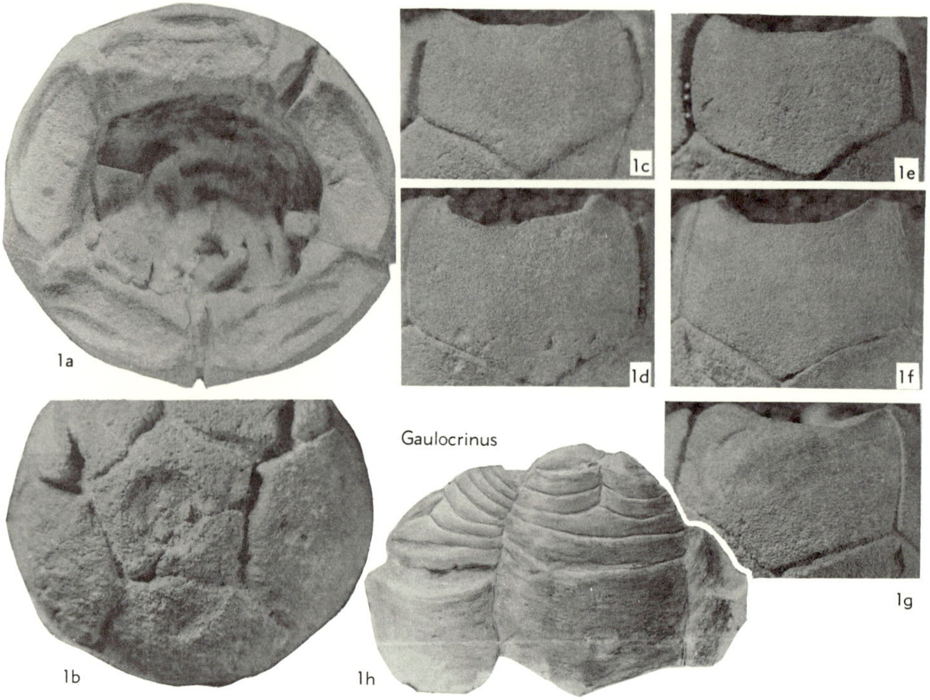

FIG. 527. Gaulocrinidae (p. T786).

[WANNER (1930b, p. 43) has described and figured an abnormal specimen classed as belonging to *P. cuspidatus* WANNER (1916a) differing from others in having both a rhombic radianal and pentagonal anal *X*. The radianal occurs directly beneath the *C* radial, as in *Mysticocrinus*, and accordingly the specimen was compared by WANNER with this genus, which he interpreted as ancestral to *Prophyllocrinus*.] *U.Perm.*, Indon. (Timor, Basleo).——FIG. 528,*2a*; 529,*3a,b*. *P. cuspidatus* WANNER; 528,*2a*, plate diagram of specimen classed as abnormal by WANNER because of presence of small plates seemingly representing radianal and anal *X* (radials black, radianal cross ruled, *X* stippled) (Wanner, 1930b, mod.); 529, *3a,b*, lat. and dorsal views of cup, ×1.5 (Wanner, 1916a).——FIG. 528,*2b*; 529,*3c-e*. **P. dentatus*; 528,*2b*, plate diagram (radials black), ray orientation not identifiable (Wanner, 1916a, mod.); 529, *3c-e*, ventral, dorsal, and lat. views of holotype crown, ×2 (Wanner, 1916a).

Ancistrocrinus WANNER, 1924, p. 119 [**A. vermistriatus*; M]. Generally very similar to *Prophyllocrinus*, but cup plates more massive and externally bulbous, infrabasal circlet in distinct concavity, mostly concealed by proximal columnals, and chiefly distinguished by very asymmetrical radials with distal left portion of each strongly elevated as spikelike projection and rounded articular facet confined to right half of plate. Arms composed of taller than average brachials, branching isotomously on primibrachs *2*. *U.Perm.*, Indon.(Timor, Basleo).——FIG. 528,*3*. *A. depressus* WANNER; plate diagram (radials black), orientation of rays not identifiable (Wanner, 1930b, mod.).——FIG. 529,*2*. **A. vermistriatus; 2a,b*, lat. and dorsal views of holotype cup, ×1.5 (Wanner, 1924).

Proapsidocrinus WANNER, 1924, p. 116 [**P. permicus*; OD]. Like *Ancistrocrinus*, but narrowly rounded articular facets of radials submedial, bordered on right by moderate upward projection and on left by very strong one arched inward to meet those of other radials at summit of cup, thus forming deep niches for reception of arms, which branch isotomously on primibrachs *2* and secundibrachs *2*, latter supporting 7 or more tertibrachs on either side. *U.Perm.*, Indon.(Timor, Basleo).——FIG. 528,*1*; 529,*1*. **P. permicus*; 528,*1*, plate diagram (radials black), orientation of rays not identifiable (Moore, n); 529,*1a-c*, lat., dorsal, and ventral views of holotype incomplete crown, ×1 (Wanner, 1924).

Family PALAEOHOLOPODIDAE Wanner, 1916

[*nom. correct.* MOORE, herein (*pro* Palaeoholopidae WANNER, 1916a, p. 279)]

Crown gumdrop shaped, with subvertical, longitudinally nearly straight sides and evenly rounded, very low summit formed by wide and extremely short arms curved inward to meet centrally. Cup consisting of five quadrangular radials surrounding tall cylindrical base presumably formed by fused basal and infrabasal circlets, cup attached to featureless low or moderately tall peduncle, no true stem. Arms unbranched, composed of three to five brachials which tend to be inclined clockwise. *U. Perm.*

Palaeoholopus WANNER, 1916, p. 279 [**P. pretiosus*; M]. Crown with nearly vertical sides and flatly rounded summit, 0.7 of height formed by 5 quadrangular radials of cup, which has very even summit formed by wide straight articular facets of radials, base of cup consisting of flat circular disc thought to be formed by fused basal and infrabasal circlets; cup attached to stemlike peduncle, which is cylindrical sutureless tube with diameter equal to that of cup and sides confluent with cup. Arms formed by very wide short brachials which are longitudinally convex, arms unbranched, with pointed tips meeting at summit of crown. *U.Perm.*, Indon.(Timor).——FIG. 525,*1*; 526,*1*. **P. pretiosus*; 525,*1*, plate diagram (radials black) (Wanner, 1930b, mod.); 526,*1a,b*, ventral and lat. views of crown, ×1 (Wanner, 1916a).

Permobrachypus MOORE & STRIMPLE, 1973 [*nom. subst. pro Brachypus* WANNER, 1929a, p. 320 (*non* VON MEYER, 1814; *nec* SWAINSON, 1824; *nec* MEIGEN, 1824; *nec* SCHOENHERR, 1825; *nec* GRAY, 1825; *nec* FITZINGER, 1826; *nec* GUILDING, 1828)] [**Brachypus adhaerens* WANNER, 1929a, p. 320; M]. Essentially similar to *Palaeoholopus* but radials proportionally wider and shorter and arms forming relatively higher more evenly rounded summit of crown. Peduncle very short and with sides tending to flare outward from base of cup. Arms composed of only 3 brachials which progressively decrease in width upward. *U.Perm.*, Indon. (Timor).——FIG. 525,*4*; 526,*2*. **P. adhaerens* (WANNER); 525,*4*, plate diagram (radials black) (Wanner, 1930b, mod.); 526,*2a-c*, 2 lat. and ventral views of holotype crown, ×2 (Wanner, 1930b).

Superfamily ICTHYOCRINACEA Angelin, 1878

[*nom. transl.* MOORE & STRIMPLE, 1973 (*ex* Icthyocrinidae

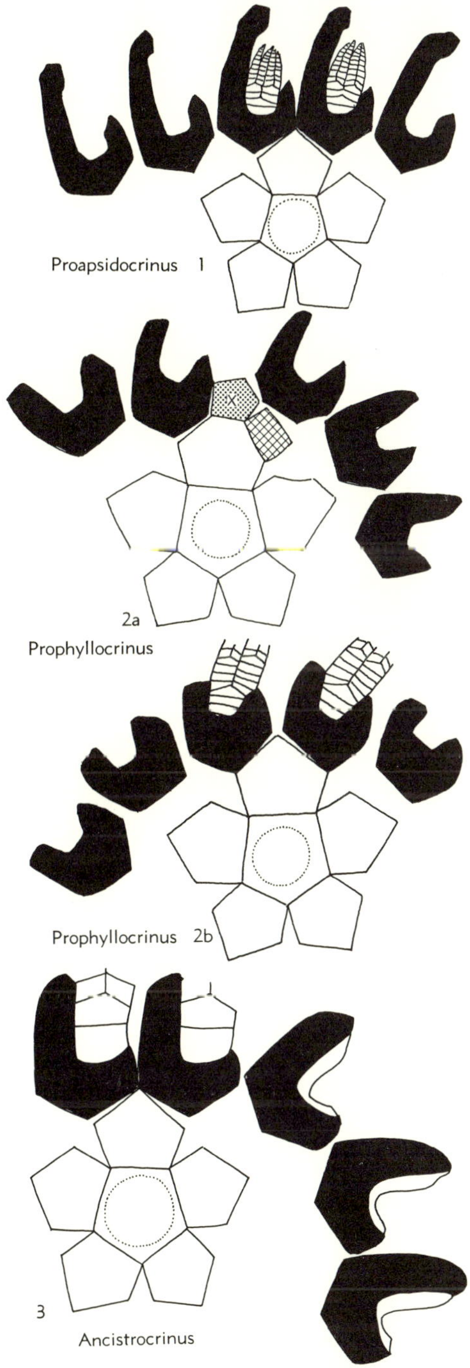

FIG. 528. Prophyllocrinidae (radials black, anal *X* stippled, radianal cross ruled) (p. *T*787-*T*788).

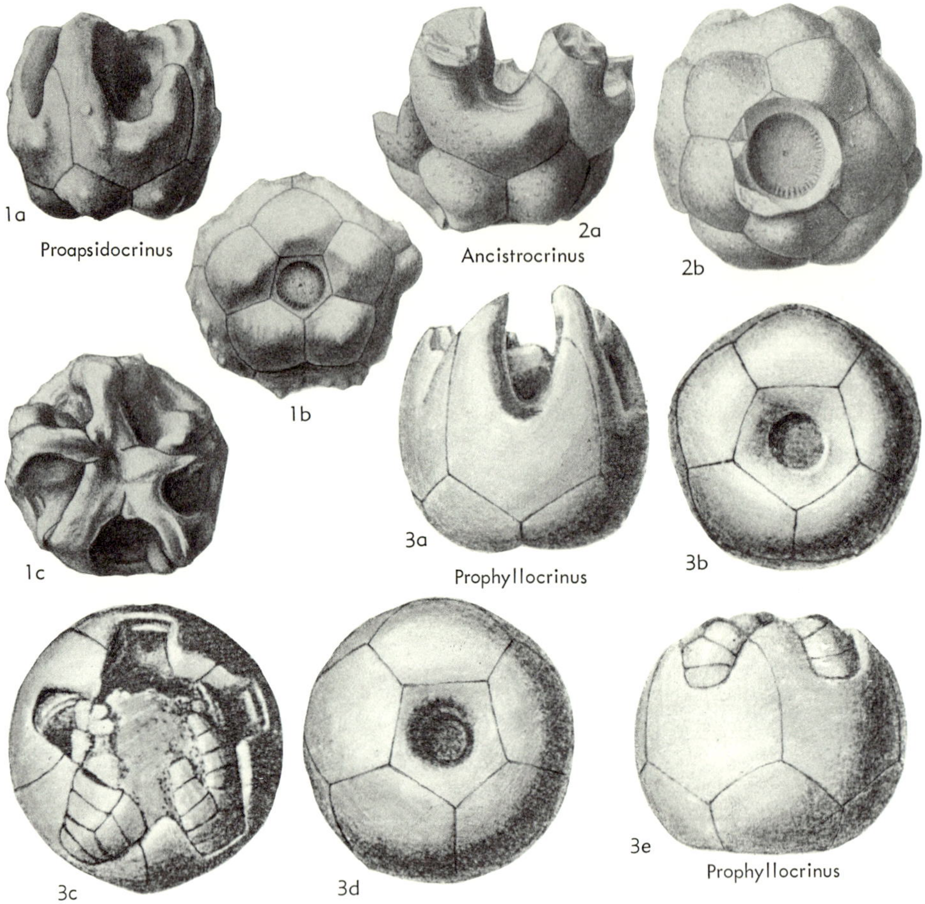

FIG. 529. Prophyllocrinidae (p. *T787-T788*).

ANGELIN, 1878, p. 13; *nom. correct.* MOORE & STRIMPLE, 1973, *pro* Ichthyocrinidae ANGELIN, 1878, p. 13)] [The ICZN *Code*, 1964 (art. 11e), requires that "a family-group name must, when first published, be based on the name then considered to be valid for a contained genus . . . ." *Ichthyocrinus* HALL, 1852 (erroneously attributed to CONRAD, 1842) for erection of Ichthyocrinidae by ANGELIN (1878) is not such a valid name and consequently the family name given by ANGELIN must be rejected. The fact that CONRAD transliterated *ichthys* from the Greek as *icthy-* does not prevent *Icthyocrinus* as published by him (1842, p. 279) from being construed as a "correct original spelling" (*Code*, art. 32,a,ii) and therefore *Ichthyocrinus* cannot legally replace *Icthyocrinus*]

Crown globose to elongate ovoid, many-armed, with cup not separated from fixed proximal brachials which form much or most of calyx, arms joined together closely, without associated interbrachials, cup anal plates lacking or restricted to radianal or anal *X*. Arms branching isotomously. Stem transversely circular. *L.Sil.-L.Miss.* or *low. L.Carb.*

Family ICTHYOCRINIDAE Angelin, 1878

[*nom. correct.* MOORE & STRIMPLE, 1973 (*pro Ichthyocrinidae* ANGELIN, 1878, p. 13)] [See nomenclatural note under Icthyocrinacea]

Characters of superfamily. [The unimportant, even rudimentary nature of the infrabasal circlet was considered by SPRINGER to be a diagnostic feature of this assemblage, but the dominance of fixed ray plates unassociated with interprimibrachs in forming a stoutly built calyx seems more significant.] *L.Sil.(Medinan)-L.Miss.(Osag.)*.

Icthyocrinus CONRAD, 1842, p. 279 [**I. laevis*; M] [=*Ichthyocrinus* HALL, 1852, p. 195 *(nom. van.)* (obj.)]. Crown ovoid to pear shaped, expanding upward from base of calyx which is formed by small radials and subjacent cup plates; greatly larger part of calyx composed of interlocking, firmly sutured ray plates; small infrabasal circlet

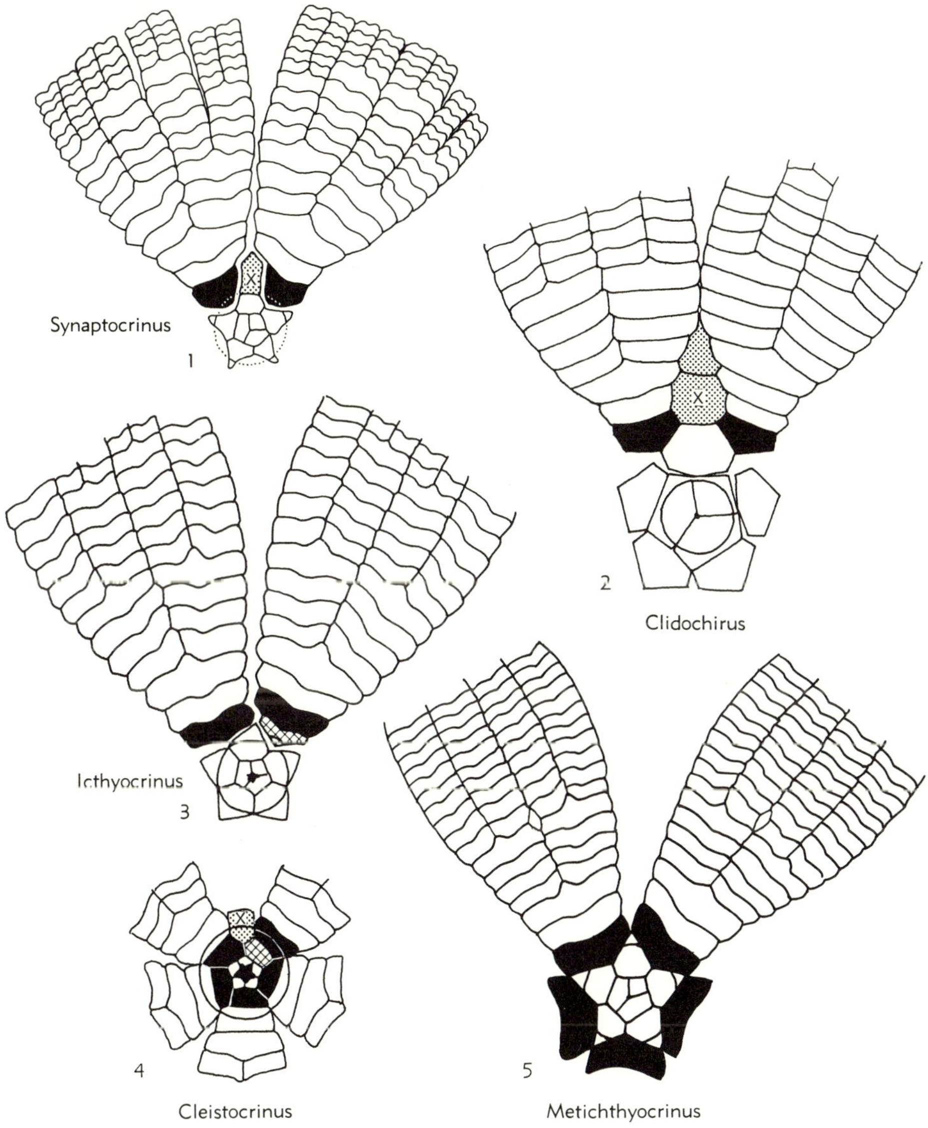

FIG. 530. Icthyocrinidae (radials black, anals stippled, radianal cross ruled) (p. *T790-T794*).

entirely concealed by stem or resorbed; posterior basal not differentiated, orientation of crown being determined by position of small infrabasal in *C* ray and occurrence of primitive radianal in this ray directly below radial; no anal *X* or interradial plates. Arms in calyx and above it branching isotomously on primibrachs 2, secundibrachs 3 or more commonly 4. Stem composed of thin, circular columnals tapering abruptly beneath cup or gradually reduced in width, axial canal small, circular or bluntly pentagonal. *M.Sil.(Niagaran)*, USA(N.Y.-Ind.-Ill.)-Eu.(Sweden-Eng.); *L.Dev.* USA(Tenn.)-Eu.(Czech.).——FIG. 530,*3*; 531, *3a-c*. **I. laevis*, Lockport Ls., USA(N.Y.)-Can. (Ont.); 530,*3*, plate diagram of part of calyx (radials black, radianal cross ruled) and arms of *C* and *D* rays (mod. from Springer, 1920); 531, *3a,b*, dorsal view of crown (*CD* interray directed upward) and *B*-ray view of same, specimen from Lockport, N.Y., ×2 (Springer, 1920); 531,*3c*, *AB*-interray view of crown from Grimsby, Ont., ×2 (Springer, 1920).——FIG. 531,*3d*. *I. pyri-*

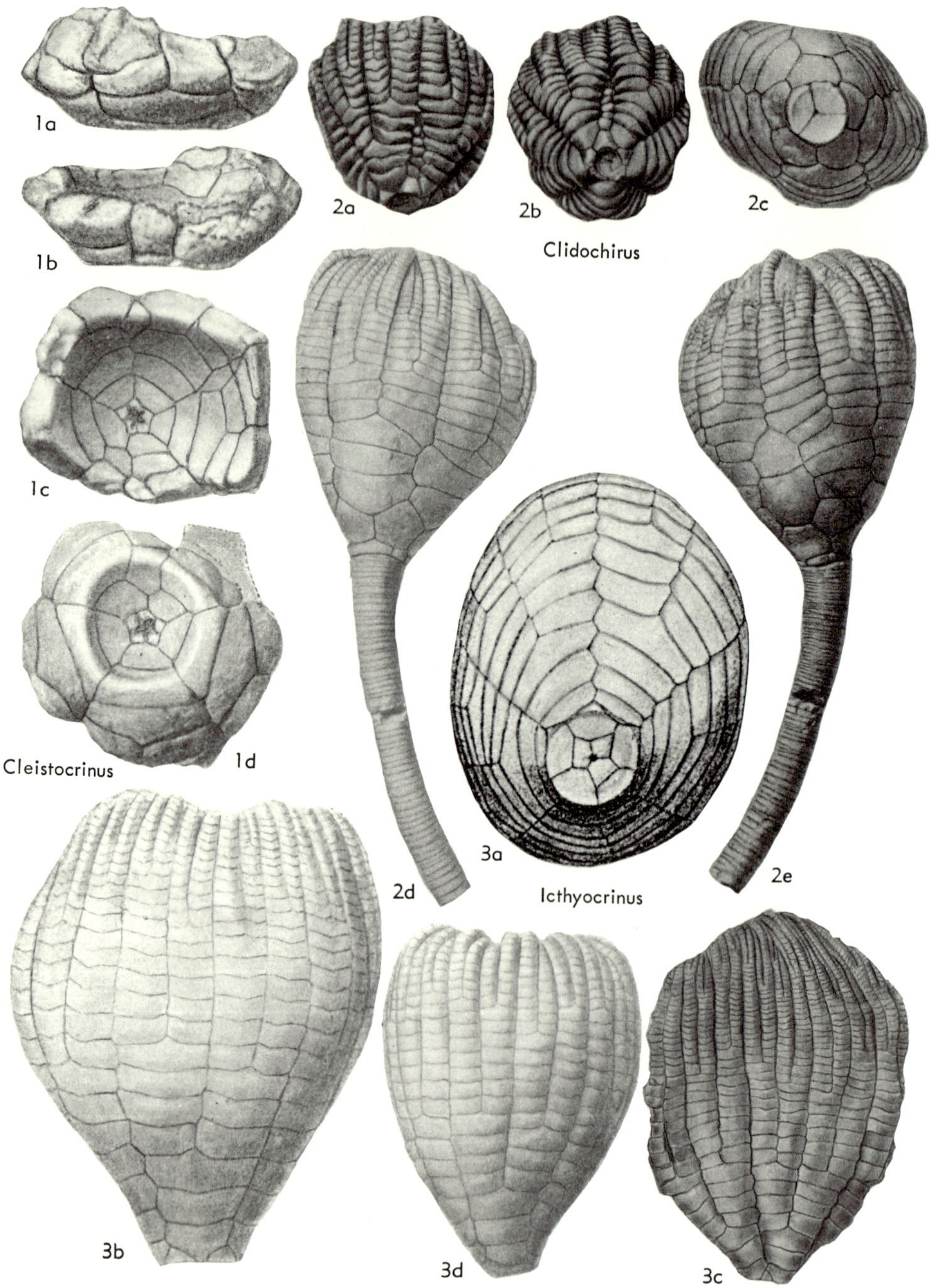

FIG. 531. Icthyocrinidae (p. T790-T793).

formis (PHILLIPS), Visby F., Sweden (Gotl.); *D*-ray view of undistorted crown, ×2 (Springer, 1920).

Cleistocrinus SPRINGER, 1920, p. 292 [**Calpiocrinus humilis* ANGELIN, 1878, p. 12; OD]. Lower part of calyx strongly built, shallow bowl shaped, with gently sloping sides; infrabasal circlet very diminutive or lacking; basals very small, entirely covered by stem, which likewise conceals all radials except in *C* ray where lower half of radial and radianal plate directly beneath it are included in stem impression; proximal part of primibrachs concealed by stem except in *C* ray; ray branching isotomously on primibrachs *2*, parts of calyx above secundibrachs *2* and free arms unknown; pentagonal anal *X* in *CD* interray rests on beveled edges of the *D* radial and radianal. Stem large, transversely circular, shown by impression on calyx to have facets with long fine crenulae; axial canal quinquestellate. *U.Sil.*, Eu.(Sweden).——FIG. 530,*4*; 531,*1*. **C. humilis* (ANGELIN), Wenlock Ls., Gotl.; 530,*4*, plate diagram of calyx (radials black, radianal cross ruled, anal *X* stippled) and proximal parts of arms (mod. from Springer, 1920); 531,*1a-d*, ant., post., ventral, and dorsal views of syntype, *CD* interray directed upward in *1d* and downward in *1c*, ×2 (Springer, 1920).

Clidochirus ANGELIN, 1878, p. 12 [**C. pyrum*; M] [=*Clidocrinus* WANNER, 1916a, p. 251 *(nom. null.)* (obj.)]. Crown elongate ovoid to pear shaped, generally resembling *Icthyocrinus* in appearance, but distinguished by larger plates in lower part of calyx, infrabasals sloping upward and visible from side (but concealed in one species); *CD* basal larger than others and broadly truncate distally for support of large anal *X*, which is followed by additional anal. Arms laterally interlocked, divided isotomously on primibrachs *2*, secundibrachs *4* and at higher levels. Stem very little tapered next below calyx and subequal in diameter throughout its observed length, formed of low columnals in proximal region but taller ones distally, articular facets distinctly crenulate. *L.Sil.*, USA(Ohio); *U.Sil.*, Eu. (Sweden); *L.Dev.*, USA(W.Va.-N.Y.); *U.Dev.*, USA(Iowa); *L.Miss.*, USA(Ky.).——FIG. 530,*2*; 531,*2c-e*. **C. pyrum*, U.Sil.(Wenlock), Gotl.; 530,*2*, plate diagram of lower part of calyx (radials black, anals stippled), with proximal part of *C* and *D* rays separated by anal plates (mod. from Springer, 1920); 531,*2c*, dorsal view of crown, *CD* interray upward, ×1 (Springer, 1920); 531,*2d,e*, ant. and post. views of complete crown with attached stem, ×1 (Springer, 1920).——FIG. 531,*2a,b*. *C. keyserensis* SPRINGER, L.Dev. (Helderberg.), USA(W.Va.); *2a,b*, post. and dorsal views of complete crown, *2b* with *CD* interray upward, ×1 (Springer, 1920).

Metichthyocrinus SPRINGER, 1906, p. 517 [**Ichthyocrinus burlingtonensis* HALL, 1858b, p. 557; M]. Crown subglobular, with smoothly sloping sides of calyx and free arms well inturned at summit; stem concealing subhorizontal infrabasals and basal circlet except for their distal extremities. Arms expanding upward, closely abutting with brachials interlocking. Stem tapering somewhat next below calyx but thereafter nearly uniform in diameter, columnals very thin proximally but much thicker distally with tendency in some species toward heteromorphy. [The type species of this genus is not the form recorded by SPRINGER (1920, p. 320), given as *Cyathocrinites tiaraeformis* TROOST *(nom. nud.)* (=*Ichthyocrinus tiaraeformis* TROOST in HALL, 1858b, p. 558). This is because in his original publication SPRINGER cited a single species as *M. burling-*

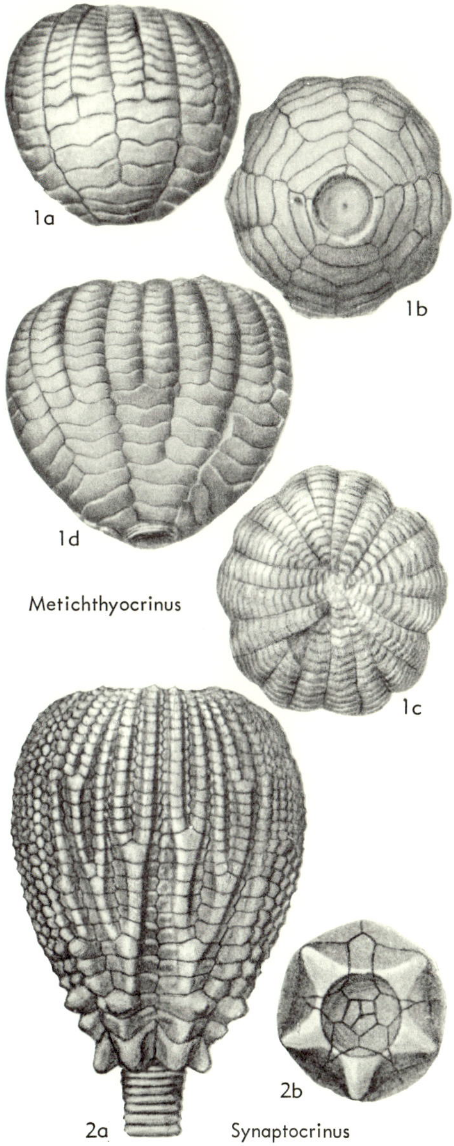

FIG. 532. Icthyocrinidae (p. *T793-T794*).

tonensis Hall (Springer, 1906, p. 522, pl. 6, fig. 15) (=*Ichthyocrinus burlingtonensis* Hall, 1858, p. 557). Hall's species is thus fixed as type of *Metichthyocrinus* by monotypy.] *L.Miss.*, USA(Iowa-Ind.-Ky.-Tenn.).——Fig. 530,*5*; 532, *1a-c*. *M. tiaraeformis* (Troost), L.Miss.(Osag., New Providence F.), USA(Tenn.); 530,*5*, plate diagram of cup (radials black) and major part of *C* and *D* ray (mod. from Springer, 1920); 531,*1a-c*, lat., basal, and summit views of complete crown, ×1 (Springer, 1920).——Fig. 532, *1d*. *M. clarkensis* (Miller & Gurley), L.Miss. (Osag., New Providence F.), USA(Tenn.); lat. view of crown, identity of rays unspecified, ×1 (Springer, 1920).

Synaptocrinus Springer, 1920, p. 301 [**Forbesiocrinus nuntius* Hall, 1862, p. 124; OD]. Crown pear shaped, widest near summit; calyx includes cup plates largely concealed by stem, *CD* basal notably elongate and pointed distally; ray plates bifurcating isotomously, brachials strongly interlocked; type species distinguished by rounded elevation of the median part of each brachial, producing appearance of raised branching arms entirely foreign to typical icthyocrinids. Stem heteromorphic, not appreciably reduced in diameter distally. *M.Dev.*, USA(N.Y.); *U.Dev.*, NW. Can.——Fig. 530,*1*; 532,*2*. **S. nuntius* (Hall), Hamilton Gr., USA(N.Y.); 530,*1*, plate diagram of part of cup (radials black, anal *X* stippled) and arms of *C* and *D* rays (mod. from Springer, 1920); 532,*2a,b*, post. view of crown (reconstr.) and diagrammatic view of base with *CD* interray directed upward, ×2 (Springer, 1920).

Superfamily SAGENOCRINITACEA Roemer, 1854

[*nom. transl.* Moore & Strimple, 1973 (*ex* Sagenocrinitidae Bassler, 1938, p. 23; *nom. subst. pro* Sagenocrinidae Roemer in Bronn, 1854, p. 228)]

Crown globose to elongate ovoid, with many laterally abutting arms and joined together also by interradial and interbrachial plates, arms curved inward at top of crown above tegmen, which therefore generally cannot be seen. Stoutly built calyx comprising lower one-half or more of crown includes one or several anal plates in posterior interray or uncommonly lacks them entirely. Stem mostly large, transversely circular. *U.Sil.-U.Perm.*

Family HOMALOCRINIDAE Angelin, 1878

[Homalocrinidae Angelin, 1878, p. 11]

Crown small, rotund, with cup (defined as including plates below summit of radials) not distinctly separated from higher part of crown, in this respect clearly differentiated from Lecanocrinidae and Nipterocrinidae; infrabasals extending beyond stem impression; radianal in primitive position directly beneath *C* radial, anal *X* extending above articular facets of adjoining radials. Arms gently to somewhat strongly rounded transversely, composed mostly of brachials wider than high, with only few of them in each ray joined together by firm sutures to form fixed brachials, branching of arms isotomous up to *Y* level at upper limit of fixed brachials and heterotomous above *Y* line; one or two large interprimibrachs between contiguous rays or with larger number of small, irregularly arranged and shaped interprimibrachs; interbrachials very few or lacking. Stem transversely circular, composed of low columnals or in intermediate and distal regions of columnals with height nearly equal to diameter; axial canal circular, very tiny. [Endotomous branching of arms in upper part of crown is a most diagnostic attribute of the family.] *U.Sil.*

Homalocrinus Angelin, 1878, p. 11 [**H. parabasalis* (=*Taxocrinus nanus* Salter, 1873, p. 126, *nom. nud.*; *H. dudleyensis* Springer, 1906, p. 482); M] [=*Leiocrinus* Springer, 1902, p. 95 (type, *Calpiocrinus ovatus* Angelin, 1878; OD) (*non* d'Orbigny, 1850, p. 180)]. Large anal *X* resting on truncated distal margin of posterior basal and extending upward so as to separate primibrachs of *C* and *D* rays; other characters as for family. *U.Sil.*, Eu.(Sweden-Eng.).——Fig. 533,*3*; 534,*2a-h*. **H. parabasalis*; 533,*3*, plate diagram (radials black, radianal cross ruled, *X* and higher anals stippled), *CD* interray upward (Springer, 1920, mod.); 534,*2a-c,h*, *E*-ray, *CD*-interray, *C*-ray oblique from below, and dorsal (*CD* interray up) views of specimen from Gotland figured by Angelin, ×1.5 (Springer, 1920); 534,*2d-f*, *A*-ray, dorsal (*A* ray down), and *D*-ray views of specimens from Dudley, Eng., ×2 (Springer, 1920); 534,*2g*, *CD*-interray view of juvenile crown from Gotland, ×2 (Springer, 1920).——Fig. 534,*2i-k*. *H. liljevalli* Springer, Sweden(Gotl.); *2i,j*, dorsal (*CD* interray up) and *B*-ray views of holotype crown, ×2; *2k*, under side of finely plated tegmen of another specimen showing 5 centrally placed orals and ambulacra passing between them to arms, ×2 (all Springer, 1920).

Anisocrinus Angelin, 1878, p. 13 [**A. interradiatus*; M]. Radianal directly or somewhat obliquely at left below *C* radial; interrays with very large plates resting on shoulders of radials and arched over by fixed brachials, anal *X* largest of these, distinguished by its position on truncated distal margin of posterior basal and by its contact with radianal; large proximal plates of interrays may be followed by 1 or 2 smaller ones above; *Y* line at summit of axillary secundibrachs *4* or *5*. *U.Sil.*, Eu.(Sweden)-N.Am.(Ill.-Ind.-Ky.-Tenn.).——Fig. 533,*1*; 534,*1a-c*. **A. interradiatus*, Sweden(Gotl.); 533,*1*, plate diagram (radials black, radianal cross ruled, anal *X* stippled; *y*, upper line of fixed brachials), *CD* interray upward (Springer, 1920, mod.); 534,*1a*, *A*-ray view of essentially perfect crown showing upflared infrabasal circlet and truncate summit formed by infolded arms, ×2; 534,*1b,c*, *CD*-interray and dorsal (*CD* interray up) views of incomplete crown, ×2 (all Springer, 1920).

Asaphocrinus Springer, 1920, p. 174 [**A. bassleri*; OD]. Crown expanding upward, with transversely well-rounded arms not abutting laterally; interray areas with numerous irregularly shaped and arranged plates; infrabasals visible from side, *CD* basal larger than others and firmly joined to rhombic radianal and large anal *X*, latter supporting numerous smaller anals in form of sac or tube interlocking laterally with contiguous brachials. Each ray with several isotomous bifurcations of arms at different levels. *U.Sil.*, USA(N.Y.-Ind.-Tenn.).——Fig. 533,*2*; 534,*3a-c*. **A. bassleri*, Brownsport Ls., Tenn.; 533,*2*, plate diagram (radials black, radianal cross ruled, *X* and higher anals stippled), *CD* interray upward (Springer, 1920, mod.); 534,*3a-c*, *D*-ray, *A*-ray, and *DE*-interray views of three crowns, ×1 (Springer, 1920).——Fig. 534,*3d*. *A. ornatus* (Hall), Rochester Sh., N.Y.; *CD*-interray view of crown showing tubelike appearance of anal plate series, ×1.5 (Springer, 1920).

Family SAGENOCRINITIDAE Roemer, 1854

[*nom. correct*. Bassler, 1938, p. 23 (*pro* Sagenocrinidae Roemer in Bronn, 1854, p. 228)] [incl. Forbesiocrinidae Jaekel, 1918, p. 78]

Crown ovoid to pear shaped, elongate cup (radials, anals, and lower plate circlets) not at all set off from bulk of large stout calyx composed of numerous fixed brachials in all rays and prominent many-plated interradial and interbrachial areas; free arms mostly well differentiated from calyx, transversely rounded, separated from one another and bifurcating several times isotomously, distal extremities of arms curved strongly inward at top of crown; infrabasal circlet subhorizontal, largely, or entirely concealed by column; basals and radials commonly not larger than fixed brachials and interbrachials, all massive thick plates firmly joined together; radianal and anal *X* present in all genera. Stem homeomorphic or distinctly heteromorphic, composed of

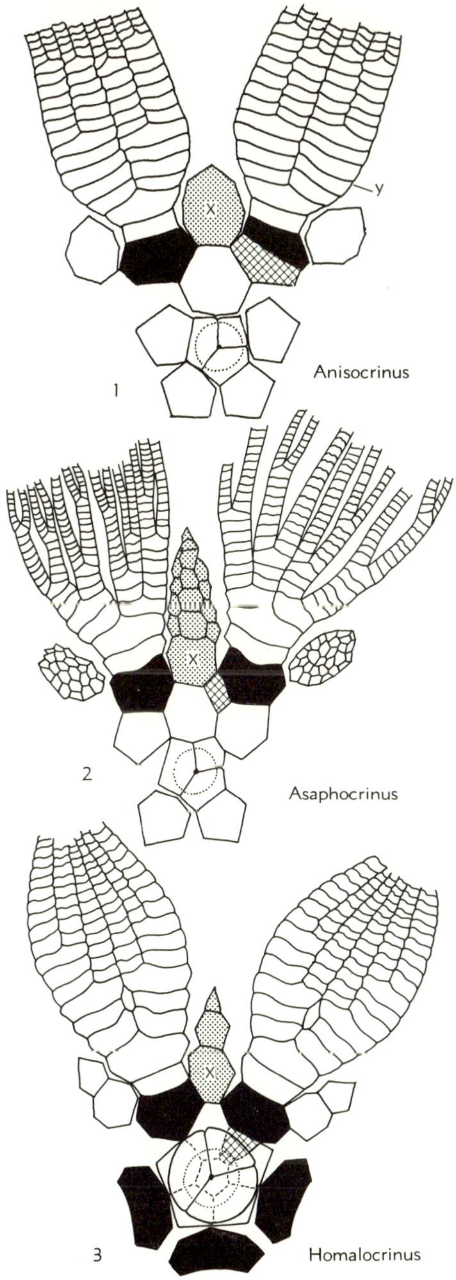

Fig. 533. Homalocrinidae (radials black, anals stippled, radianal cross ruled) (p. *T794-T795*).

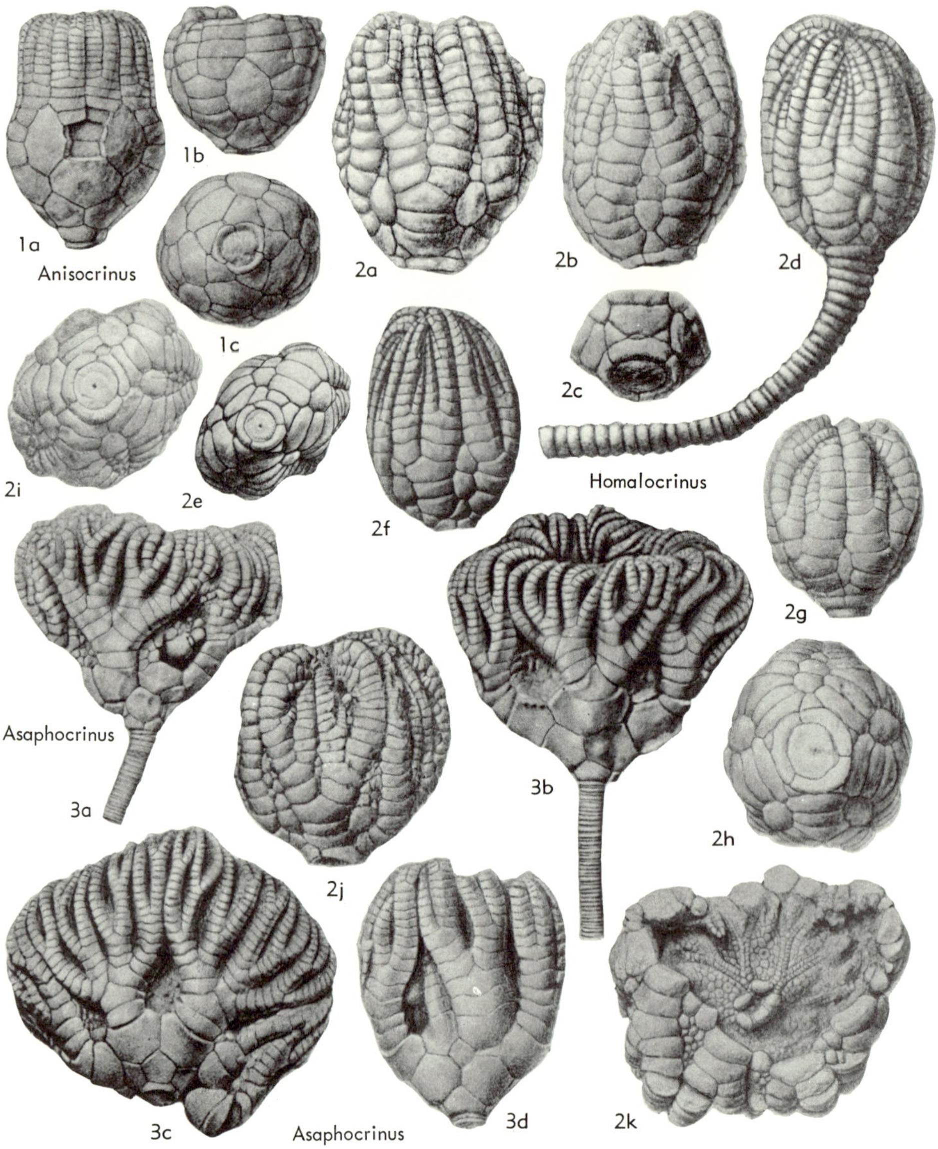

FIG. 534. Homalocrinidae (p. *T794-T795*).

relatively low columnals which commonly are perforated by quinquestellate axial canal. *U.Sil.; Miss.; L.Perm.*

Sagenocrinites AUSTIN & AUSTIN, 1842, p. 110 [**Actinocrinites? expansus* PHILLIPS in MURCHISON, 1839, p. 674; M] [=*Sagenocrinus* MORRIS, 1843, p. 58 *(nom. van.)* (obj.); *Periechocrinites* AUSTIN & AUSTIN, 1843, p. 203 (type, *P. articulosus*; =*Actinocrinites? expansus*, obj.; OD) (*non* BASSLER, 1938, p. 149); *Sageniocrinus* ZITTEL, 1895, p. 130 *(nom. null.)*]. Crown ovoid to biturbinate, small to moderately large; infrabasals somewhat recumbent, not entirely covered by stem impression; basal circlet commonly interrupted in *C*-ray position by radianal which occurs in primitive position directly beneath *C* radial and touches small infrabasal; *CD* basal not larger than others or truncated distally; anal, interradial and interbrachial areas filled by numerous polygonal solid plates in more than one series and characterized by appreciable breadth of these areas; fixed and

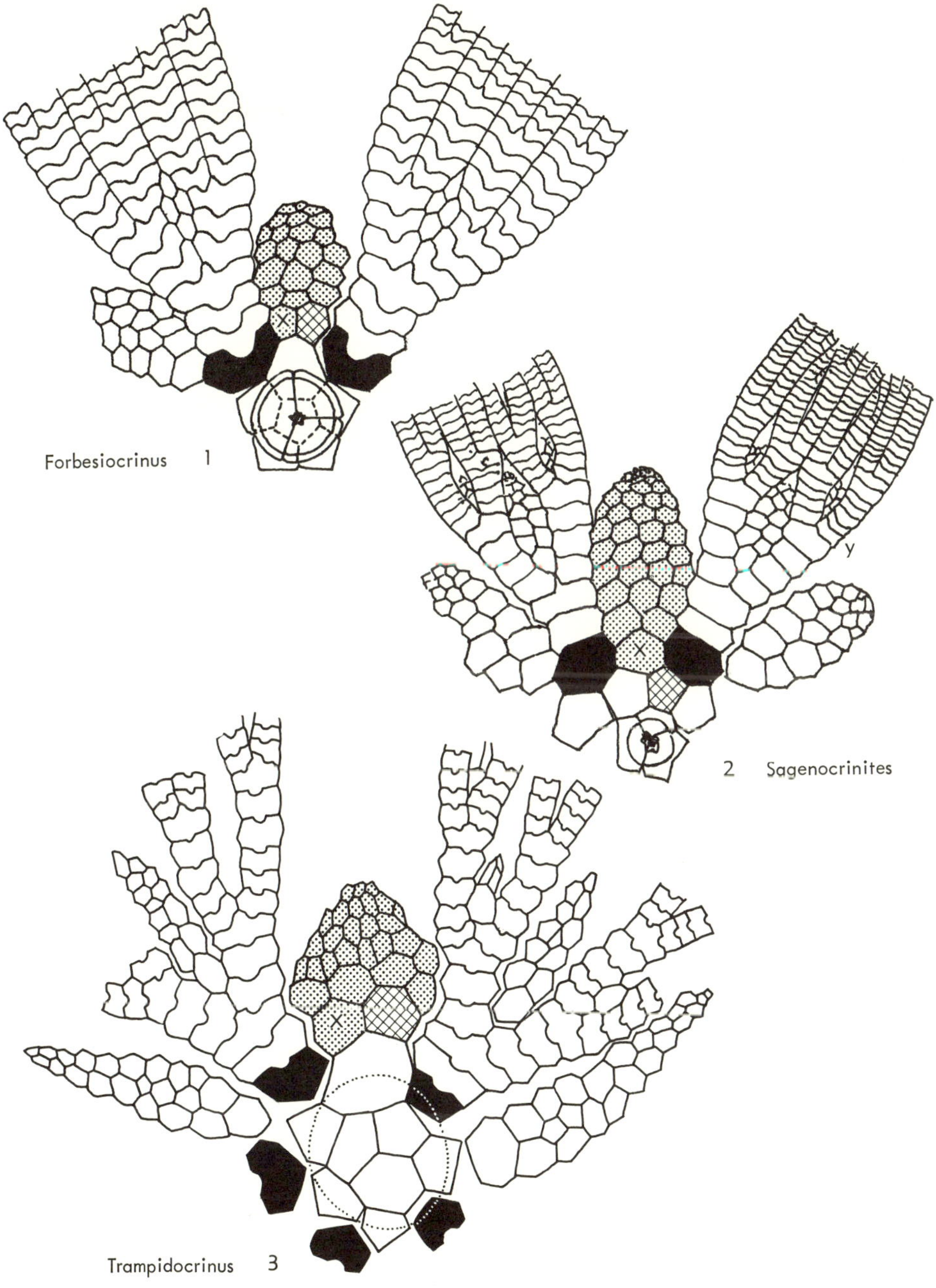

FIG. 535. Sagenocrinitidae (p. *T*796-*T*800).

free brachials in isotomously branching successions, narrowing upward and with straight to distinctly sinuous interbrachial sutures; each ray or arm branch with 2 primibrachs, 3 or 4 secundibrachs, 5 to 8 tertibrachs, and indefinite number of plates in higher brachitaxes. Stem transversely circular, heteromorphic, not expanding at calyx; columnals penetrated by quinquestellate axial canal. *U.Sil.,* Eu.(Eng.-Sweden)-USA(Ind.-Tenn.).——FIG. 535,*2*; 536,*1a,b*. **S. expansus* (PHILLIPS),

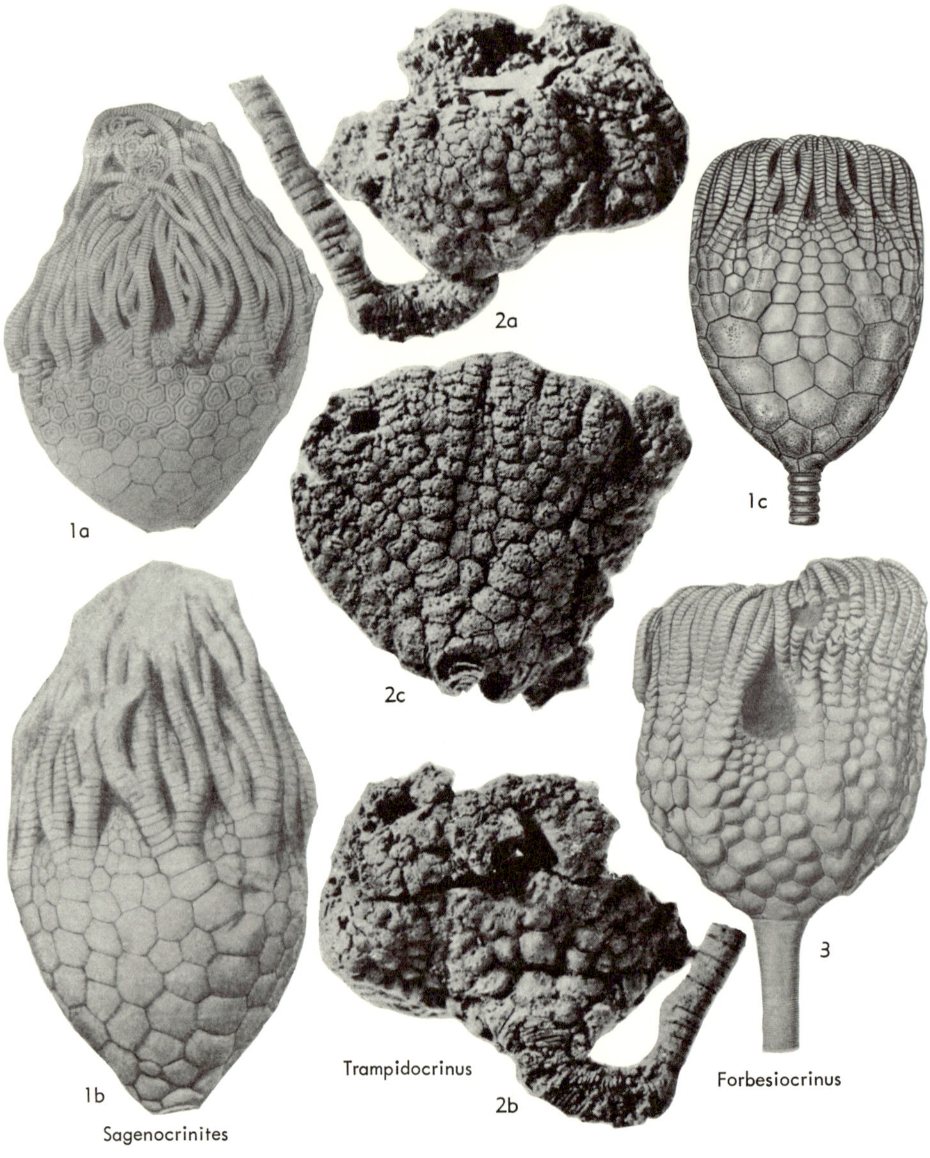

FIG. 536. Sagenocrinitidae (radials black, anals stippled) (p. T796-T800).

Wenlock Ls., Eng.; 535,*2,* plate diagram of calyx (radials black, radianal cross ruled, *X* and higher anal plates stippled; *y,* upper line of fixed brachials) and post. rays with adjoining interrays (mod. from Springer, 1920); 536,*1a,b,* side view of incomplete crown (rays not specified) and post. view of nearly complete specimen, both from Dudley, Eng., ×0.7, ×1.5 (Springer, 1920).——FIG. 536,*1c*. *S. clarki* SPRINGER, Brownsport F., USA(Tenn.); post. side of crown, ×0.7 (Springer, 1920).

Forbesiocrinus DE KONINCK & LE HON, 1854, p. 118 [**F. nobilis*; M] [=*Proguettardicrinus* STEINMANN, 1908, p. 153 (type, *Forbesiocrinus greenei* MILLER & GURLEY, 1896b, p. 57; OD)]. Crown globular to elongate ovoid, medium sized to unusually large, with stout calyx formed by firmly sutured thick cup and ray plates joined laterally with large areas of interradial and interbrachial plates; transversely rounded free arms in upper

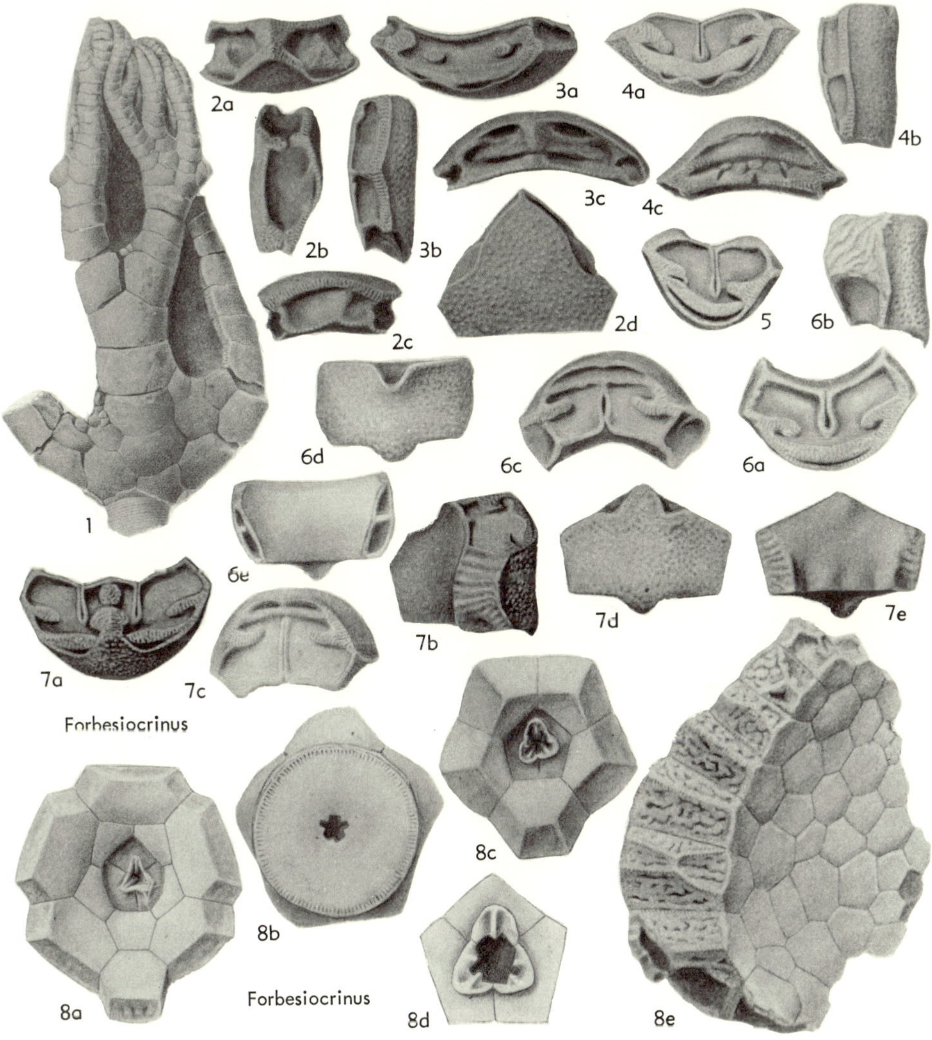

FIG. 537. Sagenocrinitidae (p. *T*798-*T*800).

part of crown branching isotomously and separated, incurved distally; infrabasal circlet barely visible beyond large stem impression externally, very thick, and narrowing internally to small pentagon; only distal extremities of basals exposed, *CD* largest, with distal margin beveled for contact with anal *X* at left and upwardly displaced radianal at right, these plates followed by numerous other medium-sized polygonal anals in posterior interray; radials approximately equal in size and shape to proximal primibrachs, laterally united with one another except where separated by anals. Both fixed and free arms branching evenly isotomously, composed of brachials with strongly sinuous sutures produced by prominent patelloid processes and sockets on outer parts of their articular facets. Large stem composed of low circular columnals which decrease gradually in diameter for short or considerable distance beneath calyx but after initial taper maintain fairly constant diameter, with or without distinct increase in height of columnals; stem penetrated by quinquestellate axial canal. *L.Carb.(Tournais.)*, Eu.(Belg.-Eng.); *L.Miss.(Osag.)-U.Miss.(Meramec.)*, USA(Iowa-Mo.-Ill.-Ind.-Ky.-Tenn.-Ohio).——FIG. 535,*1*; 536,*3*; 537,*1-7*. **F. nobilis*, Tournais., Belg.; 535,*1*, plate diagram of calyx (symbols as in Fig. 535,*2*), *C* and *D* lower part of arms, and adjoining interrays (mod. from Springer, 1920); 536,*3*, post. view of crown from

Burlington Ls.(Osag.) at Burlington, Iowa, showing very numerous anal plates, outer ones joined firmly to brachials of *C* and *D* rays, ×0.5 (Springer, 1920); 537,*1*, *C*-ray view of somewhat crushed crown from Tournais., Belg., ×0.7 (Springer, 1920); 537,*2-7*, dissociated calyx and arm plates from Tournais., Belg.; *2a-d*, distal, side, prox., and ext. views of basal; *3a-c*, distal, side, and prox. views of radial; *4a-c*, distal, side, and prox. views of first primibrach; 5, distal view of 2nd secundibrach; *6a-e*, distal, side, prox., ext., and int. views of 2nd primibrach; *7a-e*, distal, side, prox., ext., and int. views of primaxil; all ×2 (Springer, 1920).——FIG. 537,*8*. *F. saffordi* HALL, U.Miss.(Meramec.), USA(Ind., Spergen); *8a*, int. view of basal part of calyx with radials (*A* radial upward) and anal, ×2; *8b*, ext. view of basal circlet (post. basal upward) and stem impression, ×2; *8c*, int. view of infrabasal and basal circlets (post. side downward), ×2; *8d*, int. view of infrabasal circlet showing subtriangular rim surrounding axial canal (small infrabasal in *A* ray), ×4; *8e*, part of interior showing corrugated edges of thick brachials for attachment to interbrachials, ×1.5 (all Springer, 1920).

Trampidocrinus LANE & WEBSTER, 1966, p. 17 [**T. phiala*; OD]. Crown medium sized to large, compact, sides of calyx flaring rather evenly outward from stem with free arms above calyx curved strongly inward; notably similar to *Forbesiocrinus* in general features, including arrangement of large anal *X* and radianal plates above large *CD* basal but differing radically in broad contact of *B* and *C* radials with infrabasals. Isotomous branching of fixed and free arms well defined, as is strongly sinuous interbrachial sutures produced by patelloid processes. Stem very unlike that of *Forbesiocrinus* and corresponding closely to *Nevadacrinus* (dactylocrinid flexible crinoid) in having a curved, gently tapered basillarid region composed of 25 to 40 very low columnals with strongly crenulate articula ending distally at a bulbous wedge shaped columnal (cuneus) with facetal planes disposed at angle of 45° to 55° to each other; beyond cuneus stem is distinctly heteromorphic, composed of moderately tall columnals with longitudinally convex profile alternating with much thinner, straight-sided columnals; curvature and abrupt flexure of stem is in direction of *CD* interray, thus differing from *Nevadacrinus* in which curvature and flexure appear to be toward *AB* interray. *L.Perm.*, USA(Nev.).——FIG. 535,*3*; 536,*2a-c*. **T. phiala*; 535,*3*, plate diagram of calyx (symbols as in Fig. 535,*2*), lower part of *C* and *D* arms, and adjoining interrays (mod. from Lane & Webster, 1966); 536,*2a,b*, views of crown with part of attached stem from *B*-ray side (distal arms of *A* ray at right) and from *DE*-interray side (*CD* interray at right), ×1 (Lane & Webster, 1966); 536,*2c*, *DE*-interray view of holotype (*D* ray at right), ×1 (Lane & Webster, 1966).

Family DACTYLOCRINIDAE Bather, 1899

[Dactylocrinidae BATHER, 1899b, p. 923] [incl. Calpiocrinidae JAEKEL, 1918, p. 79]

Crown ovoid, moderately elevated, with virtually no distinction between fixed brachials forming upper part of calyx and cup composed of radials and lower plate circlets; radianal normally lacking but anal *X* commonly distinct and relatively large, followed by other anal plates. Proximal columnal concealing infrabasal circlet in most genera, as well as much or all of basal plates and even portions of radials. First and second branching of arms confined to calyx in which brachitaxes and lower plates are fixed, is isotomous; higher branching involves free brachitaxes, which are separated laterally from one another and are strongly heterotomous. Within this family free arms and their free brachitaxes considerably exceed length and number of arms incorporated in the calyx; accordingly, heterotomy prevails over isotomy in arm structure. Interbrachials few or may be absent. *U.Sil.-L.Miss.; L.Penn.-U.Perm.*

Dactylocrinus QUENSTEDT, 1876, p. 520 [**Dimerocrinites oligoptilus* PACHT, 1852, p. 339; M] [*non Dactylocrinus* SLADEN, 1878, p. 245] [=*Dimerocrinites* PACHT, 1852 (*non* PHILLIPS in MURCHISON, 1839) (obj.); *Aristocrinus* ROWLEY, 1895, p. 218 (type, *A. concavus*; M); *Callawaycrinus* ROWLEY, 1895, p. 219 (=*Aristocrinus*, obj.)]. Crown rotund to elongate, widening upward; base flat, broadly concave, or deeply excavated, with part of radials and all lower plates involved in bowl for reception of stem; infrabasal circlet entirely surrounded by basals, which are relatively narrow and notably elongated, *CD* basal more than others and truncated distally for support of large polygonal anal *X*; radials with large articular facets and beveled shoulders for sutural contact with large interradial plate and with anal *X* in posterior interray; no radianal, *X* followed by several anal plates which are suturally joined laterally to brachials. Arms branching isotomously in each ray on primibrachs 2 and secundibrachs *3* to *5*, in more distal portions characterized by strong heterotomy of bi-endotomous type, with main arm trunks subequal or with inner ones somewhat smaller than outer. Column enlarging proximally. [Name derived from fingerlike appearance of ramules.] *M.Dev.-L.Miss.*, Eu.(Belg.-Ger.-W.USSR)-USA (Ohio-Mich.-Mo.-Iowa).——FIG. 538,*3a,b*; 539,*1a,b*. **D. oligoptilus* (PACHT), U.Dev., USSR; 538,*3a,b*, plate diagram (radials black, *X* and other anals stippled), *CD* interray

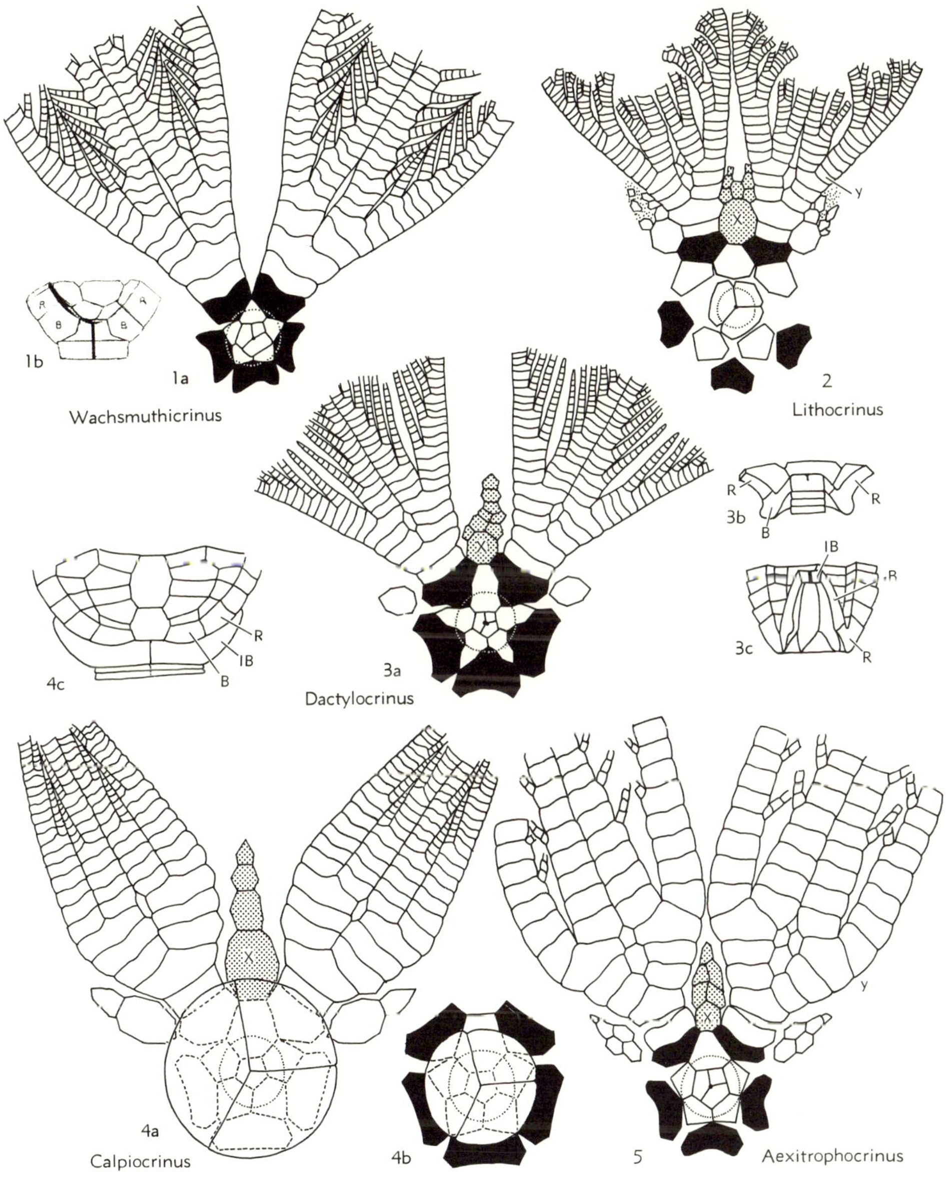

FIG. 538. Dactylocrinidae (p. *T*800-*T*804, *T*806).

upward, and section of basal part of cup (Springer, 1920, mod.); 539,*1a,* post. side of crown showing isotomous divisions of main branches of *C* and *D* rays with ramules on facing sides (bi-endotomous), also arrangement of anal plates, ×1 (Springer, 1920); 539,*1b,* dorsal side of cup showing basal concavity, basals entirely concealing infrabasals, ×1 (*B,* basal; *R,* radial) (Springer, 1920).——FIG. 538,*3c*. *D. excavatus* (SCHULTZ), M.Dev., Ger.; vert. sec. of lower part of calyx (*B,* basal; *IB,* infrabasal; *R,* radial) (Springer, 1920).——FIG. 539,*1c-e*. *D. concavus* (ROWLEY), M.Dev.(Hamilton), USA(Mo.); *D*-ray, *CD*-interray, and dorsal views, ×1.5 (Springer, 1920).

Aexitrophocrinus STRIMPLE & WATKINS, 1969, p. 229 [**Synerocrinus formosus* MOORE & PLUMMER, 1940, p. 94; OD] [=*Trautscholdicrinus* MOORE & PLUMMER in MOORE, 1939c, p. 195 (*non* YAKOVLEV & IVANOV. 1939, p. 66) (type, *T. jaekeli*; OD); *Synerocrinus* MOORE & PLUMMER in MOORE,

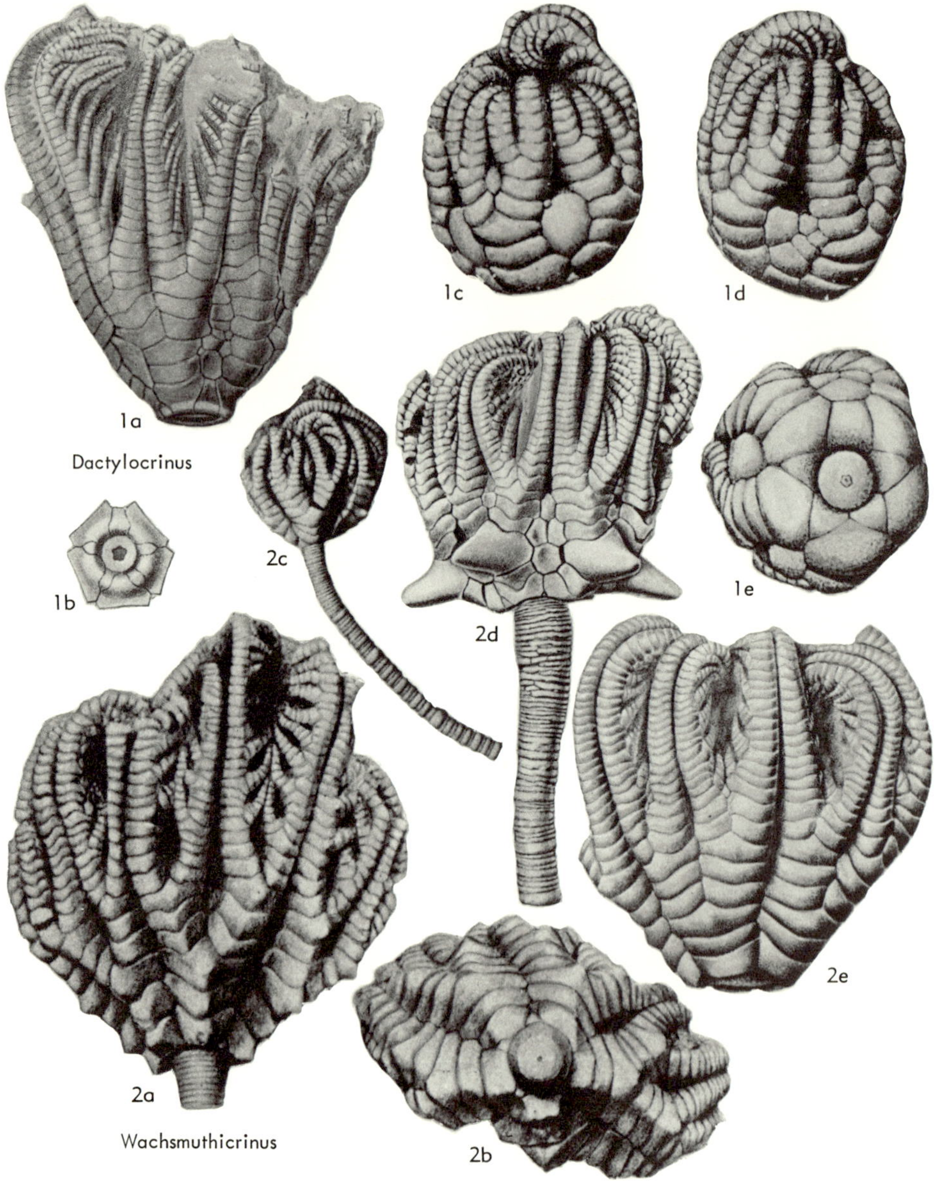

FIG. 539. Dactylocrinidae (p. T800-T801, T806).

1939c, p. 195 (*non* JAEKEL, 1898a); *Talanterocrinus* MOORE & PLUMMER, 1940, p. 96 (type, *T. jaekeli*, =*Trautscholdicrinus jaekeli*, =*Forbesiocrinus incurvus* TRAUTSCHOLD, 1867, =*Synerocrinus incurvus* JAEKEL, 1898a; OD)]. Differs from *Dactylocrinus* chiefly in much smaller number and diminutive size of short ramules, presence of 4 or more interradial plates between each contiguous pair of rays, and 2 primibrachs, 3 secundibrachs, and 9 or more tertibrachs in each brachitaxis; large anal *X* with oblique distal facets supporting double series of additional anal plates which are suturally joined to *C*- and *D*-ray brachials. Stem medium sized to large, narrowing somewhat beneath proximal region. [In appearance this genus corresponds rather closely to *Synerocrinus* but differs in having anal plates suturally interlocked with adjacent brachials. *Synerocrinus* has a tubular anal series not so connected. MOORE & PLUMMER (1940, p. 90)

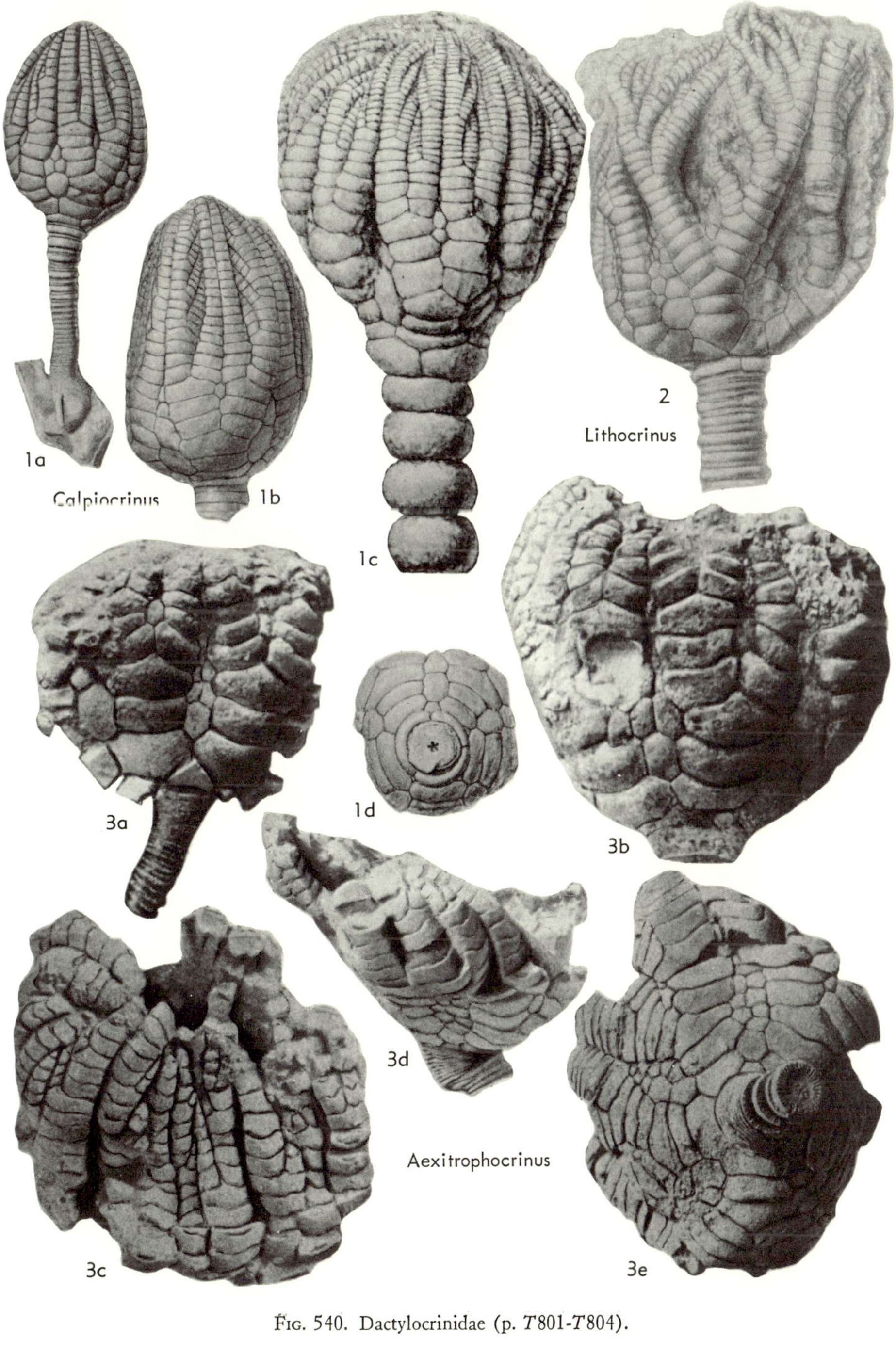

Fig. 540. Dactylocrinidae (p. *T*801-*T*804).

overlooked this distinction and failed to recognize the heterotomous nature of distal parts of the arm branches. They also misconstrued the nature of the type species of *Synerocrinus* and needlessly introduced a new generic name called *Trautscholdicrinus,* later changed without explanation to *Talanterocrinus.*] *M.Penn.-U.Penn.,* USA(Okla.-Colo.-Texas).——FIG. 538,*5*; 540,*3a,b*. **A. formosus* (MOORE & PLUMMER), M.Penn.(Desmoines., Millsap Lake F.), USA(Texas, Parker Co.); 538,*5*, plate diagram (radials black, *X* and higher anals stippled; *y,* upper line of fixed brachials), *CD* interray upward (Moore, n); 540, *3a,b, C*-ray view of paratype and *D*-ray view of holotype, ×1 (Moore & Plummer, 1940).——FIG. 540, *3c-e*. *A. mcguirei* (MOORE), U.Penn. (Virgil., Brownsville Ls.), USA(Okla., Osage Co.); *3c-e,* holotype from above, side, and below (with *CD* interray upward), ×1.5 (Moore, 1939c).

Apodactylocrinus STRIMPLE & LEVORSON, 1971, p. 9 [**A. keithi*; OD]. Crown small, subglobular, like *Dactylocrinus,* except posterior basal short, anal plate in shallow notch between *C* and *D* radials. *U.Dev.,* USA(Iowa).

Calpiocrinus ANGELIN, 1878, p. 12 [**C. fimbriatus*; SD SPRINGER, 1920, p. 145]. Crown ovoid; chiefly characterized by its highly specialized base with very large infrabasals enveloping basals and to variable extent even radials and part of primibrachs; large anal *X* resting on truncated extremity of *CD* basal and followed by other anals; interbrachials consisting of single large plate between adjacent rays or by such plate followed by smaller ones in series. Free arms heterotomous of bi-endotomous type, with small ramules confined to distal region. Column large, short, tapering gradually to encrusting root. [Corresponds to *Homalocrinus* in enlargement of infrabasals externally, but in *Homalocrinus* this is much less extreme. Also, heterotomy of latter genus is endotomous with much larger ramules.] *U.Sil.,* Eu.(Sweden-Eng.-Czech.).——FIG. 538,*4b*; 540, *1c,d*. **C. fimbriatus,* Wenlock Ls.; 538,*4b,* plate diagram (radials black), *CD* interray upward (Moore, n); 540,*1c, B*-ray view of syntype from Gotland (Fara), ×1.5 (Springer, 1920); 540,*1d,* dorsal view of specimen from Eng.(Wisby), *CD* interray upward, showing stellate axial canal of stem, ×1.5 (Springer, 1920).——FIG. 540,*1a*. *C. intermedius* SPRINGER, Wenlock Ls., Eng.(Dudley); *CD*-interray view of complete crown with attached stem, ×1.5 (Springer, 1920).——FIG. 540,*1b*. *C. ovatus* (ANGELIN), Wenlock Ls., Sweden(Gotl.); *C*-ray view of holotype, ×1.5 (Springer, 1920).——FIG. 538,*4a,c*. *C. rotundatus* SPRINGER, Wenlock Ls., Sweden(Gotl.); plate diagram (radial and basal outlines dashed, *X* and higher anals stippled), *CD* interray upward, and vert. sec. lower part of calyx (*B,* basal; *IB,* infrabasal; *R,* radial) (mod. from Springer, 1920).

Lithocrinus WACHSMUTH & SPRINGER, 1880, p. 275 [**Forbesiocrinus divaricatus* ANGELIN, 1878, p. 9; SD BATHER, 1899b, p. 923]. Crown elongate, expanding upward; infrabasals somewhat recumbent, with parts exposed beyond stem impression; moderately large basals and radials subequal in size, without well-marked distinction from upper part of calyx formed by fixed brachials and interbrachials; large anal *X* followed by 3 anals and additional higher ones, border anals suturally joined to brachials at left and right, small plate on upper right shoulder of anal *X* and interpreted by SPRINGER as radianal in upper oblique position, but this is doubtful. Arms branching isotomously in calyx on primibrachs *2* and secundibrachs *3,* with similar bifurcations of free arms at 2 or more levels in different arm branches, heterotomy in distal part of crown very irregular, without clearly marked pattern. Stem large, transversely circular columnals distinctly heteromorphic. *U.Sil.,* Eu.(Sweden).——FIG. 538,*2*; 540,*2*. **C. divaricatus* (ANGELIN), Wenlock Ls., Sweden(Gotl.); 538,*2,* plate diagram (radials black, *X* and higher anals stippled; *y,* upper line of fixed brachials), *CD* interray upward (mod. from Springer, 1920); 540,*2, D*-ray view of holotype showing crown and attached stem, ×1.5 (Springer, 1920).

Nevadacrinus LANE & WEBSTER, 1966, p. 15 [**N. geniculatus*; OD]. Crown small, expanding upward from low bowl-shaped calyx, free arms separated, not touching one another laterally and not suturally united with interbrachials above primibrachs; infrabasal circlet relatively large, visible from side, with small plate directly in contact with *C* radial, as well as large *CD* basal, which has oblique distal margins for support of 2 anal plates interpreted by authors as anal *X* at left and upwardly displaced radianal at right (latter here considered very doubtful), these plates followed by additional ones above; anal plates and interprimibrachs joined laterally to brachials by firm sutures. Arms branching isotomously on primibrachs *4* (rarely on primibrach *5*), with additional bifurcations on secundibrachs *2* and possibly higher; articular facets of brachials with patelloid processes. Transversely circular stem composed in proximal (basillarid) region by 14 to 16 very low columnals with strongly crenulate articular facets, moderately curved in *A*-ray direction, joined distally to very prominent bulbous thick columnal (cuneus) with angularly divergent facets, followed distally by columnals thicker than those in the basillarid region and moderately heteromorphic. [LANE & WEBSTER's assignment of this genus to the Taxocrinidea is not accepted here, for the rather wide group of anal plates, firmly sutured laterally to radials and brachials is far removed from the linear anal series in form of a tube found in typical Taxocrinida.] *L.Perm.,* USA(Nev.).——FIG. 541,*4*; 542,*1*. **N. geniculatus*; 541,*4,* plate diagram (radials black, radianal cross ruled, *X* and other anals stippled), with part of *C* and *D* rays and adjoining interrays (Moore, n); 542,*1a,b, AB*-interray and *D*-ray

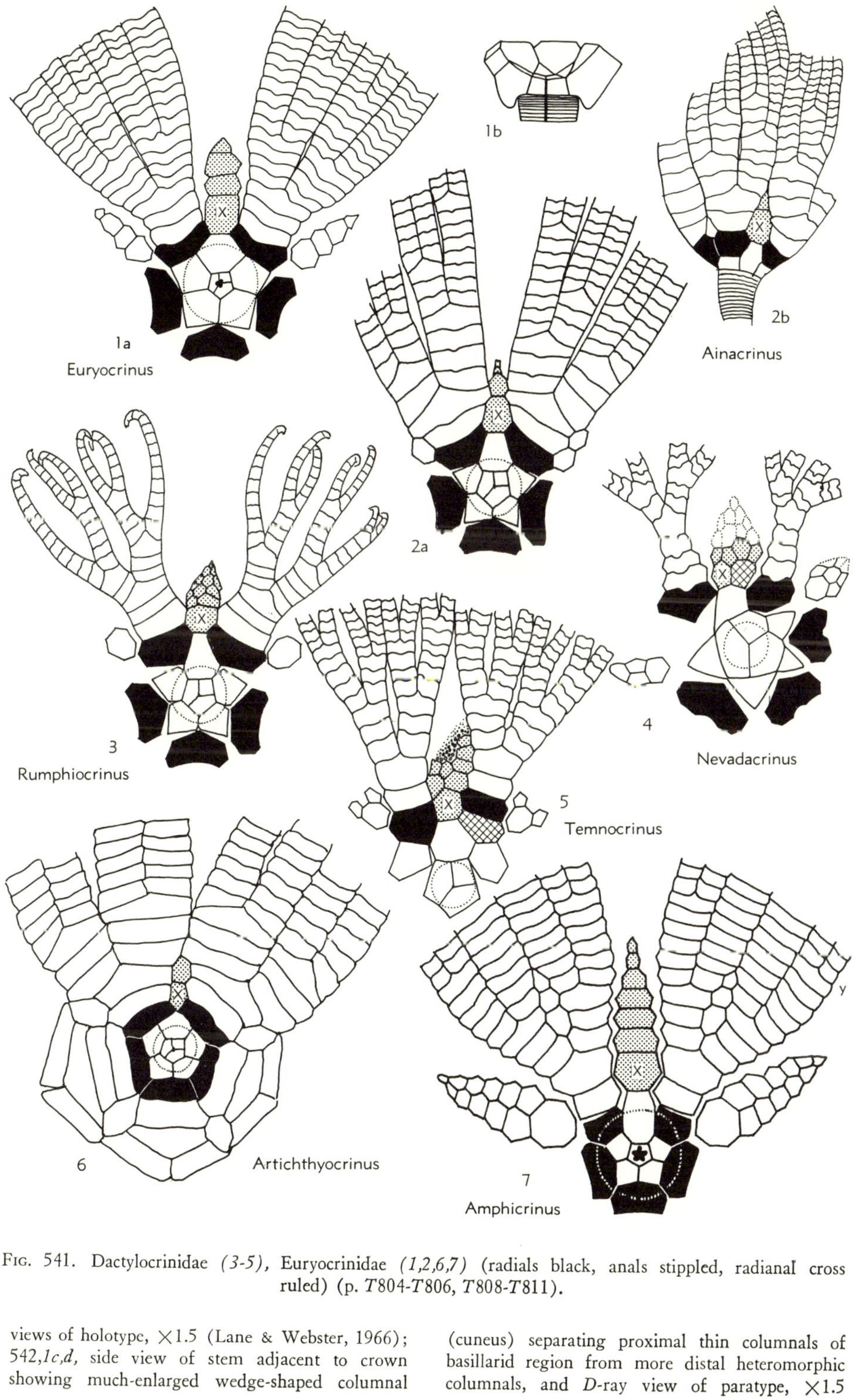

FIG. 541. Dactylocrinidae *(3-5)*, Euryocrinidae *(1,2,6,7)* (radials black, anals stippled, radianal cross ruled) (p. *T*804-*T*806, *T*808-*T*811).

views of holotype, ×1.5 (Lane & Webster, 1966); 542,*1c,d*, side view of stem adjacent to crown showing much-enlarged wedge-shaped columnal (cuneus) separating proximal thin columnals of basillarid region from more distal heteromorphic columnals, and *D*-ray view of paratype, ×1.5

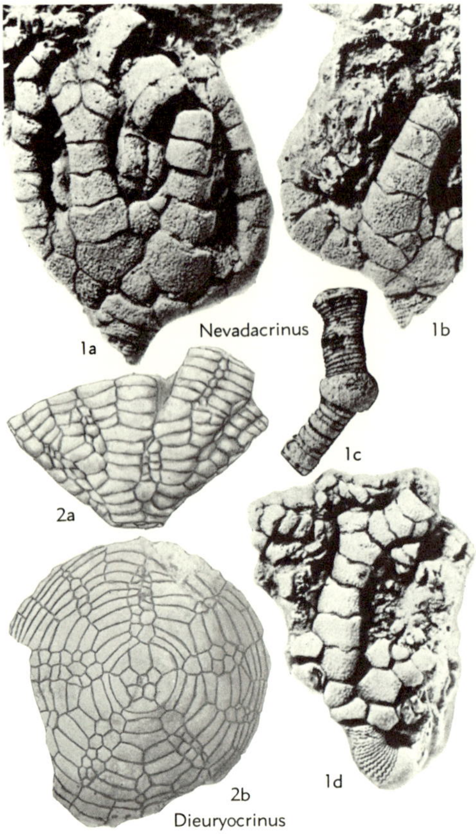

FIG. 542. Dactylocrinidae *(1)*, Euryocrinidae *(2)* (p. *T*804-*T*805, *T*811).

(Lane & Webster, 1966).

Rumphiocrinus WANNER, 1924, p. 122 [**R. singularis*; OD]. Crown ovoid with height slightly greater than width, transversely well-rounded arms incurved at summit; infrabasal circlet and proximal parts of basals covered by stem, *CD* basal distally truncate for support of large anal *X*, which is followed by several additional irregularly arranged anals laterally joined by firm sutures to brachials; calyx formed by radials and lower plate circlets combined with fixed brachials consisting of 3 or 4 primibrachs and 2 or 3 secundibrachs in each ray, large interprimibrach followed by 1 or 2 small ones in each interray, including previously mentioned anals. Arms branching in calyx isotomous, separated above first secundibrach, in most arm branches with additional bifurcation above axillary secundibrachs. Stem formed by moderately tall strongly crenulate columnals with gradual reduction in diameter distally and showing moderate curvature toward *CD* interray. [Relatively large plate above right distal margin of anal *X* corresponding to similarly placed plates in *Nevadacrinus, Dactylocrinus, Aexitrophocrinus,* and *Lithocrinus,* doubtfully interpreted as radianal displaced upward.] *U.Perm.,* Indon.(Timor).——FIG. 541,*3*; 543,*3a,b*. **R. singularis,* Basleo; 541,*3,* plate diagram of calyx (radials black, *X* and other anals stippled), *C* and *D* rays, and adjacent interrays (mod. from Wanner, 1924); 543,*3a,b,* post. and ant. views of holotype, ×1.5 (Wanner, 1924).

Temnocrinus SPRINGER, 1902, p. 94 [**Cyathocrinites tuberculatus* MILLER, 1821, p. 88; M]. Crown elongate, expanding upward; infrabasals slightly recumbent, almost entirely covered by stem impression, *CD* basal larger than others and truncate distally for support of large anal *X,* which is followed by 2 or 3 rows of polygonal plates which interlock suturally with brachials at sides; proximal plate of *C* ray identified as lower element of compound radial and defined as radianal in primitive position; all radials followed by fixed brachials which include 2 primibrachs, 3 secundibrachs, and 2 to 5 tertibrachs for each arm branch, sutures between brachials sinuous; interprimibrachs few, followed by perisome passing into tegmen. Proximal part of stem composed of low circular columnals which decrease in diameter downward along with gradual increase in height. *U.Sil.,* Eu.(Eng.).——FIG. 541,*5*; 543, *6a-c*. **T. tuberculatus* (MILLER), Wenlock Ls., Dudley; 541,*5,* plate diagram part of calyx (radials black, radianal cross ruled, *X* and higher anals stippled), *C* and *D* rays, and adjacent interrays (mod. from Springer, 1920); 543,*6a-c,* post. views of two crowns and *D*-ray view of third, ×1, ×1, ×1.5 (Springer, 1920).

Wachsmuthicrinus SPRINGER, 1902, p. 95 [**Forbesiocrinus thiemei* HALL, 1861a, p. 317; M]. Crown elongate, expanding upward from radials; infrabasals in central part of stem impression surrounded by basals of subequal size and shape, *CD* basal identified only by its relation to azygous small infrabasal in *C* ray, neither radianal nor anal *X* present. Arms branching isotomously in calyx, on primibrachs *2* and secundibrachs *3,* above which strongly marked heterotomy of biendotomous type closely resembles that of *Dactylocrinus*; interprimibrachs none to 4 or more and interbrachials none to 1 or 2; outer edges of brachials and main arms may interlock laterally even in distal parts of crown as well as in calyx formed by fixed brachials and lower plates. Stem large, expanding proximally. *Dev.,* USSR; *L.Miss. (Osag.),* USA(Ky.-Ind.-Mo.-Iowa-N.Mex.); *low. L.Carb.,* Eu.(Scot.).——FIG. 538,*1*; 539,*2a-c*. **W. thiemei* (HALL), L.Miss.(Osag., Burlington Ls.), USA(Iowa); 538,*1a,b,* plate diagram (radials black) with *D* and *C* rays at left and right above, and vert. sec. of cup (mod. from Springer, 1920); 539,*2a,b,* side and dorsal views of holotype, rays not identified, ×1.5 (Springer, 1920); 539,*2c,* side view of specimen with part of attached stem,

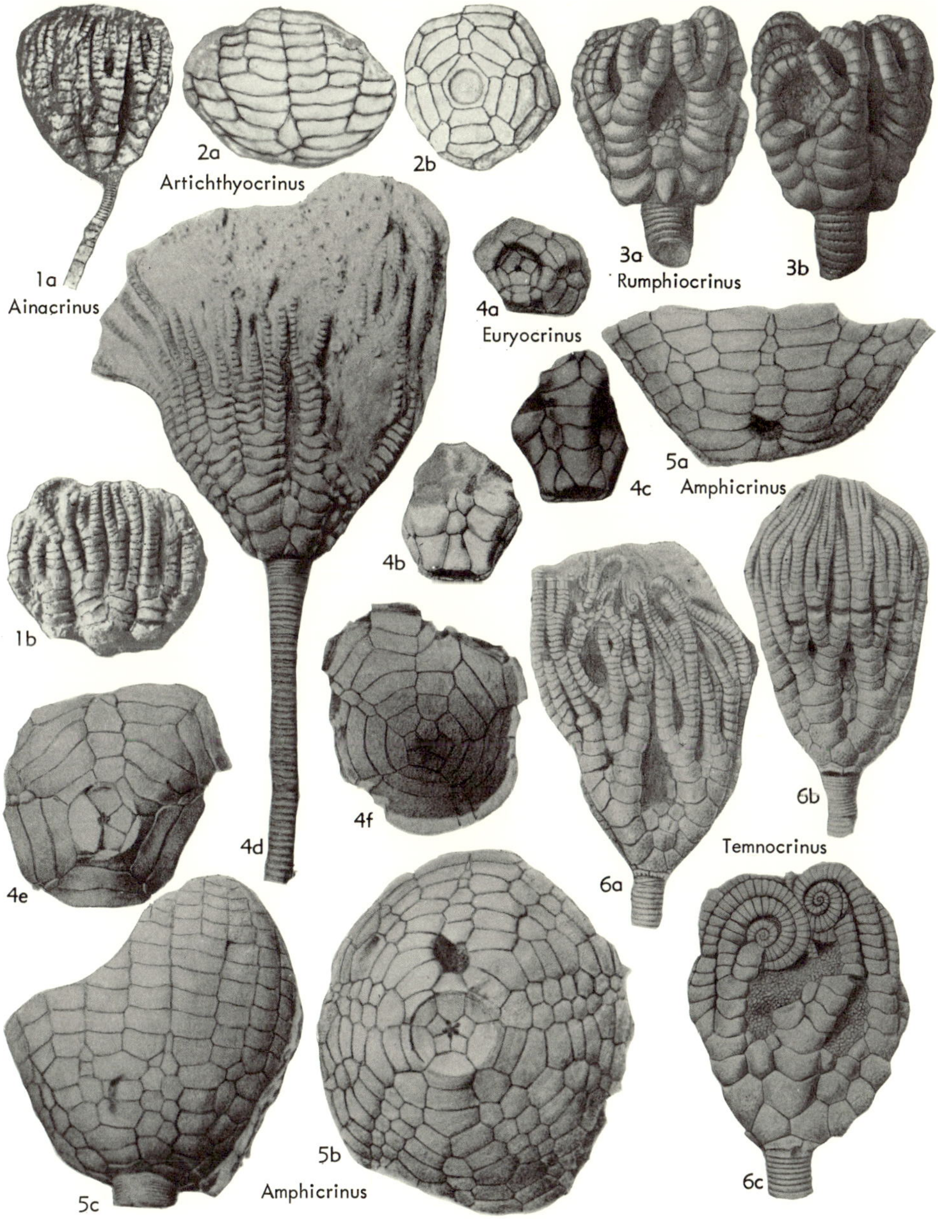

FIG. 543. Dactylocrinidae *(3,6)*, Euryocrinidae *(1,2,4,5)* (p. *T*806, *T*808-*T*811).

×1 (Springer, 1920).——FIG. 539,*2d*. *W. spinifer* (HALL), L.Miss.(Osag., Burlington Ls.), USA (Iowa); side view of crown with part of attached stem, ×1 (Springer, 1920).——FIG. 539,*2e*. *W. spinulosus* (MILLER & GURLEY), L.Miss.(Osag., New Providence Sh.), USA(Ky., Lincoln Co.); side view of crown showing brachials interlocked to height of 2nd arm bifurcations, ×1 (Springer, 1920).

Zenocrinus MOORE & STRIMPLE, 1973, p. 36 [**Z. zeus*; OD]. Small, low crown with wide shallow calyx rounded steeply upward at sides in fixed brachials of proximal parts of arms which branch isotomously on primibrachs *2* and secundibrachs *2* (or exceptionally *3*), arms abutting but free above axillary secundibrachs, with strongly heterotomous

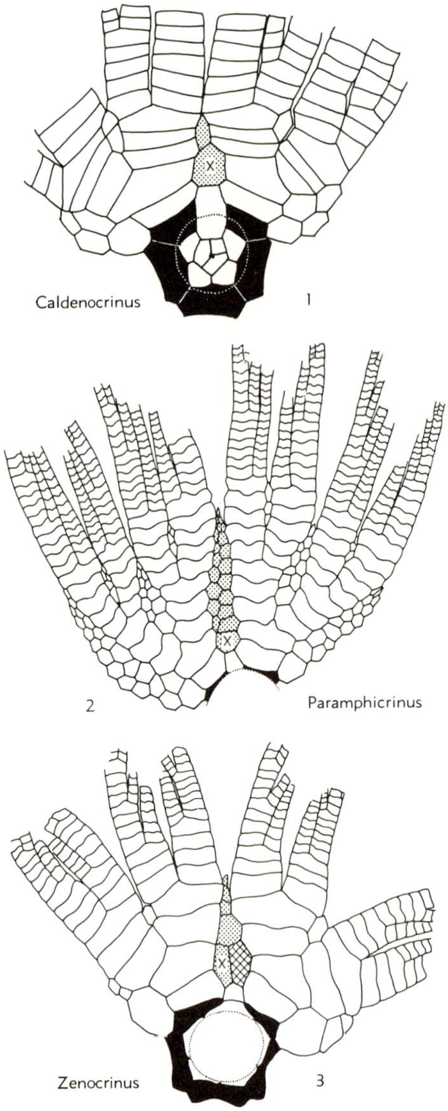

FIG. 544. Dactylocrinidae *(3)*, Euryocrinidae *(1,2)* (radials black, anals stippled, radianal cross ruled) (p. *T*807-*T*808, *T*811-*T*812).

branching of bi-endotomous type in upper part of crown, each complete ray with 16 slender arm tips bent inward (indicating 80 or more arm tips in perfect crown); infrabasals and most of basal circlet concealed by large proximal columnals; radials exposed all around, laterally abutting except where separated by *CD* basal or (?abnormally) arching over this plate to allow *C* and *D* radials to meet; *CD* interray variable, with anal *X* followed by 6 additional anals in holotype of type species but with only 2 plates (lower one very large) in paratype; other interrays with single large plate touching radials and separating fixed brachials, with or without 1 to 3 small plates above them; single intersecundibrach in each ray. Stem circular in section, composed of very low columnals with peripherally finely crenulate facets, tapering away from calyx and curved strongly toward *A* ray. *L.Penn.(Morrow.)*, USA(Okla.).——FIG. 544,*3*; 545,*2*. **Z. zeus*, Wapanucka Ls., Pontotoc Co., Okla.; 544,*3*, plate diagram (radials black, radianal cross ruled, *X* and higher anals stippled), *CD* interray at midline (Moore & Strimple, n); 545,*2a*, oblique dorsal view of paratype from post. side showing curved and tapered proximal part of stem, ×1.5 (Moore & Strimple, 1973); 545,*2b,c*, *D*-ray and dorsal views of holotype, latter with *CD* interray centrally placed above stem, ×1.5 (Moore & Strimple, 1973).

Family EURYOCRINIDAE Moore & Strimple, 1973

[Euryocrinidae MOORE & STRIMPLE, 1973, p. 16]

Crown rotund to elongate with arm branches in upper part closely appressed or separated; cup, defined to include plates below *Z* line at summit of radials not at all distinct from remainder of calyx formed by firmly sutured brachials, interbrachials, and anals; chief distinctive feature of the family is prominence of the stoutly built calyx which may have thick plates and considerably exceed upper part of crown formed by relatively short, stubby free arms which tend to branch isotomously rather than heterotomously. Infrabasal circlet surrounded by basals entirely concealed by proximal columnal; *CD* basal larger than others and typically truncate distally for support of anal *X*, which may be followed by additional anal plates. Arms in calyx branching isotomously on primibrachs *2*, less commonly on *3*. Stem transversely circular. [Main distinguishing features of this assemblage are stout nature of the calyx, isotomous branching of the arms, and general absence of interbrachials (exception, *Amphicrinus*).] *M.Dev.-U.Penn.*

Euryocrinus PHILLIPS, 1836, p. 205 [**E. concavus*; M] [=*Euryocrinites* AUSTIN & AUSTIN, 1842, p. 109 *(nom. van.)* (obj.)]. Crown expanding upward from radials, its lower part comprising stout bowl-shaped calyx formed of anals, and fixed brachials and interprimibrachs in addition to cup plates; infrabasal circlet diminutive or resorbed, surrounding basals moderately large with *CD* basal truncate distally for support of large anal

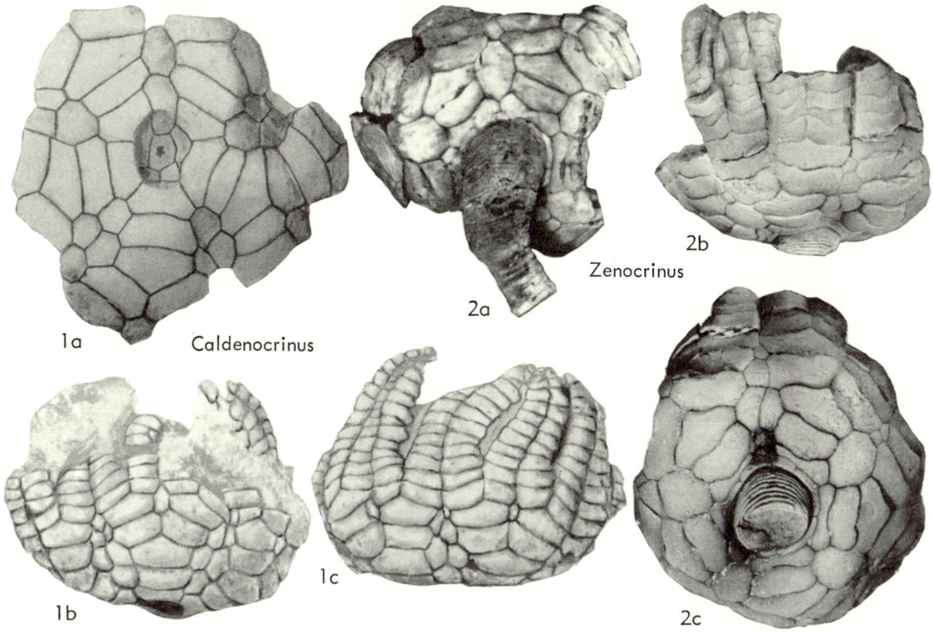

FIG. 545. Dactylocrinidae *(2)*, Euryocrinidae *(1)* (p. *T*807-*T*808, *T*811).

X, which is followed by others mostly in single series filling *CD* interray and firmly interlocked laterally with brachials. Arms branching isotomously on primibrach *3* and thereafter at higher levels, with branches laterally abutting and tending to be joined firmly; interprimibrachs few, usually limited to single series; interbrachials uncommon but may be present. Column large, expanding toward calyx. *M.Dev.*, USA(Iowa); *U. Dev.*, USA(Ohio-Mich.); *L.Miss.*, USA(Ky.-Ind.-Tenn.); *L.Carb.(Visean)*, Eu.(Eng.).——FIG. 543, *4a-d*. *E. barrisi* SPRINGER, M.Dev.(Hamilton), USA(Mich.-Iowa); *4a-c*, dorsal, post., and *B*-ray views of specimens from Michigan, ×1 (Springer, 1920); *4d*, *B*-ray side of crown with attached stem, from Iowa, ×1 (Springer, 1920).——FIG. 541,*1*; 543,*4e,f*. **E. concavus*, L.Carb., Eng.; 541,*1a,b*, plate diagram (radials black, *X* and higher anals stippled), with *CD* interray centrally placed above, and vert. sec. of cup (mod. from Springer, 1920); 543,*4e,f*, dorsal and ventral views of lower part of calyx, *CD* interray directed upward in each, ×1 (Springer, 1920).

Ainacrinus WRIGHT, 1939, p. 56 [**Synerocrinus? smithi* WRIGHT, 1934, p. 253; OD]. Crown subglobose to ovoid, with arms infolded at summit; infrabasal circlet moderately large but entirely covered by proximal columnal; *CD* basal broadly truncate for support of anal *X*, followed by one or more other anals in series, laterally joined firmly to brachials. Arms branching isotomously on primibrachs *2*, followed by 2 or more additional bifurcations at different levels in various arm branches; interprimibrachs few, normally only 1 or 2 is each interray but may be as many as 4. Stem composed of low circular columnals which expand toward base of calyx. *L.Carb. (Visean)*, Eu.(Scot.).——FIG. 541,*2*; 543,*1*. **A. smithi* (WRIGHT), Ayrshire; 541,*2a*, diagram of plate arrangement in cup (radials black, *X* and higher anals stippled) and *C* and *D* rays; 541,*2b*, *D*-ray view of crown showing anal *X* (stippled) at right; 543,*1a,b*, side views of crown with attached stem and another crown, ×1 (Wright, 1950).

Amphicrinus SPRINGER, 1906, p. 518 [**A. scoticus* SPRINGER in WRIGHT, 1914, p. 161; SM]. Crown subglobular, consisting mostly of strongly built calyx, small free arms in upper part of crown laterally abutting but not joined together; stem impression covering diminutive tripartite (or ?fused) infrabasals, basals (except distal extremity of *CD* basal) and most of radials; *CD* interray identified by single plate defined as anal *X* resting on truncated posterior basal, proximal plates of other interrays not touching basals but resting on oblique shoulders of radials; radianal lacking. Rays with 2 primibrachs in calyx followed by isotomously branching arms with 3 to 4 or more bifurcations, with tendency toward heterotomy observed in summit portion of crown; sutures between brachials sinuous and above primibrachs commonly with patelloid processes. Stem large, next to crown composed of very low columnals

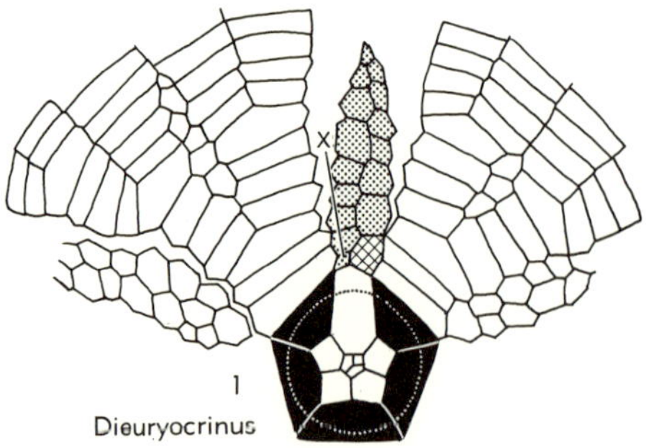

FIG. 546. Euryocrinidae (p. T811).

with crenulate facets, tapering gradually and curving gently away from cup; columnals pierced by quinquestellate axial canal. *L.Carb.(Visean)*, Eu.(Scot.)-USSR-USA(Okla.).——FIG. 541,7; 543, *5a-c*. **A. scoticus*; 541,*7*, plate diagram (radials black, *X* and higher anals stippled; *y*, upper line of fixed brachials), with *C* and *D* rays accompanied by adjoining interbrachials (mod. from Springer, 1920); 543,*5a,b*, post. and dorsal views of somewhat flattened calyx, Eng.(Roscobie), ×1 (Springer, 1920); 543,*5c*, *AB*-interray view of crown, Scot.(Hurlet Ls.), ×1 (Springer, 1920).

Artichthyocrinus WRIGHT, 1923, p. 481 [**A. springeri*; OD]. Crown globose to ovoid, generally wider than high; broad shallow calyx formed by rigidly sutured plates; small infrabasal circlet sur-

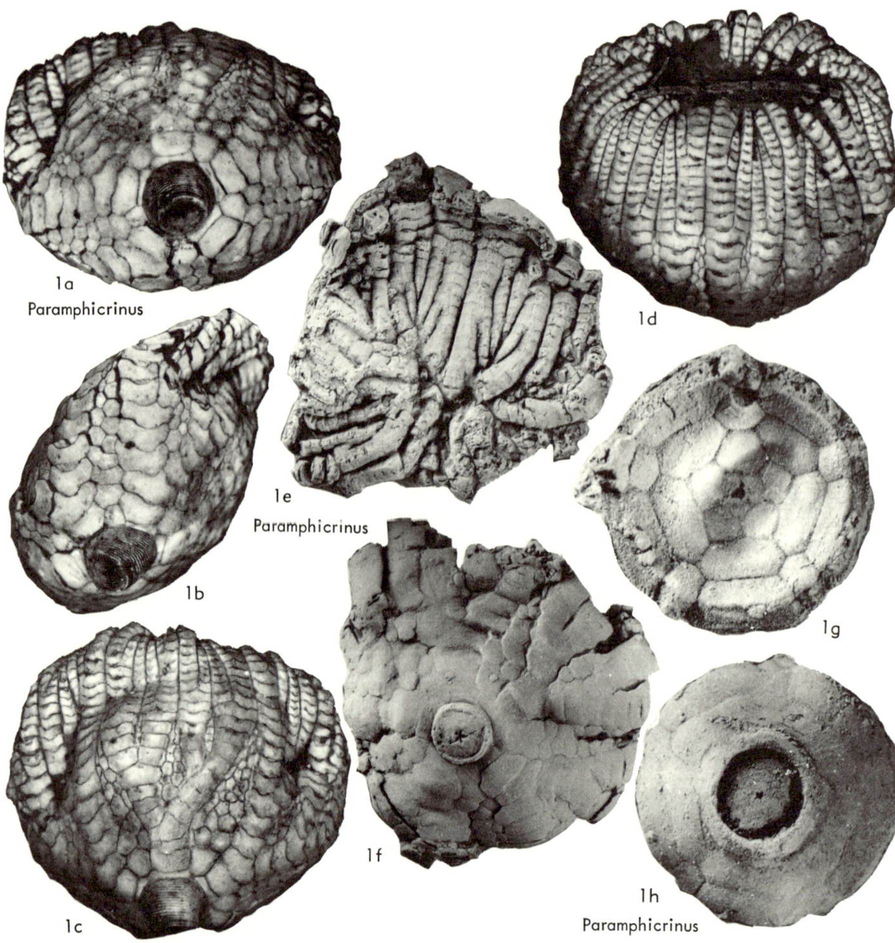

FIG. 547. Euryocrinidae (p. T811-T812).

rounded by basals and entirely concealed by proximal columnal, present in sharply indented faceted calyx. Arms with nearly flat outer surfaces with brachials fitted smoothly together laterally or interlocked, articular facets somewhat arcuate; interradials commonly 1 in each interray; small anal *X* followed by additional plate in posterior interray, orientation of calyx mainly determinable from *C*-position of small infrabasal. *L.Carb.(Visean)*, Eu.(Scot.).——Fig. 541,*6*; 543,*2*. **A. springeri*; 541,*6*, plate diagram (radials black, *X* and higher anal stippled), with part of *C* and *D* rays (Moore, n); 543,*2a,b*, *AB*-interray and dorsal views of specimen from Fife, ×1.5 (Wright, 1950).

Caldenocrinus Wright, 1946, p. 33 [**C. curtus*; OD]. Crown low, rounded, with shallow bowl-shaped calyx of firmly sutured plates; infrabasals concealed by stem, which also covers basal circlet except distal part of *CD* basal and encroaches on proximal tips of radials; large anal *X* followed by one or more additional anals; 4 or 5 moderately large interprimibrachs in other interrays. Arms branching heterotomously above axillary primibrachs *2* but not strongly so. *L.Carb.(Visean)*, Eu.(Scot.).——Fig. 544,*1*; 545,*1a-c*. **C. curtus*; 544,*1*, plate diagram (radials black, *X* and higher anals stippled), *CD*-interray upward (Moore, n); 545,*1a*, paratype from Roscobie, Fife, dorsal view with *CD* interray upward, ×1 (Wright, 1954); 545,*1b,c*, holotype from Seafield Tower Ls., near Kirkcaldy, Fife, *B*-ray and *DE*-interray views, ×1 (Wright, 1954).

Dieuryocrinus Wright, 1954, p. 165 [**Euryocrinus duplex* Wright, 1942, p. 280; OD]. Lower part of broadly flaring calyx formed by stout sutured plates, rather closely resembling *Euryocrinus* and *Amphicrinus*, but with double series of anal plates in *CD* interray, in which proximal small plate at left is interpreted as anal *X* and that at right as upwardly displaced radianal. Primibrachs 3, as in *Euryocrinus*, rather than 2 as in *Amphicrinus*; free arms of upper crown unknown. Stem impression not indented. *L.Carb.(Visean)*, Eu.(Eng.).——Fig. 542,*2*; 546,*1*. **D. duplex* (Wright); 542,*2a,b*, *AB*-interray and dorsal views of large calyx, *CD* interray directed upward in *2b*, ×0.75 (Wright, 1954); 546,*1*, plate diagram (radials black, radianal cross ruled, *X* and higher anals stippled), with part of *C* and *D* rays and adjoining interrays (Moore, n).

Paramphicrinus Strimple & Moore, 1971, p. 40 [**Amphicrinus oklahomaensis* Strimple, 1939a, p. 4 (364); OD]. Crown rotund, medium-sized, with stoutly built bowl-shaped calyx forming lower half of crown and closely parallel free arms upper half, curving inward at summit so that arm tips nearly meet; infrabasal circlet and all except posterior plate of basal circlet covered by stem, subhorizontal in attitude; only distal edges of radials exposed, those of *C* and *D* rays slightly smaller than others and visible parts asymmetrical; no

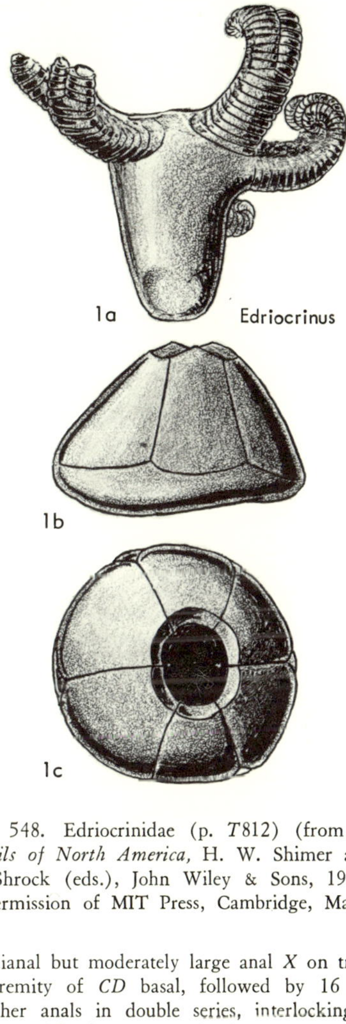

Fig. 548. Edriocrinidae (p. *T*812) (from *Index Fossils of North America*, H. W. Shimer and R. R. Shrock (eds.), John Wiley & Sons, 1944, by permission of MIT Press, Cambridge, Mass.).

radianal but moderately large anal *X* on truncate extremity of *CD* basal, followed by 16 to 20 higher anals in double series, interlocking sideward with fixed brachials from primibrach to tertibrach series. Arms broad in proximal portion, branching isotomously in calyx on primibrachs *2* and secundibrachs *3* in each ray, thereafter displaying well-marked heterotomy of bi-endotomous type, distal parts of all arms narrowing upward, with 80 to 100 arm tips at crown summit, each complete half-ray with 20 arms, interbrachial sutures mostly sinuous; interradial polygonal plates 16 to 20 beginning between first primibrachs with large plate, much smaller distal ones suturally joined to tertibrachs; interbrachials beginning between proximal secundibrachs and extending upward nearly to distal tertibrachs; 2 small areas of 3 to 5 intertertibrachs in each ray. Stem large, composed of very low crenulate columnals, with quinquelobate axial canal, proximal part tapering gradually and strongly curved toward *DE* interray or *A* ray. *L.Penn.(Morrow.)-U.Penn.(Missour.)*,

USA(Okla.-Ark.-Ill.-Kans.).——FIG. 544,2; 547,*1*. **P. oklahomaensis* (STRIMPLE), U.Penn.(Missour.), USA(Okla.-Ill.); 544,*2*, plate diagram of post. side of crown (radials black, *X* and higher anals stippled), showing anal plates of *CD* interray (center) flanked by *C* and *D* rays on right and left, based on specimen from LaSalle Ls., Ill.; 547,*1a-d*, views of crown from Missour.(LaSalle Ls.), Ill., from below (*B* ray upward), post. side, *B*-ray side, and obliquely from above (*DE* interray located centrally), all ×0.9 (Strimple & Moore, 1971); 547,*1e,f*, holotype from Missour. (Stanton Ls.), near Bartlesville, Okla., viewed from above (*C* and *D* rays centrally placed at left and right of midline above) and dorsal view (*A* ray directed upward) showing proximal columnal with stellate axial canal, ×0.9 (Strimple, 1939a).——FIG. 547,*1g,h*. *P. poundi* (STRIMPLE), Missour.(Stanton Ls.), USA(Kans., Montgomery Co.); ventral and dorsal views of holotype, *CD* interray directed upward, ×3.5 (Strimple, 1939a).

Order UNCERTAIN

Family EDRIOCRINIDAE S. A. Miller, 1889

[Edriocrinidae S. A. MILLER, 1889, p. 214] [Materials for this family prepared by H. L. STRIMPLE]

Stalkless crinoids attached by base of cup, at least during youthful stages. Basals four, commonly fused, anal *X* in line with radials; arms broad, incurving distally, formed of very short brachials. *L.Dev.-M. Dev.*

Edriocrinus HALL, 1858, p. 278 [**E. pocilliformis* HALL, 1859, p. 121; SD MILLER, 1889, p. 242]. Cup thought to be pseudomonocyclic, i.e., infrabasals probably present in ontogeny but fused or resolved in process of producing base for attachment to foreign objects. Arms strikingly similar to those of flexible crinoids. [Some evidence indicates that some individuals, or even species, became free in full maturity and used the rigid arms for limited locomotion. Some species (e.g., *E. sacculus*) developed a long fused basal cone that could have been buried in a soft marl of the substrate, and yet others (e.g., *E. dispansus*, Fig. 548,*1b*) have a broad flat base suitable for resting on soft substrate without becoming mired.] *L.Dev.(Helderberg.)*, USA(N.Y.-Md.-Va.-W.Va.-Tenn.-Mo.-Okla.); *L.Dev.-M.Dev.*, Eu.(Boh.); *M. Dev.*, Afr.(Alg.).——FIG. 548,*1a*. *E. sacculus* HALL, L.Dev.(Oriskany), Md.; adult showing attachment scar, ×1.3 (Moore & Laudon, 1944).——FIG. 548,*1b,c*. *E. dispansus* KIRK, L.Dev. (Linden), Tenn.; *1b,c*, lat. and ventral view of cup, ×1.3 (Moore & Laudon, 1944).